Biomolecular Thermodynamics

From Theory to Application

Foundations of Biochemistry and Biophysics

Founding Series Editor: John J. (Jack) Correia

Currently available

Quantitative Understanding of Biosystems: An Introduction to Biophysics
Thomas M. Nordlund

Biomolecular Thermodynamics: From Theory to Application
Douglas E. Barrick

Biomolecular Kinetics: A Step-by-Step Guide
Clive R. Bagshaw

Forthcoming

Physical Principles in Nucleic Acid Chemistry
David E. Draper

RNA Biophysics
Kathleen B. Hall

Biomolecular Thermodynamics
From Theory to Application

Douglas E. Barrick

CRC Press
Taylor & Francis Group
Boca Raton London New York

CRC Press is an imprint of the
Taylor & Francis Group, an **informa** business

CRC Press
Taylor & Francis Group
6000 Broken Sound Parkway NW, Suite 300
Boca Raton, FL 33487-2742

© 2018 by Taylor & Francis Group, LLC
CRC Press is an imprint of Taylor & Francis Group, an Informa business

No claim to original U.S. Government works

Printed on acid-free paper

International Standard Book Number-13: 978-1-4398-0019-5 (Paperback)
978-1-138-06884-1 (Hardback)

Library of Congress Cataloging-in-Publication Data

Names: Barrick, Douglas, author.
Title: Biomolecular thermodynamics : from theory to application / Douglas Barrick.
Other titles: Foundations of biochemistry and biophysics.
Description: Boca Raton : Taylor & Francis, 2017. | Series: Foundations of
biochemistry and biophysics
Identifiers: LCCN 2017005294| ISBN 9781439800195 (pbk. : alk. paper) | ISBN
9781138068841 (hardback)
Subjects: | MESH: Thermodynamics | Biochemical Phenomena | Models, Theoretical
Classification: LCC QP517.T48 | NLM QU 34 | DDC 572/.436--dc23
LC record available at https://lccn.loc.gov/2017005294

Visit the Taylor & Francis Web site at
http://www.taylorandfrancis.com

and the CRC Press Web site at
http://www.crcpress.com

Printed and bound in the United States of America by Sheridan

To Debbie and Timmy, who together are my role models for science, love, and life.

Contents

Detailed Contents

Chapter 4
The Second Law and Entropy   131

Series Preface

Biophysics encompasses the application of the principles, tools, and techniques of the physical sciences to problems in biology, including determination and analysis of structures, energetics, dynamics, and interactions of biological molecules. Biochemistry addresses the mechanisms underlying the complex reactions driving life, from enzyme catalysis and regulation to the structure and function of molecules. Research in these two areas has a huge impact in pharmaceutical sciences and medicine.

These two highly interconnected fields are the focus of this book series. It covers both the use of traditional tools from physical chemistry such as nuclear magnetic resonance (NMR), x-ray crystallography, and neutron diffraction, as well as novel techniques including scanning probe microscopy, laser tweezers, ultrafast laser spectroscopy, and computational approaches. A major goal of this series is to facilitate interdisciplinary research by training biologists and biochemists in quantitative aspects of modern biomedical research and teaching core biological principles to students in physical sciences and engineering.

Proposals for new volumes in the series may be directed to Lu Han, senior publishing editor at CRC Press, Taylor & Francis Group (lu.han@taylorandfrancis.com).

Preface

This book introduces students to the concepts and skills necessary to understand the behavior of molecular systems in the biological and chemical sciences at a quantitative level, through the formalisms of classical and statistical thermodynamics. Although the application of thermodynamics to biomolecular and chemical problems is not new, most existing books written for traditional courses focus on the underlying theory and equations but do not provide adequate understanding of how to apply these equations in the analysis of experimental measurements. Thus, it is often difficult for students to properly leverage the knowledge obtained from traditional (bio)physical chemistry courses to "real world" physical chemical research. Moreover, many treatments of classical and statistical thermodynamics are complete and rigorous, but do not provide a lot of intuition either to newcomers or returning customers.

Thus, my goals with this book are twofold: provide a deep and intuitive understanding of the ideas and equations of thermodynamics, and provide the computational and statistical skills to analyze data using the thermodynamic framework. To achieve the first goal, I have developed a presentation that I hope is both intuitive and rigorous. In my experience, many of the great texts in physical chemistry, both at the undergraduate and graduate level (e.g., Hill, McQuarrie, Callan), are better at presenting to those who already understand thermodynamics. The equations are concise, the narrative is logical, and the endpoint is advanced. In reading these texts for the first (and second, and third) time, I learned to do the math, but I did not understand *why* I was doing the math. For example, why does analysis of simple heat engines lead to the most important equation in thermodynamics, the Clausius inequality? What is the statistical basis for spontaneous heat flow from hot to cold? Why does the Lagrange method take us from the rather arcane-seeming ensemble construct to the equilibrium (Boltzmann) distribution? What do the binding polynomials of Wyman and Gill represent, beyond the arithmetic?

To develop this intuition, I have emphasized on illustrations and examples, following the lead of Dill and Bromberg's groundbreaking book on driving forces in physical chemistry. Using modern three-dimensional plotting software, the results of manipulations using multivariable calculus (such as the Legendre transform and Lagrange method in classical and statistical thermodynamics) can be directly visualized in space, rather than simply as equations. By combining step by step pictures of the math, it is hoped that students will come away with the concepts, rather than just learning derivations (such as the Boltzmann distribution) as a memorized progression of equations. At the same time, I have made efforts not to sacrifice the rigor of the equations, but to expand them, especially in areas that are typically left off. Though for some audiences, some of the details can be skipped without losing a working understanding, it is hoped that those with a deep interest in fundamentals will find what they are looking for.

My second major goal with this book is to teach students how to use computers to think about and solve problems in biological and chemical thermodynamics. Most physical chemistry books present equations, and draw plots of equations as continuous curves. I emphasize to my students at Hopkins that we do not interact with nature by measuring curves and lines, rather, we measure points. That is, our data are discrete. It is what we do with our measurements that tell us about our system, how it behaves, where equilibrium lies, and why. In this book, readers will learn to extract thermodynamic quantities from a broad spectrum of experimental measurements using computers to analyze data, test models, and extract thermodynamic parameters. It is my opinion that students learn more, for example, from fitting a set of p-V measurements for a gas with various models than from memorizing ideal and van der Waals equations of state. As an important additional step, several methods are presented to help students think about and quantify uncertainties in fitted parameters, to avoid the common pitfall of treating fitted quantities as error-free. Another emphasis with computer-aided learning is to use random number generation and simulation to generate statistical distributions and to simulate and analyze the dynamics of molecular systems. This approach greatly complements the equations and logic of statistical thermodynamics. When students simulate the energy distributions of a molecular system and compare these to theoretical distributions from statistical thermodynamics, the theory becomes practice.

To achieve this second goal, the material in this book has been designed to be taught with the aid of a high-level interpreted mathematics program such as Mathematica or MATLAB®. Having taught semester-long biophysical chemistry courses with both programs, I find Mathematica to be easier for students to both learn and use, and as a result, their efforts are directed more toward learning and doing physical chemistry than to debugging code. For this reason, examples in this book are based on Mathematica. Other software packages that could be used in conjunction with this book include python (and its modules numpy, scipy, simpy, matplotlib), especially when run in the Jupyter notebook platform, which has much of the look and functionality of a Mathematica notebook. A third package that could be used here is R, which lacks much of the symbolic math and visualization, but is strong in statistical analysis. A major advantage of both python (and its add-ons) and R is that they are available at no cost.

The first two chapters of this book focus on background and mathematical tools that are used throughout. Chapter 1 focuses on probabilities and statistics, including how probabilities for events combine, how outcomes build up over many events to give discrete distributions of various types (with emphasis on binomial and multinomial distributions), and how these distributions relate to continuous distributions (with emphasis on the Gaussian and exponential distributions). Chapter 2 reviews multivariable calculus, emphasizing partial differentiation, maximization of multivariable functions, exact differentials and their properties, and path integration. Chapter 2 then describes curve fitting using linear and nonlinear least squares, and presents methods (relating to the covariance matrix, bootstrap technique, chi-squared statistics, and the f-test) to estimate errors in fitted parameters, their correlations with each other, and the goodness of fit of different models.

The main part of this book progresses from mostly classical thermodynamic to mostly statistical thermodynamic material. Chapter 3 introduces some fundamental concepts relating to system, surroundings (and the "thermodynamic reservoir"), equilibrium, and reversibility. Next, the first law is presented and discussed, along with analysis of heat and work. Emphasis is placed on gasses to take advantage of their simple equations of state. After discussing reversible changes, irreversible heat transfers and expansions are analyzed for some simple systems. This section on irreversible thermodynamics can easily be skipped by beginning students.

Chapter 4 introduces the second law and entropy. Though emphasis is placed on a classical derivation and analysis, using heat engines to derive the Clausius inequality, a few statistical models are developed in parallel, to show how spontaneous change (and entropy increase) corresponds to increasing the number of configurations available. One of these models, which describes heat transfer between two subsystems, is used throughout this book, especially when developing statistical thermodynamics in Chapters 9 and 10. As with Chapter 3, entropy changes are calculated for some irreversible transformations, and can be skipped by beginners.

Chapter 5 takes the idea of the entropy as a thermodynamic potential for an isolated system and transforms it into potentials for nonisolated systems. The most important of these is the Gibbs free energy potential at constant temperature and pressure. These transforms are done in a simple approach using differentials, but are also developed in a mathematically more rigorous way using Legendre transforms. This rigor is essential for those seeking a solid understanding of the fundamentals, but again it can be skipped by beginning students. Chapter 5 also introduces the concept of molar (intensive) quantities rather than the extensive quantities discussed in Chapters 3 and 4. Using these molar quantities, the thermodynamics of mixtures (and partial molar quantities) are introduced. Emphasis is placed on the partial molar free energies (chemical potentials) and their relationships (leading to the Gibbs–Duhem equation).

Chapter 6 uses the concept of chemical potential to discuss phase equilibrium in simple systems. Using experimental enthalpy, heat capacity, and volume data for water, a phase diagram is constructed using chemical potentials and their pressure and temperature dependences. Following this exercise, the Gibbs–Duhem equation is applied to different phases of material, and evaluated graphically. This graphical analysis provides a visual picture that leads directly to the Gibbs phase rule and the Clausius–Clapeyron equation.

Chapter 7 explores the concentration dependence of the chemical potential, progressing from the ideal gas to the ideal solution to ideal dilute solutions. Along the way, the concept of the standard state is developed. Using the mole-fraction standard state, thermodynamics of mixing is developed for an ideal solution (following Raoult's law). Using a lattice mixing model, nonideality is built in, and the mixing thermodynamics of the regular (or mean-field) solution is developed. In addition to showing how complexity can be built into a simple model, the regular solution reveals liquid–liquid phase separation behavior. Finally, the thermodynamics of chemical reactions is developed using chemical potentials on a molar scale. Two approaches are taken. First, a difference approach is used to develop the familiar relationships between reaction free energies, concentrations, and equilibrium constants. Second, a calculus-based approach is developed to show the relation between the reaction free energy, the Gibbs energy, and the position of chemical equilibrium. Finally, the temperature and pressure dependence of chemical reactions are discussed.

Chapter 8 applies the reaction thermodynamics developed in Chapter 7 to a fundamental process in biology: the protein folding reaction. This "reaction" is spectacular in its structural complexity, yet in many cases, it can be modeled as a very simple reaction involving just two thermodynamic states. Emphasis in this chapter is placed on methods for studying the reaction, and on the importance of obtaining data (i.e., "melts") where there is a measurable equilibrium between folded and unfolded conformations. Thermal and chemical denaturation approaches are described in which equilibrium is monitored with optical probes, and calorimetric methods are also described. Key thermodynamic signatures of the folding reaction are discussed, including large entropy and enthalpy change, and a significant heat capacity change.

Chapters 9 through 11 present a formal development of statistical thermodynamics. In Chapter 9, the motivations for the ensemble method (including the challenges of constructing detailed dynamic models, and the ergodic hypothesis) are discussed. Concepts and variables of ensembles are developed and illustrated for simple models. Using the heat flow model from Chapter 4, an ensemble is developed to model an isolated system (the "microcanonical" ensemble). A method is presented to determine the equilibrium ensemble distribution in which the number of ensemble configurations is maximized, subject to a constraint of fixed ensemble size, using the Lagrange method. In this derivation, considerable use is made of visuals, showing how constraints interact with configurational variables, and how multivariable calculus provides the equilibrium solution. The resulting distribution defines a microcanonical "partition function," which is fundamentally related to the entropy—the thermodynamic potential for the isolated system.

Chapter 10 extends the ensemble approach to systems that can exchange thermal energy (but not volume) with their surroundings. The equilibrium position (i.e., the Boltzmann distribution) for this "canonical" ensemble is again found by Lagrange's method with an additional constraint on the total energy of the ensemble. The resulting canonical partition function is logarithmically proportional to the Helmholtz free energy—again, the thermodynamic potential for the constant volume system. This approach is then extended to systems that can expand and contract, leading to the equilibrium distribution for a system at constant temperature and pressure, and providing a direct connection to the Gibbs free energy potential.

With the fundamental ensembles in place, Chapter 11 makes some key modifications to the statistical thermodynamic approach. First, the ensemble approach is applied to a system with just one molecule, leading to molecular partition functions. We show how, as long as molecules do not interact, these simple molecular partition functions can be combined to give a partition function for systems with many molecules, and how they can also be used to extract thermodynamic quantities. One place where these molecular partition functions are particularly useful is in analyzing chemical reactions. By building an ensemble of reactive molecules, we generate a "reaction" partition function that is used in one form or another in the remaining chapters.

Chapter 12 applies the reaction partition function to the "helix–coil transition" in polypeptides. Unlike the folding of globular proteins described in Chapter 8, the helix–coil transition involves many partly folded states. We develop statistical thermodynamic models of increasing complexity, progressing from a model in which all residues are independent and identical to a model in which residues are not identical, to models where residues are energetically coupled to their nearest-neighbor. The nearest-neighbor or "Ising" approach is a workhorse model for describing cooperativity in chemical and biological systems. In addition to learning about this important type of reaction, this chapter shows how to build statistical thermodynamic models, starting simple and moving toward complex, as is demanded by experimental observations.

Chapters 13 and 14 describe the binding of multiple ligands to macromolecules. To describe such data, we develop a binding partition function referred to as a "binding polynomial." In Chapter 13, we develop a macroscopic scheme for ligand binding which counts the number of ligands bound but not their arrangement among binding sites. Two different phrasings of binding constants are introduced (stepwise and overall), and their relative advantages and disadvantages are discussed. The relationship between the binding polynomial and various experimental representations of binding data are described, including the fraction of ligand bound, the binding capacity, and the (somewhat outdated but still useful) Hill plot, as are the effects of positive and negative cooperativity on these representations. Chapter 13 concludes by treating the binding of multiple different ligands to a macromolecule, and introduces the concept of thermodynamic linkage.

Chapter 14 analyzes binding from a microscopic perspective. Although the microscopic approach can be more complicated (in terms of parameters and equations), it provides a more mechanistic and more fundamental way of thinking about binding. Again, binding constants are phrased in terms stepwise and overall reactions, and the relationships between these constants (and the macroscopic constants) are analyzed using basic topology and graph theory. Unlike our approach with the helix–coil transition, we start here with the most complex and general models, and then simplify these models to include macromolecular structure and symmetry. The result is a lean set of models for noncooperative and cooperative ligand binding. We conclude by combining conformational changes (ideas from Chapters 8 and 12) with ligand binding both at the macroscopic and microscopic level, leading to a thermodynamic description of macromolecular allostery.

MATLAB® is a registered trademark of The MathWorks, Inc. For product information, please contact:

The MathWorks, Inc.
3 Apple Hill Drive
Natick, MA 01760-2098 USA Tel: 508 647 7000
Fax: 508-647-7001
E-mail: HYPERLINK "mailto:info@mathworks.com"info@mathworks.com
Web: HYPERLINK "http://www.mathworks.com/"www.mathworks.com

Acknowledgments

Many people have taught me physical chemistry, and helped me understand it, think about how to apply it to biological problems, and teach me how to teach it to students. I have been very fortunate to have had the luck to have taken classes from, researched with, taught with, taught to, or just crossed path with these individuals.

At the University of Colorado, I had the blind luck of wandering into Larry Gold's lab, and learning how to think about biology using physical chemistry with Larry, and with Gary Stormo and Tom Schneider. These three profoundly shaped my interest in entropy (and information theory) from a statistical viewpoint. Their application of this material to molecular biology was truly ahead of its time. Also at Colorado, I got my first exposure to thermodynamics from Fred Stein (a visiting professor from Western State) and quantum mechanics from Carl Lineberger. From Fred and Carl I fell in love with the subject.

At Stanford I learned much about physical chemistry from Buzz Baldwin, my graduate mentor, and in particular, learned about protein folding and the helix-coil transition. I sincerely hope that Chapters 8 and 12 of this book gives a satisfactory retelling what Buzz attempted to teach me. Also in Buzz's lab, I learned to apply physical chemistry to biopolymers from an amazing group of graduate students and postdocs; some of these include Susan Marqusee, Fred Hughson, Jayant Udgaonkar, Janette Carey, Marty Scholtz, Andy Robertson, Thomas Kiefhaber, and Steve Mayo.

At the University of Oregon, I owe much to Rick Dahlquist, my postdoctoral mentor. Among other things, Rick taught me to speak fluently in equations. I remember many times when I asked Rick a question, either about theory or about data, and he would say "let's go to the board for a minute, shall we?" What followed was a derivation that answered my question with precision and clarity. I also benefitted greatly from interactions with others at the Institute of Molecular Biology in Eugene, including John Schellman, Pete von Hippel, and Brian Matthews. Eugene in the 1990s was a magical place for biomolecular physical chemistry. During this time, I also learned much about cooperativity and allostery from Chien Ho at Carnegie Mellon, and was able to do some physical chemistry on hemoglobin (a rite-of-passage system for biophysical chemists of my era) with Chien and with Nancy Ho.

Through another series of lucky events, I moved from one magical place of biophysical chemistry to another when I took a position at Johns Hopkins. When I arrived at Hopkins, I was fortunate to do science with Jeremy Berg, who made me think it was not a crazy idea to write a textbook (which I now know is untrue). I also benefitted from regular discussions about protein folding with David Shortle and Mario Amzel at the School of Medicine.

At my home department, the T.C. Jenkins Department of Biophysics, I have been lucky to work with many great colleagues and discuss thermodynamics of biopolymers, including David Draper, Ed Lattman, Juliette Lecomte, George Rose, and Richard Cone. George and Richard were particularly important, setting an example of what it means to be great teachers; my style in the classroom is an attempt to emulate them both, and I hope I have captured some of their essence in this text. I thank Sarah Woodson for encouraging me to include statistical examples in the early part of this text to complement the dry (but elegant) subject of heat engines. Above all, I have benefitted from my constant scientific interactions with Bertrand Garcia-Moreno, who has a wealth of knowledge of biophysics, a deep and intuitive understanding of biothermodynamics, a keen supporter of education, a great mentor, and a patient chair. Bertrand has *always* been willing to help me with all things within and outside of science.

Outside of my department, I have benefitted greatly from interactions with many Hopkins colleagues, including Vince Hilser, Peter Privalov, Ernesto Freire, Doug Poland, Jan Hoh, and Ananya Majumdar. I have also learned much from colleagues at Washington University in St. Louis, including Tim Lohman, the late Gary Ackers, and in particular, Rohit Pappu, who has taught me everything I know about polymer physics, and has been a great source of knowledge and ideas in statistical thermodynamics. I have also learned a tremendous amount of thermodynamics through the Gibbs conference on biothermodynamics and regular faculty participants, including Madeline Shea, Dorothy Beckett, Wayne Bolen, and in particular, Jack Correia. Finally, I have benefitted from interactions with faculty in the Protein Folding Consortium, including Ken Dill, Tobin Sosnick, Walter Englander, Heinrich Roder, Josh Wand, Susan Marqusee, Vijay Pande, Cathy Royer, and in particular, Bob Matthews. In addition to learning about the folding reaction through our group discussions, they have all been a source of support in finishing this text.

Finally I would like to acknowledge authors of key texts that I have learned from over the years. They include Peter Atkins, Don McQuarrie, and Terell Hill, Herbert Callen, Ken Dill, and Sarina Bromberg, all authors of general texts that I have read until their bindings gave out. With these I include the three-volume set by Charles Cantor and Paul Schimmel, the helix-coil bible by Doug Poland and Harold Scheraga, and the beautiful binding and linkage text by Jeffries Wyman and Stan Gill. None of my copies of these books is in very good shape any more, and I am better for it.

Note to Instructors

This book is designed for courses at the undergraduate and graduate level. To illustrate the flexibility of the book, I describe two separate courses that differ in emphasis and order of presentation. The outline for the undergraduate course begins by focusing on classical thermodynamics and uses the classical as a vehicle to teach the statistical. The outline for the graduate course begins by focusing on essential mathematics and statistics. These differences are motivated, in part, by the expectation that most undergraduates will have familiarity with classical (but not statistical) thermodynamic descriptions of both materials and (bio)chemical reactions, although this familiarity is often shallow and lacking in intuition. In contrast, graduate students in molecular biosciences programs often have a better understanding of classical thermodynamics and at least some exposure to statistical thermodynamics and have generally committed to biomolecular studies.

Example of an undergraduate course in biothermodynamics. At the undergraduate level, the book serves as a primary reading/teaching source for an upper-division course in biothermodynamics to biosciences majors. I have been teaching and refining such a course at Johns Hopkins for 20 years. This course is essential for students majoring in biophysics, but it also greatly benefits chemistry, biochemistry, molecular biology, and cell biology majors as well. One of the goals of the course is to present the theory that is normally provided in a traditional first-semester junior-level physical chemistry course in a context that provides direct application to complex and important problems in biological systems. After a few lectures on statistics and mathematics, the course develops classical thermodynamics, from laws to potentials, and from simple gasses to mixtures and reactions. Statistical examples are kept to a minimum at first, and are used to complement the classical equations. Some of the more advanced topics (irreversible processes, Legendre transforms, and Gibbs–Duhem analysis) can be skipped without loss of generality.

With classical thermodynamics in hand, students tackle the protein folding problem, and learn how to go from data to energies and the like, using a simple two-state model. To analyze more complex equilibria, the concepts, tools, and equations of statistical thermodynamics are introduced, focusing on the concepts behind the equations. After learning to write and analyze partition functions, students analyze complex reactions including the helix–coil transition and binding of multiple ligands to macromolecules. These applications give the students the confidence to analyze complex problems, and teach them how to model-build. Computer-aided learning is a key component of this course, as are the regular use of problem sets. In my course at Hopkins, I complement three 1-hour lectures per week (for 13 weeks) with a 1-hour "recitation" in the computer lab, where students learn to do computer-aided math, simulate random distributions, and analyze data, as well as engage in real-world activities such as throwing coins, tossing dice, and playing statistical-thermodynamics card games. Instructors who are interested in such hands-on teaching aids should feel free to contact me.

Example of a graduate course in biological thermodynamics (or first semester of a two-semester molecular biophysics series). At the graduate level, the book serves both as a reading resource and a problem-solving guide for a graduate course in biological thermodynamics. Such a course is a cornerstone in quantitative biology programs, including molecular biophysics and biochemistry, which are likely to see a significant growth in the next decade. In addition, the book has value in more traditional graduate programs in chemistry, chemical engineering, biological physics, and computational biology, and serves as a reading/problems source for independent study. The graduate course has two phases. In phase one, which covers the first third of the course, students receive parallel training in three disciplines: (1) partition function formalism of statistical thermodynamics, (2) simulation as a learning tool, and (3) statistical analysis of data. This phase develops the skills necessary to formulate equilibrium expressions, simulate data, and analyze the data to extract the relevant thermodynamic functions. In phase two, students use these skills to address a range of phenomena relevant to understanding biological function. These phenomena include two-state, three-state, and multistate conformational equilibrium, single-site ligand binding to a two-state equilibrium, binding of multiple independent ligands, binding of coupled ligands, chemical denaturation, proton titration, helix–coil transition theory, allostery, catalysis, and higher-order assembly reactions.

Author

Douglas E. Barrick is the T.C. Jenkins Professor in the Department of Biophysics at Johns Hopkins University. He earned his PhD in biochemistry at Stanford University (1993) and carried out postoctoral research in biophysics and structural biology at the University of Oregon (1993–1996). He has been honored as the recipient of the Beckman Young Investigator Award, the Helen Hay Whitney Postdoctoral fellowship, and Howard Hughes Medical Institute Predoctoral Fellowship. He has been an editorial board member of the journals *Protein Science* and *Biophysical Journal*, and he has been an organizer of the Gibbs Conference on Biothermodynamics. Research in his lab focuses on the study of protein physical chemistry, protein, folding, and macromolecular structure and assembly.

CHAPTER 1

Probabilities and Statistics in Chemical and Biothermodynamics

Goals and Summary

The goals of this chapter are to familiarize students with the concepts of probabilities and statistics to understand physical chemical principles and to analyze complex problems in chemical and biothermodynamics. The first part of the chapter will focus on events, outcomes, and their combinations. We will develop the concept of the probability distribution as a collection of probabilities for all possible outcomes. We will emphasize the differences between specific sequences of elementary events, and collections composed of specific sequences that share some overall property. This distinction is essential to statistical thermodynamics, where we are often limited to measurements of overall composition, where each composition is consistent with many different arrangements of molecules. The "indistinguishability" of these different arrangements is directly related to important thermodynamic concepts such as entropy.

In the process we will introduce a number of key probability distributions, including both discrete distributions (most importantly the binomial and multinomial distributions) and continuous distributions (emphasizing Gaussian and decaying exponential distributions). We will describe how to derive various average quantities from probability distributions; in subsequent chapters, such derivations provide a means to directly test and refine statistical thermodynamic models and learn about molecular systems.

One of the most important subjects in chemistry and biology is the reaction of molecules to form new molecules of different sizes, shapes, and types. New covalent bonds are formed, new configurations can be adopted, new complexes and assemblies are built, and old ones are taken apart and rearranged. The application of physical chemistry allows these kinds of transformations to be described quantitatively, providing access to underlying forces and mechanisms, and providing predictive power to describe how complex systems of molecules will react as conditions change.

Biochemical systems almost always involve very large numbers of molecules. Rather than describing the behavior of each individual molecule, which is difficult even for small systems (and yields much more information than is of practical value for large systems), we will seek to understand reactions and material transformations in terms of "*distributions*" of atoms and molecules into different reactive species. As examples, what fraction of an enzyme is bound with substrate, what fraction of the hemoglobin tetramer binds to oxygen (and over what arrangement of subunits), and what is the distribution of bound transcription factors over a collection of promoters? These examples are all linked—through "fraction" and "distribution" to the statistical concepts of populations and probabilities. To understand

Figure 1.1 Elementary and composite outcomes and their probabilities. In this example, four elementary events are shown, each with three exclusive outcomes (A, B, and C). Probabilities for each elementary outcome are given as, for example, $p(B_2)$, where B indicates the outcome, and the subscript 2 indicates the event. For the independent events considered in this book, probabilities of the same outcome occurring in different events are the same, so the elementary event subscript can be dropped. However, such event markers are useful for representing a specific sequence for a composite outcome (often called a "permutation") and its probability. For probabilities of sequence-independent composite outcomes (often called "combinations"), the numeral following the semicolon represents the total number of elementary events that are combined ($3 + 1 + 0$ in this example).

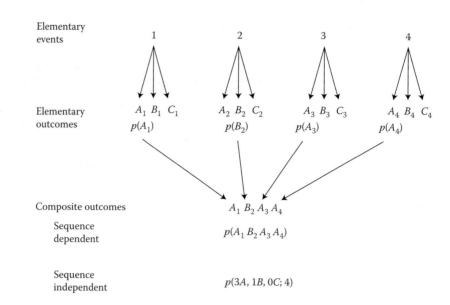

complex reaction thermodynamics, especially at the level of biological complexity, we must start with a good understanding of probabilities and statistics.

We will begin by reviewing basic concepts of probabilities and the distributions that govern them. Instead of starting with molecular probabilities, we will develop probability theory using familiar, everyday examples (games, weather, coins, and dice), because these examples are easy to think about. We will first develop ideas for "elementary events," which result in one of a number of mutually exclusive outcomes (**Figure 1.1**), and will then discuss "composite outcomes" in which outcomes of elementary events are combined. By taking into account all the possible outcomes, we come to the idea of the probability distribution. Though we will start with examples of outcomes are "discrete" (i.e., there are a collection of precise outcomes but no values in between), there are important distributions that are "continuous," allowing all possible values, and we will adjust our picture of distributions accordingly.

ELEMENTARY EVENTS

As described above, elementary events are random events in which a single outcome from a list of two or more possibilities occurs. Because each elementary event yields only a single outcome, these outcomes are referred to as "mutually exclusive." Examples of elementary events and their outcomes are whether or not it will rain tomorrow (here, "tomorrow" is the event, and the outcomes are "rain" and "no rain"), and whether or not a team will win a football game (here the game is the event, winning and losing are mutually exclusive outcomes). Other examples of elementary events, which we will use to develop equations for probabilities, include die rolls and coin tosses (which can be reasonably approximated to occur at an exact point in time). In probability theory, these elementary events are also referred to as "trials."

Relationship between probabilities, random samples, and populations

Returning to the examples involving a football game and the weather, the relationship between the probability and the outcome deserves a bit more explanation. Although in both examples a particular outcome may be described as likely or unlikely (it is likely to rain; the team is unlikely to win), neither outcome can be predicted with certainty. Tomorrow may be sunny all day; the underdog team may win. The only two certainties for each statement are that both "outcomes" will not come true (because of mutual exclusion), but one of them will (the outcomes are "collectively exhaustive"; e.g., the team will win or lose[†]).

In principle, probabilities can be described with very high levels of precision. However, the discussion above demonstrates that when we apply probabilities to single "real-world" events, we often have little quantitative confidence about outcome ahead of time, especially when outcomes have similar probabilities. When we record a single event, the distribution is precise, but our sampling of it is random.

Given this uncertainty of outcomes, one might question the usefulness of probability-based "predictions" for descriptions of the real world. Though it is better to have a probability to estimate a single event than not to have a probability,[‡] the real value of probabilities comes not from prediction of single events, but from prediction of many events in aggregate. If rainfall was monitored over some number (N_{tot}) of different days, we can define a fractional population of rainy days as

$$\frac{N_{rain}}{N_{rain} + N_{no\ rain}} = \frac{N_{rain}}{N_{tot}} = f_{rain} \qquad (1.1)$$

where N_{rain} and $N_{no\ rain}$ are the number of rainy and rain-free days in N_{tot} total days. In the framework in Figure 1.1, each day is an elementary event, and the fraction in Equation 1.1 describes a sequence-independent composite outcome. This kind of "fractional population," which we will use throughout this text, should not be confused with everyday usage, where population refers to a number greater than one (e.g., the number of people living on Earth at the same time). Our fractional populations will range from zero to one.

For modest values of N_{tot}, the fractional population also depends on the specific collection of elementary events we chose to sample (e.g., days sampled for rain). We can eliminate this uncertainty by making the number of days in our sample ($N_{rain} + N_{sun}$) very large. In this case, the fractional population converges to the actual probability:

$$\lim_{N_{tot} \to \infty} \frac{N_{rain}}{N_{tot}} = \lim_{n \to \infty} f_{rain} = p_{rain} = 0.6 \qquad (1.2)$$

Although collecting large averages for a single football game or a certain rainy day isn't practical, large averages are the norm in studies of molecular systems. Almost all experimental studies of molecules (especially thermodynamic studies) involve collections of large numbers (typically 10^{12} or more)[§] of molecules. Quantities of interest, such as the energy of simple physical systems, position of reaction for chemical systems, degree of assembly, extent of structure formation, and distribution of ligands over binding sites for biomolecular systems are all determined very precisely by the averaging concept expressed in Equation 1.2. Thus, when we measure populations within a large collection of molecules, it is equivalent to determining underlying probabilities for different molecular states.

This connection between probabilities and populations provides a bridge between theory and molecular models (which define probabilities) and experiment (which measures populations). To take advantage of this connection, we need to understand the basic features of probabilities, and how to combine elementary

---

[†] Barring a tie.

[‡] A gambler that placed only random bets, without taking into account the probability of various outcomes, would lose money very quickly.

[§] To appreciate the impracticality of applying this kind of large-number averaging to football games (or rainy days), it is worth writing out 10^{12} longhand: 1,000,000,000,000 games (days). On a schedule of one game per week, our two teams would have had to start playing each other the week the universe began.

probabilities to describe composite outcomes. Before developing equations to combine probabilities, we will show some pictures from "set theory" that help illustrate the process and types of combinations.

Set theory diagrams depict elementary and composite outcomes

We will begin a formal discussion of probability and statistics by focusing on simple elementary events that have two or more outcomes satisfying the "mutually exclusive" and "collectively exhaustive" criteria above (Figure 1.1, top level). Mutual exclusivity means that no more than one outcome can result from a single elementary event. Collective exhaustivity means no less than one of the specified outcomes can occur. Outcomes of elementary events can be depicted using diagrams found in set theory (**Figures 1.2, 1.3, and 1.4**). In these diagrams, outcomes are represented as areas inside shapes such as circles and wedges. Mutually exclusive outcomes are typically depicted by nonoverlapping shapes (see the "pie" and "Euler" diagrams in Figure 1.2), and the probabilities of the outcomes are often represented by the area bounded by each shape.

In addition to depicting the relationships between outcomes of elementary events, these set theory diagrams can be useful to motivate simple "combination" concepts (Figure 1.1, bottom). Combinations of mutually exclusive events are often depicted using "Venn" diagrams (Figure 1.4), which are useful for showing overlap in nonexclusive outcomes (see Figure 1.4); these diagrams sometimes sacrifice probability information (encoded as areas in Figure 1.2) in favor of generality and symmetry.

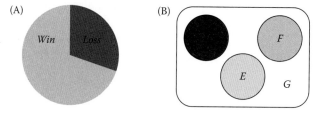

Figure 1.2 Depicting elementary and outcomes diagrammatically. (A) A pie chart and (B) a Euler diagram. Panel A depicts an event having two outcomes, panel B depicts an event with four outcomes. Colored areas are outcomes of a particular type, within the "sample space" of the elementary event. Relative areas are proportional to probability. In (A), the team has a higher probability of a win than a loss ($p_{win} = 0.7$; $p_{loss} = 0.3$). In (B), the relative probabilities of outcomes D, E, and F are the same, whereas the probability of the fourth outcome (G the white area outside the circles but bounded by the rectangle) is larger. In the pie diagram, there is no "outside" probability; the outcomes bounded by the circle are collectively exhaustive.

Figure 1.3 How exclusive outcomes of a single elementary event combine. The pie chart (A) and Euler diagram (B) are the same as in Figure 1.2. The "or" combination, in this case given by the union (∪), is designated with light shading. The "and" combination, which would be determined by intersection (∩), is absent from both representations; that is, it is zero (Equation 1.4).

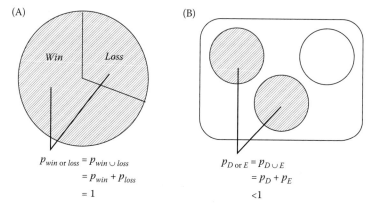

$$p_{win\ or\ loss} = p_{win\ \cup\ loss}$$
$$= p_{win} + p_{loss}$$
$$= 1$$

$$p_{D\ or\ E} = p_{D\ \cup\ E}$$
$$= p_D + p_E$$
$$< 1$$

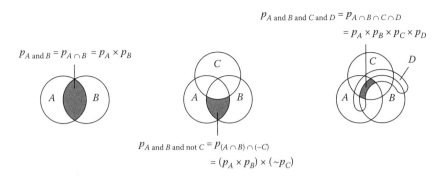

$$p_{A \text{ and } B \text{ and } C \text{ and } D} = p_{A \cap B \cap C \cap D}$$
$$= p_A \times p_B \times p_C \times p_D$$

$$p_{A \text{ and } B} = p_{A \cap B} = p_A \times p_B$$

$$p_{A \text{ and } B \text{ and not } C} = p_{(A \cap B) \cap (\sim C)}$$
$$= (p_A \times p_B) \times (\sim p_C)$$

Figure 1.4 Three "and" combinations of independent outcomes from different events, depicted as intersections in Venn diagrams. The areas bounded by each circle (and by the crescent on the right) represent outcomes of separate events. Outcomes can be of different types (e.g., a win and a rainy day, Equation 1.7) or they can be of the same type from different elementary events (e.g., rain on two different days, Equation 1.4, or heads on independent coin tosses). The "and" combination is represented as the intersection of the two outcomes (shaded; left, center), and is calculated as the product of the two probabilities. The product rule easily generalizes to more than two outcomes (right, although depicting a large number of outcomes is rather impractical) and Venn diagrams can also depict "*not*" conjunctions (referred to in set theory as the "complement" of an outcome; $\sim C$, center). Unlike the Euler diagram shown in Figure 1.2, the Venn diagram does not typically depict an external boundary, although the region outside of each outcome is considered to be nonzero. For example, the area outside of A represents outcomes where A did not occur ($\sim A$). For a detailed description of Venn diagrams and a comparison with those from Euler, see Edwards, 2004.

In addition to providing nice pictures, set theory gives a concise way to write equations for the probabilities for various combinations of elementary events. Both the underlying concepts and formal equations for such combinations are useful for analyzing large collections of objects, such as complex biomolecules, where drawing diagrams like those in Figures 1.2 through 1.4 would be impractical.

HOW PROBABILITIES COMBINE

The rules for combining different outcomes and events and calculating the probabilities of these combinations falls under a branch of mathematics called "combinatorics." A key question in both combinatorics and thermodynamics is "given the probabilities of elementary outcomes, what is the probability of a specific combination of outcomes?"

The two most common ways that outcomes are combined is the "*and*" and the "*or*" conjunctions. The meaning of the "*and*" conjunction is that both outcomes occur, and is equivalent to the idea of "intersection" in set theory (represented by the symbol $\cap$). The meaning of the "*or*" conjunction, which is related to the idea of "union" in set theory (represented by the symbol $\cup$), depends on its context. In English usage, "*or*" typically means one or the other outcome occurs, but not both,[†] whereas in logic, "or" often includes both outcomes. Here, we will stick with the English usage.

To calculate probabilities for combinations of outcomes, it is generally the case that the "*and*" conjunction multiplies the probabilities of individual outcomes, whereas the "*or*" conjunction adds the probabilities of individual outcomes. However, this generality depends on whether outcomes are mutually exclusive, as

---

[†] An exception in English is when *or* is used to mean *that is* (for example, "Arachnophobia, *or* fear of spiders, decreases one's enthusiasm to sleep in a cave.")

with outcomes from a single elementary event (Figure 1.1, top), or are independent (as is often the case for combinations from separate events).[†] We first consider the simple combination rules for exclusive outcomes within single events, and then use these principles to calculate probabilities for combined outcomes of different events.

Combining probabilities for mutually exclusive outcomes within a single events

For mutually exclusive outcomes from the same elementary event, the "or" combination is very simple: probabilities of exclusive outcomes add. For the examples in Figure 1.2, we can represent "or" combinations as follows:

$$p_{win\,or\,loss} = p_{win \cup loss} = p_{win} + p_{loss} = 0.7 + 0.3 = 1$$
$$p_{D\,or\,E} = p_{D \cup E} = p_D + p_E < 1 \tag{1.3}$$

Because winning and losing in Figure 1.2A are collectively exhaustive, the "or" combination gives a probability of one—nothing else can happen. In the example in Figure 1.2B, the "or" combination applied to outcomes D and E gives a probability less than one (but greater than zero), because there are two additional outcomes. For the win–loss example, the logical and English "or" combinations are equivalent because there is no overlap (the outcomes are mutually exclusive).[‡] These two unions are shown using gray shading in Figure 1.3.

In contrast, the "and" combination for mutually exclusive outcomes of a single elementary event requires no formula to remember, other than this simple identity:

$$p_{win\,and\,loss} = p_{win \cap loss} = 0$$
$$p_{d\,and\,e} = p_{D \cap E} = 0 \tag{1.4}$$

Example 1.1: A single coin toss as an elementary event

A normal coin toss has two possible mutually exclusive outcomes: "heads" and "tails." We assume that the coin is "fair;" that is, the chances of heads and tails are equal ($p_H = p_T$). Moreover, because these two outcomes are collectively exhaustive (we assume the chance of the coin landing on its side is zero), $p_H = p_T = 1/2$. By applying the rule for "and" combinations to mutually exclusive outcomes (Equation 1.4), the probability of observing both heads and tails in a single event (toss) is

$$p_{H\,and\,T} = p_{H \cap T} \equiv 0 \tag{1.5}$$

By applying the rule for "or" combinations to mutually exclusive outcomes (Equation 1.3), the probability of observing heads or tails in a single event (toss) is

$$p_{H\,or\,T} = H \cup T = p_H + p_T = \frac{1}{2} + \frac{1}{2} = 1.0 \tag{1.6}$$

consistent with our assertion that these two outcomes are collectively exhaustive.

[†] The details of combination also depend on whether or not the outcomes to be combined influence one another's probabilities. Here we will assume they do not.

[‡] If there were overlap, the union would include the common area (or "intersection") twice, and a correction would be required.

Example 1.2: A single site for binding ligands on a protein as an elementary event

Consider a protein molecule with a single site that can bind two different types of ligands (*A*, black circle, and *B*, gray circle):

The scheme above assumes that only one ligand can bind at a time; that is, the binding of ligands *A* and *B* are exclusive. The strength of interaction of ligands *A* and *B* can be represented by numerical binding constants K_A and K_B, respectively. As we will show in Chapter 13, the probabilities of binding of ligands *A* and *B* are

$$p_A = \frac{K_A[A]}{1 + K_A[A] + K_B[B]}; \quad p_B = \frac{K_B[B]}{1 + K_A[A] + K_B[B]}$$

Assuming values of $K_A = 10^6 \ M^{-1}$ and $K_B = 10^7 \ M^{-1}$, respectively, and free ligand concentrations $[A] = [B] = 3 \times 10^{-7} \ M$ (0.3 μM), we can calculate numerical values for the probability of binding to each ligand as

$$p_A = \frac{10^6[3 \times 10^{-7}]}{1 + 10^6[3 \times 10^{-7}] + 10^7[3 \times 10^{-7}]} = \frac{0.3}{1 + 0.3 + 3} \approx 0.07$$

$$p_B = \frac{10^7[3 \times 10^{-7}]}{1 + 10^6[3 \times 10^{-7}] + 10^7[3 \times 10^{-7}]} = \frac{3}{1 + 0.3 + 3} \approx 0.70$$

The probability of having no ligand bound can be obtained by evaluating all the collectively exhaustive outcomes together:

$$p_A + p_B + p_O = 1;$$
$$p_O = 1 - p_A - p_B = 1 - 0.07 - 0.7 = 0.23$$

The probability of having ligand *A* *or* ligand *B* bound is

$$p_{A \ or \ B} = p_{A \cup B} = 0.70 + 0.07 = 0.77$$

The probability of having ligand *A* and ligand *B* bound is

$$p_{A \ and \ B} = p_{A \cap B} = 0$$

In general, exclusive binding is likely to result when there is "steric" overlap between bound ligands (as is the case when ligands bind a single site), because the energetic cost for such overlap is untenable.

The probability that both mutually exclusive outcomes occur is zero. Consistent with this, there is no overlap between exclusive outcomes, as is depicted in Figure 1.3. This differs from the "and" combination for *independent outcomes of separate events*, which involves multiplication of individual probabilities, described below.

Two examples of elementary events that can be used to illustrate these probability combinations are the coin toss and the six-sided die roll.

Combining probabilities for outcomes from multiple independent events

In thermodynamic analysis of biological molecules, we typically focus on large collections of molecules. A key component of such analysis is to take a known (or postulated) set of probabilities for individual molecules to be in different configurations, and calculate the probability that the entire collection will be distributed in various ways. Although there are often a few important constraints on the collection of molecules as a whole, each molecule can often be treated as an "independent event," and the various available configurations as "outcomes," each with an associated probability. Though a collection of molecules and the configurations they adopt differ conceptually from the time-dependent events we have described above (as with red and white balls in a jar, molecules exist whether we analyze them or not, whereas coin tosses are specific events in time, initiated by us), the underlying statistics are the same. Thus, we will continue with the familiar examples above (weather, coins, dice) to develop rules for combination of independent events.

"And" combinations of independent outcomes of separate events involve multiplication

For independent outcomes from two different events, the probability of an "*and*" combination is obtained by a simple multiplication of the probabilities describing each outcome. In terms of set theory, this amounts to the intersection of the two sets of outcomes ($\cap$, Figure 1.4). These events may either be of the same type (the weather on two consecutive days, or two different coin tosses) or they may be of different types (the weather and a football game, or a coin toss and a die roll). Using the example above, the probability of the team winning and the day being rainy is

$$p_{win\,and\,rain} = p_{win\cap rain} = p_{win}\times p_{rain} = 0.6\times 0.7 = 0.42 \tag{1.7}$$

If the probability of rain tomorrow is the same as that for today,

$$\begin{aligned} p_{rain\,today\,and\,tomorrow} &= p_{rain\,today\cap rain\,tomorrow} \\ &= p_{rain\,today}\times p_{rain\,tomorrow} = 0.6\times 0.6 = 0.36 \end{aligned} \tag{1.8}$$

As with combinations described above, we can depict combinations involving independent outcomes using diagrams from set theory. For independent events, the Venn diagram is particularly useful, illustrating all possible intersections of the different outcomes (Figure 1.4). As shown for the intersection on the right side of Figure 1.4, the probability of an "and" combination can easily be extended to more than two independent outcomes, simply by multiplying the probabilities of each elementary outcome.

Extending to N independent events, if the probability of getting outcome O_1 on the first event is p_{o_1}, the probability of O_2 on the second event is p_{o_2}, and so on, then the probability of getting O_1 "and" O_2 "and" so on is the product of the individual probabilities, that is,

$$p_{O_1\,and\,O_2\,and...and\,O_N} = p_{O_1}\times p_{O_2}\times\cdots\times p_{O_N} = \prod_{i=1}^{N}p_{O_i} \tag{1.9}$$

In Equation 1.9, all outcomes of the N events are explicitly represented. If some outcomes are the same (e.g., N_A outcomes of type A, N_B total outcomes of type B, and so on), Equation 1.9 simplifies to

$$p_{A_1\,and\,B_2\,and\cdots and\,A_n} = (p_A)^{N_A}\times(p_B)^{N_B}\times\cdots\times(p_N)^{N_N} \tag{1.10}$$

The numbers are left in place on the left-hand side to emphasize that we are talking about one *particular sequence* here. If all the outcomes are the same (e.g., A), the expression simplifies further to

$$p_{A_1 \text{ and } A_2 \text{ and}\dots A_n} = (p_A)^n \tag{1.11}$$

This multiplication rule is obviously different from the situation for mutually exclusive events, where the intersection was zero by assertion. In between the independent and mutually exclusive limits, outcomes may influence one another. For example, a prediction of a high chance of rain may influence the chance a person on the street has an umbrella. The rules for such "conditional probabilities" and their combinations are slightly more complex, and will not be developed here. Interested readers should see Blom, 1989.

"Or" combinations of independent outcomes for separate events involve addition and multiplication

As described above, the English conjunction "*or*" combines only the probabilities corresponding to the occurrence of one or the other outcome, but not both, whereas set theory and formal logic (the union, ∪) includes both. We adhere to the more common usage (not both). For exclusive outcomes of a single event, these two representations of "or" are the same (zero intersection), and the corresponding probability is calculated simply by adding the probabilities directly, that is, as the union of the outcomes (see Equation 1.3, Figure 1.3). For independent outcomes of different events, there is a nonzero chance that both outcomes can occur (shaded regions, Figure 1.4). A simple addition formula of the form of Equation 1.3 would include this intersection, so we must remove it by subtraction. In fact, to get the English "or" (not both) combination we must subtract the intersection *each time* we add a probability from an independent event.

For the example involving the weather and the game, to calculate the probability the team will win "or" it will be rainy, we only want to include the probability of a win on a dry day, and combine it with the probability of a loss and rain. We can apply these restrictions by subtracting the intersection ($p_{win \cap rain} = p_{win} \times p_{rain}$) away from each outcome before combining:

$$p_{win \text{ or } rain} = \{p_{win} - (p_{win} \times p_{rain})\} + \{p_{rain} - (p_{win} \times p_{rain})\}$$
$$= p_{win} + p_{rain} - 2(p_{win} \times p_{rain}) \tag{1.12}$$

The English logical "or" (the union) for these same outcomes is (Problem 1.1)

$$p_{(win \cup rain)} = p_{win} + p_{rain} - 1(p_{win} \times p_{rain})$$
$$= p_{win \text{ or } rain} + (p_{win} \times p_{rain}) \tag{1.13}$$

Alternatively, we could multiply using "not" statements (Problem 1.2). We will represent an event in which an outcome is not obtained (but another one is) using the "∼" symbol (e.g., ∼*rain* represents an outcome other than *rain*). For only two weather outcomes, the probability of sun is equal to the probability of not having rain ($p_{sun} = p_{\sim rain} = 1 - p_{rain}$). Working with "not" statements is particularly useful in "or" combinations where there are more than two possible outcomes, where the corrections that eliminates the overlap of outcomes become rather complex. Although additional outcomes add complexity to the expression of collective exhaustiveness ($1 = p_A + p_B + p_C + \cdots$), the "not" statement remains simple: $p_{\sim A} = 1 - p_A$ (the right-hand side of this equality represents all the other outcomes implicitly, no matter how many outcomes there are). To get the common "or (but not both)" combination, we combine the probability of getting one outcome in one event and not the others (independent events, allowing a simple multiplication)

Table 1.1 Rules for combining probabilities

	And	**Or**
Mutually exclusive outcomes of a single event	$p_{A \text{ and } B} = p_{A \cap B} = 0$	$p_{A \text{ or } B} = p_{A \cup B} = p_A + p_B$
Independent outcomes of two events	$p_{A \text{ and } B} = p_{A \cap B} = p_A \times p_B$	$p_{A \text{ or } B} = p_{A \cup B} - p_{A \cap B}$ $= p_A + p_B - 2p_A \times p_B$
Independent outcomes of n events	$p_{O_1 \text{ and } O_2 \text{ and...and } O_n} = p_{O_1} \times p_{O_2} \times \cdots \times p_{O_n} = \prod_{i=1}^{n} p_{O_i}$	$p_{O_1 \text{ or } O_2 \text{ or... or } O_n} = \sum_{i=1}^{N} \left(p_{O_i} \prod_{j \neq i}^{N} (1 - p_{O_j}) \right)$

The rules for combination differ significantly for exclusive versus independent events. The *and* combination can easily be generalized to multiple independent outcomes, although the *or* combination involves multiple corrections. Note that the *or* combination for separate events also describes the *or* combination for exclusive outcomes of a single event, since the overlap term is zero. For the *and* combination independent outcomes of n events, the notation on the first line, right-hand side, represents a set intersection using an index i in analogy with a summation. Likewise second line is the multiplication equivalent to a summation, meaning "multiply all terms" from $i = 1$ to n.

"or" the probability of getting the next outcome in its event and not the others (these composite outcomes are mutually exclusive, allowing simple addition) (**Table 1.1**).

The combination rules developed here can easily be applied to simple trials such as coin tosses and die rolls, where probabilities of different outcomes can be stated precisely. Direct application of the rules above comes when the outcomes can be *distinguished* from one another, either by making unique marks on each coin or die, or by conducting the trials in sequence and noting the order, that is, "first toss, second toss...." (Problem 1.2a). Another simple example of independent distinguishable trials is die rolls (Example 1.3).

Example 1.3: Combining outcomes for the multiple die rolls

If we roll two separate dice that we can distinguish, either by a marking or by keeping track of the sequence, we can calculate the probability of getting a particular sequence of two outcomes (the *and* combination) by the same multiplication formula. For example, the probability of getting a one on the first role *and* a six on the second is

$$p_{1_1 6_2} = p_{1_1} \times p_{6_2} = \frac{1}{6} \times \frac{1}{6} = \frac{1}{36} \tag{1.14}$$

In general, the probability of obtaining a sequence of N particular outcomes is $(1/6)^N$. Likewise, we can calculate the probability of getting one value *or* the other value (e.g., a one on the first roll *or* a six on the second roll, but not both) as

$$p_1^1 \text{ or } p_6^2 = p_1^1 + p_6^2 - 2(p_1^1 \times p_6^2)$$
$$= \frac{1}{6} + \frac{1}{6} - 2\left(\frac{1}{6} \times \frac{1}{6} \right) \tag{1.15}$$
$$= \frac{10}{36}$$

The factor of two in front of the subtracted term reflects the fact that there were two rolls, and on each roll where the outcome was achieved (e.g., a one on the first roll, with 1/6 chance), there was a 1/6 chance the second roll would also match, which would violate the "both" exclusion.

PERMUTATION VERSUS COMPOSITION

The examples above describe how to calculate probabilities of getting a sequence of outcomes from multiple independent events using the *and* operator. The fact that we are discussing "sequences" means that we can tell the difference between getting the same "set" of outcomes in different orders. For example, the sequence of three coin tosses involving two heads and one tail, $T_1H_2H_3$ is different from $H_1T_2H_3$, which is different from $H_1H_2T_3$ (although each sequence has an equal probability of $(1/2)^3$). The number of sequences for a particular set of outcomes is equivalent to the number of "*permutations*" of the outcomes; that is, the number of orders in which the outcomes can occur.

But what if we cannot tell the events apart? This situation when unmarked coins flipped at the same time, and when identical dice thrown together. In this situation, we can only count how many of each outcome we observe, which we refer to as a "composition." Combinations that lack sequence (or permutation) information are common in studying the properties of collections of molecules. In such collections, the identities of the individual molecules cannot be distinguished from one another. Although we can measure how many molecules of a particular covalent structure equilibrate in different ways (e.g., how the molecules are distributed in terms of their energies, their extent of binding to ligands, their extent of folding, or their extent of assembly) most experiments don't distinguish individual molecules from each other. Fortunately, information about composition is sufficient to extract most of the thermodynamic parameters we will be interested in, although we must account for the underlying permutation statistics to connect experimental studies with thermodynamics (and quantities related to the entropy in particular).

How do we calculate the probabilities of observing a given composition of independent outcomes? The solution to this problem uses the statistical ideas we have discussed above. We simply sum all of the different probabilities for each sequence that is consistent with the specified composition (we sum because different sequences are mutually exclusive). If, as is often the case, each sequence has the same probability, this is equivalent to multiplying the probability of one of the sequences with the number of different sequences that have the specified composition. Here we will analyze coin tosses to derive the probability of compositions of just two outcomes. For two outcomes, the number of sequences for each composition is given by the "binomial coefficients." Later, when we consider more than two outcomes (e.g., rolling multiple die), the number of sequences consistent with a particular composition will be given by closely analogous "multinomial coefficients."

The key idea is that the probability of observing a particular composition is equal to the probability of any particular sequence that is consistent with the composition times the total number of consistent outcomes. This is because for independent outcomes, each sequence is equally probable. When three coins are tossed, the sequence $T_1H_2H_3$ is as likely as $H_1T_2H_3$. Moreover, each sequence is mutually exclusive. That is, it is not possible to get the sequence $T_1H_2H_3$ and $H_1T_2H_3$. We can use the rule above for *or* combinations of mutually exclusive events (Equation 1.3) to represent the probability of a particular composition as the sum of the probabilities of all consistent sequences. Since the sequences all have the same probabilities, this is equivalent to the probability of any one of the sequences times the number of consistent sequences. For coin tosses, this is equal to

$$p(N_H, N_T; N_{tot}) = \sum_{\substack{consistent \\ sequences}} p_{seq} = \sum_{\substack{consistent \\ sequences}} p_H{}^{N_H} \times p_T{}^{N_T}$$

$$= W(N_H, N_T; N_{tot}) \times p_H{}^{N_H} \times p_T{}^{N_T}$$

$$= W(N_H, N_T; N_{tot}) \times p_H{}^{N_H} \times (1 - p_H)^{(N_{tot} - N_H)} \qquad (1.16)$$

$$= W(N_H, N_T; N_{tot}) \times 0.5^{N_H} \times (0.5)^{(N_{tot} - N_H)}$$

$$= W(N_H, N_T; N_{tot}) \times 0.5^{N_{tot}}$$

The summand on the right-hand side is the probability of each and every sequence with N_H heads and N_T tails. $W(N_H, N_T; N_{tot})$ is the number of sequences (or ways[†]) that N_H heads can be obtained in N_{tot} tosses, with the remainder being tails. The last equality in Equation 1.16 uses the constraint that $N_H + N_T = N_{tot}$ to include the total number of tosses explicitly.[‡]

Calculating the number of permutations

To obtain the probability of a particular coin-toss composition, ignoring sequence, we need to know how many unique permutations of N_H heads and $N_T = N_{tot} - N_H$ tails exist. Here is a step-by-step solution to this problem (see **Figure 1.5**):

1. Arrange all of the N_H heads into a single stack, separate from the tails.
2. Build a long coin holder that has N_{tot} slots, each of which can hold a single coin. Label each slot in the coin holder from 1 to N_{tot}.
3. Take one of the heads from the pile and count out all the ways (slots) it can be placed in the holder. There are N_{tot} different slots, so the number of ways the first coin can be placed is N_{tot}. Note that if there were only one head and $N_{tot} - 1$ tails, this would be the answer to our calculation:

$$W(1; N_{tot}) = N_{tot} \tag{1.17}$$

(The tails can only be put in the remaining slots one way, so they do not contribute explicitly to the formula.)

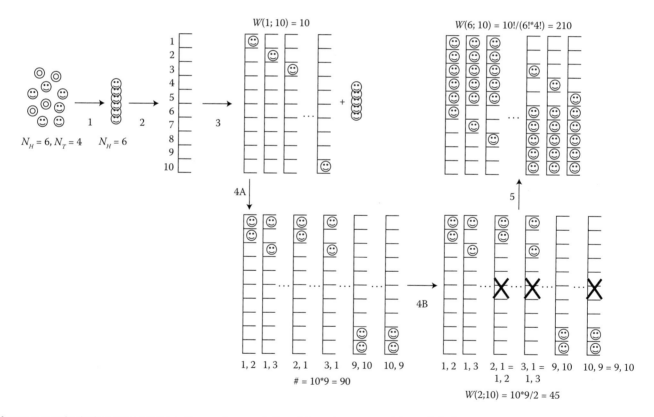

Figure 1.5 Calculating the number of ways (permutations) to get a composition of six heads in ten tosses. As described in the text, the heads are collected in a group (step 1). The first head can be placed in ten possible slots in a coin holder (step 3). For each of these, the next head in the group can be placed in all of the nine remaining slots (step 4A), giving ninety possibilities. However, each of these possibilities will be identical to another arrangement where the coins were placed in the opposite order. Eliminating the duplicates (step 4B) gives 45 arrangements. Steps 4A and 4B are repeated until the last head has been placed in the coin holder (step 5).

[†] W goes by many names. In addition to permutations and sequences, which have been introduced here, W is often referred to as "ways," "multiplicity," and "degeneracy," depending on the context.

[‡] And to eliminate N_T and p_T from the expression. Note that this substitution only works when $p_H = p_T$.

4A. Now for each of the N_{tot} placements of the first coin, take in the next heads on the stack and count all the number of remaining ways it can be placed. Since one of the N_{tot} slots will be taken, there will be only $N_{tot} - 1$ slots remaining. Since this is true for each of the N_{tot} arrangements with the first coin, placing the second heads can be done in $N_{tot}*(N_{tot} - 1)$ ways.

4B. However, there is a problem with our calculation. Because we cannot tell heads apart from one another, each of the $N_{tot}*(N_{tot} - 1)$ placements will be identical to one of the others. We have counted the placement of the first heads at slot i and the second at slot j separately from placement of the first heads at slot j and the second at slot i, although we cannot tell these two placements apart. This double-counting will occur for every i,j pair (Figure 1.5); that is, we counted every unique pair *twice*. Thus, we must divide by two:

$$W(2; N_{tot}) = \frac{N_{tot} \times (N_{tot} - 1)}{2} \tag{1.18}$$

5. Now repeat steps 4A and 4B for all of the remaining heads. The next (third) heads can be placed in $(N_{tot} - 2)$ positions (step 4A), but its placement generates *three* identical arrangements for every triplet of i,j,k slots, so we must correct by a factor of three. Continuing the placement of the remaining $N_H - 3$ heads, and correcting each step for overcounting up to a factor N_H gives

$$\begin{aligned} W(N_H; N_{tot}) &= \frac{N_{tot} \times (N_{tot} - 1) \times (N_{tot} - 2) \times \cdots \times (N_{tot} - N_H + 1)}{1 \times 2 \times 3 \times \cdots \times N_H} \\ &= \frac{N_{tot} \times (N_{tot} - 1) \times (N_{tot} - 2) \times \cdots \times (N_{tot} - N_H + 1)}{N_H!} \end{aligned} \tag{1.19}$$

where the notation $N!$ is referred to as the "factorial of N.[†]" We can simplify this equation by expressing the numerator as a factorial as well:

$$\begin{aligned} N_{tot} &\times (N_{tot} - 1) \times (N_{tot} - 2) \times \cdots \times (N_{tot} - N_H + 1) \\ &= N_{tot} \times (N_{tot} - 1) \times (N_{tot} - 2) \times \cdots (N_{tot} - N_H + 1) \\ &\quad \times \frac{(N_{tot} - N_H) \times (N_{tot} - N_H - 1) \times \cdots \times 3 \times 2 \times 1}{(N_{tot} - N_H) \times (N_{tot} - N_H - 1) \times \cdots \times 3 \times 2 \times 1} \\ &= \frac{N_{tot}!}{(N_{tot} - N_H)!} = \frac{N_{tot}!}{N_T!} \end{aligned} \tag{1.20}$$

Combining Equations 1.23 and 1.24, we get the expression

$$\begin{aligned} W(N_H; N_{tot}) &= \frac{N_{tot}!}{N_H! \times (N_{tot} - N_H)!} = \frac{N_{tot}!}{N_H! \times N_T!} \\ &= \binom{N_{tot}}{N_H} = W(N_H, N_T; N_{tot}) \end{aligned} \tag{1.21}$$

The second line in Equation 1.21 shows a shorthand notation for the permutation expression, which is often read "N_{tot} *choose* N_H". These coefficients are used in the binomial theorem and Pascal's Triangle (**Box 1.1**).

Finally, note that we have not explicitly included the number of tails (N_T) in our expression for the number of permutations, nor do we need to. For each unique arrangement of heads, there are N_T remaining slots, and there is only one way to add the N_T tails to fill those slots, since like heads, the tails are all the same. If it pleases you to include N_T in the number of permutations, you can do so, but it does not add anything more than specifying the total number of throws (N_{tot}; see the various versions of Equation 1.25). N_T in the expression on the RHS of Equation 1.19.

---

† $N!$ is the product of the integer N and all smaller integers down to one: $N! = 1 \times 2 \times 3 \times \cdots \times N$. Although it is not obvious from this definition, the quantity $0!$ is taken to be one (as is $1!$).

Box 1.1: Permutations, binomial coefficients, and Pascal's triangle

The formula for the number of permutations of two mutually exclusive and collectively exhaustive outcomes has applications in mathematics and physics, in addition to calculating probabilities for coin-toss statistics. One important example is the *"binomial theorem,"* which states that an expression of the type $(a + b)^N$, where a and b are numbers and N is a positive integer, can be expressed using the sum

$$(a+b)^N = \sum_{i=0}^{N} \frac{N!}{i!(N-i)!} a^{N-i}b^i = \sum_{i=0}^{N} \binom{N}{i} a^{N-i}b^i \qquad (1.22)$$

You are probably familiar with these expressions for small values of N (0,1,2,3):

$$
\begin{aligned}
N=0: &\quad (a+b)^N = 1 \\
N=1: &\quad (a+b)^N = a+b \\
N=2: &\quad (a+b)^N = a^2 + 2ab + b^2 \\
N=3: &\quad (a+b)^N = a^3 + 3a^2b + 3ab^2 + b^3
\end{aligned}
\qquad (1.23)
$$

The coefficients from the binomial expansions on the right-hand side of Equation 1.23 can be arranged in a simple, easy-to-calculate pattern called "Pascal's Triangle" (**Figure 1.6**).

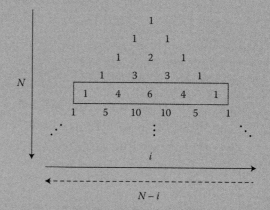

Figure 1.6 Pascal's triangle as a way to generate binomial coefficients. The outer edges of the triangle are always ones, and entries inside the triangle are the sum of the two flanking elements in the row above. Applied to coin tosses, rows down the page correspond to increasing numbers (N_{tot}) of tosses, whereas entries from right to left correspond to increasing numbers of heads in combination. For example, the red boxed row corresponds to combinations for $N_{tot} = 4$ tosses, with from zero to four heads (four to zero tails) from left to right.

Discrete probability distributions

The examples involved events where the outcomes can have only a few possible values. We can roll a die and get a 3 or a 4, but we can't get the value 3.5 in any single roll. Likewise, when we toss a coin single coin, we can get an H or a T, but nothing in between. These types of outcomes are called discrete. And although going from elementary to composite events increases the number of values possible (e.g., 101 values for 100 coin tosses), the outcomes remain discrete.

In this section, we will think about discrete outcomes in a more holisitic way. Instead of thinking about the probability of getting *one particular combination* (e.g., six heads in ten tosses), we will think about the arrangement of all of the possible combinations (e.g., from zero to 101 heads for 100 tosses). A complete set of such outcomes is called a "distribution," and when outcomes are discrete, we have a "discrete distribution."

In addition to games of chance, discrete distributions are often used in describing experiments where individual events are "counted," like radioactive decay measurements, and in single-molecule measurements. Discrete distributions are also important when distributions of molecules are studied, and the numbers of molecules with different types of properties are grouped together. For example, when different species equilibrate by chemical reaction, we can treat the reactants and products as discrete and distinct. We will encounter a number of biomolecular reactions (binding, folding, and assembly) that have the same statistics as distributions that govern games of chance, such as the coin and die examples above. Here we discuss some common discrete distributions, starting with one that we have already discussed, the binomial distribution.

The binomial distribution

Probably the most important discrete distribution in statistical thermodynamics is the binomial distribution.[†] When N_{tot} events that have two mutually exclusive outcomes (A and B) are combined without regard for sequence, the probabilities of different compositions (N_A and N_B, respectively) are given by the binomial distribution. The binomial distribution is calculated as

$$p(N_A; N_{tot}) = W(N_A; N_{tot}) \times p_A^{N_A} \times (1 - p_A)^{(N_{tot} - N_A)}$$
$$= \frac{N_{tot}!}{N_A!(N_{tot} - N_A)!} \times p_A^{N_A} \times (1 - p_A)^{(N_{tot} - N_A)} \qquad (1.24)$$

p_A is the probability for outcome A. p_B and N_B are not included explicitly in Equation 1.24; thus, the distribution is given using a single composition variable. However, p_B and N_B are implied by the relationships $p_B = 1 - p_A$ and $N_B = N_{tot} - N_A$.

The binomial distribution (Equation 1.24) is identical to the coin toss expression developed above (Equations 1.16 and 1.21). Although the two elementary outcomes in the coin toss have equal probability, the binomial distribution also applies to outcomes with unequal probabilities (as long as they sum to one). Two binomial distributions are shown in **Figure 1.7**, with equal (black) and unequal (white) probabilities.

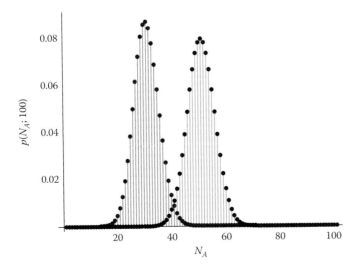

Figure 1.7 Binomial distributions with equal (black) and unequal (red; $p_A = 0.3$, $p_B = 0.7$) probabilities. The total number of events is 100. Each point corresponds to one of the 101 combinations of N_A and $N_B = 100 - N_A$ outcomes.

[†] And its big brother, the multinomial distribution (see below).

One notable feature of the binomial distribution is that its variance, $\sigma_{N_A}^2$, is given as

$$\begin{aligned} \sigma_{N_A}^2 &= N_{tot} \times p_A (1 - p_A) \\ &= N_{tot} \times p_A \times p_B \end{aligned}$$

(1.25)

where the symbols are as in Equation 1.24 (see Bevington, 1969). Thus, the width of the binomial distribution, which is proportional to the square root of the variance (σ_{N_A}, sometimes referred to as the "root-mean-squared deviation," or RMSD), increases with the square root of N_{tot} (Problem 1.20). An important consequence of this square root dependence is that if binomial distributions with different N_{tot} values are plotted over their whole range of possible values (from $N_A = 0$ to $N_A = N_{tot}$), the apparent spread *decreases* with increasing N_{tot} (Problem 1.6). This square root dependence is a central feature of random molecular processes such as diffusion and the spatial separation of subunits of disordered polymer chains.

Another implication of Equation 1.25 that has important thermodynamic implications is that for a fixed number of events N_{tot}, the width of the distribution is maximized when $p_A = p_B = 0.5$, and goes to zero when either p_A or p_B goes to zero (Problem 1.11). As will be developed in Chapters 8 and 10, this principle is closely connected to thermodynamic properties such as heat capacity, which is proportional to the spread among states with different energies.[†]

The Poisson distribution

Another discrete distribution that is commonly observed in molecular (and non-molecular) systems is the Poisson distribution. This distribution describes the probability that a number N_c of random, independent occurrences, or "counts" will happen in a fixed interval (often a period of time). Examples include the arrival of cosmic rays per unit time, the number of mutations in a fragment of irradiated DNA, the number of receptor proteins in a small patch of cell's membrane, and the number of text messages received by a college student per minute.[‡]

The Poisson distribution has many features in common with the binomial distribution. Both distributions are discrete, and both can be evaluated in terms the probability of success of achieving some type of outcome. The binomial distribution could be used to describe the probability associated with these types of observations, as long as the elementary events are independent. The main difference is that there is no limit to the number of outcomes that can be obtained in a Poisson trial ($N_c \to \infty$), whereas a binomial trial is bounded by N_{tot}. The Poisson distribution can be derived as an approximation to the binomial distribution by treating N_{tot} as very large, and p_c (and thus, N_c) as very small (see Bevington, 1969, for a derivation[§]). This treatment results in the formula

$$p(N_c; \mu) = \frac{\mu^{N_c} e^{-\mu}}{N_c!}$$

(1.26)

where μ is the mean of the number of outcomes within the sampling interval. Poisson distributions with different averages are shown in **Figure 1.8**. One notable feature of the Poisson distribution is its asymmetry, which is most obvious for small values of μ (see black and red bars, Figure 1.8). For large average values, the distribution loses this asymmetry, and can be closely approximated by the normal (Gaussian) distribution (see below).

Another noteworthy feature of the Poisson distribution is that its variance ($\sigma_{N_c}^2$, see Equation 1.33 below) is equal to the average (μ). This leads to a square-root

---

[†] Other thermodynamic parameters sensitive to fluctuations are compressibility and binding capacity, which are maximized through expressions analogous to Equation 1.25.

[‡] At suitably low protein density so that proteins are not interacting with each other.

[§] Because the probability of success for the elementary outcome is taken to be very small, the Poisson distribution is sometimes referred to as the "law of rare events," or "law of small numbers."

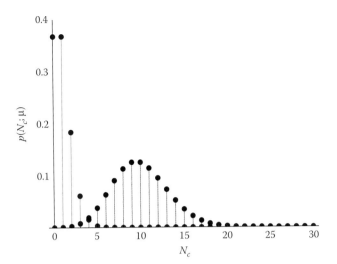

Figure 1.8 Poisson distributions with averages (μ, Equation 1.26) of one (black) and ten (red). The Poisson distribution gives the probability that some number of uncorrelated occurrences (N_c, for "counts") will occur within a fixed interval. Notice that the width of the two distributions are proportional to the averages.

relationship between the mean of the distribution and its width, similar to the binomial distribution (Equation 1.25). In electronic circuits and optical devices that are sensitive enough to detect individual electrons and photons, respectively, the intrinsic "noise" follows a Poisson distribution, and is referred to as "shot" noise.[†]

The FFT ("for the first time") and geometric distributions

When there is some (usually low) probability p_O that a given outcome (O) will be achieved in a single trial, then the probability that it will take N independent trials to achieve the outcome is given by the FFT[‡] (sometimes called a "shifted geometric") distribution:

$$p(N) = (1 - p_O)^{N-1} \times p_O \qquad (1.27)$$

Since the outcome cannot be achieved *before* the first trial, the FFT distribution differs from the binomial and Poisson distributions in that it begins at $N = 1$ rather than $N = 0$. Equation 1.27 can be rationalized by recognizing that if an outcome is obtained at trial N (with probability p_O), it *did not* happen in any of the $N - 1$ trials (with probability $p_{\sim O} = 1 - p_O$). Assuming that each trial is independent, the probability from each trial multiplies (the *and* combination). This discrete distribution applies to time-dependent processes that are independent of one another,[§] such as first-order chemical reactions where the individual molecules each have independent reaction probabilities. Two FFT probability distributions are given in **Figure 1.9**. The distribution with the greater probability per unit time (black bars) decays more sharply, because the likelihood of success per trial is greater.

The FFT distribution is closely related to the geometric distribution, which has a slightly different formula (Problem 1.13). The geometric distribution gives the probability that there will be N trials *before* the successful outcome, and thus begins at $N = 0$ (corresponding to success on the first trial, thus preceded by $N = 0$ unsuccessful trials).

The multinomial distribution

The discrete distributions above give the probability as a function of a single variable. The Poisson distribution, for example, is defined as a function of the number of counts. The binomial and FFT distributions are also described in terms of a single outcome variable (e.g., number of heads), with the other "variable" (e.g., number of tails) fixed by N_{tot}.

[†] Apparently the name "shot" was inspired by the sound of pouring lead shot onto a plate. Perhaps coincidentally, this was first described by Walter Schottky.

[‡] Do not confuse this with the fast Fourier transform, which is often given the same abbreviation.

[§] Processes of this type are referred to as "Markov chains."

Figure 1.9 FFT distributions with probabilities of 0.2 (black) and 0.04 (red). The higher probability for the black curve increases the likelihood of an event in an early trial. Because the FFT distribution is normalized, the likelihood of an event at late trial is decreased.

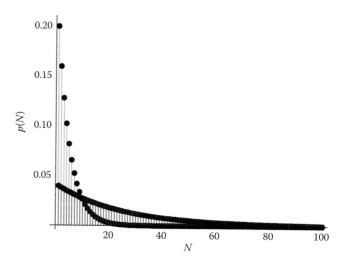

What if we have more than two possible outcomes per elementary event, as dice rolls? The distribution describes the probability of obtaining a collection $(N_1, N_2,...,$ N_j, with sum N_{tot}) of j mutually exclusive and collectively exhaustive outcomes, each with probability $p_1, p_2,..., p_j$. As we will see, this distribution is appropriate for describing the number of ways that N_{tot} molecules can be distributed among j different states, a key problem in statistical thermodynamics. The j states can be of different energy (in which case, additional criteria will be applied that modify the distribution, as described in Chapter 10) or they can have the same energy.

To develop the equations that describe the multinomial distribution, we will use a familiar example: rolling dice. The derivation for dice rolls is analogous to the binomial distribution for tossing coins (a die can be considered a multisided coin, or alternatively, a coin can be considered a two-sided die). For the familiar six-sided die (Example 1.3; see also Problem 1.0) there are $j = 6$ mutually exclusive and collectively exhaustive outcomes. For a die that is "fair," each outcome has an equal probability of 1/6. When N_{tot} dice are thrown and the sequence of throws is recorded (or the dice can be distinguished from one another by a mark or color), the probability of a particular sequence (e.g., 2,3,4,1,1,6) is

$$p_{2_1 3_2 4_3 1_4 1_5 6_6} = p_2 \times p_3 \times p_4 \times p_1 \times p_1 \times p_6$$
$$= \left(\frac{1}{6}\right)^6 \tag{1.28}$$

More generally, for any sequence with a distribution $N_1, N_2,..., N_j$, the probability of a particular sequence (there are likely to be many such sequences, but all sequences have the same probability if the dice are true) is

$$p_{sequence} = p_1^{N_1} \times p_2^{N_2} \times p_3^{N_3} \times p_4^{N_4} \times p_5^{N_5} \times p_6^{N_6}$$
$$= \left(\frac{1}{6}\right)^{N_{tot}} \tag{1.29}$$

where N_1 is the number of ones in the sequence, etc.

Lacking sequence information, we can still calculate the probability of getting a composition of $\{N_1, N_2, N_3, N_4, N_5, N_6\}$ as

$$p(N_1, N_2, N_3, N_4, N_5, N_6; N_{tot}) = W(N_1, N_2,..., N_6; N_{tot}) \times p_{sequence}$$
$$= W(N_1, N_2,..., N_6; N_{tot}) \times p_1^{N_1} \times p_2^{N_2} \times \cdots \times p_6^{N_6}$$
$$= W(N_1, N_2,..., N_6; N_{tot}) \times \left(\frac{1}{6}\right)^{N_{tot}} \tag{1.30}$$

The same method that was used to calculate $W(N_H, N_T; N)$ for coin tosses (Figure 1.5) can be used to calculate the number of sequences $W(N_1, N_2, N_3, N_4, N_5, N_6; N)$ for die rolls of a given composition. Divide the dice into six piles based on their outcomes (1–6). Build a die holder with N_{tot} numbered slots. Start distributing the ones first and calculate the ways the dice can be distributed: N_{tot} positions for the die on the top of the ones pile, $(N_{tot} - 1)$ for the next,..., $(N_{tot} - N_1 + 1)$ for the last. Then correct for the overcounting: a factor of two for the second die, a further factor of three for the third,..., and a further factor of N_1 for the last (for a total of $N_1!$). Then start distributing the twos. There are $(N_{tot} - N_1)$ ways to distribute the first two, $(N_{tot} - N_1 - 1)$ for the next,..., and $(N_{tot} - N_1 - N_2 + 1)$ for the last. Correcting for overcounting the twos introduces an additional factor $N_2!$ in the denominator. This is continued for the remaining four piles (threes though sixes). This leads to an expression for the multinomial coefficient of

$$W(N_1, N_2, N_3, N_4, N_5, N_6; N_{tot}) = \frac{N_{tot}!}{N_1! \times N_2! \times N_3! \times N_4! \times N_5! \times N_6!}$$
$$= \frac{N_{tot}!}{\prod_{i=1}^{6} N_i!} \tag{1.31}$$

Inserting Equation 1.35 into 1.34 gives the multinomial probability distribution for six elementary outcomes. This distribution is easy to generalize to any other number of outcomes (note that if we evaluate the multinomial distribution for two collectively exhaustive outcomes, we recover the binomial distribution).

Although in general it is not easy to depict the multinomial probability distribution (a combination of Equations 1.30 and 1.31) because $j - 1$ independent axes are needed for the outcomes and an additional axis is needed for the probability, it can be depicted for $j =$ three outcomes using N_1 and N_2 to give the combinations of outcomes (with N_3 implied as $N_{tot} - N_1 - N_2$). **Figure 1.10** shows such a distribution, which could be applied to a collection of three-sided dice, or the distribution of molecules among three distinct states.

As with the binomial distribution, the multinomial distribution becomes sharply peaked as the number of events (or molecules in a system) increases.

Average values for discrete outcomes

The discrete probability distributions provide a fundamental description of a system with one or more random variables: how likely is it that a particular observation will be made. On a molecular level, this can correspond to how molecules populate different energy levels, react to form different species, bind different ligands, adopt different configurations, etc. But in many cases, these different distributions are not directly measurable in experiments. We often don't have instruments that can count molecules in each different states and provide us with populations. Instead, we study molecular features by measuring quantities, or "observables" (we will refer to such observables as Y_{obs}), that are sensitive to different molecular states.

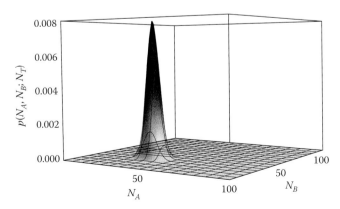

Figure 1.10 Multinomial distribution for three distinct outcomes with equal probabilities ($p_A = p_B = p_C = 1/3$). The total number of events (N_{tot}) is 100. Although the surface plotted here appears to be a continuous surface, this is a discrete distribution, defined only for integer values of the variables N_A, N_B, and (implicitly here) N_C.

What we get back from such a measurement is an average value, reflecting the probability of each state (p_i) and its corresponding contribution to the observable (Y_{obs}). More precisely, we say that Y_{obs} is a "population-weighted average" over all the states:

$$Y_{obs} = \langle Y \rangle = \sum_{i=1}^{N} Y_i p_i \tag{1.32}$$

The angle brackets ($<\ >$) will be used as a short-hand for average value. Equation 1.36 is one of the most important ways to connect experimental measurements to the mechanistic and thermodynamic descriptions of molecular behavior.

A related equation provides a measure of the average spread among different combined outcomes. This spread is referred to as the "variance" of the distribution:

$$\sigma_Y^2 = \langle (Y - \langle Y \rangle)^2 \rangle = \sum_{i=1}^{N} (Y_i - \langle \langle Y \rangle \rangle)^2 p_i \tag{1.33}$$

Variances are important for calculating spreads of energies in molecular systems, which can be directly connected to experimentally measured heat capacities, as described in Chapter 10. The square ($Y_i - \langle Y \rangle$) in Equation 1.33, which is referred to as the "residual" or "deviation" of Y from its average, can be factored into a useful form:

$$
\begin{aligned}
\sigma_Y^2 &= \sum_{i=1}^{N} (Y_i - \langle Y \rangle) \times (Y_i - \langle Y \rangle) p_i \\
&= \sum_{i=1}^{N} (Y_i^2 - 2Y_i \langle Y \rangle + \langle Y \rangle^2) p_i \\
&= \sum_{i=1}^{N} Y_i^2 p_i - 2\langle Y \rangle \sum_{i=1}^{N} Y_i p_i + \langle Y \rangle^2 \sum_{i=1}^{N} p_i \\
&= \langle Y^2 \rangle - 2\langle Y \rangle^2 + \langle Y \rangle^2 \\
&= \langle Y^2 \rangle - \langle Y \rangle^2
\end{aligned} \tag{1.34}
$$

Often it is preferable to represent the spread of a distribution as the square root of the variance, or RMSD, since it has the same dimension as the mean, and thus gives a better intuitive understanding in terms of the variable of interest.

CONTINUOUS DISTRIBUTIONS

The distributions described above give probabilities for discrete values, and can be applied to games of chance, counting experiments, and to distributions of molecules among different discrete molecular states. There are, however, a number of distributions that are not restricted to discrete values. These "continuous distribution functions" (i.e., functions that can be used to obtain a probability at any value within a specified range) are appropriate to describe speeds in collections of molecules,[†] lifetimes of molecular complexes, and distances between groups in polymeric molecules.

[†] There are practical limits to how fast gas molecules can move. Specifically, gas molecules are unlikely to move faster than a few multiples of $(3kT/m)^{1/2}$, where k is Boltzmann's constant (1.38×10^{-23} J K^{-1}), T is temperature on the Kelvin scale, and m is the molecular mass in kg.

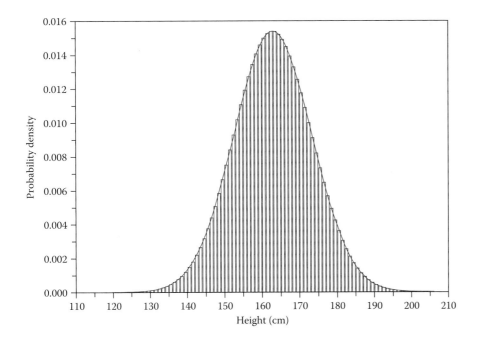

Figure 1.11 Going from a continuous to a discrete distribution. A probability density for the height of college-age women in the United States (data from 1999 to 2002; mean = 162.9 cm, standard deviation = 10.3 cm) is plotted assuming a Gaussian distribution (solid line). This distribution was converted to a discrete distribution, with a bin width of $\Delta h = 1$ cm, by taking the probability of being within a particular bin as in Equation 1.35.

There are a number of obvious similarities between discrete and continuous distribution functions, and some very important differences. In a crude sense, both types of distribution functions give a picture of how likely a particular value is to be obtained. High values in each distribution correspond to high probabilities, low values to low probabilities. Moreover, the two types of distributions can often be converted. Continuous distribution functions can be converted to discrete probability functions by averaging values over fixed intervals (or bins) to give a bar plot (**Figure 1.11**). Likewise, when "sampling" from a population that is distributed continuously (i.e., collecting a bunch of measurements of a particular variable), the easiest way to represent the results is with a discrete histogram. For example, the heights of students at a university can be lumped together into the numbers of students from 169 to 170 cm, 170 to 171 cm,..., and can be plotted as a histogram (Figure 1.11). And as described below, discrete probability distributions often closely approximate continuous distributions, especially when the average becomes large.

However, there is a fundamental difference between discrete and continuous distributions. For discrete distributions such as the binomial distribution, the value at a particular outcome ($p(n_H; N)$ for N_H heads out of N coin tosses) can be interpreted directly as a probability of that outcome. The probability of obtaining a combination of two heads and one tail in three tosses of a coin is 3/8, as given by the binomial distribution. However, for a variable that is continuous, the probability of obtaining a particular value in a single trial is effectively zero.[†] This is because there are an infinite number of values in any interval of a continuous variable.

Our experiences with making measurements appear to be at odds with the picture of zero probability because we (and often our instruments) round off numbers. For example, if we measure a continuous variable like the heights of our classmates, we would surely find some number of students with "the same" height, such as 170 cm. This is because we round off numbers (to the nearest centimeter in this example), which naturally groups our measurements into bins (Δh). In effect, we are taking a continuous distribution and making it discrete. In a simple measurement like this,

---

[†] Unless the continuous distribution is infinitely sharp, though this would not really be much of a "distribution."

we don't have the ability to obtain values like 170.3839201... cm, (nor, typically, would we have need of this much information).

Nonetheless, continuous distributions are very useful for the analysis of discrete data, for making models, and for analyzing the "goodness of fit" of a model to data. To accommodate the fact that the probability of obtaining a particular value in a continuous probability distribution is zero, we need to modify the nature of our distributions. In our example with heights, where we unintentionally constructed a discrete histogram, we can relate the probability of getting heights within a particular range (a bin of width $\Delta h = 1$ cm) to the continuous distribution function we seek with the approximation

$$p_{170 \text{ to } 171} = \frac{N_{170 \text{ to } 171}}{\sum N}$$

$$\approx \rho(170.5) \times \Delta h$$

$$= \rho(170.5) \times 1 \,\text{cm}$$

(1.35)

The continuous distribution function $\rho(h)$ is often referred to as a "probability density."

Equation 1.35 can be made more and more accurate by making Δh smaller and smaller. By replacing the bin width Δh with an infinitesimal width dh, we get the expression

$$dp(170.5) = \rho(170.5)dh$$

(1.36)

As with the probability expression in Equation 1.35, the left-hand side of Equation 1.36 describes the probability of falling within a range, this time from 170.5 cm to $170.5 + dH$ cm. Because this range is infinitely narrow, we represent this range as a differential probability. We can use Equation 1.36 to calculate the exact probability of heights falling within a certain range using the expression

$$p(170 \text{ to } 171) = \int_{170}^{171} \rho(h)dh$$

(1.37)

The reason Equation 1.37 is exact but Equation 1.35 is approximate in that Equation 1.37 accounts for the variation in $\rho(h)$ through the interval, whereas Equation 1.35 uses a single-value approximation, treating the probability within the interval as a rectangle.

One common range to integrate over is from the lowest possible value (for height, the lowest "possible" value would be zero cm,[†] but for distributions that can range over negative values, a lower integration limit of negative infinity is used) up to a value of interest. This gives a "cumulative" distribution function Q[‡]

$$Q(h') = \int_{0}^{h'} \rho(h)dh$$

(1.38)

$$Q(x') = \int_{-\infty}^{x'} \rho(x)dx$$

(1.39)

Equation 1.38 is for our height example, Equation 1.39 is for a general continuous random variable x. The primes in Equation 1.30 indicate integration to a particular

---

[†] People cannot have negative heights.
[‡] In Mathematica, cumulative distribution functions are referred to as CDFs.

value from the lower limit. Because probability densities are always positive, Q always increases with the random variable. And in analogy with the "collectively exhaustive" criterion we introduced for discrete probability distributions, the cumulative distribution calculated over all values is unity:

$$Q(\infty) = \int_{-\infty}^{+\infty} \rho(x)dx = 1 \tag{1.40}$$

This results from appropriately scaling $\rho(x)$; probability densities satisfying Equation 1.40 are said to be "normalized."

A final comment on the difference between discrete probabilities and continuous probability densities relates to their dimensionality. Discrete probabilities are dimensionless, as are the differential and integrated probabilities of continuous distributions (Equations 1.35 through 1.40). That these dimensionless probabilities are related to the product of $\rho(x)$ and dx means that when dx has dimensions, $\rho(x)$ has the inverse of those dimensions. Thus for our example of student heights, if the units of dh are chosen to be cm, the units of $\rho(h)$ must be inverse cm.

A few of the most important continuous probability densities in biophysical chemistry are the exponential and Gaussian distributions, which are described below.

The exponential decay distribution

The exponential decay distribution is used to describe energy distributions in molecular systems with a large number of molecules. As will be described in Chapter 10, the exponential decay distribution describes the probability that molecules have a particular energy value that decays inversely with the energy. The exponential decay distribution is also used to give lifetimes of complexes that decompose through random fluctuations. This distribution is the continuous analog of the geometric distribution. The exponential decay distribution has the form

$$\rho_E(x) = \alpha e^{-\alpha x} \tag{1.41}$$

Here, α is a positive constant that describes how sharply the distribution decays with increasing values of x. The dimensions of α are the inverse of the dimensions of x so that the exponent remains dimensionless. The parameter α also appears as a prefactor that scales the initial value of the probability distribution. This compensates the sharpness of the decay, and ensures that the area under the $\rho_E(x)$ is unity (satisfying the condition for normalization, Equation 1.40). Some examples of exponential probability densities are shown in **Figure 1.12**.

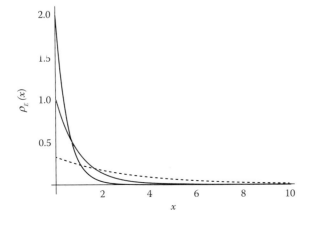

Figure 1.12 Exponential probability densities with different decay constants. The black solid curve corresponds to $\alpha = 2.0$, the red solid curve corresponds to $\alpha = 1.0$, and the dashed curve corresponds to $\alpha = 0.33$. Note that the distribution that decays rapidly starts from a higher value, a requirement of normalization. Unlike the discrete distributions above, smooth lines are used to emphasize the continuous nature of the exponential probability distribution.

Figure 1.13 Gaussian distributions functions with the same average value ($\bar{x} = 5$) but different spreads. The black solid curve corresponds to $\sigma = 0.5$, the red solid curve corresponds to $\sigma = 1.5$, and the dashed curve corresponds to $\sigma = 5.0$. Note that as the spreads get wider, the peak value gets shorter to maintain a constant area of unity.

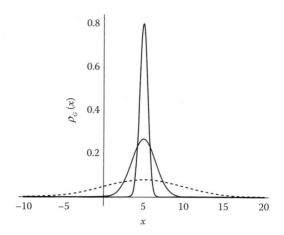

The Gaussian distribution

One of the most common and well-known continuous distributions is the Gaussian, or normal distribution.[†] This distribution describes the probability of obtaining particular value for a continuous variable that is distributed around an average value $\bar{x}$:

$$\rho_G(x) = \frac{1}{\sigma\sqrt{2\pi}} e^{-(x-\bar{x})^2/2\sigma^2} \tag{1.42}$$

Here, σ describes the width of the distribution, with a large value corresponding to a wide distribution. As with the exponential distribution, σ appears as a prefactor (in this case inversely) to scale the height of the distribution, ensuring that $\rho_G(x)$ is normalized. Some examples of Gaussian distributions are given in **Figure 1.13.**

The Gaussian distribution describes distributions resulting from the sum of a large number of independent but identically distributed random values. One situation we have already discussed that involves sums of independent, identically distributed random values is the tossing of multiple coins. When we toss N_{tot} coins and add the number of heads (n_H), each single coin toss has the same probability distribution ($p_H = p_T = 1/2$), and subsequent tosses are independent of one another. When the total number of tosses is low, the binomial distribution describes the discrete probabilities for each outcome, and the Gaussian is a poor approximation. However, when N_{tot} becomes very large, the distribution is well-approximated by a Gaussian distribution, and it looks more and more continous. This is explored in Problem 1.8. The same is true for the Poisson distribution as the average μ becomes large.

One example where the Gaussian distribution comes into play at the molecular scale is in the diffusion of molecules on the molecular scale. Diffusion of molecules can be viewed as a process in which molecules take random steps in space, sometimes forward and sometimes backward. After a period of time, the position diffusing molecules, relative to the starting point, can be approximated as the sum of a large number of random steps, sometimes forward, and sometimes backward. The resulting distribution of molecules will be Gaussian. The ends of an unstructured polymer can also be approximated in this simple way.

[†] Not to be confused with the property of being "normalized", which means that the area under the probability density function is one. Although the Gaussian distribution is indeed normalized, so are all other probability distributions, both continuous and discrete.

Another important situation involving the Gaussian distribution is the experimental determination of a quantity for which there is random measurement error. In addition to the real variation in height from student to student, there is variation associated with the measurement itself. This type of error would be revealed by measuring the height of the *same* student over and over. In such cases, the Gaussian distribution describes the probability of making a measurement that has a particular value of x. The peak $\bar{x}$ in the Gaussian can be thought of as the "true" value of the variable, and σ represents the uncertainty, or error, associated with any single measurement of x.

Average values for continuous distributions

As described above for discrete distributions, we will often be interested in calculating average values from continuous distributions, especially when we are interested in connecting molecular-level descriptions with observations made in experiments. Averages are calculated much the same way as with discrete probabilities, but we will use the infinitesimal probability element $dp = \rho_x(dx)$ to multiply the quantity being averaged, and will use integration instead of summation. The average value for the quantity x described by the distribution is

$$\langle x \rangle = \int_{-\infty}^{+\infty} x \rho(x) dx \tag{1.43}$$

Likewise, the average value of x^2 can be calculated as

$$\langle x^2 \rangle = \int_{-\infty}^{+\infty} x^2 \rho(x) dx \tag{1.44}$$

This average is often called the "second moment" of x; higher order moments ($<x^n>$) are calculated the same way. By using the factorization in Equation 1.34 (this also holds for integration of continuous probability densities; see Problem 1.21), the variance of a continuous distribution can easily be calculated from the second moment (Equation 1.44) and from the square of the first moment (Equation 1.43).

Problems

1.0 When rolling a single six-sided die,[†] what is the probability of rolling a six? When single rolling a 20-sided die (a d20, with icosahedral shape), what is the probability of rolling a 15 or higher? What is the probability of rolling either a 1 or a 20 on a single d20 roll?

1.1 Consider independent events that have the same set of (multiple) outcomes (*A*, *B*, *C*,...) and associated probabilities (p_A, p_B, p_C,...). Calculate the probability of getting one or the other (including both). How does this differ from the "or" (not both) calculation below (see Problem 1.2)?

1.2 Again, consider a collection of independent events that have the same set of (multiple) outcomes (*A*, *B*, *C*,...) and associated probabilities (p_A, p_B, p_C,...). Using the *not* conjunction (e.g., $\sim p_A$, $\sim p_B$), calculate the probability of getting A on the first event or B on the second event, but not both *A* and *B*. Show that your answer is equivalent to the *or* formula for two independent outcomes of different events in Figure 1.3A. Use Venn diagrams to illustrate your thinking.

[†] Somewhat confusingly, die is the singular form of the noun describing a cube with dots on the faces; dice is the plural.

(a) What is the probability of tossing two coins in sequence and getting a head on the first toss and a tail on the second? What are the probabilities each of the other three permutations? What is the probability of getting a head on the first toss or a tail on the second toss, but not both?

(b) When tossing N coins, what is the probability of getting a head on every toss? Is this a permutation or a composition?

1.3 Assume that you know 2000 people who live in your country, which has a population of 300,000,000. Assume that they are randomly distributed geographically. If you get on an airplane that holds 201 people, what are the chances that you know someone on the plane (excluding yourself, i.e., among the 200 other passengers)?

1.4 A pitcher in baseball has a probability p_s of throwing a strike and $p_b = 1 - p_s$ of throwing a ball. Assuming the batter adopts a strategy of never swinging at a pitch, and that each pitch is independent of each other pitch (and the strike count and identity of the batter), what is the probability that the pitcher will "walk" the batter (any combination of four balls with zero, one, or two strikes)? At what value of p_b would the strategy of not swinging the bat yield more walks than strike-outs (three strikes in combination with three or fewer balls)? What is the probability of getting on base with a walk if the $p_b = p_s = 0.5$?

1.5 Starting with the probabilities of *specific sequences* of coin tosses that contain N_H heads and N_T tails, use the "or" conjunction to show that the probability of getting a particular *composition* is given by Equation 1.16.

1.6 Using Mathematica (or another package of your choice), generate a random coin toss simulator using the random number generator. Throw 10 *coins at once, and count the total number of heads, N_H; 10. Repeat with the ten* coins 100 times, and plot the distribution as a histogram. Generate another distributions where 1000 coins are tossed at once, and this is repeated 10,000 times, and plot this distribution as a separate histogram. Scale the x-axis of each plot from 0 to N_{tot} for (i.e., $N_H = 0$ to 10 for the first plot, 0 to 1000 for the second plot). How do the standard deviations σ of your distributions ($N_{tot} = 10$ vs. 1000 tosses) compare? Does this match what you expect for the dependence of σ on N_{tot}?

How does the sharpness of the distribution (which can be quantified as σ/N_{tot}) compare for your two distributions?

Note that you can do this in one of two ways. Either generate N_{tot} (10, 1000) random binary numbers (1 and 0, each with equal probability) to get the number of heads, and then repeat this 100 and 10,000 times, respectively, or you can sample randomly from the appropriate distribution. A useful Mathematica command for generating i random numbers from a distribution "dist" is "RandomVariate[dist, i]." And to plot, a useful command is "Histogram[list,{Nmin, Nmax, dN}]."

1.7 Generate binomial distributions (not from random trials but from the actual formula for the distribution for the two coin-toss combinations in Problem 1.6, that is, 10 and 1000 tosses of a fair coin). Plot the binomial distributions along with the simulated tosses, vertically scaling the former so that it matches the latter (what scale factor do you need for each and why?).

1.8 For the two histograms of 100 and 10,000 trials in Problem 1.6, use Mathematica to fit a Gaussian distribution to each toss (using nonlinear least squares; see Chapter 2). You will need to find a command that counts the number of elements in each bin, and convert to $x–y$ data. What are the fitted parameters for each (note you will need three parameters for each: mean, standard deviation, and a prefactor that scales the distribution to the right height). Generate plots that show

(a) The random distributions from Problem 1.6

(b) The true binomials (scaled) from Problem 1.7 and

(c) The fitted Gaussians

1.9 Demonstrate the assertion that the absolute width of the binomial distribution (e.g., in "total number of heads") increases with the square root of the total number of trials (N_{tot}), by plotting the two distributions from Problem 1.7 at the same absolute scale (0–600 total tosses).

1.10 Now demonstrate that the relative width of the binomial distribution decreases on a relative scale, by plotting both distributions from Problem 1.7 from 0 to N_{tot} (1–10 and 0–1000, respectively).

1.11 Use the derivative test to show that for a fixed total number of events N, the variance of the binomial distribution is maximal at $p_A = p_B = 0.5$. For $N_{tot} = 100$, plot the variance and the square root of the variance as a function of p_A over the range $0 < p_A < 1$.

1.12 Use Equation 1.32 to show that the average of the Poisson distribution is equal to its mean.

$$\langle n \rangle = \sum_{n=0}^{\infty} n p(n; \mu) = \sum_{n=1}^{\infty} n \frac{\mu^n e^{-\mu}}{n!}$$

$$= e^{-\mu} \sum_{n=1}^{\infty} n \frac{\mu^n}{n!} = e^{-\mu} \sum_{n=1}^{\infty} \frac{\mu^n}{(n-1)!}$$

$$= e^{-\mu} \mu \sum_{n=1}^{\infty} \frac{\mu^{n-1}}{(n-1)!}$$

1.13 Use the *and* combination to show that the geometric distribution is given as

$$p_{geo} = (1 - p_0)^j \times p_0$$

where j is the number of trials before a successful outcome and p_0 is the probability of a successful outcome in each independent trial.

1.14 Use the binomial theorem to show that the binomial distribution is normalized.

1.15 Use the series expansion of the exponential function, that is,

$$e^x = \sum_{i=0}^{\infty} \frac{x^i}{i!}$$

to show that the Poisson distribution is normalized.

1.16 Using the result that

$$\int_{-\infty}^{\infty} e^{-x^2} dx = \sqrt{\pi}$$

show that the Gaussian distribution is normalized.

1.17 For the following discrete and continuous distributions, calculate the average (mean) values.

    i. A binomial distribution with $p_A = 0.2$, $p_B = 0.8$.

    ii. A Poisson distribution with an average of $\mu = 2.3$ counts/second.

    iii. An exponential decay distribution with a decay constant of $\alpha = 0.26$ sec^{-1}.

1.18 Consider a collection of molecules that can adopt discrete energy values of $\varepsilon = 1, 2, 3, 4, \ldots,$ *infinity*. As we will develop in Chapter 10, collections of molecules will adopt equilibrium distributions that decay exponentially with the energy. Using a "for the first time" (FFT) distribution (a discrete version of an exponential distribution) with $p_o = 0.1$, use Mathematica to plot such a distribution from the $\varepsilon = 0$ to the $\varepsilon = 50$ energy level. What is the average energy in this distribution?

1.19 Show that for the Gaussian distribution, the first moment is equal to $\bar{x}$ (the mean value).

1.20 Show that for the binomial distribution, the RMSD increases with the square root of N_{tot}.

1.21 Show that for a distribution with a continuous probability density $\rho(x)$, the variance of x factors in the same way as for the discrete distribution (Equation 1.34), namely,

$$\sigma_x^2 = \langle x^2 \rangle - \langle x \rangle^2$$

1.22 Use the results of Problems 1.19 and 1.20 to show that the mean-square deviation {i.e., the average of $(x - \bar{x})^2$} for the Gaussian distribution is equal to σ_x^2.

CHAPTER 2

Mathematical Tools in Thermodynamics

Goals and Summary

The goal of this chapter is to provide students with the mathematical background to understand physical chemical principles and to analyze experiments in biothermodynamics. Assuming a basic knowledge of calculus, we will develop key concepts for thermodynamic analysis. After a quick review of single-variable calculus, we will describe results from multivariable calculus needed for classical and statistical thermodynamics, most notably, maximizing (and minimizing) multivariable functions, expressing and testing exact (and inexact) differentials, integration of multivariable functions along paths, and understanding factors that result in path independence of integration.

We will then describe methods for analyzing data using least-squares methods, providing an analytical framework for model testing, and for determination of physical parameters from experimental measurements. We will develop exact methods for fitting linear models to data, and describe the iterative approach used for fitting nonlinear models. Some common pitfalls encountered during fitting will be described. The chapter will conclude with a discussion of methods for analyzing the uncertainties in fitted parameter values. Nonlinear least squares is an important tool for analyzing physical chemical data, and will be used throughout this text, from analysis of simple gasses to complex conformational and allosteric transitions.

CALCULUS IN THERMODYNAMICS

Thermodynamics uses equations based on physical relationships and empirical formulas. For reasons outlined below, these equations are most easily manipulated using calculus, which connects observable properties of materials to tiny (more precisely, infinitesimal) changes, through derivatives and integrals. Here we review some important properties related to differentiation and integration. Although much of the calculus discussed here will be familiar to most readers, some of the underlying ideas may not be. It is hoped that the graphical illustrations provided below will give readers an intuitive understanding of these ideas, which are critical to the logic of thermodynamic analysis.

Derivatives of single-variable functions

The most familiar representation of the derivative is a single quantity that gives the slope of a function. Derivatives are often represented with the notation $df(x)/dx$ (or df/dx for short[†]), but are also represented using a "prime," for example, $f'(x)$ when

---

[†] This "d" nomenclature is attributed to Gottfried Leibniz. The "prime" notation is attributed to Joseph-Louis Lagrange.

the original function is $f(x)$. In this book, we will mostly use the "d" notation, although the f' notation will sometimes be used when we plan on integrating. Methods for calculating derivatives of single-variable functions can be found in the first few chapters of any calculus book (see Problem 2.1 for practice).

One important application of derivatives is finding extrema of functions, that is, minima and maxima of $f(x)$. For a single variable function, the x values of extrema (often denoted as x^*) have first derivative values of zero,[†] that is,

$$\frac{df(x^*)}{dx} = 0 \qquad (2.1)$$

All extrema of smooth, continuous functions are critical points.[‡]

For the polynomial function shown in **Figure 2.1**, that is,

$$f(x) = x^2 - 3x \qquad (2.2)$$

the first derivative is[§]

$$\frac{df(x)}{dx} = 2x - 3 \qquad (2.3)$$

Setting the first derivative equal to zero and solving for x gives

$$2x^* - 3 = 0$$

$$x^* = \frac{3}{2} \qquad (2.4)$$

suggesting that the function has a minimum or maximum at $x^* = 3/2$. Indeed, when the first derivative is plotted (red line, Figure 2.1), it crosses the horizontal axis is at $x = 3/2$, further suggesting a minimum or a maximum.[¶]

Although the "first-derivative test" (i.e., finding critical points) allows us to locate extrema, this test alone does not allow us to say whether an extremum is a minimum or a maximum. The second derivative test resolves this ambiguity.[**] If the first derivative is zero and the second derivative is positive (upward concavity), the extremum is a minimum. If the second derivative is negative (downward concavity), the

Figure 2.1 The derivative of a single variable function and its relation to slope and critical points. The upward facing parabola in black (Equation 2.2), and its derivative in red (plotted on the same numerical scale as y). The value of the derivative increases linearly with x, consistent with the increasing slope of the parabola. At the minimum of the parabola, at $x^* = 3/2$, the derivative is zero.

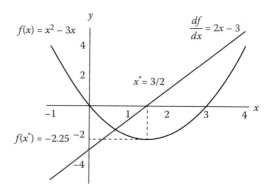

[†] Alternatively, minima and maxima can occur at points where the first derivative is undefined, although this does not happen in equilibrium thermodynamics except for phase transitions.

[‡] Although the converse is not always true: not all critical points are minima or maxima (see Problem 2.2).

[§] This expression comes from the power rule for differentiation, where for the polynomial $y = cx^n$, $dy/dx = ncx^{n-1}$.

[¶] If the first derivative becomes zero at x^* but does not cross the x-axis, it means the function $f(x)$ has an inflection point rather than a minimum or maximum.

[**] Another less elegant but perfectly reasonable way to determine whether a critical point is a minimum or a maximum is to plot the function and look at it. Clearly, the critical point at $x^* = 3.2$ for Equation 2.1 is a minimum. This graphical approach also identifies rare cases where a critical point is neither a minimum or a maximum (Problem 2.2).

extremum is a maximum. In the example in Figure 2.1, the second derivative is positive:

$$\frac{d^2 f}{dx^2} = 2 \qquad (2.5)$$

Thus, the parabola is a minimum at $x^* = 3/2$.

Differentials involving single-variable functions

In many treatments of elementary calculus, the derivative is treated as a single entity rather than a ratio. That is, the derivative itself is either a number, when evaluated at a single value of x, or is a well-defined function that gives slope as function of x. However, as the df/dx notation suggests, the derivative has properties of a ratio of the two quantities: df and dx. Each of these quantities, called "*differentials*," are infinitesimally small changes in f and x.

In thermodynamic formulas, differentials are often separated instead of treating them as a ratio that gives a single slope. For the parabola in the previous section (Equation 2.2), separation gives

$$df = (2x - 3)dx \qquad (2.6)$$

This relationship describes how much an infinitesimal variation in x (dx) changes the value of the function $f(x)$ (df). The quantity $2x - 3$ serves as a "sensitivity coefficient": the amount that f responds to a change in x depends on the value of x itself. At large positive values of x, an increase in x brings about a large increase in $f(x)$, whereas at values of x near $3/2$, it brings about a small increase in $f(x)$. At x values less than $3/2$, an increase in x brings about a *decrease* in $f(x)$.

Although df and dx are infinitesimally small, it is important to remember that they are neither zero, nor are they equal to each other, in general. If this seems puzzling, it may be helpful to think of dx and df as increments Δx and Δf (with really small Δ's). Note that because the sensitivity coefficient is equal to the derivative of f (Equation 2.3), the differential form (Equation 2.6) can be written

$$df = \frac{df}{dx} dx \qquad (2.7)$$

Although Equation 2.7 seems rather trivial, we will encounter a similar (but less trivial) form of this relationship when we calculate differentials of multiple variables.

Differentials are useful in classical thermodynamics for two reasons. First, for many thermodynamic quantities (most importantly, energy and entropy), we don't know *absolute* values. Instead, we focus on the *change* in those quantities in going from one set of measurable variables (e.g., pressure, temperature, concentration) to another. The differential is a precise expression of such change. Second, in thermodynamics we focus on reversible changes.[†] When changes are made reversibly, a system remains at equilibrium at all times (or very near to it). To maintain equilibrium, these changes must be tiny.[‡] The tiny changes we apply to a system to keep it at equilibrium can be represented by differentials, which are infinitesimal. Although each differential change would be too small to notice individually, they can easily be added together mathematically by integration.

[†] As will be described, many of the formulas we use to analyze thermodynamic changes can only be applied to reversible changes. In most cases, equations for irreversible changes are too complex to be useful.

[‡] It is not uncommon for students of thermodynamics to be confused about how an equilibrium system can undergo a change. If this confusion becomes overwhelming, skip ahead to the section in Chapter 3 entitled "Equilibrium, Change of State, and Reversibility," and then come back once you are clear.

Integrals of single-variable functions

Most readers are familiar with a geometric picture of the integral as the "area under the curve." Another familiar description is that the integral undoes what the derivative does (i.e., the integral is the "anti-derivative"). We will tend toward the second description, undoing derivatives (and differentials). However, the area picture will be important for analyzing work (**Box 2.1**), for example, the work associated with the change of volume of a gas against an external pressure.

We begin our review of the single-variable integral by noting an important asymmetry between the derivative and the anti-derivative. The derivative of a function derivative $f'(x)$ is uniquely determined by the starting function $f(x)$ (i.e., a function uniquely determines its slope, see Figure 2.1. However, the integral is, in general, not uniquely determined starting from $f'(x)$ from the derivative, the general (indefinite) integral is not uniquely determined (**Figure 2.2**). Information is lost in the process of differentiation. There are many other functions (an infinite number) that have the same slope, but are shifted up or down along y (Figure 2.2B). Starting from the slope, the most general form of the integral (the "indefinite" integral) is only known to an arbitrary constant (often represented by c, corresponding to the amount of shift). In analytical form, the indefinite integral is expressed as[†]

$$\int f'(x)\,dx = f(x) + c \tag{2.8}$$

The situation is simpler if the limits of integration are specified, which results in what is referred to as a "definite" integral. The definite (limit-bounded) integral is the quantity most connected with the closely connected to the area under the curve. Specifically, when a function (here, $f'(x)$) is integrated from $x = a$ to $x = b$, the integral is the area under the $f'(x)$ curve from a to b (**Figure 2.3**). Because the curve being integrated (the "integrand") is a derivative, and there is only one derivative curve regardless of the value of an indefinite integral offset c, the area under the derivative curve does not depend on the value of c. That is, specific limits of integration uniquely

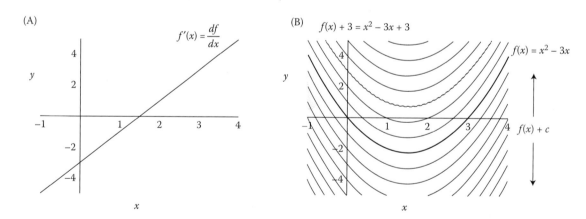

Figure 2.2 The indefinite integral of a single-variable function. (A) Starting from the derivative from Figure 2.1A, indefinite integration leads to the family of curves (red in panel B) of the same slope but different y offsets (given by the constant c). The original function from Figure 2.1A (solid black) can be viewed as a member of this family of curves with c is equal to zero. The dashed curve is offset in the y direction by $c = 3$.

[†] Another notation for integration starts with a function $f(x)$, and expresses the integral of $f(x)$ using an upper case F, that is,

$$\int f(x)\,dx = F(x) + c$$

This notation is useful when we are interested in both the derivative and the integral of a single function. Here we stick with the prime notation to highlight the relationship between the integral and derivative.

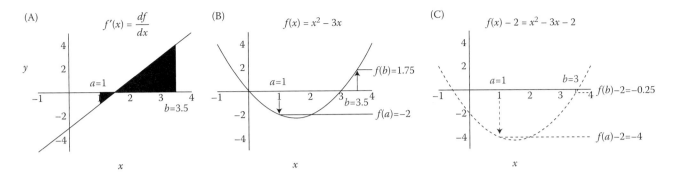

Figure 2.3 The definite integral of a single-variable function. (A) Integrating the function $f'(x) = 2x - 3$ from the limits $a = 1$ to $b = 3.5$ gives a single value (3.75), which corresponds to the area between the curve $f'(x)$ and the x-axis. This is true even though the indefinite integral of $f'(x)$ does not produce a single curve but a family of curves (see Figure 2.2). Numerically, the definite integral is simply the difference between the integrated function at the two limits. In panel B, this is $f(b) - f(a)$. In panel C, this is $f(b) + c - (f(a) + c) = f(b) - f(a)$. That is, the constant of integration (e.g., $c = -2$ in panel C) cancels in the difference.

determine the value of the definite integral. Whereas the indefinite integral produces a family of functions, the definite integral produces a single number.

Represented as an equation, the definite integral is similar in form to the indefinite integral (Equation 2.8), namely,

$$\int_a^b f'(x)\,dx = \left\{f(x)+c\right\}\Big|_a^b \tag{2.9}$$
$$= f(b)+c-\left\{f(a)+c\right\}$$
$$= f(b)-f(a)$$

The first line of the right-hand side of Equation 2.9 is a common way to represent the second, that is, the difference of the integrated function at the two limits. The integration constant c cancels in the definite integral regardless of its value. This cancellation can be seen in Figure 2.3: though the two functions in panels (B) and (C) differ by an offset of two, the differences defined by the direct integral are the same.

In thermodynamics, we often use a type of integration that is in between indefinite and definite. Rather than specifying both limits of integration, we specify a single limit as a reference point along x (for now, we will refer to it as x_r).[†] Analytically, single-limit integration looks like this[‡]:

$$\int_{x_r} f'(x)\,dx = f(x)-f(x_r) \tag{2.10}$$

In this type of integration we recover a *function* of the variable x, rather than a family of functions or a single number. If the value of $f(x)$ at the limit x_r (designated $f(x_r)$) is known, single-limit integration selects a single function from the family of functions produced by indefinite integration. The selected function passes through the value $f(x_r)$ at $x = x_r$. This is shown in **Figure 2.4A** for the integration of Equation 2.3, with a limit of $x_r = 1$, and a value at that limit of $f(x_r) = 2$. If $f(x_r)$ is not known,

---

[†] In Chapter 7, this reference point will define a "standard state."

[‡] Another way to write this type of single-point integral, which may be less offensive to mathematicians, is

$$\int_{x_r}^x f'(t)\,dt = f(x)-f(x_r)$$

That is, the upper limit of the integration is the variable x. The use of the dummy variable t instead of x inside the integral avoids confusion of the variable of integration with the variable x. For our purposes, our intentions will be clear enough using equations that look like Equation 2.10 that we won't bother with the more mathematically proper change of variable.

Box 2.1: Using differentials and integration to calculate work

As will be developed in Chapter 3, when a system changes its shape or volume, work is typically done, either *on* the system or *by* the system. At the bulk (macroscopic) level, examples include expansion of a gas, increasing the surface area of a liquid, and stretching a wire. At the molecular (microscopic) level, examples include stretching polymers, bringing charged groups together, and removing molecules from a crystal lattice. In the simplest case, work done by a force during an infinitesimal displacement of a object in the x direction is given by

$$dw = -\vec{F} \bullet dx = -F_x dx \tag{2.11}$$

$\vec{F}$ is a force vector that opposes (or promotes) displacement, and F_x is the component of F in the x direction. To calculate the work associated with a finite change, Equation 2.11 is integrated:

$$w = -\int_{x_i}^{x_f} F_x dx \tag{2.12}$$

Where the subscripts i and f indicate initial and final states.[†]

[†] Though these subscripts make it tempting to write the left-hand side of Equation 2.12 as w_f-w_i, since work is only defined as a property associated with change. This will be discussed extensively in Chapter 3.

---

the integrated function can be regarded as giving the difference $f(x) - f(x_r)$ (**Figure 2.4B**). This second approach is common in thermodynamic analysis. Note that this single-limit approach applies equally well to integration of multivariable functions (but the reference "point" must include coordinates for all variables, e.g., x_r, y_r for a function of x and y).

This single-limit integration is quite similar to the approach taken to solve differential equations, where an "initial condition" or "boundary condition" is given (Problem 2.5). Whether the value of $f(x_r)$ is known or not, it can be viewed as a constant of integration.[†]

Derivatives of multivariable functions

The properties of chemical and biological systems usually depend on more than one variable. These can be physical variables (e.g., temperature and pressure) or chemical variables (material states and molecular concentrations). Thus, the formulas that describe chemical and biological systems depend on multiple variables.

Functions of two independent variables (often written $f(x,y)$) can be visualized in three dimensions, using the third dimension, z, to plot $f(x,y)$. Such plots take the form of a surface. For example, the function is plotted in **Figure 2.5**.

$$f(x,y) = (3x - x^2)ye^{-y} \tag{2.13}$$

Derivatives play important roles in analysis of multivariable functions, but are more complex than those of single-variable functions. As with single-variable functions, derivatives are used to identify critical points (e.g., the maximum in Figure 2.5) by setting the derivatives to zero and solving. However, with multivariable functions can be differentiated with respect to each variable. With two independent varia-

---

[†] It might seem that x_r acts as a second integration constant (with $f(x_r)$ acting as the first constant), however this is not the case for the following reason: the value of x_r is arbitrary, and we are free to select any value we want (though there are practical limits that are important when choosing a standard state; see Chapter 7).

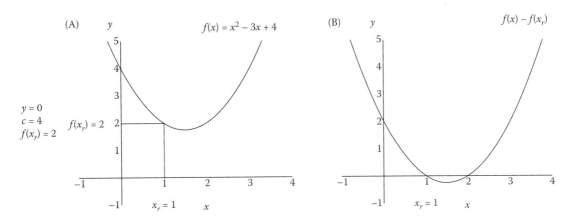

Figure 2.4 Single-limit integration of a single-variable function. (A) When a single limit of integration (x_r, in this case 1) is specified and the value of the function is known at that reference point ($f(x_r)$, in this case $f(x_r) = 2$), a single curve is selected from the family of integrals in Figure 2.2B (in this case, the selected curve turns corresponds to the indefinite integral with the constant $c = 4$). (B) If the value $f(x_r)$ is not known, then by selecting the curve from the family of integrals that passes through $y = 0$ at x_r, a difference from the reference (or standard state) value is obtained.

bles, as in Figure 2.5, we can differentiate f with respect to x, with respect to y, or with respect to a combination of the two. These can be thought of as derivatives of f in different directions. We will generally differentiate with respect to each independent variable (here x, y) one at a time.[†]

Derivatives with respect to a single independent variable, where other variables are fixed, are referred to as "partial derivatives." A partial derivative of $f(x,y)$ in Equation 2.13 with respect to x is written (and evaluated) as

$$\left(\frac{\partial f}{\partial x}\right)_y = \left(\frac{\partial\{(3x - x^2)ye^{-y}\}}{\partial x}\right)_y$$
$$= ye^{-y}\left(\frac{\partial(3x - x^2)}{\partial x}\right)_y \qquad (2.14)$$
$$= ye^{-y}(3 - 2x)$$

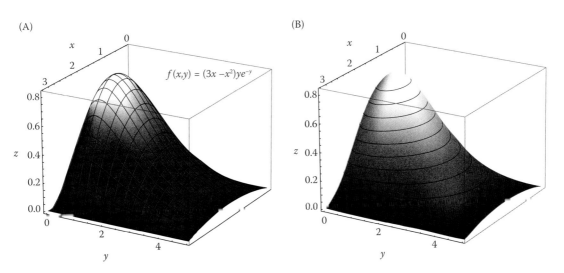

Figure 2.5 A function of two variables, plotted in three dimensions. Equation 2.13 is plotted with mesh lines at (A) constant x and y values, and (B) constant z values. The flat lines in panel (B) are often referred to as "level-curves," and play an important role in constrained maximization (Chapter 10).

[†] The one at a time approach differs from differentiating with respect to a combination of x and y, for example, along the diagonal line $y = x$. To explore this kind of directional derivative, see a multivariable calculus text.

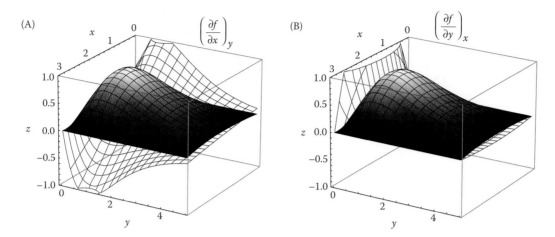

Figure 2.6 Partial derivatives of a multivariable function. The function in Figure 2.5 (Equation 2.13) is reproduced (red surface) along with its partial derivatives (white surfaces) with respect to x and y (A and B, respectively). The value of the partial derivatives is equal to the slope of the tangent line in the direction of the derivative.

The symbol ∂ is used in place of Latin d's to indicate that the derivative is partial. The right-subscripted variable (y in this case) indicates the independent variable (or sometimes variables[†]) held constant. The key to computing a partial derivative is that all variables other than the differentiation variable (ye^{-y} in Equation 2.14) are treated as constant, that is, they can be taken outside the derivative. Similarly, the partial of $f(x,y)$ with respect to y (Problem 2.6) is

$$\left(\frac{\partial f}{\partial y}\right)_x = (3x - x^2)(e^{-y} - ye^{-y}) \tag{2.15}$$

Figure 2.6 shows the partial derivatives of the surface shown in Figure 2.5 with respect to x and y.

Maximizing (and minimizing) multivariable functions

As with single variable functions, partial derivatives of multivariable functions give the slope of the tangent line in the direction of the derivative (e.g., a partial derivative with respect to x gives the tangent in the x-direction). When a partial derivative is equal to zero, the surfaces is likely to have a maximum or minimum *in the direction of the partial derivative*, although in other directions the need not be critical points. Global maxima and minima require that *all* partial derivatives are zero.[‡] For example, for the maximum in the surface in Figure 2.5 the two partial derivative equations below are simultaneously satisfied:

$$\left(\frac{\partial f}{\partial x}\right)_y = 0 = ye^{-y}(3 - 2x^*) \tag{2.16}$$

$$\left(\frac{\partial f}{\partial y}\right)_x = 0 = (3x - x^2)(e^{-y^*} - y^* e^{-y^*}) \tag{2.17}$$

---

† In general, for n independent variables, there will be $n-1$ variables held constant. Sometimes it is convenient to represent these as "$j \neq x$," rather than listing them all.

‡ Note that as with single variable functions, there are special cases where all partial derivatives equal to zero but the function is not a global maximum or minimum. One example is a "saddle-point" (maximum in one direction, minimum in another). For more on this, and how to resolve minima from maxima, consult a calculus text.

Again, the asterisk indicates values of x and y at the maximum. Equations 2.16 and 2.17 compose a system of two equations with two unknowns (values for x^* and y^*). Thus, they should be adequate to determine one (or more) solutions (i.e., critical points), if they exist. If Equations 2.16 and 2.17 were linear, a simple matrix manipulation would be enough to find a solution. However, their nonlinearity makes solution a little trickier.

In principle, the pair of Equations 2.16 and 2.17 could be solved by expressing one of the functions in terms of just x or just y (to the first power), and substituting into the other function. Unfortunately, neither function can be rearranged in this way. However, each Equation (2.16 and 2.17) is expressed as a function of only of x times a function of only of y, that is, $g(x)h(y) = 0$. Solutions to such equations are given when either (or both) $g(x)$ or $h(y)$ are zero. For example, Equation 2.16 is satisfied when

$$g(x) = (2x - 3) = 0 \qquad (2.18)$$

or when $x=3/2$. This solution is a line in the x–y plane parallel to the y-axis. Another set of solutions to Equation 2.16 come from the relation

$$h(y) = ye^{-y} = 0 \qquad (2.19)$$

or when $y=0$. This solution is a line along the x-axis. These two lines are shown in **Figure 2.7A**.

Likewise, Equation 2.17 is satisfied when $x = 0$, $x = 3$, $y = 0$, and $y = \infty$ (Problem 2.7). These lines of zeros are shown in **Figure 2.7B**. The requirement that *both* derivatives must be zero at a critical point occurs when zero lines of the two partial derivatives intersect, as shown in **Figure 2.7C**. The intersection of two such lines occurs at the conspicuous maximum at $x^* = 3/2$, $y^* = 1$. Note that based on the criterion of intersecting lines, there are other critical points in the region shown, though they are less obvious from a plot of the surface (see Problem 2.8).

As with single-variable functions, determining whether a critical point on a multi-variable function is a minimum or a maximum (or a saddle-point, that is, a minimum in one direction and a maximum in another) involves computing the signs of the second derivatives. However, this test is a bit more complicated, involving not only evaluation of each partial second derivative, but also cross-derivatives. For the thermodynamic functions we will be working with, it is usually fairly obvious from the context whether a critical point is a minimum or maximum. Thus, we will not

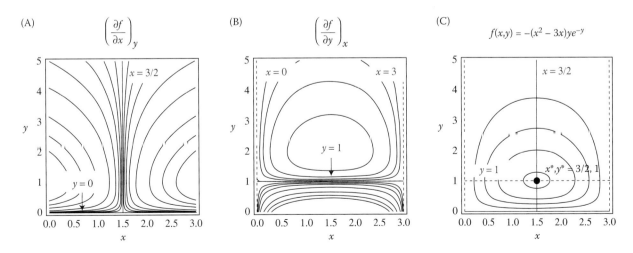

Figure 2.7 The relationship between the critical points of a surface and its partial derivatives. (A) The partial derivative with respect to x of the surface given by Equation 2.13, (B) the partial derivative with respect to y, and (C) the surface itself. Red solid and dashed lines show solutions obtained by setting x and y partial derivatives equal to zero. Contour lines are shown in black. In (C), the intersection of solid and dotted red zero gives the critical point. The maximum at ($x = 3/2$, $y = 1$) is shown.

elaborate on this extended second derivative test, though treatments can be found in any multivariable calculus text.

Differentials of multivariable functions. As described above, differentials are useful for analysis of how thermodynamic functions change when the system's variables change. In the previous section, we reviewed derivatives of multivariable functions. Here we will develop expressions to represent differentials of multivariable functions.

For functions with more than one independent variable, there will be more than one variation that contributes to the differential relationship. For example, for the two-variable function $f(x,y)$, the differential $df(x,y)$ should generally respond to changes in both x and y (dx and dy). In analogy with the single variable differential above, the sensitivity of a multivariable function to variation in each of its variables is proportional to sensitivity coefficients for each variable. Variation in x (by dx) should lead to a change in $f(x,y)$ of

$$d_x f = \left(\frac{\partial f}{\partial x}\right)_y dx \tag{2.20}$$

We have attached the x subscript on the left side of Equation 2.20 to make it clear that this change was brought about by variation in x but not y.[†] Likewise, variation in y (by dy) produces a change

$$d_y f = \left(\frac{\partial f}{\partial y}\right)_x dy \tag{2.21}$$

What if we vary x and y at the same time? It seems reasonable to expect that we can simply add the two changes in $f(x,y)$ resulting from these two variations, that is,

$$\begin{aligned} d_{tot} f &= d_x f + d_y f \\ &= \left(\frac{\partial f}{\partial x}\right)_y dx + \left(\frac{\partial f}{\partial y}\right)_x dy \end{aligned} \tag{2.22}$$

If df can be expressed in this way, it is said to be an ***exact differential***. The concept of exact differential can be generalized to functions of an arbitrary numbers of variables. For a function of n variables, that is, $f(x_1, x_2, ..., x_n)$, an exact differential of all of the variables is written

$$\begin{aligned} df &= \left(\frac{\partial f}{\partial x_1}\right)_{x_2,...,x_n} dx_1 + \left(\frac{\partial f}{\partial x_2}\right)_{x_1,...,x_n} dx_2 + \cdots + \left(\frac{\partial f}{\partial x_2}\right)_{x_1,x_2,...} dx_n \\ &= \sum_{i=1}^{n} \left(\frac{\partial f}{\partial x_i}\right)_{x_{j\neq i}} dx_i \end{aligned} \tag{2.23}$$

As with the single-variable differential, the derivatives in Equation 2.23 should be thought of as sensitivity coefficients. If the function has high sensitivity to a particular variable, the corresponding partial derivative amplifies the change in that variable.

In Equations 2.22 and 2.23, variations dx_1, dx_2, ... associated with each independent variable are independent of each other and are additive. Independence means that each independent variable can be adjusted on its own without invalidating the equality (as long as the variation is infinitesimal), although such changes do change

---

[†] Though making this connection explicitly is useful for the present derivation, we will not use the subscript notation of Equation 2.20 in general.

the *value* of *df*. That is, we can arbitrarily choose the sign and magnitude[†] of variation of a particular variation (dx_i) without affecting contributions to *df* from the other terms. We will use this "independence of variation" in many contexts in the coming chapters.

In addition, there is another less obvious implication of the independent additivity of exact differentials formulas. Based on additivity, each sensitivity coefficient might be expected to be independent of the other independent variables (and their differential variations). However, this is *not* typically the case. Rather, the sensitivity coefficients of most multivariable functions are also multivariable functions, showing sensitivity to multiple variables. This mutual sensitivity can be computed using mixed partial derivatives (see Problem 2.9).[‡] Given the interdependence of the sensitivity coefficients, why don't mixed partial derivatives show up in our formulas (Equations 2.22 and 2.23) for the differential?

The simplicity of the exact differential formula comes from the fact that the variations associated with the differential are infinitesimal. This can be seen by making variations to independent variables in a specific order. For example, in Equation 2.22, we could first make a perturbation in *x*, from x_i to $x_f = x_i + dx$, while holding *y* constant, and then make our perturbation in y (by *dy*) at the new x_f value. The first step makes the following contribution to the differential:

$$d_1 f = \left(\frac{\partial f}{\partial x}\right)_y dx \tag{2.24}$$

This is the same as in Equation 2.20. In the second step, we might expect the sensitivity coefficient associated with *y* to have changed as a result of the change in *x* that occurred in the first step. The new sensitivity coefficient can be represented using the old value and a term proportional to the size of the *x* perturbation in the first step:

$$\left(\frac{\partial f}{\partial y}\right)_{x_f = x_i + dx} = \left(\frac{\partial f}{\partial y}\right)_{x_i} + \left(\frac{\partial^2 f}{\partial x \partial y}\right) dx \tag{2.25}$$

With this perturbed sensitivity coefficient, the second step makes the following contribution to the differential:

$$\begin{aligned} d_2 f &= \left(\frac{\partial f}{\partial y}\right)_{x_f} dy \\ &= \left[\left(\frac{\partial f}{\partial y}\right)_{x_i} + \left(\frac{\partial^2 f}{\partial x \partial y}\right) dx\right] dy \\ &= \left(\frac{\partial f}{\partial y}\right)_{x_i} dy + \left(\frac{\partial^2 f}{\partial x \partial y}\right) dx dy \end{aligned} \tag{2.26}$$

[†] This includes setting $dx_i = 0$.

[‡] The degree to which sensitivity coefficients depend on other variables can be quantified using cross-derivatives, which have the form

$$\left(\frac{\partial}{\partial y}\left(\frac{\partial f}{\partial x}\right)_y\right)_x = \frac{\partial^2 f}{\partial y \partial x}$$

The right-hand side of the equality is a short-hand notation where the constant subscripted variables are implied. Later in this chapter, we will use the fact that for well-behaved (continuous, smooth) functions, the order of cross-differentiation does not matter, that is,

$$\frac{\partial^2 f}{\partial y \partial x} = \frac{\partial^2 f}{\partial x \partial y}$$

Summing the differentials from the two steps gives

$$df = d_1f + d_2f$$
$$= \left(\frac{\partial f}{\partial x}\right)_{y_i} dx + \left(\frac{\partial f}{\partial y}\right)_{x_i} dy + \left(\frac{\partial^2 f}{\partial x \partial y}\right) dx dy \tag{2.27}$$

The first two terms in the bottom expression are the same as for the exact differential, implying that the third term does not contribute. This is because the product of two differentials makes the third term infinitesimally small compared to the first two terms.[†] Since all cross-derivative terms will have similar products, they can all be safely ignored. A concrete example of this vanishing cross-term is given in **Box 2.2**.

Box 2.2: The area of a rectangle as an illustration of the exact differential formula

As a visual justification of the exact differential relation, consider the area of a rectangle A, as a function of length and width (x and y):

$$A = xy \tag{2.28}$$

The question is, how does the area change with respect to x and y? To represent this pictorially, we will start with finite perturbations to the length and width (by Δx and Δy). The resulting change in area, ΔA, is shown pictorially in **Figure 2.8**.

There are three new pieces to the area. We can calculate these pieces in the following way. First, if we increment x (by Δx) but do not change y, the change in area is $\Delta A_1 = y \Delta x$. If we next we increment y (by Δy), the change in area becomes $\Delta A = (x + \Delta x)\Delta y$. This new area contains to the second and third pieces or area in Figure 2.8: $\Delta A_2 = x \Delta y$ and $\Delta A_3 = \Delta x \Delta y$. Adding these three pieces gives the total change in area:

$$\Delta A = y \Delta x + x \Delta y + \Delta x \Delta y \tag{2.29}$$

If we let the increments in x and y become infinitesimally small ($\Delta x, \Delta y \to dx, dy$), then the third term in Equation 2.29 can be ignored, because it multiplies two infinitesimals, whereas the first and second only multiply one:

$$dA = y dx + x dy + dx dy \cong y dx + x dy \tag{2.30}$$

Note that the coefficients multiplying the differentials in Equation 2.30 are equal to the partial derivatives of the area function (Equation 2.6), that is,

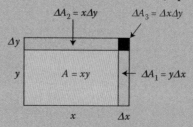

Figure 2.8 How the area of a rectangle responds to increments in its independent variables, length, and width. Starting at area A (shaded gray), the length and width can be incremented by Δx and Δy, making three new rectangular contributions to area. As the increments in x and y get smaller, the contribution made by $\Delta x \Delta y$ (red) goes to zero.

(Continued)

[†]   Keeping in mind that the first two terms are infinitesimal, the third must indeed be *really* tiny!

Box 2.2 (*Continued*): The area of a rectangle as an illustration of the exact differential formula

$$A = xy; \quad \left(\frac{\partial A}{\partial x}\right)_y = y; \quad \left(\frac{\partial A}{\partial y}\right)_x = x \tag{2.31}$$

Substitution of these partial derivatives into Equation 2.30 gives

$$dA = \left(\frac{\partial A}{\partial x}\right)_y dx + \left(\frac{\partial A}{\partial y}\right)_x dy \tag{2.32}$$

which is of the same form as the exact differential for two variables (Equation 2.22).

Integration of exact differentials of multivariable functions. The fact that the cross-terms do not contribute to the differentials of multivariable functions means the differentials of multivariable functions are independent of the order in which each independent variable is evaluated. In the example above, we would have gotten the same differential df (Equation 2.22) if we evaluated the differential in reverse order, (perturb y first, then x). More importantly, and perhaps more surprisingly, exact differentials can be integrated over their variables in a way that is independent of path. For example, integrating the exact differential Equation 2.22 from (x_i, y_i) to (x_f, y_f) gives

$$\Delta f = \int_{f(x_i,y_i)}^{f(x_f,y_f)} df = \int_{x_i,y_i}^{x_f,y_f} \left[\left(\frac{\partial f}{\partial x}\right)_y dx + \left(\frac{\partial f}{\partial y}\right)_x dy\right] \tag{2.33}$$

The integral on the left-hand side of Equation 2.33 is simply the difference $f_f - f_i$. Because this difference Δf does not depend on the path connecting the initial and final states, but simply depends on the value of f at these two end points, f is referred to as a "state function." The importance of the state function in thermodynamics will be discussed in the next section. Here, we will explore the mathematical relationships between exact differentials and state functions.

Though it is perhaps not a surprise that differential df of a well-behaved function $f(x,y)$ integrates simply,[†] as shown on the left-hand side of Equation 2.33, it is not obvious that the integral on the right-hand side should independent of the path.[‡] This path independence is particularly surprising since an explicit path must be specified in order to evaluate this multivariable integral analytically. What features of this integral contribute to this paradoxical relation to the path?

To be sure, the integral on the right-hand side of Equation 2.33 is in a rather unconventional form. Though it has multiple variables of integration (dx and dy), it is *not* a double integral over both x and y (which would lead to a volume under the $f(x,y)$ surface). Rather, the right-hand side of Equation 2.33 is more like a "line integral,"[§] also encountered in multivariable calculus. Both types of integral operate on a path, and can be viewed as representing the area of a "curtain" draping from the function (the integrand) down to the independent variable plane (here x, y) along the integration path (**Figure 2.9**). However, ordinary line integrals have a single multivariable integrand function, whereas integrals of exact differentials (e.g., the right-hand side

---

[†] Note that this is the same behavior as for definite integration of single-variable functions, described previously.

[‡] Though the right-hand side of Equation 2.33 *has* to be independent of the path, since it is equal to Δf, which itself is independent of the path.

[§] Line integrals use as their integration variable the arc length along the path of integration. Despite their names, the path does not need to be a straight line but can be curved (and can even close on itself). To learn more about line-integrals, see a multivariable calculus text.

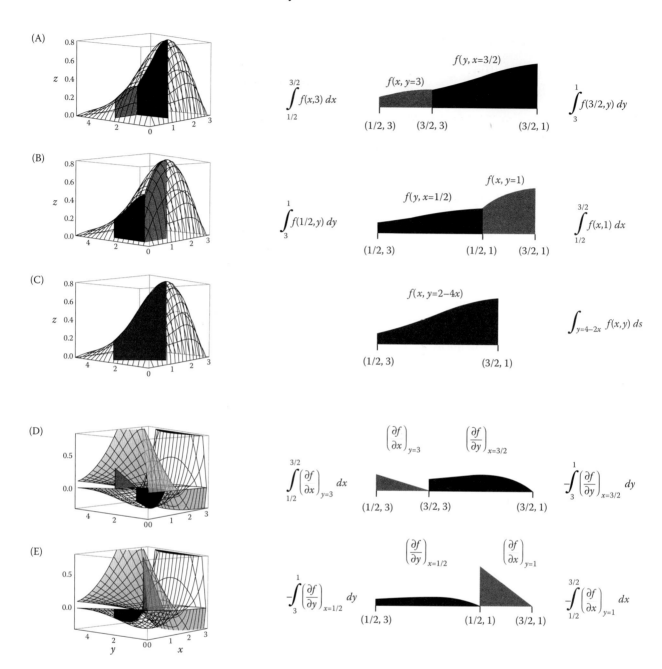

Figure 2.9 Similarities and differences between line integrals of functions and integrals of exact differentials. (A–C) Line integrals of the multivariable function in Figure 2.5 (Equation 2.13), and (D,E) integrals of the exact differential of the multivariable function. The white surfaces in A–C show the multivariable function. The white and gray surfaces in D and E show its partial derivatives. The red and black panels show the result of each step in integration (through their areas), both in the space of the integrated function (left), and flattened out in profile (right) to allow direct comparison of areas in different paths. Note that for the steps along y in (D, E), the area on the right (black) is inverted, because the integration variable decreases along the path (from $y = 3$ to $y = 1$, making the integral positive-valued). In (C), the path involves variation in x and y along the line $y = 4 - 2x$ (dark red). In the other panels, x and y are varied in separate steps (red and black, respectively). All paths connect the same initial (1/2, 3) and final (3/2, 1) states. The line integrals in (A–C) have different values along the three different paths (different summed areas, Problem 2.10). The exact integrals in (D, E) have the same value regardless of the path (Problem 2.11).

of Equation 2.33; we will refer to these as "exact integrals" for brevity) involve different integrands for each independent integration variable (here dx and dy). Another key difference is that, unlike exact differentials, line integrals usually depend on the path.[†]

[†]   An important exception is the line integral of a gradient of a scalar field. The gradient of a scalar field (which is a conservative vector field) has the same form as our exact differential expression, and not surprisingly, integrates in a path-dependent way. This result underlies physical processes such as work in a gravitational or electrical potential.

Figure 2.9 compares line integrals and exact integrals along different paths. The values of the line integrals are simply the sum of the areas bounded by the integrand and the path on the x–y plane (red, note for the single-step path (C) there is only one area), because they integrate over arc-length (which steadily increases along the path, for obvious reasons). The values of the exact integrals (D, E) result from a similar sum of areas, but each term depends in sign on whether the integrating variable increases (here, x from 1/2 to 3/2) or decreases (here, y from 3 to 1). Thus, the negative of the black areas in Figure 2.9 are added to the red areas in integration of the exact differential (D, E). Though the value of the line integral depends on the path (compare Figures 2.1A [A–C]; Problem 2.10), the value of the exact integral does not (compare Figures 2.1A [D,E]; Problem 2.11).

The importance of state functions in thermodynamics. As the name implies, state functions depend only on the current state of the system (in thermodynamics, the state of the system is given by pressure, volume, temperature, and composition), but not how the system has been prepared. That is, state functions are independent of the "history" of the system. The use of state functions has clear advantages for thermodynamic analysis, since we can calculate such quantities (energies, heat capacities, and entropies, and also positions of reaction and levels of assembly) without knowledge of the past. And when a system changes from one state to another, we can calculate the change in the associated state functions by any path we choose, even if the actual change was by a different path.[†]

As described above, the integration of exact differentials is path independent. Thus, if a differential df is exact (Equation 2.22), then we can be sure the function it came from is a state function, and that the integral of df gives Δf on any path. This relationship goes both ways. Given a state function $f(x,y)$, we can be certain its differential is exact.[‡] As described above, the conditions necessary to ensure an exact differential relationship (e.g., df in Equation 2.22) is that the function $f(x,y)$ must be well-behaved (continuous and smooth). The logical consequence that state functions are well-behaved is entirely consistent with the idea above that the value of a state function is uniquely determined by its current conditions (e.g., temperature, volume, and composition) and not on its past.

It is likely that all of the functions you have analyzed in a multivariable calculus course are state functions.[§] Like these, most (but not all) functions we work with in thermodynamics are state functions; thus they have exact differentials. One example is the internal energy (U).[¶] The internal energy is unambiguously defined by the state of a system, regardless of what happened to it in the past, or how it was prepared in its current state. For a system of fixed composition, U is uniquely determined by specifying a pair of variables, for example, the volume V and the temperature T. We can represent this functional relation by the expression $U = f(V, T)$, or simply $U(V, T)$.[**]

Based on our discussion above, we can expect the differential of internal energy to be exact. If we make perturbations to T and V of dT and dV, the differential of internal energy internal (dU) that results is given by

$$dU = \left(\frac{\partial U}{\partial V}\right)_T dV + \left(\frac{\partial U}{\partial T}\right)_V dT \qquad (2.34)$$

[†] This is not to say that we do not have to specify *a* path when calculating changes in thermodynamic variables such as energy. Rather, we can use *any* path for calculation, allowing us to select a path where the calculations are simplified. Most importantly, we can use reversible paths to calculate state function changes (like energy) for systems that changed state irreversibly (typically, irreversible calculations are extremely difficult to carry out).

[‡] If we asserted otherwise, that is, if a state function had an inexact differential, its integration would depend on the path, which would be inconsistent with the idea that the starting function depends only on state.

[§] That is, they are well-behaved, meaning continuous and smooth.

[¶] The internal energy will be defined more precisely in Chapter 3.

[**] In Chapter 5, we will argue that S and V are more useful as independent variables for U. But for now we will use T here because readers at this stage will be more familiar with T than S.

If a system undergoes large (i.e., noninfinitesimal) changes in temperature and volume ($\Delta T = T_f - T_i$ and $\Delta V = V_f - V_i$, where the subscripts i and f designate initial and final states), the change in internal energy is simply

$$\Delta U = U_f - U_i \tag{2.35}$$

This is the integral of the left-hand side of Equation 2.34. Equation 2.35 holds for all paths that connect states i and f, even irreversible paths.

Unfortunately, not all thermodynamic quantities are state functions.[†] Heat and work are two such quantities.[‡] Following the logic above, when a system changes state, the heat flow and the work done depend on the path. We refer to such functions as "**path functions.**" Unlike state functions, this path dependence means that heat and work can't be associated with a particular state of a system. Though we can say a system has a specific amount of a certain amount of internal energy, we can't say that it possesses a certain amount of heat or work. Rather, we use such path functions only when talking about specific *changes* to a system, and only along a specified path. In such changes, it is still valid (and useful) to represent heat flow and work exchange as a sum of many small heat flows and work exchanges, and in the infinitesimal limit, we integrate as before:

$$q = \int_{i \to f} dq \tag{2.36}$$

$$w = \int_{i \to f} dw \tag{2.37}$$

There are two differences in integrating path functions compared to state functions (Equation 2.33). First, we don't use deltas on the left-hand side of Equations 2.36 and 2.37 to represent change in heat or work; we simply use q and w. Second, a path needs to be specified. In Equations 2.36 and 2.37, this is given with the subscripted arrow; although this nomenclature will be omitted for simplicity in subsequent chapters, a particular path is implied in Equations 2.36 and 2.37, and must be specified for analytical calculations.

The Euler test for exact differentials. Though all state functions have exact differentials, and all exact differentials integrate to state functions, *not all differential relationships are exact*. For such "inexact" differentials, there is no corresponding integrated function analogous a state function. Is there some way to tell whether a differential is exact?

Consider the two-variable differential relationship

$$df = S(x,y)dx + T(x,y)dy \tag{2.38}$$

where $S(x,y)$ and $T(x,y)$ are functions of x and y. If df is exact, then we can be sure there exists be a well-behaved state function $f(x,y)$. If we can find such a function, then we know the differential is exact. According to Equation 2.22, if df is exact, then

$$S(x,y) = \left(\frac{\partial f}{\partial x}\right)_y \quad \text{and} \quad T(x,y) = \left(\frac{\partial f}{\partial y}\right)_x \tag{2.39}$$

One way to verify that a state function $f(x, y)$ exists is to make educated guesses for f based on $S(x, y)$ and $T(x, y)$. If we can find a function that differentiates to give Equation 2.38, then we have shown df to be exact, and have simultaneously found a state-function representation. The problem with this approach is that if we fail to identify such a function, there is no guarantee that we were just not clever enough in our educated guesswork.

[†] If they were, we would not need to provide this lengthy development of state functions.
[‡] Heat and work will be defined in Chapter 3, along with their many associated subtleties.

It turns out there is a less haphazard way to test whether a differential is exact. This method is based on the fact that for a well-behaved function $f(x,y)$,[†] cross-differentiation of $f(x,y)$ separately with respect to x and to y is independent of the order of differentiation. This is the Euler relationship:

$$\left(\frac{\partial^2 f}{\partial y \partial x}\right) = \left(\frac{\partial^2 f}{\partial x \partial y}\right), \text{ or } \left(\frac{\partial}{\partial y}\left(\frac{\partial f}{\partial x}\right)_y\right) = \left(\frac{\partial}{\partial x}\left(\frac{\partial f}{\partial y}\right)_x\right) \tag{2.40}$$

If df in Equation 2.38 is exact, then Equation 2.39 can be combined with the Euler relationship to yield

$$\left(\frac{\partial S(x,y)}{\partial y}\right)_x = \left(\frac{\partial T(x,y)}{\partial x}\right)_y \tag{2.41}$$

The derivatives in Equation 2.41 can be computed and compared; equality of the two derivatives demonstrates both that the differential is exact, and that a differentiable function $f(x,y)$ exists.

In the area example above, the exactness of dA, based on the Euler test (Equation 2.41) is easy to verify. In this case, $S(x,y) = y$, and $T(x,y) = x$. Applying the Euler test gives

$$\left(\frac{\partial S(x,y)}{\partial y}\right)_x = \left(\frac{\partial y}{\partial y}\right)_x = 1; \quad \left(\frac{\partial T(x,y)}{\partial x}\right)_y = \left(\frac{\partial x}{\partial x}\right)_y = 1 = \left(\frac{\partial S(x,y)}{\partial y}\right)_x \tag{2.42}$$

Using differentials to represent minima and maxima in multivariable functions. As described above, the critical points in a multivariable function occur at points where each partial derivative[‡] is equal to zero. Instead of representing this as a series of equations (such as Equations 2.16 and 2.17), we can combine these derivative relationships using the exact differential expression, setting df equal to zero. For example, for a function of two independent variables, df takes the following form at a critical point:

$$df = \left(\frac{\partial f}{\partial x}\right)_y dx + \left(\frac{\partial f}{\partial y}\right)_x dy = 0 \tag{2.43}$$

This expression extends to n independent variables by adding an analogous term for each variable:

$$df = \sum_n \left(\frac{\partial f}{\partial x_i}\right)_{x_{j\neq i}} dx_i = 0 \tag{2.44}$$

There are three ways that the middle part of Equation 2.43 (and 2.44) can be equal to zero. Though we are only interested in one of these types of solutions, discussion of the other two will help clarify our intended solution. First, we are not interested in the solution in which the variations dx and dy (and the additional terms in Equation 2.44) are all zero. The point of Equations 2.43 and 2.44 in representing critical points is to describe locations where x and y can be changed without affecting f (i.e., $df = 0$ for $dx, dy \neq 0$).

Second, we are not interested in the specific solutions to Equations 2.43 and 2.44 that result from variations dx and dy that are chosen (by adjusting the size and sign) to offset contributions from the various sensitivity coefficients. This type of solution can be seen by rearranging Equation 2.43:

[†] Again, continuous and smooth.
[‡] That is, the partial of f with respect to each of n independent variables.

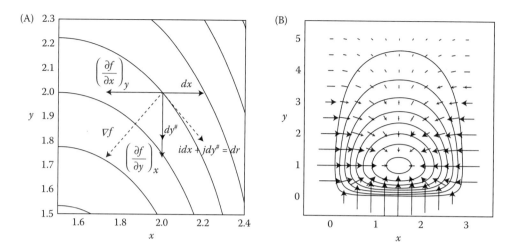

Figure 2.10 A solution to the exact differential expression that does not correspond to a critical point. (A) Starting at an arbitrary point (here, $x = y = 2$) away from the maximum in Equation 2.13, variations in x and y are chosen according to the condition given by Equation 2.45, which ensures that $df = 0$. Starting with an arbitrary value of $dx = 0.2$ (black arrow), Equation 2.45 gives a (constrained) value $dy^{\#}$ (in this case, -0.2). The resulting variation in the x–y plane, which is the vector sum of dx and $dy^{\#}$ ($dr^{\#}$, dashed black line) follows the level curve—the line where $f(x,y)$ remains constant (and thus $df = 0$). Note that this constrained variation $d\vec{r}$ is perpendicular to the vector sum of the partial derivatives of $f(x,y)$, which is equal to the gradient of $f(x,y)$ (Equation 2.46). This is shown more generally in (B), where gradient vectors at different points (red arrows) is perpendicular to the level curves.

$$dy^{\#} = -\frac{(\partial f/\partial x)_y}{(\partial f/\partial y)_x}dx \tag{2.45}$$

For a given value of dx, there will be a specific $dy^{\#}$ value, scaled by the sensitivity coefficients in the x and y directions that solves the equation $df = 0$ (Equation 2.43). The hash-mark indicates that $dy^{\#}$ in Equation 2.45 is not free to vary, but is fixed by the chosen value dx and the partial derivatives. This solution does not identify a maximum or minimum; instead, it corresponds to variation along the level curve of $f(x,y)$, as shown in **Figure 2.10A**.

Rather, the solutions we are after (maxima and minima) are level in all directions. That is, regardless of the combinations of perturbations to the independent variables (dx, dy, ...), $df = 0$. We can vary x but not y, y but not x, or any combination of the two, and f remains constant. This only happens when each partial derivative is zero.

The gradient of multivariable functions

One partial derivative quantity that is important in maximizing multivariable functions is the gradient, which provides a measure of the slope in multiple directions. Numerically, the gradient is a vector sum of the partial derivatives of each independent variable. For $f(x,y)$, the gradient is

$$\nabla f(x,y) = \vec{i}\left(\frac{\partial f}{\partial x}\right)_y + \vec{j}\left(\frac{\partial f}{\partial y}\right)_x$$
$$= \left\langle \left(\frac{\partial f}{\partial x}\right)_y, \left(\frac{\partial f}{\partial y}\right)_x \right\rangle \tag{2.46}$$

The symbol ∇ (sometimes called "del") can be thought of as a partial derivative operator:

$$\nabla = \vec{i}\frac{\partial}{\partial x} + \vec{j}\frac{\partial}{\partial y} \tag{2.47}$$

(with additional terms for additional independent variables). The notation ∇f is read as "the gradient of f." The quantities $\vec{i}$ and $\vec{j}$ are unit vectors in the x and y direction to keep the components of the vectors (the partial derivatives) separate. We will mostly use the format in the second equality of Equation 2.46, which is most convenient for manipulation of equations. The second equality provides a nice reminder that the gradient is a pair[†] of numbers rather than a single scalar value. The gradient gives the steepness of a function, pointing in the direction of maximal slope. Though we will not demonstrate it here, the gradient at any given point is perpendicular to the level curve passing through that point.[‡] **Figure 2.10B** shows this, where the gradient vectors are drawn as red arrows, and run perpendicular to the level curves (black lines).

Note that the exact differential can be thought of as the dot product of the variation vector $d\vec{r}$ (where $d\vec{r} = \vec{i}dx + \vec{j}dy$) and the gradient vector. That is,

$$df = \nabla f \bullet d\vec{r} \tag{2.48}$$

(see Problem 2.19). If we set the exact differential to zero, the solution obtained at critical points results when $\nabla f = 0$. In other words, maxima and minima in multivariable functions have gradients of zero. As discussed above, there are two other ways that the vector form of the exact differential equation (2.48) can equal zero that don't map out critical points. One is the trivial solution that $dr = 0$. Since we are interested in varying x and y (and r), we are not interested in this solution. The third way, which is more interesting (but more subtle) is that neither ∇f nor $d\vec{r}$ are zero. But the two are perpendicular to each other. Because ∇f is perpendicular to the level curve, this constrained variation $d\vec{r}^{\#}$ must be parallel to the level curve. The important implication of this somewhat tricky train of reasoning is that that movement along the movement along x and y in a direction tangent to a level curve identifies a constrained maximum (or minimum). This idea is a key concept that we will return to in Chapters 9 and 10, when we develop the distribution functions of statistical thermodynamics.

FITTING CONTINUOUS CURVES TO DISCRETE DATA

In most physical chemistry textbooks (including this one), equations are developed to approximate the behavior of molecules and collections of molecules. These are either empirical relationships that approximate the behavior of molecules and materials, or they are developed from statistical thermodynamics models. Then the features of these equations are illustrated through graphing them as continuous curves. However, when we *measure* how molecules behave, we don't measure lines or curves. Experimental data consist of collections of discrete points, no matter how much data we collect. To compare equations to real systems, we need a way to compare curves to points. There are two reasons this kind of comparison is important. First, we can use it to test whether our equation (and often the underlying model used to generate it) is consistent with the data (if not, we disprove the model and construct a better one). Second, once we have an equation that adequately describes our data, we can use the statistical tools to determine (or "fit") unknown numerical parameters. This allows us to measure fundamental quantities such as binding constants, energies, and rates of reaction. This type of comparison (equations with discrete data) is referred to as "curve fitting," and is the subject of the remainder of this chapter.

As an example of a curve-fitting problem, consider an experimental investigation between the pressure and volume of a gas. We could fix the volume at specific value, V_1, and record the pressure, p_1. Then we would change the volume to a differ-

---

[†] Or in general an n-tuple, where n is the number of independent variables (and the number of partial derivatives).

[‡] You can find details in your calculus book.

ent value, V_2 and record the pressure p_2, holding the temperature constant.[†] We would repeat this until we had sampled the volume (and pressure) range we are interested in at enough different values to stringently test our model (or models) and fit our unknown parameters. Obviously, the resulting data set is a discrete set of p,V points, not a smooth curve.

What does this discrete set of points tell us about the behavior of the gas? We might like to know if the gas behaves ideally, or if behavior is more complex? If behavior is complex, we might want to test whether the gas is nonideal because the molecules attract each other, or if they repel each other. To do this, we need to compare our data to the ideal gas equation, and to various nonideal equations (like the van der Waals equation, see Chapter 3), and see which equation best approximates the data. If the data are best fitted by a nonideal curve, we might like to know how strongly the gas molecules attract or repel. Model testing and parameter determination must be done simultaneously, because with poor parameter values, a good model will compare poorly to the data. A complicating factor is that all data have random error. Because of random error, no curve will match perfectly, so we must compare among different amounts of imperfection, and decide how much imperfection we will tolerate in the end.

The least-squares approach to compare discrete data and with a continuous curve

In this section we will develop a quantitative metric to compare a function with a given a set of parameters to a set of discrete data, both graphically and numerically. In the following section, we will use these concepts to adjust the curve (by adjusting the parameter) to optimize the fit (and the parameters).

We will represent a generic continuous "fitting" function as

$$f(x; p_1, p_2, ..., p_m)$$
$$= f(x; \vec{p}) \tag{2.49}$$

where x represents the independent variable, and the p_i are the adjustable parameters. The second line uses a more compact vector notation for parameter values:

$$\vec{p} = p_1, p_2, ..., p_m \tag{2.50}$$

We will also use vector notation to represent the data set:

$$\vec{x}_{obs} = x_1, x_2, ..., x_n \tag{2.51}$$

and

$$\vec{y}_{obs} = y_1, y_2, ..., y_n \tag{2.52}$$

To compare the function to the data, we will make a discrete version of the function, by evaluating the function at each value of $\vec{x}_{obs}$, that is, $f(x_i; \vec{p})$. If the function closely approximates the data, the differences between the discrete function values and the observed y_i values, that is,

$$r_i = y_i - f(x_i; \vec{p}) \tag{2.53}$$

will be small. We will refer to each of the n r_i values as "residuals." A close match between the measured y_{obs} values and the fitting function means will result in small residuals. Large residuals mean that the fitting function does not match data. Thus, it might seem like we could select the function that best fits the data simply by minimiz-

---

[†] We could choose to let the temperature vary as, but for now it will be simpler to develop a model with just one independent variable.

ing the sum of the residuals. However, since residuals can be positive or negative, a small sum can be obtained by an offsetting collection of large positive and negative residuals. Thus, the sum of residuals is an imperfect measure of the goodness of fit.

The *square* of each of the residuals does not suffer this problem, since each square has a positive value, even if the residual is negative. Thus, we can find a good fit by minimizing the sum of the *square* of the residuals (*SSR*),

$$SSR = \sum_{i=1}^{n} r_i^2$$

$$= \sum_{i=1}^{n} \{y_i - f(x_i; \vec{p})\}^2$$

(2.54)

We will take the best fit to correspond to the equation $f(x;\vec{p})$ that minimizes SSR.[†] For obvious reasons, this approach is referred to as a "least-squares" method.

A second way that residuals can be helpful in curve-fitting is to make a plot of the distribution of residuals as a function of x values. In such a plot, residuals should be randomly distributed, uniformly scattering above and below the $r = 0$ value. This type of random residual distribution indicates that the equation gives a good representation of the data. A nonrandom distribution of residuals indicates a poor fit, and can be used to determine how the fitting equation (and often the underlying model) should be modified. Though these residual plots require more effort to create, they should be considered an essential part of curve fitting.

As an example of least-squares fitting (and analysis of residual distributions), consider the discrete data set shown in **Figure 2.11** (black circles). The data are clearly nonlinear, and have the overall shape of an upward opening parabola. Thus, we might try to fit the data by least-squares fitting the function

$$f(x) = p_1 x^2$$

(2.55)

where p_1 is a fitting parameter. Although the parameter p_1 can be adjusted so that the function roughly approximates the data (red curve, Figure 2.11B), there is a clear deviation between the best fitted function and the data—at negative values of x, the function $p_1 x^2$ has lower values than the corresponding y values, and at positive values of x, the function has higher values. This deviation results in systematic (i.e., nonrandom) distribution of residuals (red points, Figure 2.11C).

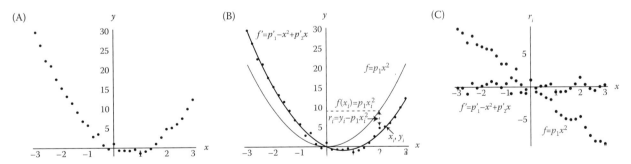

Figure 2.11 A quadratic fit to a data set. (A) A set of discrete data points ($n = 31$ x, y pairs), which show a clear nonlinear relation. (B) Least-squares fits of two different quadratic equations to the data. The equation that includes a linear offset (black, Equation 2.56) gives a good approximation to the data, whereas the equation with a single quadratic term (red, Equation 2.55) deviates significantly. This can be seen as a nonrandom distribution of residuals (C) for the single parameter fitting function. Note, least-squares solutions for the two models in (B) were found using nonlinear least squares, though the equations are both linear in their parameters, allowing analytical solution of the least-squares parameters p_1 and p_2 (see below).

[†] As a justification for this approach, it can be shown that as long as the errors in the y-values are randomly distributed over each of the points, the least-squares solution has the "maximum likelihood" of representing the true equation from which the y values were taken. For justification and further details of least-squares methods, see Bevington (1969), Johnson (1992), and Hughes and Hase (2010).

As described above, a nonrandom distribution of residuals can be used to determine how a function should be modified. The residuals resulting from fitting Equation 2.55 to the data in Figure 2.11A are linearly distributed, with a negative slope (of about $-2/5$) and a zero y-intercept. Thus, we would expect that adding a linear term with an adjustable slope parameter to the fitting function, that is,

$$f'(x) = p_1' x^2 + p_2' x \qquad (2.56)$$

should improve the fit, and indeed it does.[†] The black curve in Figure 2.11B, which corresponds to Equation 2.56 fits the data very well (Figure 2.11B), and the residuals (black points, Figure 2.11C) are both small and randomly distributed. Though it is always possible to include additional terms to "improve" a fit further (i.e., decrease SSR), adding more terms to Equation 2.56 would almost certainly do little more than attempt to fit the random "noise," and is often regarded as "overfitting" (not a good thing; see Silver (2012) for an informal but compelling discussion of overfitting).

Solving the least-squares optimization problem

Now that we have established that we will use the SSR as a metric for curve-fitting, we need to describe *how* the least-squares procedure works. What does it mean to minimize SSR in a least-squares fitting problem? Here, we will answer this question in two ways. We will first represent the minimization problem, pictorially, which provides an intuitive understanding of least-squares optimization. We will then apply the classical analytical approach of minimization, namely, differentiating and setting the result to zero. Though this approach works for a class of fitting functions that are linear in their parameters p_i, it is not solvable for functions that depend nonlinearly on their parameters. For nonlinear functions, we will return to the conceptual picture to outline commonly used "search" methods to find least-squares equations and their parameters.

A visual picture of least-squares

Before developing analytical methods for least-squares solutions, it is important to start with a clear understanding of which variables are being minimized. The key concept is that unlike the relationships that fill calculus books, we don't differentiate x or y in the least-squares approach. Instead, we take the derivative of the SSR, and we differentiate with respect to the parameters $p_1, p_2, ..., p_m$ (more compactly, $\vec{p}$). Writing SSR as a multivariable function of its parameters,

$$SSR = SSR(p_1, p_2, ..., p_m) = SSR(\vec{p}) \qquad (2.57)$$

we can visualize (at least in principle) Equation 2.57 as a surface in a space with axes $p_1, p_2, ..., p_m$, along with a vertical SSR axis. The best fit is obtained at the point in this space (which we will call $\vec{p}_{best}$) where $SSR(\vec{p})$ is a minimum.

For the example shown in Figure 2.11, SSR values for the two-term parabola (Equation 2.56) are shown as a function of the two parameters, p_1 and p_2 (**Figure 2.12**). There is a clear minimum in this space (indicated with the black arrow), which corresponds to the least-squares solution $\vec{p}_{best} = (p_{1best}, p_{2best}) = (2.33, -2.97)$.

An analytical approach: Linear least squares

As mentioned above, one approach to this minimization is to apply the methods of calculus, differentiating SSR with respect to each parameter in $\vec{p}$, setting the deriv-

---

[†] Alternatively, we might choose to modify our fit simply by comparing the deviation between the data and the fitted curve (red curve, Figure 2.11), as it is clear that the fitted curve is too low at negative values of x, and too high at positive values. However, the functional form of the deviation (i.e., linear) is much clearer from the residual plot (Figure 2.11C) than from the data itself (Figure 2.11B).

(A)

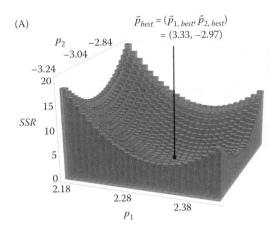

(B)

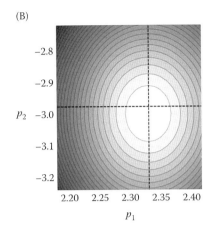

Figure 2.12 A pictorial view of least-squares minimization in the $SSR(\vec{p})$ parameter space for the quadratic fitting example.
(A) For the data set shown in Figure 2.11, SSR is calculated for different sets of parameter values p_1 and p_2 from the two-parameter Equation 2.56. The minimum on this SSR versus $\vec{p}$ surface (black arrow) marks the least-squares solution to the fitting problem. (B) The same surface is in a contour plot, where the dashed lines indicate least-squares parameter values. Note that because the SSR surface is asymmetric, the minimum appears slightly off-center from the lowest contour circle.

ative to zero, and solving for $\vec{p}_{best}$. Because there is almost always more than one adjustable parameter, this minimization is a multivariable problem. Thus, the SSR minimum must be found simultaneously over all of the parameters. This requires the expression of the *total* differential of SSR with respect to simultaneous variations in all the parameter values (dp_i):

$$
\begin{aligned}
d(SSR) &= \sum_{j=1}^{m} \left(\frac{\partial SSR}{\partial p_j} \right)_{p_{k \neq j}} dp_j \\
&= \sum_{j=1}^{m} \frac{\partial}{\partial p_j} \left(\sum_{i=1}^{n} \{ y_i - f(x_i; \vec{p}) \}^2 \right) dp_j \\
&= -\sum_{j=1}^{m} \left(\sum_{i=1}^{n} 2 \{ y_i - f(x_i; \vec{p}) \} \frac{\partial f(x_i; \vec{p})}{\partial p_j} \right)_{p_{k \neq j}} dp_j
\end{aligned}
\tag{2.58}
$$

The inside sum is over the n data pairs (x_i, y_i), whereas the outside sum is over the m parameters. In the third line of Equation 2.58, the first term in the sum is simply the residual at each point, r_i. The second term gives the sensitivity of the fitting function to the *j*th parameter, at each data point. For some fitting equations (linear equations, see below), setting $d(SSR)$ to zero provides a means to finding $\vec{p}_{best}$:

$$
\begin{aligned}
0 &= \sum_{j=1}^{m} \left(\sum_{i=1}^{n} \{ y_i - f(x_i, \vec{p}) \} \frac{\partial f(x_i; \vec{p})}{\partial p_j} \right)^*_{p_{k \neq j}} dp_j \\
&= \sum_{j=1}^{m} \left(\sum_{i=1}^{n} \{ y_i - f(x_i, \vec{p}_{best}) \} \frac{\partial f(x_i; \vec{p}_{best})}{\partial p_j} \right)_{p_{k \neq j}} dp_j
\end{aligned}
\tag{2.59}
$$

The asterisk and the "*best*" subscripts in Equation 2.59 indicates sets of parameter values where SSR is a minimum (as described above for critical points identified by derivative tests). Note that the factor -2 in Equation 2.58 can be dropped without affecting Equation 2.59.

Although Equation 2.58 appears rather formidable, it does have an analytical solution if the fitting function is *linear* in each of the parameters, that is, if $f(x, \vec{p})$ has the form

$$
f(x, \vec{p}) = p_1 g_1(x) + p_2 g_2(x) + \ldots + p_m g_m(x)
\tag{2.60}
$$

Box 2.3: Linear and nonlinear fitting equations

Whether or not a model is "linear" or "nonlinear" is an important distinction in curve-fitting, determining whether a model can be fitted analytically by differentiation, or approximately, using a search method. Perhaps counterintuitively, this distinction is not whether the fitting function $f(x;\vec{p})$ is a straight or curved line when plotted versus x, but whether it is straight or curved when plotted as a function of the p_j parameters,

Here are some examples of linear fitting functions:

$$f(x,\vec{p}) = p_1 + p_2 x \tag{2.61}$$

$$f(x,\vec{p}) = p_1 + p_2 x^2 \tag{2.62}$$

$$f(x,\vec{p}) = p_1 + p_2 e^{-x} \tag{2.63}$$

For these functions, least-squares parameter values can be found analytically by solving a set of equations such as that in (2.71) below:

In contrast, the following example functions are nonlinear in their parameters:

$$f(x,\vec{p}) = p_1 + x^{p_2} \tag{2.64}$$

$$f(x,\vec{p}) = p_1 + e^{-p_2 x} \tag{2.65}$$

$$f(x,\vec{p}) = p_1 \frac{p_2 x}{1 + p_2 x} \tag{2.66}$$

For these equations least-squares parameter values are found approximately, using search methods.

where the $g_j(x)$ terms do not contain any p_i parameters (though they can depend nonlinearly on x, as does the quadratic Equation 2.56). For examples of linear and nonlinear fitting functions, see **Box 2.3**.

For equations that are linear in the parameters p_j, the sensitivity coefficients in Equation 2.59 have a particularly simple form:

$$\left(\frac{\partial f(x_i; \vec{p})}{\partial p_j}\right)_{p_{k \neq j}} = g_j(x_i) \tag{2.67}$$

where the $g_j(x_i)$ expressions are from Equation 2.60. This simplifies Equation 2.59 to

$$\sum_{j=1}^{m}\left(\sum_{i=1}^{n}\{y_i - f(x_i, \vec{p}_{best})\}g_j(x_i)\right)dp_j = 0 \tag{2.68}$$

Because the variations in p_j (i.e., dp_j) are independent of one another,[†] Equation 2.68 can be treated as a set m of independent equations (one for each of the m p_j parameters):

[†] In problems like the linear least-squares optimization here, you should think dp_j terms as variations that are under our control. Though we don't adjust the data (Equations 2.51 and 2.52), we do get to ask how much *SSR* changes when we change one, two, or all of the p_j values (our choice as to which ones and how many). The same rationale is used when we enforce nonzero values for dp_i (Equation 2.69). We will use nearly the same logic to treat variation in particle number and energy in deriving the Boltzmann distribution in Chapter 10, although we will need a couple of key constraints to conserve total mass and energy.

$$\left(\sum_{i=1}^{n}\{y_i - f(x_i, \vec{p}_{best})\}g_1(x_i)\right)dp_1 = 0$$

$$\left(\sum_{i=1}^{n}\{y_i - f(x_i, \vec{p}_{best})\}g_2(x_i)\right)dp_2 = 0 \qquad (2.69)$$

$$\vdots$$

$$\left(\sum_{i=1}^{n}\{y_i - f(x_i, \vec{p}_{best})\}g_m(x_i)\right)dp_m = 0$$

Because the dp_i are nonzero, the zeros in Equations 2.69 must come from the sums in the parentheses. That is,

$$\sum_{i=1}^{n}(y_i - f(x_i, \vec{p}_{best}))g_1(x_i) = 0$$

$$\sum_{i=1}^{n}(y_i - f(x_i, \vec{p}_{best}))g_2(x_i) = 0 \qquad (2.70)$$

$$\vdots$$

$$\sum_{i=1}^{n}(y_i - f(x_i, \vec{p}_{best}))g_m(x_i) = 0$$

Rearranging and substituting the linear representation of the fitting function (Equation 2.60) gives

$$p_{1,best}\sum_{i=1}^{n}g_1(x_i)^2 + p_{2,best}\sum_{i=1}^{n}g_2(x_i)g_1(x_i) + \cdots + p_{m,best}\sum_{i=1}^{n}g_m(x_i)g_1(x_i) = \sum_{i=1}^{n}y_ig_1(x_i)$$

$$p_{1,best}\sum_{i=1}^{n}g_1(x_i)g_2(x_i) + p_{2,best}\sum_{i=1}^{n}g_2(x_i)^2 + \cdots + p_{m,best}\sum_{i=1}^{n}g_m(x_i)g_2(x_i) = \sum_{i=1}^{n}y_ig_2(x_i)$$

$$\vdots$$

$$p_{1,best}\sum_{i=1}^{n}g_1(x_i)g_m(x_i) + p_{2,best}\sum_{i=1}^{n}g_2(x_i)g_m(x_i) + \cdots + p_{m,best}\sum_{i=1}^{n}g_m(x_i)^2 = \sum_{i=1}^{n}y_ig_m(x_i)$$

$$(2.71)$$

Representing this set of equations in matrix form (which separates the unknowns parameters from their coefficients) gives

$$\begin{bmatrix} \sum_{i=1}^{n}g_1(x_i)^2 & \sum_{i=1}^{n}g_2(x_i)g_1(x_i) & \cdots & \sum_{i=1}^{n}g_m(x_i)g_1(x_i) \\ \sum_{i=1}^{n}g_1(x_i)g_2(x_i) & \sum_{i=1}^{n}g_2(x_i)^2 & \cdots & \sum_{i=1}^{n}g_m(x_i)g_2(x_i) \\ \vdots & & & \\ \sum_{i=1}^{n}g_1(x_i)g_m(x_i) & \sum_{i=1}^{n}g_2(x_i)g_m(x_i) & \cdots & \sum_{i=1}^{n}g_m(x_i)^2 \end{bmatrix} \begin{bmatrix} p_{1,best} \\ p_{2,best} \\ \vdots \\ p_{m,best} \end{bmatrix} = \begin{bmatrix} \sum_{i=1}^{n}y_ig_1(x_i) \\ \sum_{i=1}^{n}y_ig_2(x_i) \\ \vdots \\ \sum_{i=1}^{n}y_ig_m(x_i) \end{bmatrix}$$

$$(2.72)$$

The sums on the left and right sides of (2.72) are nothing more than different parts of the fitting function evaluated at different values of x_i and y_i. Most importantly, each of these sums is a constant, dependent on the data (the values of x_i and y_i) but not the parameter values. Thus, Equation 2.72 is simply a set of m equations that are linear in the m optimal (but unknown) parameter values. Solution by substitution and elimination gives numerical values for each of the best fitting parameter values, $\vec{p}_{best}$ (Problem 2.20).

A search approximation: Nonlinear least squares

Many equations used to model thermodynamic behavior are nonlinear in their unknown parameters (see Box 2.3). As a result, the derivative of the fitting function (Equation 2.67) has a more complicated form, where some terms include the parameter values themselves. This makes the corresponding system of equations (those analogous to Equation 2.71) nonlinear. Thus, they cannot be solved by simple algebra methods (i.e., substitution and elimination).

Instead, numerical search methods can be used for nonlinear least-squares problems. Although the numerical search methods are less direct than solving a system of equations analytically, the algorithms used to approximate nonlinear parameters are fast and efficient, even with the slowest personal computers available.[†] Here we will describe the general ideas behind nonlinear least-squares fitting (NLLS). Appendix at the end of this book shows how NLLS is implemented in Mathematica (using the command NonlinearModelFit).

The idea behind the search methods in nonlinear least-squares fitting is really quite simple (**Figure 2.13**). Starting with an arbitrary set of parameters $\vec{p}$, one considers variations to each of the m p_i parameters (in vector notation, a "shift vector" $\Delta\vec{p}$), and looks for parameter changes that decrease the value of SSR. If SSR can be decreased, the new parameter values are considered "better," than the starting parameters (they give the model is closer to the data, in a least-squares sense), and are adopted in place of the starting parameters. This procedure is repeated (or "iterated") until there is no further decrease in SSR. At this point ("iteration z" in Figure 2.13), the SSR value is at a minimum (see Figure 2.12), and current parameter set is taken as the best set of parameters (i.e., $\vec{p}_{best} \equiv \vec{p}_z$).

However, there is one complication that relates to how to "start" a nonlinear least-squares search. To determine how SSR changes when parameter values are modified, there must be a starting set of parameters to modify. Once a fit is underway, these values come from the previous iteration (the middle section of Figure 2.13). But there is no previous iteration at the start. Thus, to get the whole thing going, "initial guesses" for each parameter must be provided prior to fitting. It may seem a bit backward to need parameter values at the start of a fit, since a major goal of fitting the data is to determine (unknown) parameter values.

Fortunately, in most fitting problems chemistry, initial guesses can be quite far from the (unknown) best-fit values[‡] without negatively impacting the final fit. To make reasonable initial guesses, characteristic data points can be identified that are closely related to model parameters, such as y-limits at low and high x values, midpoints, and inflections. Parameters representing y-intercepts can be guessed at from the y values, either by measurement or extrapolation to $x = 0$ (e.g., p_1 in Equations 2.61 through 2.65). Saturation and steady-state values can be approximated from y values at high values of x, (e.g., p_1 in Equation 2.66). Parameters reflecting folding and binding transitions can be guessed at from midpoints, inflections, and peaks (e.g., p_2 in Equations 2.65 and 2.66).

[†] And note that although parameter values obtained from indirect search methods are correctly referred to as "approximate," in most cases (with a reasonable model) they are determined to very high precision, and are no worse than parameters determined by linear least squares. Rather, the uncertainty in fitted parameters typically comes from errors in the data.

[‡] For most models, being within an order of magnitude is sufficient.

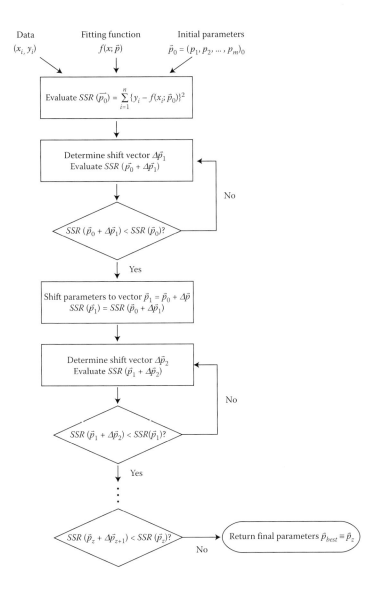

Data
(x_i, y_i)

Fitting function
$f(x; \vec{p})$

Initial parameters
$\vec{p}_0 = (p_1, p_2, \ldots, p_m)_0$

Evaluate $SSR(\vec{p}_0) = \sum_{i=1}^{n} \{y_i - f(x_i; \vec{p}_0)\}^2$

Determine shift vector $\Delta\vec{p}_1$
Evaluate $SSR(\vec{p}_0 + \Delta\vec{p}_1)$

$SSR(\vec{p}_0 + \Delta\vec{p}_1) < SSR(\vec{p}_0)$?   No   Yes

Shift parameters to vector $\vec{p}_1 = \vec{p}_0 + \Delta\vec{p}$
$SSR(\vec{p}_1) = SSR(\vec{p}_0 + \Delta\vec{p}_1)$

Determine shift vector $\Delta\vec{p}_2$
Evaluate $SSR(\vec{p}_1 + \Delta\vec{p}_2)$

$SSR(\vec{p}_1 + \Delta\vec{p}_2) < SSR(\vec{p}_1)$?   No   Yes

$SSR(\vec{p}_z + \Delta\vec{p}_{z+1}) < SSR(\vec{p}_z)$?   No   Return final parameters $\vec{p}_{best} \equiv \vec{p}_z$

Figure 2.13 Schematic flow-chart for NLLS fitting. Fitting begins with a set of data set (Equations 2.51 and 2.52), a parameter-dependent fitting function (Equation 2.49), and a set of initial parameter guesses (Equation 2.50). After determining a starting SSR value using these initial parameters, a shift vector $\Delta\vec{p}_1$ is determined (either randomly or by evaluating the local features of $SSR(\vec{p}_0)$), and a new SSR value at $\vec{p}_0 + \Delta\vec{p}_1$ computed and is compared to the starting value. If the new SSR value is lower than the starting value, parameters are shifted by $\Delta\vec{p}_1$ to $\vec{p}_1$. Then the process repeats itself (a new parameter shift is identified, SSR values are compared, and the new shift is applied to the current parameters, if SSR decreases). The process stops when no shifts can be identified that significantly decrease SSR, and the final parameters are returned.

With initial parameters in hand, a second important issue is how to change parameters during the fit.[†] In parameter space, this can be viewed as asking which direction to change parameters, and by how much (**Figure 2.14**)? The easiest (but not the most efficient) way is to make a random change to parameter values. This is sometimes referred to as a "Monte Carlo" search. Following the flow chart in Figure 2.13, if SSR decreases, then the old parameters are replaced with the new ones, otherwise the old parameters are kept. This kind of search does have some advantages, but it is slow. A lot of computation is spent on parameter changes that will be rejected (because they increase SSR).[‡] A more informed (and usually more efficient) search can be done by evaluating the slope of the $SSR(\vec{p})$ function at the current parameter locus, and changing the parameters in the direction of the largest negative slope. This is done by calculating the gradient ($\nabla SSR(\vec{p})$); as described above, the gradient is a vector that points in the direction of maximal slope; in this case, the vector is in the plane of fitted parameters; see Problem 2.21). Once the direction of shift is selected, the distance of the shift must also be determined. This can come from trial and error, or it can be informed by the ratio of the gradient to the curvature of the $SSR(\vec{p})$ surface.

[†] Though the method used to choose shift vectors is important, it is often either hidden from the user or it is ignored. For simple fitting problems, this is fine, but for large data sets or difficult models, the method used matters a lot.

[‡] Alternatively, the opposite direction in parameter space can be tried; if SSR increases going one direction, it will probably decrease in the opposite direction.

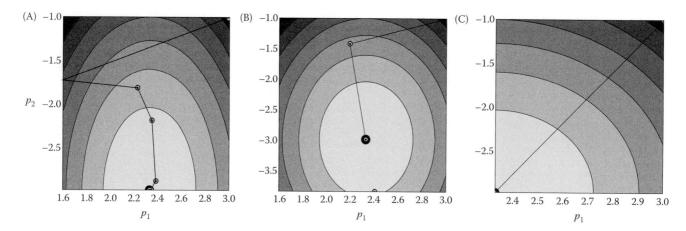

Figure 2.14 Different search methods for NLLS fitting of a two-parameter parabola. The data from Figure 2.11A were fitted with the two-parameter Equation 2.56; individual iterations (black lines) are drawn on the $SSR(p_1,p_2)$ surface defined by the fitting function 2.56. In all three searches, the initial guesses were $p_1 = 3.0$, $p_2 = -1.0$. Red and gray circles show points where SSR and its gradient were evaluated, respectively. (A) Quasi-Newton search, with 10 steps, and 11 evaluations of the SSR and its gradient. (B) Conjugate gradient search, with two steps, and five evaluations of the SSR and its gradient. (C) Levenberg–Marquardt search, with a single step and two evaluations of SSR. In all three cases, the same minimum was found (to at least six significant figures for each parameter).

There are several nonlinear least-squares algorithms that use different methods to select a shift vector. Some of these algorithms are shown in Figure 2.14. These algorithms differ in the number of steps it takes them to find the minimum, and the "radius of convergence," that is, how far the initial guesses can be from the best-fitted parameter values and still find minimum. The Levenberg–Marquardt method provides a good compromise between efficiency and sensitivity, and is used in examples throughout this book. More details about these methods can be found in the Mathematica documentation center, and in Bevington (1969).

A final issue that must be considered is when to end a search. Remember, these search methods are approximate, so there is always a chance that another iteration could further decrease SSR, Very close to a minimum in parameter space, these additional changes in SSR will be tiny, and associated parameter changes will be insignificant, given the noise level in the original data. To avoid a protracted search among (nearly) equally good parameter values, a "convergence criterion" is usually specified that stops the search. This is often achieved by supplying a threshold minimum decrease in SSR: if successive iterations decrease SSR by less than some small percentage of the total value, the search is terminated, and the most recent set of parameters are returned. In practice, for most nonlinear fits, the default convergence threshold value is generally adequate for a good fit. Thus the user is often not aware of its value, though there is sometimes a risk that if data are really good, the fitting algorithm might terminate too easily.

Pitfalls in NLLS

Although nonlinear least-squares fitting is quite robust when it is applied appropriately, there are a number of pitfalls to avoid. Some pitfalls are obvious; others are rather subtle. The most obvious pitfalls will prevent the user from executing the search. Others will search but will lead to a very poor fit. Though such failures can be annoying, these are actually good problems in that they alert the user to problems that can be fixed. For the most subtle pitfalls, the search appears to succeed, but there are problems associated with the resulting parameters. It is up to the user to watch out for these problems, fix them if possible, and if not, be aware of the limitations of the fit. The following section describes some of the common errors encountered in least-squares fitting.

1. ***Too few data points, or too many parameters.*** Regardless of whether the fitted model is linear or nonlinear, there must be more data points than parameters. Viewing the fitting problem as the solution of a set of equations (which, for the linear

case, is given by Equations 2.71 and 2.72), this requirement ensures that there are more equations (data points) than unknowns (parameters). Otherwise, there are likely to be an infinite number of solutions (parameter sets) that fit the data.

2. ***Poor initial guesses.*** Though a successful fit can be obtained using initial guesses that differ substantially from best-fit parameters, there are limits. In such cases, NLLS algorithms run into numerical difficulties (either producing a memory overflow, or simply failing to find changes that produce a systematic decrease in *SSR*). Depending on the details, this can either produce error messages, or can produce a very bad fit, and nonsensical parameters. Thus, it is very important to *view* the fit and the data together, and to think about whether the parameters are numerically reasonable.

   Sometimes initial guesses are off because they are in different units than a scaling constant in the equation. For example, if an equation uses a value of the gas constant, *R*, of 8.314 joule per mole, and guesses for enthalpies and free energies are in kilojoules per mole, initial guess-associated problems will likely be encountered. One way to test for mistakes in initial guesses is to plot the function to be fitted, using the initial guesses as constants (a "guess function"), along with the observed data. If there is a big difference between the guess function and the data, then the initial guesses need to be adjusted.[†] Making adjustments to the guess function and monitoring the response provides a systematic way to find better guesses (or to finding them in the first place).

3. ***Nonuniform errors in the observed $\vec{y}_{obs}$ values.*** One of the assumptions of the least-squares fitting procedure is that each measured y_i value has the same error.[‡] This assumption is often met when all observed values are obtained with a single instrument without adjusting settings that influence sensitivity, and the raw data are fitted. But when observations from different instruments are combined, or more commonly, when data are processed nonlinearly prior to the *NLLS* analysis, errors may be different from one value of y_{obs} to the next. Common examples of such nonlinear transforms are taking ratios of concentration-based measurements into equilibrium constants, taking the log of concentration data in kinetic studies, and "linearizing" data in Scatchard plots and Lineweaver–Burke plots (in binding and enzyme activity measurements). In such cases, the fitting algorithm is likely to adjust parameters so that the large residuals associated with points with magnified uncertainty are minimized at the expense of residuals that are well-determined.

   A logarithmic transformation of data containing error is shown in **Figure 2.15**. Data are generated from the exponential decay equation

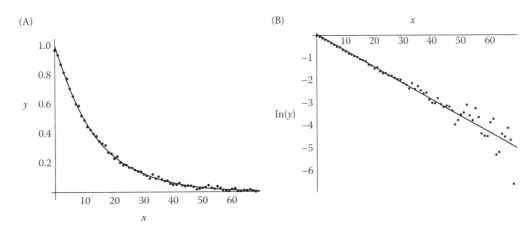

Figure 2.15 A logarithmic transform and effects on errors. (A) An exponential decay (from Equation 2.73, with $y_0 = 1.0$, and $k = 0.07$, with Gaussian errors with a standard deviation of 0.01). The solid line is a two-parameter nonlinear fit to the data, with fitted parameters $y_0 = 0.99991$, and $k = 0.06971$. (B) A logarithmic transformation, which linearizes the exponential decay, but amplifies the error at large values of x (small values of y). The solid line is a two-parameter unweighted fit, with fitted parameters $\ln(y_0) = 0.05714$, and $k = 0.07252$.

[†] Though it is always possible that the shape of the function cannot be made to match the data. For example, a linear function cannot be made to match data that shows curvature (although this sort of problem with the fitting function typically does converge on a fit, but a poor one).

[‡] Also, the error in the y_i values are assumed to come from a Gaussian distribution, and the x_i values are assumed to have no error.

$$y = y_0 e^{-kx} \tag{2.73}$$

Each of the y_i points has an added random error term taken from a Gaussian distribution with a standard deviation of 0.01. The points in Figure 2.15B are a logarithmic transformation of the data. This transformation magnifies the scatter on the right side of the logarithmic plot, generating large residuals. As a result, fitting to the linearized data (solid line, panel B) overemphasizes these noisy transformed points, and compromises the fit. For the direct fit to the data (panel A), the decay constant k is within 0.4% of the "true" value (0.0697 vs. 0.07); in contrast, for the transformed data, the fitted decay constant k differs from the true value by 3.6% (0.0725 vs. 0.07).

One way to fix this problem is to "weight" each of the n data points according to the uncertainty associated with each point during the least-squares minimization (Problem 2.22). Instead of minimizing the sum of the squares of the residuals (Equation 2.58), a "weighted sum of squares" is minimized:

$$\sum_{j=1}^{m} \frac{\partial}{\partial p_j} \left(\sum_{i=1}^{n} W_i \{ y_i - f(x_i; \vec{p}) \}^2 \right) dp_j = 0 \tag{2.74}$$

Typically,[†] the weighting factor W_i is the reciprocal of square of the uncertainty:

$$W_i = \frac{1}{\sigma_{y_i}^2} \tag{2.75}$$

A weighted fit to the linearized decay data in Figure 2.15 gives a significantly closer fit to the true parameters (the fitted k parameter is within 1.1%) than the unweighted fit, although it is not as close as the direct fit to the exponential decay.

4. **The effects of a large outlier on the fit.** If each observed y_i value has some error associated with it, then the occasional y_i value is expected to deviate substantially from the true value. If enough data points are collected, there are likely to be both positive and negative deviations, and these will tend to cancel. However, on occasion, something goes wrong with the experiment, such as an error in preparing a particular sample, or a temporary variation in instrument power or temperature. These nonrandom errors have the potential to produce a point (or a few) that differs significantly from the trend followed by the rest of the data. Because the *NLLS* algorithm minimizes the *square* of the residuals, the fitted curve will be strongly influenced by such an outlier.

The question is, "what does one do with an outlier?" Two extreme options are to do nothing (keep the data point), and to remove the data point. Statisticians do not agree in which is most appropriate (or least offensive), and we will not pick between these two. A third option (which seems less offensive), is to repeat the entire set of measurements and see if the suspected point remains deviant. If so, it is probably real. If not, it was probably a result of systematic error, giving some sense that the second data set can be interpreted with confidence.[‡] With careful attention paid to the second set of measurements, the source of the deviation may be isolated. But at the very least, if the "do nothing" extreme is used, it should be supplemented by an awareness of how the outlier may be affecting the fit: if the point were to be dropped from the data, how much would the fitted parameters change?

5. **The wrong model.** One of the main goals of *NLLS* fitting is to determine the values of parameters of interest. To determine meaningful parameters, a close agreement between the model and the data are a must—a poor fit will almost certainly mean questionable parameter values. Poor fitting may be in the form of large systematic residuals, which indicate that the fitting function does not correctly describe the data. In this case, parameters will not only have large numerical uncertainties, they

---

† A description of uncertainty weighting can be found in Bevington (1969).

‡ However, even this simple strategy can also be viewed as problematic, as it may license to collect data sets until one comes along that "looks good."

are likely to have questionable meaning. If the fitting function is not capable of reproducing the data, the underlying physical model is probably wrong; thus, fitted parameters will lack physical meaning (or at least, they won't mean what the modeler thinks they mean).

6. ***Problems with the data.*** Another way that a low-quality fit can be obtained is if the data are of low quality. Even with a great model, large errors in the $\{y_i\}$ values will result in high parameter uncertainties. The result of large $\{y_i\}$ errors will be large *uniform* residuals (and note that with high residuals, it is difficult to identify subtle deviations between the model and the measurements). Although parameter uncertainties can be decreased to some extent by collecting more data points (i.e., increasing i in the fitting equations above), decreasing the measurement error is often a more practical way to decrease parameter uncertainties.[†]

   However, even a good fit (low *SSR* values) is no guarantee that parameters are well-determined. A set of $\{x_i\}$ may not sample the region of x over which a parameter influences the fitting function. For example, if a time series is collected to determine an exponential decay parameter, but the rate of decay is much faster than the time steps (the differences between the x_i measurements), then the decay constant can at best be considered to be larger than a limiting rate (Problem 2.21). This would result in a large uncertainty in the fitted value of the decay constant.

7. ***Parameter correlation.*** Even when data are collected over an appropriate region of x, subsets of parameters can exhibit large uncertainties through correlation with each other. Correlated parameters have a similar influence on the shape of the fitting function, and as a result, a large value of one parameter can be compensated by a small value of the other. As an extreme example, consider the equation describing the time-dependence of a small displacement to a pendulum

$$\theta(t) = \theta_0 \cos\left(2\pi\sqrt{\frac{g}{\ell}}\,t + \delta\right) \tag{2.76}$$

From a careful measurement of the displacement angle as a function of time ($\theta(t)$), one could fit the amplitude (θ_0), phase δ, and the pendulum length ℓ, because all three of these affect different features of the equation. However, one cannot simultaneously fit ℓ and the gravitational constant g, because these two parameters affect the equation the same way: large fitted values of ℓ are compensated by large fitted values of g. As a result, these two parameters have infinite correlation, and cannot be simultaneously fitted.

The ratio of parameters in the pendulum example (Equation 2.76) is a rather obvious form of parameter correlation. However, significant parameter correlation can occur even when parameters are associated with different mathematical features of a fitting function. As described above, such correlations increase parameter uncertainty. In a linear fit, for example, the slope and intercept are correlated, even though they clearly have different effects on the equation of a line. Uncertainties due to correlation are often enhanced by long extrapolations. For a line, uncertainties in intercept are enhanced when data are collected over a narrow range that is far from the y-axis (**Figure 2.16A**).[‡] The *SSR* surface (i.e., *SSR* as a function of slope and intercept values) of the extrapolated linear fit has the shape of a broad diagonal trough (**Figure 2.16B**). A better view of the *SSR* surface is obtained in a contour plot, where curves of constant *SSR* are highly eccentric (i.e., flattened) ellipses that are tilted with respect to the parameter axes (**Figure 2.16C**).

[†] Like many other statistically averaged quantities, the uncertainties in fitted parameters decrease with the square root of the number of measurements ($n^{1/2}$; see Tellinghuisin, 2008). To decrease fitted parameter uncertainty by a factor of two, n must be increased by a factor of four. While collecting four times as much data may be tolerable, it is likely that collecting one hundred times more data (to decrease parameter uncertainties by a factor of ten) is beyond the patience of most experimentalists.

[‡] As we will describe in Chapter 8, this kind of long extrapolation will be encountered in analysis of the sharp conformational transitions of biological macromolecules.

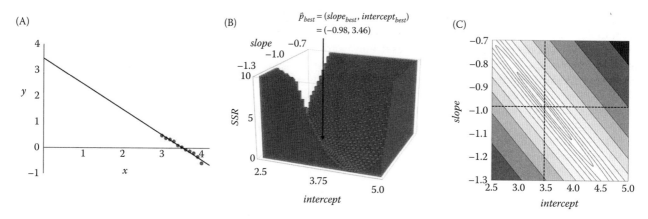

Figure 2.16 Fitting a function with strongly correlated parameters. (A) Data with noise (gray points, taken from a line $y = -1x + 3.5$ with 5% random error added) is limited to a narrow range of x ($x = 3$ to $x = 4$). The line is the best fit of the line $y = mx + b$, with best-fitted parameters of -0.98 and 3.46, respectively. (B) The SSR surface as a function of the two fitted parameters shows a narrow diagonal trough. (C) The shape of this trough is seen more clearly as a SSR contour plot, in which contour lines are highly eccentric diagonal ellipses. Because the major axes of these ellipses are diagonal in the parameter space, the two parameters have strong negative correlation: large intercept values can be accommodated by large (negative) slopes. This type of correlation is typical with parameters extrapolated over large distances, as is seen in equilibrium protein folding studies (Chapter 8).

This tilting means that the two parameters are highly correlated. In this case, a high slope can be balanced by a low intercept, and vice versa. The result of this correlation is large parameter uncertainties.

One way to decrease this correlation (and thus decrease parameter uncertainty) is to minimize the extrapolation. For example, in the linear fit shown in Figure 2.16A, fitting as a function of slope and the value of y at $x = 3.5$ (the mean of the measured x_i values), this correlation can be greatly reduced (Problem 2.24). However, it is not always possible to eliminate correlation by reparameterization, especially for complicated, nonlinear models. In such cases, it is critical that we come up with quantitative measures of parameter uncertainty, and the underlying correlations that contribute to this uncertainty.

Putting quantitative error bars on fitted parameters

Although fitted parameters are derived quantities that are usually measured after data have been collected, such parameters should be considered to be experimentally determined; thus, parameters have uncertainties. When a parameter value is estimated from a fit, its value is unlikely to match the "true" value, but will differ from the true value by some (unknown) amount. In the section above, several potential sources of parameter uncertainty were discussed, including high errors in measured y_i values and strong correlations between parameters.

Here we will discuss some methods to estimate parameter uncertainties. One way to do this is to collect multiple, independent data sets, fit each data set independently, and compare parameters from different sets. This approach will allow us to calculate a mean and a standard deviation of each measured parameter. The more times we do this, the closer our parameter means will be to their true values. Although this approach works in principle, it is often not practical to collect multiple data sets, especially if each data set is large, if samples are precious, or if instrument time is limited. Such cases call for a method to estimate parameter uncertainties from a single data set. If this sounds like a difficult proposition since we have one value for each fitted parameter, keep in mind that we made many measurements (the y_i values) to determine this single value. In the methods described below, we will combine error estimates from these measurements with the structural features of our fitting equation to determine parameter uncertainties.

For *linear* least squares, there is an analytical method for obtaining uncertainties in each fitted parameter. This method requires that errors in the measured y_i values have Gaussian distributions, and either that each of these Gaussian distributions is the same or that differences have been accounted for by weighting the fit (Equations 2.74 and 2.75). This method takes the equations used to determine the best fitted parameters and extracts an expression for the variance of each parameter as well as the covariance between parameter pairs. For m parameters, these quantities are represented by the "covariance matrix, V"[†]:

$$V = \begin{bmatrix} v_{1,1} & v_{2,1} & \cdots & v_{m,1} \\ v_{1,2} & v_{2,2} & \cdots & v_{m,2} \\ \vdots & \vdots & \ddots & \vdots \\ v_{1,m} & v_{2,m} & \cdots & v_{m,m} \end{bmatrix} \tag{2.77}$$

The square root of each diagonal element of V give estimates of the standard deviation for the corresponding parameter. For example, for the *ith parameter*,

$$s_{p_i} = v_{i,i}^{1/2} \tag{2.78}$$

For details on calculating V, see Appendix 2.1.

Although Equations 2.77 and 2.78 are not exact for nonlinear least-squares, as long as parameters are reasonably well-determined, these equations provide an adequate representation of parameter uncertainty (see Tellinghuisin, 2008). Most least-squares fitting packages use Equation 2.78 to give parameter errors. In Mathematica, the V matrix can be obtained using the command `fit["CovarianceMatrix"]`, where `fit` is the variable name associated with the `NonlinearModelFit[…]` command.

To illustrate the estimation of parameter uncertainties from the covariance matrix, and to compare them to more statistically rigorous approaches (described below), we will perform nonlinear least squares to model exponential decay (**Figure 2.17**). This type of decay is commonly observed in chemical kinetics, where a reactant is converted to product by a first-order mechanism. Figure 2.17 shows a data set that decays exponentially to a nonzero endpoint, with random error (from a Gaussian distribution with a standard deviation of 0.01) added to each y_i value. This data is well-fitted by the exponential decay equation,

$$f(x) = (y_0 - y_\infty)e^{-kx} + y_\infty \tag{2.79}$$

Here, y_0 is the initial y value at $x = 0$, y_∞ is the final y value at $x = \infty$, and k is the decay constant.

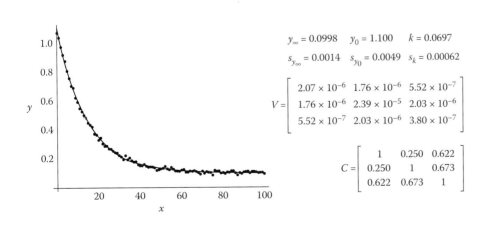

$$y_\infty = 0.0998 \quad y_0 = 1.100 \quad k = 0.0697$$
$$s_{y_\infty} = 0.0014 \quad s_{y_0} = 0.0049 \quad s_k = 0.00062$$

$$V = \begin{bmatrix} 2.07 \times 10^{-6} & 1.76 \times 10^{-6} & 5.52 \times 10^{-7} \\ 1.76 \times 10^{-6} & 2.39 \times 10^{-5} & 2.03 \times 10^{-6} \\ 5.52 \times 10^{-7} & 2.03 \times 10^{-6} & 3.80 \times 10^{-7} \end{bmatrix}$$

$$C = \begin{bmatrix} 1 & 0.250 & 0.622 \\ 0.250 & 1 & 0.673 \\ 0.622 & 0.673 & 1 \end{bmatrix}$$

Figure 2.17 A nonlinear least-squares fit of exponential decay, and the resulting covariance and correlation matrices. Left, 101 data points with random Gaussian error of $\sigma_y = 0.01$ (approximately 1%), fitted with equation 2.79. The sum of the square of residuals (SSR) for this fit is 0.0902. Right, fitted parameters along with the covariance matrix V, and derived parameter uncertainties (s_{p_i}, Equation 2.78). The correlation matrix C is calculated from the elements of V using Equation 2.81. The ordering of the columns and rows of V and C from left to right and top to bottom is y_∞, y_0, k.

[†] *V is sometimes referred to as a "variance-covariance matrix," because it contains both variances (diagonal terms) and covariances (off-diagonal terms).*

Figure 2.17 also lists the V matrix from the fit. The diagonal elements of V can be used to calculate error estimates for each of the fitted parameters. For example, for the decay constant, the error estimate can be approximated as[†]

$$s_k \approx \sqrt{V_{k,k}} = \sqrt{3.80 \times 10^{-7}} = 6.16 \times 10^{-4} \tag{2.80}$$

As long as the rate constant estimates have a Gaussian distribution about the mean, s_k can be interpreted as a standard deviation, meaning that if multiple independent data sets like that in Figure 2.17 were collected and fitted, we could expect 67% of our fitted k values to lie in an interval that is $\pm 6.16 \times 10^{-4}$ of the true mean value of k.

In addition to providing information on parameter uncertainties, we can use the V matrix to calculate correlations between parameters. This provides a quantitative measure of the extent of parameter correlation described above (Figure. 2.17). Correlation is quantified using the values of the off-diagonal covariance parameters $V_{i,j}$, which are proportional to the correlation between parameters i and j. One shortcoming of the raw $V_{i,j}$ parameters is that they depend on the scale of the two parameters as well as the correlation.[‡] To account for parameter scale, we can compute "coefficients of correlation" ($C_{i,j}$) for each parameter pair:

$$C_{i,j} = \frac{V_{i,j}}{V_{i,i}^{1/2} \times V_{j,j}^{1/2}} \tag{2.81}$$

For each pair, the correlation coefficient is simply the covariance between parameters i and j divided by the standard deviations for the two parameters (the square root variances). Each of these three quantities can be found in the matrix V. Correlation coefficients between parameters range from minus one to plus one. A $C_{i,j}$ value of zero means that parameters i and j are completely uncorrelated, whereas values of mean that the parameters are completely correlated (positively or negatively). Although there is no "threshold" value at which parameters should be considered to be correlated,[§] absolute values of $C_{i,j}$ greater than 0.95 are cause for concern.

Values of $C_{i,j}$ are often organized into an $m \times m$ matrix, just like the covariance matrix.

$$C = \begin{bmatrix} C_{1,1} & C_{2,1} & \cdots & C_{m,1} \\ C_{1,2} & C_{2,2} & \cdots & C_{m,2} \\ \vdots & \vdots & \ddots & \vdots \\ C_{1,m} & C_{2,m} & \cdots & C_{m,m} \end{bmatrix} \tag{2.82}$$

In addition to being square, V and C are both symmetric, that is, $C_{i,j} = C_{j,i}$. In words, this relates to the truism that if parameter i is correlated with parameter j, then j is correlated with i to the same degree.

Analysis of parameter distributions using the "bootstrap" method

Though the use of the covariance matrix to estimate parameter uncertainty is common, as described above, for nonlinear models these estimates should be regarded as approximate. Here we will describe a statistical method that explores the nonlinear least-squares parameter distribution itself. This method uses the results of the least-squares fit, including the fitted parameters and residuals, to generate many

---

[†] We take Equation 2.80 as an approximation since Equation 2.78 is not strictly valid for nonlinear least squares.

[‡] As an illustration of "scale," if parameters i and j have units of joules, the will have significantly larger numerical covariances than if kilojoules were used instead.

[§] In principle, any nonzero value of $C_{i,j}$ implies correlation between parameters i and j, and most parameters show some correlation.

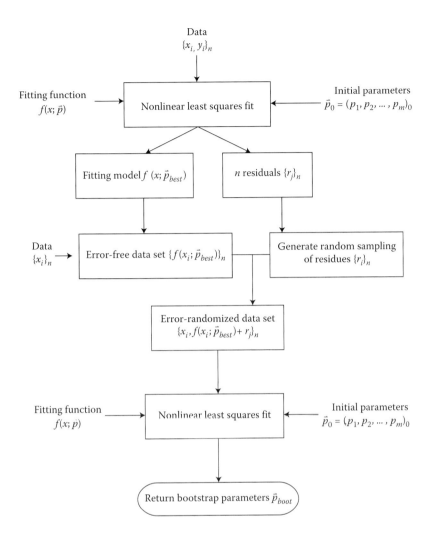

Figure 2.18 The bootstrap method for estimating uncertainties in fitted parameters. In the first step, nonlinear least squares is performed on the experimental data set $\{x_i, y_i\}_n$, as illustrated in Figure 2.13. Using the best-fit parameters, the fitting function is used to generate error-free data for each value x_i in the original data set. The resulting error-free set $\{f(x_i); \vec{p}_{best}\}_i$ is combined with residuals that are randomly sampled with replacement from the original fit. This error-randomized data set is fitted using the same fitting function as used in the original fit. The resulting "bootstrapped" parameters are saved, and the second round of fitting (red arrows and boxes) is repeated (many times) using a different randomized residual set. Through repetition, a large number of bootstrapped parameters are obtained, which can be used for statistical analysis of parameter uncertainty.

simulated data sets that have randomized errors. These "resampled" data sets are then refit using the original model, generating a distribution of fitted (resampled) parameters. A flow-chart for this "bootstrap" resampling approach is shown in **Figure 2.18**.

In the first step, the original fitted function $f(x_i; \vec{p})$ is used to generate "error-free" y values at each of the n original x_i values. Note that some implementations of this test randomly select x_i values with replacement, changing the structure of the data set. This approach, which changes the structure of the data set somewhat, is valuable for large data sets that are sampled with high density (such as velocity sedimentation data in analytical ultracentrifugation). This alternative approach is more closely related to the "jack-knife" test for parameter uncertainty. Using the same x_i values means that the resampled data sets have the same "structure" as the original data set.

In the second step, error values are randomly selected from the n-long list of residuals generated in the original fit, and are added to each of the error-free values. This results in n synthetic y_i values with random errors corresponding to each of the original x_i points.[†] An important aspect of this random error selection is that

---

[†] Note that some implementations of this test randomly select x_i values with replacement, changing the structure of the data set. This approach, which changes the structure of the data set somewhat, is valuable for large data sets that are sampled with high density (such as velocity sedimentation data in analytical ultracentrifugation). This alternative approach is more closely related to the "jack-knife" test for parameter uncertainty.

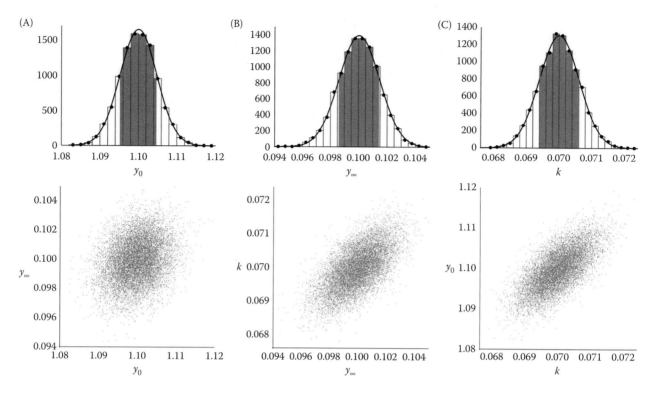

Figure 2.19 Bootstrap analysis of the exponential decay fit. A 10,000 iteration bootstrap procedure outlined in Figure 2.18 was applied to the fit in Figure 2.17. Upper plots (A–C) show histograms of for each of the three fitted parameters. The solid black curves are Gaussian fits to the parameter distributions (the bin-heights at the bin centers). The gray filled portion of the histogram bars indicates the 67% confidence intervals (Table 2.1). Lower plots show two-dimensional scatter plots between pairs of fitted bootstrap parameter values. Each gray point is one of 10,000 bootstrap pairs. Parameter correlation results in a tilted elliptical distribution of points (panels B and C).

residuals are chosen "with replacement."[†] Sampling-with-replacement methods like this are often referred to as "bootstrap" methods.

In the third step, these simulated data sets are fitted using the original model, and fitted bootstrapped parameters are recorded. This procedure is repeated many (thousands of) times, building up a distribution of values for each fitted parameter. From these distributions, measurements of parameter spread can be calculated, such as root-mean-square deviation. If the simulated parameters are normally distributed,[‡] the root-mean-square deviation provides a good approximation of the error each parameter.

Another means to represent parameter uncertainties, which is more appropriate if the parameter distribution is asymmetric, is to determine upper and lower percentiled confidence intervals. For example, 95% confidence intervals include 95% of the parameters in the distribution. These are usually chosen so that 2.5% of the bootstrapped parameters are below the lower bound, and 2.5% are above the upper bound.[§]

[†] The meaning of the term "with replacement" can be illustrated using an "urn" problem (Chapter 1). Imagine an urn with r red balls and $n-r$ white balls, and you want to randomly sample the color of multiple balls, one at a time. For your first sample, you reach into the urn, grab a ball, and note its color. For your second sample, you can either replace the first ball (and mix them up—this is "sampling with replacement"), or you can leave the first one out and grab a second ball from the $n-1$ remaining balls ("sampling without replacement").

[‡] Because the bootstrap method gives us a distribution of parameters, the question of whether the distribution is approximately Gaussian normal is easy to test (by inspection and/or fitting), as will be shown in the example below.

[§] The 67% confidence limits, which are similarly calculated (with 16.5% of the bootstrap parameters below and 16.5% above), should provide good approximations of the standard deviation of bootstrap distribution, as long as the distribution is close to Gaussian.

Table 2.1 Estimates of parameter uncertainty from the exponential decay fit

Parameter	Best fit value	Uncertainty from covariance matrix	Bootstrap standard deviation	Bootstrap 67% confidence intervals	f-statistic 67% confidence intervals
y_0	1.0999	0.00489	0.00479	1.0953, 1.1046	1.084, 1.116
y_∞	0.0998	0.00144	0.00142	0.0986, 0.1014	0.0984, 0.1012
k	0.0697	0.000616	0.00061	0.06941, 0.07061	0.0677, 0.0717

Note: The data set and fit is from Figure 2.17. The uncertainty from the covariance matrix is from Equation 2.80. Bootstrap analysis was performed as in Figure 2.18, with 10,000 iterations (Figure 2.19). Confidence intervals from f-statistics were determined as illustrated in Figures 2.21 and 2.22.

The bootstrap method is illustrated for the exponential decay fit in **Figure 2.19**. The residuals and parameters from the fit in Figure 2.17 were used to generate and fit synthetic data with randomized errors 10,000 times. The resulting parameter distributions for y_0, y_∞, and k from Equation 2.79) are shown in the top panels Figure 2.19. To good approximation, these three distributions are Gaussian (see black solid lines). The standard deviations for these three parameter distributions and the 67% confidence intervals calculated from the bootstrap are very similar, and are quite close to the simpler covariance matrix estimates of parameter uncertainty (see **Table 2.1**).

However, the upper and lower confidence intervals suggest that there is a slight asymmetry in the parameter distributions, based on the fact that the lower and upper confidence intervals differ slightly. Such differences (which are often much larger than in the present example) are not unexpected when using NLLS. As long as asymmetry is modest, the simpler covariance matrix estimate is good enough. For high asymmetry, the upper and lower confidence intervals should be reported instead. And to reiterate, obtaining multiple independent data sets can be obtained through repetition and separate parameters can be fitted for each set should be preferred overall of these computed estimates of parameter uncertainties.

In addition to providing a measure of uncertainties, the bootstrap method provides a means to visualize parameter correlation. One way to do this is to calculate scatter plots in which pairs of parameter values from each bootstrap iteration are plotted as single points in parameter space (Figure 2.19). For the three-parameter model, there are three unique pairs for which correlations can be examined. The two amplitude parameters (y_∞ and y_0) show relatively little correlation (Figure 2.19A, lower). In contrast, the rate parameter k shows some correlation with both of the amplitude parameters, as evidenced by the diagonal, ellipsoidal distribution of parameter pairs in $y_\infty - k$ and $k - y_0$ space (Figure 2.19B, C, lower panels). These observations are consistent with the numerical values in the correlation matrix (Figure 2.17): C_{y_∞, y_0} is smaller (0.25) than $C_{y_\infty, k}$ and $C_{y_0, k}$ (0.622 and 0.673).

In Figure 2.19 the abscissas for the one-dimensional parameter histograms (upper) have been scaled to match those on the corresponding pairwise correlation plots (lower), allowing the overall spread of each parameter to be compared to its correlated spread. Such comparison shows that the overall distributions (i.e., the single-parameter histograms) include all the variance, even when it is strongly correlated. In fact, the one-dimensional histograms are simply projections of the parameter of interest from the correlation plots onto their respective axes.

Analysis of parameter uncertainties and comparing models using the f test

Another way to estimating parameter uncertainties is based on looking at the how the *SSR* function changes with parameter values near the minimum identified by NLLS. Since least squares minimizes the *SSR*, shifting parameters away from best-

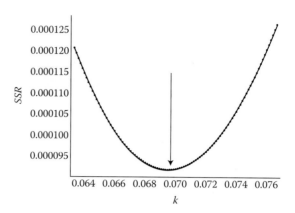

Figure 2.20 The variation in the sum of square residuals as a function of a test parameter. The decay data from Figure 2.17 is fitted with the exponential decay Equation 2.79 with adjustable parameters y_0 and y_∞, with the k parameter fixed at various values around the optimized value ($k_{best} = 0.06969$, determined by the full three-parameter fit in Figure 2.17). The SSR value is minimized at the optimal value of k.

fitted values will increase the *SSR*. An example of this is shown in **Figure 2.20**, where the decay parameter k from the exponential fit (Figure 2.17) is treated as a constant at values above and below the best-fit value, and new *SSR* values are determined by optimizing the remaining parameters. The further we go from the best fit value of k, the greater the *SSR* becomes, corresponding to an increasingly poor fit.

At some point, the fit will be poor enough, and the *SSR* will be large enough, that we should be able to reject the corresponding parameter value. Such rejection points would give us a means to assign confidence intervals. The question is, what threshold *SSR* values should we use to place our confidence intervals? It turns out that the *SSR* is an imperfect metric for placing such thresholds, because in addition to providing sensitivity to the quality of the fit, it depends on some specific details of the data set from which it was calculated. First, since the *SSR* is a sum, it increases linearly with the number of data points in our fit. If we measure extra points, our *SSR* should increase. An easy way to correct this is to divide the *SSR* by the number of data points. An even better correction is to divide by the number of "degrees of freedom." The number of degrees of freedom (represented with the symbol ν) is equal to the number of data points n minus the number of fitted parameters (m). This correction leads to a quantity that is often the "mean square error" (*MSE*):

$$MSE = \frac{SSR}{\nu} = \frac{\sum_{i=1}^{n}(y_i - f(x_i))^2}{n-m} \tag{2.83}$$

A second issue with the *SSR* is that it depends on the amount of error associated with each measurement. If measurements of y_i are very precise, the value of *MSE* should be small, but if the measurements of y_i are imprecise, the *MSE* will be large. If we know the error in each y_i point *a priori*, we can correct for the effects of error on *MSE*:

$$\chi^2 = \sum_{i=1}^{n} \frac{(y_i - f(x_i))^2}{\sigma_{y_i}^2} \tag{2.84}$$

The sum in Equation 2.84 is known as "chi-squared." It is similar to *SSR*, but each square residual is divided by the variance for the measurement of y_i. If each y_i value has the same uncertainty,[†] then the variance can be taken outside the sum:

$$\chi^2 = \frac{1}{\sigma_y^2} \sum_{i=1}^{n} (y_i - f(x_i))^2 = \frac{SSR}{\sigma_y^2} \tag{2.85}$$

---

† As is commonly assumed with least-squares fitting.

Making both corrections leads to a quantity called the "reduced" chi-squared:

$$\chi_r^2 = \frac{\chi^2}{\nu} = \frac{1}{(n-m)\sigma_y^2} \sum_{i=1}^{n} (y_i - f(x_i))^2 = \frac{MSR}{\sigma_y^2} \qquad (2.86)$$

As described in Appendix 2.2, there is a well-established statistical test (the "chi-squared test") to determine if an experimentally determined χ^2 value is consistent with the random error of the data, or if it reflects a shortcoming of the model. If the uncertainties in the y_i values are known, χ^2 can be calculated and compared with a theoretical distribution for χ^2 values. By making this comparison for different values of the test parameter, confidence intervals can be determined (see Appendix 2.2).

A practical limitation of the chi-squared test is that we typically don't have accurate measurements of $\sigma_{y_i}^2$. If our estimate of $\sigma_{y_i}^2$ is too large, we will underestimate χ^2, and overestimate confidence intervals on our test parameter. Conversely, if our estimate of $\sigma_{y_i}^2$ is too small, we will underestimate confidence intervals, and believe that our test parameter is better determined than it is.

The solution to this problem is to evaluate ratios of chi-squared values for different "models." When we search for confidence limits for fitted parameters, these models consist of a function where we fix one parameter (the parameter we are testing for confidence; in Figure 2.20 this "test" parameter is k), and the function in which all parameters are adjustable (i.e., $f(x; \vec{p})$).[†] Because both models are applied to the same data set, both have the same value of σ_y^2. Thus, in a ratio of these two χ^2 values, σ_y^2 cancels.

The ratio of χ^2 values, corrected by the number of degrees of freedom for each model, is called an "f-ratio":

$$f \equiv \frac{\chi_1^2/\nu_1}{\chi_2^2/\nu_2} = \frac{SSR_1/\nu_1}{SSR_2/\nu_2} \qquad (2.87)$$

The definition on the left is true for all models. The equality on the right is true as long as there is a single value for σ_y^2. In testing confidence intervals, the restricted model will be called model 1 (in the numerator), and the full model will be called model 2 (in the denominator).

The f-ratio is useful both because it can be easily calculated from fitting, and (more importantly) because there is an analytical formula for the probability $p_f(\nu_1, \nu_2)$ of getting different values of the f-ratio. This distribution (Appendix 2.2) gives the probability of getting a ratio of χ^2 values[‡] for models 1 and 2, assuming the two models are statistically equivalent.[§] A plot of $p_f(\nu_1, \nu_2)$ for the exponential fitting example in Figure 2.17 is shown in **Figure 2.21**.

As we described for χ^2 values (see Appendix 2.2), we can ask whether our observed f-ratio is probable or improbable. If the restricted model fits poorly, the associated f-ratio will be high; comparison to the f-ratio probability distribution will provide a means to reject the hypothesis that the restricted model is statistically equivalent to the full model. By calculating f-ratios for different values of the test parameter, we can find the points where the f-ratio exceeds a specified probability threshold. Parameters that lead to f-ratios outside this range result in significantly poor fits, and be disregarded on statistical grounds. This "f test" is shown in **Figure 2.22** for the test parameter k in exponential decay fitting (Figure 2.17).

[†] Sometimes the terms "restricted" and "unrestricted" are used to describe the fixed-parameter model and the full model. The fixed-parameter model is also referred to as a "nested" model, since it is contained within the full model.

[‡] or SSR values for f-tests involving equal σ_y^2.

[§] Other than having different degrees of freedom ν_1 and ν_2.

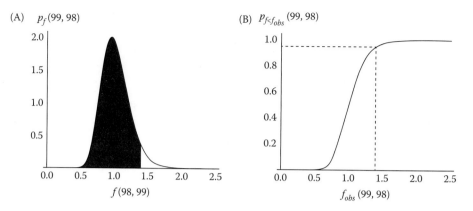

(A) $p_f(99, 98)$

(B) $p_{f<f_{obs}}(99, 98)$

$f(98, 99)$

$f_{obs}(99, 98)$

Figure 2.21 The *f* probability distribution and its integral. (A) The *f* probability distribution for $\nu_1 = 99$ and $\nu_2 = 98$ degrees of freedom. This corresponds to the situation for determining the parameter confidence intervals for the exponential fit example (Figures 2.18 and 2.20), with $n = 101$ data points, three adjustable parameters for the full model (model 2), and two adjustable parameters for the restricted model (model 1). (B) The integrated *f*-ratio probability distribution function. The dashed horizontal line at a value of 0.95 marks the threshold value (at $f_{95\%} = 1.3955$, dashed vertical line); this corresponds to the red shaded area under the p_f curve in (A). If models 1 and 2 are statistically indistinguishable, resulting *f*-ratios will be below the limiting (1.3955) value 95% of the time. Models with *f*-values above this threshold are 95% likely to be significantly different.

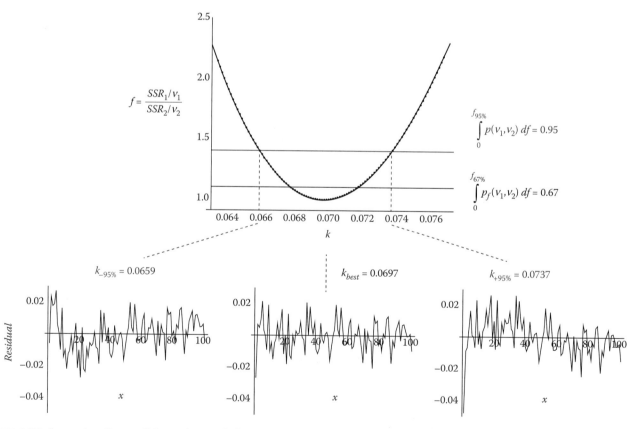

$$f = \frac{SSR_1/\nu_1}{SSR_2/\nu_2}$$

$$\int_0^{f_{95\%}} p(\nu_1, \nu_2)\, df = 0.95$$

$$\int_0^{f_{67\%}} p_f(\nu_1, \nu_2)\, df = 0.67$$

k

$k_{-95\%} = 0.0659$

$k_{best} = 0.0697$

$k_{+95\%} = 0.0737$

Residual

x

x

x

Figure 2.22 Assessing the confidence intervals for a parameter using *f*-statistics. The variation in the *f*-ratio (top) for the exponential fit with different fixed values of *k*. The horizontal lines show limiting *f*-values at 67 and 95% confidence levels ($f_{67\%} = 1.093$ and $f_{95\%} = 1.395$ in red and black, respectively). Values of *k* that produce these limiting *f*-ratios (dashed vertical lines, only shown here for $f_{95\%}$) define the lower and upper 95% confidence intervals on *k*. For the full three-parameter fit (model 2, resulting in a *k* value of 0.0697 and $\chi_2^2/\nu_2 = 0.00902/98 = 9.21 \times 10^{-5}$) the residuals are randomly distributed (lower, middle). With *k* fixed at the lower and upper 95% confidence limits (model 1, $\chi_1^2/\nu_1 = 0.01272/99 = 1.285 \times 10^{-4}$), residuals show a nonrandom distribution (lower, left, and right).

Tolerance limits for the f test in Figure 2.22 are set to 67% and 95% (black and red lines). Randomly sampling the f-ratio probability distribution would give f-ratio values that exceed these limits 33% and 5% of the time. The intersection points between these tolerance values ($f_{67\%}$ and $f_{95\%}$) and the f-ratio curve determined by variation of the test parameter k give the 67% and 95% confidence limits for this parameter. Values of 67% confidence intervals for k are quite similar to those obtained by bootstrap analysis (Table 2.1). An advantage of the f-stat method is that requires many fewer rounds of fitting than the bootstrap This computational savings is particularly important when fitting large data sets with complicated models.

Problems

2.1 Use a pencil (or pen) and paper to calculate the derivatives (df/dx) of the following functions:

$$f(x) = 3x^4 + 2x + 5$$
$$f(x) = Kx/1 + Kx$$
$$f(x) = ln(x) + e^{-Kx}$$

In the second and third equations, K and k are constants.

2.2 For the function $f(x) = x^3$, identify a critical point. Use the second derivative test to determine whether this critical point is a maximum or a minimum. Plot $f(x)$ versus x to check your answer.

2.3 Use a pencil (or pen) and paper to solve the following integrals. Show any steps you needed to take:

$$\int 3x \, dx$$

$$\int_1^4 (x^3 + 2) \, dx$$

$$\int_1 y^2 e^{-2x} dx$$

2.4 Find the integrated function $F(x)$ by solving the single limit integral

$$\int_1^x t^{-1} \, dt$$

using the boundary condition that at $x=1$, the integrated function has a value of $F(1)=2$.

2.5 For the simple unimolecular reaction of species A to species B, that is,

$$A \xrightarrow{k} B$$

solve (i.e., integrate) the differential equation for the rate of change of concentration of species A as a function of time:

$$\frac{d[A]}{dt} = -k[A]$$

using the initial (or "boundary") condition that at time $t = 0$, the concentration

[A] = 1 Molar. Your answer should have no derivative, but instead should have [A] as a function of time. Begin by separating the derivative into a differential, and collecting the [A] variables on one side of the equation. This single boundary problem is equivalent to single-point integration.

2.6 Confirm the partial derivative given in Equation 2.15, namely,

$$\left(\frac{\partial z}{\partial y}\right)_x = -(x^2 - 3x)(e^{-y} - ye^{-y})$$

where

$$f(x, y) = (3x - x^2)ye^{-y}$$

2.7 Find solutions to the partial-derivative equation given by Equation 2.17, namely,

$$\left(\frac{\partial z}{\partial y}\right)_x = 0 = -(x^2 - 3x)(e^{-y} - ye^{-y})$$

(Note, solve the right-hand side set to zero—forget about the derivative).

2.8 Use the intersection of zero lines in Figure 2.7C identify the other critical points. Plot the surface (Equation 2.13) in the vicinity of these points to determine whether these points are minima, maxima, or saddle-points. Alternatively, you can use the extended second derivative test found in calculus texts).

2.9 Calculate the cross derivatives of the equation for the surface in Figure 2.5 (Equation 2.13)

$$\frac{\partial^2 f}{\partial x \partial y} \quad \text{and} \quad \frac{\partial^2 f}{\partial y \partial x}$$

and verify that the same value is obtained regardless the order of cross-differentiation.

2.10 Calculate values of the line integral in Figure 2.9 by the paths shown in Panels A (from [3/2, 3] to [1/2, 3] and then to [1/2, 1]) and B (same endpoints, but through point [3/2, 1]). Do the two paths produce the same value of the line integral?

2.11 Calculate the integral of the exact differential in Figure 2.9 by the paths shown in panels D (from [3/2, 3] to [1/2, 3] and then to [1/2, 1]) and E (same endpoints, but through point [3/2, 1]), and show that the value is independent of the path.

2.12 Calculate the exact (or "total") differentials of the following functions.

$$f(x, y) = x^3 e^{-2y}$$

$$f(x, y) = \frac{2}{x} + \cos(3y)$$

2.13 Identify whether the following two differentials are exact, and if possible, give an expression for f(x,y) in each case.

Differential 1 $df(x, y) = 4x^3 y \, dx + x^4 \, dy$

Differential 2 $df(x, y) = \frac{x}{y} dx + \frac{y}{x} dy$

2.14 Calculate the differential dz of the paraboloid $z = x^2 + y^2$:

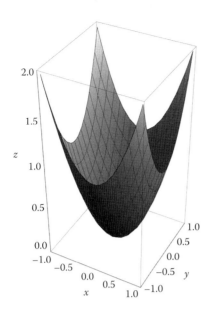

2.15 Use the Euler test to show that your answer to Question 1.1 is an exact differential.

2.16 Starting with the differential you determined in Problem 2.14 above, integrate to calculate Δz starting at $(x,y = 1,1)$ and ending at $(x,y = 2,2)$. Do this in two parts by first changing x from 1 to 2 at $y = 1$, and then changing y from 1 to 2 at $x = 2$ as shown below:

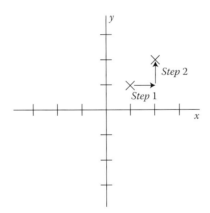

2.17 In a polar coordinate system, the paraboloid in Problem 2.14 is given as

$$z = r^2$$

That is, given the axial symmetry of the paraboloid, there is no angular dependence. Calculate dz using the polar coordinate system.

2.18 Now demonstrate the path independence of the integral

$$\Delta z = \int dz$$

by repeating the integral along a direct line from $(x,y = 1,1)$ to $(x,y = 2,2)$ shown below, using your differential expressed in polar coordinates, and comparing to your answer in Problem 2.16.

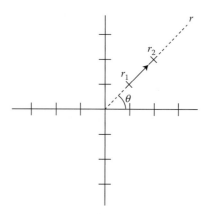

2.19 Using vector notation, show that the dot product in Equation 2.48 produces the equality in Equation 2.43. As a reminder, the dot product for two vectors $\vec{a}$ and $\vec{b}$ is given by,

$$\vec{a} \bullet \vec{b} = a_1 b_1 + a_2 b_2$$

where

$$\vec{a} = i a_1 + j a_2 = \langle a_1, a_2 \rangle$$

and

$$\vec{b} = i b_1 + j b_2 = \langle b_1, b_2 \rangle$$

2.20 Using the analytical approach to linear least squares, find the values for best-fitting p_1 and p_2 values by Equation 2.71 (or of you prefer the notation, Equation 2.72), assuming a quadratic fitting function with a linear offset (Equation 2.56). Use the following data set for the fit:

x_i	y_i	x_i	y_i
−3.0	29.755	0.2	−0.682
−2.8	26.442	0.4	−0.719
−2.6	22.434	0.6	−0.806
−2.4	21.116	0.8	−0.464
−2.2	17.692	1.0	−1.435
−2.0	15.398	1.2	−0.964
−1.8	13.306	1.4	−0.306
−1.6	11.561	1.6	0.969
−1.4	9.2473	1.8	2.614
−1.2	6.8393	2.0	4.776
−1.0	5.7989	2.2	5.081
−0.8	2.8783	2.4	5.994
−0.6	3.5373	2.6	7.454
−0.4	0.7149	2.8	9.947
−0.2	−0.315	3.0	12.314
0.0	1.0311		

2.21 Using the data file from Problem 2.20, use Mathematica to do a linear (or nonlinear) least-squares to fit the parameters p_1 and p_2. Do you get the same results as in Problem 2.20?

2.22 For the transformed exponential decay data in Figure 2.15B, which is reproduced below, use a weighted fit to determine the decay constant k and the starting amplitude $\ln(y_0)$. Weighting can be implemented in the Mathematica command `NonlinearModelFit` (and the command `LinearModelFit`) using the option "`Weighting->{W_i}`" where {W_i} is a list of weights for each of the n y_i points. For this problem, you will need to calculate the weights according to Equation 2.75. Use the standard propagation of error formula, in this case,

$$\sigma^2_{\ln(y_i)} = \left(\frac{\partial \ln(y_i)}{\partial y_i}\right)^2 \sigma^2_{y_i}$$

x_i	y_i	x_i	y_i	x_i	y_i	x_i	y_i	x_i	y_i
0	-0.0272	14	-0.9558	28	-1.9415	42	-2.7882		
1	-0.0621	15	-1.0528	29	-1.9585	43	-3.0469	56	-3.6318
2	-0.1337	16	-1.0925	30	-2.1015	44	-3.1767	57	-4.3536
3	-0.1972	17	-1.1328	31	-2.1857	45	-3.1177	58	-4.4543
4	-0.2509	18	-1.2971	32	-2.3561	46	-3.1609	59	-4.4693
5	-0.3442	19	-1.3250	33	-2.1123	47	-3.3425	60	-3.8567
6	-0.4188	20	-1.4697	34	-2.3703	48	-3.9485	61	-3.7058
7	-0.5031	21	-1.4069	35	-2.2499	49	-3.7557	62	-5.2884
8	-0.5246	22	-1.5939	36	-2.4373	50	-3.5093	63	-5.1662
9	-0.6467	23	-1.6792	37	-2.5710	51	-3.4199	65	-4.4883
10	-0.7056	24	-1.6728	38	-2.5408	52	-3.0795	66	-4.1035
11	-0.8001	25	-1.7769	39	-2.8594	53	-3.5515	67	-4.6371
12	-0.8482	26	-1.7939	40	-3.0003	54	-3.7592	68	-6.5767
13	-0.9112	27	-1.8780	41	-3.0083	55	-3.2394		

2.23 The following two data sets are derived from a two-parameter exponential decay equation (Figure 2.15) with the same parameters but a different set of {x_i} values. Use NLLS to fit the two data sets. Based on the covariance matrix, which data set gives a better determination of the decay constant k? In words, rationalize your answer.

Data set 1

x_i	y_i	x_i	y_i	x_i	y_i	x_i	y_i	x_i	y_i
0	0.973	20	0.230	40	0.050	60	0.021	80	0.001
1	0.940	21	0.245	41	0.049	61	0.025	81	-0.006
2	0.875	22	0.203	42	0.062	62	0.005	82	0.005
3	0.821	23	0.187	43	0.048	63	0.006	83	0.002
4	0.778	24	0.188	44	0.042	64	0.013	84	0.013
5	0.709	25	0.169	45	0.044	65	0.011	85	0.015
6	0.658	26	0.166	46	0.042	66	0.017	86	-0.001

Continued

x_i	y_i	x_i	y_i	x_i	y_i	x_i	y_i	x_i	y_i
7	0.605	27	0.153	47	0.035	67	0.010	87	−0.015
8	0.592	28	0.143	48	0.019	68	0.001	88	−0.001
9	0.524	29	0.141	49	0.023	69	−0.009	89	−0.005
10	0.494	30	0.122	50	0.030	70	0.012	90	0.005
11	0.449	31	0.112	51	0.033	71	−0.004	91	0.009
12	0.428	32	0.095	52	0.046	72	0.009	92	0.011
13	0.402	33	0.121	53	0.029	73	0.003	93	−0.002
14	0.384	34	0.093	54	0.023	74	0.002	94	0.008
15	0.349	35	0.105	55	0.039	75	−0.015	95	0.008
16	0.335	36	0.087	56	0.026	76	0.017	96	0.005
17	0.322	37	0.076	57	0.013	77	0.012	97	0.001
18	0.273	38	0.079	58	0.012	78	0.014	98	0.001
19	0.266	39	0.057	59	0.011	79	−0.002	99	0.002
								100	−0.009

Data set 2

x_i	y_i	x_i	y_i	x_i	y_i	x_i	y_i	x_i	y_i
0	0.973	2000	−0.017	4000	−0.011	6000	0.006	8000	−0.002
100	0.008	2100	0.015	4100	−0.007	6100	0.011	8100	−0.010
200	0.005	2200	−0.011	4200	0.009	6200	−0.008	8200	0.002
300	0.010	2300	−0.013	4300	−0.002	6300	−0.006	8300	−0.001
400	0.022	2400	0.001	4400	−0.004	6400	0.001	8400	0.011
500	0.004	2500	−0.005	4500	0.001	6500	0.001	8500	0.012
600	0.001	2600	0.004	4600	0.002	6600	0.007	8600	−0.003
700	−0.008	2700	0.002	4700	−0.002	6700	0.000	8700	−0.017
800	0.021	2800	0.003	4800	−0.015	6800	−0.007	8800	−0.003
900	−0.009	2900	0.010	4900	−0.009	6900	−0.017	8900	−0.007
1000	−0.003	3000	0.000	5000	0.000	7000	0.005	9000	0.003
1100	−0.014	3100	−0.002	5100	0.005	7100	−0.011	9100	0.007
1200	−0.004	3200	−0.012	5200	0.020	7200	0.002	9200	0.010
1300	0.000	3300	0.022	5300	0.004	7300	−0.003	9300	−0.003
1400	0.009	3400	0.001	5400	0.000	7400	−0.003	9400	0.007
1500	−0.001	3500	0.019	5500	0.018	7500	−0.021	9500	0.007
1600	0.009	3600	0.007	5600	0.007	7600	0.012	9600	0.004
1700	0.018	3700	0.001	5700	−0.006	7700	0.008	9700	−0.001
1800	−0.010	3800	0.009	5800	−0.006	7800	0.009	9800	0.000
1900	0.001	3900	−0.008	5900	−0.005	7900	−0.006	9900	0.001
								10000	−0.009

2.24 Use nonlinear (or linear) least squares to fit the data in Figure 2.16A (below), using a model with a slope and the y value at $x = 3.5$ ($y_{3.5}$):

$$y = m(x - 3.5) - y_{3.5}$$

Make a contour plot of the *SSR* surface like in Figure 2.16C. Compare the extent of parameter correlation to the fit in Figure 2.16.

x_i	y_i
3.0	0.488
3.1	0.388
3.2	0.339
3.3	0.264
3.4	0.107
3.5	0.038
3.6	-0.059
3.7	-0.122
3.8	-0.194
3.9	-0.335
4.0	-0.582

2.25 Calculate the correlation between the two fitted parameters in Problem 2.24 (m and $y_{3.5}$), and compare these to correlation between the fitted parameters in Figure 2.16 (m and the intercept b; to do this, you will need to fit the data from Problem 2.24 using a simple line).

2.26 Using the standard formula for propagation of errors, that is

$$\sigma_f = \sqrt{\left(\frac{\partial f}{\partial p_1}\right)^2 \sigma_{p_1}{}^2 + \left(\frac{\partial f}{\partial p_2}\right)^2 \sigma_{p_2}{}^2}$$

where the function f is a function of parameters p_1 and p_2, and the variances in p_1 and p_2 (σ_{p_1} and σ_{p_2}) are known, calculate the uncertainty in the value of the fitted line in Problem 2.24 as a function of x, given the parameter uncertainties you determined in Problem 2.24. What is the estimated uncertainty in y at $x=0$? What your results should suggest is that although you can sometimes reparameterize a model to minimize parameter correlation, the old parameters remain correlated.

2.27 Using the `NonlinearModelFit` command of Mathematica, determine the parameter errors and correlation for the data in Figure 2.11, using a fitting model

$$f(x) = p_1 x^2 + p_2 x$$

The primes from the equation in Figure 2.11 have been omitted for simplicity. The data are given in the table below:

x_i	y_i	x_i	y_i
-3.0	29.75	0.2	-0.68
-2.8	26.44	0.4	-0.72
-2.6	22.43	0.6	-0.81
-2.4	21.16	0.8	-0.46
-2.2	17.69	1.0	-1.43
-2.0	15.40	1.2	-0.96
-1.8	13.31	1.4	-0.31
-1.6	11.56	1.6	0.97
-1.4	9.25	1.8	2.61
-1.2	6.84	2.0	4.77
-1.0	5.80	2.2	5.08
-.08	2.88	2.4	5.99
-0.6	3.54	2.6	7.45
-0.4	0.71	2.8	9.95
-0.2	-0.31	3.0	2.31
0.0	1.03		

2.28 For a two parameter fit with n data points, write matrices F and $F^T F = A$. Invert A to generate $\sigma_y^2 V$.

2.29 Use your results from Problem 2.28 to calculate numerical results for the parabola fit to the data set in Problem 2.27.

2.30 For a three parameter fit with n data points, write matrices F and $F^T F = A$. Invert A to generate $\sigma_y^2 V$.

2.31 For the "urn problem," use probabilities so show that a two-step sampling without replacement protocol has the same probability as grabbing two balls at once.

2.32 Show that for the χ^2 probability distribution (Appendix 2.2), the value of the mode is $\nu - 2$.

2.33 Show that if the χ^2 probability distribution is normalized, the two right-hand sides of Equation A2.1.1 are equal, that is,

$$\int_{\chi_{obs}^2}^{\infty} P_{\chi^2} d\chi^2 = 1 - \int_{0}^{\chi_{obs}^2} P_{\chi^2} d\chi^2$$

2.34 For the decay data in Figure 2.17, which is reproduced below, use the f test to compare the three-parameter exponential decay model (Equation 2.79) and a five parameter quartic model, that is,

$$f(x) = a + bx + cx^2 + dx^3 + ex^4$$

Since, *a priori*, either model could fit better, you should test both f-ratios. *Note, although $f_{obs}(v_1, v_2) = 1/f_{obs}(v_2, v_1)$, since they are just ratios of χ_{obs}^2, there is no such reciprocal relationship for probability distributions for f-ratios. That* *is, $p_f(v_1, v_2) = 1/p_f(v_2, v_2)$. Rather, both $p_f(v_1, v_2)$ and $p_f(v_2, v_1)$ must be calculated separately.*

x_i	y_i	x_i	y_i	x_i	y_i	x_i	y_i	x_i	y_i
0	1.073	21	0.345	42	0.162	63	0.106	84	0.113
1	1.040	22	0.303	43	0.148	64	0.113	85	0.115
2	0.975	23	0.287	44	0.142	65	0.111	86	0.099
3	0.921	24	0.288	45	0.144	66	0.117	87	0.085
4	0.878	25	0.269	46	0.142	67	0.110	88	0.099
5	0.809	26	0.266	47	0.135	68	0.101	89	0.095
6	0.758	27	0.253	48	0.119	69	0.091	90	0.105
7	0.705	28	0.243	49	0.123	70	0.112	91	0.109
8	0.692	29	0.241	50	0.130	71	0.096	92	0.111
9	0.624	30	0.222	51	0.133	72	0.109	93	0.098
10	0.594	31	0.212	52	0.146	73	0.103	94	0.108
11	0.549	32	0.195	53	0.129	74	0.102	95	0.108
12	0.528	33	0.221	54	0.123	75	0.085	96	0.105
13	0.502	34	0.193	55	0.139	76	0.117	97	0.101
14	0.484	35	0.205	56	0.126	77	0.112	98	0.101
15	0.449	36	0.187	57	0.113	78	0.114	99	0.102
16	0.435	37	0.176	58	0.112	79	0.098	100	0.091
17	0.422	38	0.179	59	0.111	80	0.101		
18	0.373	39	0.157	60	0.121	81	0.094		
19	0.366	40	0.150	61	0.125	82	0.105		
20	0.330	41	0.149	62	0.105	83	0.102		

Appendix 2.1: Determining the Covariance Matrix in Least-Squares Fitting[†]

For linear least-squares fitting, matrix Equation 2.72 provides a direct means to obtain the best parameters for the model, given the data. It turns out that this equation, and in particular, the matrix on the left-hand side that multiplies the parameter vector, that is,

$$
\begin{bmatrix}
\sum\limits_{i=1}^{n} g_1(x_i)^2 & \sum\limits_{i=1}^{n} g_2(x_i)g_1(x_i) & \cdots & \sum\limits_{i=1}^{n} g_m(x_i)g_1(x_i) \\
\sum\limits_{i=1}^{n} g_1(x_i)g_2(x_i) & \sum\limits_{i=1}^{n} g_2(x_i)^2 & \cdots & \sum\limits_{i=1}^{n} g_m(x_i)g_2(x_i) \\
\vdots & & & \\
\sum\limits_{i=1}^{n} g_1(x_i)g_m(x_i) & \sum\limits_{i=1}^{n} g_2(x_i)g_m(x_i) & \cdots & \sum\limits_{i=1}^{n} g_m(x_i)^2
\end{bmatrix} \equiv A
$$

(A2.1.1)

provides a direct route to calculating the covariance matrix V, as well. To investigate this relationship, we will analyze this matrix (which Equation A2.1.1 defines as A) in some detail. First, notice that each term in the A matrix is a sum over n x_i values. This suggests that it results from multiplication of two simpler matrices.[‡] To get sums of length n for each term, the left matrix would need to have n rows, and the right would have to have n columns. Moreover, to get the correct dimensions (m by m) for A, the two simpler matrices would need to have m rows (left matrix) and m columns (right matrix). One factorization is

$$
A = \begin{bmatrix}
g_1(x_1) & g_1(x_2) & \cdots & g_1(x_n) \\
g_2(x_1) & g_2(x_2) & \cdots & g_2(x_n) \\
\vdots & \vdots & \ddots & \vdots \\
g_m(x_1) & g_m(x_2) & \cdots & g_m(x_n)
\end{bmatrix}
\times
\begin{bmatrix}
g_1(x_1) & g_2(x_1) & \cdots & g_m(x_1) \\
g_1(x_2) & g_2(x_2) & \cdots & g_m(x_2) \\
\vdots & \vdots & \ddots & \vdots \\
g_1(x_n) & g_2(x_n) & \cdots & g_m(x_n)
\end{bmatrix}
$$

(A2.1.2)

Note that the columns of the left matrix are identical to the rows of the right matrix in Equation A2.1.2. This means that the two matrices are transposes of one another:

$$
A = \begin{bmatrix}
g_1(x_1) & g_2(x_1) & \cdots & g_m(x_1) \\
g_1(x_2) & g_2(x_2) & \cdots & g_m(x_2) \\
\vdots & \vdots & \ddots & \vdots \\
g_1(x_n) & g_2(x_n) & \cdots & g_m(x_n)
\end{bmatrix}^{T}
\times
\begin{bmatrix}
g_1(x_1) & g_2(x_1) & \cdots & g_m(x_1) \\
g_1(x_2) & g_2(x_2) & \cdots & g_m(x_2) \\
\vdots & \vdots & \ddots & \vdots \\
g_1(x_n) & g_2(x_n) & \cdots & g_m(x_n)
\end{bmatrix} \equiv F^{T}F
$$

(A2.1.3)

[†] This appendix requires some familiarity with linear algebra, including matrix inversion, determinants, and cofactors. Though neither this nor the subsequent Appendix 2.2 are required for any portion of this text, it is hoped that it will provide students who wish to apply modeling to their research a deeper understanding of how to interpret results from least-squares fitting. Both appendices can be skipped by students encountering this material for the first time.

[‡] Or, if you prefer, it can be factored into two matrices.

where we define matrix **F** as the left matrix in Equation A2.1.3.

Matrix **F** has two interesting features. First, it appears in the set of linear equations that relate the x_i and y_i values through the function $f(x; \vec{p})$:

$$F\vec{p} + E = Y \tag{A2.1.4}$$

where E is a vector of error values (e_i) for each data point, and **Y** is a vector of observed y values. Written explicitly, Equation A2.1.4 is

$$\begin{bmatrix} g_1(x_1) & g_2(x_1) & \cdots & g_m(x_1) \\ g_1(x_2) & g_2(x_2) & \cdots & g_m(x_2) \\ \vdots & \vdots & \ddots & \vdots \\ g_1(x_n) & g_2(x_n) & \cdots & g_m(x_n) \end{bmatrix} \begin{bmatrix} p_1 \\ p_2 \\ \vdots \\ p_m \end{bmatrix} + \begin{bmatrix} e_1 \\ e_2 \\ \vdots \\ e_m \end{bmatrix} = \begin{bmatrix} y_1 \\ y_2 \\ \vdots \\ y_m \end{bmatrix} \tag{A2.1.5}$$

Second, **F** can be regarded as series of derivatives of the fitting function with respect to each of the parameters, evaluated at each of the n x_i values (see Equation 2.67):

$$F = \begin{bmatrix} \dfrac{\partial f(x_1)}{\partial p_1} & \dfrac{\partial f(x_1)}{\partial p_2} & \cdots & \dfrac{\partial f(x_n)}{\partial p_1} \\ \dfrac{\partial f(x_2)}{\partial p_1} & \dfrac{\partial f(x_2)}{\partial p_2} & \cdots & \dfrac{\partial f(x_n)}{\partial p_2} \\ \vdots & \vdots & \ddots & \vdots \\ \dfrac{\partial f(x_n)}{\partial p_1} & \dfrac{\partial f(x_n)}{\partial p_2} & \cdots & \dfrac{\partial f(x_n)}{\partial p_m} \end{bmatrix} \tag{A2.1.6}$$

The covariance matrix **V** is obtained by taking the inverse of the matrix $A = F^TF$. The inverse of this $m \times m$ matrix is also an $m \times m$ matrix, with elements proportional to the elements of the covariance matrix.

$$V \propto (F^TF)^{-1} \tag{A2.1.7}$$

Although matrix inversion is something that is best left to computers, and will not be attempted here, it is worth thinking about what inversion does to generate the elements of V. One way to represent each element in V is through the use of determinants and "cofactors."[†] The elements of an inverted matrix are equal to the the cofactor of the transpose divided by the determinant of the matrix:

$$V_{i,j} \propto \frac{cof_{j,i}(A)}{|A|} = \frac{cof_{j,i}(F^TF)}{|F^TF|} \tag{A2.1.8}$$

Notice that the cofactor in the numerator is for the element in row j, column i. We can keep consistent subscripts (ij) between v and the cofactor if we first transpose matrix A:

$$V_{i,j} \propto \frac{cof_{i,j}(A^T)}{|A|} = \frac{cof_{i,j}(FF^T)}{|F^TF|} = \frac{cof_{i,j}(FF^T)}{|FF^T|} \tag{A2.1.9}$$

The second equality comes from the fact that the transpose of a product of two matrices is the product of the two transposed matrices multiplied in reverse order,[‡] and the fact that transposing a transpose of a matrix gives back the original matrix. The third equality takes advantage of the fact that the determinant of a matrix and its transpose are equal.[§]

[†] A cofactor of the $a_{i,j}$ element of matrix A is the determinant formed from a smaller matrix where row i and column j of the A matrix are left out. See Strang, 2008.

[‡] That is, $(CB)^T = B^TC^T$.

[§] That is, $|A^T| = |A|$. Then again, since A is symmetric, $A^T = A$.

We can get a sense of what the proportionality constant should be in Equations A2.1.7 through A2.1.9 using dimensional analysis. For the diagonal $v_{i,i}$, the numerator has units of $y^{2(m-1)}$ (from the squared ∂f terms) divided by $p_1^2 \times p_2^2 \times \ldots \times p_m^2 / p_i^2$. The denominator has units of y^{2m} divided by $p_1^2 \times p_2^2 \times \ldots \times p_m^2$. As a result, the quotient in Equation A2.1.9 has units of p_i^2 divided by y^2. As defined in Equation 2.78, $v_{i,i}$ has units of p_i^2. Thus, the proportionality constant needs units of y^2. Because the uncertainty in the fitted parameter values should increase with the uncertainty of the measured y_i values, we will take the variance in the measured y values (σ_y^2, with units y^2) as the proportionality constant:

$$v_{i,j} = \sigma_y^2 \, \frac{cof_{i,j}(A^T)}{|A|} \tag{A2.1.10}$$

An example of the matrix manipulations that produce the $v_{i,j}$ elements is given in Problems 2.28 and 2.29 for a two-parameter fitting model. The results are

$$v_{1,1} = \frac{1}{\sigma_y^2} \times \frac{\displaystyle\sum_{i=1}^{n} (\partial f(x_i)/\partial p_2)^2}{\left[\displaystyle\sum_{i=1}^{n} (\partial f(x_i)/\partial p_1)^2 \sum_{i=1}^{n} (\partial f(x_i)/\partial p_2)^2 - \left\{ \displaystyle\sum_{i=1}^{n} (\partial f(x_i)/\partial p_1)(\partial f(x_i)/\partial p_2) \right\}^2 \right]} \tag{A2.1.11}$$

$$v_{2,2} = \frac{1}{\sigma_y^2} \times \frac{\displaystyle\sum_{i=1}^{n} (\partial f(x_i)/\partial p_1)^2}{\left[\displaystyle\sum_{i=1}^{n} (\partial f(x_i)/\partial p_1)^2 \sum_{i=1}^{n} (\partial f(x_i)/\partial p_2)^2 - \left\{ \displaystyle\sum_{i=1}^{n} (\partial f(x_i)/\partial p_1)(\partial f(x_i)/\partial p_2) \right\}^2 \right]} \tag{A2.1.12}$$

$$v_{1,2} = v_{2,1} = \frac{1}{\sigma_y^2} \times \frac{\displaystyle\sum_{i=1}^{n} (\partial f(x_i)/\partial p_1)(\partial f(x_i)/\partial p_2)}{\left[\displaystyle\sum_{i=1}^{n} (\partial f(x_i)/\partial p_1)^2 \sum_{i=1}^{n} (\partial f(x_i)/\partial p_2)^2 - \left\{ \displaystyle\sum_{i=1}^{n} (\partial f(x_i)/\partial p_1)(\partial f(x_i)/\partial p_2) \right\}^2 \right]} \tag{A2.1.13}$$

Even though the two-parameter fit is about the simplest type of least squares that can be performed, the resulting equations for $v_{i,j}$ (Equations A2.1.11 through A2.1.13) are not simple. Nonetheless, for this simple case, a little bit of thought about the structure of these equations can provide some intuition about their relation to parameter uncertainty. Each summations can be thought of a measure of the sensitivity of the fitting function to each parameter (in this case, p_1 and p_2), averaged over all the points in the fit (the $\{x_i\}$). Focusing first on the numerators, each of the variance parameters is large when the fitting function has high sensitivity to the *other* fitting parameter. If $f(x)$ has high sensitivity to p_2, but low sensitivity to p_1, then p_1 will have high uncertainty compared to p_2.

If at each point in the fit, the $f(x)$ sensitivities are the same for p_1 and p_2, then the covariance parameter $v_{1,2}$ will be large (Equation A2.1.13). High covariance decreases the individual parameter variance values through the denominators of Equations

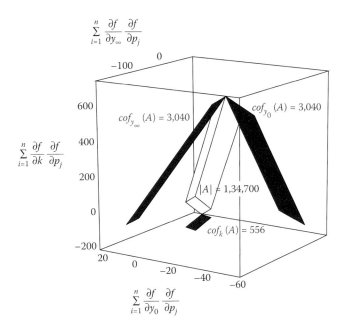

Figure A2.1.1 A geometric representation of parameter variance. For the three-parameter exponential decay (Figure 2.17), the determinant of the covariance matrix A is equal to the volume of the parallelepiped shown in white. The cofactors for each of the three parameters are equal to the areas of the parallelograms projected onto the three faces of the coordinate system. The ratio of the parallelogram areas to the parallelepiped volume gives the variances of each parameter (Equation A2.1.10).

A2.1.11 and A2.1.12. The denominators of all three terms (equal to $|A|$) express the extent to which the sensitivities of $f(x)$ to the parameters p_i are independent of one another. The first product of sums represents the total sensitivity, and the second sum represents the extent to which that sensitivity is coupled between parameters. If there is no joint sensitivity, the difference will be maximized, decreasing parameter uncertainties. As the second term increases, the denominator approaches zero, increasing parameter uncertainties increase.

A more succinct way to think about how the $v_{i,j}$ equations work is to return to the more general cofactor/determinant picture (Equation A2.1.10).[†] The cofactor of matrix element $v_{i,j}$ includes sensitivities of $f(x)$ to all the parameters except parameters i and j (or just i, for diagonal variance elements). The determinant of A includes all sensitivities including those parameters. Geometrically, the ratios of the cofactor to the determinant can be interpreted in terms of areas and volumes. The numerical value of a determinant is equal to the volume of an m-dimensional parallelepiped (here, m is the number of parameters), with sides given by the row vectors of the corresponding matrix (here, the A matrix). Cofactors are also determinants, thus their numerical values are equal to parallelepipeds of dimension $m-1$. The three-dimensional parallelepiped corresponding to the determinant of the A matrix for the exponential decay fit in Figure 2.17 is shown in **Figure A2.1.1** (white), along with the two-dimensional cofactors (i.e., parallelograms, red).

Appendix 2.2: Testing parameters and models with the χ^2 and f-ratio probability distributions

Estimation of confidence intervals using the f-ratio test relies on an analytical probability distribution (the f-distribution) that expresses the ratio of χ^2 values for different models (Equation 2.86). This f probability distribution is used to find values

---

[†] The benefits of this more general picture are particularly important as the number of parameters increases. For m parameters, the number of terms in the numerators of the $v_{i,j}$ expressions such as Equtions A2.1.11 through A2.1.13 goes as $(m-1)!$, and the number in the denominator as $m!$

of f at various confidence limits (e.g., 67% and 95%). In addition to finding confidence limits for model parameters (see Figures 2.21 and 2.22 for the exponential fit), the f-ratio can be used to compare different models to determine whether subtle differences between fits are statistically significant.

Here we will describe the probability distributions that underlie these statistical tests. Their derivations are lengthy and will not be presented. Rather, our goal is to provide an understanding of what these distributions mean, and how to use them to quantitatively evaluate the quality of a fit.

The χ^2 distribution

An essential concept in understanding these distributions is that when we perform a fit on a finite number (n) of measurements $\{y_n\}$, and each of these measurements has error, the collection of measurements is a random sample.[†] As a result, quantities derived from the fit can also be thought of as random samples (from some unknown "parent distribution"). This is true both for fitted parameter values (as shown in the bootstrap section above, Figure 2.19) and for χ^2 values that result from fitting. There is some chance that we will get a low χ^2 value (by good fortune, each measurement in the data set $\{y_n\}$ might fall very close to the true values), and some chance we will get a high χ^2 value (through no fault of our own, the data set might contain a large number of outliers).

Whereas the $\{y_n\}$ values (and to a good approximation, each residual) are distributed according to Gaussian statistics, the squares of residuals (and thus χ^2) are not.[‡] This can be seen in **Figure A2.2.1**, which shows a histogram of χ^2 values from fitting the exponential decay data with different random errors. The χ^2 histogram is peaked at *approximately*[§] $\nu = 98$, but as described above, there is a good chance of obtaining lower and higher values.

It turns out that there is an analytical probability distribution for obtaining a particular value of χ^2, given ν degrees of freedom[¶]:

$$p_{\chi^2} = \frac{(\chi^2)^{(\nu-2)/2}e^{-\chi^2/2}}{2^{\nu/2}\,\Gamma(\nu/2)} \tag{A2.2.1}$$

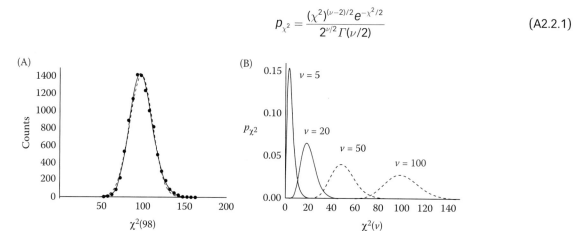

Figure A2.2.1 The χ^2 distribution. (A) Values of χ^2 for the three-parameter exponential decay model fit to 10,000 data sets. Each data set has the same 101 $\{x_i\}$ values (Figure 2.17), but the $\{y_i\}$ values have independent random Gaussian error (with a standard deviation of 0.01). Black dots show the resulting counts for bin widths of five units. The solid black line shows the theoretical χ^2 probability distribution with 98 degrees of freedom (scaled by the product of the bin width and the number of fitted χ^2 values to match the histogram). The red dashed line is a fitted (and scaled) Gaussian probability distribution. Compared to the symmetrical Gaussian distribution, the χ^2 probability distribution is asymmetric, being sharper on the low side, and broader on the high side. (B) Four χ^2 probability distributions with different degrees of freedom. Asymmetry is most obvious for small values of ν.

[†] This remains true even though the various y_i values are typically collected under different conditions, that is, at different values of x_i.

[‡] Note that the sum of residuals resulting from a least-squares fit is *not* Gaussian distributed, but is equal to zero, with positive and negative residuals offsetting.

[§] Actually, as described below, the histogram should peak at $\nu - 2 = 96$.

[¶] See Pugh and Winslow (1966) for a derivation.

This distribution is gives an excellent match to the histogram of values in Figure A2.2.1A (black curve). Compared to the Gaussian distribution (red dashed curve, Figure A2.2.1A), the χ^2 distribution is asymmetric (skewed right). This asymmetry can also be recognized by comparing the mean and the peak (or "mode") of the χ^2 distribution: whereas the mean and mode are the same value for the Gaussian distribution, the mode of the χ^2 distribution ($\nu-2$) is lower than the mean (ν).

One of the pieces of Equation A2.2.1 may be unfamiliar to some readers. The "gamma function" (Γ) shows up a lot in statistics, especially in distributions of squares of random variables (including distributions for the correlation coefficient and the f-statistic). The gamma function can be thought of as a continuous extension of the factorial function. Although the Γ function is generally defined by an integral,[†] we only need to evaluate Γ at whole- and half-integer values (given by $\nu/2$), which are

$$\Gamma(\nu) = \begin{cases} (\nu-1)! = \dfrac{\nu!}{\nu} & \nu = 0,\ 1,\ 2,\ \ldots \\[2em] (\nu-1)(\nu-2)\ldots\left(\dfrac{3}{2}\right)\left(\dfrac{\sqrt{\pi}}{2}\right) & \nu = \dfrac{1}{2},\ \dfrac{3}{2},\ \dfrac{5}{2},\ \ldots \end{cases}$$

(A2.2.2)

If the χ^2 function seems unpleasantly complicated, you can take comfort in the fact that most statistics texts have tables full of χ^2 values for different degrees of freedom. Even better, the χ^2 probability distribution is built into Mathematica.[‡]

Figure A2.2.1B shows four different χ^2 distributions with different degrees of freedom ν. As ν increases, the peak shifts right, becomes broader (there is more possibility for spread in measured χ^2 values), and the asymmetry becomes less noticeable. One way to evaluate the "goodness of fit" for a particular model and data set $\{x_i, y_i\}$ would be to compare the observed χ^2 value to the distribution with the same number of degrees of freedom. If the χ^2 value is near the peak of the distribution, the fit is about as good as should be expected, given the error each y_i value. For the exponential fit in Figure 2.17, the value of χ^2 is 90.2, which is near to (but in fact less than) the peak of the theoretical $\chi^2(98)$ distribution (at 96, see Figure A2.2.2A). Thus, the exponential decay in Figure 2.17 can be considered to be quite good.

If instead the χ^2 value is significantly larger than the theoretical peak value, the fit should be regarded as poor. The question is, what value of χ^2 should be regarded as "significantly larger?" A fairly standard statistical approach to answer this question is to evaluate the chance that a set $\{y_n\}$ of randomly selected measurements would give a χ^2 value that is as large as (or larger than) the observed value. This requires integration of the χ^2 probability distribution:

$$p_{\chi^2 \geq \chi^2_{obs}} = \int_{\chi^2_{obs}}^{\infty} p_{\chi^2}\, d\chi^2$$

$$= 1 - \int_{0}^{\chi^2_{obs}} p_{\chi^2}\, d\chi^2$$

(A2.2.3)

where p_{χ^2} is as given in Equation A2.2.1. In the second line (which results from the fact that p_{χ^2} is normalized, see Problem 2.33), the integral is referred to as the "cumulative distribution function ('CDF')" from χ^2. Values of this integral are tabulated in most statistics texts, and can be generated using the "CDF" command in Mathematica.

[†] Specifically,

$$\Gamma(\nu) = \int_{0}^{\infty} x^{\nu-1} e^{-x}\, dx$$

[‡] The Mathematica command is "ChiSquareDistribution[v]," and can be used along with the commands "PDF" and "CDF" to get probabilities, probability distributions, and cumulative distributions.

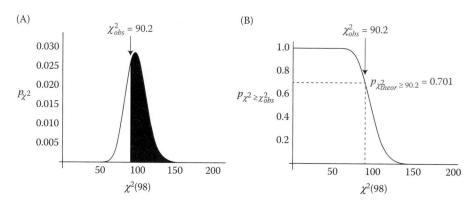

Figure A2.2.2 The integral distribution function of χ^2 as a means to test goodness of fit. (A) The χ^2 probability function (p_{χ^2}) for $\nu = 98$ degrees of freedom (Equation A2.2.1). The observed χ^2 value (90.2) for the exponential fit in Figure 2.17 is indicated. (B) The integral of the χ^2 distribution function, obtained by integrating p_{χ^2} from χ^2_{obs} to infinity (Equation A2.2.3). The dashed line gives the probability (0.701) that a randomly selected χ^2 value would be greater than 90.2. This probability corresponds to the red shaded section in panel A.

Integration of the χ^2 probability function with $\nu=98$ degrees of freedom is shown in **Figure A2.2.2**. The integral of p_{χ^2} (i.e., the CDF) increases sigmoidally from zero to one as χ^2 increases; its "complement" ($p_{\chi^2 \geq \chi^2_{obs}}$, Equation A2.2.3) decreases sigmoidally from one to zero (Figure A2.2.2B). For the exponential decay fit in Figure 2.17, the value of $p_{\chi^2 \geq \chi^2_{obs}=90.2}$ is 0.701. This means there is a 70% chance a random set of $\{y_i\}$ values that conform to the exponential decay model would give a larger value of χ^2 (i.e., a worse fit) than the observed value ($\chi^2=90.2$). This can be taken as strong statistical evidence that the data in Figure 2.17 are consistent with the exponential decay model.

The *f*-distribution

The χ^2 distribution provides a good way to evaluate whether a particular model adequately describes a set of data. In contrast, the *f*-distribution allows different models to be compared to determine if one is better than the other. This is particularly useful when comparing a simple model and a more complicated model. It is generally believed that simpler models should be favored, but not at the expense of a poor fit. If a more complicated model significantly improves the fit, the more complicated one should be favored.

However, more complicated models (i.e., models with a larger number of parameters) can usually be expected to fit a little bit better, since they have more freedom to "soak up" the residuals. When should a modest improvement in fit be considered significant? The *f*-distribution provides a statistical test for comparison of two models, based both on the differences in the quality of the fit (i.e., the *SSR* values) and the complexity of the two models (i.e., differences in numbers of degrees of freedom).

In the application described in the main text, we applied the *f*-test to evaluate the confidence limits for fitted parameters. In that context, we can think of our more complicated model as the full model with *m* adjustable parameters, and the simpler model as one where the parameter we are testing is fixed. Using the definition of the *f*-ratio given in the text, that is,

$$f(\nu_1, \nu_2) = \frac{\chi_1^2/\nu_1}{\chi_2^2/\nu_2}$$

(A2.2.4)

the full model is model 2 with ν_2 degrees of freedom, and the fixed-parameter model is model 1, with $\nu_1 = \nu_2-1$ degrees of freedom.[†]

[†] Note that when finding confidence intervals for a single parameter, the relationship $\nu_2 = \nu_1 - 1$ is true regardless of the number of data points *n*.

One nice feature of f-ratios is that the variance in y values (i.e., σ_y^2) cancels (as long as the variance is the same for all points x_i). Thus, rather than having to know the absolute values of the variances *a priori* to calculate χ^2 (or guessing them after the fit, based on residuals), we can calculate the f-ratio by simply taking a ratio of the sum of the square residuals (SSRs):

$$f(\nu_1, \nu_2) = \frac{SSR_1/\nu_1}{SSR_2/\nu_2} \tag{A2.2.5}$$

This formula, which can be calculated directly from the fits of models 1 and 2, was used to calculate f-values in Figure 2.22 for measuring the confidence limits on the exponential decay parameter k.

To calculate threshold values, we need a probability distribution for f, that is, $p_f(\nu_1, \nu_2)$ (p). Because f-values are ratios of χ^2 values (for which we have a probability distribution, Equation A2.2.1), we can calculate $p_f(\nu_1, \nu_2)$ explicitly (for a derivation, see Pugh and Winslow, 1966):

$$p_f(\nu_1, \nu_2) = \frac{\Gamma[(\nu_1 + \nu_2)/2]}{\Gamma[\nu_1/2]\Gamma[\nu_2/2]}\left(\frac{\nu_1}{\nu_2}\right)^{\nu_1/2} \frac{f^{(\nu_1-1)/2}}{(1 - f\nu_1/\nu_2)^{(\nu_1+\nu_2)/2}} \tag{A2.2.6}$$

Although probability distribution in Equation A2.2.6 is somewhat formidable, tables of $p_f(\nu_1, \nu_2)$ values can be found in statistics texts, and the probability distribution is built into Mathematica.[†] Some f distributions for different ν_1, ν_2 pairs are shown in **Figure A2.2.3**.

As with the analysis using χ^2, we can calculate the probability that an f-ratio selected randomly from the f probability distribution exceeds a particular value (in this case the value f_{obs} obtained from a data set), from two statistically equivalent models. This is achieved by integrating p_f from f_{obs} to infinity:

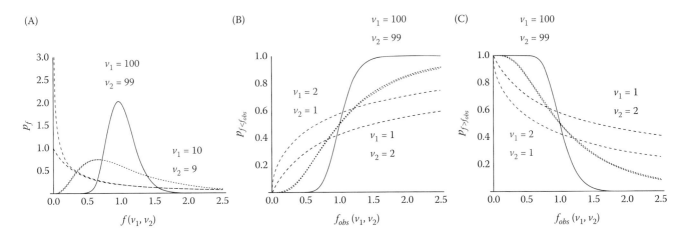

Figure A2.2.3 Probability distributions for the f-ratio for different ν_1, ν_2 pairs. (A) Probability distribution of f-ratios for different degrees of freedom. Dashed curves give probabilities for models with $\nu_1 = 1$, $\nu_2 = 2$ degrees of freedom (black, $p_f(1, 2)$), and the reverse ($p_f(2, 1)$, red). Dotted curves show $p_f(9, 10)$ (black) and $p_f(10, 9)$ (red). Solid curves show $p_f(99, 100)$ and $p_f(100, 99)$. For models with a large number of degrees of freedom, there is very little difference between $p_f(\nu_1, \nu_2)$ and $p_f(\nu_2, \nu_1)$ when the two models differ by a single degree of freedom, although these differences are significant for small values of ν. (B) Cumulative (i.e., integrated) f-ratio probability distributions from (A). These distributions give the probability that a random draw from the f-ratio distribution will be smaller than a given value of f_{obs}, that is, $p_{f<f_{obs}}(\nu_1, \nu_2)$. (C) The probability that a value drawn from the f-ratio distribution exceeds a given value of f_{obs} ($p_{f>f_{obs}}(\nu_1, \nu_2)$, Equation A2.2.7). This probability, which is equal to one minus the probabilities from (B), is the basis for the f test.

[†] The Mathematica command is "FRatioDistribution[v1, v2]."

$$p_{f>f_{obs}} = \int_{f_{obs}}^{\infty} p_f df$$

(A2.2.7)

$$= 1 - \int_{0}^{f_{obs}} p_f df$$

As with χ^2, the integral in the second line is the cumulative probability distribution for f.[†] Fortunately, analytical expressions of these integrated p_f functions are available in statistics texts and in mathematical software packages.[‡] Integrated f-distributions are shown in Figure A2.2.3.

Using Equation A2.2.7, we can compare the goodness of fit of two arbitrary models. If we find an experimentally determined f-ratio to be improbably large,[§] compared to values from the f-ratio distribution, model 2 should be favored. For example, given the data in Figure 2.17, we could compare the exponential decay model with a polynomial equation with five parameters using the f test to determine whether there are significant differences in fitting between the two models (see Problem 2.34).

When determining confidence limits for a fitted parameter, we can use the f-ratio distribution to explore confidence limits. In this case, we set the "test" parameter to various fixed values, optimize the remaining free parameters, and determine the resulting SSR. We repeat this procedure, and approximate what values of the test parameter lead to a threshold probability for the f-ratio. For example, the 95% confidence limits are calculated as

$$p_{f \geq f_{95\%}} = 0.05 = \int_{f_{95\%}}^{\infty} p_f df$$

(A2.2.8)

The analytical expression for the integral in Equation A2.2.8 can be used to solve for $f_{95\%}$ (or for other confidence limits, by substituting the appropriate value for 0.05 in the middle equality).

[†] Sometimes referred to as p_f, where the capital F indicates integration of f.
[‡] The Mathematica command is "`CDF[FRatioDistribution[v1, v2]]`."
[§] Or improbably small. In general, either model 1 or model 2 could provide a better fit, and both f and $1/f$ should be tested (evaluating against $p_f(v_1, v_2)$ and $p_f(v_2, v_1)$, respectively). For tests of parameter confidence intervals, we don't need to worry about model 1 (the restricted, nested model) fitting better than model 2.

CHAPTER 3

The Framework of Thermodynamics and the First Law

Goal and Summary

The goals of this chapter are twofold. First, to give students an understanding of the logic of thermodynamic analysis, key concepts are described, including the differences between macroscopic and microscopic (or "statistical") approaches to thermodynamics, the partitioning between "system" and "surroundings," the concepts of reversibility and irreversibility, and the importance of the equilibrium state and corresponding equations of state. These concepts are key to understanding the first and second law (Chapter 4), the connection to free energy (Chapters 5 through 8), and the different types of statistical ensembles that will be analyzed in Chapters 9 through 11.

The second goal of this chapter is to introduce the first law of thermodynamics, re-enforcing the differences between path and state functions, and highlighting the use of simple state functions to calculate thermodynamic changes along complicated paths. General concepts of work will be introduced, focusing on expansion and compression of ideal gases. Heat flow will be described in terms of temperature changes and heat capacities. Considerable attention will be given to analysis of adiabatic expansion, to illustrate various approaches to thermodynamic analysis (empirical observation, comparison of alternative of paths, and differential approaches). We will conclude with a discussion of the van der Waals gas, to provide a picture of nonideality and its origins.

WHAT IS THERMODYNAMICS AND WHAT DOES IT TREAT?

Thermodynamics describes the properties (especially energetic properties) of matter. As suggested from the first part of its name, thermodynamics describes properties that are "*thermal*" in nature, such as temperature, heat capacity, various forms of energy, and also energetic processes like heat flow and work. These last two quantities played in important role in the early development of thermodynamics in the 1800s, which was motivated by advances in machine building and the desire to increase the efficiency of machines in industry, transportation, agriculture, and mining. With the subsequent understanding of atomic and molecular structures, developments in the thermodynamics of machines were adapted to describe the thermodynamic properties of materials and chemical reactions.

The second part of the word "thermodynamics" suggests a description of the changing or "*dynamic*" properties of matter. To be sure, *some* aspects of thermodynamics deal with dynamic, changing properties. For example, the second law of thermodynamics describes the direction of spontaneous change for bulk or

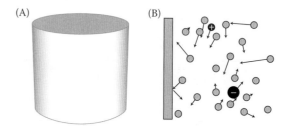

Figure 3.1 Bulk and molecular views of equilibrium systems. (A) A bulk, macroscopic view of an aqueous salt solution is shown. (B) A molecular snapshot of the solution is depicted, with motions indicated with arrows. A microscopic view would show ceaseless and chaotic thermal jostling, and would require a lot of information to specify each microstate (in principle, the positions and velocities of each atom). In contrast, very little detail is needed to depict the bulk equilibrium state, which generally appears constant over time.[†]

"macroscopic"[‡] properties. In addition, statistical thermodynamics often provides a direct connection between energy and motions on the molecular or "microscopic" scale. However, some of the strongest (and by far the simplest) analytical statements of thermodynamics (especially classical thermodynamics) apply to systems that do not appear to be changing when viewed in bulk (**Figure 3.1**). Such "equilibrium states" are central to thermodynamic analysis. Figure 3.1 illustrates that although equilibrium states appear to be static at the bulk level, there is constant change associated with the ceaseless motions on the molecular level. Indeed, if one were restricted to a microscopic view, it would be difficult to conclude (without extensive averaging) that a system is in an equilibrium state. Equilibrium thermodynamics is remarkable in that it captures the bulk behavior of materials and reactions that depend on these incredibly detailed underlying dynamics. This is done either by effectively ignoring most of the molecular details (in "classical" thermodynamics), or by cleverly organizing them into a framework involving probabilities (in "statistical thermodynamics").

Classical and statistical thermodynamics

Within the field of thermodynamics, an important distinction can be made between "classical" and "statistical" approaches (**Figure 3.2**). These two approaches to thermodynamics have different assumptions and viewpoints, different methods of analysis, and provide complementary results. To a large extent, classical thermodynamics is free from models (especially molecular models), whereas statistical thermodynamics depends nearly entirely on molecular models. Deep insights can often be extracted from the models of statistical thermodynamics, often leading to the prediction of complex macroscopic behavior.

However, the dependence of statistical thermodynamics on models can sometimes be a drawback. In some cases we don't have good models. In other cases, we can come up with appropriate models, but the they end up being too complicated to be useful. Especially in such cases, the generality of classical thermodynamics can be a strength. When classical thermodynamics is used with the right measurements, it provides a useful macroscopic description of a system under a variety of conditions. In short, there are unique strengths to both the classical and statistical approaches.

The differences between classical and statistical thermodynamics can be understood by considering the context in which they were developed. Much of classical

---

[†] One notable exception is simple substances at their critical points, where coexisting gas and liquid phases show large fluctuations in bulk properties.

[‡] The adjectives "macroscopic" and "microscopic" can be thought of as low- and high-magnification views of material. A macroscopic view is too coarse to see the molecules (and their medium sized aggregates) that display the random jostlings of thermal energy (see Figure 3.1).

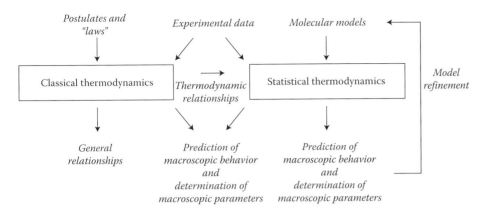

Figure 3.2 The relationship between classical and statistical thermodynamics. These two branches of thermodynamics differ in their inputs and outputs. Classical thermodynamics rests on some very general assumptions and laws, whereas statistical thermodynamics depends on molecular models as well as relationships from classical thermodynamics. Both branches are used to analyze experimental data and extract thermodynamic quantities (e.g., energies, binding constants, and entropies) and to make predictions regarding behavior on the macroscopic level. Statistical thermodynamics also provides insights and makes predictions on the molecular level. By comparing statistical thermodynamic predictions with experimental measurement, molecular parameters can be determined, and models can be improved, if warranted.

thermodynamics was developed before the discovery of atoms and molecules, and thus focuses on macroscopic systems. In contrast, statistical thermodynamics was developed with recognition of atoms and molecules,[†] with a major goal of connecting the microscopic properties of atoms and molecules to bulk properties of materials. Though these approaches differ significantly, the divisions between the classical and statistical approach are often blurry. Key derivations of statistical thermodynamics require relations from the classical approach (Figure 3.2). Likewise, classical thermodynamic analysis of chemical reactions must begin with descriptions of relevant molecular species; for complex reactions involving many species, the classical approach starts to look statistical.

The six chapters (Chapters 3–8) will focus mostly on classical, bulk thermodynamics. The statistical formalism will be developed starting in Chapter 9. Beginning with the classical approach allows the laws of thermodynamics to be developed and applied to simple, familiar systems, both physical and chemical. However, the classical thermodynamic description of macroscopic systems in Chapters 3–8 will sometimes be augmented with molecular models along with statistical considerations. Though outside the domain of classical thermodynamics, these models help in understanding the origins of quantities such as energy, heat capacity, and in particular, entropy (Chapter 4).

DIVIDING UP THE UNIVERSE: SYSTEM AND SURROUNDINGS

The motivation for thermodynamic studies in chemistry and biology is often to understand the properties of a well-defined sample (be it solid, liquid, gas, or a complex mixture) or a mechanical or electrical device. In most cases, such samples will be able to exchange one or more forms of energy and/or matter with the environment. Thus, to develop an understanding of thermodynamics and its applications, we need precise definitions that specify our sample (which we refer to as "the system"), the environment (which we refer to as "the surroundings"), and the

---

[†] Ludwig Boltzmann's early contributions to statistical thermodynamics were made before the concept of atoms and molecules was accepted, and provided early support for the atomic view that emerged. See "Boltzmann's Atom" for a readable account of this important period in the development of chemistry and physics.

extent to which the system and surroundings interact (through thermal, mechanical, and material transfer).

*The **system*** is defined as the part of the universe we choose to study. In biothermodynamics, the system is usually an aqueous solution with various macromolecules, assemblies, and small molecules. However, systems can also be simpler, such as a vessel containing a gas, a pure liquid, or a solid, or can be more complex, for example, a living cell, a whole multicellular organism, or a planet.

*The **surroundings*** is defined as rest of the universe. Fortunately, from a practical point of view, we do not need to concern ourselves with the entire universe. Rather, we focus on that part of the universe that exchanges matter and energy with the system (i.e., the immediate vicinity), although in some cases, distant sources of energy must be included.[†]

*The **boundary*** is defined as border between the system and surroundings. The boundary's job is to permit (or prevent) interaction between the system and surroundings in terms of exchange of material, heat flow, and volume change (for now we will ignore external fields and exotic kinds of work such as that resulting from stirrers and electrical resistors). The boundary is typically regarded as being thermodynamically negligible, that is, it takes up no volume, has no heat capacity, and does not release or adsorb material. In most real situations, the "negligible thermodynamics" of the boundary is an approximation.[‡]

One important thing to keep in mind when partitioning of the universe into system and surroundings is that these divisions are artificial, and are created by us to simplify thermodynamic analysis. Thus, *the thermodynamics we develop must be insensitive to this partitioning*. If we flip labels, renaming the system as surroundings and vice versa, all analysis should lead to the same results as with the original (unflipped) labeling. With this qualification in mind, we will further classify systems based on their interaction with the surroundings[§] (**Figure 3.3**).

*The **open system*** can exchange matter (through its boundaries) with the surroundings. Defining the number of moles of material in the system as n (or if we have a mixture of species, n_1, n_2, ...),[¶] we can associate the open system with the differential $dn \neq 0$. Through the use of a differential, we can express both a finite change (integrating over n), and small material fluctuations around an equilibrium point.[**]

*The **closed system*** cannot exchange matter with the surroundings. For the closed system, $dn = 0$.

Figure 3.3 Different types of systems and their properties. The systems are composed of a simple macroscopic fluid (liquid or gas, gray) that can in principle expand (dV), exchange material (dn), or exchange thermal energy (heat flow, dq) with the surroundings. Moving from left to right, the boundary becomes more restrictive, preventing the flow of material, heat, and extent (i.e., expansion and compression) between the system and surroundings.

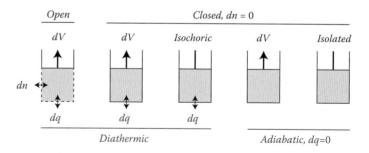

[†] For example, the surroundings of photosynthesizing plants must include the sun (or at least its light energy).

[‡] Though for some simple systems like a block of metal submerged in a heat bath, the boundary has no material extent.

[§] This classification scheme is really a scheme about differences in boundaries, but it is more common usage to refer, for example, to an adiabatic system than an adiabatic boundary.

[¶] In this text, we will use lower-case n to represent the number of moles, and upper-case N to represent the number of molecules (or other fundamental units, including the coins and dice in Chapter 1).

[**] Some texts represent infinitesimal fluctuations about equilibrium using the symbol δ in place of the differential; thus, a fluctuation about equilibrium would be represented as $\delta n \neq 0$.

The **diathermic system** exchanges energy with the surroundings in the form of heat flow. Using q to represent heat, $dq \neq 0$ for a diathermic system.[†] There are two types of diathermic systems that are easy to represent analytically: (1) isobaric systems, with constant pressure but variable volume ($dp = 0$, $dV \neq 0$) and (2) isochoric systems, with constant volumes but variable pressure ($dV = 0$, $dp \neq 0$). The isobaric, diathermic system ($dV \neq 0$, $dq \neq 0$) is the most important one for biological thermodynamics, and we will discuss it a lot, especially by fixing p and T to be constant ($dp = dT = 0$). This special type of system is often referred to an **isothermal system**.

The **adiabatic system** does not exchange heat energy with the surroundings, that is, it is thermally insulated. Common examples that are approximately adiabatic are Dewar flasks from the laboratory, and thermos bottles and ice coolers from picnics. For adiabatic systems, $dq = 0$.

The **isolated system** exchanges nothing with the surroundings. It is adiabatic ($dq = 0$), isochoric ($dV = 0$) and closed ($dn = 0$). One important example of an isolated system is the universe.[‡] As we shall see from the first and second laws of thermodynamics, isolated systems have constant energy (the first law) and increasing entropy (the second law). Extending these limits to the universe has important practical and philosophical implications.

EQUILIBRIUM, CHANGES OF STATE, AND REVERSIBILITY

The meaning of equilibrium in classical thermodynamics

Equilibrium states are much easier to analyze with thermodynamics than non-equilibrium states. This is because when equilibrium is achieved, bulk thermodynamic variables are well defined and are uniform throughout. As a result, the variables that describe the properties of the material are well defined, and are uniform on a macroscopic length scale. The most important of these are the "intensive" variables, which do not depend on the size of the system.[§] Intensive variables include pressure (p), temperature (T), and concentrations of different chemical components. We will typically give concentrations on the molarity scale, moles per liter, $M_i = n_i/V$ (or simply $[i]$), where i indicates a particular species.[¶]

Equilibrium behavior can either be described within a system or between a system and its surroundings (**Figure 3.4**). Equilibrium between a system and its

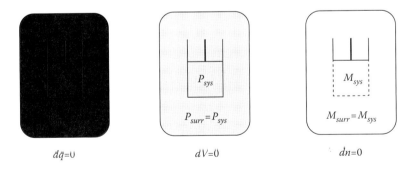

$$dq=0 \qquad dV=0 \qquad dn=0$$

Figure 3.4 Simple systems in equilibrium with their surroundings. For each of the three systems, the relevant intensive variable (X_{sys}) is matched to that of the surroundings (X_{surr}). For the diathermal system (left) the temperature is matched, so there is no heat flow. For the expandable system (middle) the pressure is matched, so there is no volume change. For the open system (right) the concentrations are matched (only one concentration is shown for simplicity) so there is no flow of material.

[†] Heat and work will be defined more precisely in the discussion of the first law later in this chapter.

[‡] Although this statement is somewhat difficult to make with conviction, since we don't have a good picture of the boundary of the universe, and the concept itself is more in the realm of philosophy than chemistry and biology.

[§] Variables that depend on the size of the system are called "extensive." To determine whether a variable is intensive or extensive, simply ask "if the system is doubled, does the value of the variable double?" For volume and mass (extensive variables), the answer is clearly "yes," whereas for pressure, temperature, and concentration (intensive variables), the answer is "no."

[¶] Although we use molarity in this book, other concentration scales can be used equally well. (e.g., gram/milliliter, mole fraction; see Chapter 7).

surroundings has a number of important implications. Specifically, a diathermic system in equilibrium with its surroundings has the same temperature as its surroundings ($T_{sys} = T_{surr}$), a variable-volume system in equilibrium with its surroundings has the same pressure as its surroundings ($p_{sys} = p_{surr}$), and an open system in equilibrium with its surroundings has the same concentrations as its surroundings ($M_{i,sys} = M_{i,surr}$). This uniformity suggests that there will be no heat flow ($dq = 0$), no volume change ($dV = 0$), and no material flow ($dn_i = 0$). If left alone, most[†] systems in equilibrium with their surroundings should remain unchanged.

Although systems in equilibrium have the advantage of possessing well-defined thermodynamic properties, the fact that these systems do not change their state even when permitted by the boundary would seem to limit their use in classical thermodynamic analysis. This is because several important thermodynamic quantities (like the entropy) can't be measured directly (or calculated) using classical thermodynamics alone. Rather, classical thermodynamics is restricted to the analysis of how these quantities *change* when a system changes state. Thus, our goal is to analyze the thermodynamics associated with changes of state, while retaining the simplicity of equilibrium systems. So-called "reversible changes" will allow us to achieve this goal.

Irreversible and reversible changes of state

The most obvious way to bring about a change of state to a system is to significantly modify the surroundings so that it is no longer in equilibrium with the system (**Figure 3.5**). Away from equilibrium, there is appreciable heat flow, volume change, and material flow between the system and surroundings. Though such changes are easy to bring about, they are hard to analyze. The problem is that in most nonequilibrium changes, T_{sys}, p_{sys}, and $M_{i,sys}$ are heterogeneous across the system; as a result, they cannot be defined with a single system-wide parameter (Figure 3.5); without well-defined values for these quantities, thermodynamic

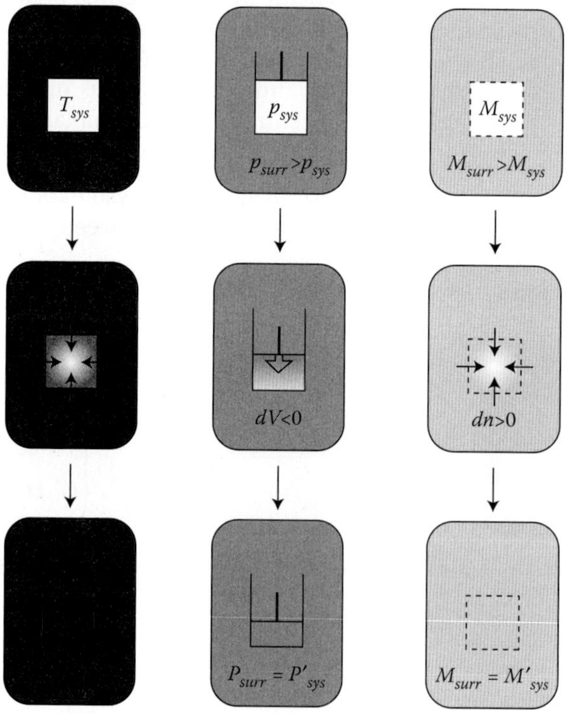

Figure 3.5 Irreversible changes between equilibrium states. At the start of each transformation (top), there is a difference between intensive variables of the system and that of the surroundings. On the left, a large temperature difference results in a rapid and irreversible heat flow into the system, which produces a temperature gradient in the system. As a result, the temperature of the system cannot be defined as a single value during the transformation. In the center, a pressure difference results in a rapid and irreversible volume decrease, which results in a density gradient, making the pressure in the system undefined. On the right, a concentration difference results in a rapid and irreversible material flow into the system, which results in a composition gradient, making the concentration in the system undefined. Eventually the gradients within the system are dissipated (bottom), and a new equilibrium is reestablished, with the intensive variables of the system matching those of the surroundings. Although each system has converted between well-defined equilibrium states, the paths taken are not easily described.

[†] In addition to critical-point example noted above, systems in less exotic phase equilibrium can vary in the relative amount of each phase present, provided volume change and heat flow to the surroundings are permitted. For example, a diathermic system with water and ice at fixed values of $T = 0°C$ and $p = 1$ *atmosphere* can, over time, accumulate more ice at the expense of water (or vice versa). This situation is sometimes referred to as "neutral equilibrium" (see Adkins, 1984).

calculations become very complicated. Thus, we have a dilemma: *equilibrium systems have well-defined properties, but they don't appear to change their state; in contrast, systems undergoing typical irreversible changes of state are out of equilibrium, and don't have well-defined properties.*

Thus, a very basic but important question for classical thermodynamic analysis is "*can a system be manipulated to change its macroscopic properties while maintaining equilibrium?*"[†] Fortunately (but perhaps counter-intuitively), the answer is "yes." As long as changes are made very slowly,[‡] the system can be assumed to have uniform, well-defined properties such as T, p, and M. This slow approach to making "equilibrium changes" can be brought about by creating an infinitesimal imbalance between the system and the surroundings. For example, a tiny amount of heat flow (dq) can be achieved by adjusting the temperature of the surroundings to be only slightly different than that of the system (i.e., $T_{surr} = T_{sys} + dT$). Although the result of this heat flow leaves the system very close to its starting temperature, the process can be repeated by incrementing (or decrementing) T_{surr} by dT, and allowing another infinitesimal heat flow, dq. Repeating this procedure many times will result in a finite heat flow q, and a finite change in temperature from the initial temperature T_i to the final temperature T_f (**Figure 3.6**).

Using this differential approach to change the system, there is never more than an infinitesimal difference between the intensive variables of the system and the surroundings (dT in Figure 3.6). As a result, the direction of change can be "reversed" by making an infinitesimal change to either the system or the surroundings. Thus, these types of changes are called "**reversible**" changes of state.[§] In contrast, the changes of state in Figure 3.5 cannot be reversed by an infinitesimal change; thus, these processes are called "**irreversible**" changes of state.

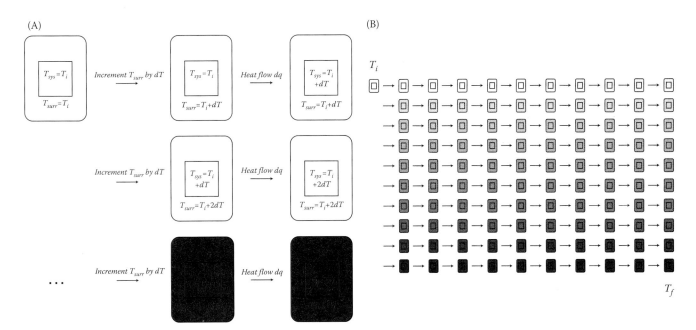

Figure 3.6 A reversible change of state involving heat flow. The temperature of the system is raised from T_i to T_f by a series of infinitesimal temperature changes to the surroundings by $dT > 0$. (A) A few of the many steps in between T_i and T_f are shown, including the infinitesimal temperature imbalances that produce reversible heat flow (dq), in this case into the system. (B) A very large number of these tiny steps produces a finite temperature change, $\Delta T = T_f - T_i$, and a finite heat flow q.

[†] This question will become particularly relevant when we discuss the second law of thermodynamics, which asserts that all spontaneous changes of state are "irreversible" and thus are not at equilibrium.

[‡] The question is, "how slow is slowly"? Although the answer depends on the system, the uniformity of T, p, and M will be maintained if the rates of heat flow, volume change, and material flux are slower than thermal diffusion, acoustic propagation, and molecular diffusion within the system.

[§] Rather than the rather awkward expression "equilibrium changes of state."

In the same spirit as the reversible heat flow, the differential approach can be used to make reversible changes in volume and in composition by making infinitesimal changes in pressure and concentration to the surroundings, respectively.

The surroundings as a "reservoir"

Although the reversible transformation in Figure 3.6 facilitates thermodynamic calculations by maintaining equilibrium, this type of change of state is somewhat contrived, particularly with regard to the surroundings. It is often impractical to smoothly adjust the temperature, pressure, and concentration of the surroundings. This is particularly true when the surroundings is much larger than the system, such as a chemical reactions in a small beaker (system) in a laboratory (surroundings). If the reaction gives off heat, the system will likely increase its temperature above the laboratory temperature, and it is probably impractical to change the temperature of the laboratory to match the reaction. The same can be said for more complicated "systems" like internal combustion engines.

Not being able to manipulate large surroundings is a limitation in some contexts, its constancy provides a significant simplification for thermodynamic analysis. Large surroundings absorb (or provide) large quantities of heat from the system without changing the temperature.[†] Likewise, large surroundings accommodate substantial volume change from the system without changing pressure. Further, large surroundings exchange significant amounts of material with a system without changing concentration. Thus, thermodynamic changes to the reservoir can be calculated using constant values for T_{surr}, p_{surr}, and M_{surr} (see Figure 3.5). We will refer to this type of large, invariant surroundings as a "*reservoir*."[‡]

Though it is easy to calculate thermodynamic changes for reservoir-type surroundings, calculating irreversible thermodynamic changes for the system remains a challenge. The intensive variables of the system (T_{sys}, p_{sys}, and M_{sys}) are *not* constant, and as described above, they are likely to become undefined. The problem is that the the rates of exchange (heat, shape, and material) between the system and surroundings are fast compared equilibration within the system. One way to avoid this problem is to slow down the rate of exchange, by modifying the boundary. **Figure 3.7** illustrates this approach. Heat flow can be limited by a boundary that is adiabatic except for a very narrow heat conduit. As long as the conduit is of sufficiently low thermal conductance that heat transfer between the system and surroundings is substantially slower than system equilibration, the system will

Figure 3.7 Irreversible quasi-static changes of state. The top row shows a quasistatic heat flow from a hot surroundings to a cold system through a very narrow (low conductance) heat pipe. Heat flow can be thought of as the result of a large number (n total) of microscopic infinitesimals (dq) as indicated by the summations. At all intermediate stages, the system maintains internal equilibrium (middle, with a well-defined temperature T′). The bottom row shows an analogous flow of material from surroundings to system. The two are connected by a single narrow hole (top) that only occasionally allows passage of a single molecule (spheres).

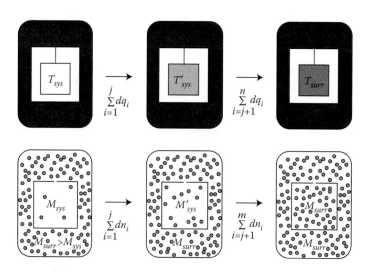

[†] Jumping ahead a little bit, we would describe large surroundings as having a very high heat capacity.
[‡] More specifically, a "thermal reservoir" for heat exchange, a "mechanical reservoir" for volume change, and a "material reservoir" for material exchange.

remain in equilibrium states throughout the entire process.[†] Likewise, material transfer can be limited by making a single narrow passageway between the system and surroundings that permits molecules through in single file. These types of very slow changes of state, which are often referred to as "**quasistatic**,"[‡] maintain both the system and the surroundings at (different) equilibrium states throughout the transformation facilitating thermodynamic calculations.

THERMODYNAMIC VARIABLES AND EQUATIONS OF STATE

Variables in thermodynamics

As mentioned above, if a simple system is left alone, it tends toward an equilibrium state in which T, p, and M_i can be described with just a single value that applies to all regions.[§] Moreover, we can define an equilibrium state using just a few of these parameters, using an "equation of state." Two questions naturally arise from this simple statement: (1) how many variables are sufficient to "specify" the thermodynamic state of the system,[¶] and (2) which variables should be used?

For single-phase systems,[**] the number of variables required to specify the thermodynamic state is equal to the number of components in the system (molecules that cannot be interconverted by chemical reaction) and two additional variables. Thus, for a system with one component, specifying the number of moles of the component (n) along with two other system variables (such as pressure and temperature) is enough to uniquely specify the state of the system. If we specify n, T, and p, then the volume should be uniquely defined. One shorthand notation that captures this functional dependence is $V = V(n, T, p)$. This choice of variables (n, T, p) is not the only way the thermodynamic state of a one-component system can be specified. Instead, we might specify the thermodynamic state using n, V, and T, which would allow us to uniquely define the pressure, that is, $p = p(n, V, T)$. Choosing variables that best describe a particular system depends in part on the features of the system, and in part on how the system interacts with its surroundings. An explanation of this second point will be postponed until we introduce the second law as a means to develop thermodynamic potential functions.

Functions that have the form given above (e.g., $p = p(n, V, T)$) are called "***equations of state***." These equations are extremely useful for thermodynamic calculation. A well-known example of an equation of state is the "ideal gas law," which relates the pressure, temperature, and volume of a fixed amount of n moles of gas as follows:

$$pV = nRT \qquad (3.1)$$

[†] This does not mean that the system and the surroundings are in equilibrium, which would be inconsistent with the irreversible nature of such a transformation.

[‡] Note that the term "quasi-static" is used differently by different authors. For example, Adkins uses the term to describe the slow reversible transformations like that depicted in Figure 3.6, rather than the slow irreversible transformations in Figure 3.7. The key idea—that the change of state is so slow that the system and surroundings can be regarded as being in equilibrium states the whole time—is the same in both usages, but the relationship between the system and surroundings differs.

[§] As long as we don't "zoom in" too much to very tiny regions. The issue here is that we need a certain amount of averaging, either over space, time, or both, to describe the intensive variables p, T, and M_i. This will be discussed further when we develop statistical thermodynamics (Chapters 9–11). Note that if a system has multiple phases of component i, the concentration within different phases will be different.

[¶] By "specify," we mean that all of the bulk thermodynamic variables are uniquely defined. In a particular thermodynamic state, we can have a single value for each measurable variables (e.g., T, p, V) and also for state functions that are only measurable by through changes (e.g., U, S, and G).

[**] We will discuss phases in detail in Chapter 6. For now, just consider phases to be different states of matter such as solid, liquid, and gas.

R is a fundamental constant called the gas constant, and has a value of 8.31447 J mol^{-1} K^{-1}. This law applies to a number of simple gases at low pressures and high temperatures, although substantial deviations occur away from these limits. Thus, Equation 3.1 can be thought of as a "limiting law." The ideal gas law can of course be rearranged to give equations of state that isolate p, V, or T, as described above.

Because of its simplicity, the ideal gas law is a good equation of state to learn how to use the laws of thermodynamics to analyze processes like heat flow and expansion work below, and entropy change in Chapter 4. The parameters of the ideal gas law (p, T, and V) are easy to measure, and the relationships between them are very simple.

Ways to derive the ideal gas equation of state

Before we start using the ideal gas law to illustrate the workings of classical thermodynamics, we will describe the origins and derivations of the ideal gas law (**Figure 3.8**). The ideal gas law can be derived in several ways, either through empirical observation of how the pressure of simple gases changes with n, V, and T (the classical thermodynamic approach), or to start with a model (either classical or quantum mechanical), and determine what the average macroscopic properties should be (the statistical thermodynamic approach).

The first, and perhaps the simplest derivation of the ideal gas equation of state comes from measurements (empirical observations; Figure 3.8A) of the relationship between pairs of the four variables in the equation, holding the others constant. For example, the relationship between the pressure and volume of a fixed quantity of a gas at a single temperature was found by to follow a simple inverse law by Robert Boyle in 1660:

$$p = \frac{c}{V} \tag{3.2}$$

where c is a constant at a fixed temperature and mole number. Simple relationships can also be derived between pressure and temperature, volume, and temperature, and volume and particle number (Figure 3.8). Combining these relationships leads to the ideal gas equation of state (Equation 3.1, see Problem 3.1).

A second classical approach uses differentials as a starting (Figure 3.8B). This approach is more general, serving as a starting point for deriving equations of state for nonideal gases and other phases of matter. As described in Chapter 2, the dif-

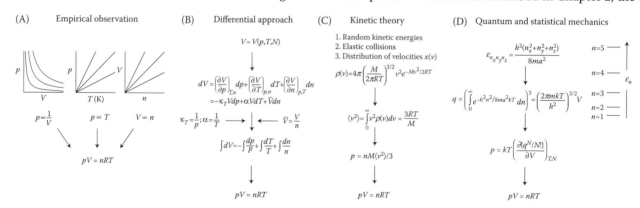

Figure 3.8 Different ways to arrive at the ideal gas law. The earliest formulation involved experimental measurement of how p, T, V, and n are interrelated for gases at low pressures and high temperatures. Alternatively, using the properties of exact differentials of state functions, the equation of state can be obtained by integration, using measured relations for compressibility and thermal expansion coefficients. Alternatively, using kinetic theory with the assumptions that the only energy present is kinetic energy and that it is randomly distributed, a density function for molecular speeds can be generated (the Maxwell distribution), and can be used to calculate pressure through the rate of momentum transfer to the walls, assuming elastic collisions. Finally, by assuming that gas molecules act as noninteracting particles restrained to a three dimensional box, quantum mechanical energy levels can be derived and can be used to construct a partition function, which can be manipulated to calculate the pressure.

ferential approach describes change. In this approach, experimental information is introduced in the form of differential coefficients (i.e., partial derivatives), which often correspond to known (or measurable) quantities (compare the first and second lines of Equation 3.4 below).

The differential approach begins with a general equation of state that is expressed using the variables of interest, without knowledge or explicit statement of the functional form

$$V = V(p, T, n) \tag{3.3}$$

This general equation of state implies an exact differential relationship as described in Chapter 2:

$$\begin{aligned}
dV &= \left(\frac{\partial V}{\partial p}\right)_{T,n} dp + \left(\frac{\partial V}{Tp}\right)_{p,n} dT + \left(\frac{\partial V}{Np}\right)_{p,T} dn \\
&= -V\kappa_T dp + V\alpha dT + \bar{V} dn
\end{aligned} \tag{3.4}$$

In the second line, κ_T is the "isothermal compressibility," α is the "coefficient of thermal expansion," and $\bar{V}$ is the molar volume.[†]

A third way to derive the ideal gas equation of state comes from "kinetic theory" (Figure 3.8C).[‡] This derivation starts with a classical model in which the gas molecules do not interact with one another, and interact with the boundary of their vessel through elastic collisions. The only modes of energy storage in this model are those associated with kinetics energy. For monatomic gases this means translation in each of three dimensions. By assuming a random energy distribution among molecules, the average velocity can be calculated and related to pressure. In addition to leading to the ideal gas equation of state (Figure 3.8), this approach provides a number of simple relations involving the thermodynamic properties of "classical" ideal gases that will be used in our development of the laws of thermodynamics (**Table 3.1**).

Finally, the ideal gas equation of state can be obtained by combining quantum and statistical thermodynamics (Figure 3.8D). By analyzing an isolated particle in a container using quantum mechanics, discrete energy values are obtained.[§] In one of the more spectacular derivations in physical chemistry, statistical thermodynamic averaging of this somewhat abstract model yields the ideal gas law (Equation 3.1).[¶]

A graphical representation of the ideal gas law

Because the ideal gas law will be used to develop and understand the first and second law of thermodynamics, it is worth visualizing about how the variables relate to one another (**Figure 3.9**). At constant mole number, the ideal gas law connects three variables (p, V, and T). One particularly useful way to represent these three variables on a two-dimensional plot is to plot pressure–volume curves at different fixed temperatures (Figure 3.9A). Such plots are often referred to as "indicator diagrams," and the constant temperature curves are referred to as "isotherms." For the ideal gas law, these isotherms are simple inverse decays. Alternatively, other pairs of variables can be depicted (p vs. T at constant V; T vs. V at constant p). A more general way to depict the gas law graphically is as a surface in all three dimensions (p, V, and T; Figure 3.9B).

[†] Molar quantities are discussed in detail in Chapter 5.
[‡] This approach, which was developed in the 1800s by Ludwig Boltzmann, was one of the first quantitative analyses based on a molecular model. The success of Boltzmann's approach strongly supported the existence of molecules as the constituents of matter, and provided a picture of thermal energy as random motions of these molecular constituents.
[§] From the "particle in a box" model.
[¶] See McQuarrie, 1984 for a full derivation.

Table 3.1 Thermodynamic relationships for ideal gasses that derive from classical kinetic theory

Mechanical variables		Monatomic gas	Diatomic gas
Pressure	p	nRT/V	nRT/V
Volume	V	nRT/p	nRT/p
Coefficient of thermal expansion	$\alpha = \dfrac{1}{V}\left(\dfrac{\partial V}{\partial T}\right)$	$nR/(pV) = T^{-1}$	$nR/(pV) = T^{-1}$
Coefficient of isothermal compressibility	$\kappa_T = -\dfrac{1}{V}\left(\dfrac{\partial V}{\partial p}\right)_T$	$nRT/(p^2V) = p^{-1}$	$nRT/(p^2V) = p^{-1}$
Heat and energy variables			
Internal energy	U	$3/2nRT$	$5/2nRT$
Heat capacity at constant volume	$C_V = \left(\dfrac{\partial U}{\partial T}\right)_V$	$3/2nR$	$5/2nR$
Heat capacity at constant pressure	$C_p = \left(\dfrac{\partial\{U + PV\}}{\partial T}\right)_p$	$5/2nR$	$7/2nR$
Heat capacity ratio	$\gamma = C_p/C_V$	$5/3$	$7/5$
Internal pressure	$\pi_T = \left(\dfrac{\partial U}{\partial V}\right)_T$	0	0

Note: Monatomic (e.g., He, Ne) and diatomic gasses (e.g., N_2, O_2) have the same mechanical properties (parameters connecting pressure and volume), but they differ in energy and energy storage parameters. This is because diatomic gasses can store energy in ways other than linear translation, including tumbling rotations.

Figure 3.9 Graphical depiction of the ideal gas equation of state. Equation 3.1 is plotted for $n = 1$ mole. (A) When temperature is fixed, pressure shows a simple inverse decay curve with volume. Because of their constant temperature, these curves are referred to as isotherms. The family of isotherms includes temperatures from 50 (solid black) to 800 K (dotted black). (B) In three dimensions, pressure is given by a surface of inverse decays (with volume) that increase linearly with temperature.

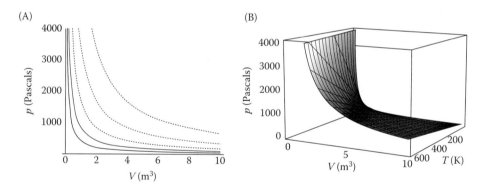

THE FIRST LAW OF THERMODYNAMICS

There are two main laws of thermodynamics, which are sensibly referred to as the "first" and "second" laws.[†] These laws put restrictions on how thermodynamic variables change when a system (and its surroundings) changes. These restrictions include the flow of energy and the distribution of temperature and material. In chemical reactions, the first and second laws determine the equilibrium positions of reactions. These laws also put restrictions on more complicated processes, such as the accuracy of replication in living organisms, and the efficiencies of mechanical engines in converting energy.

[†] In addition, there are two accessory laws, often referred to as the "zeroth law" and "third law." Although they tidy up a number of thermodynamic subtleties, we won't concern ourselves with them here.

One confusing thing about the laws of thermodynamics is that each law can be stated in different ways. On one hand, the first and second laws can each be stated analytically through a simple equation.[†] On the other hand, each law can be stated descriptively, that is, with words rather than equations. For the descriptive versions, there are multiple ways each law can be stated, and these different statements seem loosely connected, at best. To get a full understanding of thermodynamics, it is important to understand both the descriptive and the analytical statements of the laws of thermodynamics, and to see how the different statements are related.

The first law in words

One popular descriptive statement of the first law is *"energy is conserved."* Conservation means that energy is neither created nor destroyed, but instead remains constant. This is obviously true for an isolated system. However, for systems that are not isolated, if we focus only on the system, the internal energy can go up or down. In this case, the conservation statement applies when we consider the system and the surroundings together: any change in the internal energy of the system will be matched by an equal and opposite change in the energy of the surroundings.[‡]

If we lump the system and surroundings together, we can regard the combination as an isolated system. As described above, we can consider the universe to be an isolated system, since there should be no heat flow and no work that comes from outside the universe. Thus, the first law says that the internal energy of the universe is constant, although it can be redistributed among its parts.

Returning to focus on the system, the first law is sometimes represented with a statement that energy of the system can be changed either by heat or by work, and different combinations of heat and work can bring about the same change of state. Although this may seem obvious now, it was a surprise in the early 1800s. At the time, heat was viewed (incorrectly) as a mysterious fluid that flowed from hot to cold. The idea that two very different processes—performing organized work on a system, and pouring "caloric fluid" into a system—can lead to the same energy change seemed unlikely. Fortunately, replacement of the caloric fluid idea with an understanding of molecules and their internal modes of energy storage makes the equivalence of heat and work more tenable.[§]

The first law in equation form

The statement of the first law in terms of work and heat paints a picture of the first law as an energy balance sheet. This balance can be represented as an equation:

$$dU = dq + \sum_{all\ forms} dw_i \tag{3.5}$$

[†] Especially for the second law, the equation is *deceptively* simple—that is, the equation has broad implications that are not obvious at first glance.

[‡] Here we avoid using the phrase "internal energy of the surroundings" since the word internal seems closely connected to the system. It should be kept in mind that the labels "system" and "surroundings" are our own creation, and the quantities and laws of thermodynamics must remain valid no matter how we label.

[§] In fact, the association of the adjective "internal" with the energy of a system reflects an appreciation that there must be modes of energy storage within materials that are invisible on the macroscopic scale. Such invisible "internal" modes of energy storage provided support for the existence of atoms and molecules, and vice versa.

where U is the internal energy of the system.[†] As written above, dq represents an infinitesimal heat flow into the system (from the surroundings). The dw_i terms represent infinitesimal amounts of work done *on* the system by different processes (see Table 3.2). We will usually represent the sum of different types of work using a single dw term[‡]:

$$dU = dq + dw \tag{3.6}$$

Note that Equations 3.5 and 3.6 differ from some first law formulas in terms of the sign of the work term. In this alternative formulation, the work is taken as that done *by the system on the surroundings*, which can be a useful viewpoint for applications involving heat engines, where heat is input, and work is output. I prefer the form in Equation 3.6 because all forms of energy exchange have the same effect: "positive" heat flow and all forms of work increase the internal energy of the system.

There is an important feature of the first law that is easy to miss in the simplicity of Equation 3.6. Although the internal energy is a state function, and thus dU is an exact differential (see Chapter 2), neither dq nor dw are exact. Thus, neither heat nor work is a state function. A system at equilibrium possesses a uniquely determined amount of internal energy, but it makes no sense to say that a system possesses a certain amount of heat or a certain amount of work. Instead, we regard heat and work as *processes* by which a system changes its energy. This distinction has important implications for finite changes of state, and the paths of finite change.

The importance of pathways in first law calculations

In addition its infinitesimal form (Equation 3.6), the first law is useful in integrated form, where it is used to analyze finite changes of state. Starting with Equation 3.6,

$$\int_j^f dU = \int_p dq + \int_p dw \tag{3.7}$$

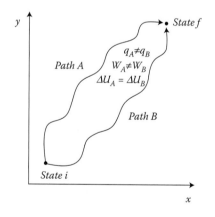

Although we did not worry about paths for the differential form of the first law (because state is not significantly changed by differential variation; if we are not changing the state of the system, we don't need to worry about path), path is critical for finite state changes involving heat and work. Since U is a state function, the left-hand side of Equation 3.7 does not depend on the path of integration. The internal energy integral can be represented by a single value, $\Delta U = U_f - U_i$. In contrast, the heat and work integrals depend on path, indicated by the p subscripts in Equation 3.7.

It is customary to represent the integrated form of the first law (Equation 3.7) as

$$\Delta U = q + w \tag{3.8}$$

Unlike internal energy, Δ's are not associated with the q and w terms. Doing so would imply that we could take the difference $q_f - q_i$ (and $w_f - w_i$); since the system does not posses a defined amount of heat (or work), such differences are not meaningful.

An illustration of how the path differences play out in the first law is shown in **Figure 3.10**. When a system changes from an initial to a final state, many paths can be taken (an infinite number, two of which are shown). Although the change in

Figure 3.10 Different paths (A and B) connecting two states (i and f), and the corresponding first-law quantities. In this general picture, x and y are variables that determine the state of the system (e.g., pressure and temperature). The values of x and y (e.g., x_i, y_i) uniquely determine state functions like internal energy (e.g., U_i), so that the same value ΔU is obtained regardless of the path. In contrast, the work done and heat flow depend on path.

[†] Note that some textbooks use the symbol E for internal energy.

[‡] In most of the calculations done in this chapter, work will only be of one form: expansion and compression against an opposing force.

internal energy is the same regardless of path, different paths will almost always have different heat and work exchanges.[†] Though we are often interested in changes that take complicated paths (where thermodynamic calculations are difficult), the first law tells us we can calculate the internal energy change along any path that connects the initial and final states, including simple paths where we can easily determine the work and heat. The first law guarantees that the sum of the heat and work along the simple path will provide the internal energy change along the complicated path.

From the discussion above, it should be obvious that state functions have favorable properties for thermodynamic analysis compared to path functions like heat and work. One might then ask: "why bother with heat and work?" Why not leave heat and work to the engineers and steamship captains, and focus directly for functions like the internal energy (and, as developed in Chapters 4 and 5, entropy and free energy)? One answer is that we don't have meters that directly measure changes in internal energy, entropy, and free energy.[‡] However, we can directly measure heat flow by calorimetry, and can measure work by coupling to a mechanical device.[§] Thus, in addition to serving as historical relics of the development of thermodynamics, heat and work retain their importance in thermodynamic studies. But before we can use heat and work in thermodynamic calculations, we need to develop formal definitions of these two quantities.

WORK

That is, work is any type of energy transfer between the system and surroundings that is not heat flow. This definition, which comes from our formulation of the first law (Equations 3.5 and 3.8), is complete, it is also vague. A little more detail can be added with the following distinction: for heat flow, energy enters (or leaves) the system in a disorganized form, especially when viewed at a molecular level. For work, energy enters (or leaves) in an organized form (although it may rapidly become disorganized). So, work can be considered to be an "organized energy flow" between system and surroundings. Though this distinction gives a nice visual understanding, it is hard to make a quantitative law that measures the extent of organization of energy (we will return to this point in Chapter 4 when we discuss entropy).

A quantitative expression for work on an object was given in Chapter 2, relating force and displacement. In differential form,

$$dw = -F_x dx \qquad (3.9)$$

where dx is a displacement, and F_x is the component of a force in the x direction. A simple example is the displacement of an object in the upward x direction, against the downward force of gravity. The negative sign in Equation 3.9 ensures that if the object is raised, work is positive, and can be this energy can be recovered later by lowering the object. Although the coordinates of thermodynamic systems are typically not simple linear displacements, and the interacting "forces" are often complex, the work done in thermodynamic changes of state can usually be expressed in a differential form similar to Equation:

$$dw = -\phi d\xi \qquad (3.10)$$

[†] In fact, Equation 3.8 demands that if q depends on path, w must also depend on path (in an offsetting way), because their sum does not depend on path.

[‡] In simple cases, the molecular models of statistical thermodynamics can be used to connect internal energy with measurable quantities like temperature (as is true for an ideal gas). But this model-dependent approach is outside the scope of classical thermodynamics.

[§] In simplest form, a device that lifts a weight against gravity.

where ξ is a system coordinate that changes in a work process. The variable ϕ can be thought of as a general force opposing (or driving) the process.[†] When the system undergoes a finite change of state, the work can be calculated by integration (typically along a specified path):

$$w = \int_i^f \phi \, d\xi \qquad (3.11)$$

Some examples of work processes in thermodynamic systems are given in **Table 3.2**. One general type of work involves a change in the system's physical dimensions, which we will refer to here as expansion work (though our discussion is equally valid in the reverse direction, i.e., compression). Expansion work along a single coordinate is relevant to describe the work of linear deformation of polymers such as a proteins and nucleic acids (see Problem 3.9). Expansion work in two dimensions is relevant to materials in which surface properties differ from interior properties, such as droplets, bubbles, membranes, and macromolecules with large surfaces. In such cases, a generalized force can be represented with a surface tension γ. Expansion work in three dimensions is important for systems that change their volumes, and will be discussed extensively in the next few chapters.

Work can also result from system changes in an external field. Examples include polarization against an electrical field, magnetization against a magnetic field, and mass displacement in a gravitational field (Table 3.2). "Chemical work," involving either addition of a component to an open system, or change in the concentrations of reactive species, is very important in the thermodynamics of phase equilibrium and chemical reactions. We will discuss this type of work in detail starting in Chapter 5, where we will develop expressions for the driving force μ, which is referred to as the "chemical potential."

Table 3.2 Types of work and expressions for thermodynamic calculations

Shape change		
Stretching	$F_x dx$	$F_x = x$-component of force
		$x =$ distance
Surface area change	γdA	$\gamma =$ surface tension
		$A =$ area
Volume change	$-p dV$	$p =$ hydrostatic pressure
		$V =$ volume
Polarization		
Electrical polarization	$E \cdot dp$	$E =$ electric field strength
		$p =$ total electric dipole moment
Magnetization	$B \cdot dm$	$B =$ magnetic induction
		$m =$ total magnetic dipole moment
Chemical Work	$\mu_i dn_i$	$\mu_i =$ chemical potential of species i
		$n_i =$ moles of species i

[†] Although ϕ is analogous to as a "force," it need not have the same dimension as a true force (i.e., Newton's). Rather, the dimensions of ϕ must combine with those of ξ to give units of energy, as is the case with F and x.

The work associated with expansion of a gas

To develop the first law, we will analyze work resulting from reversible expansion and compression of a simple gas.[†] Although we do not often pay attention to the small volume changes in biological systems, developing work in terms of volume has several advantages. First, volume change (i.e., expansion and compression) is easy to visualize. Second, although the work associated with gas expansion can be analyzed without knowing any molecular details, we can often extract molecular details by using of an equation of state (e.g., the van der Waals equation described below) to analyze data. And ultimately, once we combine the constraints of the second law with the first law, we will return to gas expansion to develop relationships between concentration, free energy, and ultimately, reaction thermodynamics (Chapter 7).

Our analysis of expansion work begins with a gas in a box with a single movable wall (a piston, **Figure 3.11**). Reversible volume change requires the pressure of the system and surroundings to be equal.[‡] This ensures equilibrium both within the system, and between the system and surroundings. Equilibrium within the system means we can use an equation of state to represent the pressure of the system, p_{sys}. Equilibrium between the system and the surroundings means we can also use the equation of state of the *system* to represent the pressure of the *surroundings* ($p_{sys} = p_{surr}$).

To reversibly compress the gas, we can increase the pressure of the surroundings by a tiny amount (by $dp_{surr} > 0$), or we can decrease the pressure of the gas a little bit (by $dp_{sys} < 0$).[§] This will allow the gas to compress by an infinitesimal amount (dV) until its pressure rises to match the external pressure. Repeating this over and over many times will produce a finite decrease in volume, but the change will be through a series of equilibrium states, and each can be described by the equation of state of the gas.

For compression, work can be calculated by considering the internal force F_{in} opposing the compression. If we define a coordinate, z, to be positive in the direction of expansion (Figure 3.11), the upward internal force exerted by the system is positive. Writing the work integral against this force gives

$$w = -\int_{z_i}^{z_f} F_{in} dz \qquad (3.12)$$

This work is that done on the gas (i.e., the system), which is consistent with the first-law sign convention introduced above.

Recognizing that force per unit area is pressure, and can substitute F_{sys} by the pressure of the gas (force per unit area) times the area (A) of the moveable wall:

$$w = -\int_{z_i}^{z_f} p_{sys} A dz = -\int_{V_i}^{V_f} p_{sys} dV \qquad (3.13)$$

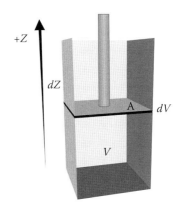

Figure 3.11 The work associated with volume change. A gas is enclosed in a box of volume V. The gas is bounded on the bottom and all four sides by fixed walls (here the front wall is not shown). The top wall, with area A, is movable; here it is drawn as a piston, with a shaft to control its position. If the pressure of the gas is equal to that of the surroundings ($p_{sys} = p_{surr}$), the piston will not move. If the pressure of the surroundings is changed by dp_{surr}, the piston will be displaced by dz (with z parallel to the direction of movement) until the equilibrium is reestablished (at $p_{surr} + dp_{surr}$). This movement results in an infinitesimal volume change for the system (red, dV), which is equal to A times dz.

[†] More generally, we could discuss the compression and expression of a "fluid." Both liquids and gases are considered to be fluids, since both can flow freely to accommodate a deformation, and don't maintain steady-state stresses or strains. But because gases often have very simple equations of state, and have much larger changes in volume in response to pressure change, they make better models for introducing expansion work than do liquids.

[‡] This is analogous to our reversible heat-flow example, which requires the temperature of the system and surroundings to be equal.

[§] Similarly, compression of the gas can be brought about by increasing the pressure of the surroundings (by $dp_{surr} > 0$), or by decreasing the pressure of the system (by $dp_{sys} < 0$).

The second equality results from the fact that the volume change dV is equal to A times dz (Figure 3.11). This substitution replaces parameters that depend on details of the system's geometry (area of the moveable wall, its coordinate z, and the force on the wall) with the more general parameters p and V. This substitution also allows for more complicated volume changes than perpendicular, one-dimensional expansion of a single wall.[†]

The key result of this analysis (Equation 3.13) is that $dw = -p_{sys}dV$. We will make extensive use of this formula throughout this book. Note that this expression is consistent with our first-law sign convention: when the gas expands ($dV > 0$), work is negative (p is always positive for our gas).

The integral of $-pdV$ (Equation 3.13) has an important graphical interpretation. If pressure is plotted versus volume during an expansion, the work is given by the (negative) area under the pressure–volume curve (see Figure 3.13). This area representation is particularly useful for evaluating the work done in closed mechanical cycles, providing insight into the efficiency of heat engines and forms the basis of the second law (Chapter 4).

THE REVERSIBLE WORK ASSOCIATED WITH FOUR FUNDAMENTAL CHANGES OF STATE

To calculate work by integrating $-pdV$, we need to specify a path for volume change.[‡] For any finite volume change, there are many paths that connect initial and final states (as in Figure 3.10), but not all paths lead to simple work calculations. Here we will focus on four reversible paths where one variable is held constant. This allows us to take the constant variable outside the work integral, simplifying the solution. Four paths that achieve this are

1. Isobaric change of state (constant pressure)
2. Isochoric change of state (constant volume)
3. Isothermal change of state (constant temperature)
4. Adiabatic change of state (no heat flow[§])

For the ideal gas, the four fundamental transformations are illustrated on a p–V diagram in **Figure 3.12**. In the following section we will calculate (by integration) the work done in reversible expansion of an ideal gas by these four fundamental processes; you should confirm that the results of integration match what you would expect from considerations of area under the p–V curve in Figure 3.12.

Work done in reversible constant-pressure (isobaric) expansion

Reversible constant-pressure volume change leads to a very simple calculation of work in which the constant pressure term can be taken outside of the integral:

$$w = -p_{sys} \int_{V_i}^{V_f} dV = -p_{sys}(V_f - V_i) = -p_{sys}\Delta V \tag{3.14}$$

Interpreted as the area under the p–V curve, the work done in isobaric expansion is a simple rectangle bounded by the pressure and the volume limits. Because

---

[†] For a nice demonstration that $dw = -pdV$ holds for a general volume change involving irregular surfaces, see Fermi (1936).

[‡] Remember, work is a path function, and it depends on how the change of state was carried out.

[§] As described in Chapter 4, *reversible* adiabatic changes of state are constant entropy.

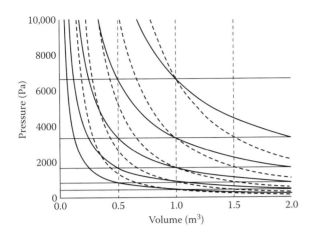

Figure 3.12 Four simple changes of state involving an ideal gas. Reversible transformations are shown for one mole of a monatomic ideal gas. Solid black: constant temperature volume changes. Dashed black: adiabatic volume change. Solid red: constant pressure volume change. Dashed red: constant volume pressure change. The isobars, isochores, and adiabats all cut through the isotherms, consistent with changes in temperature for each (for example, the adiabats cut down through the isotherms as volume is increased, indicating that temperature falls). Unlike the other three types of curves, the shapes of the adiabats depend on the number of degrees of freedom (energy storage modes) available to the gas; the adiabats shown here are for a monatomic gas.

expansion involves a positive ΔV term, the work calculated from Equation 3.14 is negative, decreasing the internal energy of the system. This is consistent with our earlier definition of work, where positive work is that which is done *on the system* (increasing its internal energy).

In practical terms, reversibility can be brought about by infinitesimally incrementing the pressure of the system by dp_{sys},[†] which allows the gas to expand by dV, restoring the pressure of the system to dp_{sys} so that equilibrium with the surroundings is maintained (and pressure remains constant, within an infinitesimal deviation). By repeating this process many times, the system can be expanded from V_i to V_f. Reversible isothermal compression of the system can be brought about in a similar way, by infinitesimally decrementing the pressure of the system, allowing the gas to compress by dV, restoring the system pressure to match the surroundings. In this case, $\Delta V < 0$; thus, work done on the system is positive.

Comparing reversible isobaric expansion with isothermal expansion (Figure 3.12), it should be clear that the temperature increases for the former, since isobars cut

Example 3.1

Calculate the work done for the reversible constant pressure expansion of 0.05 moles of helium at 2 atmospheres, from an initial volume of 0.5 L to a final volume of 2.0 L.

For the isobaric expansion, the work done is calculated from Equation 3.14. Using *SI* units,

$$w = -p\Delta V = -2.27 \times 10^5 \, \text{Pa} \times (0.002 \, \text{m}^3 - 0.0005 \, \text{m}^3)$$
$$= -341 \, \text{J}$$

The calculation does not depend on the gas behaving ideally. However, calculation of the initial and final temperatures from the data given above would require the use of the ideal gas law (Problem 3.5) or another equation of state.

[†] Most likely bin an infinitesimal heat flow into the system.

through higher-temperature isotherms on expansion. A molecular picture is helpful to understand this temperature increase. Because n is constant, at the larger volume V_f, the gas density is lower. To maintain constant pressure, the collisions must be made more frequent and more energetic, which requires higher average velocities, corresponding to a higher temperature.

Example 3.2

Calculate the work done for the reversible isothermal expansion of 0.05 moles of helium at an initial volume of 0.5 L and a temperature of 273.15 K (0°C) to a final volume of 2.0 L, assuming ideal gas behavior. How does this compare to the work done for the constant pressure expansion (Example 3.1)?

For the isothermal gas, work is calculated according to Equation 3.16:

$$w = -nRT \ln\left(\frac{V_f}{V_i}\right) = -0.05\,\text{mol} \times \frac{8.314\,\text{J}}{\text{mol}\cdot\text{K}} \times 273.15\,\text{K} \times \ln(4)$$

$$= -157\,\text{J}$$

This is considerably smaller in magnitude (less negative) than the work in the isobaric expansion at $p_i = 2.27 \times 10^5$ Pa ($w = -341$ J). In other words, more work is done *by the system on the surroundings* for expansion at constant pressure. This is reflected in the differences in the areas under the p-V curves (**Figure 3.13**).

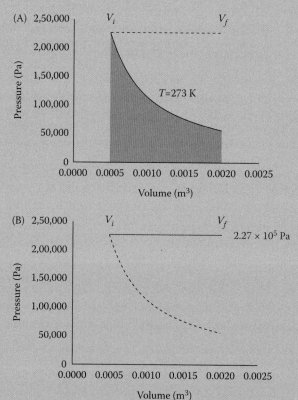

Figure 3.13 The work of isothermal versus isobaric expansion of an ideal gas. In the two panels, 0.5 mol of an ideal gas is reversibly expanded from the same initial conditions ($T_i = 273.15$ K, $V_i = 0.0005$ m³), to the same final volume $V_f = 0.002$ m³. (A) Reversible isothermal expansion. The work done *on the system* is the negative of the gray shaded area under the black isotherm, limited by V_i and V_f (Equation 3.16). (B) Reversible isobaric expansion. The work done is the negative of the red shaded area under the red isobar. As can be seen from the differences in areas, there is a larger (but more negative) work term for the isobaric expansion. In each panel, the other process is shown as a dashed line for comparison.

Work done in constant volume pressure change

When the pressure of a gas is increased at fixed volume (by heat flow), no work is done. This is consistent with the picture of work as area under the curve. Since there is no change in volume, the "curve" is a vertical line (see the red dashed lines in Figure 3.12) with no width and thus no area. If we try to evaluate the work integral (Equation 3.13), $dV = 0$. As with isobaric expansion, this result holds for fluids that have nonideal behavior (and thus more complex equations of state).

Work done in reversible isothermal expansion of an ideal gas

When a gas changes its volume isothermally, the pressure changes (Figure 3.12). Thus, unlike our approach in for isobaric expansion (Equation 3.14), we cannot take pressure outside of the work integral. Instead, we can eliminate pressure by substitution with the ideal gas law:

$$w = -\int_{V_i}^{V_f} p_{sys} dV = -\int_{V_i}^{V_f} \frac{nRT}{V} dV = -nRT \int_{V_i}^{V_f} \frac{1}{V} dV \tag{3.15}$$

R can always be taken outside the integral; n can be taken out because the system is closed (and we are not considering chemical reaction), and T can be taken out since the volume change is isothermal. The integral of V^{-1} should be easily recognizable as the natural log of V, giving

$$w = -nRT(\ln V_f - \ln V_i) = -nRT \ln \frac{V_f}{V_i} \tag{3.16}$$

Expansion of the system means that V_f is larger than V_i, thus the ratio of the volumes is larger than one, making a positive value for the log of the volume ratio. Since n, R, and T are all positive quantities (T is on the absolute scale), the work for isothermal expansion is negative. Again, this is consistent with our convention for the sign for work. Compression work calculated using Equation 3.16 comes out positive.

By comparing the curve for isothermal expansion of an ideal gas with that for an adiabatic expansion (Figure 3.12), it should be clear that heat must flow into the system during the isothermal expansion. Starting at the same initial state, the pressure falls more rapidly for the adiabat than for the isotherm. This means that if there is no heat flow, the temperature will fall during a reversible expansion. Thus, for the isotherm, heat must flow into the system to keep the temperature constant. In molecular terms, the work done corresponds to gas molecules inside the system disproportionately transferring energy to the moveable wall (and from the wall to the surroundings). This energy transfer would decrease the speed of the gas molecules and thus the temperature of the system. Thus, to maintain the temperature, heat must flow into the system to restore this lost molecular kinetic energy. We will return to this point when we calculate the heat flow associated with an isothermal expansion of an ideal gas.

Work done in reversible adiabatic expansion of an ideal gas

Of the four fundamental state changes, adiabatic expansion is the most challenging to analyze. In addition to the volume change, both the pressure and the temperature change. Thus, we cannot move pressure outside the work integral, in contrast to Equation 3.14, and if we substitute it with the ideal gas law, we can't move the temperature outside the work integral.

What we need to solve the work integral (Equation 3.14) is a functional expression of the form $p = p(V)$. The decrease in pressure for adiabatic expansion of an ideal gas is similar to that for isothermal expansion. However, the decrease is sharper than the inverse dependence (Figure 3.12), because the heat flow that is required to maintain constant temperature in the isothermal expansion is prevented by the adiabatic boundary. This sharp dependence suggests a decay of the form

$$p = \frac{\lambda}{V^\gamma} \tag{3.17}$$

where γ is a constant larger than one, and λ is a proportionality constant. In the next section, we will show that $\gamma = C_p/C_V$, where C_p and C_V are heat capacities at constant pressure and volume, respectively. The fact that C_p is larger than C_V for all simple systems ensures that $\gamma > 1$. For monatomic and diatomic ideal gases, $\gamma = 5/3$ and $7/5$, respectively.

The value of λ can be obtained by rearranging Equation 3.17, and recognizing that for a particular adiabat, the value of λ remains unchanged at all (p, V) points on the curve, including the starting point (p_i, V_i). That is,

$$\lambda = pV^\gamma = p_i V_i^\gamma \tag{3.18}$$

Combining Equations 3.17 and 3.18 gives

$$p = \frac{p_i V_i^\gamma}{V^\gamma} \tag{3.19}$$

Substituting Equation 3.19 into the integral (Equation 3.13) gives the work for reversible adiabatic expansion (and compression), we can substitute Equation 3.17 into the work integral (Equation 3.13) and solve:

$$w = -\int_{V_i}^{V_f} \frac{p_i V_i^\gamma}{V^\gamma} dV = \frac{p_i V_i^\gamma}{(\gamma-1)V^{\gamma-1}}\bigg|_{V_i}^{V_f} = \frac{p_i V_i^\gamma}{(\gamma-1)}\left\{ \frac{1}{V_f^{\gamma-1}} - \frac{1}{V_i^{\gamma-1}} \right\} \tag{3.20}$$

Equation 3.20 can be used to directly determine the work of reversible adiabatic expansion, either by assuming a specific value of γ based the type of gas (Problem 3.6), or by determining the value of γ by fitting experimental data measurement (Problem 3.7).

Although Equation 3.20 is adequate for calculation of the adiabatic expansion work for the ideal gas, there is a much simpler form that can be derived, which takes advantage of the difference between C_V and C_p. As derived below, for the ideal gas,

$$C_p - C_V = nR \tag{3.21}$$

Dividing by C_V gives

$$\gamma - 1 = \frac{nR}{C_V} \tag{3.22}$$

Substituting Equation 3.22 into Equation 3.20 gives

$$w = \frac{p_i V_i \times V_i^{nR/C_V}}{nR/C_V}\left\{ \frac{1}{V_f^{nR/C_V}} - \frac{1}{V_i^{nR/C_V}} \right\}$$

$$= \frac{p_i V_i C_V}{nR}\left\{ \left(\frac{V_i}{V_f}\right)^{nR/C_V} - 1 \right\} \tag{3.23}$$

The ideal gas law can be used to replace pressure with temperature, giving

$$w = \frac{nRT_iC_V}{nR}\left\{\left(\frac{V_i}{V_f}\right)^{nR/C_V} - 1\right\}$$

$$= T_iC_V\left\{\left(\frac{V_i}{V_f}\right)^{nR/C_V} - 1\right\} \qquad (3.24)$$

$$= C_V\left\{T_i\left(\frac{V_i}{V_f}\right)^{nR/C_V} - T_i\right\}$$

By recognizing that $T_i(V_i/V_f)^{nR/C_V} = T_f$ (see Problem 3.8), Equation 3.24 simplifies to

$$w - C_V\{T_f - T_i\} = C_V\Delta T \qquad (3.25)$$

Though this is a rather lengthy derivation, the simplicity of Equation 3.25 is worth the effort. At this point, we should review the assumptions that went into this derivation. In addition to assuming an ideal gas, we assumed that p varies inversely with V^{C_p/C_V}. When we include heat and internal energy in our analysis, we will see an alternative (and simpler) derivations for Equation 3.25.

HEAT

When we introduced work in the section above, we stated that we can use the first law to define work as energy change that does not involve heat flow. Not surprisingly, we can also do the reverse. Heat flow can be defined as any process that changes the internal energy of a system that isn't work. Now that we have a way to calculate work, this "definition by exclusion" has some value, and we will use this definition later. But for now, we will need a direct and more practical way to express heat flow.

We will define heat as *the amount of energy that flows as a result of a difference in temperature between system and surroundings*. Using this definition, we can calculate heat flow from the temperature change in the system. In differential form, a tiny bit of heat flow dq into the system results in a tiny increase in temperature dT.[†] By introducing a proportionality constant C, referred to as a heat capacity, we can make this relationship into an equality:

$$dq = CdT \qquad (3.26)$$

The heat capacity, which has dimensions of energy per unit of temperature,[‡] can be thought of as a measure of how effectively a system can store energy that is introduced in the form of heat. A system with a high heat capacity can adsorb a lot of and not change its temperature very much. In this regard, the heat capacity is like buffer capacity β in acid–base equilibria (**Figure 3.14**). This analogy is particularly clear when heat and buffer capacity are represented as derivatives:

[†] Likewise, a tiny heat flow out of the system ($dq < 0$ by our convention) results in a tiny temperature decrease ($dT < 0$). It is never the case that dq and dT have opposite signs. Based on Equation 3.26, this means that the heat capacity is *always positive*.

[‡] Note that the heat capacity, as defined in Equation 3.26, is an extensive quantity. In most applications (starting in Chapter 5), we will find it more convenient to divide this extensive quantity by the amount of material present. This results in an intensive heat capacity (often referred to as a "molar" heat capacity) that represents the extent of heat storage of the material itself. We will return to molar quantities in Chapter 5.

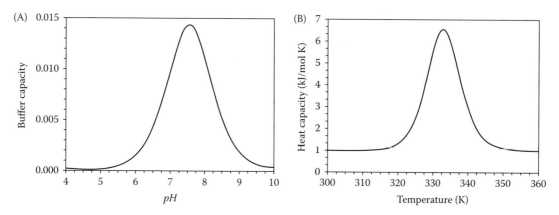

Figure 3.14 Analogy between buffer and heat capacity. (A) The ability of a 25 mM solution of a buffer with a pK_a of 7.55 (the value for HEPES, a commonly used biological buffer) to resist pH change as acid is added. The buffer solution has maximum capacity to absorb acid when the pH matches the pK_a of the buffer. At this maximum, the concentrations of the protonated and unprotonated forms of the buffer are equal. (B) The heat capacity of a solution containing a small protein that unfolds at a midpoint temperature (T_m) of 333 K. The protein solution has a maximum capacity to absorb heat, per degree of temperature increase, at $T = T_m$, where the concentrations of the folded and unfolded forms of the protein are equal. The maximum seen at both midpoints is a result of maximizing the sensitivity of the underlying reaction (acid–base equilibrium, protein folding) to the variable driving the reaction (pH and temperature).

$$C = \frac{dq}{dT} \tag{3.27}$$

$$\beta = -\frac{dn}{dpH} \tag{3.28}$$

Equation 3.27 is a simple rearrangement of Equation 3.26. Derivation of Equation 3.28 can be found in most *Biochemistry* and *Analytical Chemistry* text-books. The quantity n in Equation 3.28 represents the number of moles of a "strong" acid added to a buffer solution. The minus sign keeps β from having a negative sign, since pH decreases as acid is added. A high buffer capacity means that a lot of acid must be added to bring about a unit change in pH, just as a high heat capacity means a large amount of heat must be added to bring about a unit change in temperature.

In Figure 3.14, the buffer and heat capacities are shown for a simple single-site buffer, and for a reaction in which a protein is driven to unfold with temperature. Both quantities show a peak, which occurs when the two major forms (protonated and unprotonated for the buffer; folded and unfolded for the protein are at the same concentration). For the buffer, this peak occurs at a pH equal to the buffer pK_a. For the protein, the peak occurs at a temperature equal to T_m. The protein folding reaction will be discussed extensively in Chapter 8.

Types of heat capacities

Heat capacities are determined by molecular interactions and their associated energies, and thus, provide an important link to understanding behavior in complex molecular systems. Like internal energy, the heat capacity depends on the state of a system—the number and type of molecules present, their interaction energies, the volume, and often the temperature. However, since heat capacity is defined in terms of heat flow, its value depends on the path (or at least the direction of the path) heat flow. The heat capacity of a system heated at constant volume differs from that at constant pressure. Thus, when we discuss heat capacity and its relation to heat flow, we must be specific about how the system is manipulated (or constrained) during heating. The two most important varieties of heat capacities are those at constant volume and constant pressure, and are abbreviated are C_V and C_p, respectively. Both C_p and C_V are state functions, but they represent two different state functions.

We have already encountered C_p and C_V (and their ratio) when we discussed the work done during adiabatic expansion. Starting from the definition leading to Equation 3.27, we can use partial derivatives to indicate that volume and pressure are being held constant:

$$C_V = \left(\frac{\partial q}{\partial T}\right)_V \tag{3.29}$$

$$C_p = \left(\frac{\partial q}{\partial T}\right)_p \tag{3.30}$$

In Figure 3.12, C_V is associated with the red dashed vertical lines. This corresponds to a transformation where heat flow into the system increases pressure and temperature (or conversely, heat flow out decreases the pressure and temperature). In contrast, C_p is associated with the red horizontal lines in Figure 3.12, where heat flow into the system increases both temperature and volume (or the converse).

THE HEAT FLOW ASSOCIATED WITH THE FOUR FUNDAMENTAL CHANGES IN AN IDEAL GAS

Here we will use heat capacities to determine heat flows for the four fundamental changes described above. For the three changes that permit heat flow (adiabatic change prevents heat flow), one variable is held constant: pressure, volume, and temperature. Here we will calculate the heat flow associated with these four changes, again analyzing reversible transformations of an ideal gas. Although we have only defined two path-dependent heat capacities (Equations 3.29 and 3.30), by using the first law of thermodynamics along with results from our work calculations, we will be able to treat all four fundamental changes of state.

Heat flow for reversible constant-pressure (isobaric) expansion of an ideal gas

To calculate the heat flow when an ideal gas expands at constant pressure, the appropriate heat capacity is C_p. When an ideal gas expands at constant pressure, it cuts upward through isotherms, reflecting an increase in temperature (Figure 3.12). Rearranging Equation 3.30 to a form resembling Equation 3.26 provides an expression relating infinitesimal temperature change to an infinitesimal heat flow:

$$dq = C_p dT \tag{3.31}$$

Integrating this expression from the initial to the final temperature,

$$q = \int_{T_i}^{T_f} C_p dT \tag{3.32}$$

The solution of this integral depends on how the heat capacity varies with temperature. Over very large temperature ranges (especially at low temperatures on the absolute scale), even simple systems have temperature dependent heat capacities.[†] However, over the narrow temperature ranges typically accessed in the laboratory (especially in studies of molecules in aqueous solution), the heat

---

[†] This is because energies translation, rotation, and vibration are quantized, and as the available thermal energy (i.e., temperature) falls below the quantum energy spacings, these modes of energy storage become inaccessible, and the heat capacity falls. This effect is most evident for diatomic and polyatomic molecules.

capacity can often be treated as independent of temperature, making Equation 3.32 easy to integrate:

$$q = C_p \int_{T_i}^{T_f} dT = C_p(T_f - T_i) = C_p \Delta T \tag{3.33}$$

For classical ideal gases, the value of C_p is $5nR/2$ and $7nR/2$ for monatomic and diatomic gases, respectively (see Table 3.1).

Using the first law, we can combine the heat and work of reversible isobaric expansion of an ideal gas to calculate the internal energy change for expansion:

$$\Delta U = q + w = C_p \Delta T - p \Delta V \tag{3.34}$$

This result can be made more compact by using the ideal gas law to express the work term in terms of temperature, and by expressing the heat capacity using the classical formulas given in Table 3.1. These substitutions lead to ΔU values of $3nR\Delta T/2$ and $5nR\Delta T/2$ for isobaric expansion of monatomic and diatomic ideal gases, respectively (Problem 3.10). Importantly, because internal energy is a state function, these values of ΔU (and more generally, Equation 3.34) also apply to gases expanding *irreversibly* between the same starting and ending points (from T_i, V_i, p_i to T_f, V_f, $p_f = p_i$), although the irreversible heat and work values would be different.

Heat flow for reversible constant-volume (isochoric) heating of an ideal gas

The heat flow for constant volume transformation in which temperature and pressure change can be calculated using C_V, in the same way we analyzed to constant pressure heat flow. The differential relation between heat flow and temperature change is analogous to Equation 3.31, and can be integrated to give the heat flow:

$$q = \int_{T_i}^{T_f} C_V dT = C_V \Delta T \tag{3.35}$$

The second equality holds as long as C_V is independent of temperature.

The internal energy change for constant volume heating can be calculated using the first law,

$$\Delta U = q + w = C_V \Delta T + 0 \tag{3.36}$$

That is, since there is no work done in a simple constant volume process, the change in internal energy is equal to the heat flow. Substituting values of C_V for classical monatomic and diatomic ideal gases gives the *same internal energy changes as for the isobaric expansion,* as long as the two processes have the same temperature limits. But it is important to recognize that these are not the same transformations. One is an expansion, one is not. We could start the transformations at the same thermodynamic state (same n, p_i, V_i, and T_i), but they would separate to different end states (to p_i, V_f, and T_f for the isobaric expansion, and to p_f, V_i, and T_f for the isochoric heating), moving in perpendicular directions on the p-V diagram. The reason ΔU is the same for the two processes is that for an ideal gas, *internal energy depends only on temperature* (Table 3.1). The fact that the internal energy of an ideal gas depends only on temperature is a fact worth remembering.

Heat flow for reversible isothermal expansion of an ideal gas

When we discussed work done in isothermal expansion, we anticipated there would be heat flow. Unfortunately, the simple method we have developed to calculate heat flow depends on temperature change, and for the isothermal transformation, there is no temperature change.[†] This does not mean that there is no heat flow; rather, the heat that enters the system is removed as (negative) expansion work. Again, because the internal energy of an ideal gas depends only on temperature, we can state that the internal energy remains constant during isothermal expansion of an ideal gas:

$$\Delta U = q + w = 0 \tag{3.37}$$

$$q = -w = +nRT\ln\frac{V_f}{V_i} \tag{3.38}$$

The second expression on the right-hand side of Equation 3.38 is obtained by substituting the result of our work calculation for the reversible isothermal expansion of an ideal gas (Equation 3.16).

Heat flow for reversible adiabatic expansion of an ideal gas

By definition, the heat flow for an adiabatic process is zero. Thus, we can calculate the change in internal energy as

$$\Delta U = w = C_V \Delta T \tag{3.39}$$

where the term on the right-hand side is the result of our work calculation above (Equation 3.25). Although Equation 3.39 gives us everything we need to know about this expansion, it depends on Equation 3.17 ($p = \lambda V^{-\gamma}$), which we gave without justification. Here we will demonstrate the validity of Equations 3.39 and 3.25 using the path independence of state functions, along with the results obtained above for isothermal and constant volume heat flows.

Since internal energy is a state function, the internal energy change in an adiabatic expansion (and thus the work in the adiabatic process) can be obtained by constructing a nonadiabatic path that connects the initial and final states of an adiabatic expansion. One simple way to do this is a two-step path in which the first step is isothermal expansion from V_i to V_f (the volume limits of the adiabatic expansion), and the second step isochoric cooling from T_i to T_f by letting heat escape into the surroundings (**Figure 3.15**).

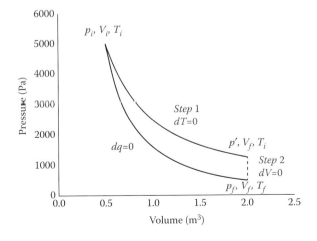

Figure 3.15 A two-step process that connects the end points of an adiabatic expansion. The black curve shows reversible adiabatic expansion of one mole of an ideal monatomic gas from p_i, V_i, T_i to p_f, V_f, T_f. The red curves show a two-step path connecting the initial and final states. The first step is a reversible isothermal expansion at T_i from V_i to V_f, and a constant volume cooling from T_i to T_f. Although the work and heat flow are different along the two paths (both these quantities are path functions), the change in internal energy is the same.

[†] This is one reason it does not make sense to define heat capacity at constant temperature, C_T.

With these two paths connecting the initial and final states, we can equate the internal energy changes:

$$\Delta U_{ad} = \Delta U_1 + \Delta U_2 \tag{3.40}$$

where subscripts 1 and 2 refer to the isothermal and isochoric steps. Reiterating, the internal energy change for an isothermal process involving an ideal gas is zero. Thus, all of the internal energy change along the nonadiabatic path is produced in the constant-volume cooling step (i.e., $\Delta U_{ad} - \Delta U_2$). Because there is no work done in this second step, the internal energy change for the second step is equal to the heat flow:

$$\Delta U_{ad} = w_{ad} = C_V \Delta T \tag{3.41}$$

This is same result as our previous calculation for adiabatic work (Equation 2.46), but was much easier to derive. Moreover, it does not depend on an unjustified p-V relation.

But now that we have obtained Equation 3.41, we should be able to use it to demonstrate that $p = p_i(V_i/V)^\gamma$ (Equation 3.17). One way to do this would be to work backward from Equation 3.25 to Equation 3.19. A similar, but slightly more elegant approach will be taken here. We begin by recognizing that the analysis leading to Equation 3.41 applies to an ideal gas adiabat of any size (i.e., any volume change). We can make the expansion smaller and smaller, and this relationship still holds. In the limit that V_f is only infinitesimally larger than V_i ($V_f = V_i + dV$), Equation 3.41 is obtained in differential form:

$$dU = dw = C_V dT = -p dV \tag{3.42}$$

Equation 3.42 can also be obtained in by calculating the exact differential of U using T and V as variables (**Box 3.1**). Substituting the ideal gas law into the right-hand side of Equation 3.42 gives

$$C_V dT = -\frac{nRT}{V} dV \tag{3.43}$$

Separating variables and integrating leads to an expression for a finite reversible adiabatic volume change for an ideal gas:

$$-\int_{V_i}^{V_f} \frac{nR}{V} dV = \int_{T_i}^{T_f} \frac{C_V}{T} dT \tag{3.44}$$

$$-nR\ln\left(\frac{V_f}{V_i}\right) = C_V \ln\left(\frac{T_f}{T_i}\right) \tag{3.45}$$

$$\ln\left(\frac{V_i}{V_f}\right)^{nR/C_V} = \ln\left(\frac{T_f}{T_i}\right) \tag{3.46}$$

$$\left(\frac{V_i}{V_f}\right)^{nR/C_V} = \left(\frac{T_f}{T_i}\right) \tag{3.47}$$

Substituting the ideal gas law into the right-hand side of Equation 3.47 replaces the temperature ratio with a pressure–volume ratio:

$$\left(\frac{V_i}{V_f}\right)^{nR/C_V} = \left(\frac{p_f V_f / nR}{p_i V_i / nR}\right) \tag{3.48}$$

$$\left(\frac{V_i}{V_f}\right)^{1+\frac{nR}{C_V}} = \frac{p_f}{p_i} \tag{3.49}$$

$$\left(\frac{V_i}{V_f}\right)^{\frac{C_V + nR}{C_V}} = \frac{p_f}{p_i} \tag{3.50}$$

Inspection of the heat capacities in Table 3.1 shows regardless of the gas, C_p is larger than C_V by nR, that is, $C_p = C_V + nR$. Substituting this relationship into Equation 3.50 and rearranging gives

$$p_f = \frac{1}{V_f^{C_p/C_V}} \times p_i V_i^{C_p/C_V} \tag{3.51}$$

$$= \frac{1}{V_f^{\gamma}} \times p_i V_i^{\gamma}$$

where we substituted $\gamma = C_p/C_V$ in the second line.

The integration that led to Equation 3.51 was carried out as a definite integral from state i to state f (Equation 3.44). However, Equation 3.51 would remain valid had we integrated from state i to any other final state (p_3, V_3, or p_4, V_4, or ...) by appropriately modifying the 2 subscript. We can represent the generality of the final state by with the *variables* p and V (rather than specific values p_i and V_i; this is the single-limit integral described in Chapter 2):

$$p = \frac{1}{V^{\gamma}} \times p_1 V_1^{\gamma} = \frac{\lambda}{V^{\gamma}} \tag{3.52}$$

Box 3.1: An approach to representing adiabatic energy change using exact differentials

One way to derive expressions for the work and internal energy change for an ideal gas is to use exact differentials, as described in Chapter 2. For a single-component gas, the internal energy can be expressed using the variables V and T:

$$U = U(V,T) \tag{3.53}$$
$$= U(T)$$

The second line in Equation 3.53 reflects the fact that for an ideal gas, the internal energy depends only on temperature, that is, it has no volume dependence (Table 3.1). Choosing T and V as state variables is reasonable because both of these quantities vary in adiabatic expansion, and because for a simple single-phase system with fixed n, two variables are sufficient to specify the thermodynamic state (and its variation).

Calculation of the total differential from Equation 3.53 (see Chapter 2) gives

$$dU = \left(\frac{\partial U}{\partial T}\right)_V dT + \left(\frac{\partial U}{\partial V}\right)_T dV = \left(\frac{\partial U}{\partial T}\right)_V dT = C_V dT \tag{3.54}$$

(Continued)

Box 3.1 (*Continued*): An approach to representing adiabatic energy change using exact differentials

Again, the simpler right-hand side reflects the fact that for an ideal gas, internal energy is a function only of the temperature, that is,

$$\left(\frac{\partial U}{\partial V}\right)_T = 0 \tag{3.55}$$

For real gases, the derivative on the left hand side of Equation 3.55 (called the "internal pressure," π_T) is nonzero. A negative value of π_T indicates repulsive interactions, whereas a positive value of π_T indicates attractive interactions. Experimental determination of π_T provides a means to quantify the interactions between gas molecules (Problem 3.25).

As with Equation 3.42, Equation 3.54 can be used to relate the heat capacity to the work done in an adiabatic expansion,

$$dU = C_V dT = -pdV \tag{3.56}$$

which can be integrated (Equations 3.44 through 3.51) to yield the pressure-volume relationship for the reversible adiabatic expansion of an ideal gas (Equation 3.17).

This relationship is identical to Equations 3.17 and 3.19. The shorthand λ serves as a reminder that $p_1 V_1^\gamma$ is a constant and can be thought of as a constant of integration. An equivalent expression would be obtained by indefinite integration of Equation 3.43, which would introduce an additive constant (in logarithmic form) defined by the initial state of the system (Problem 3.11).

THE WORK ASSOCIATED WITH THE IRREVERSIBLE EXPANSION OF AN IDEAL GAS

So far, we have only analyzed reversible transformations. As discussed at the beginning of the chapter, analysis of irreversible changes is challenging, because variables that are important for calculations tend not to be well defined. These include pressure and temperature. One approach to analyzing irreversible thermodynamic changes is to assume that the surroundings are very large, and adjust quickly in response to changes in the system. Although this "reservoir" case is a limiting approximation, it can highlight the differences between reversible and irreversible changes. Here we will use the reservoir approach to analyze irreversible gas expansions. We will postpone discussion of irreversible heat flow until Chapter 4, where we develop entropy and its relation to reversible and irreversible processes.

Adiabatic irreversible expansion of a small system against a mechanical reservoir

Irreversible expansion occurs when the pressure of the surroundings is significantly lower[†] than the pressure of the system. If the surroundings is vast (compared to the system), and if its pressure equilibrates quickly in the surroundings, then the pressure of the surroundings can be regarded as constant (i.e., it acts as a mechanical reservoir). The work done *on the surroundings* is given by

$$w_{surr} = \int_{V_i}^{V_f} p_{surr}dV = p_{surr}\Delta V \tag{3.57}$$

[†] By "significantly," we mean *non infinitesimally* lower.

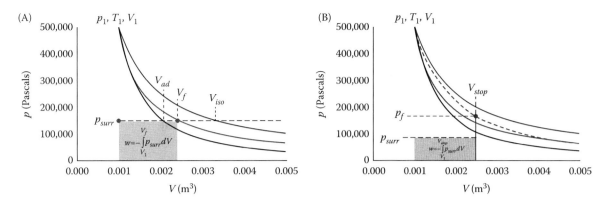

Figure 3.16 Irreversible adiabatic expansions against a mechanical reservoir at constant pressure p_{surr}. (A) Expansion of 0.2 mol of a monatomic ideal gas against a constant external pressure $p_{surr} = 150{,}000$ Pa, to a final equilibrium state where $p_{sys} = p_{surr}$ (gray curve). (B) Expansion to a fixed volume of $V_{stop} = 0.0025$ m³, against a constant external pressure of $p_{surr} = 83{,}140$ Pa (dashed line). Following equilibration, $p_{sys} = p_f - 188{,}280$ Pa $> p_{surr}$. In both expansions, the system starts at an equilibrium state $p_i = 0.5$ MPa, $V_i = 0.001$ m³. Reversible isothermal (red) and adiabatic (black) expansions from this starting point are shown for an ideal monatomic gas for comparison. The constant pressures of the surroundings, represented using dashed horizontal lines, are not meant to depict equilibrium states of the system, but rather define the work done on the surroundings as rectangular shaded areas. In (A), the final volume of the system, V_f, is given by Equation 3.59 (0.00239 m³); the gray curve between the isotherm and the adiabat shows the relationship between the final volume and the pressure of the mechanical reservoir (Equation 3.60). Like the dashed constant pressure lines, the gray line is *not* meant to depict states of the system during equilibrium, but instead depicts a series of endstates resulting from different expansions. In (B), the dashed line shows the final pressure that would be obtained for various stopping volumes, expanding against a constant pressure of 83,140 Pa. At this pressure, the system will not expand past a volume of 0.004 m³ (where the dashed curve intersects gray curve line in A).

where ΔV refers to the *system* volume change. If this volume change is adiabatic, $dw = dU$. Combined with the fact that energy is conserved, this adiabatic constraint allows us to write Equation 3.57 as

$$\Delta U_{sys} = -\Delta U_{surr} = -p_{surr}\Delta V \qquad (3.58)$$

As expected, the work on the surroundings is equal but opposite that on the system.

An important practical consideration in analysis of irreversible expansion is the endpoint volume, V_f, and relatedly, how the irreversible expansion is "stopped." For reversible expansion, this is not a consideration, because p_{sys} is always matched with p_{surr}. Thus, reversible expansion is easy to stop at any point.[†] For irreversible expansion, we can either allow the system to expand to a volume where p_{sys} matches p_{surr} (once the system relaxes to equilibrium[‡]), or we can physically stop the expansion (with a barrier) at a specific volume where p_{sys} remains larger than p_{surr} (again, once the system equilibrates). These two expansions are illustrated in **Figure 3.16**.

Expansion to mechanical equilibrium with the surroundings

To analyze expansion to an equilibrium state in which $p_{sys} = p_{surr}$, we need to determine the final volume in order to evaluate the work and the change in energy using Equation 3.58. By assuming the system to be an ideal gas, some limits on the final volume can be obtained. First, the system cannot end up at a

---

[†] In a sense, the reversible expansion is always stopped.

[‡] Although the system is out of equilibrium during the expansion, equilibrium will reestablish shortly after the expansion is complete. By analyzing the energy change for the reequilibrated system using Equation 3.58, we can determine the final volume (for the "unstopped" expansion, Figure 3.16A) or the final pressure (for the stopped expansion, Figure 3.16B).

volume greater than that of the reversible isothermal reference curve[†] to p_{surr} (i.e., $V_f < V_{iso}$, Figure 3.16A). This is because the reversible isothermal ideal gas expansion occurs without internal energy change. Expanding past this isothermal reference would require the system to *increase* its energy, which is inconsistent with the negative work and the lack of heat flow. Second, the gas cannot end up at a volume *less* than for reversible adiabatic expansion to p_{surr} (i.e., $V_f > V_{ad}$, Figure 3.16A). This is because the system will have done less work than the corresponding reversible adiabatic expansion, and thus it will have lost less energy and will end up at a higher temperature (i.e., on a warmer isotherm). Thus, the expansion is limited to a final volume on the $p = p_{surr}$ segment bounded by the reversible isotherm and the adiabat (Figure 3.16A).

The exact volume at which the system will come to rest (V_f) can be calculated by equating the energy change (i.e., work) for irreversible expansion with that for a reversible path involving an isothermal expansion to V_f, followed by a constant volume cooling to p_{surr} (Problem 3.18). For a monoatomic ideal gas, the volume change is

$$V_f = \frac{V_1}{5}\left(\frac{3p_1}{p_{surr}} + 2\right) \tag{3.59}$$

The lower the pressure of the surroundings, the greater the final volume. By rearranging Equation 3.59, a relationship can be obtained that gives the pressure of the surroundings explicitly as a function of the final volume:

$$p_{surr} = \frac{3p_1 v_1}{(5V_f - 2V_1)} \tag{3.60}$$

Equation 3.61 can be plotted on a p-V diagram to visualize the relationship between the pressure of the reservoir and the extent of expansion of the system. It should be understood that this equation does not define a curve that the system travels along; rather, the curve defined by this equation sets work limits for irreversible expansion, and can be compared to its reversible adiabatic counterpart. Though the work done for both types of expansion (reversible and irreversible) increases as the extent of expansion increases, work increases less steeply for the irreversible case (**Figure 3.17A,B**).

The ratio of irreversible to irreversible work can be expressed as (Problems 3.20 and 3.21)

$$\frac{w_{irr,ad}}{w_{rev,ad}} = \frac{-p_{surr}(V_f - V_1)}{C_V \Delta T}$$

$$= \frac{-2p_{surr}\left(\left[\frac{1}{5}\left(\frac{3p_1}{p_{surr}} + 2\right)\right] - 1\right)}{3p_1\left(\left[\frac{1}{5}\left(\frac{3p_1}{p_{surr}} + 2\right)\right]^{\gamma-1} - 1\right)} \tag{3.61}$$

$$= \frac{-2(V_f - V_1)}{(5V_f - 2V_1)\left\{\left(\frac{V_f}{V_1}\right)^{1-\gamma} - 1\right\}}$$

[†] Note that the reversible isothermal reference curve included in these examples is *nonadiabatic*. Rather it serves as an upper bounds for the volume and pressure increase for unrestricted and stopped irreversible expansions, respectively.

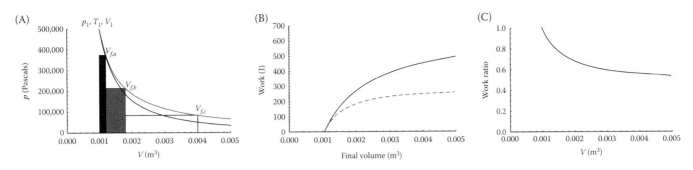

Figure 3.17 The work done in irreversible adiabatic expansion against a mechanical reservoir, compared with reversible expansion. (A) Irreversible expansion of an ideal monatomic gas (0.2 mol, starting at 0.5 MPa, 0.001 m³; the same as in Figure 3.16) against a constant pressure reservoir at constant pressure p_{surr}. Expansion is completed when the pressure of the gas (once returned to an equilibrium state) is equal to p_{surr}. The black line shows a reversible adiabatic expansion gray line shows the dependence of the final system pressure on different reservoir pressures (which leads to expansion to different final volumes, $V_{f,a}$, $V_{f,b}$, and $V_{f,c}$). The three rectangles show how much work would be done by expanding against different external pressures. For small expansion to high p_{surr}, most of the work that would be obtained in a reversible is recovered (corresponding to the area of the red rectangle). At lower p_{surr} values (i.e., for larger expansions), a smaller fraction of the available (reversible) work is recovered from irreversible expansion (compare red with gray and white rectangles). (B) Direct comparison of the work done for reversible (black) and irreversible (dashed) adiabatic expansions to different final volumes. Different limiting work values are obtained for large expansions to low pressure. (C) The ratio of irreversible to reversible work falls from a value of 1.0 for very small expansions (where p_{ex} is close to $p_1 = 0.5$ MPa) to a limiting value of 0.004 m³ for large expansions.

Though for tiny expansions (where p_{sys} is not very much less than p_1), nearly all of the work is recovered (Figure 3.17C), the irreversible to reversible work ratio falls steadily to a limiting value of 0.4.

Expansion to a stopping volume where $p_{sys} > p_{surr}$

If instead of letting the system expand until the pressure matches that of the mechanical reservoir, we stop the expansion at a volume of our choosing (V_f), we need to determine the final pressure, p_f, of the system. Because less work is done than for reversible adiabatic expansion to the same volume (see Figure 3.17A), the irreversibly expanded system will have a higher final energy than the reversibly expanded adiabatic system at the same final volume. However, the irreversible expansion does lose *some* energy, and a result, is lower in energy than the isothermally expanded system at the same final volume. Thus, the adiabat and isotherm again provide lower and upper limits to the final state, but in this case, it is the pressure that lies between these two limiting values (Figure 3.16B).

The internal energy decrease for irreversible adiabatic expansion, $-p_{ex}\Delta V$, can be set equal to the internal energy change calculated along a reversible path connecting the initial and final states. A convenient reversible path for this connection is an isothermal expansion to V_f followed by constant volume cooling to p_f. The internal energy change along this path is $C_v\Delta T$. By setting these two energy changes equal, and using the ideal gas law, p_f can be determined:

$$
\begin{aligned}
-p_{surr}(V_f - V_1) &= C_V(T_f - T_1) \\
&= \frac{3}{2}nR(T_f - T_1) \\
&= \frac{3}{2}(p_f V_f - p_1 V_1)
\end{aligned}
\qquad (3.62)
$$

Rearranging Equation 3.62 leads to an expression for p_f in terms of known quantities (starting conditions and the stopping volume):

$$\frac{3}{2}p_f V_f = -p_{surr}(V_f - V_1) + \frac{3}{2}p_1 V_1, \text{ or}$$

$$p_f = \frac{-2p_{surr}(V_f - V_1) + 3p_1 V_1}{3V_f} \tag{3.63}$$

$$= \frac{V_1(3p_1 + 2p_{surr})}{3V_f} - \frac{2p_{surr}}{3}$$

In Equation 3.63, if $V_f = V_1$, then $p_f = p_1$. As V_f increases, p_f decreases from p_1, and approaches p_{surr}. Note, for a given value of p_{surr}, V_1 cannot exceed that given by Equation 3.59. Another way to say this is that the system cannot come to rest at a pressure below p_{surr}. Thus, the irreversible expansion curve given by Equation 3.60 provides a lower limit to the extent to which stopped expansion can proceed. This is shown in Figure 3.16B where the stopped expansion curve (dashed line, given by Equation 3.63) achieves its maximal V_f value when it intersects the curve for irreversible expansion to mechanical equilibrium (gray curve).

Expansion against vacuum

Expansion against a vacuum can be regarded as expansion against a mechanical reservoir at constant pressure $p_{surr} = 0$. Based on Equation 3.56,

$$w_{sys} = -\int_{V_i}^{V_f} p_{surr}dV = 0\Delta V = 0 \tag{3.64}$$

As in Figure 3.16A, we will take V_f as a fixed at a value greater than V_i, set by an external stop (**Figure 3.18**). If the vacuum expansion is adiabatic, then the internal energy change for the system is zero.

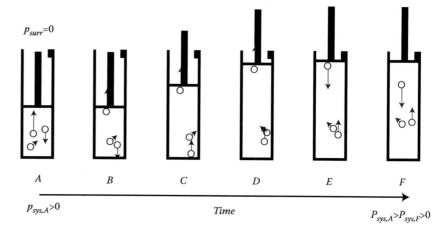

Figure 3.18 A molecular picture of expansion of an ideal gas against a vacuum.
The container is taken to be adiabatic and thermodynamically idealized (i.e., with a massless, frictionless piston). In state A, the piston is at rest, but the gas is moving with an average velocity (and temperature), and thus exerts some pressure against the walls (in a time-averaged sense). At time B, a gas molecule that was moving upward contacts the piston. Because the piston has no mass, it moves along with the colliding molecule at the same speed. Because there is no friction, the piston does not slow down until it hits the stop at time E. At this point, the gas molecule is elastically reflected off the now stationary piston. Since the gas molecules all retain their velocity, the temperature remains constant. If the gas is ideal, the internal energy of the gas does not change, which is consistent with the observation that with adiabatic walls, $\Delta U = 0$.

The constancy of energy for adiabatic vacuum expansion seems surprising for two reasons. First, we have a tendency to think that an expansion in which a wall moves should slow down the gas molecules within the system, thereby lowering the temperature. This is because it is easy to overlook the idealizations that are often made in thermodynamic calculations of expansion. One idealization is that the moveable wall is frictionless, so no energy is dissipated at the junctions of the moving and fixed walls. Another idealization is that the wall is massless. This means that once expansion commences, the first gas molecule that strikes the wall will continue to move along at the same velocity, along with the massless wall (Figure 3.18). Unless the wall is overtaken by a faster gas molecule, the wall and the contacting gas molecule will continue to move until the stop is reached. Upon stopping, the gas molecule will be reflected off the wall as if the wall was never moving. Thus, no gas molecules were slowed down in the process of moving the wall. For an ideal gas, this means that neither the temperature nor the internal energy changes. Surprisingly, the constancy of temperature means that the vacuum expanded state ends up on its own isotherm. In terms of the initial and final states of the *system*, it is as if the gas expanded reversibly *along its isotherm*, balancing work with heat flow (but with a balance of zero and zero).

Note that this does not mean the system traveled along its isotherm. Also note that if wall had mass, kinetic energy would be transferred to the external stop, creating an additional mechanism for energy flow (albeit a small energy transfer) from system to surroundings. The idealized massless wall means that the moving wall has no energy to transfer.

The constancy of energy (more specifically, the statement that work is zero) for vacuum expansion also seems at odds with the work calculated for irreversible expansion against a constant pressure reservoir above (See Figure 3.17B, dashed curve, and Equation 3.61), in which the system comes to pressure equilibrium with the surroundings. In the limit that $p_{surr} = 0$, the constant pressure treatment suggests that work done by the system plateaus at a positive value (0.4 for a monatomic gas). The reason for this apparent discrepancy is that as p_{surr} goes to zero, the volume of the system goes to infinity upon expansion to equilibrium. Thus, the work integral for vacuum expansion to $p_{sys} = p_{surr} = 0$ is of indeterminate form:

$$w_{sys} = -\int_{V_1}^{\infty} p_{surr} dV = -(0 \times V)|_{V_1}^{\infty} \tag{3.65}$$

This integral can be represented rectangle of infinitesimal height, but infinite width, which converges to a finite (negative), nonzero value. It is only when expansion against $p_{surr} = 0$ is stopped at a finite value (Figure 3.18) that $w_{sys} = 0$.

THE CONNECTION BETWEEN HEAT CAPACITIES AND STATE FUNCTIONS

Although we use different values of heat capacity (C_V vs. C_p) depending on how a system changes its state, each of these values is uniquely defined by the state of the system; that is, both C_p and C_v are state functions. Here we will develop some relationships between C_V and C_p and some other state functions. These relationships will be useful in analysis of thermodynamics of phase transitions and reactions in subsequent chapters.

Returning to Equation 3.27, the connection of C_V to state variables is quite direct, and can be obtained by a simple rearrangement of the first law. If volume is constant, no work is done.[†] From the first law, $dU = dq$. Substituting into Equation 3.27 gives

$$C_V = \left(\frac{\partial U}{\partial T}\right)_V \qquad (3.66)$$

The connection of C_p to state variables is a little more complicated. Almost all materials expand when heated at constant pressure ($\alpha > 0$, Table 3.1), so there is expansion work associated with heating ($dw = -pdV < 0$). Thus, substitution of the first law into Equation 3.30 gives

$$C_p = \left(\frac{\partial q}{\partial T}\right)_p = \left(\frac{\partial U + p\partial V}{\partial T}\right)_p \qquad (3.67)$$

We could simplify this relationship if we had a function with a differential equal to $dU + pdV$ at constant pressure. A function that satisfies this criterion is

$$H \equiv U + pV \qquad (3.68)$$

H, which is referred to as the "enthalpy," can be thought of as the internal energy plus the cost of pushing back the universe to make a hole in which to place (or expand) the system. H is useful for analyzing the thermodynamics of systems at constant pressure. To see that the differential of H has the form we are seeking (that is, $dH = dU + pdV$), we can calculate the total differential of $H = H(U, P, V)$ as described in Chapter 2:

$$
\begin{aligned}
dH &= \left(\frac{\partial H}{\partial U}\right)_{p,V} dU + \left(\frac{\partial H}{\partial p}\right)_{U,V} dp + \left(\frac{\partial H}{\partial V}\right)_{U,p} dV \\
&= \left(\frac{\partial \{U + pV\}}{\partial U}\right)_{p,V} dU + \left(\frac{\partial \{U + pV\}}{\partial p}\right)_{U,V} dp + \left(\frac{\partial \{U + pV\}}{\partial V}\right)_{U,p} dV \\
&= \left(\frac{\partial U}{\partial U}\right)_{p,V} dU + V\left(\frac{\partial p}{\partial p}\right)_{U,V} dp + p\left(\frac{\partial V}{\partial V}\right)_{U,p} dV \\
&= dU + Vdp + pdV
\end{aligned}
\qquad (3.68)
$$

At constant pressure, the second term in the differential is zero, giving $dH = dU + pdV$. Thus, dH simplifies Equation 3.66, providing a connection between H and C_p analogous to that between U and C_V:

$$C_p = \left(\frac{\partial H}{\partial T}\right)_p \qquad (3.70)$$

One final note on the enthalpy function. By combining Equations 3.30 and 3.69, that is,

$$\left(\frac{\partial q}{\partial T}\right)_p = \left(\frac{\partial H}{\partial T}\right)_p \qquad (3.71)$$

we see that at constant pressure, $dq = dH$. In other words, the heat flow into a system at constant pressure is equal to the change in the system's enthalpy.

[†] Assuming there is no nonexpansion work (e.g., stirring or polarization).

The relationship between C_V and C_p

In our analysis of adiabatic expansion, the relationship between the heat capacities at constant volume and constant pressure played an important role. Here we will explore the relationship between these two quantities, using the first law of thermodynamics. Because heat capacities can be measured directly using calorimetry (and their ratios can be inferred from adiabatic expansion), these relationships can be tested, and if needed, refined.

When a fixed amount of heat flows into a system at constant volume, all of the energy associated with the flow remains in the system (because no work can be done). However, when the same amount of heat flows into a system at constant pressure, there is no volume restriction. Thus, the constant pressure system is likely to expand,[†] resulting in loss of some of the heat energy to work. The internal energy will increase *less* at constant pressure than at constant volume, which corresponds to a lower final temperature. Thus, an amount of heat flow dq will result in a larger temperature change (dT) at constant volume than at constant pressure, corresponding to a smaller heat capacity. More concisely,

$$C_V = \left(\frac{\partial q}{\partial T}\right)_V < \left(\frac{\partial q}{\partial T}\right)_p = C_p \qquad (3.72)$$

For an ideal gas, we can determine the numerical difference between C_V and C_p (nR, Equation 3.21) using a p-V diagram (**Figure 3.19**). From a common initial state at p_i, V_i, and T_i, we can compare two ways to increase the temperature to a T_f value: constant volume heating and constant pressure expansion (vertical and horizontal lines, Figure 3.19). These two processes end on the same isotherm, but at different pressures and volumes.

Because internal energy is constant along the T_f isotherm (dashed line, Figure 3.19), the two final states must have the same internal energy, U_f. Since both changes started at the same initial state, the internal energy changes must be the same for the two changes of state, that is,

$$\Delta U_V = \Delta U_p \qquad (3.73)$$

Using the first law, Equation 3.73 can be written as

$$C_V(T_f - T_i) = \Delta C_p(T_f - T_i) - p_i(V_f - V_i) \qquad (3.74)$$

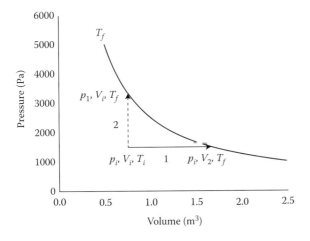

Figure 3.19 Using an isotherm to compare C_p and C_V for an ideal gas. Two changes of state are compared that take an ideal gas from a single starting point (p_i, V_i, T_i) to an isotherm with $T_f > T_i$ (red). In transformation 1, the pressure is held constant (horizontal arrow), in transformation 2 the volume is held constant (vertical dashed arrow). Since the internal energy change is the same for both transformations (because temperature changes are the same and the gas is ideal), extra heat must flow at constant pressure to make up for the energy lost to expansion work (area under the horizontal arrow). Thus, C_p must be larger than at C_V.

[†] The requirement for expansion is that $\alpha = V^{-1}(\partial V/\partial T)_p > 0$. This is the case for most but not all systems. Water is a notable exception, both upon melting (liquid water is more dense than ice) and upon heating just above melting temperature, where density *decreases* slightly with temperature.

where V_f is the for the constant pressure expansion. Rearranging Equation 3.74 gives

$$
\begin{aligned}
C_p - C_V &= \frac{p_i V_f - p_i V_i}{T_f - T_i} \\
&= \frac{nRT_f - nRT_i}{T_f - T_i} \\
&= \frac{nR(T_f - T_i)}{T_f - T_i} \\
&= nR
\end{aligned}
\tag{3.75}
$$

The gas-law substitution of nRT_f for $p_i V_f$ in the second line of Equation 3.76 can be justified by recognizing that $p_i V_f$ is the final coordinate in the constant pressure expansion to T_f (Figure 3.19). This result ing nR value is independent of whether the gas is monatomic ($C_p - C_V = 5nR/2 - 3nR/2$) or diatomic ($C_p - C_V = 5nR/2 - 3nR/2$; Table 3.1), it just requires that the gas behaves ideally.

A NONIDEAL MODEL: THE VAN DER WALLS EQUATION OF STATE

As a model for understanding the basic concepts of thermodynamics, the ideal gas has great appeal. It allows the kinetic energy of molecules to be related to a variety of macroscopic properties, such as pressure, heat capacity, and internal energy. And as shown above, the simplicity of the ideal gas equation of state can be used to develop and quantify the concepts of work and heat.

Though many real gases adhere to the ideal gas law at low density and high temperature, they often show significant deviations at high density and low. A fundamental feature of the ideal gas model is that molecules *don't* interact (except via elastic collisions that allow energy to redistribute), either through forces that attract molecules to each other, or forces that repel molecules from each other. However, most molecules "feel" each other to some extent. This is obviously true in liquids and solids,[†] where the close proximity of molecules promotes interaction. However, interactions also occur in gases at high densities, where interactions between molecules are likely to occur. Low temperature promotes these weak interactions because of entropy considerations, which we will consider later.

One model that retains much of the simplicity of the ideal gas, but includes attractive and repulsive interactions is the "van der Waals" gas. The equation of state for the van der Waals gas is

$$
p = \frac{nRT}{V - nb} - \frac{an^2}{V^2}
\tag{3.76}
$$

Both a and b are positive constants that have different values for different gases (**Table 3.3**). The first term in Equation 3.76 has similar form to the ideal gas law, but the volume is diminished by an amount that is proportional to the number of gas molecules present. This volume decrement can be thought of as an "excluded volume" term, reflecting the fact that real atoms and molecules are not point masses but instead occupy space that other molecules cannot penetrate (Problem 3.22). This exclusion can be thought of as a repulsive interaction, albeit very short ranged. The excluded volume modification serves to increase the pressure over what it would be

---

[†] Without intermolecular attractions, there would be no liquids or solids. And without repulsions, we would all fall through the floor.

Table 3.3 Parameters for the van der Waals equation of state for several gasses

	A (10^5 Pa L^2 mol^{-2})	B (L mol^{-1})
He	0.03457	0.0237
Ne	0.2135	0.01709
Ar	1.363	0.03219
Kr	2.349	0.03978
Xe	4.250	0.05105
N_2	1.408	0.03913
O_2	1.378	0.03183
CO_2	3.640	0.04267
H_2O	5.536	0.03049
Benzene	18.24	0.1145

Source: Weast, R.C. 1979. *The Handbook of Chemistry and Physics*, 60th Edition. CRC Press, Boca Raton, FL.

for an ideal gas, because with less volume available, the gas molecules will spend more time hitting the walls of the container.

The second term in Equation 3.75 diminishes the pressure of the gas in proportion to the square of the density of molecules.[†] This pressure decrement can be thought of as reflecting an attractive interaction between gas molecules. Its dependence on the square of the molecular density reflects the fact that in a random distribution of molecules, the probability that two molecules will be close enough to favorably interact depends on the probability that one will be there times the probability that the other will be there.

Of the two interaction terms in the van der Waals equation of state, only the attractive term contributes to the internal energy. The excluded volume term simply defines a region of space that cannot be occupied, but does not contribute an interaction energy.[‡] Taking the attractive term into account gives an expression for the internal energy of a monatomic van der Waals gas of

$$U = \frac{3}{2}nRT - \frac{an^2}{V}$$

(3.77)

The first term on the right-hand side is the same as for a monatomic ideal gas, and has to do with kinetic energy. The second term accounts for the favorable pairwise interaction energy between gas molecules, and decreases the internal energy compared to the ideal gas at the same temperature. Note that as a result of this second term, the energy of a van der Waals gas is dependent on volume. Thus, unlike the ideal gas, the internal energy of a van der Waals gas varies in an isothermal volume change (see Problem 3.21).

Values for the a and b constants of the van der Waals equations have been estimated for a number of real gases by analyzing p–V measurements at high densities and low temperatures (Table 3.3; see Problem 3.27). Inspection of these values provides insight into molecular properties. For example, the b terms for the noble gases (Ne, Ar, Kr, and Xe) increase with atomic number, consistent with increasing

---

[†] The term n/V can be thought of as a number density or concentration term.

[‡] Outside the excluded volume region the gas molecules don't feel any repulsion, and they are not allowed inside the excluded volume region.

excluded volume. Moreover, a significant increase in the attractive term is seen for the same series (by more than 100-fold), which reflects a significant increase in polarizability (and thus, intermolecular interaction) with atomic number (and number of electrons). Among the di- and tri-atomics in Table 3.3, the molecules with a permanent dipole (CO_2 and H_2O) show a stronger attractive term; for H_2O, this attraction may be further enhanced by hydrogen bonding. Benzene shows a very strong attractive term, which may be result of favorable aromatic stacking interactions.

Problems

3.1 The following three data sets give p as a function of V, T, and n for a perfectly ideal gas. Specifically, data set 1 gives p as a function of V at fixed values $T = 1$ K and $n = 1$ mol. Data set 2 gives p as a function of T at fixed values $V = 1$ m^3 and $n = 1$ mol. Data set 3 gives p as a function of n at fixed values of $V = 1$ m^3 and $T = 1$ K. Plot each of these data sets separately and fit each to an equation that seems appropriate. What is the value for the fitted constant in each plot?

3.2 Using the data from Problem 3.2.abcd, globally fit a single equation to an appropriate equation that defines pressure in terms of the three relevant variables, that is, $p \equiv p(n,V,T)$. What is the value for the single fitted constant?

3.3 (This problem involves some somewhat advanced graphing). Using Mathematica or a similar program, construct a three-dimensional plot to show the data sets 1 and 2, and your fitted curve. How well does your model fit the data? Try to come up with a way to include data set 3 in your plot (you might consider using color to encode information on this plot). Can you also illustrate how data set 3 is fitted in this plot? In Mathematica, useful commands might include `Graphics[Line[...]]` or `ContourPlot3D[...]`.

3.4 This problem derives the ideal gas law from the general differential equation of state. For ideal gases with a single component, experiments show that $\kappa_T = p^{-1}$, and $\alpha = T^{-1}$. Moreover, for a single component, $\bar{V} = V/n$ (see Chapter 5). Starting with Equation 3.4, substitute the three results above and integrate. Within a multiplicative constant, you should be able to recover the ideal gas law. Hint: although you should be able to separate variables, integration will require multiple steps. Use the fact that p, T, V, and n are state functions, and integrate from a single starting point with variables p_i, V_i, T_i, n_i to general values p, V, T, n.

3.5 For the constant pressure expansion in Example 3.1, assuming that the gas behaves ideally, calculate the initial and final temperatures for the expansion.

3.6 Consider the adiabatic expansion of 0.05 moles of helium, a monatomic gas, from initial volume and temperature $V_i = 0.1$ m^3, $T_i = 273.15$ K, to $V_f = 2$ m^3. Use Equation 3.20 to calculate the work done. Confirm that this work value is the same as that obtained using Equation 3.27.

3.7 For the following adiabatic expansion data for an ideal gas, use nonlinear least-squares to determine γ. Based on your results, is the gas monatomic or diatomic?

3.8 Demonstrate the relationship

$$T_i \left(\frac{V_i}{V_f} \right)^{nR/C_V} = T_f$$

for the adiabatic expansion of an ideal gas.

3.9 Calculate the work required to pull the ends of a linear polymer (such as DNA and unfolded polypeptide chains) apart. Assume that the restoring force opposing separation can be expressed as

$$F = -\frac{k_B T}{L_p}\left\{\frac{1}{4(1-x/L_c)^2} - \frac{1}{4} + \frac{x}{L_c}\right\}$$

This force model is often associated with a statistical polymer model called the "worm-like chain." The variable x is the separation between ends. The variable L_c is the "contour length," you can think of this as a fixed value describing the total length of the polymer chain, whether it is stretched out or not. Because the polymer will generally curve and fold back on itself, the end-to-end separation is almost always shorter (and never longer) than L_c. The parameter L_p is the "persistence length," which you can think of as the average distance over which the polymer changes direction. L_p is a measure of polymer stiffness.

3.10 Starting with Equation 3.34, demonstrate that for an isobaric expansion of a monatomic ideal gas, the internal energy change ΔU has a value of $3/2nR\Delta T$. Demonstrate that for a diatomic ideal gas, $\Delta U = 5/2nR\Delta T$.

3.11 Show that for a reversible adiabatic expansion of an ideal gas, an expression of the same functional form as Equation 3.52 is obtained by indefinite integration of Equation 3.43. From your comparison, what is the value of the integration constant?

3.12 For both the ideal gas, demonstrate the thermodynamic relationship

$$\pi_T = T\left(\frac{\partial p}{\partial T}\right)_V - p$$

3.13 Based on what you know about heat capacities of ideal gases, come up with an expression for the constant volume heat capacity change, C_V, for CO_2 (a linear triatomic molecule) at room temperature? Assume it behaves as an ideal gas. Justify your answer.

3.14 As you did in Problem 3.13, come up with an expression for C_V for H_2O, a bent triatomic molecule. Again, justify your answer.

3.15 For the following *reversible processes for an ideal gas*, give the heat flow, work done, and internal energy change for the following processes from the initial state i (with variables p_i, V_i, and T_i) to the final state f (p_f, V_f, and T_f). Assume that these two states differ by a finite amount in at least one variable (probably two).

	q	w	ΔU
Constant Pressure expansion	$C_p(T_f-T_i) = C_p\Delta T$	$-p(V_f-V_i) = -p\Delta V$	$C_p\Delta T-p\Delta V$
Constant volume heating	$C_V(T_f-T_i) = C_V\Delta T$	zero	$C_V\Delta T$
Vacuum expansion	0	0	0
Isothermal expansion	$nRT\ln(V_f/V_i)$	$-nRT\ln(V_f/V_i)$	0
Adiabatic expansion	0	$C_V(T_f-T_i) = C_V\Delta T$	$C_V\Delta T$

3.16 On the *p-V* diagram below, draw two curves, one for the adiabatic *compression* of a monatomic ideal gas from state i to V_f, and the other for isothermal *compression* from the same starting point to the same final volume, V_f. Using the curves you drew, illustrate the work done for the two compressions. Based on our sign convention, rank the numerical size of each work term relative to zero and to each other (using greater-than and less-than signs, i.e., $a < b < c$).

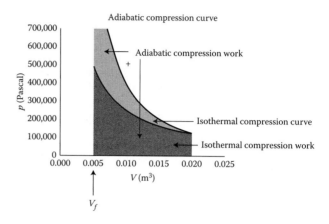

3.17 For the following reversible adiabatic ideal gas expansion data, determine the value of γ using curve-fitting. Based on your fitted value, what can you say about the gas?

V (m³)	p (Pa)	V (m³)	p (Pa)	V (m³)	p (Pa)	V (m³)	p (Pa)
0.0010	9,981,600	0.0060	504,100	0.0110	185,100	0.0160	92,520
0.0015	5,100,500	0.0065	442,300	0.0115	167,700	0.0165	92,800
0.0020	3,143,600	0.0070	384,400	0.0120	165,000	0.0170	98,400
0.0025	2,184,800	0.0075	354,000	0.0125	147,800	0.0175	82,500
0.0030	1,597,400	0.0080	330,800	0.0130	145,600	0.0180	97,000
0.0035	1,227,300	0.0085	284,900	0.0135	142,100	0.0185	84,500
0.0040	1,001,200	0.0090	240,300	0.0140	103,100	0.0190	76,100
0.0045	825,400	0.0095	217,500	0.0145	128,300	0.0195	62,400
0.0050	686,600	0.0100	218,000	0.0150	102,300	0.0200	55,000
0.0055	593,200	0.0105	192,200	0.0155	99,190		

3.18 Show that for an ideal monatomic gas expanding adiabatically against the constant external pressure of a mechanical reservoir (p_{surr}), the final volume (V_f, at which $p_{sys} = p_{surr}$) is given by Equation 3.58, that is,

$$V_f = \frac{V_1}{5}\left(\frac{3p_1}{p_{surr}} + 2\right)$$

where p_1 and $V1$ are the starting pressures and volume (see Figure 3.16A).

3.19 Rearrange your answer from Problem 3.17 to generate an expression that relates the pressure of the mechanical reservoir, p_{surr}, to the final volume for adiabatic irreversible expansion of a monatomic ideal gas (Equation 3.59).

3.20 Show that for an ideal monatomic gas expanding adiabatically against the constant external pressure of a mechanical reservoir (p_{surr}) to a final equilibrium state where $p_{sys} = p_{surr}$, the ratio of work compared with reversible adiabatic expansion over the same volume limits is given by the first equality in Equation 3.60, that is,

$$\frac{W_{irr,ad}}{W_{rev,ad}} = \frac{-2p_{surr}\left(\left\{\frac{1}{5}\left(\frac{3p_1}{p_{surr}}\right)+2\right\}-1\right)}{3p_1\left(\left\{\frac{1}{5}\left(\frac{3p_1}{p_{surr}}\right)+2\right\}^{\gamma-1}-1\right)}$$

Hint: for this and the next problem, you will need to substitute the pressure–volume relationship for adiabatic expansion of an ideal gas.

3.21 Show that for an ideal monatomic gas expanding adiabatically against the constant external pressure of a mechanical reservoir (p_{surr}) to a final equilibrium state where $p_{sys} = p_{surr}$, the ratio of work compared with reversible adiabatic expansion over the same volume limits decays to zero as extent of expansion (V_f) increases (i.e., as p_{surr} decreases) according to the second equality in Equation 3.60:

$$\frac{W_{irr,ad}}{W_{rev,ad}} = \frac{-2(V_f - V_1)}{(5V_f - 2V_1)\{(V_f / V_1)^{1-\gamma} - 1\}}$$

3.22 Treating the energy of an arbitrary system (i.e., make no assumptions about ideality) as a function of temperature and volume, give an expression for the exact differential of the energy function, dU.

3.23 Using the result that for a van der Waals gas, the internal energy is

$$U = \frac{3}{2}nRT - a\frac{n^2}{V}$$

give the exact differential of dU, based on your answer to Problem 3.18.

3.24 Use the Euler test to show that the exact differential for the van der Waals gas from Problem 3.19 is exact.

3.25 Derive an expression for the internal pressure π_T for a van der Waals gas, using an equation, assuming that the van der Waals coefficients (a and b) are independent of temperature. Given that a and b are always positive, what can you say about the internal pressure of the van der Waals gas?

3.26 Estimate the excluded volume term (in liters per mole) you would expect for argon gas, using the accepted "radius" of 1.8 Å. How does this compare with the experimentally determined van der Waals b coefficient of 0.03219 L per mole?

3.27 Using Mathematica, plot six isotherms (at the six temperatures 200 K, 230 K, 270 K, 300 K, 400 K, and 800 K) for the expansion of one mole of CO_2, which behaves as a van der Waals gas. Use the van der Waals coefficients in Table 3.3.

For your plot, use the following limits:

p *(on the y axis)* from -1×10^7 Pa to 4×10^7 Pa.

V *(on the x axis)* from 0 to 0.001 m³.

Describe the overall shapes of the curves, how they relate to isotherms for the ideal gas, and what might be going on at low, high, and intermediate temperatures.

3.28 Consider a *reversible isothermal* expansion at 300 K of one mole of CO_2, using the van der Waals equation of state, from $V_i = 0.5$ L to $V_f = 2.0$ L, as plotted below:

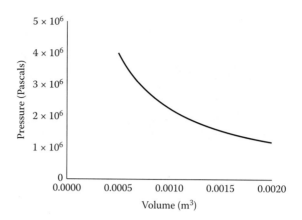

Calculate the work done, w, in this expansion, using van der Waals coefficients in Table 3.3. Is this quantity positive or negative? Indicate this work on the plot above.

3.29 Calculate the heat flow for the isothermal expansion of CO_2 described in Problem 3.28.

3.30 Use nonlinear least squares to determine if the following isothermal (100 K) expansion data (p vs. V) for an unknown gas (0.1 mole) behaves ideally. Show the fitted curve along with the data, as well as an appropriately scaled residual plot to test for ideality. If it does not behave ideally, fit the data with the van der Waals equation to determine the a and b coefficients.

3.31 Use nonlinear least squares to globally fit the series of isotherms (100 K, 200 K, 300 K, 400 K, and 500 K) to the van der Waals equation (again, 0.1 mole gas), and determine the a and b coefficients. In global fitting, you should write a single equation that is a function of both temperature and pressure, and in your NonlinearModelFit command, you data file should have have three columns {pressure, volume, temperature}. Plot your fit along with the data, either as a series of curves on a 2D plot, or on a 3D plot.

CHAPTER 4

The Second Law and Entropy

Goals and Summary

The goal of this chapter is to give students both an analytical and an intuitive understanding of the second law of thermodynamics. The second law is beautiful in its predictive power, but compared to its counterpart in the first law, it can appear to be rather untidy, and mastering its subtleties can be difficult. Students will learn two equally important approaches to the second law, a classical approach using heat engines, and a statistical, counting-based approach. Both approaches are essential to master the second law, and understanding each will help to reinforce the other. Students will learn the importance of reversibility and irreversibility in the second law, and how the entropy acts as a potential for spontaneous change and for determining stable equilibrium configurations.

We will begin by comparing processes that occur spontaneously with those that do not, even though they comply with the first law. Analysis of these types of processes shows multiplicity (the number of ways a system can be configured) to be a key indicator of whether or not a process is spontaneous, and allows us to identify both the direction of spontaneous change, and the position at which a system will "come to rest," that is, the position of equilibrium. A central quantity in this analysis is entropy, which will be developed both in terms of classical macroscopic variables (heat flow and temperature), and in terms of microscopic variables (multiplicity of molecular configurations). Using the classical approach, calculations of entropy change will be made for reversible transformations in simple systems, and will be compared to approximations for irreversible (i.e., spontaneous) processes. The use of entropy as a thermodynamic potential (giving the direction for spontaneous change, and marking the position of equilibrium) for isolated systems will be developed. Using the statistical approach, entropy will be calculated from analysis of molecular configurations using simple models. Although we will postpone a detailed statistical treatment until Chapter 9, comparisons here shows good agreement with between the classical and statistical models, showing a direct connection between the classical, heat-based approach and the statistical approach to the second law.

The first law tells us that energy is conserved during all processes, and that both heat and work contribute to this conservation, without favor to one or the other form of transfer. But the first law says nothing about *which way* a process will go (whether the process goes forward or backward). Clearly, if energy is conserved when a process moves forward, it is also conserved when the process moves backward—both directions are consistent with the first law. But most processes show a preference to move in one direction. Buildings fall down, hot coffee cools, and chemical reactions progress toward equilibrium. In each of these processes, energy is conserved but is redistributed. Whereas the first law guarantees the conservation of energy, the second law identifies the preferred distribution, and puts this "preference" into quantitative terms through a quantity called the "entropy."

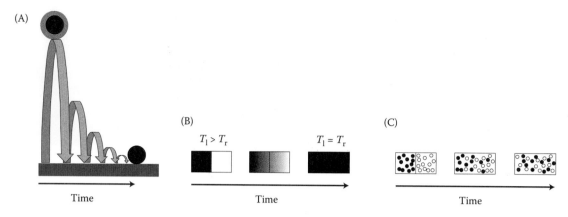

Figure 4.1 Spontaneous, irreversible processes. (A) A ball bouncing on the floor comes to rest, but does not begin hopping on its own. (B) Two bodies at different temperatures come to the same temperature, but bodies do not form thermal gradients on their own. (C) Simple fluids made of distinguishable types of molecules mix with each other to uniformity, but do not spontaneously form concentration gradients. In all three cases, the reverse (unobserved) processes do not violate the first law (energy is conserved), but they do violate the second law.

SOME FAMILIAR EXAMPLES OF SPONTANEOUS CHANGE

Three familiar examples of spontaneous change[†] are shown in **Figure 4.1**. Spontaneous mechanical change is illustrated by a bouncing ball (Figure 4.1A). Though the ball continues to bounce for a little while, each bounce is smaller than the previous one, and after several bounces the ball comes to rest on the floor. We never observe the reverse process, in which a ball at rest on the floor starts bouncing on its own.

Spontaneous thermal change is seen when two bodies at different temperatures are connected by a heat-conducting boundary (Figure 4.1B). Heat flows until the two bodies have the same temperature; thereafter, the bodies remain at this temperature.[‡] We never observe the reverse process, in which two bodies at the same temperature develop a temperature difference spontaneously.

Spontaneous material change is seen when two fluids composed of different (distinguishable) species are brought into contact through an open boundary (Figure 4.1C). In the absence of strong interactions, both species spread from high concentration to low concentration, until they are uniformly distributed.[§] For simple mixtures, we never observe the reverse process, in which concentration gradients develop spontaneously.[¶]

To understand why these processes are directional, we need to analyze the fine details of the energy distributions, and how this distribution changes over time. The bouncing ball appears to have a simple energy distribution, combining potential energy (gravity), and kinetic energy exchanging in an oscillatory way (Figure 4.1A). These two modes of energy are recognizable because they are "organized" on a macroscopic scale—they can be described by just a few macroscopic coordinates (ball position and speed).

If the ball and the floor were perfectly elastic, and the air was removed, bouncing would continue at the same height forever. That is, the total gravitational and kinetic

---

[†] As discussed in Chapter 3, we use the word "spontaneous" synonymously with the words "natural" and "irreversible."

[‡] Unless acted on by an outside source.

[§] For example, ideal gases, which have no intermolecular interactions, show this kind of mixing. In contrast, some strongly interacting liquids, such as oil and water, don't mix.

[¶] Simple mixtures can develop spontaneous concentration gradients, but this requires external fields or spatially inhomogeneous reactive surfaces.

energy would be constant, and would simply oscillate between these two energy forms. For real balls and real floors, both types of energy are lost over time. Where does the energy go? The first law requires the energy to be converted to new modes. These modes are less conspicuous than the two starting modes (kinetic and potential). They include thermal energy in the floor, the ball, and the air (the ball makes a bouncing sound through traveling pressure waves, and also loses energy to friction against the air). With a very precise thermometer, a slight rise in temperature of these materials (ball, table, air) might be detected.

The key point as far as the second law is concerned is that the process described above (the bouncing ball coming to rest) is spontaneous; the reverse process is not. The new distribution of energy (into many thermal modes) is favored over the starting distribution (of just two modes). The spreading of energy into all available modes is a key feature of the second law.

To convince yourself of the improbability of the reverse process (i.e., for the ball to start bouncing spontaneously), think of what would have to happen to the hidden disorganized energy. A significant amount of this thermal energy would need to come rushing back to the ball, all at once, from the motions of the molecules in the floor, the air, and the ball itself, in an organized (upward) direction. As a result, the ball would get a little upward kick. Just when the ball returned to the floor, another upward kick would need to appear in the same place and direction, to drive the ball further into the air. After repeating several times, the ball would be back where it started. Although this kind of synchronous behavior is not *impossible*, it is *highly improbable*. There are so many more ways that the molecules can move that don't involve kicking the ball in just the right way that we can consider the process in Figure 4.1A to be irreversible.

SPONTANEOUS CHANGE AND STATISTICS

As the example with the bouncing ball illustrates, there is an important connection between spontaneity, probability, and statistics, especially the statistics of energy distributions. This connection is a central feature of the second law. Most students with a little background in chemistry have loose, qualitative familiarity with this connection through the concept of "randomness." But a quantitative analysis of randomness requires some careful statistical analysis.

To help with this analysis, we will consider models for the processes shown in Figure 4.1B and C (heat flow and mixing), where probabilities can be calculated using simple statistics. The simplification in both models comes from treating configurations of the system as "discrete" (and therefore countable). Though this analysis approaches the realm of statistical thermodynamics, we include it here to complement the somewhat abstract classical thermodynamic development of the second law.

Spontaneous change and the mixing of simple fluids

To calculate the probability of different spatial distributions of two different fluids (*A* and *B*), we will partition the space in our container into molecule-sized boxes (**Figure 4.2A**). This type of model is often referred to as a "*lattice model*." We will construct the partitions so the number of lattice cells, N_c, is the same as the total number of *A* and *B* molecules, that is, $N_c = N_A + N_B$.[†] We will prescribe two rules for the placement of molecules. First, no more than one molecule can occupy a given cell, regardless of type. This rule captures the concept of excluded volume—molecules cannot overlap. Second, no preferential interactions exist between the *A* and *B* molecules. That is, placing *A* and *B* molecules in adjacent cells is no better (and no worse) than placing *A* and *A* (or *B* and *B*) molecules in adjacent cells. This is

---

[†] Throughout this text, we will use N to count the number of molecules, and n to count the number of moles. Thus, $n = N/L_0$, where L_0 is Avogadro's number ($6.02 \ldots \times 10^{23}$ molecules per mole).

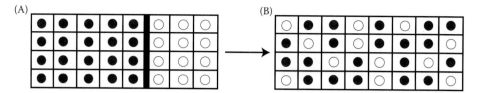

Figure 4.2 Mixing of two distinguishable, noninteracting species on a two-dimensional lattice. $N_A = 20$ red molecules and $N_B = 12$ white molecules are partitioned over a lattice of $N_c = 32$ cells. (A) Initially, the two types of molecules are separated from each other by a barrier that prevents exchange (thick line). In this starting state, there is only one arrangement the molecules can adopt. (B) After the barrier is removed, the molecules can exchange by a process that maintains one and only one molecule per cell; one such arrangement is shown. Because there are many "well-mixed" configurations like that shown in (B), conversion from the unmixed configuration (A) to the collection of mixed configurations is spontaneous. (Note that by itself, the *specific* arrangement in panel B is no more likely than that in panel A.)

equivalent to saying that there is no energy of interaction (other than the excluded volume term).[†]

With these rules, we can regard mixing as the conversion from a state where the A and B molecules are all partitioned to separate sides of the lattice by an impermeable barrier (Figure 4.2A) to a state where the barrier is removed and the molecules are free to exchange lattice sites (**Figure 4.2B**).

With this model, we will calculate the number of ways the A and B molecules can be arranged when they are free to mix, $W(N_A, N_B; N_c)$, and compare it to the number of ways the molecules can be arranged before the partition is removed. Assuming each arrangement, or "configuration," to be equally likely, we will represent the direction of spontaneous change, in terms of the distribution of A and B molecules, as that which increases the number of configurations.

One way to count the number of mixed configurations is to start with an empty lattice and ask how many ways there are to distribute the A molecules (which we cannot distinguish from one another) over the N_c lattice sites. The first molecule can be placed at N_c different sites, the second can be placed at $N_c - 1$ sites, ..., and the N_Ath molecule can be placed at $N_c - (N_A - 1)$ sites. Because each of these placements is independent, the total number of placements is related to the product of the number for each step.

However, this procedure counts some arrangements multiple times.[‡] For example, we counted the configuration where the first A molecule was placed in lattice site 1 and the second A in site 2 as distinct from the configuration where the first A molecule was placed in site 2, and the second in site 1. Since we cannot tell the A molecules apart, we cannot define an "order" (first, second, ...) in which they were placed. You should convince yourself that each placement of pairs of A molecules occurs twice, but should be counted once. Thus, we correct by a factor of two. Then, when we place the third A molecule in with each (statistically corrected) A pair, each distinguishable triple occurs three times, requiring correction by a factor of three. The fourth requires a correction by a further factor of four, and so on. Correcting by the appropriate factor for each of the N_A molecules leads to

$$
\begin{aligned}
W(N_A; N_c) &= \frac{N_c \times (N_c - 1) \times \ldots \times (N_c - N_A + 1)}{1 \times 2 \times \ldots \times N_A} \\
&= \frac{N_c!}{N_A! \times (N_c - N_A) \times (N_c - N_A - 1) \times \ldots 1} \\
&= \frac{N_c!}{N_A! \times (N_c - N_A)!} = \frac{N_c!}{N_A! \times N_B!}
\end{aligned}
\tag{4.1}
$$

[†] The effects of relaxing the first rule (excluded volume) are examined in Problem 4.2. The effects of relaxing the second rule (interaction between molecules in adjacent lattice cells) are examined in Chapter 7.

[‡] In the same way that we over-counted when we analyzed coin tosses and dice throws in Chapter 1.

The notation $W(N_A; N_c)$ on the left-hand side of Equation 4.1 reminds us that so far, we have only explicitly worried about distribution of the N_A molecules over the N_c sites. However, it turns out that including the B molecules in our calculation does not change the results. This is because for each arrangement of A molecules, there is only one way to arrange the B molecules over remaining $N_B = N_C - N_A$ sites. Thus, the total number of ways to arrange both the A and B molecules is given by Equation 4.1, that is, $W(N_A, N_B; N_c) = W(N_A; N_c)$. This expression is simply the binomial coefficient, the same statistical expression that gives the number of combination of N_H heads and N_T tails when N_{tot} coins are tossed (Chapter 1). Though we could have argued for binomial statistics and skip the analysis above, it is good to spend some effort thinking about the specific problem at hand, rather than just a particular distribution.

In contrast, there is only one way to arrange the A and B molecules prior to mixing (Figure 4.2A). The A molecules are restricted to N_A cells, and the binomial coefficient for placing N_A molecules in N_A cells ($W(N_A; N_A)$) is one (Problem 4.1). The same is true for placement of N_B molecules in N_B cells. Thus, mixing *greatly* increases the number of ways the A and B molecules can be arranged. Even for the small system shown in Figure 4.2, $W(N_A, N_B; N_c) = 32!/((20!)(12!)) \approx 2 \times 10^8$. Since each *arrangement* is equally probable (they all have the same energies), the likelihood the system will repartition to the starting (unmixed) state is less than one in 100 million.

It should be emphasized that the configuration shown in Figure 4.2B is just one of the many mixed configurations. On its own, this particular configuration is no more likely to occur than the unmixed configuration—both have multiplicities of one.[†] A system like the one in Figure 4.2 seems to mix because that there are *many* configurations that appear to be well mixed, and comparatively few that appear to be unmixed.

Spontaneous change and the dissipation of temperature differences

As with the mixing example above, we will use a discrete model to provide a statistical picture of why temperature differences between materials in thermal contact dissipate spontaneously (Figure 4.1B). Here, rather than discretizing space, we will discretize energy, using a model that has discrete molecular energy levels.[‡] To make things simple, we will assume these energy levels are evenly spaced in a "ladder," having uniform energy differences between adjacent states (**Figure 4.3**). We will use total energy as a proxy for the temperature of each subsystem (as is the case for an ideal gas, where $U = C_V T$; Chapter 3).

Figure 4.3 shows a discrete energy model with two separate subsystems, each containing four molecules and four energy levels (ε).[§] The two subsystems are insulated from the surroundings, so that together, they behave as an isolated system. The number of molecules in different energy levels (N_i) is indicated by placing circles in different rungs of the energy ladder. The sum of the energies of the molecules gives the total energy for each subsystem (e.g., E_L). On the left, the subsystems have different total energies ($E_L = 10$ units $> E_R = 4$ units), suggesting different temperatures ($T_L > T_R$). Upon establishing thermal contact (middle), the systems are allowed to exchange energy in the form of heat flow, as long as the total energy remains constant at 14 units (because the combined system is isolated). On the

---

[†] And in fact, the "unmixed" configuration is included in calculation of the "mixed" configurations.

[‡] Meaning that individual molecules can only have specific amounts of energy. This is a common situation when molecules are analyzed using quantum mechanics (e.g., harmonic oscillators), but is also useful for describing molecular interactions using classical models and in describing chemical reactions.

[§] Note that for some systems, such as electrons filling the orbitals of an atom, quantum mechanical factors (like Pauli exclusion) can influence the occupancies of energy levels. Here we will concern ourselves with collections of distinct molecules, where we can ignore quantum statistics.

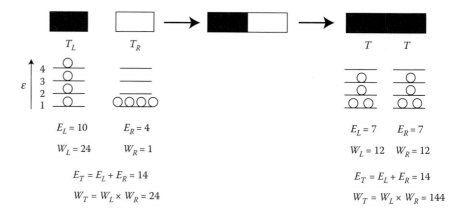

Figure 4.3 The statistics of heat flow in an energy ladder model. Two bodies at different temperatures (left) are composed of four indistinguishable particles that occupy discrete, evenly spaced energy levels. The higher temperature of the hot body (dark red) means greater energy (E_L, left panel) than for the cold body (white). Bringing the two bodies together (middle) allows transfer of energy; assuming the two bodies are isolated from the surroundings, the total combined energy is constant (at 14 units). The number of ways a particular configuration of particles can be achieved is given by the multinomial coefficient, Equation 1.23. Because the particle arrangements within each of the two bodies are independent, the total number of unique arrangements is given as the product of that for each body ($W_T = W_L*W_R$). Accordingly, there are more ways that the thermally contacting system can be arranged in the configuration on the right ($W_T = 144$), where the temperature is the same for both bodies, than for the configuration in the left ($W_T = 24$), where there is a temperature difference. This calculation is consistent with spontaneous decay of temperature differences. Note that there are other possible configurations consistent with the total energy of 14 units (we will return to this point later in this chapter, see Figure 4.17).

right, the two subsystems are drawn with equal energies (7 units each, conserving total energy at 14 units), suggesting equal temperatures (in between T_L and T_R).

Starting with an assumption that each configuration is equally likely, the direction the combined system goes spontaneously (either dissipating or creating an energy difference between the subsystems) depends on the number of configurations that are consistent with different energy (loosely, temperature) distributions. For *each subsystem*, the number of ways of achieving a specific configuration is equivalent to the number of ways we can place the four different molecules into the different energy levels, where we cannot tell different molecules apart when they are in the same energy level, but we *can* when they are in different energy levels. This problem is equivalent to the number of ways four-sided dice[†] (corresponding to the number of molecules) can be thrown to give a particular composition of ones, twos, threes, and fours (corresponding to the four distinct energy levels). For this type of problem, W is given by multinomial coefficients for each configuration (Equation 1.23). W is greatest when all molecules are in different levels (the $E_L = 10$ distribution of molecules on the left side of Figure 4.3). W is lowest when all molecules are in the same level ($E_R = 4$; equivalent to rolling all ones with a set of dice). Other temperature distributions (not shown, but see Figure 4.17) have intermediate values of W.

The values of W for each subsystem determine the direction of spontaneous change through their combined probabilities. Since the two subsystems are independent with respect to particle configuration (so long as the total energy adds up to 14 units), we can get the total number of ways for the combined system (W_T) by multiplying the W value for each subsystem. In the example in Figure 4.3, W_T is maximized when both systems have the same energy, consistent with our everyday experience that temperature differences dissipate spontaneously.

---

† Four-sided dice are regular tetrahedra, as will be familiar to old gamers.

THE DIRECTIONALITY OF HEAT FLOW AT THE MACROSCOPIC (CLASSICAL) LEVEL

In the thermal equilibration model (Figure 4.1B), there are more configurations when the energy (and thus, temperature) is evenly distributed than when it is concentrated in one subsystem. This is consistent with the familiar experience that heat spontaneously flows from hot to cold. In this section, we return to classical thermodynamics to generalize this statement in a quantitative, model-independent way. This generalization is arrived at through the analysis of a special type of heat engine called the Carnot engine.[†]

Though this idealized heat engine seems unrelated to biological and chemical reactions, it provides an elegant route to an analytical form of the second law, and it produces a quantitative definition of the entropy in terms of classical thermodynamic variables. Entropy is *the key quantity* for analyzing chemical and biological equilibria. Our analysis of Carnot engines will lead to entropy-based rules for determining the direction of spontaneous change. All that is required is a few empirical observations relating heat, work, and temperature. Because of their centrality, these observations are often given as definitions of the second law, in the form of the "Clausius" and "Kelvin" statements.

The Clausius and Kelvin statements of the second law

Like the first law, the second law can be expressed in words, as well as in analytical formulas. The Clausius and Kelvin statements give the direction of spontaneous change, either through heat flow or through work done by heat engines. Here we will start with the words of Clausius and Kelvin regarding heat engines, and will derive an expression for entropy from them.

The Clausius statement of the second law is

> No process is possible where the sole result is transfer of heat from a body at lower temperature to a body at higher temperature.

The "Clausius statement," which is consistent with our everyday experiences, says that heat flows from hot to cold, but not from cold to hot. Hot bodies cool; cold bodies warm (Figure 4.1B).

The Kelvin[‡] statement of the second law is

> No process is possible where the sole result is absorption of heat from a reservoir and its complete conversion into work.

Although the Clausius and Kelvin statements seem quite different, it can be shown that if one of the two statements is true, then the other statement must be true. To understand the Kelvin statement and develop a connection to entropy, we need to discuss the properties of heat engines.

Heat engines

An "*engine*" is a device that uses a cycle to convert energy from one form to another, but returns to its original state after the energy conversion process is complete. A heat engine receives heat (q_{in}) from a hot thermal reservoir (at temperature T_h) and

---

[†] The development of thermodynamics and the second law emphasized heat engines because engineers were behind its development. One of their major concerns, which was important during the industrial revolution of the 1800s, was in increasing the efficiency of machines.

[‡] Attributing scientific accomplishments to Kelvin can be confusing, in part because he had two names. He started out as William Thompson, but later in life he was Knighted and then named Baron "Kelvin." He is broadly remembered by the second name, which was given to the absolute temperature scale in his honor. In addition to thermodynamics, Kelvin made significant contributions to the fields of electricity, magnetism, and engineering. For a biography of Kelvin's life, see Lindley (2004).

Figure 4.4 Schematic representation of a heat engine. The engine takes an amount of heat, q_{in}, from a heat source, and converts some of it to work (taken as positive as drawn—see footnote 14). According to the Kelvin statement of the second law, not all of this heat can be converted to work; some must be ejected to a second thermal reservoir as heat (q_{out}, see Problem 4.5). According to the Clausius statement, the temperature of the reservoir supplying the heat to the engine must be greater than the temperature of the reservoir accepting the unconverted heat from the engine (i.e., $T_h > T_c$). The Carnot theorem (below) specifies the maximum amount of work that can be done, based on the temperatures of the two reservoirs.

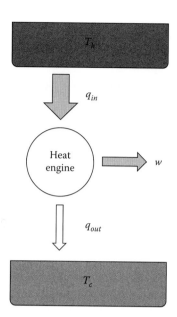

converts some of it into mechanical work $w^{\dagger}$ (**Figure 4.4**). The Kelvin statement says that not all of the received heat can be converted into work, that is, $w < q_{in}$. Since the first law tells us the energy of the unconverted heat ($q_{in} - w$) cannot be destroyed, and since the engine cannot be changed by a cycle, the unconverted heat energy must leave the engine in the form of heat (q_{out}, Figure 4.4).

In analysis of heat engines, it is useful to define the "efficiency," ε. The efficiency is simply the ratio of the work done to the heat supplied, that is,

$$\varepsilon = \frac{w}{q_{in}} = \frac{q_{in} - q_{out}}{q_{in}} = 1 - \frac{q_{out}}{q_{in}} \tag{4.2}$$

Using the first law along with the statements of Clausius and Kelvin, we can obtain some limits on the maximum efficiency that can be achieved. The first law requires the efficiency to be no greater than one. The Kelvin statement of the second law requires the efficiency of less than one.[‡] The Clausius statement requires that the temperature of the reservoir supplying the heat be greater than that of the reservoir receiving the unconverted heat (i.e., $T_h > T_c$).

Together, these restrictions limit the efficiency to $0 \le \varepsilon \le 1$. Between these two limits, the precise value of ε depends on the temperature of the two reservoirs. The numerical value of ε can be obtained through analysis of a special type of heat engine called the "Carnot" heat engine. In addition to providing a value for ε, analysis of the Carnot engine provides a definition of a state function, which we will define as the entropy, which gives a more general definition of the second law.

Carnot heat engines and efficiencies

The Carnot engine is a very simple engine that converts heat into work in four reversible steps (**Figure 4.5**). These steps are (1) a reversible isothermal expansion at a hot temperature T_h, (2) a reversible adiabatic expansion to a cold temperature T_c,

---

[†] For the heat engine, we will take work done *by* the engine to be positive, and will take *both* heat flows (q_{in} and q_{out}) to be positive when they are in the direction depicted in Figure 4.4. Although this differs from our convention in Chapter 3, it matches the treatments typically used to describe engine operations (Equation 4.2).

[‡] In principle, an efficiency of one can be achieved, but this requires a cold reservoir temperature of absolute zero. The third law of thermodynamics states that materials cannot reach absolute zero, thus keeping the Kelvin statement intact.

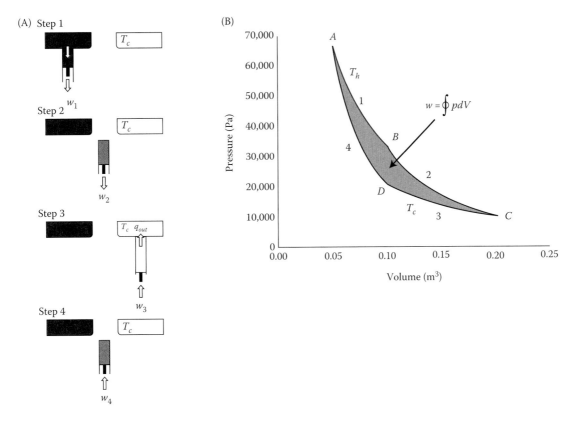

Figure 4.5 The Carnot engine. (A) Schematic, showing the engine as a compressible gas that cycles between two thermal reservoirs at temperatures T_h and T_c in four steps. (B) Pressure–volume relationship along the Carnot cycle. In the first step, the gas expands isothermally at T_h, extracting heat (q_{in}) from the hot reservoir. In the second step the gas expands adiabatically until its temperature falls to T_c. In the third step, the gas is compressed isothermally at T_c, expelling heat into the cold reservoir (q_{out}). In the fourth step, the gas is compressed adiabatically until its temperature rises to T_h. At this point, the engine has returned to its original state. All steps are reversible. The total work output of a single cycle is the area bounded inside the four curves (gray; note that the circle over the integral symbol indicates integration over a closed path).

(3) a reversible isothermal compression at temperature T_c, and (4) a reversible adiabatic compression to temperature T_h. This fourth step completes the cycle, returning the engine to its initial state. Such an engine can be constructed using a piston filled with a gas (Figure 4.5A).[†] For all steps, the piston is in mechanical equilibrium[‡] with an external opposing force that matches the pressure of the gas in the piston, maintaining reversibility.[§]

Because steps 2 and 4 are adiabatic, heat flows only in the isothermal steps 1 and 3 (**Table 4.1**). In contrast, work is performed in all four steps. Work is done *by* the engine in steps 1 and 2. Work is done *on* the engine in steps 3 and 4, to restore the engine to its original state.[¶] Starting with the concept that work is the area under the p–V curve,[**] you should be able to convince yourself that the work output of one

---

[†] The gas does not have to be ideal, but it makes analysis easier.

[‡] And thermal equilibrium for in isothermal steps 1 and 3.

[§] Since pressure in the surroundings must match that of the Carnot engine, such engines cannot be run against a mechanical reservoir (i.e., constant pressure). This is unfortunate, as it means that our real-world engines must run at lower efficiency than the Carnot engine.

[¶] Note that if we only have reversible isothermal and adiabatic steps available to us, we require two steps, one of each kind, to return the engine to its original state. In Figure 4.5B, neither an isotherm nor an adiabat alone gets us from point C back to starting point A. Further, there is only unique pair of steps that gets from C to A. This fact relates to yet another way to state the second law of thermodynamics, which is "For a given state, there are nearby states that are adiabatically inaccessible." This rather abstract second-law statement was made by Constantin Caratheodory, in an attempt to derive the second law axiomatically, without appeal to heat engines.

[**] But the sign of the work depends on which direction the piston moves (expanding or compressing), that is, the sign of dV.

Table 4.1 Heat flow and work in an ideal gas Carnot cycle

Process	Heat flow into engine[a]	Work done by engine[b]
Step 1: Isothermal expansion, T_h	$q_1 = q_{in} = nRT_h \ln V_B/V_A$	$w_1 = nRT_h \ln V_B/V_A$
Step 2: Adiabatic expansion $T_h \rightarrow T_c$	0	$w_2 = -C_v(T_c - T_h)$
Step 3: Isothermal compression, T_c	$q_3 = -q_{out} = nRT_c \ln V_D/V_C$	$w_3 = nRT_c \ln V_D/V_C$
Step 4: Adiabatic compression, $T_c \rightarrow T_h$	0	$w_4 = -C_v(T_h - T_c)$

[a] The heat that enters the engine in step 3 is opposite in sign from q_{out} (the heat *leaving* the engine).

[b] The sign of the work is opposite to our convention for Chapter 2, if the engine is regarded as the system.

cycle of the Carnot engine is the area inside the closed loop on the *p–V* diagram (gray, Figure 4.5B).

If the working material of a Carnot engine is an ideal gas, the heat and the work for each step can easily be calculated (Problem 4.11). Since the heat capacity of an ideal gas is independent of temperature, the work done in the two adiabatic steps is equal in magnitude but of opposite sign (because ΔT reverses), so this part of the work cancels, leaving only the first and third (isothermal) steps to contribute to the total work. In the first step, a significant amount of work is done by the engine, but the work output is reduced by the third step, where work is done on the engine. To maximize the work done per cycle, the compression work in the third step should be minimized. This is achieved by making the adiabat in step 2 (and in step 4) as long as possible, to connect to as low (temperature) an isotherm as possible (**Figure 4.6, top**). If the adiabatic expansion (step 2) could extend to absolute zero,[†] the work on the system that would be required in step 3 would be zero.[‡] Thus, an efficiency of one would be achieved for a cold reservoir at absolute zero. In contrast, when the cold reservoir is at the same temperature as the hot reservoir (corresponding to an adiabat of zero length), isotherms 1 and 3 are identical. In such a situation, no network would be done, leading to an efficiency of zero (**Figure 4.6, bottom**).

The relationship between efficiency and separation between isotherms means that efficiency is maximized when the temperature difference between the two thermal reservoirs is maximized. This observation is consistent[§] with the following expression relating the efficiency of the Carnot engine to temperature:

$$\varepsilon = \frac{T_h - T_c}{T_h} = 1 - \frac{T_c}{T_h} \tag{4.3}$$

Comparing Equations 4.2 and 4.3 shows that for a Carnot engine,

$$\frac{T_c}{T_h} = \frac{q_{out}}{q_{in}} \tag{4.4}$$

[†]　This would require an adiabatic expansion to infinite volume.

[‡]　Another way to understand this is that an ideal gas at zero Kelvin exerts no pressure anywhere along its isotherm, so the reversible compression step would require no work. In addition to violating the third law, which states that we can't get to absolute zero, we note here that the concept of an ideal gas at zero Kelvin is pretty untenable.

[§]　It is consistent with the Equation 4.3, but it does not prove it. For a proof, see Fermi (1936) or Adkins (1983).

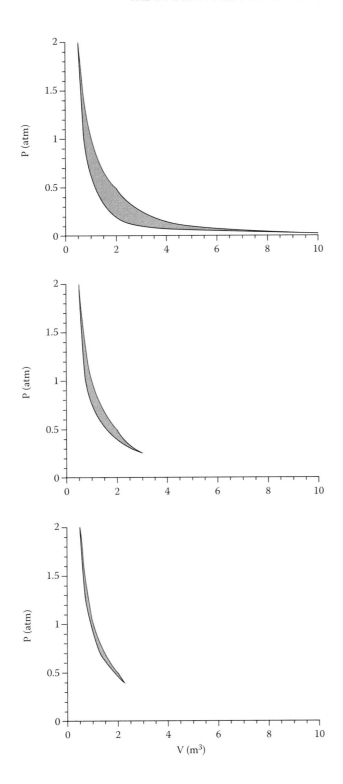

Figure 4.6 Three different Carnot cycles with different efficiencies. All three cycles start at the same place ($p_A = 2$ atm, $V_A = 0.1$ m³) and undergo the same isothermal expansion (to $p_B = 0.5$ atm, $V_B = 2$ m³), absorbing the same amount of heat (q_{in}) from the same hot reservoir at a constant temperature, T_h. The cycles differ in that they inject heat into cold reservoirs at different temperatures. The temperature of the cold reservoir is determined by the length of the adiabatic expansion. In the top cycle, heat is injected into a cold reservoir at very low temperature (resulting in a low-pressure compression isotherm). In the middle cycle, the cold reservoir is at an intermediate temperature. In the bottom cycle the cold reservoir is nearly the same temperature as the hot reservoir. As the compression isotherm approaches the expansion isotherm (i.e., T_c approaches T_h), less work is done (gray area inside each cycle), and efficiency goes to zero.

Rearranging gives the expression

$$\frac{q_{in}}{T_h} = \frac{q_{out}}{T_c} \qquad (4.5)$$

Expressing Equation 4.5 in terms of heat flow entering the system (here the engine),

$$\frac{q_1}{T_h} = -\frac{q_3}{T_c} \qquad (4.6)$$

Rearranging Equation 4.6 gives

$$\frac{q_1}{T_1} + \frac{q_3}{T_3} = 0 \tag{4.7}$$

Remember, the Carnot engine travels in cycles, finishing in the same state that it starts. Thus, as far as the engine is concerned, all functions of state must be unchanged after each cycle. Equation 4.7 suggests that even though q is not a state function, the quantity q/T acts as a state function in the Carnot cycle (as long as changes are made reversibly).

FROM THE CARNOT CYCLE TO MORE GENERAL REVERSIBLE PROCESSES

Although the development above is based on a single Carnot cycle, and thus seems rather specialized (and thus somewhat limiting), two generalizations can be made that expand the usefulness of these results. First, *any* reversible cycle that only exchanges heat at two fixed (but different) temperatures can be regarded as a Carnot engine. The two heat transfer steps must be isotherms (because heat flow is reversible). Since isotherms don't cross,[†] the system must get from one isotherm to the other adiabatically (we asserted that heat flow is restricted to the isothermal processes). Since it can be shown that adiabats are unique (see Problem 4.9), there is only one way to specify a reversible cycle, given two thermal reservoirs, and that is the Carnot cycle.

Second, any reversible cycle that exchanges heat over a range of temperatures can be represented as a series of Carnot cycles, and thus, the statements above about Carnot efficiency apply to the general reversible cycle. This is true even when there are no adiabatic expansion steps. The idea behind this generalization is shown in **Figure 4.7**. For an arbitrary reversible cycle depicted on a p–V diagram, a collection of isotherms and adiabats can be drawn that completely contain the cycle. The general cycle can be approximated using segments from whichever isotherm or adiabat is closest to the arbitrary cycle. Although this approximation has jagged corners as a result of the intersections between the isotherms and adiabats (Figure 4.7A), by using more isotherms and adiabats of closer spacing, this jagged character can be smoothed out (Figure 4.7B, C), giving a very good approximation of the arbitrary cycle.

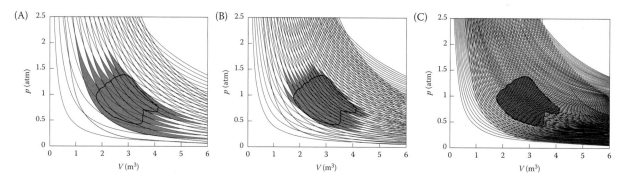

Figure 4.7 Approximating an arbitrary reversible cycle as a set of small Carnot cycles. The thick black line circle an arbitrary reversible cycle on a p–V plane. The thin red and black lines show reversible isotherms and adiabats for an ideal monoatomic gas. The cycle in black can be approximated by the array of Carnot cycles that contain the original cycle (gray shaded diamonds). Although the approximation is not so good when isotherms and adiabats are spread out (A), the approximation improves with increased isotherm and adiabat density (B and C).

[†] If isotherms with different temperatures, T_a and T_b, crossed on the p–V diagram, there would be a point on the diagram (the hypothetical pressure and volume of intersection) that has two temperatures (T_a and T_b), and thus, temperature would not behave as a state function.

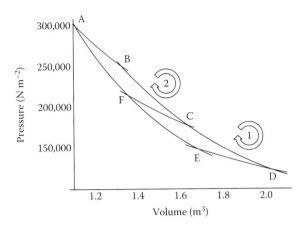

Figure 4.8 Simplification of Carnot cycles with shared isotherms. The isotherms (red) and adiabats (black) define two Carnot cycles. In the clockwise direction, cycle 1 connects points FCDEF; cycle 2 connects points ABCFA. In cycle 1, the shared isotherm is traced in the F to C direction, whereas in cycle 2, it is traced from C to F. Thus, the system (the working material of the engine) accepts an amount of heat $nRT\ln V_C/V_F$ (assuming an ideal gas) on the common isotherm in cycle 1 (a positive heat flow to the system since $V_C > V_F$), and the amount $nRT\ln V_F/V_C$ in cycle 2. Thus, the two heat flows are equal but opposite, so they cancel when both cycles are run. Consistent with this analysis, note that running these two cycles are run together is equivalent to a single large Carnot cycle ABDEA, in which the FC isotherm plays no role.

The connection between the general reversible cycle and the Carnot cycle goes beyond the "tiling" approximation shown in Figure 4.7. Each turn of the arbitrary cycle is equivalent to the sum of a single turn of each of the little Carnot cycles that tile the original cycle, all in the same direction.[†] This is clearly true for the work, which is equivalent to the area inside the cycle: the area bounded by the general cycle is approximated by the sum of the areas of the little Carnot cycles (gray shading, Figure 4.7). In addition, the heat exchange by the collection of little cycles approximates the heat flow in the original cycle. This is easier to see for the "outside" isothermal edges of the Carnot cycle approximation, but the presence of many "inside" isothermal edges would seem to complicate things. However, the inside isotherms don't contribute to heat flow, because they are transited in opposite directions by adjacent Carnot cycles, and thus the heat flows associated with these inside steps cancels (**Figure 4.8**).[‡]

Because the little Carnot cycles are reversible, an expression equivalent to Equation 4.7 applies to each of the little cycles. Since each of these Equation 4.7 equivalents is equal to zero, the sum of these equations is also equal to zero, or

$$\sum_{i=1}^{\#C}\left(\frac{q_h}{T_h}+\frac{q_c}{T_c}\right)_i=0 \qquad (4.8)$$

where the index i refers to the individual Carnot cycles, and $\#C$ is the total number of cycles. Again, due to the cancellation of heat flows from "inside isothermal segments" described above, all of the terms in Equation 4.8 cancel except for the outside terms (which correspond to the isothermal segments on the outside edge of the approximating cycle). Thus, the sum can be written as

$$\sum_{i=1}^{n}\frac{q_i}{T_i}=0 \qquad (4.9)$$

[†] "In the same direction" means that if, on the p–V diagram, the original cycle turns in the clockwise direction, then each little Carnot cycle turns in the clockwise direction.

[‡] Though the same cancellation argument can be applied to inside adiabats, it is not necessary because there is no heat flow.

where n is the number of outside isothermal segments. In words, Equation 4.9 says that for an arbitrary reversible cycle approximated using n reversible isothermal steps, the sum of the ratio of heat exchanged to the temperature at which the heat is exchanged is zero. Because the best approximation to an arbitrary reversible cycle is obtained when the Carnot cycles increase in number and shrink in size, it is natural to make each step infinitesimal, and replace the sum with an integral:

$$\oint \frac{dq}{T} = 0 \qquad (4.10)$$

As in Figure 4.5B, the circle in the integral symbol indicates integration over a closed (and reversible) cycle.

Importantly, Equation 4.10 implies that the quantity dq/T behaves an exact differential (even though dq does not). This means that there is a state function that can be associated with the ratio of the reversible heat flow to the temperature at which it flows, and if the system is reversibly cycled to its original state, this quantity is unchanged. This important state function is defined as the entropy (represented by the letter S), with a differential given by

$$dS \equiv \frac{dq_{rev}}{T} \qquad (4.11)$$

The subscript "*rev*" indicates that the equality holds for *reversible* heat flow. Since heat has units of energy (joules), entropy has dimensions of energy per unit temperature (joule per kelvin). This is why we encounter entropy multiplied by temperature in thermodynamic equations (e.g., TdS, SdT, and $T\Delta S$).

One way to interpret the definition of entropy in Equation 4.11 is that the magnitude of entropy produced by heat injection into a system is modulated by the temperature. Clearly, as long as T is positive-valued, injection of heat increases entropy. However, if the system temperature is high, heat injection increases the entropy by a small amount. At low temperature, the same heat injection produces a large entropy increase. From the mainstream perspective of entropy as a measure of disorder, this inverse temperature relationship would suggest that heat flow into a system generates more disorder at low temperature than at high temperature.

Combining Equations 4.9 and 4.10 gives the entropy change of a reversible cycle:

$$\oint \frac{dq}{T} = \oint dS = \Delta S = 0 \qquad (4.12)$$

As we will show below, if the cycle is not reversible, the left-hand equality (involving dq/T) is invalidated; the right-hand equality is valid, but it must be accompanied by an entropy increase in the surroundings.

ENTROPY CALCULATIONS FOR SOME SIMPLE REVERSIBLE PROCESSES

For noncyclical reversible transformations from an initial state to different final state, the change in entropy of the system is given by

$$\Delta S = S_f - S_i = \int_i^f dS = \int_i^f \frac{dq}{T} \qquad (4.13)$$

Because S is a state function, the path from i to f need not be specified, other than that it must be a reversible one for the right-most equality to hold. As described in

Chapter 3 for internal energy, the path independence expressed in Equation 4.13 allows entropy change for a complicated transition to be determined by selecting a simple path for calculation. Here we will calculate entropy changes for the four simple state changes discussed in Chapter 3 (constant p, constant V, constant T, and adiabatic change; see Figure 3.12).

1. **Entropy change for reversible constant pressure expansion.** As described in Chapter 3, reversible isobaric expansion involves substantial heat flow into the system, in part to keep the pressure of the system matched to the constant pressure of the surroundings. Starting with Equation 4.13, we can calculate the entropy change as

$$\Delta S = S_f - S_i = \int_i^f dS = \int_i^f \frac{dq}{T} = \int_i^f \frac{C_p dT}{T} \tag{4.14}$$

If the heat capacity is independent of temperature, the integral is easily solved:

$$\Delta S = C_p \int_i^f \frac{dT}{T} = C_p \ln\left(\frac{T_f}{T_i}\right) \tag{4.15}$$

For a monatomic ideal gas (where C_p is independent of T), this can be written as

$$\Delta S = \frac{5}{2} nR \ln\left(\frac{T_f}{T_i}\right) \tag{4.16}$$

Figure 4.9A compares the temperature dependence of the entropy change to that of the heat flow. Whereas heat flow increases linearly with T_f for the constant pressure expansion,[†] entropy increases logarithmically. As anticipated above, at low temperature, a unit of heat flow produces a large increase in entropy, whereas at higher temperatures, the same unit of heat flow produces a small increase in entropy.

This logarithmic temperature dependence of the entropy connects to a useful heuristic classical view of the entropy as a measure of the energy in a system that is unavailable to do work (i.e., energy that is not "free").[‡] The thermal energy in a hot system (e.g., a unit of heat flow) is more available for work than in a cold system. In terms of the Carnot cycle (Equation 4.3), efficiency (and thus, work) is maximized when the reservoir delivering the heat is at a high temperature, and the reservoir receiving the wasted heat is at a low temperature. **Figure 4.9B** compares q and ΔS as

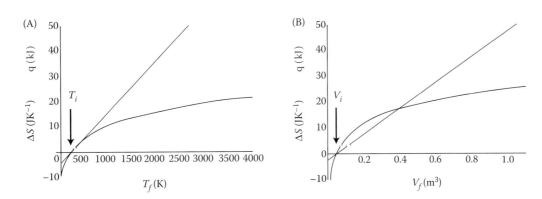

Figure 4.9 Entropy change (black) and heat flow (red; note the different dimensions) for reversible constant pressure expansion of one mole of an ideal monatomic gas as a function of (A) temperature and (B) volume. Both quantities are calculated from an initial state at $T_i = 300$ K, and $V_i = 0.5$ m³. The heat flow scales linearly with temperature and volume, but the entropy scales logarithmically (see Problem 4.13).

[†] Because C_p is independent of T for the ideal gas.
[‡] Similar descriptions are of "isothermally unavailable energy."

a function of volume, which is perhaps a more appropriate coordinate for expansion than temperature. For an ideal gas, the entropy change is (Problem 4.13)

$$\Delta S = C_p \ln\left(\frac{V_f}{V_i}\right) \tag{4.17}$$

For a constant pressure expansion, entropy increases with volume, but the increase is nonlinear (as with temperature, Figure 4.9A). The largest entropy increase per unit volume occurs when a system starts in a highly compressed (i.e., small volume) state.

2. **Entropy change for reversible constant volume heating.** Here, volume is fixed, and reversibility is achieved by raising the temperature of the surroundings slowly, allowing heat to flow into the system to maintain thermal equilibrium. The entropy calculation is nearly the same as for the constant pressure example above, with C_V replacing C_p:

$$\Delta S = \int_i^f \frac{dq}{T} = \int_i^f \frac{C_V dT}{T}$$
$$= C_V \ln\left(\frac{T_f}{T_i}\right) \tag{4.18}$$

The equality on the second line applies when C_V is independent of T. For a monatomic gas,

$$\Delta S = \frac{3}{2} nR \ln\left(\frac{T_f}{T_i}\right) \tag{4.19}$$

Like the constant pressure transformation (Figure 4.9), the constant volume transformation involves linear heat flow and logarithmic entropy change (**Figure 4.10**). However, the proportionality constants are different for constant volume and constant pressure transformations (C_V vs. C_p). A given temperature increase at constant pressure results in a larger entropy increase than at constant volume.

3. **Entropy change for reversible constant temperature expansion of an ideal gas.** In the two preceding examples, the constraints of constant pressure and volume expansion permit the entropy to be calculated through using heat capacities. For constant temperature expansion, there is no simple analogous expression that can be inserted into the integrand of Equation 4.13. Instead, we can obtain an expression for dq by recalling that for an ideal gas, the internal energy is constant in an isothermal expansion. Thus, $dq = -dw$, allowing us to express the entropy change from the work:

$$\Delta S = \int_i^f \frac{dq}{T} = \int_i^f \frac{-dw}{T}$$
$$= \int_i^f \frac{pdV}{T} = \int_i^f \frac{nRTdV}{VT} = nR \ln\left(\frac{V_f}{V_i}\right) \tag{4.20}$$

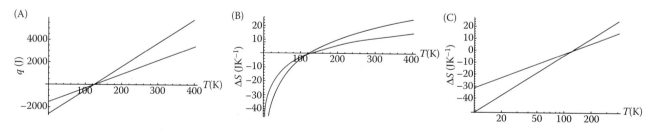

Figure 4.10 Comparison of heat flow and entropy changes in an ideal monoatomic gas in constant pressure expansion (black) and constant volume heating (red). For all curves, one mole of gas at an initial volume of 10 L and a pressure of 1 atm (and thus, by the ideal gas law, $T_i \sim 121.8$ K). (A) The amount of heat that flows in reversibly changing the temperature from T_i to T). (B,C) The entropy change ($\Delta S = S(T) - S_i$)) in reversibly changing the temperature from T_i to T, on a linear (B) and a logarithmic (C) temperature scale.

This indicates that the entropy increases for isothermal expansion (because for expansion, $V_f > V_i$). As with the constant pressure expansion, the volume dependence is logarithmic.

4. **Entropy change for reversible adiabatic expansion of an ideal gas.** As long as adiabatic expansion is reversible, Equation 4.13 requires that there is no entropy change for an adiabatic expansion, since $dq = 0$ along the entire adiabat. As with the "constant volume work calculation" in Chapter 3, you can think of the integral on the right-hand side of Equation 4.13 as summing up a bunch of zeros ($dq/T = 0/T = 0$). Even though this "sum" has an infinite number of terms, each term contributes nothing; thus, the sum remains zero. Because entropy remains constant, reversible adiabatic expansion (and compression) can be regarded as "isentropic."

COMPARISON OF REVERSIBLE AND IRREVERSIBLE CYCLES

The development of entropy above was for reversible transformations. In part this is because (as with first law calculations) analysis of irreversible processes is not easy, since variables such as T and p are not well defined. But the difference between reversibility and irreversibility is more profound than simply determining how difficult the calculations are. Reversible changes maintain constant total entropy (system plus surroundings), whereas irreversible changes increase total entropy.[†] The "reverse" of an irreversible transformation, which would *decrease* the total entropy, does not happen.

To consider the effects of irreversibility on entropy and its relation to heat flow, we return to Carnot cycle efficiency. The reversible Carnot cycle acts as an upper limit on efficiency, as stated in the "***Carnot theorem***":

> No heat engine operating between two temperatures can be more efficient than a reversible Carnot engine operating between the same two temperatures.
>
> —Sadi Carnot.

The phrase "No heat engine" includes irreversible engines.[‡] Expressing the Carnot theorem in terms of efficiencies, making use of Equations 4.2 and 4.3:

$$\varepsilon_C > \varepsilon_{irr}$$
$$1 - \frac{T_C}{T_h} > 1 - \frac{q_{out,irr}}{q_{in,irr}}$$
$$\frac{T_C}{T_h} < \frac{q_{out,irr}}{q_{in,irr}}$$

(4.21)

Note the change in the direction of the inequality ("greater than" to "less than") from line 2 to line 3 that accompanies the change of sign. The subscripts C and *irr* indicate Carnot and irreversible engines. The use of Equation 4.3 for the efficiency of the Carnot engine on the left side is justified because the temperature is well defined throughout the cycle. Although this is not the case for the irreversible engine, the heat flows are still well defined, and can be used on the right-hand side. Rearranging Equation 4.21 gives

$$\frac{q_{in,irr}}{T_h} - \frac{q_{out,irr}}{T_c} < 0$$

(4.22)

[†] Emphasis must be placed on *total* entropy (system plus surroundings). There is nothing preventing a spontaneous (i.e., irreversible) processes that decrease the system's entropy, but it must be accompanied by a larger increase the entropy of the surroundings.

[‡] By "irreversible engine" we mean an engine that transforms through one or more irreversible heat flow and/or work steps, but still returns to its original state after a cycle. Although the entropy of the engine is unchanged if it returned to its original state, the entropy of the surroundings (the reservoirs plus the work repository) must be increased.

Substituting q_{in} and q_{out} by q_1 and $-q_3$, as per our sign convention (Chapter 3) gives

$$\frac{q_{1,irr}}{T_h} + \frac{q_{3,irr}}{T_c} < 0 \qquad (4.23)$$

Compare this equation with the equation for the reversible Carnot cycle (4.7). In the reversible cycle, the sum of these two terms is equal to zero, whereas in the irreversible cycle the sum exceeds zero. These two Equations 4.7 and 4.21 can be combined into a single equation:

$$\frac{q_1}{T_h} + \frac{q_3}{T_c} \leq 0 \qquad (4.24)$$

Though we are not specifying which step (or steps) in the cycle is irreversible, we are assuming that heat is being exchanged with hot and cold *reservoirs*, which define the temperatures in Equation 4.24.

As in Figure 4.7, we can use a collection of little cycles to approximate an irreversible general cycle that exchanges heat over a range of temperatures. In this approximation, we will make all of the inside cycles reversible, so that the edges all cancel. Applying the same reasoning as for the reversible general cycle, for the irreversible general cycle we can write

$$\sum_{i=1}^{n} \frac{q_i}{T_i} < 0 \qquad (4.25)$$

where again, the sum is over the outside isothermal edges of the little Carnot cycles. In the limit that the number of cycles is very large, we can replace the sum in Equation 4.25 with an integral:

$$\oint_{irr} \frac{dq}{T} < 0 \qquad (4.26)$$

Given the inequality in the irreversible path integral, it is worth remembering that Equation 4.26 refers to the *system* (i.e., the engine, not the reservoirs and work repository). From the reversible cycle (Equation 4.10), we identified the quantity dq/T as a differential of a state function, and named this state function the "entropy" (Equation 4.11). Regardless of whether or not the cycle was reversible, if the engine is returned to its original state as a result of the cycle, all the state functions for the engine must remain unchanged by the cycle. This includes the entropy, even for the irreversible. The inequality in Equation 4.26 should be attributed to the left-hand side, that is, to the dq/T terms.

A GENERAL FORM FOR IRREVERSIBLE ENTROPY CHANGES

Since entropy is a state function, the change in entropy when a system changes from an initial to a final state is the same regardless of whether transformation was reversible or irreversible. For reversible change, the system entropy change is obtained by integrating dq/T (Equation 4.13). Combining with the irreversible entropy change gives

$$\Delta S_{irr} = \Delta S_{rev} = \int_{i}^{f} \frac{dq_{rev}}{T} \qquad (4.27)$$

Comparing to the irreversible heat flow (the noncyclic version of Equation 4.25), we introduce an inequality to the above equation:

$$\int\limits_{\substack{i \\ irr}}^{d} \frac{dq}{T} < \Delta S_{irr} = \Delta S_{rev} = \int\limits_{i}^{f} \frac{dq_{rev}}{T} \qquad (4.28)$$

On the left-hand side of Equation 4.28, the irreversible label is associated with the integral in recognition of the fact that only part of the transformation need be irreversible. For the right-hand side, heat transfer must be reversible along the entire pathway.

Equation 4.28 and its differential analog are often condensed to give two closely related relations that are often regarded as the second law in equation form:

$$\Delta S \geq \int\limits_{i}^{f} \frac{dq}{T} \qquad (4.29)$$

and

$$dS \geq \frac{dq}{T} \qquad (4.30)$$

In Equation 4.29 and 4.30, the equality applies to reversible transformations, and the inequality applies to irreversible transformations.[†] These two equations are often referred to as the "Clausius inequalities." These inequalities are particularly simple when there is no heat flow:

$$\Delta S \geq 0 \qquad (4.31)$$

and

$$dS \geq 0 \qquad (4.32)$$

In addition to describing the entropy change of a system undergoing adiabatic expansion, Equation 4.31 has important applications to isolated systems. For example, if we view a heat engine and its reservoirs to be an isolated system, Equation 4.31 says that reversible processes (including cycles) do not change entropy, whereas irreversible processes (and cycles) always increase the entropy. This applies to any type of isolated system, no matter how complex, *including the entire universe.* These concepts give rise to another statement of the second law:

> The entropy of an isolated system (the universe) increases during any spontaneous process.

Here, "spontaneous" is synonymous with irreversible. Since all processes that occur in isolation (i.e., without an external driving force) must be spontaneous, this means that all natural processes increase entropy. Although reversible processes do not increase entropy, they do not decrease it. Rather, the second law states that process that decrease entropy do not occur in isolation.[‡]

ENTROPY AS A THERMODYNAMIC POTENTIAL

A key feature of the inequality developed in the previous section is that it makes the entropy into a thermodynamic potential. By potential, we mean that when an

---

[†] For reversible heat flow, there is no temperature gradient, and thus, T is well-defined throughout the system and surroundings. However, for irreversible flow, the system and surroundings have to be carefully manipulated to generate T values that can be used in calculation of dS.

[‡] These would be the reverse of irreversible processes, which can't happen if the process is irreversible.

Figure 4.11 The entropy potential in an isolated system. The direction of spontaneous (irreversible) change is that which increases S. At $\xi = a$, spontaneous change is to the right (toward increasing values of ξ, arrow). At $\xi = c$, spontaneous change is to the left (toward decreasing values of ξ). At $\xi = b$, S is maximized, and thus the isolated system remains at this coordinate. This maximum defines the equilibrium position for the system.

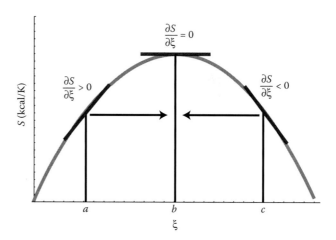

isolated system is away from equilibrium, entropy identifies the direction of spontaneous change and provides a driving force for change. Moreover, as a potential, entropy locates the position of equilibrium. Identifying the position of equilibrium is extremely important in analyzing the thermodynamics of chemical reactions.

Though readers may be unfamiliar with the concept of a thermodynamic potential, everyone is familiar with potentials associated with simple mechanical systems with just a few degrees of freedom. For a ball on a hill, gravity provides a potential, determining which direction the ball will move (downhill), and where it will come to rest (an equilibrium position, at the bottom of the hill). Both pieces of information can be quantified by taking derivatives, and for the equilibrium position, using the derivative to find a minimum.

Potential energy is also important on the molecular scale for determining favorable bonding patterns, and these interactions clearly influence bulk equilibrium properties of a system of molecules. However, translating all this molecular information into a bulk potential is difficult for simple systems, and is impossible for complicated systems. The beauty of the classical entropy is that it ignores all the molecular detail. It acts as a potential for the bulk properties of a system, implicitly averaging over these important (but often inaccessible) molecular details.

The entropy acts as a potential because of Equation 4.29, the Clausius inequality for the isolated system. Equilibrium is achieved when the isolated system is in a reversible configuration, and at that configuration the Clausius relationship acts as an equality (i.e., $dS = 0$), and entropy is at a maximum.[†] To help visualize this, it helps to think about some sort of internal configuration variable for the isolated system, and evaluate how entropy changes as this variable is changed. We will call his variable ξ (the greek lower case Xi, pronounced "*ksi*"). ξ could describe volume distributions within parts of the system, distribution of materials among phases, extent of mixing, or importantly, position of a chemical reaction. For such a system, S can be plotted as a function of ξ (**Figure 4.11**). When the configuration is away from the entropy maximum, the system moves toward the maximum, either by increasing or by decreasing ξ (at points a and c, respectively). When S is maximum (point b), the system remains in the same configuration (i.e., at equilibrium), just like a ball that is free to roll stays at rest in a valley.

The potential nature of entropy can also be expressed using differentials. Entropy is a state function, and if it depends on our new variable ξ (as we have drawn in Figure 4.11), we can write $S = S(\xi)$. The differential of entropy will be exact, and can be written as

---

[†] Unlike familiar potentials like gravitation and electrical potential where the system moves spontaneously down the potential, an isolated the system moves *up* the entropy potential. Though we could multiply S by -1 to get a potential that minimizes, this is not the convention.

Table 4.2 How the entropy gradient sets the direction of spontaneous change and the position of equilibrium

Value of ξ	$d\xi$ (by assertion)	$\left(\dfrac{\partial S}{\partial \xi}\right)$	$dS = \left(\dfrac{\partial S}{\partial \xi}\right) d\xi$	Direction of spontaneous change
a	$+$	$+$	$+ (dS = + \times +)$	Increasing ξ (right)
b	$+$	0	$0 (dS = 0 \times +)$	None
c	$+$	$-$	$- (dS = - \times +)$	Decreasing ξ (left)

Note: Here, $+$ and $-$ indicate changes that are greater than and less than zero, respectively. The sign of the partial derivative is taken from Figure 4.11. The sign of dS is determined by that of the product of the derivative and the variation $d\xi$, taken by assertion here to be positive. At point b, the absence of a direction for spontaneous change is consistent with an equilibrium state.

$$dS = \left(\frac{\partial S}{\partial \xi}\right) d\xi \qquad (4.33)$$

(see Chapter 2).[†] Using Equation 4.33, we can evaluate the entropy variation (dS) that result from variation in the internal coordinate ($d\xi$). Depending on the sign of $d\xi$ (and thus the direction of movement), the sign of the partial derivative determines whether S will increase, decrease, or remain unchanged (**Table 4.2**). At point a, increasing ξ increases S (and is thus the direction of spontaneous change), because $dS/d\xi$ is positive. Conversely, at point c, increasing ξ decreases S because $dS/d\xi$ is *negative*. Thus, spontaneous change at point c is brought about by *decreasing* ξ ($d\xi < 0$, i.e., going *backward*). At point b, increasing ξ neither increases nor decreases S because $dS/d\xi$ is zero (S is at a maximum), marking point b as the equilibrium point.

ENTROPY CALCULATIONS FOR SOME IRREVERSIBLE PROCESSES

In general, it is difficult to calculate entropy changes for irreversible processes, because the Clausius relationship becomes an inequality, and because intensive variables (T, p, and density) are not uniform, and thus cannot be defined by a single number. Nonetheless, it is possible to construct some systems for which irreversible entropy calculations for the system and surroundings can be made. These calculations take advantage of the fact that entropy is a state function; thus, for a given change, ΔS is the same regardless of whether the change was reversible or not. Here we give two related examples of irreversible heat flow, followed by two examples of irreversible expansion. Although some of examples may seem rather contrived, their analysis provides a concrete example of how entropy increases for spontaneous changes.

Heat flow between two identical bodies at different temperatures

One of the simplest irreversible processes for which entropy change can be analyzed is the flow of heat between two bodies initially at a different temperature (A and B at T_A and T_B, **Figure 4.12**). To simplify the calculation, we will assume the two bodies are otherwise the same, and their heat capacities are independent of

---

[†] If multiple internal coordinates can be varied independently, additional terms will appear on the right-hand side of Equation 4.33 that correspond to variation from each coordinate. In this situation, the driving force and direction of equilibrium would be given by the gradient of the entropy, and the equilibrium position would be the coordinate where $\nabla S = 0$.

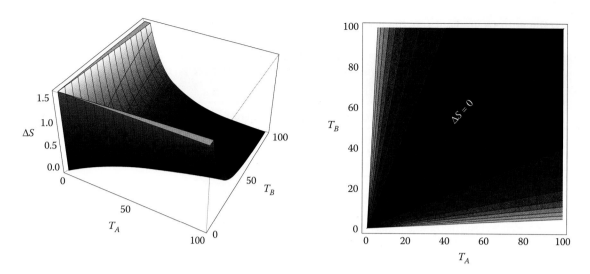

Figure 4.12 Entropy change for irreversible heat flow between two bodies at different starting temperatures T_A and T_B. The heat capacities of the two bodies are taken to be equal and independent of temperature; here, a value $C_V = 1$ was used for simplicity. Left, a three-dimensional plot showing that ΔS becomes large when the difference between T_A and T_B is large. When $T_A = T_B$ (the equilibrium state), there is no entropy change; this corresponds to the diagonal on the contour plot (right, contours increase by steps of 0.2 from dark to light).

temperature. Based on the common experience, such bodies come to a common temperature T_f (see Figure 2.1). From our assumptions of heat capacities above, it is straightforward to show that

$$T_f = T_A + \frac{(T_B - T_A)}{2} = \frac{T_B + T_A}{2} \tag{4.34}$$

Although the heat flow is irreversible, the fact that entropy is a state function means that the irreversible entropy change is the same as that for a reversible transformation between the initial temperatures T_A and T_B and the final temperature, T_f. Assuming volume is constant, the entropy change for reversible heat flow to (or from) body A can be calculated using Equation 4.18:

$$\begin{aligned} \Delta S_A &= C_V \ln\left(\frac{T_f}{T_A}\right) \\ &= C_V \ln\left(\frac{T_A + T_B}{2T_A}\right) \end{aligned} \tag{4.35}$$

The expression for ΔS_B has the same form as Equation 4.35, but with T_B in the denominator instead of T_A. The total entropy change is

$$\begin{aligned} \Delta S_{tot} &= \Delta S_A + \Delta S_B \\ &= C_V \left\{ \ln\left(\frac{T_A + T_B}{2T_A}\right) + \ln\left(\frac{T_A + T_B}{2T_B}\right) \right\} \\ &= C_V \left\{ \ln\left(\frac{\{T_A + T_B\}^2}{4T_A T_B}\right) \right\} \end{aligned} \tag{4.36}$$

When $T_A = T_B$, $\{T_A + T_B\}^2 = 4T_A T_B$, and thus $\Delta S = 0$. This corresponds to the condition required for reversibility, that is, the equilibrium condition. When $T_A \neq T_B$, $\{T_A + T_B\}^2 > 4T_A T_B$ (regardless of whether T_A is greater than T_B, or vice versa), and thus, $\Delta S_{tot} > 0$. In other words, heat flows spontaneously from the hot body to the cold body, and ceases when the two temperatures are equal.

Spontaneous development of a temperature gradient on a large scale would result in an entropy decrease. Note that because entropy is a state function, the total entropy change given in Equation 4.36 also applies when heat flow is irreversible (which it would be, for if the two subsystems start at different temperatures).

Heat flow between a small body and a thermal reservoir

As discussed in Chapter 3, we often analyze systems that are small compared to their surroundings. In such cases, reversible heat flow requires careful adjustment of the temperature of the surroundings to match the system. For most processes in nature, gradual adjustment of the temperature of the surroundings does not occur. Rather, the surroundings for many natural processes more closely approximates a thermal reservoir, remaining at a fixed temperature. If the surroundings equilibrates fast compared to the system, the entropy change for the reservoir is easy to calculate and combine with that of the system.

As in the previous example, we will consider a system that is held at constant volume, with a temperature-independent heat capacity, C_V. We will calculate the entropy change associated with a heat flow that results in a temperature change from an initial temperature T_i to a final temperature equal to that of the surroundings, T_{surr}. From Equation 4.18, $\Delta S_{sys} = C_V \ln(T_{surr}/T_i)$. Again, this equality holds regardless of whether the heat flow was reversible or not, because entropy is a state function. For the surroundings, the entropy change is

$$\Delta S_{surr} = \int \frac{dq_{surr}}{T_{surr}} = \frac{1}{T_{surr}} \int dq_{surr} = \frac{q_{surr}}{T_{surr}}$$
$$= \frac{-q_{sys}}{T_{surr}} = \frac{C_V(T_i - T_{surr})}{T_{surr}} \tag{4.37}$$

In the first line, T_{surr} is taken outside the integral because it is constant. In the second line, we invoke the fact that heat flow into the surroundings is equal but opposite that into the system. The entropy change of the system is the same as in the previous example (Equation 4.34), and can be combined with Equation 4.36 to give the total entropy change:

$$\Delta S_{tot} = \Delta S_{sys} + \Delta S_{surr}$$
$$= C_V \left\{ \ln\left(\frac{T_{surr}}{T_i}\right) - \frac{(T_{surr} - T_i)}{T_{surr}} \right\} \tag{4.38}$$

This entropy change is shown as a function of T_i and T_{surr} in **Figure 4.13**. As with entropy flow between two blocks, when T_i and T_{surr} are equal, $\Delta S_{tot} = 0$ (Equation 4.38). When T_i and T_{surr} are different (in either direction), the total entropy change increases (Figure 4.13, Problem 4.16).

Although the entropy surface for heat flow between two identical blocks looks quite similar to that for heat flow into the reservoir (compare Figures 4.11 and 4.12), the asymmetry between the small system and infinite thermal reservoir results in an asymmetric total entropy change. The entropy change is greater when heat is transferred from a hot system to a cold reservoir ($T_i \gg T_{surr}$) than from a hot reservoir to a cold system ($T_i \ll T_{surr}$). This is because when $T_i \gg T_{surr}$, all of the transferred energy distributes over a low temperature surroundings, where it produces a large entropy increase (remember, $dS = dq/T$). When $T_i \ll T_{surr}$, the transferred energy is removed from high temperature surroundings, where it produces a small entropy decrease.

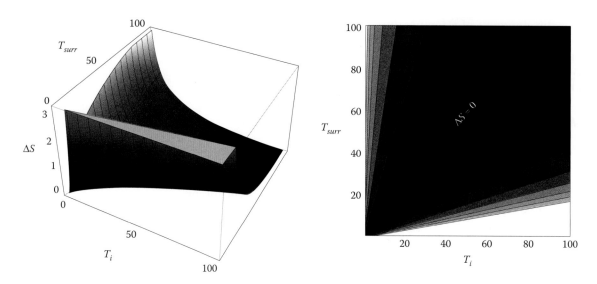

Figure 4.13 Entropy change for irreversible heat flow between a body at initial temperature T_i and a thermal reservoir at T_{surr}. The heat capacity of the system is set to 1 for simplicity. The heat capacity of the surroundings is taken to be infinite. Left, the entropy surface as a function of the initial temperature of the system (T_i) and the temperature of the thermal reservoir. When $T_i = T_{sys}$, the entropy change is zero (right, contour plot). For all other values of T_i and T_{sys}, the total entropy change is positive, reflecting the spontaneous nature of heat flow over a finite temperature difference.

ENTROPY CHANGES FOR IRREVERSIBLE EXPANSIONS

In Chapter 3, we discussed some irreversible gas expansions, invoking a mechanical reservoir as surroundings, with $p_{sys} < p_{surr}$. Since these expansions are spontaneous and irreversible, we should expect entropy to increase.

Entropy change for irreversible expansion to mechanical equilibrium with a constant pressure surroundings

To calculate the entropy change in this adiabatic expansion to equilibrium against a constant pressure surroundings (Figure 3.16A), we will evaluate the change for the system and surroundings separately. Because the walls are adiabatic, there is no heat flow. It is tempting to conclude from the Clausius relationship (Equation 4.29) that because $q = 0$, there is no entropy change in the system. However, the system is clearly out of equilibrium through the expansion so we cannot use the Clausius equality to calculate entropy. Instead, we are forced to use the inequality, which only gives us a limiting value:

$$\Delta S_{sys} = \int_{S1}^{Sf} dS > \int_{1}^{f} \frac{dq}{T} = 0 \tag{4.39}$$

or

$$\Delta S_{sys} > 0 \tag{4.40}$$

Although this may be a comforting result, we already knew as much. To obtain a numerical estimate of ΔS_{sys} we can take advantage of the fact that entropy is a state function, and calculate ΔS_{sys} along a reversible path from state 1 to state f. One simple path for this is a reversible isothermal expansion from T_1, V_1 to T_1, V_f (call it "step A"), followed by a constant volume cooling from T_1 to T_f ("step B"). The entropy change along the isotherm, step A, can be calculated from Equation 4.20, using Equation 3.60 to obtain a value for V_f:

$$\Delta S_A = \int_1^f \frac{pdV}{T} = nR \ln\left(\frac{V_f}{V_1}\right)$$

$$= nR \ln\left\{\frac{\dfrac{V_1}{5}\left(\dfrac{3p_1}{p_{surr}}+2\right)}{V_1}\right\} \tag{4.41}$$

$$= nR \ln\left(\frac{3p_1}{5p_{surr}}+\frac{2}{5}\right)$$

The entropy in the second step of the reversible transformation, "step B," comes from constant volume heat flow (Equation 4.18):

$$\Delta S_B = C_V \ln\left(\frac{T_f}{T_1}\right)$$

$$= \frac{3}{2}nR \ln\left(\frac{T_f}{T_1}\right) \tag{4.42}$$

To combine this with the entropy change for step A, it is helpful to express T in terms of pressure and volume, again using Equation 3.60:

$$\Delta S_B = \frac{3}{2}nR \ln\left(\frac{p_f V_f/nR}{p_1 V_1/nR}\right)$$

$$= \frac{3}{2}nR \ln\left(\frac{p_{surr} V_f}{p_1 V_1}\right)$$

$$= \frac{3}{2}nR \ln\left(\frac{p_{surr}\times\dfrac{V_1}{5}\left(\dfrac{3p_1}{p_{surr}}+2\right)}{p_1 V_1}\right) \tag{4.43}$$

$$= \frac{3}{2}nR \ln\left(\frac{3}{5}+\frac{2p_{surr}}{5p_1}\right)$$

Combining with the entropy change from step A gives the total system entropy change for the reversible (and the irreversible) transformation:

$$\Delta S_{sys} = \Delta S_A + \Delta S_B$$

$$= nR\left[\ln\left(\frac{3p_1}{5p_{surr}}+\frac{2}{5}\right)+\frac{3}{2}\ln\left(\frac{3}{5}+\frac{2p_{surr}}{5p_1}\right)\right] \tag{4.44}$$

This entropy change is shown in **Figure 4.14**.

To calculate the total entropy change (ΔS_{tot}), which determines the spontaneity of the expansion, we also need to determine the entropy change for the surroundings. For the reversible adiabatic expansion, there is no heat flow, and thus no entropy change in the surroundings, due to Clausius' equality (Equation 4.29). For the irreversible adiabatic expansion, we are treating the surroundings as rapidly equilibrating throughout the expansion.[†] Since no heat flows into the surroundings from the

---

[†] This equilibrium might be referred to as "external"" equilibration to remind us that the system is *not* equilibrated during the expansion.

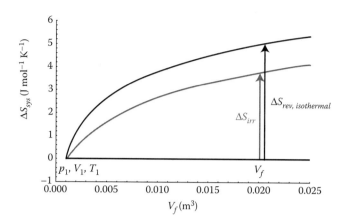

Figure 4.14 System entropy changes associated with irreversible adiabatic expansion of an ideal gas. 0.2 mol of an ideal monatomic gas is expanded adiabatically and irreversibly from $V_1 = 0.001$ m³, $T_1 = 300$ K (the same initial conditions as in Figures 4.11 and 4.12) against a constant external pressure, p_{surr}, to an equilibrium state where $p_{sys} = p_{surr}$ (gray curve, given by Equation 4.44). The red curve and black horizontal line correspond to reversible isothermal and reversible adiabatic expansions curves starting from the same initial state. The gray curve is not meant to depict the entropy variation through a single expansion, but rather entropy changes for different expansions against different constant p_{surr} values (and thus, different final volume values). For $V_f = 0.02$ m³, entropy changes are shown for irreversible (gray) and reversible isothermal expansions (red). For the adiabatic expansions (both reversible and irreversible), because the surroundings are assumed to act as a mechanical reservoir, there is zero entropy change for the surroundings. Thus, for both adiabatic expansions, $\Delta S_{sys} = \Delta S_{tot}$.

system owing to the adiabatic boundary, the entropy change for the surroundings is zero for the irreversible expansion. Thus, the system entropy increase depicted in Figure 4.14 is also equal to ΔS_{tot}. Note that for the reversible isothermal expansion, this is not the case, since heat flows from the surroundings into the system.

The entropy change for this irreversible adiabatic expansion is compared to reversible expansions in Figure 4.16. Combined with the entropy change of the surroundings, these changes reflect the degree of spontaneity of the expansion. Clearly, the entropy increase ($\Delta S_{tot} > 0$) for the irreversible expansion is consistent with the assertion that this process is spontaneous. In contrast, the reversible adiabatic expansion is not driven by entropy ($\Delta S_{tot} = 0$), and can be considered to be in a thermodynamically neutral. This reflects the fact that a reversible transformation is in equilibrium throughout the process. In total, reversible changes occur at constant entropy. By this reasoning, we would expect the total entropy to remain constant for the reversible isothermal expansion as well. Because the entropy of the system increased in the expansion (in fact, by *more* than for the irreversible expansion), the entropy of the surroundings should decrease by the same amount. This can be demonstrated quantitatively by integrating Clausius' equality, recognizing that $dq_{surr} = -dq_{sys}$ (Problem 4.19). Second, Figure 4.16 illustrates that for irreversible adiabatic expansion, the more the system expands, the greater the entropy increase. Although the volume variation shown in the curve in Figure 4.16 involves different external pressures; this the same type of entropy change is seen for expansions at the same external pressure to different stopping volumes (Problem 4.20).

Entropy change for an irreversible expansion against vacuum

It is instructive to evaluate the entropy change for adiabatic expansion against vacuum, and to compare results to isothermal and irreversible adiabatic expansions. Our intuition tells us that an expansion against vacuum is spontaneous, and thus, should produce an overall entropy increase ($\Delta S_{tot} > 0$). Although the reverse of expansion against vacuum (which would be equivalent to a spontaneous contraction of the material within the system to occupy only one region of space) may happen for molecules with strong attractive interactions (e.g., condensation of solutions), for an ideal gas such a contraction would be so highly improbable that we can safely assume it never happens.

To simplify analysis, we will stipulate that expansion is stopped at V_f, and that the boundary is adiabatic (although adiabatic walls are not strictly necessary, see Problem 4.23). When we evaluated expansion work in Chapter 3, we argued that because the internal energy of the ideal gas is unchanged in vacuum expansion (since both work and heat flow are zero), the expanded system comes to rest on its starting isotherm, with pressure determined by the stopping volume, V_f. Thus, the entropy change of the system is the same as for reversible isothermal expansion (Equation 4.20), because entropy is a state function. This entropy increase is shown in red in Figure 4.14. However, unlike reversible isothermal expansion, there is no entropy change for the surroundings to offset this increase.[†] Thus, the total entropy change is given by

$$\Delta S_{tot} = nR \ln\left(\frac{V_f}{V_i}\right) \tag{4.45}$$

This result is consistent with our intuition: entropy increases as a result of expansion. This result also demonstrates that although the Clausius relationship connects heat flow and entropy change, heat does not have to flow for entropy to change.

ENTROPY AND MOLECULAR STATISTICS

Our development from classical thermodynamics of the potential for spontaneous change (i.e., the entropy) is agnostic to molecular details such as multiplicity. Although a rigorous connection between the multiplicities of molecular distributions and entropy requires the tools of statistical thermodynamics (Chapters 9–11), considerable insight on the entropy–multiplicity relationship can be obtained by applying a few simple equations to simple molecular models, like those described in the beginning of this chapter (Figures 4.2 and 4.3). The derivations in the remainder of this chapter use discrete models to provide insight into how entropy relates to the number of states available.[‡]

Statistical entropy and multiplicity

The central formula for calculating entropy from statistics is the "Boltzmann entropy formula":

$$S = k \ln W \tag{4.46}$$

where W is the number of configurations. The constant k is Boltzmann's constant, $1.38 \ldots \times 10^{-23}$ J K^{-1}.[§] Boltzmann's constant can be thought of as a gas constant for a single molecule (rather than for a mole of molecules), that is, $k = R/N_A$.[¶] As with the Clausius formulation of entropy (Equation 4.29), the statistical entropy in Equation 4.46 is an extensive quantity.

[†] This can be appreciated by recognizing that for adiabatic expansion against vacuum, there is no surroundings. The system expanding against vacuum can be regarded as an isolated system.

[‡] The extent to which a system distributes among different microstates is often loosely connected to the extent of disorder (Callen, 1985), or relatedly, to the information content (or more precisely, the information that is "missing" from a system). A relationship between statistical entropy and information was developed by Claude Shannon around 75 years ago (Shannon, 1948) elegantly described; for a lucid description, see Ben-Naim, 2008.

[§] Equation 4.45 remains permanently associated with Ludwig Boltzmann, in part by inscription on his tomb in Vienna, Austria.

[¶] L_0 is Avogado's number, that is, $6.02 \ldots \times 10^{23}$mol^{-2}.

Equation 4.46 can either be applied to subsets of microstates available in a model, or it can be applied to all the available microstates. Application to subsets of microstates is useful for calculating the entropy change associated with a change in some sort of model variable, for example, the heat flow from left to right, depicted in Figure 4.3, or the extent of reaction developed below. Application to all available microstates is useful for calculating the total entropy of a model, and provides insight into the overal equilibrium distribution. However, calculation of total entropy using Equation 4.46 is only appropriate when each microstate represented in *W has the same energy*! If some microstates have high energies, they will be populated less frequently, and will contribute less to the entropy than that given by Equation 4.45. At the end of the chapter, we will derive another statistical entropy formula (the Gibbs entropy formula) that accommodates energy variation, and; in Chapter 10 we will develop equations to calculate the equilibrium distribution of a model with microstates that have different energies.

Although we have not derived the Boltzmann entropy formula (derivation will be postponed until we develop the necessary tools in Chapter 9), there are several features of Equation 4.46 that match our expectations for the entropy. Most importantly, as noted above, the Boltzmann entropy is extensive. Since W increases (sharply) with the number of particles in the system, the statistical entropy increases with the size of the system. The logarithmic relationship between S and W makes the size dependence additive[†]: doubling the size of the system doubles the entropy (Problems 4.26 through 4.28), as is the requirement for an extensive function. Below, we will analyze three simple models to show that the statistical entropy calculated using the Boltzmann formula matches results from classical thermodynamics.

These include models for expansion of a gas (Equation 4.20) and unimolecular chemical reaction. We will also apply the Boltzmann entropy formula to the heat flow model developed at the beginning of this chapter (Figure 4.3). But before we analyze the statistical entropy for these three examples, we need to define a few relationships relating to molecular multiplicities, distinguishable and indistinguishable configurations, and bulk properties.

Counting the microstates

Applying the Boltzmann equation to determine entropy from a model requires that we count things (represented by W). Here, we will try to precisely define what we should be counting, and how counting relates to model configurations and to bulk properties. In Chapter 2, we counted sequences and permutations, and referred to the count as a multiplicity. Now that we are shifting to models involving tiny components (molecules), we will refer to the things we should be counting as "***microstates.***" You can think of this term as synonymous with "sequence" or "permutation" in the examples in Chapter 2. It is the number of microstates that is used in Boltzmann's formula.

A key concept for counting in molecular models is that different microstates are often indistinguishable. Said another way, a single distinguishable configuration may comprise multiple (indistinguishable) microstates. We encountered an example of this in our heat flow model (see **Table 4.3**), and analogous examples with coins and dice in Chapter 2. Whether there is a one-to-one correspondence between distinguishable configurations and microstates really depends on how much detail we build into our model. Although we could always construct a model with enough

---

[†] Or *nearly* additive. For some statistical models (see Problems 4.28 and 4.29), it appears that doubling the size of the system does not quite double the entropy. Such behavior would appear to provide a route to spontaneously decrease the entropy of a system without work or heat exchange, which would be in violation of the second law. This apparent violation is referred to as the "Gibbs Paradox"; it arises from a subtle counting error in the paradoxical model.

Table 4.3 Relationships between macrostates, configurations, and microstates for some countable model systems

Model	Relevant macrostate variable	Fixed model parameters	Model variable	Examples of distinguishable configurations	Number of distinguishable configurations	Multiplicity of configurations	Total number of microstates W
Lattice expansion	Volume	$N = 15,\ N_c = 50$	N_c	For $N_c = 50$:	$\dfrac{50!}{15! \times 35!}$	1	$\dfrac{50!}{15! \times 35!}$
Chemical reaction	Concentration	$N_c = N_A + N_B = 100$	$N_A\ (= N_c - N_B)$	For $N_A = N_B = 50$:	$\dfrac{32!}{20! \times 12!}$	1	$\dfrac{32!}{20! \times 12!}$
Heat flow[a]	Energy per subsystem[b]	$E_T = 14$	$E_L\ (= E_T - E_R)$	For $E_L = E_R = 7$: {3,0,0,1} {3,0,0,1}, {3,0,0,1} {2,1,1,0}, {3,0,0,1} {1,3,0,0}, ...	9	16; 48; 16; ...	400[c]

Note: Table entries to the left of the double line refer to general properties of the model and the bulk variable represented by the model. Entries to the right give microscopic details, and are illustrated for only one of many values of the model variable (e.g., $N_c = 50$ cells for the lattice expansion model.

[a] The heat-flow example is from Figure 4.3.

[b] The energy of each subsystem is closely related to the temperature of each subsystem.

[c] The degeneracies for the heat-flow model are doubled compared to the number of entries in Figure 4.17, which only shows half the total microstates for simplicity. Each value comes from the product of multinomial coefficients for the left and right subsystems.

detail to distinguish all the microstates,[†] such highly detailed models often require much more bookkeeping than models that include indistinguishable microstates.

Table 4.3 shows the relationship between microstates and configurations for the three models we will discuss below, and how the model parameters and microstate details determine bulk or macroscopic behavior. For the models of expansion and the chemical reaction, each microstate is a distinguishable molecular configuration. For the heat flow model, there are multiple indistinguishable configurations for each distinguishable configuration.

Importantly, as long as each microstate has the same energy, each makes an equal contribution to the properties of the system, including the entropy, regardless of whether they are distinguishable or indistinguishable. This is closely connected to the concept of "equal *a priori* probability." Though the calculation of entropy using Boltzmann's formula must be made for microstates of the same energy, it is perfectly valid to compare entropies of two groups of microstates that have different energies.[‡] Where Equation 4.46 runs into trouble is in its attempts to calculate the total entropies of systems composed of microstates with different energies, because the low-energy microstates contribute more than the high-energy microstates.

The statistical entropy of a simple chemical reaction

Consider the following reaction:

$$A \rightleftharpoons B \tag{4.47}$$

In Scheme 4.47, molecules can each adopt one of two mutually exclusive and distinguishable conformations through a chemical reaction. We will refer to this type of system as "two-state" for obvious reasons. Such reactions can be simple interconversions of spin-up and spin-down states of a collection of hydrogen atoms in a magnetic field, or can be complex conformational changes in a macromolecule involving many atoms, such as the cooperative folding of a protein. To aid in counting the number of states for a given extent of reaction, we can array the reacting A and B molecules on a lattice with the number of cells equal to the number of molecules ($N_c = N_A + N_B$, **Figure 4.15**).[§]

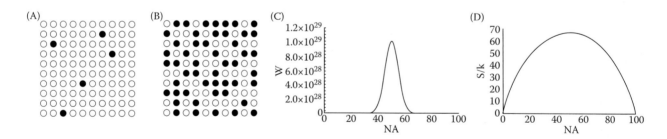

Figure 4.15 A two-state lattice reaction model with $N = 100$ molecules. The A and B configurations are colored black and white, respectively. (A) A microstate from a low multiplicity ($W = 3,921,225$), low entropy ($S = 15.2$ k) composition ($N_A = 5$). (B) A microstate from a high multiplicity ($W = 1 \times 10^{29}$), high entropy ($S = 66.8$ k) composition ($N_A = N_B = 50$). (C), (D) The multiplicity and entropy as a function of composition (represented as N_A, the number of molecules in the A conformation).

[†] For example, in the heat flow model, we could restrict the four particles to four lattice sites and prevent them from exchanging position (but allow energy exchange), which would allow us to distinguish each microstate. One downside of this approach is that such a model doesn't really correspond to a system we are likely to study (a gas or liquid, e.g., where molecules are not distinguishable to us—in principle, a crystal would be an exception).

[‡] That is, if group 1 is composed of microstates with identical energies ε_1, and group 2 is composed of different microstates identical energies ε_2 (where $\varepsilon_1 \neq \varepsilon_2$), we can apply Boltzmann's entropy formula separately to group 1 and to group 2, and use the results to calculate the entropy of converting group 1 to group 2, $\Delta S = S_2 - S_1$.

[§] Note, however, that we are not considering volume explicitly, so the lattice geometry is not needed. Here it is used for easy visualization and counting.

Though A and B molecules can be distinguished from each other, A's cannot be distinguished from other A's, and B's cannot be distinguished from other B's. For a given composition, the number of distinct configurations can be calculated as the product of all the ways the A molecules can be distributed over the N_c sites, multiplied by the number of ways the B molecules can be placed in the remaining $N_c - N_A = N_B$ sites. But for the second placement (of B's in a lattice with the A's already positioned), there is only one arrangement. And for the first placement (A's in an empty lattice) the number of configurations is given by the binomial distribution (Equation 1.15), namely,

$$W(N_A; N_c) = \frac{N_c!}{N_A!(N - N_A)!}$$
$$= \frac{N_c!}{N_A!N_B!}$$

(4.48)

This is identical (both the logic and the resulting expression for W) to the mixing problem at the beginning of the chapter. Indeed, mixing of reactants and products plays a key role in reaction thermodynamics (see Chapter 7). Note that for each configuration, there are no indistinguishable conformations that need to be considered. Thus the number of microstates is equal to the number of configurations (Equation 4.48). The Boltzmann statistical entropy at a particular reaction composition is

$$S = k \ln W(N_A; N_c)$$
$$= k \ln \frac{N_c!}{N_A!N_B!}$$
$$= k \ln N_c! - k \ln N_A! - k \ln N_B!$$

(4.49)

Equation 4.49 can be further simplified using Stirling's approximation (**Box 4.1**):

Box 4.1: Stirling's approximation for dealing with large factorials

Statistical counting formulas (e.g., Equation 4.48) often involve factorials. One problem with factorials is that as the argument of a factorial increases, the factorial gets *really large*. Though computers can work with big numbers, there are practical limits. For example, as far as the program Matlab® is concerned (at the time of this writing), the factorial of 171 is infinity.[†] We will be interested in numbers of particles *much* larger than 171 (on the order of Avogadro's number), for which computing factorials is not possible. Often we take the log of these factorials, which results in a much smaller number, but if we are required to calculate the factorial and then take its log, the calculation problem remains.

Fortunately, there is a good way to approximate the log of a factorial, known as "Stirling's approximation." When x gets reasonably large (see Problem 4.32), we can approximate the log of $x!$ as

$$\ln x! \approx +x \ln x - x + \frac{1}{2} \ln(2\pi x)$$

(4.51)

For really large values of x, the last term makes a negligible contribution, and we can further simplify the approximation:

$$\ln x! \approx x \ln x - x$$

(4.52)

[†] Obviously, this is not the case. The number 171! is a mere 171 times larger than 170!, which Matlab estimates to be 7.26×10^{306}. Mathematica has better luck calculating the factorial of large numbers, though working with numbers as large as 10,000! is rather cumbersome.

$$S = k\left(N_c\ln N_c - N_c - N_A\ln N_A + N_A - N_B\ln N_B + N_B\right)$$
$$= k\left(N_c\ln N_c - N_A\ln N_A - N_B\ln N_B\right)$$
$$= k\left(\ln \frac{N_c^{N_c}}{N_A^{N_A}N_B^{N_B}}\right)$$

(4.50)

Because Equation 4.50 does not contain factorials of large numbers, it is much easier to work with than Equation 4.59. Figure 4.15 shows W and S (Equations 4.48 and 4.49) as a function of the composition. Because the logarithmic function increases monotonically with its argument (here W), the composition with the maximum number of arrangements will maximize the statistical entropy. As you might expect, maximum entropy is achieved when the number of A's and B's are equal.

The statistical entropy of an expanding two-dimensional lattice gas

Above, we calculated the entropy of expansion of a gas based on classical considerations (Equation 4.44). Here we will rationalize the spontaneity of vacuum expansion using a lattice model (**Figure 4.16**) to evaluate W, and will evaluate S from W with Equation 4.46.

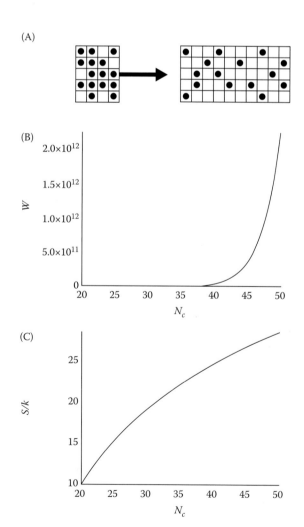

Figure 4.16 Expansion of a two-dimensional lattice gas. The $N = 15$ gas molecules are black circles; unfilled lattice cells are empty. (A) Expansion from $N_c = 20$ cells ($W = 15,504$, $S = 9.65$ k) to $N_c = 50$ cells ($W = 2.25 \times 10^{12}$, 28.4 k). (B) The multiplicity and (C) entropy are plotted as a function of volume (represented as the number of lattice cells, N_c).

Since volume is the relevant macroscopic variable for gas expansion, we should cast the lattice model in terms of volume.[†] The total volume is simply the number of lattice cells times the volume per cell, $V = N_c \times V_c$ (although this linear scaling by V_c affects neither W nor S). We will model $N \leq N_c$ molecules to be free to distribute among the lattice points, where again, only one molecule is allowed per lattice site, and again we will exclude interaction energies among gas molecules (that is, molecules do not feel each other, even when they are in adjacent lattice sites).

At each volume (i.e., at each value of $N_c \geq N$), the number of ways the N identical molecules can be arranged over N_c distinct sites is given by a binomial coefficient:

$$W(N; N_c) = \frac{N_c!}{N! \times (N_c - N)!} \tag{4.53}$$

Each of these $W(N; N_c)$ configurations is distinguishable, and gets counted once in the microstate count. The statistical entropy is given by combining Equations 4.53 and 4.46:

$$S = k \ln \frac{N_c!}{N!(N_c - N)!} \tag{4.54}$$

Although this equation appears to the same as the entropy equation for the two-state chemical reaction (Equation 4.49), the independent variable is different. The variable for the chemical reaction is the number of molecules of a particular type (N_A, in the denominator of W), and is bounded from above by the total number of molecules ($N_A \leq N = N_c$). The variable for the lattice expansion is the total number of lattice cells (N_c, in the numerator of W), and is bounded from below by the number of gas molecules ($N_c \geq N$). In analogy to coin toss statistics, calculating the statistical entropy of the chemical reaction is like counting the number of arrangements of heads for different compositions (N_h) in a fixed number of tosses ($N = N_h + N_t$), whereas the lattice expansion is like counting the number of arrangements of a fixed number of heads for a variable number of coins tosses.

We previously calculated the classical entropy of expansion of an ideal gas, and can now compare the classical and statistical formulas. For the isothermal expansion, our classical formula (Equation 4.21, and for the identical formula for vacuum expansion, Equation 4.44) gave ΔS in terms of volume change from V_i to V_f. For comparison, we will calculate the statistical entropy change in going from N_i to N_f lattice cells using Equation 4.54:

$$\begin{aligned} \Delta S = S_f - S_i &= k \left(\ln \frac{N_f!}{N!(N_f - N)!} - \ln \frac{N_i!}{N!(N_i - N)!} \right) \\ &= k \ln \left(\frac{N_f!(N_i - N)!}{N_i!(N_f - N)!} \right) \end{aligned} \tag{4.55}$$

Although there are some similarities between Equation 4.55 and the classical expansion entropy (both increase logarithmically with their expansion variable), the statistical expression involves a factorial, and the classical expression does not. As with the lattice reaction, the factorial can be eliminated using Stirling's approximation[‡]:

[†] Although this two-dimensional example may seem more like an area problem than a volume problem, each lattice cell can be expected to have some thickness. Thus, the volume goes as the number of lattice points, regardless of whether we arrange the cells in a two-dimensional plane or a three-dimensional box (or for that matter, any irregular 3-d lattice shape).

[‡] As long as N_i, N_f and N are large.

$$\Delta S = k\left\{\ln N_f! + \ln(N_i - N)! - \ln N_i! - \ln(N_f - N)!\right\}$$
$$= k\left\{\begin{array}{l} N_f\ln N_f - N_f + (N_i - N)\ln(N_i - N) - (N_i - N) \\ -N_{ci}\ln N_i + N_i - (N_f - N)\ln(N_f - N) + (N_f - N) \end{array}\right\}$$
$$= k\left\{N_f\ln N_f + (N_i - N)\ln(N_i - N) - N_i\ln N_i - (N_f - N)\ln(N_f - N)\right\}$$
$$= k\left\{N_f\ln\frac{N_f}{(N_f - N)} - N_{c1}\ln\frac{N_i}{(N_i - N)} - N\ln\frac{(N_i - N)}{(N_f - N)}\right\} \tag{4.56}$$

When the lattice becomes large relative to the number of gas molecules (i.e., $N \ll N_c$), Equation 4.56 simplifies to

$$\Delta S = k\left(N_f\ln\frac{N_f}{N_f} - N_i\ln\frac{N_i}{N_i}\right) - kN\ln\frac{N_i}{N_f}$$
$$= kN\ln\frac{N_f}{N_i} \tag{4.57}$$

To relate this expression to the volume change, the numerator and denominator of 4.57 can each be multiplied by V_c:

$$\Delta S = kN\ln\left(\frac{N_f V_c}{N_i V_c}\right) = Nk\ln\left(\frac{V_f}{V_i}\right)$$
$$= nR\ln\left(\frac{V_f}{V_i}\right) \tag{4.58}$$

The second line uses the relationships $R = kL_0$ and $n = N/L_0$. Equation 4.58 is the same as the expression obtained classically for isothermal expansion of an ideal gas. The agreement between the statistical and the classical approaches supports the validity of each.[†]

The statistical entropy of heat flow

At the start of this chapter, we constructed a model for heat flow in which multiplicity is increased when energy distributes equally among different parts of a system (Figure 4.3), consistent with the observation that heat flows in the direction that leads to uniform temperature distribution. The statistical formula for entropy (Equation 4.46) can be applied to this model to demonstrate that this direction of change corresponds to an increase in entropy. In our heat-flow model, there are multiple indistinguishable arrangements for each distinguishable configuration, given by the multinomial distribution (Equation 1.31; Figure 4.3).

One way to apply the Boltzmann entropy formula to the heat-flow model is to calculate and compare entropies of different distinguishable configurations, using the multiplicities of each configuration for W in the Boltzmann entropy formula. For the configuration on the left-hand hand side of Figure 4.3 (we will represent this configuration with the occupancy numbers {1,1,1,1} {4,0,0,0}, **Figure 4.17**), which corresponds to a large bulk energy difference, the number of microstates is $W_T = W_L \times W_R = 24$; thus, $S = k\ln 24 \approx 3.18\ k$.[‡] For the configuration on the right-

---

[†] Note that although the lattice gas is not an ideal gas (the excluded volume V_c acts as a very strong, short-range repulsion), in the limit that $N \ll N_c$ in the analysis above, double occupancies become very unlikely even if they could occur. This is equivalent to the low-density (ideal gas) limit.

[‡] It might seem that the application of the Boltzmann entropy formula to the heat flow model is inappropriate, since different molecular energy levels are involved, and different distributions of temperature have different energies. The key point here is that the total energy the combined system (left and right) is constant. Using language to be developed in Chapters 9 and 10, the combined system can be analyzed as a microcanonical ensemble, although the subsystems and individual particles cannot.

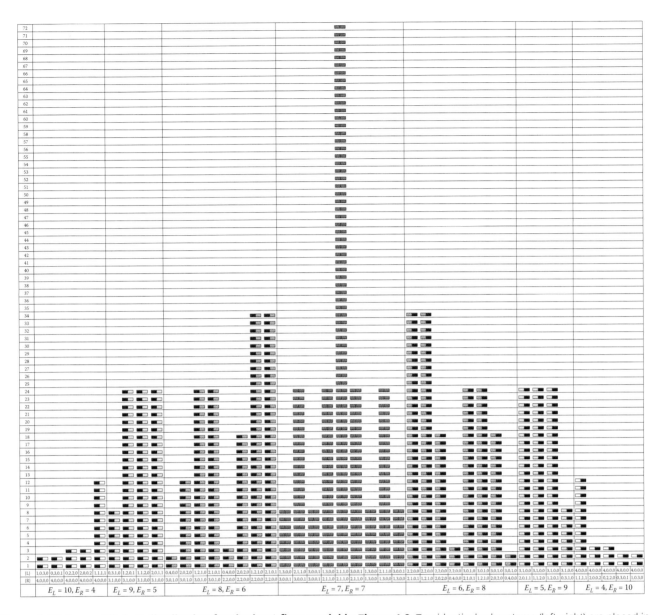

Figure 4.17 An equilibrium distribution for the heat flow model in Figure 4.3. Two identical subsystems (left, right) are placed in thermal contact, with a fixed total energy of 14 units. The temperature (and energy) of each subsystem is indicated by red color saturation, and by the pair of energies in the bottom row. Seven bulk energy distributions are consistent with the fixed total energy restriction ($N_T = 14$). For each energy distribution, there are a number of distinguishable molecular configurations that can be achieved. For example, the {1,0,3,0}, {4,0,0,0} configuration (left) has the same has the same energy distribution ($E_L = 10$, $E_R = 4$) as the {0,3,0,1}, {4,0,0,0} configuration. For each distinguishable conformation, indistinguishable microstates are stacked vertically in proportion to their multiplicity (divided by two for simplicity). Each of these microstates is equiprobable. The collection of 718 microstates shown here corresponds to this equiprobable distribution (only half of 1436 microstates consistent with 14 total units of energy are shown).

hand side of Figure 4.3, where bulk temperature is uniform (with occupancy numbers {2,1,1,0} {2,1,1,0}), the number of microstates is $W_T = W_L \times W_R = 144$; thus, $S = k \ln 144 \approx 4.97\, k$. Thus, the entropy change in going from left to right in Figure 4.3 is

$$\Delta S \approx (4.97 - 3.18)k = 1.79k \qquad (4.59)$$

This increase in the statistical entropy is consistent with spontaneous temperature equilibration.

With a little more generality, we could use Boltzmann's entropy formula to calculate the entropy change from the $E_L = 10$, $E_R = 4$ energy profile to the $E_L = 7$, $E_R = 7$ profile, including all distinguishable configurations. For the $E_L = 10$, $E_R = 4$ profile, there are two distinguishable configurations, $\{1,1,1,1\}\{4,0,0,0\}$ and $\{2,0,0,2\}\{4,0,0,0\}$, with 24 and 6 microstates, respectively (Figure 4.17),[†] These are mutually exclusive, so they add up to give $W_{10,4} = 30$. For the $E_L = 7$, $E_R = 7$ profile there are nine distinguishable configurations (three independent configurations per side), each comprising multiple indistinguishable microstates, which sum to give $W_{7,7} = 400$. Using these numbers in the Boltzmann entropy formula gives

$$\Delta S = S_{7,7} - S_{10,4}$$
$$= k(\ln 400 - \ln 30)$$
$$= k \ln\left(\frac{40}{3}\right) \tag{4.60}$$
$$\approx 2.590k$$

This entropy increase is larger than that depicted in Figure 4.3 (Equation 4.6), which leaves out many $E_L = 7$, $E_R = 7$ microstates.

Before leaving the heat-flow model (for now[‡]), we will consider its equilibrium distribution and use the Boltzmann formula to calculate the total entropy. As described above, because each microstate has the same energy, each has the same probability. If we consider the average (equilibrated) properties of our heat flow model,[§] we should see the same numbers of each microstate. Even though the $E_L = 10$, $E_R = 4$ configurations are rare (and collectively have relatively low entropy), they would occur *some* of the time,[¶] as would their mirror image configurations ($E_L = 4$, $E_R = 10$). Configurations from other distributions (e.g., $E_L = 9$, $E_R = 5$) would also occur.

Based on the principal of equal *a priori* probability, the probability of different configurations can be obtained by dividing the number of microstates consistent with the configuration by the total number of microstates:

$$p_{\{1,1,1,1\}\{4,0,0,0\}} = \frac{W_{\{1,1,1,1\}\{4,0,0,0\}}}{\sum\limits_{i=1}^{\text{all configs}} W_i} = \frac{24}{1436} \approx 0.0167 \tag{4.61}$$

The denominator of Equation 4.61 is often referred to as the "microcanonical partition function" (abbreviated Ω; see Chapter 9). Figure 4.17 shows a collection of heat flow models with the populations of configurations in proportion to the number of microstates for each configuration. Entries with the same bulk energy profile (degree of pink shading) are grouped by thick lines. Different distinguishable configurations with the same bulk energy profile are grouped using thin lines.

Although Figure 4.17 is rather busy, it provides an overall impression of the average temperature distribution we would expect for our heat-flow model. The most probable set of configurations have a uniform energy distribution from left to right. However, there are significant deviations; the most common deviations involve a single unit of energy transfer between the two subsystems. The configurations with the greatest energy differences are the least common. For larger systems (more par-

---

[†] A detailed accounting of the number of different configurations and indistinguishable microstates for the heat-flow model is given in Chapter 9 (see Figure 9.10).

[‡] We will return to the heat-flow model in our treatment of statistical thermodynamics and the "microcanonical partition function" in Chapter 9.

[§] We could average either by letting our model sample different configurations over time, or by creating a large number of replica heat-flow models in random configurations.

[¶] More precisely, "the indistinguishable microstates consistent with the configuration on the left would be expected to occur some of the time."

ticles), these large deviations become very rare. The population distribution represented in Figure 4.17 is investigated numerically and graphically in Problem 4.3.

As long as the energies of all microstates are the same, we can use the Boltzmann formula (Equation 4.46) to calculate the total entropy of the heat-flow model in its equilibrium distribution:

$$S = k \ln \left(\sum_{i=1}^{\text{all configs}} W_i \right) = k \ln \Omega = k \ln(1436) \approx 7.27k \qquad (4.62)$$

This is the total entropy of our heat-flow model when it allowed to equilibrate among all available microstates with 14 total units of energy. Any restriction to a smaller number of microstates, no matter how they are grouped (e.g., $E_L = 7, E_R = 7$, or $\{2,1,1,1\} \{2,1,1,1\}$, or worse still $\{1,1,1,1\} \{4,0,0,0\}$) would decrease the entropy. Stated another way, the distribution with equal partitioning among all microstates maximizes the entropy, and is thus the equilibrium distribution.

ENTROPY AND FRACTIONAL POPULATIONS

If all microstates have the same energy (as is the case in our heat-flow model), the Boltzmann entropy formula can be expressed in terms of microstate populations. Since each microstate has the same population, and since the populations sum to one, the number of microstates is given by

$$W = \frac{1}{p_j} \qquad (4.63)$$

Thus, the Boltzmann entropy formula (Equation 4.46) can be written as

$$S = -k \ln p_j \qquad (4.64)$$

Multiplying the right side of Equation 4.64 by the sum of the probabilities of all the $m = 1436$ microstates (which has a value of one, and thus does not invalidate the equality) gives

$$S = -k \ln p_j \sum_{i=1}^{m} p_i$$
$$= -k \sum_{i=1}^{m} p_i \ln p_j \qquad (4.65)$$

Note that although the two subscripts on the population terms are different (i vs. j), since each microstate population term has the same value, the j subscript can be replaced with an i, essentially chasing the index of the summation[†]:

$$S = -k \sum_{i=1}^{m} p_i \ln p_i \qquad (4.66)$$

Equation 4.66 is often referred to as the "Gibbs entropy formula." Remarkably, although we derived this equation for microstates with the same energy, it also applies when microstates differ in energy, and thus, probability (unlike the Boltzmann formula). In situations where microstate energies differ, the Gibbs entropy can be factored into two terms (see Problems 4.34 through 4.36). One term

---

[†] This index slight-of-hand *only works* if the populations are all the same!

looks like the Boltzmann entropy (Equation 4.46), except that the microstates are counted fractionally in W, in proportion to their populations. The other term is the average energy over the ground state divided by the temperature, and captures the classical entropy associated with heat storage (dq/T).

Problems

4.1 Evaluate the binomial coefficients for each compartment in Figure 4.2A, that is, $N_A A$ molecules on N_A lattice cells, and $N_B B$ molecules on N_B lattice cells.

4.2 Calculate the multiplicity W and the entropy of mixing two distinguishable species on a lattice, ignoring the excluded volume term (that is, any number of molecules can occupy the same lattice site).

4.3 For Figure 4.17, give the populations and the entropies. Make a bar graph of these two quantities, ranging from 10:4 to 4:10. What to the bar graphs indicate about the position of equilibrium?

4.4 What is the total entropy for the system shown in Figure 4.17?

4.5 The Kelvin statement says that some of the energy supplied to a heat engine (q_{in}) must be expelled as heat (q_{out}), not work, but does not say anything about the two temperatures, per se. Build an argument from the Kelvin statement that if the temperature of the supplying and accepting thermal reservoirs in a heat engine are the same, no work can be done (i.e., the efficiency goes to zero).

4.6 Show that if the Kelvin statement is false, the Clausius statement is false. This will require you to come up with hypothetical devices and think through their consequences, not merely use equations.

4.7 Show that if the Clausius statement is false, the Kelvin statement is false.

4.8 For an ideal gas, would it be possible for a single adiabatic curve to cross an isotherm more than once, as shown in the figure below (isotherm in red, and proposed adiabat in dashed black)?

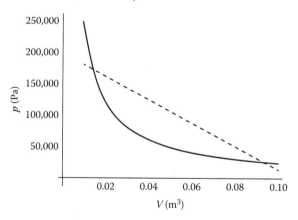

What would be the implications for the first law? What about the second law?

4.9 In discussing the reversible Carnot engine and its implications, it was stated that "adiabats are unique." This statement can be interpreted as meaning that if a system starts at a particular temperature T_h, pressure p_h, and volume V_h, and expands *reversibly* without heat flow to another temperature T_c, there is only one adiabat that the system can follow. In general, how does this statement restrict the pressure and volume when the system reaches the temperature T_c adiabatically? If this were not the case, which law of thermodynamics would be violated? Use a p–V diagram to help illustrate your answer.

4.10 As an alternative to the approximation of an arbitrary, reversible cycle in the $p–V$ plane as a net of little Carnot cycles, we could develop a similar argument using an array of tall, narrow Carnot cycles where the isotherms touch the original cycle and the adiabats are very close together. Draw this kind of approximation to an arbitrary cycle in the $p–V$ plane. What are the advantages of this type of tiling over that presented in Figure 4.7?

4.11 For a Carnot engine operating between temperatures T_h and T_c, and between points A, B, C, and D (as in Figure 3.3), calculate the heat flow into the engine, and the work done by the engine, during each step of the cycle (i.e., A->B, B->C, C->D, and D->A).

4.12 Derive expressions for the total heat flow into, and the total work done by, a Carnot engine operating between temperatures T_h and T_c, and between points A, B, C, and D (as in Figure 3.3) during a single turn of the cycle. What is the relationship between these two totals? You can assume that the working material of the Carnot engine is a monoatomic ideal gas.

4.13 Derive the volume dependence of the heat flow and entropy change of expansion of one mole of an ideal monatomic gas at constant pressure, as depicted in Figure 4.10A and B.

4.14 In Problem 4.13, the pressure appeared explicitly in the expression for heat flow, but not the number of moles of gas, whereas for pressure did not appear explicitly in the entropy expression, but the number of moles did. Thinking in molecular terms, rationalize this observation.

4.15 Plot the entropy changes for constant pressure expansion, for constant temperature expansion, and for adiabatic expansion of one mole of a monatomic gas from a common starting point ($p_i = 10^5$ Pa, $V_i = 0.02$ m^3).

4.16 Compare the volume dependence of entropy change at constant pressure to that at constant temperature, for a monatomic ideal gas, and generate a plot comparing the two as in Figure 4.9A. Does the entropy change increase more in a constant pressure or constant temperature expansion? In words, rationalize your answer.

4.17 Show that for heat flow between a system at initial temperature T_1 and a thermal reservoir at T_{surr}, the entropy change is positive for all values where $T_1 \neq T_{surr}$ (see Figure 4.13). There are two parts of this proof. First, calculate the value of ΔS on the line $T_1 = T_{surr}$. Then, calculate the derivative of ΔS with respect to the variables T_1 and T_{surr}.

4.18 Compare the heat flow and entropy change of adiabatic expansion of an ideal gas from V_i to V_f (and pressure p_f) to a two-step process involving isothermal expansion from V_i to V_f, followed by a constant volume depressurization from to pressure p_f.

4.19 For the irreversible adiabatic expansion depicted in Figure 4.16 (0.2 mol monatomic ideal gas at $p_1 = 500,000$ Pa, $V_1 = 0.001$ m^3, $T_1 = 300$ K, expanding against a constant pressure mechanical reservoir), calculate the entropy change for an expansion to a final state in mechanical equilibrium with the surroundings where $V_f = 0.02$ m^3.

4.20 Derive and plot the system entropy change for irreversible adiabatic expansion of a monatomic ideal gas. For your plot, use one mole of gas a starting pressure $p_{sys} = 10$ atm, $V_{sys} = 1$ L. Expand to a final stopped volume of 2 L, against a constant reservoir pressure of 0.1 atm.

4.21 For stopped irreversible expansion against a mechanical reservoir, how does the entropy change with external pressure? Given the same starting conditions and stopping volume, does a lower external pressure produce a bigger or smaller entropy change?

4.22 For 0.2 mol of ideal monatomic gas, calculate the entropy change for the system and the surroundings for irreversible expansion and heat flow from $p_1 = 500{,}000$ Pa, $V_1 = 0.001$ m3, $T_1 = 300$ K, to a final state that is in thermal equilibrium with the surroundings at $p_f = 24{,}942$ Pa, $T_f = 300$ K, $V_f = 0.02$ m³. To do this, assume the surroundings is both a thermal reservoir (constant $T = T_f$ with a thermal response time much shorter than the system's) and a mechanical reservoir (constant $p_{surr} = 15{,}310$ with a mechanical response time much shorter than the system's). Hint: one way to do this is to compare the irreversible expansion to a two-step process in which the first step is an irreversible adiabatic expansion to V_f, and the second step is a constant volume heating to T_f.

4.23 When we analyzed vacuum expansion of an ideal gas to a stop, we assumed that the walls were adiabatic, and found that since the internal energy remained constant, the final temperature of the system was equal to the initial temperature. Discuss whether or not this is true if the walls can allow heat flow (i.e., they are diathermal).

4.24 Show that for the lattice gas, the change in entropy for expansion is given by Equation 4.56.

$$\Delta S = k\left(N_{c2}\ln\frac{N_{c2}}{(N_{c2}-N)} - N_{c1}\ln\frac{N_{c1}}{(N_{c1}-N)}\right)$$
$$- kN\ln\frac{(N_{c1}-N)}{(N_{c2}-N)}$$

4.25 Using the lattice model (Figure 4.2) with N_c sites, calculate the statistical entropy of mixing when N_A and N_B distinguishable molecules are initially partitioned off, and are then allowed to mix freely. Assume the lattice is completely filled, that is, $N_A + N_B = N_c$.

4.26 Show that for the heat-flow model (Figure 4.3), the statistical entropy behaves extensively by considering the left and right (hot and cold) subsystems separately, and comparing the sum to that for the entire system. What property of the formula for statistical entropy (Equation 4.46) ensures that the same result is obtained when the left and right (hot and cold) subsystems are taken as a single unit?

4.27 Investigate the extensivity of the statistical entropy formula (Equation 4.46) using the lattice gas model with N_c cells and N_p particles (where $N_p \leq N_c$). To do this, calculate the entropy of a system with 50 cells and 25 particles and compare it with the entropy of a system that has 100 cells and 50 particles (that is, it is double in size). What is the relation between the entropies of these two systems?

(a) Continuing with the extensivity of the lattice gas, come up with an equation that gives the statistical entropy as a function of the number of lattice cells N_c, when 40% of the cells are occupied (i.e., $N_p = 0.4N_c$). Make a plot that compares twice this entropy to that statistical entropy of a system that has twice as many cells and twice as many particles. Also plot the *difference* between these two quantities. *Note, you should not use Stirling's approximation in this problem (the numbers are small enough for Mathematica, and we want to compare exact results).*

(b) Derive a statistical expression for entropy for a system of N molecules with t different microstates. In other words, come up with an expression for W in the formula $S = k \ln W$ in terms of the number of molecules in each of the t microstates ($N_1, N_2, ..., N_t$). Note, this problem is the same as having N indistinguishable dice with t sides, and asking for the number of ways (and entropy) for a composition with N_1 ones, N_2, twos, ...

4.28 Starting with the statistical definition of entropy you derived in Problem 4.29, derive the expression for entropy in terms of fractional populations, that is,

$$S = -Nk \sum_{i=1}^{t} p_i \ln p_i$$

(Hint: Stirling's approximation might be helpful here.)

4.29 Consider a system below, with two subsystems (L and R, for left and right) separated by a moveable wall. Each subsystem is a two-dimensional lattice gas. L has $N_L = 10$ particles in it, and R has $N_R = 20$. Between the two subsystems, there are 60 lattice sites combined. An example of a single arrangement of particles is shown for the moveable wall position given.

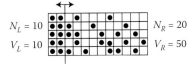

$N_L = 10$ $N_R = 20$
$V_L = 10$ $V_R = 50$

For the moveable wall position shown above (where the number of lattice cells on each side are $N_L = 10$, $N_R = 50$), calculate the multiplicity and entropy for each subsystem, and the multiplicity and entropy for the two systems combined, based on different arrangements of particles over different lattice cells.

4.30 Using statistical entropy as a potential, calculate the equilibrium position for the moveable wall in Problem 4.31. Assume that it remains vertical, *that is,* it moves integer steps of an entire column of five cells. Remember, excluded volume applies, so please, no fusion reactions).

4.31 Now use Gibbs' statistical entropy formula and calculate the entropy for each subsystem in Problem 4.31, and the total entropy, as a function of the wall position. Remember that Gibb's formula gives an intensive entropy, so you must multiply by a factor related to the size of each subsystem to get the correct value. Compare the Gibbs entropy with the Boltzmann entropy (Problem 4.32) in a plot. Note that $0*\log(0) = 0$ (you can use L'Hopital's rule if you need proof. Mathematica correctly identifies it as "indeterminate," but that is not that useful).

4.32 Use Mathematica to evaluate how close the two forms of Stirling's approximation (see Box 4.1) of $\ln(x!)$ are as a function of x. Specifically, compare $\ln(x!)$ to Equation 4.51 and to Equation 4.53 for a range of values of x, both small and large. Make this comparison both in terms of numbers (in a table) and in a plot. You may want to take the log of the factorials to facilitate comparison, either in the table, in the plot, or in both.

4.33 When we count microstates of equal energy (and populations), the count can be thought of as a sum of ones for each microstate, that is,

$$W = \sum_{i=1}^{m} 1$$

The one in the summation gives equal representation to each microstate, since each has the same population. This problem attempts to modify the Boltzmann entropy formula to include microstates with nonuniform energies and nonuniform populations.

Assume that the microstate with the lowest energy has the highest population, and label this microstate 1 (i.e., with population p_1 and energy E_1). Other microstates (2 through m) have higher energy and lower population (the relationship between population and energy will be derived in Chapter 10).

It seems appropriate to adjust our count of microstates by decreasing the contribution of high-energy microstates in proportion to their decreased populations. This can be done by dividing the population of each microstate by the population of the ground state:

$$W' = \sum_{i=1}^{m} \frac{p_i}{p_1}$$

In this representation, the ground state makes a contribution of one to and all the other states contribute less than one. This makes sense because really high-energy conformations with populations approaching zero should not make a thermodynamic contribution to the system.

Given this setup, derive an expression equivalent to the Boltzmann entropy (Equation 4.46, you can call the result S') using this modified W' expression.

4.34 Starting with your expression for S' from Problem 4.34, attempt to derive an expression for the Gibbs entropy, following the derivation from Equation 4.64 through Equation 4.66.

4.35 In the last two problems, you should have convinced yourself that the extension of the Boltzmann entropy formula to microstates with different energies (Problem 4.34) does not lead to the Gibbs entropy formula (Problem 4.35). In this problem, investigate how they differ, by subtracting your expression for S' (Problem 4.35) from the Gibbs entropy formula (i.e., the valid entropy expression for microstates of different energies). Use the relationship (which we will derive in Chapter 10) that

$$p_i/p_1 = e^{-(E_i - E_1)/kT}$$

Further Reading

Adkins, C.J. 1983. *Equilibrium Thermodynamics*, 3rd Edition. Cambridge University Press, Cambridge, UK.

Atkins, P.W. 1984. The Second Law. Scientific American Books—W. H. Freeman and Company.

Ben-Naim, A. 2008. *Entropy Demystified. The Second Law Reduced to Plain Common Sense with Seven Simulated Games*. World Scientific, Hackensack, NJ.

Callen, H. B. 1985. *Thermodynamics and Introduction to Thermostatistics*, 2nd Edition. John Wiley & Sons, New York.

Fermi, E. 1936. *Thermodynamics*. Dover Publications, Inc., Mineola, NY.

Lindley, D. 2004. *Degrees Kelvin: A Tale of Genius, Invention, and Tragedy*. Joseph Henry Press, Washington DC.

Shannon, C.E. 1948. A mathematical theory of communication. *Bell Syst. Tech. J.* 7(3), 379–423.

Zemansky, M.W. 1964. *Temperatures Very Low and Very High*. Dover Books, Mineola, NY.

CHAPTER 5

Free Energy as a Potential for the Laboratory and for Biology

Goals and Summary

The goal of this chapter is to construct thermodynamic potentials that are suited for systems that exchange energy with their surroundings, building on the second law and the Clausius inequality. These potentials will be created using the method of Legendre transforms. Of particular importance, the Gibbs free energy will be developed as a transform of the Clausius inequality, and will be shown to serve as the thermodynamic potential when pressure and temperature are held constant. Systems at constant temperature and pressure are common in the laboratory, and in much of biology. Using the calculus of exact differentials developed in Chapter 2, thermodynamic relationships will be derived from these potentials that give fundamental thermodynamic quantities such as entropy, energy, volume, and heat capacity. In many cases, these quantities would be difficult to measure directly, and the relationships developed here reveal fundamental connections between these variables between entropy, temperature, and energy.

In addition to connecting potentials to physical quantities such as temperature and pressure, these differential relationships provide a way to resolve the contributions of individual chemical species to overall thermodynamics quantities. Central to this chemical dissection is the use of "partial molar quantities," which are obtained by differentiation with respect to mole number. These partial molar quantities are intensive analogs of quantities such as energy, volume, and heat capacity, and provide a fundamental expression chemical reactivity. Of particular importance is the partial molar Gibbs free energy, referred to as the "chemical potential."

In Chapter 4, we developed entropy as a thermodynamic potential to describe the direction of spontaneous change. The entropy acts as a thermodynamic potential as long as a system is isolated, that is, when there is no work, no heat flow, and no material exchange between system and the surroundings. While this is a useful way to think about the thermodynamics of the universe (where, by definition, there is no surroundings with which to interact), systems studied in the laboratory are not isolated from their surroundings (**Figure 5.1A**).

Typically, laboratory systems are held at constant temperature and pressure (partly because such conditions are easy to achieve, requiring no thermal or mechanical insulation), which allows heat to flow and work to be done as a result of from

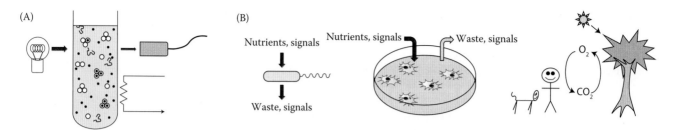

Figure 5.1 Laboratory and biological systems at constant pressure and temperature. (A) A typical laboratory sample containing a solution of large and small molecules in various states of conformation, assembly, and reaction. The sample is open to atmospheric pressure, so it can expand and contract while maintaining constant p, and has a resistive heater and/or heat sink that can maintain constant sample temperature. The distribution of various molecular species can be measured using devices such as light sources and detectors. (B) Living biological systems. Left, a bacterial cell; middle, eukaryotic cells in a culture dish; right, animals and a plant. In all three cases, temperature and pressure are usually constant (or vary slowly compared to processes of interest). In addition, material (and in one case, light) is exchanged among these living systems and their surroundings.

volume change. Even when temperature (and sometimes pressure[†]) is used as an experimental variable, the system under study is typically allowed to equilibrate with its surroundings at a series of different temperatures. Thus, variable temperature experiments are treated as a collection of measurements at fixed temperatures. Likewise, biological systems (cells, tissues, and organisms, **Figure 5.1B**) often function under conditions of constant temperature and pressure, and often exchange material with their surroundings.

How can we apply the concepts of the thermodynamic potential (predicting the position of equilibrium and the thermodynamic driving force) to laboratory and biological systems? As discussed in Chapter 3, the distinction between system and surroundings is, in a sense, arbitrary. Thus, one possibility would be to redefine the "system" to include not just the laboratory sample or the biological entity, but also the environment that exchanges heat, volume, and material. This redefined system could be considered to be isolated, allowing system entropy to serve as a potential. Though such an approach would not be incorrect, it would significantly complicate both experiment and analysis. Usually we are concerned with the behavior of a sample of interest, but we don't want to bother to measure and analyze energy and material flow in a laboratory, cell culture apparatus, or habitat. In this chapter, we will modify our entropy potential so that it implicitly keeps track of exchanges with the surroundings, while directly focusing on the variables of the system. It should be kept in mind that although these energy functions are extremely useful for analysis, their ability to serve as potentials derives entirely on the Clausius inequality, and thus, total entropy.

INTERNAL ENERGY AS A POTENTIAL: COMBINING THE FIRST AND SECOND LAWS

As a starting point to create potentials for systems that interact with their surroundings, we will develop the internal energy as a potential by combining the first and second laws. Assuming no work other than volume work, the differential form of the first law is

$$
\begin{aligned}
dU &= dq + dw \\
&= dq - pdV
\end{aligned}
\tag{5.1}
$$

To introduce the entropy, and thus, bring in the features of a thermodynamic potential, we rearrange the Clausius inequality (Equation 3.24):

[†] In practice, reactions in solution typically have very small volume changes, and unless pressure is increased significantly from 1 atmosphere, pdV work can typically be ignored (see Problem 5.1)

$$dq \leq TdS \qquad (5.2)$$

Remember, the equality applies when the system is at equilibrium, whereas the inequality applies away from equilibrium. Next, we substitute TdS (5.2) for dq in the first law (Equation 5.1), paying careful attention to how the inequality affects the relationship. If the system is away from equilibrium, we will be substituting a quantity (TdS) that is *larger* than dq on the right-hand side of Equation 5.1. Thus, dU ends up *smaller* than the substituted right-hand side:

$$dU \leq TdS - pdV \qquad (5.3)$$

The equality part of Equation 5.3 corresponds to equilibrium. The inequality corresponds to spontaneous (i.e., irreversible) change. You should spend a minute convincing yourself that the sign of the inequality in Equation 5.3 is correct.

Equation 5.3 is sort of a combination of the first and second laws in one equation. It expresses energy conservation and entropy maximization simultaneously. The fundamental variables in this differential are S and V; T and $-p$ serve as sensitivity coefficients. For pressure and volume, it makes sense that if p is high, an increase in V will produce a large decrease in U (since the system needs to do a lot of work on the surroundings to increase its volume). The analogous relationship between T, S, and U[†] is a bit less intuitive, and will be explored in Problem 5.2.

One of the key properties of Equation 5.3 is that it identifies conditions under which internal energy acts as a potential. Specifically, U acts as a potential when the right-hand side of Equation 5.3 is zero,

$$dU \leq 0 \qquad (5.4)$$

analogous to Equation 4.31. Although this condition can be achieved by setting T and p to zero, we have little interest in systems with zero temperature and pressure. Rather, U acts as a useful potential (Equation 5.4) when dS and dV are zero, that is, when S and V (the fundamental variables for U, Equation 5.3) are held constant. With S and V fixed, the direction of spontaneous change (i.e., toward equilibrium) is that which decreases the internal energy. Under these conditions, internal energy is a *minimum* at the equilibrium position (**Figure 5.2**).

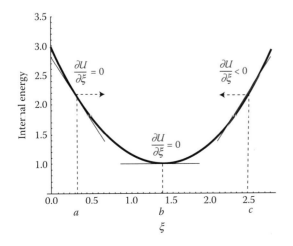

Figure 5.2 Internal energy acts as a potential in a system at constant entropy and volume. The thick black curve shows the internal energy U as a function of an internal coordinate ξ. The direction of spontaneous (irreversible) change is that which decreases U. At $\xi = a$, spontaneous change is to the right (toward increasing values of ξ, arrow). At $\xi = c$, spontaneous change is to the left (toward decreasing values of ξ). At $\xi = b$, U is minimized, and thus the system will not undergo further spontaneous change in either direction. This minimum defines the equilibrium position for the system.

[†] That is, at high temperature, a large increase in entropy corresponds to a large increase in energy.

An example of the internal energy potential $U = U(S,V)$ for an ideal gas

We will illustrate the internal energy potential by deriving an expression for the internal energy of a monatomic ideal gas in terms of S and V. To do this, we need an explicit expression that connects entropy with energy. This can be done by expressing the differential of S in terms of V and T, and substituting T with U using the energy–temperature relation for the ideal gas (Table 3.1). As was developed in Chapter 2, we can use these variables to write an exact differential S as

$$dS = \left(\frac{\partial S}{\partial T}\right)_V dT + \left(\frac{\partial S}{\partial V}\right)_T dV \tag{5.5}$$

Substituting the following two identities (Problem 5.10 and see Table 5.1 below)

$$\left(\frac{\partial S}{\partial T}\right)_V = \frac{C_V}{T} \tag{5.6}$$

$$\left(\frac{\partial S}{\partial V}\right)_T = \left(\frac{\partial p}{\partial T}\right)_V \tag{5.7}$$

into Equation 5.5 gives

$$dS = \frac{C_V}{T} dT + \left(\frac{\partial p}{\partial T}\right)_V dV \tag{5.8}$$

Inserting the (rearranged) ideal gas law ($p = nRT/V$) into the pressure derivative gives

$$dS = \frac{C_V}{T} dT + \frac{nR}{V} dV \tag{5.9}$$

This expression can be integrated along a two-step path in which T is varied with V held constant, and then V is varied with T held constant (Problem 5.10a). However, since nR is independent of $\ln T$, and for the ideal gas, C_v is independent of $\ln V$, the two differentials on the right-hand side of Equation 5.9 can be integrated over their single variables separately:

$$\int_{S_r} dS = C_V \int_{T_r} \frac{dT}{T} + nR \int_{V_r} \frac{dV}{V} \tag{5.10}$$

where the single integration limit (see Chapter 2) denotes a reference point r. Equation 5.10 can easily be solved:

$$S - S_r = C_V \ln\frac{T}{T_r} + nR \ln\frac{V}{V_r} \tag{5.11}$$

By replacing T with U in Equation 5.11 (through the relationship $U = 3nRT/2$ for a monatomic ideal gas; Table 3.1) and rearranging (Problem 5.11), we obtain the fundamental form we are looking for, namely, $U(S,V)$:

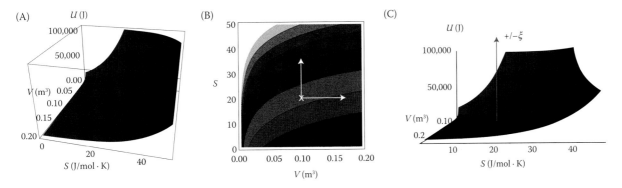

Figure 5.3 The internal energy of an ideal monatomic gas as a function of entropy and volume. In this $U(S,V)$ representation, the surface in (A) marks the equilibrium internal energy value for fixed values of entropy and volume. (B) Contour plot of the equilibrium internal energy surface versus S and V. Increasing S increases U (to lighter shades, vertical white arrow), whereas increasing V decreases U (to darker shades, horizontal white arrow). (C) Side-view, illustrating that any nonequilibrium configuration with the same value of S and V (but a different internal coordinate ξ) will have a higher value of U (gray arrow). The surface above was drawn for 1 mole, with $T_r = 273$ K, $V_r = 0.0224$ m³, and S_r arbitrarily set to a value of 10 J mol⁻¹ K⁻¹.

$$U = \frac{3}{2}nRT_r\left(\frac{V}{V_r}\right)^{2/3} e^{(S-S_r)/C_V} \tag{5.12}$$

This internal energy surface is plotted verus entropy and volume in **Figure 5.3**, for $n = 1$ mole of an ideal monatomic gas, with arbitrary reference values $T_r = 273$ K, $V_r = 22.4$ liters. The internal energy surface increases sharply with entropy at fixed volume (owing to the exponential dependence in Equation 5.12), but it decreases with increasing volume at fixed entropy (white arrows, Figure 5.3B). The increase in U with S at constant V results from the heat flow associated with this process (see Problem 5.12). The decrease in U with V at constant S (which corresponds to reversible adiabatic expansion, i.e., no heat flow), results from the volume work done by the system.

Finally, it is worth considering how the $U(S,V)$ equilibrium surface in Figure 5.3 relates to the potential plot of $U(\xi)$ at constant S and V in Figure 5.2. It is tempting to view the surface in Figure 5.3 as a potential that drives the system to lower energy values, which would occur at high volume and low entropy. However, this is *not* the correct view of the $U(S,V)$ potential. Rather, each point on the surface is an equilibrium point for a specific value of S and V. U acts as a potential when S and V are held constant, not when they are varied. According to Equation 5.3, any perturbation to the system away from the equilibrium state at constant S and V will increase U. This is illustrated by the vertical gray arrow in Figure 5.3C, where either an increase or decrease in an internal coordinate ξ increases U. The arrow is vertical (parallel to the U axis) because S and V are fixed. If we could visualize the $U(S,V)$ surface with a fourth dimension ξ, we would see an $U(\xi)$ representation such as in Figure 5.2. Points on the $U(S,V)$ surface in Figure 5.3 are minima (i.e., equilibrium values) with respect to ξ, as in Figure 5.2. An important implication is that there are no configurations that are below the $U(S,V)$ surface, though there are (nonequilibrium) configurations that are *above* the surface.

In principle, a system at constant S and V is less restrictive than an isolated system. However, in practice, it is very hard to carry out processes at constant entropy. And although constant volume systems are easy to prepare, they are not as convenient (and relevant; recall Figure 5.1) as systems at constant p and T. What we need is a differential energy function that switches both differential variables S and V in Equation 5.3 with the conjugate coefficients p and T.

OTHER ENERGY POTENTIALS FOR DIFFERENT TYPES OF SYSTEMS

To switch differential and nondifferential variables in Equation 5.3, we will introduce new energy functions that are simple combinations of U, pV, and TS. These functions are constructed so that when their differential forms are combined with Equation 5.3, the desired variable switch is obtained. We have already encountered one of these functions in Chapter 3, namely, the enthalpy:

$$H = U + pV \tag{5.13}$$

In differential form,

$$dH = dU + d(pV) = dU + pdV + Vdp \tag{5.14}$$

Substituting the inequality into Equation 5.3 for dU on the right-hand side of Equation 5.14 gives

$$\begin{aligned} dH &\leq (TdS - pdV) + pdV + Vdp \\ &\leq TdS + Vdp \end{aligned} \tag{5.15}$$

As a result of this substitution, the pdV term cancels, and is replaced with Vdp. The enthalpy inequality remains a less-than sign, because we put something into the right-hand side that is *larger* than the internal energy variation dU in Equation 5.14, if the variation is irreversible. Again, take a moment to convince yourself that the sign of the inequality is correct.

The transformation to enthalpy swaps the pressure–volume part of Equation 5.3. To swap the temperature-entropy part, we create an energy function to cancel TdS and replace it with $-SdT$:

$$G = H - TS \tag{5.16}$$

G is referred to as the "Gibbs free energy" function. In differential form,

$$dG = dH - d(TS) = dH - TdS - SdT \tag{5.17}$$

When combined with the enthalpy differential (Equation 5.15), we obtain

$$\begin{aligned} dG &\leq (TdS + VdP) - TdS - SdT \\ &\leq -SdT + VdP \end{aligned} \tag{5.18}$$

Again, the inequality results from the fact that our substitution on the right-hand side may be larger than dH in Equation 5.17.

According to Equation 5.18, the Gibbs free energy acts at a potential at constant T and p:

$$dG \leq 0 \quad (\text{constant } T, p) \tag{5.19}$$

The inequality applies away from equilibrium, where the direction of spontaneous change is that which decreases G; the equality in Equation 6.19 corresponds to equilibrium, at which point the Gibbs free energy is a minimum. We will make use of the Gibbs free energy potential extensively in the remainder of the text.

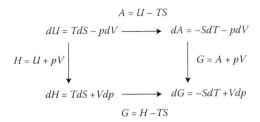

Figure 5.4 Strategy for transforming energy functions to act as potentials under various conditions. The vertical transformations swap $-pdV$ for Vdp. The horizontal transformations swap TdS for $-SdT$. Each differential defines the conditions under which the corresponding energy function acts as a potential. Here we represent transformations as equalities, which apply to reversible, equilibrium conditions.

Partly for completeness,[†] we will introduce a fourth energy function, the Helmholtz free energy function ($A = U - TS$),[‡] which swaps TdS but not pdV, giving the following result (Problem 5.3):

$$dA \leq -SdT - pdV \qquad (5.20)$$

The Helmholtz free energy acts as a potential for systems at constant temperature and volume.

The four potential energy functions described above (U, H, A, and G) and their differentials are shown in **Figure 5.4**. The inequality from the combined first-second law (Equation 5.3, which derives directly form the Clausius inequality) is retained in the variable-swapping that leads to the differentials dH, dA, and dG. Thus, the enthalpy and free energy inequalities are really just a restatement of the second law.

A more formal approach to generate *H*, *A*, and *G*: The Legendre transforms

Though the swapping procedure described above provides a simple way to generate new potentials that apply in different situations, the underlying approach is masked by the specifics. In this section, we will introduce the method underlying these manipulations, which is referred to as the method of "Legendre transforms." Legendre transforms are described in detail in Appendix 5.1.

For a single-variable function $f(x)$, the Legendre transform generates a new (but equivalent) representation, $\Lambda(D)$. The function $f(x)$ gives a point-by-point mapping of x onto y (through f). However, we could also generate an equivalent point-by-point mapping between the slope of the function (which we will represent with the variable D) and the corresponding y-intercept (which we will represent with the variable Λ). This representation contains the same information as the $f(x)$ mapping, though it maps in a different space. In this new space, we can write consider the intercept to be a function of the slope, that is, $\Lambda(D)$.

In differential form, the Legendre transform converts the single variable differential from $df = Ddx$ to $d\Lambda = xdD$. This is equivalent to our potential transforms above, for example, converting dH to dG by switching from TdS to $-SdT$.

[†] And partly because $A(T,V)$ plays a central role in statistical thermodynamics (Chapter 10).

[‡] Note that some authors use F for the Helmholtz free energy, however other authors use F for the Gibbs energy. Although it would make sense to use H for the Helmholtz energy, this symbol is already used for the enthalpy; moreover, Boltzmann (and later Shannon) used H to represent entropy.

Legendre transforms of multivariable functions

For multivariable functions, we can compute a Legendre transform of one or both (all) variables. For the energy potentials in Figure 5.4, $H(S,P)$ and $A(T,V)$ are single-variable transforms of $U(S,V)$, whereas $G(T,p)$ is a transform of both variables of $U(S,V)$. Though it is not obvious at first encounter, the approach of Legendre transforms gives $G(T,p)$ as a surface of intercepts of tangent planes with the U axis, with T and p values determined by the directional slopes (in the S and V directions, respectively) of the tangent planes. This is more easily illustrated transforming one variable at a time; here we will demonstrate starting with the $U(S,V)$ expression for a monatomic ideal gas (Equation 5.12), transforming first into enthalpy, and then into free energy.

Geometrically, the Legendre transform of $U(S,V)$ to $H(S,P)$ can be depicted by generating tangent lines to the $U(S,V)$ surface in the V-direction at various (S,V) values (**Figure 5.5A**). The intercepts of each of these lines with the U–S plane define an enthalpy value. The locus of each of these H values is determined by the S value of the original tangent in the V-direction, and the slope of the tangent, which gives the p value.

Analytically, to convert $U(S,V)$ into $H(S,P)$, transforming the fundamental variable V to p, we subtract away from U the derivative of U with respect to V multiplied by the original variable V:

$$H = U(S,V) - V\left(\frac{\partial U}{\partial V}\right)_S \tag{5.21}$$

This is analogous to Equation A5.1.7, where $\Lambda = H$. However, to obtain $H = H(S,p)$, we need to replace both $U(S,V)$ and V in Equation 5.21 with p. We can replace $U(S,V)$ using the energy–volume derivative explicitly:

$$\begin{aligned}
\left(\frac{\partial U}{\partial V}\right)_S &= -\frac{V_r^{2/3}}{V^{5/3}} nRT_r e^{(S-S_r)/C_V} \\
&= -\frac{1}{V}\left(\frac{V_r}{V}\right)^{2/3} nRT_r e^{(S-S_r)/C_V}
\end{aligned} \tag{5.22}$$

By comparing with the explicit expression for $U(S,V)$, it can easily be shown that

$$U(S,V) = -\frac{3V}{2}\left(\frac{\partial U}{\partial V}\right)_S \tag{5.23}$$

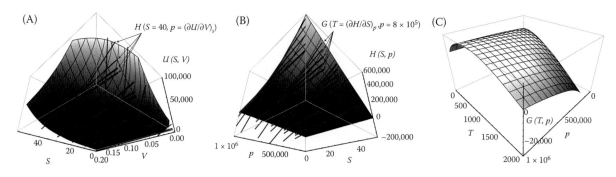

Figure 5.5 Legendre transforms of $U(S,V)$ to $H(S,p)$ to $G(T,p)$ for a monatomic ideal gas. (A) Internal energy surface from Figure 5.3, with a family of tangent lines (red) in the V direction. The intercepts of these tangents on the U–S plane (right face) give the enthalpy at S, with p given by the slope of the tangent. These tangents define the enthalpy surface, shown in (B). The intercepts to the $H(S,p)$ tangents in the S direction give the free energy at a given p value, with T given by the slope of the tangent. The resulting Gibbs free energy surface is shown in (C). The surfaces are drawn for $n = 1$ mole, either explicitly in (A), or implicitly through the chosen values of p_r, V_r, and T_r.

Substituting Equation 5.23 into 5.21 gives

$$
\begin{aligned}
H &= -\frac{3V}{2}\left(\frac{\partial U}{\partial V}\right)_S - V\left(\frac{\partial U}{\partial V}\right)_S \\
&= -\frac{5V}{2}\left(\frac{\partial U}{\partial V}\right)_S
\end{aligned}
\tag{5.24}
$$

Equation 5.24 still contains V, which can be substituted by rearranging Equation 5.22:

$$
\begin{aligned}
V^{5/3} &= \frac{-V_r^{2/3}nRT_re^{(S-S_r)/C_V}}{(\partial U/\partial V)_S} \\
V &= \left(\frac{-V_r^{2/3}nRT_re^{(S-S_r)/C_V}}{(\partial U/\partial V)_S}\right)^{3/5}
\end{aligned}
\tag{5.25}
$$

Substituting this expression for V into Equation 5.24 gives

$$
H = -\frac{5}{2}\left(\frac{\partial U}{\partial V}\right)_S\left(\frac{-V_r^{2/3}nRT_re^{(S-S_r)/C_V}}{(\partial U/\partial V)_S}\right)^{3/5}
\tag{5.26}
$$

Though Equation 5.20 is indeed the Legendre transform of U to H through V, it doesn't look like $H(S,p)$. Moreover, taking a noninteger (3/5) power of what appears to be a negative number is somewhat disconcerting, suggesting that the right-hand sided of Equation 5.26 has an imaginary component. Fortunately both problems are solved by recognizing that $(\partial U/\partial V)_S = -p$ (see Table 5.1, below). Substitution gives

$$
\begin{aligned}
H(S,p) &= \frac{5}{2}p\left(\frac{V_r^{2/3}nRT_re^{(S-S_r)/C_V}}{p}\right)^{3/5} \\
&= \frac{5}{2}p^{2/5}V_r^{2/5}(nRT_r)^{3/5}e^{3(S-S_r)/5C_V} \\
&= \frac{5}{2}p^{2/5}V_r^{2/5}(p_rV_r)^{3/5}e^{3(S-S_r)/5C_V} \\
&= \frac{5}{2}p^{2/5}p_r^{3/5}V_re^{3(S-S_r)/5C_V}
\end{aligned}
\tag{5.27}
$$

Because of the fairly extensive manipulation involved in transforming from Equations 5.21 through 5.27, it is worth checking that Equation 5.27 has the right dimensions. This is easily verified by noticing that the terms in the preexponential combine to give dimensions of pV, which is the same, dimensionally, as H.

The $H(S,p)$ surface is shown in **Figure 5.5B**. As with the internal energy, the enthalpy increases exponentially with entropy in the direction of constant pressure[†]. However, unlike the relationship between internal energy and volume, the enthalpy increases with pressure in the direction of constant entropy, although it has a soft dependence ($p^{2/5}$; see Problem 5.13).

[†] It makes sense that we have the same dependence, since we did not touch entropy in transforming $E(S,V)$ to $H(S,p)$.

From $H(S,p)$, a Legendre transform of T for S (Figure 5.5B) produces $G(T,p)$. For a monatomic ideal gas, the result of the transform is

$$G(T,p) = C_p T \ln\left(\frac{p^{2/5}T_r}{p_r^{2/5}T}\right) + T\left(C_p - S_r\right) \tag{5.28}$$

The derivation of the $G(T,p)$ surface for the monatomic ideal gas, which is shown in **Figure 5.5C**, is left as an exercise (Problem 5.16).

RELATIONSHIPS AMONG DERIVATIVES FROM THE DIFFERENTIAL FORMS OF *U*, *H*, *G*, AND *A*

As indicated in Figure 5.4, the new energy functions that result from Legendre transformation are constructed from combinations of state functions; thus, the potentials arc state functions as well. This means that their differentials are exact, and can be written in the form of Equation 1.40. For example, taking Gibbs free energy as a function of T and p, [that is, $G = G(T,p)$], we can write

$$dG = \left(\frac{\partial G}{\partial T}\right)_p dT + \left(\frac{\partial G}{\partial p}\right)_T dp \tag{5.29}$$

There are two important sets of relationships that come from these exact differential expressions. One set equates derivatives of the energy potentials (U, H, A, and G) to the variables T, p, S, and V (and later, moles and chemical potentials μ). A second set, the so-called "Maxwell relations," connects cross derivatives of T, p, S, and V (and n and μ, as described below). Both sets of relationships allow thermodynamic parameters that are difficult to measure directly to be related to quantities that can easily be measured experimentally.

Derivatives of the energy potentials

By equating exact differential expressions such as Equation 5.29 with the Legendre transforms, we can identify relationships between the derivatives of the energy potentials (U, H, G, and A) with the variables T, S, p, and V. For example, both Equations 5.29 and 5.18 represent dG, in similar (but not identical) form. By equating dG from these two expressions (using the reversible equality in Equation 5.18), we obtain

$$-SdT + Vdp = \left(\frac{\partial G}{\partial T}\right)_p dT + \left(\frac{\partial G}{\partial p}\right)_T dp \tag{5.30}$$

Because T and p are independent variables in Equation 5.30, we can set the quantities dp and dT independently without invalidating the equality.[†] This means that Equation 5.30 really defines two separate equalities, namely,

$$-SdT = \left(\frac{\partial G}{\partial T}\right)_p dT$$

$$+Vdp = \left(\frac{\partial G}{\partial p}\right)_T dp \tag{5.31}$$

[†] As an extreme (but conceptually useful) example, we could vary p but not T, or vice versa.

Since dT and dp are generally nonzero, Equation 5.31 requires that

$$S = -\left(\frac{\partial G}{\partial T}\right)_p$$

$$V = \left(\frac{\partial G}{\partial p}\right)_T$$

(5.32)

These two identities are both powerful relationships. The first identity provides a fairly simple way to determine entropy. This is important because unlike temperature and pressure,[†] there is no method that can directly measure entropy. *We will use the entropy part of* Equation 5.32 *again and again, and it should be committed to memory.* The second identity provides a sensitive way to measure volume change. Although volume is not difficult to measure in principle, volume changes during chemical reactions are usually too small to measure by conventional means (length times width times height). By determining the Gibbs free energy variation with pressure, small reaction volume changes can be measured quite accurately (Equation 5.32).

Each of the four energy differentials shown in Figure 5.4 leads to a unique pair of thermodynamic identities like those in Equation 5.32 (**Table 5.1**). The intensive variables T and p result from differentiation with respect to extensive quantities, whereas the extensive variables S and V result from differentiation with respect to intensive quantities.

One particularly fundamental (and somewhat surprising) identity comes from the derivative of internal energy with respect to entropy:

$$T = \left(\frac{\partial U}{\partial S}\right)_V$$

(5.33)

This relationship states that the temperature, a quantity that we have an intuitive (and at least some mechanistic) understanding of, is equal to the extent to which the internal energy increases as the entropy increases.[‡] At low temperature, a unit energy change corresponds to a large change in entropy. At high temperature, a unit energy change corresponds to a small change in entropy. This is consistent with the description of entropy in Chapter 4, where we stated that at low temperature, heat produces a large increase in entropy compared to that at high temperature.

Table 5.1 Energy potentials, differentials, and derivatives

Integrated form	Differential form[a]	Energy derivative identities		Maxwell Relations
U(S,V)	dU = TdS−pdV	$\left(\frac{\partial U}{\partial S}\right)_V = T,$	$\left(\frac{\partial U}{\partial V}\right)_S = -p$	$\left(\frac{\partial T}{\partial V}\right)_S = -\left(\frac{\partial p}{\partial S}\right)_V$
H(S,p) = U + PV	dH = TdS + Vdp	$\left(\frac{\partial H}{\partial S}\right)_p = T,$	$\left(\frac{\partial H}{\partial p}\right)_S = V$	$\left(\frac{\partial T}{\partial p}\right)_S = -\left(\frac{\partial V}{\partial S}\right)_p$
A(T,V) = U−TS	dA = −SdT−pdV	$\left(\frac{\partial A}{\partial T}\right)_V = -S,$	$\left(\frac{\partial A}{\partial V}\right)_T = -p$	$-\left(\frac{\partial S}{\partial V}\right)_T = -\left(\frac{\partial p}{\partial T}\right)_V$
G(T,p) = H−TS	dG = −SdT + Vdp	$\left(\frac{\partial G}{\partial T}\right)_p = -S,$	$\left(\frac{\partial G}{\partial p}\right)_T = V$	$-\left(\frac{\partial S}{\partial p}\right)_T = \left(\frac{\partial V}{\partial T}\right)_p$

[a] Differential relations are evaluated at equilibrium, and thus are equalities.

[†]   Where we have thermometers and pressure gauges.
[‡]   When volume is constant (i.e., where no pV work can be done). When pressure is constant, there is a similar relationship involving the enthalpy.

Figure 5.6 The relationship between internal energy, entropy, and temperature for an energy ladder with equal spacings. As in each subsystem in Figure 3.3, the model has four energy levels and four molecules. From left to right, internal energy (which we take as equal to the total energy) increases by one unit. Distinguishable configurations with the same total energy are shown in different rows, with multiplicities (W_i) given under each configuration. The entropy, calculated from the total multiplicity using Boltzmann's formula, increases with energy (corresponding to a positive absolute temperature, based on Equation 5.33). However, at higher energies, the increase in entropy (ΔS) per unit energy increase (ΔU) goes down, indicating an increase in temperature.

U	4	5	6	7	8	9	10
ΔU	--	1	1	1	1	1	1
	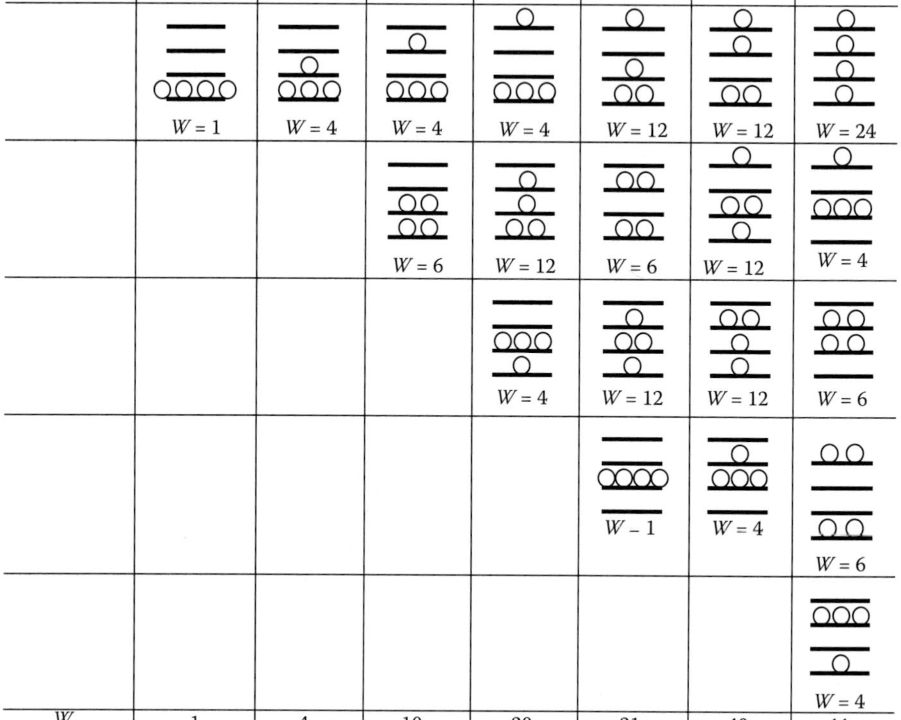 $W = 1$	$W = 4$	$W = 4$	$W = 4$	$W = 12$	$W = 12$	$W = 24$
			$W = 6$	$W = 12$	$W = 6$	$W = 12$	$W = 4$
				$W = 4$	$W = 12$	$W = 12$	$W = 6$
					$W = 1$	$W = 4$	$W = 6$
							$W = 4$
W_{tot}	1	4	10	20	31	40	44
$S/k = \ln W$	0	1.39	2.30	2.99	3.43	3.69	3.78
$\Delta S/k = \Delta\ln W$	--	1.39	0.91	0.69	0.44	0.26	0.09
$T = \Delta U/\Delta S$	--	0.72	1.10	1.45	2.25	3.84	11.1

This relationship can be illustrated with the four-particle energy ladder from Chapter 4 (**Figure 5.6**). Configurations are arranged in terms of energy (from left to right), with energy increasing by one unit from one column to the next. For each of these energy increases, the total number of microscopic configurations goes up, consistent with an entropy increase, and thus, from Equation 5.33, a positive temperature value. However, as the total energy increase, the entropy increase (ΔS, calculated using the Boltzmann entropy formula from Chapter 4) per unit energy increase (ΔU) goes down, consistent with a higher temperature. This relationship is shown in the bottom line of Figure 5.6.

Another important implication of Equation 5.33 is that as long as temperature is positive, entropy *must* increase energy is increased (and vice versa). Any situation where entropy decreased with increasing energy would require a negative temperature scale,[†] a concept that is explored in Problem 5.2.

Maxwell relations

In addition to the thermodynamic identities above, a second set of derivative relations, termed the Maxwell relations,[‡] can be obtained from cross derivatives from each of the four Legendre-transformed potentials. Recall that from Chapter 2, the Euler theorem states that for differentials of state functions, the cross-derivatives of

---

[†] Although apparently negative temperatures can be generated for systems with a finite number of energy levels, such as that in Figure 5.6 (see Chapter 4 of Zemansky [1964]), systems with an infinite number of molecular energy levels will not show negative temperature phenomena, which would require an infinite amount of energy (or more!). Most systems we encounter have an infinite number of energy levels (such as ideal gases, where the gas molecules can steadily increase their velocities (below the speed of light).

[‡] Do not confuse these with the famous Maxwell equations from electromagnetic theory. Both sets of equations are attributed to James Clerk Maxwell, but they are not the same.

the sensitivity coefficients in Figure 5.3 are equal (that is, order of differentiation does not matter). Focusing on Equation 5.29, this means that

$$\left(\frac{\partial}{\partial p}\left(\frac{\partial G}{\partial T}\right)_p\right)_T = \left(\frac{\partial}{\partial T}\left(\frac{\partial G}{\partial p}\right)_T\right)_p \tag{5.34}$$

But the "inside" derivatives in Equation 5.34 are, from the thermodynamic identities defined in Equation 5.32, equal to $-S$ and V, respectively. Substituting gives the expression

$$-\left(\frac{\partial S}{\partial p}\right)_T = \left(\frac{\partial V}{\partial T}\right)_p \tag{5.35}$$

This expression, which is referred to as a "Maxwell relation," allows a quantity that would be very difficult to measure directly (the pressure derivative of entropy) to be determined from another, much simpler measurement (the temperature dependence of volume, which is equal to the volume times the coefficient of thermal expansion, $V\alpha$). Each energy differential in Figure 5.3 defines a unique Maxwell relation that connects $p, T, S,$ and V (Table 5.1). Importantly, these derivative relationships in Table 5.1 are general. They do not depend on the type of system or material within it, nor do they depend on whether a system behaves ideally or has strong internal interactions.

Manipulation of thermodynamic derivative relations

Table 5.1 has a large number of derivatives to keep track of, and when we include concentration as a variable (below), we will have even more. However, many of these derivatives can be related to each other (see Callen, 1985), and to simple measurable quantities. In fact, all of these relationships can be expressed by just three measurable coefficients: C_p, α, and κ_T. To get to such expressions, some general rules that relate partial derivatives are required (see **Box 5.1**).

Here, we show an example of such a manipulation that provides an important connection between Gibbs free energy and heat capacity. Starting with the second temperature derivative of the Gibbs free energy, we can substitute in the entropy identity (Table 5.1):

$$\left(\frac{\partial^2 G}{\partial T^2}\right)_p = \frac{\partial}{\partial T}\left(\frac{\partial G}{\partial T}\right)_p = -\left(\frac{\partial S}{\partial T}\right)_p \tag{5.36}$$

Expanding $(\partial S/\partial T)_p$ using the chain rule (Problem 5.1) using H,

$$\left(\frac{\partial^2 G}{\partial T^2}\right)_p - \left(\frac{\partial S}{\partial T}\right)_p = -\left(\frac{\partial S}{\partial H}\right)_p\left(\frac{\partial H}{\partial T}\right)_p \tag{5.37}$$

Equation 5.37 can be simplified by applying the reciprocal rule to $(\partial S/\partial H)_p$, and by recognizing that $(\partial H/\partial T)_p = C_p$:

$$\left(\frac{\partial^2 G}{\partial T^2}\right)_p - \left(\frac{\partial S}{\partial H}\right)_p = -\frac{C_p}{\left(\dfrac{\partial H}{\partial S}\right)_p} \tag{5.38}$$

Finally, recognizing that $(\partial H/\partial S)_p = T$, the second partial temperature derivative of G can be written using the familiar variables

$$\left(\frac{\partial^2 G}{\partial T^2}\right)_p = -\frac{C_p}{T} \tag{5.39}$$

Box 5.1: Manipulation of partial derivatives

Rule 1. The chain rule. If there is a function $f(x)$ that maps x values onto y values, and another function $g(y)$ that maps y values onto z values, we can combine f and g to map x values to z values: $g(x) = g(f(x))$. In such cases, the derivative of g with respect to x is given by

$$\left(\frac{\partial g}{\partial x}\right) = \left(\frac{\partial g}{\partial f}\right)\left(\frac{\partial f}{\partial x}\right) \tag{5.40}$$

This rule is useful when it is necessary to introduce a variable (in this case, $f(x)$) that is not currently expressed in the differential equation, as we did with H in Equation 5.37.

Rule 2. The reciprocal rule. A partial derivative of variable y with respect to variable x is equal the inverse of the derivative of x with respect to y, that is

$$\left(\frac{\partial y}{\partial x}\right)_z = \frac{1}{\left(\dfrac{\partial x}{\partial y}\right)_z} \tag{5.41}$$

We used this rule to derive Equation 5.38.

Rule 3. The reciprocity theorem. If we take permute the order of partial differentiation, along with the variable held constant, the product of the three permutations is

$$\left(\frac{\partial x}{\partial y}\right)_z\left(\frac{\partial z}{\partial x}\right)_y\left(\frac{\partial y}{\partial z}\right)_x = -1 \tag{5.42}$$

This rule is useful when a parameter that is held constant needs to be exchanged with a variation parameter (i.e., a parameter within the derivative). This is really a combination of rules 1 and 2.

The behavior of energy derivatives in changes of state

As described in Chapter 3, although we can derive formal relationships in classical thermodynamics in terms of total energy, entropy, etc., we cannot apply such relationships to experimental measurements. Instead, we focus on how these functions change for a finite change of state. For a thermodynamic variable X, this means that for initial and final states i and f, we measure ΔX (where $\Delta X = X_f) - X_i$, but not X_f or X_i alone.

Although the energy derivatives in Table 5.1 involve total energies, the distributive property of differentiation[†] (i.e., the derivative of a sum (or difference) is equal to a sum (or difference) of derivatives) means that the derivatives in Table 5.1 can be applied to finite changes in energies and entropies (i.e., quantities of the form ΔX).

[†] Integration of differences in thermodynamic functions is also distributive, which will be useful in analyzing reaction thermodynamics.

For example, for the free energy–entropy relationship (Equation 5.33), we can write equations for initial and final states:

$$S_f = -\left(\frac{\partial G_f}{\partial T}\right)_p$$
$$S_i = -\left(\frac{\partial G_i}{\partial T}\right)_p$$

(5.43)

The initial and final states can differ in physical properties (e.g., phase, Chapter 6), or their extent of mixing or chemical reaction (Chapter 7). As long as the derivatives are taken at the same temperature (and pressure), we can subtract these two equations to get the change in entropy from the initial to final state:

$$S_f - S_i = -\left(\frac{\partial G_f}{\partial T}\right)_p + \left(\frac{\partial G_i}{\partial T}\right)_p = -\left(\frac{\partial \{G_f - G_i\}}{\partial T}\right)_p$$

(5.44)

Using the standard delta nomenclature, we can write Equation 5.44 as

$$\Delta S = -\left(\frac{\partial \Delta G}{\partial T}\right)$$

(5.45)

In short, deltas can be inserted into the numerators of the derivatives in Table 5.1 and into the left-hand sides of the thermodynamic identities, giving an analogous set of relations that represent changes in the thermodynamic functions from initial to final state.

CONTRIBUTIONS OF DIFFERENT CHEMICAL SPECIES TO THERMODYNAMIC STATE FUNCTIONS—MOLAR QUANTITIES

So far, our analysis of thermodynamic systems has largely ignored concentrations of different chemical components. The closest we have come is to specify the *total* number of moles in gas laws (Chapter 3). However, to analyze the thermodynamics of complex mixtures (and reactions in particular), we need to be able to represent the contribution of each component to thermodynamic functions such as the Gibbs free energy, enthalpy, and volume. Here we will develop a formalism that represents the thermodynamic contributions of each component, and combines each of these contributions to give the thermodynamics of the whole system. This approach not only provides a basis for analyzing discrete reactions and phase transitions, it generates additional species-based Maxwell relations that form the basis of thermodynamic linkage and energetic coupling. Such linkages are important for describing cooperative effects, as are seen in ligand binding equilibria and signal transduction.

Given the strong interaction between molecules in solution, especially when chemical reactions are involved, separating overall thermodynamic quantities such as free energy into contributions from individual species might seem a daunting task. Fortunately, the nonmechanistic approach of classical thermodynamics is a strength in this context. By taking partial derivatives of the quantity of interest (free energy, etc.) with respect to the number of moles each species, we define "partial molar" quantities. This approach does not require detailed structural knowledge of the species or their modes of interaction.

Table 5.2 Representation of a general system variable _X_ in total, partial, and molar terms

	Total	Partial quantity
System	X	X_i
Molar quantity	$\bar{X}$	$\bar{X}_i$

Molar quantities

A molar quantity is the amount of a particular extensive thermodynamic quantity (X, which could be energy, entropy, heat capacity, etc.) contributed per mole[†] (n for total moles, n_i, for moles of species i in a mixture). To indicate the molar version of a quantity, we will use an overbar (i.e., $\bar{X}$). To indicate the species that is contributing to the quantity (important in mixtures of different species), we will use an i subscript (i.e., X_i, $\bar{X}_i$). We will refer to such quantities as "partial" with respect to species i, because it is the part of the total $\bar{X}$ contributed by species i.[‡] This nomenclature is summarized in **Table 5.2**.

One important feature of the conversion to molar quantities is that whereas X is extensive, $\bar{X}$ is an intensive quantity. Typically, intensive quantities provide a more fundamental description of the chemistry of the system than their extensive counterparts. For example, if we wish to compare the reactivity of different chemical species such as adenosine triphosphate (ATP, a major store of chemical energy in the cell) with other organic phosphate compounds (e.g., creatine phosphate), the free energy *per mole* for the two compounds ($\bar{G}_{ATP}$ versus $\bar{G}_{CP}$) provides an immediate and clear comparison. In contrast, the total free energies of these two compounds in arbitrary systems would depend on details such as total system sizes and concentrations of each.

Here we will first develop molar quantities for a system with only a single species, to illustrate the formalism. We will do this both for molar volume and for molar Gibbs free energy. Though free energy analysis will be of greater interest to us in the long run than volume analysis, the latter provides a good starting place because it is an intuitive quantity. Since the development of molar free energies parallels that of molar volumes, intuition associated with analysis of molar volume will aid in understanding molar free energies. Following analysis of single-species systems, we will develop "partial" molar quantities for individual species in mixtures. Combining partial molar free energies with the differentials of energy potential described above (Figure 5.3) provides a powerful set of "linkage relationships" that connect species thermodynamically.

Molar volumes for systems made of only one species

For single-species systems, such as container of pure liquid water, molar quantities can be determined simply by dividing the total quantity X by the number of moles n:

$$\bar{X}_i^* = \frac{X}{n_i} \tag{5.46}$$

The asterisk in Equation 5.46 is used to indicate that the system consists of only one species, that is, the system is "pure." Although the i subscript is unnecessary for a single-species system, it serves as a useful point of comparison to mixtures.

[†] Be sure not to confuse the term "molar" in the present context with "molarity" (moles per liter, which is an intensive property). Though it would make more sense to refer to "mole quantities" (e.g. mole volume, partial mole free energy), the molar expression is widely used, and changing nomenclature here would cause more confusion than clarity.

[‡] Also, it is related to the partial derivative with respect to moles of species i.

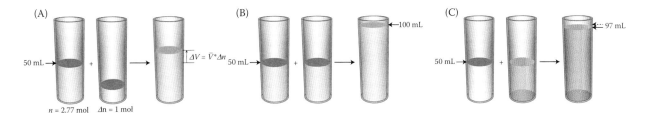

Figure 5.7 Relationship between total and molar volumes for a pure, single-component liquid (water, A and B), and a liquid of two component (water and ethanol, C). (A) For a pure liquid, the molar volume $\bar{V}^*$ is simply the change in volume (ΔV) that results when one mole is added ($\Delta n = 1$). (B) Because total and molar volumes are extensive and intensive quantities, respectively, doubling the size of the system (from n to $2n$ moles) doubles the total volume, but $\bar{V}^*$ remains unchanged. Here, combining two 50 mL volumes produces 100 mL total. However, in (C), when 50 mL of water is added to 50 mL of ethanol, the resulting mixture "shrinks" to a total volume of 97 mL. This is because the partial molar volumes of the two species are less in the mixture than in their pure liquids. Unlike pure water (B), mixing is not a true doubling of an otherwise identical system—the compositions of the uncombined pure liquids differ from the combined mixture (and from each other). Thus, extensive properties (such as total volume) need not double, and intensive properties such as $\bar{V}_i$ (in this case, $\bar{V}_{H_2O}$ and $\bar{V}_{EtOH}$) need not remain constant.

As an example, in **Figure 5.7A**, we can calculate the molar volume of pure water in the left-most cylinder by dividing the volume (50 mL total volume at 285 K) by the number of moles of water (2.77 mol)[†]:

$$\bar{V}^*_{H_2O} = \frac{V}{n_{H_2O}} = \frac{50\,\text{mL}}{2.773\,\text{mol}} = 18.03\,\text{mL/mol} \tag{5.47}$$

Rearranging Equation 5.46 gives an expression for the total system volume:

$$V = n_{H_2O} \times \bar{V}^*_{H_2O} \tag{5.48}$$

Although Equation 5.48 almost seems too obvious to write, the same form will be used (with greater impact) in mixtures, and will be particularly important when adding partial molar free energies to obtain the total Gibbs free energy of a system.

In addition to describing the volume contributed by a particular species, $\bar{V}$ acts as coefficient that describes how total volume changes when a given amount of a species is added. In the example shown in Figure 5.7A, where 1 mol of water (Δn) is added to 50 mL of pure water, the molar volume can be represented as the ratio of the volume increment ($\Delta V = 18.05$ mol) to the mole increment:

$$\bar{V}^*_{H_2O} = \frac{\Delta V}{\Delta n_{H_2O}} = \frac{18.03\,\text{mL}}{1\,\text{mol}} = 18.03\,\text{mL/mol} \tag{5.49}$$

Of course, Equation 5.49 is valid for mole increments other than $\Delta n = 1$.

Of particular interest are very small values of Δn, for which we can replace increments with differentials. For pure species i, we can write the differential form of the molar volume as

$$\bar{V}^*_i = \left(\frac{\partial V}{\partial n_i}\right) \tag{5.50}$$

In words, the molar volume of a pure component is the derivative of the total (extensive) volume with respect to the mole number. The partial derivative notation is used because some external variables (likely p and T) will be held constant,

---

[†] The units given in Equation 5.47 are not SI units (*mL/mol*), but are often used out of convenience (there are 10^6 mL in a cubic meter, which would make the molar volume of pure water 1.805×10^{-5} m^3/mol).

and importantly, for mixtures (below), the mole numbers of other species will be held fixed (see Equation 5.53 below). Finally, note that Equation 5.50 can be rearranged to give the total volume differential (analogous to the rearrangement in Chapter 2, Equations 2.3 and 2.6):

$$dV = \bar{V}_i^* dn_i = \left(\frac{\partial V}{\partial n_i}\right) dn_i \tag{5.51}$$

Molar volumes in mixtures

For mixtures of multiple species, many of the concepts introduced above still apply, although there is a major complication: partial molar quantities in mixtures depend on composition. For example, the partial molar volume of water at a mole fraction[†] of 0.5 with ethanol (a two-component or "binary" mixture) differs from the molar volume of water in pure water ($\bar{V}_{H_2O}^*$). This is illustrated in **Figure 5.7C**, where 50 mL of water is mixed with 50 mL of ethanol. Unlike the pure water example in **Figure 5.7B**, the total volume of the resulting mixture ($V = 97$ mL) is not the sum of the two starting volumes ($50 + 50$ mL). Clearly, barring chemical reaction (and sloppy pouring), the number of moles of water and ethanol (n_{H2O}, n_{EtOH}) remains unchanged. The -3 mL discrepancy reflects the dependence of partial molar volumes on composition.

This is shown quantitatively in **Figure 5.8A**, where the measured volumes of water:ethanol mixtures in different proportions (red circles) are lower than would be expected from a simple mole-fraction weighted average of pure molar volumes (black line). This mole-fraction weighted average, which provides a useful comparison with real solutions, will be referred to as showing "mixing volume ideality." Here, this ideality is represented by the equation

$$\bar{V}_{ideal} = x_{H_2O}\bar{V}_{H_2O}^* + x_{EtOH}\bar{V}_{EtOH}^* \tag{5.52}$$

The difference between the real and ideal volumes of water:ethanol mixtures, ΔV, is shown in **Figure 5.8B**.

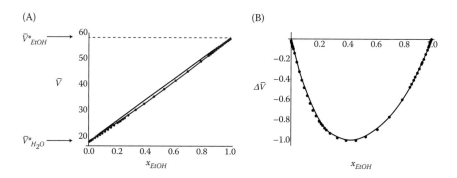

Figure 5.8 The volume of mixtures of liquid water and ethanol. (A) The total molar volume (in mL/mol, obtained by dividing the bottom line of Equation 5.60 by n_T) at different compositions. If molar volume were independent of composition ("volume ideality"), $\bar{V}$ would be a straight line (black, Equation 5.52) connecting $\bar{V}_{H_2O}^*$ and $\bar{V}_{EtOH}^*$. However, mixtures of water and ethanol have a lower measured volume (red circles; the red line is a fifth order polynomial fit to the data to guide the eye). (B) The volume difference between the ideal and real mixture ($\Delta\bar{V}$). At $x_{EtOH} = 0.2$ (the mole fraction of a 50:50 volume mixture), $\Delta\bar{V}$ is about -8 mL per mole, consistent with the 3 mL contraction described in Figure 5.7C. Data are from Marsh and Richards (1980).

[†] Mole fractions are good compositional variables because they uniquely specify molar quantities, but unlike a list of total mole numbers (an extensive quantity), they have the generality of an intensive variable. In addition, because the sum of the mole fractions must add to one, there is one fewer variable required to uniquely specify the composition than with straight mole numbers.

To capture deviations like that shown in Figure 5.8B, we need a framework for measurement (and representation) of partial molar volumes in real mixtures, and the total volumes of such mixtures. Unlike our approach to measuring partial volumes of pure species, we cannot add a finite number of moles of species i to a finite mixture and simply read off the volume change, because the composition of the mixture will change. Because partial molar quantities depend on composition, the resulting volume increment will depend on both the molar volumes of the initial and final mixtures. Thus, an equation of the form of 5.49 will not work for mixing experiments such as that shown in Figure 5.7C.

One way to minimize this change composition is to make the starting mixture very large, so that a finite addition of species i only modestly changes the composition of the resulting mixture. Although this "swimming pool" approach would permit $\bar{V}_i$ to be approximated by Equation 5.49, this method is impractical and instead, we will go the other direction, decreasing the mole increment of species i to an infinitesimal (dn_i). When only an infinitesimal amount of species i is added to a mixture, its composition stays constant. In analogy with Equation 5.50, we will define the partial molar volume as

$$\bar{V}_i = \left(\frac{\partial V}{\partial n_i} \right)_{n_{j \neq i}} \tag{5.53}$$

where V is the total volume. The subscript on the derivative means that the mole numbers of all other species are fixed ("bystanders" in this calculation).

As with Equations 5.50 and 5.51, $\bar{V}_i$ can be rearranged to calculate the differential change in total volume (dV) resulting from the variation dn_i. However, to be general, we should include variations in mole numbers of all the species present. For two species, the variation equation is[†]

$$dV = \bar{V}_A dn_A + \bar{V}_B dn_B$$
$$= \left(\frac{\partial V}{\partial n_A} \right)_{n_B} dn_A + \left(\frac{\partial V}{\partial n_B} \right)_{n_A} dn_B \tag{5.54}$$

Note that another route to Equation 5.54 would be to write V is a function of the amount of each component (i.e., $V = V(n_A, n_B)$), and evaluate the exact differential (as in Equation 2.22) at fixed p and T.

With Equation 5.54, we are in position to express the total volume in terms of the number of moles of species A and B by integration:[‡]

$$V_f = \int_0^{V_f} dV = \int_0^{V_f} (\bar{V}_A dn_A + \bar{V}_B dn_B) \tag{5.55}$$

Unfortunately, since there are multiple integration variables on the right-hand side of Equation 5.55, an integration path must be specified. Though all paths for integration are equally valid because volume is a state function, some paths lead to easier integration than others.

Mixing volume ideality

One conceptually simple path is to increase n_A from zero moles to its final value, while holding n_B at zero, and then to increase n_B from zero moles to its final value

---

[†] Here we use only two species to keep things general, but for complex mixtures, any number of $\bar{V}_i dn_i$ can be included.

[‡] The "f" subscripts are used to keep the actual number of moles of a and b distinct from the integration variable. The notation is *not* meant to reproduce an actual physical change from an initial to a final state, in contrast to the integrals changes of state in Chapters 2 and 3.

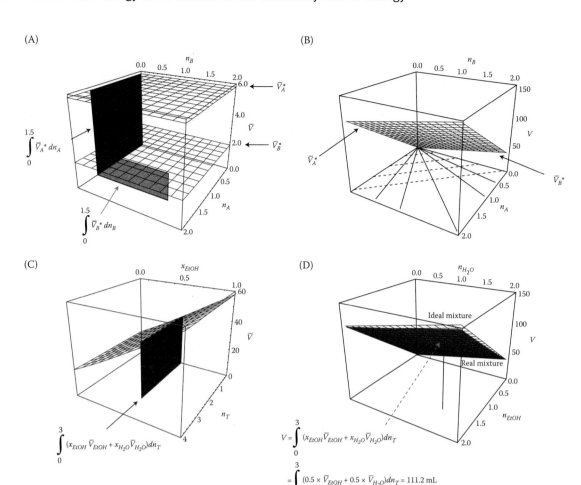

Figure 5.9 How partial molar volumes determine the volume of mixtures. (A, B) A two-component ideal mixture in which the molar volumes of each component are independent of composition (shown by the flat surfaces for $\bar{V}_A^*$ and $\bar{V}_B^*$ in A). The total volume at a given composition can be found by integrating over one component and then the other. Integration from $n_A = n_B = 0$ to n_A, n_B moles gives the total volume as a plane (B) with an intercept of zero, and slopes in the two directions (n_A, n_B) equal to partial molar volumes (Equation 5.56). For real mixtures such as water and ethanol (C, D), partial molar volumes depend on composition, and as a result, the volume of real mixtures deviates from the ideal mixture plane (red vs. white surfaces, panel D). In this case, one simple path for integration is along a line of fixed composition, using total moles as a single integration variable (C). This constant composition path corresponds to the black lines in the n_A, n_B plane (B).

while holding n_A constant (**Figure 5.9A**). For mixtures that have volume ideality (Equation 5.52), integration along this path is also analytically simple. Since the partial molar volumes are independent of concentration for volume ideality, Equation 5.55 can be simplified to

$$V_f = \int_0^{n_A} \bar{V}_A^* dn_A + \int_0^{n_B} \bar{V}_B^* dn_B$$

$$= \bar{V}_A^* n_A + \bar{V}_B^* n_B$$

(5.56)

That is, V_f is sum of the area of the two red rectangles in Figure 5.9A (Problem 5.9). The integrated form of Equation 5.56 (which is analogous to Equation 5.52) applies to all n_A, n_B values, and defines the total volume ($V = V_f$ in this case) as a plane in a space composed of the moles of each species (**Figure 5.9B**). The slopes of the plane in each direction give the partial molar volumes, and the intercept on the V axis is zero.[†]

[†] The zero intercept is consistent with the obvious fact that a system with zero moles has zero volume (at least for liquids and solids).

Mixing volume nonideality

If partial molar volumes depend on composition (as for water-ethanol mixture, Figure 5.8), most integration paths require an analytical expression of the dependence of each $\bar{V}_i$ on composition. One sensible solution to this problem is to integrate along a path where the mole fractions remain constant, by incrementing the moles of all species in the same proportion. This path is a straight line in the plane of n_A and n_B values starting at zero moles (where $V = 0$) and ending at the final composition n_A, n_B (see the black radial lines in Figure 5.9B).

Because the composition of the system is constant along these radial paths, the partial molar volumes are constant. In addition, since n_A and n_B increase in constant proportion along this path, their variations dn_A and dn_B can be represented by a single variable, the total number of moles dn_T.

$$dn_A = x_A dn_T \tag{5.57}$$

$$\begin{aligned} dn_B &= x_B dn_T \\ &= (1 - x_A)dn_T \end{aligned} \tag{5.58}$$

Substituting Equations 5.57 and 5.58 into 5.55 leads to an integral of a single variable:

$$\begin{aligned} \int_0^{V_f} dV &= \int_0^{n_A,n_b} (\bar{V}_a x_a dn_T + \bar{V}_b x_b dn_T) \\ &= \int_0^{n_T} (\bar{V}_a x_a + \bar{V}_b x_b)dn_T \end{aligned} \tag{5.59}$$

Because the terms in parentheses are constant along any given radial path, they can be taken outside the integral, leading to the solution.

$$\begin{aligned} V_f &= (\bar{V}_A x_A + \bar{V}_B x_B) \int_0^{n_T} dn_T \\ &= (\bar{V}_A x_A + \bar{V}_B x_B)n_T \end{aligned} \tag{5.60}$$

This integration is schematized in **Figure 5.9C**, where the composition is held fixed at a mole fraction of 0.5, and integration limits are zero and three total moles ($n_{H2O} = n_{EtOH} = 1.5$, given $x = 0.5$). These are the same limits as for the ideal mixture in Figure 5.9A, though the path is different. Eliminating n_T gives[†]

$$V = \bar{V}_A n_A + \bar{V}_B n_B \tag{5.61}$$

Equation 5.61 is an important (and quite sensible) result. It says that the total volume is equal to the partial molar volume of each species times the number of moles of that species. Equation 5.61 is of the same form as Equation 5.48, though it is for two components. Perhaps not surprisingly, Equation 5.61 generalizes to mixtures with any number of species. Though Equation 5.61 has a very simple form, keep in mind that the partial molar volumes depend on the composition of the system. Equation 5.61 specifies the volume of a system of *fixed* composition, given a set of partial molar volumes. Including variation in composition would require an explicit dependence of $\bar{V}_A$ and $\bar{V}_A$ on composition.

[†] To eliminate n_T, express the mole fractions in terms of moles, *that is,* $x_A = n_A/n_T$.

Table 5.3 Partial molar free energies and their relation to molar volumes

	Gibbs free energy		Volume analogy
Molar free energy of a pure (single-species) system:	$\bar{G}_i^* = \mu_i^* = \dfrac{G}{n}$	(5.62)	$\bar{V}_i^* = \dfrac{V}{n_i}$
Partial molar free energy of species i in a mixture:	$\bar{G}_i = \mu_i = \left(\dfrac{\partial G}{\partial n_i}\right)_{T,p,n_{j\neq i}}$	(5.63)	$\bar{V}_i = \left(\dfrac{\partial V}{\partial n_i}\right)_{T,p,n_{j\neq i}}$
Variation in total free energy with mole number (constant T, p):	$dG = \displaystyle\sum_{i=1}^{n\,\text{species}} \mu_i dn_i$	(5.64)	$dV = \displaystyle\sum_{i=1}^{n\,\text{species}} \bar{V}_i dn$
Total free energy:	$G = \displaystyle\sum_{i=1}^{n\,\text{species}} \mu_i n_i$	(5.65)	$V = \displaystyle\sum_{i=1}^{n\,\text{species}} \bar{V}_i n_i$

Molar free energies: The chemical potential

Our treatment of volumes provides an intuitive illustration of how to think about the contributions of each species in a mixture to a total extensive variable, and produces a clear framework for analysis of mixtures. However, the quantity of primary interest in analysis of chemical and biological reactions is not volume but Gibbs free energy. Although the decomposition of free energy into the contributions of individual species is not as intuitive as for volume, the same methods apply, and the resulting equations connecting partial molar free energies to the total have the same form as those for the volume.

Because partial molar free energy plays such an important role in thermodynamic analysis, where different species interconvert, it is given a special name, the "chemical potential," and a special symbol, μ_i. A summary of partial molar free energies their defining equations, and their analogy to molar volumes is given in **Table 5.3**. Like molar volumes, chemical potentials are the derivatives of the total Gibbs free energy with respect to mole number (Equations 5.63), and in a pure species, this derivative becomes a simple ratio (Equations 5.62). Also like volumes, the total Gibbs energy is obtained by summing the product of the chemical potential and number of moles of each species (Equations 5.65).

If variations in temperature and pressure are considered along with variations in the number of moles of each species, the chemical potentials combine through Equation 5.64 along with the physical variations (Equation 5.18, taking the equilibrium version) to give

$$dG = -SdT + Vdp + \sum_{i=1}^{i\,\text{species}} \mu_i dn_i \qquad (5.66)$$

This general variation equation for the Gibbs free energy forms the basis for a number of additional Maxwell relations (see below).

A CONSTRAINT ON THE CHEMICAL POTENTIALS: THE GIBBS–DUHEM RELATIONSHIP

Equations 5.64 and 5.65 (and the analogous volume equations) combine in a subtle but important way to restrict the thermodynamic behavior of mixtures. Starting from Equation 5.65, the total extensive Gibbs free energy appears to be a function both of the number of moles of each component and the corresponding chemical potentials. Since G is a state function, we can write the total differential in terms of both composition and chemical potentials as[†]

$$
\begin{aligned}
dG &= d\left(\sum_{i=1}^{n\,\text{species}} \mu_i n_i \right) \\
&= \sum_{i=1}^{n\,\text{species}} \mu_i dn_i + \sum_{i=1}^{n\,\text{species}} n_i d\mu_i
\end{aligned}
\tag{5.67}
$$

However, Equation 5.64 defines dG as just one of these sums. For both of these equations to be true, it must be the case that

$$
\sum_{i=1}^{n\,\text{species}} \mu_i dn_i = \sum_{i=1}^{n\,\text{species}} \mu_i dn_i + \sum_{i=1}^{n\,\text{species}} n_i d\mu_i
\tag{5.68}
$$

Subtracting the left side of Equation 5.71 from the right gives

$$
\sum_{i=1}^{n\,\text{species}} n_i d\mu_i = 0
\tag{5.69}
$$

This relationship, which is referred to as the "Gibbs-Duhem equation," states that the chemical potentials are not independent of one another. For a two-component system, if one chemical potential increases, the other decreases in proportion given by their mole ratios. For systems of just two species A and B, this coupling is particularly simple. Rearranging Equation 5.72 gives

$$
n_A d\mu_A = -n_B d\mu_B
\tag{5.70}
$$

Dividing both sides of the equation by the total number of moles $(n_T = n_A + n_B)$ gives

$$
\begin{aligned}
x_A d\mu_A &= -x_B d\mu_B \\
&= -(1 - x_A) d\mu_B
\end{aligned}
\tag{5.71}
$$

$$
d\mu_B = \frac{x_A}{(x_A - 1)} d\mu_A
\tag{5.72}
$$

This means that any variation in chemical potential of component A must be mirrored in variation of component B, in opposite sign. This connection will be developed further when we derive a relationship between the chemical potential and the concentration (and relatedly, the mole fraction).

Finally, it is worth noting that if p and T are also allowed to vary, then Equation 5.69 can be combined with Equation 5.70 to give a more general version of the Gibbs–Duhem equation:

$$
SdT - Vdp + \sum_{i=1}^{n\,\text{species}} n_i d\mu = 0
\tag{5.73}
$$

[†] In Equation 5.70, we are assuming p and T are constant. You can derive this equation using the exact differential approach described in Chapter 2.

This expression will be useful in analysis of phase diagrams, and in deriving Gibbs' famous "phase rule" in Chapter 6.

The relationship between the chemical potential and other thermodynamic potential functions

Given the relationships between the U, H, A, and G developed earlier in this chapter, we can expect corresponding relationships between U, H, A, and the chemical potential. Here, we will explore these relationships, both in differential and non-differential forms. The resulting expressions are surprisingly similar to those relating G to μ, underscoring the "potential" aspect of μ.

Starting with the enthalpy, we can rearrange Equation 5.16 to give

$$H = G + TS \tag{5.74}$$

Taking the differential leads to

$$
\begin{aligned}
dH &= dG + d(TS) \\
&= dG + TdS + SdT \\
&= \left(-SdT + Vdp + \sum_{i=1}^{n\,\text{species}} \mu_i dn_i \right) + TdS + SdT
\end{aligned}
\tag{5.75}
$$

where the term in parentheses is an expansion of dG (Equation 5.66).[†] Cancellation of SdT terms leads to

$$dH = Vdp + TdS + \sum_{i=1}^{n\,\text{species}} \mu_i dn_i \tag{5.76}$$

At constant S and p (the conditions under which H acts as a potential),

$$
\begin{aligned}
dH &= \mu_A dn_A + \mu_B dn_B + \mu_C dn_C + \ldots \\
&= \left(\frac{\partial H}{\partial n_A} \right)_{S,p,n_{j\neq A}} dn_A + \left(\frac{\partial H}{\partial n_B} \right)_{S,p,n_{j\neq B}} dn_B + \left(\frac{\partial H}{\partial n_C} \right)_{S,p,n_{j\neq C}} dn_C + \ldots
\end{aligned}
\tag{5.77}
$$

The second line of Equation 5.79 comes from the fact that H is a state function of each mole number n_i. Since each term on the right-hand side of Equation 5.79 is independent of the others, the coefficients in the two forms of Equation 5.80 must be equal, that is,

$$\mu_A = \left(\frac{\partial H}{\partial n_A} \right)_{S,p,n_{j\neq A}} , \quad \mu_B = \left(\frac{\partial H}{\partial n_B} \right)_{S,p,n_{j\neq B}} , \quad \ldots \tag{5.78}$$

Though it will not be derived here, it can be shown similarly (Problems 5.17(a) and 5.17(b)) that

$$\mu_A = \left(\frac{\partial U}{\partial n_A} \right)_{S,V,n_{j\neq A}} , \quad \mu_B = \left(\frac{\partial U}{\partial n_B} \right)_{S,V,n_{j\neq B}} , \quad \ldots \tag{5.79}$$

[†]   Here we are undoing the Legendre transform from H to G earlier in the chapter.

and that

$$\mu_A = \left(\frac{\partial A}{\partial n_A}\right)_{T,V,n_{j\neq A}} , \quad \mu_B = \left(\frac{\partial A}{\partial n_B}\right)_{T,V,n_{j\neq B}} , \ldots \tag{5.80}$$

These seemingly schizophrenic relations involving of the chemical potential can be summed up as follows:

$$\mu_i = \left(\frac{\partial G}{\partial n_i}\right)_{T,p,n_{j\neq i}} = \left(\frac{\partial H}{\partial n_i}\right)_{S,p,n_{j\neq i}} = \left(\frac{\partial U}{\partial n_i}\right)_{S,V,n_{j\neq i}} = \left(\frac{\partial A}{\partial n_i}\right)_{T,V,n_{j\neq i}} \tag{5.81}$$

In words, Equation 5.81 says that under conditions where a particular energy function serves as a potential (e.g., constant T and p for G, constant S and V for U), the chemical potential of component i is equal to the amount that the potential function changes when moles of i are added. If adding component i at constant T and p leads to a large increase in G (i.e., if species i has a high chemical potential), adding i will give the system greater potential to do various types of work. At constant S and p, H will be equally sensitive to component i, as will U at constant S and V. The name "chemical potential" nicely emphasizes this connection.

Given the nearly identical relationships between the chemical potential and derivatives of the various potential functions (Equation 5.81), one might expect similar relationships with the integrated (i.e., nondifferential) potentials. Does a direct analog of Equation 5.65 (i.e., $G = \mu_1 n_1 + \mu_2 n_2 + \cdots$) hold for H and U? In this case, the answer is "no"; μ_i is shorthand for partial molar Gibbs free energy, not partial molar enthalpy or internal energy. Combining Equation 5.65 with Equations 5.13 and 5.16 leads to

$$H = G + TS$$
$$= \sum_{i=1}^{n\,\text{species}} \mu_i n_i + TS \tag{5.82}$$

Likewise,

$$U = H - PV$$
$$= \sum_{i=1}^{n\,\text{species}} \mu_i n_i + TS - pV \tag{5.83}$$

It would be equally valid to express Equations 5.82 and 5.83 in terms of partial molar enthalpy and internal energy, which would generate simple expressions analogous to Equation 5.65. However, given our focus on thermodynamics at constant pressure and temperature, it is most convenient to define chemical potentials in terms of molar Gibbs energies (Equations 5.62 and 5.65).

Maxwell relations involving variations in composition

In the section on Maxwell relations above, second[†] (cross) derivatives of the potential functions were calculated using the variables S, T, p, and V, and based on the Euler criterion for exact differentials, were shown to be equal to one another. Now

---

[†] Note that although some of these relations do not appear to be second derivatives, it should be kept in mind that coefficients such as S and V are themselves first derivatives of potential functions such as G.

dG	$-SdT$	$+Vdp$	$+\mu_1 dn_1$	$+\mu_2 dn_2$
$-SdT$	$-\left(\dfrac{\partial S}{\partial T}\right)_{p,n_1,n_2}$	$\left(\dfrac{\partial V}{\partial T}\right)_{p,n_1,n_2}$	$\left(\dfrac{\partial \mu_1}{\partial T}\right)_{p,n_1,n_2}$	$\left(\dfrac{\partial \mu_2}{\partial T}\right)_{p,n_1,n_2}$
$+VdP$	$-\left(\dfrac{\partial S}{\partial p}\right)_{T,n_1,n_2}$	$\left(\dfrac{\partial V}{\partial p}\right)_{T,n_1,n_2}$	$\left(\dfrac{\partial \mu_1}{\partial p}\right)_{T,n_1,n_2}$	$\left(\dfrac{\partial \mu_2}{\partial p}\right)_{T,n_1,n_2}$
$+\mu_1 dn_1$	$-\left(\dfrac{\partial S}{\partial n_1}\right)_{p,T,n_2}$	$\left(\dfrac{\partial V}{\partial n_1}\right)_{p,T,n_2}$	$\left(\dfrac{\partial \mu_1}{\partial n_1}\right)_{p,T,n_2}$	$\left(\dfrac{\partial \mu_2}{\partial n_1}\right)_{p,T,n_2}$
$+\mu_2 dn_2$	$-\left(\dfrac{\partial S}{\partial n_2}\right)_{p,T,n_1}$	$\left(\dfrac{\partial V}{\partial n_2}\right)_{p,T,n_1}$	$\left(\dfrac{\partial \mu_1}{\partial n_2}\right)_{p,T,n_1}$	$\left(\dfrac{\partial \mu_2}{\partial n_2}\right)_{p,T,n_1}$

Figure 5.10 Second derivatives of terms contributing to the Gibbs energy function for a two-species system (Equation 5.84). The top row and left column express dG in terms of its fundamental variables and coefficients. For each entry, the nondifferential coefficient from the head of the column (black) is differentiated with respect to the fundamental variable from the left of the row (red). Second derivatives of G with respect to a single variable (i.e., first derivatives of S, V, and μ_i) are along the diagonal (gray shaded). Cross-derivatives are off the diagonal.

that we have expanded our expressions for these potentials (G in particular) to include contributions of different species, we can obtain second derivatives that include mole numbers and chemical potentials. Some of these new Maxwell relations define additional partial molar quantities; others form the basis of chemical linkage.

To start, consider a two component mixture.[†] We will analyze the mixture in terms of the Gibbs free energy, with fundamental variables p, T, n_1, and n_2. From Equation 5.66,

$$dG = -SdT + Vdp + \mu_1 dn_1 + \mu_2 dn_2 \tag{5.84}$$

Each of the four coefficients ($-S$, V, μ_1, and μ_2) can be differentiated with respect to each of the four corresponding fundamental variables, as shown in the array in **Figure 5.10**.

For this two-component system, four of these derivatives are really second-derivatives of the Gibbs free energy with respect to a single variable (e.g., $-(\partial S/\partial T)_{p,n} = (\partial^2 G/\partial T^2)_{p,n}$). These are on the diagonal in Figure 5.10 in shaded boxes. In general, there will be $2 + C$ of these terms; the two comes from the nonchemical variables T and p, and C corresponds is the number of components (see **Box 5.2**). From the total of 16 second derivatives for a two-component system (in general, $(2 + C)^2$), 12 will be cross derivatives (in general, $2 + 3C + C^2$). However, only half of the cross-derivatives are unique. For the two-component system, the Euler relation gives 6 Maxwell relations among the 12 cross-derivatives for the two-component system. In Figure 5.10, the Maxwell relations correspond to equalities between the off-diagonal terms that are transposed across the diagonal. That is, the derivative in row i, column j is equal to that in row j, column i. For example, the derivative at $i = 3$, $j = 1$ and its counterpart at $i = 1$, $j = 3$ define the Maxwell relation

$$-\left(\frac{\partial S}{\partial n_1}\right)_{p,T,n_2} = \left(\frac{\partial \mu_1}{\partial T}\right)_{p,n_1,n_2} \tag{5.85}$$

This equality comes from the fact that the two sides of Equation 5.85 are equal to cross derivatives of the Gibbs energy with respect to the same two variables, but taken in opposite order:

---

† The subtle but important difference between components and species will be discussed in Box 5.2.

Box 5.2: Species versus components

For systems that undergo reaction among many species, it is generally the case that the amounts of only a few different compounds need to be specified, and chemical reaction spontaneously distributes these compounds into forms that differ in bond formation, conformation, ligand binding, and degree of assembly. We will refer to this subset of defining compounds as "components," and the resulting molecules, complexes, and conformers that form as "species."

As an example, consider the reactions that can be formed by combination of human hemoglobin (a tetramer of $\alpha_2\beta_2$ stoichiometry, where each of the four subunits can bind a ligand such as dioxygen, O_2) with O_2. The primary components in such a reaction are hemoglobin and O_2. Additional components include water and whatever salts and buffer are present; while these additional components are important for determining reactivity of hemoglobin for O_2, but do not appear explicitly in the reaction stoichiometry. Although there are just two primary components in the hemoglobin–O_2 reaction, the number of species that are formed is considerably greater than two (**Figure 5.11**).

This distinction between components and species is important when evaluating the minimum number of parameters needed to specify the thermodynamic state of a system (usually two plus the number of components) and the number of phases that can coexist in equilibrium (again, the number of components is the defining variable). In the hemoglobin example above, although many species can be formed in solution, an equilibrium distribution is uniquely defined simply by specifying the number of moles of each component (n_{Hb}, n_{O2}, n_{H2O}, ...), and two other variables (p and T). Once these variables are specified, the distribution of the various species is determined by the relative Gibbs free energies of each species. Though we may not know the values of these free energies (rather, determining them is often a goal of studying such a mixture), the species distribution is uniquely defined by the amounts of each component, and the values of p and T.

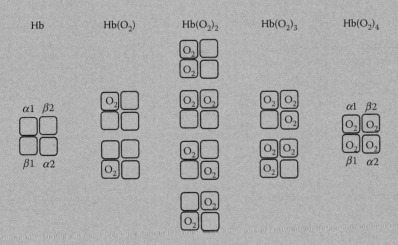

Figure 5.11 Species that can be formed from a two-component mixture of Hemoglobin (Hb) and dioxygen (O₂). The top line shows five different "macroscopic" ligation states (zero to four dioxygens bound). The diagram underneath shows unique "microscopic" ligation states (the distinction between macroscopic and microscopic states, in regard to ligand binding, will be made in Chapters 13 and 14). Because the hemoglobin tetramer is constructed of a dimer of α/β dimers, where the α- and β-subunits are similar but not identical, there are eight different partly ligated arrangements.

dG	Vdp	$\mu_1 dn_1$	$\mu_2 dn_2$
$-SdT$	$\left(\dfrac{\partial V}{\partial T}\right)_{p,n_1,n_2} = -\left(\dfrac{\partial S}{\partial p}\right)_{T,n_1,n_2}$	$\left(\dfrac{\partial \mu_1}{\partial T}\right)_{p,n_1,n_2} = -\left(\dfrac{\partial S}{\partial n_1}\right)_{p,T,n_2} = \overline{S}_1$	$\left(\dfrac{\partial \mu_2}{\partial T}\right)_{p,n_1,n_2} = -\left(\dfrac{\partial S}{\partial n_2}\right)_{p,T,n_1} = \overline{S}_2$
Vdp		$\left(\dfrac{\partial \mu_1}{\partial p}\right)_{T,n_1,n_2} = \left(\dfrac{\partial V}{\partial n_1}\right)_{p,T,n_2} = \overline{V}_1$	$\left(\dfrac{\partial \mu_2}{\partial p}\right)_{T,n_1,n_2} = \left(\dfrac{\partial V}{\partial n_2}\right)_{p,T,n_1} = \overline{V}_2$
$\mu_1 dn_1$			$\left(\dfrac{\partial \mu_2}{\partial n_1}\right)_{p,T,n_2} = \left(\dfrac{\partial \mu_1}{\partial n_2}\right)_{p,T,n_1}$

Figure 5.12 Maxwell relations in a two-species system. These Maxwell relations are equalities among cross-derivatives of the free energy. (e.g., $\partial^2 G/\partial T\partial n_i = \partial \mu_i/\partial T$), and each defines a unique Maxwell relation.

$$-\left(\frac{\partial S}{\partial n_1}\right) = \frac{1}{\partial n_1}\left(\frac{\partial G}{\partial T}\right) = \frac{1}{\partial T}\left(\frac{\partial G}{\partial n_1}\right) = \left(\frac{\partial \mu_1}{\partial T}\right) \tag{5.86}$$

For the two-component system, all of the Maxwell relations are collected in **Figure 5.12**. These Maxwell relations are of three types. The first (represented by a single equation) relates the total system variables (V, T, S, and p) already discussed (see Table 5.1). The second type (four equations for the two-component system) relate the chemical potentials to the entropy and volume. These equations define the partial molar entropies and volumes we have discussed above (Problem 5.19). The third type (one equation for the two-component system, in general $C(C-1)/2$; see Problem 5.20) relates the chemical potentials of different components to each other, and gives rise to thermodynamic linkage among species.

PARTIAL PRESSURES OF MIXTURES OF GASES

The partial molar quantities discussed above are all obtained by differentiating extensive quantities with respect to the mole number n_i to produce an intensive, per-mole quantity. It can also be useful to resolve pressure into contributions made by different species (i.e., "partial pressures"). However, since pressure is already an intensive quantity, partial pressures are not obtained by differentiation with respect to n_i. Instead, we will simply *define* the partial pressure of component i as its mole fraction of the total pressure:

$$p_i \equiv x_i p_T \tag{5.87}$$

Because mole fractions sum to one, Equation 5.87 ensures that the sum of the partial pressures is equal to the total pressure (Problem 5.21):

$$p_T = \sum_{i=1}^{c} p_i \tag{5.88}$$

If each component in the gas behaves ideally, Equation 5.88 can be simplified. Remember, for ideal gases, there are no interactions between molecules; the gas molecules only interact with the vessel walls through collision. For *mixtures* of ideal gases, this is true both for "self" interactions (among molecules from the same component), and for "nonself" interactions (among molecules from different components). If there are no interactions, the pressure exerted by any given component will be independent of the other components that happen to be in the mixture.

Thus, the partial pressure of any component will be the same as if all the other components were not there, that is,

$$p_i = \frac{n_i RT}{V} \tag{5.89}$$

Inserting this partial form of the ideal gas law into Equation 5.89 gives

$$p_T = \sum_{i=1}^{c} \frac{n_i RT}{V} = \frac{RT}{V} \sum_{i=1}^{c} n_i \tag{5.90}$$

This additivity relationship (which only applies to ideal gas mixtures) is referred to as "Dalton's law," and says that the total pressure of a mixture of ideal gases is equal to the sum of each of the pressures that would be exerted by each separate gas.

When gas molecules interact, we cannot equate partial pressure values with the pressures of the isolated components. The number of collisions that a particular component makes per unit time will depend on the concentrations of other gases present. If the other gases exclude volume through repulsive interactions, they will increase the collision frequency of component i with the wall, whereas if they attract, they will decrease the collision frequency by forming larger, slower moving clusters that include component i. Although the total pressure will still be determined by the individual collisions of the gas molecules and the wall, ascribing individual collisions to individual components is not appropriate when there are interactions. Though we can still define partial pressures of non-ideal gases using Equation 5.88, these p_i values should at best be regarded as the pressure that each component exerts in the background of the other molecules present.

Problems

5.1 The protein myoglobin (Mb) binds to a molecule of dioxygen (O_2) in a vacant cavity in the protein's interior. Assuming the O_2 to come from an ideal gas at constant temperature $T = 310$ K and constant pressure $p = 1.015 \times 10^5$ Pa, calculate the work associated with binding of one mole of O_2 to Mb, assuming a one-to-one stoichiometry. Compare the work term you calculated to the measured heat of binding of O_2 to myoglobin of 48 kJ mol^{-1}.

(a) Now calculate the work of O_2 binding to Mb under the same conditions as in Problem 5.1, but where the unbound O_2 is dissolved in solution, rather than in the gas phase. Assuming the molar volume of dioxygen (O_2) is 31 mL/mol (an estimate from the van der Waals b coefficient, Chapter 3).

5.2 This problem explores the thermodynamic identity that defines temperature as the entropy sensitivity of the energy at equilibrium, that is,

$$T = \left(\frac{\partial U}{\partial S} \right)_T$$

Consider the energy ladder shown below, with three particles and three evenly spaced energy levels (you could do the same with four particles over four levels such as in Figure 5.6, but it would be a lot to keep track of). For the given energy values, the total energy ($E_{tot} = N_1 \varepsilon_1 + N_2 \varepsilon_2 + N_3 \varepsilon_3$) can range from 3 (with occupancies shown) to 9.

$\varepsilon_3 = 3$ ___ ___ ___ ___ ___ ___ ___

$\varepsilon_2 = 2$ ___ ___ ___ ___ ___ ___ ___

$\varepsilon_1 = 1$ ●●● ___ ___ ___ ___ ___ ___

$W_{3,1} =$	$W_{4,1} =$	$W_{5,1} =$	$W_{6,1} =$	$W_{7,1} =$	$W_{8,1} =$	$W_{9,1} =$
$W_{3,2} =$	$W_{4,2} =$	$W_{5,2} =$	$W_{6,2} =$	$W_{7,2} =$	$W_{8,2} =$	$W_{9,2} =$
$W_{3,3} =$	$W_{4,3} =$	$W_{5,3} =$	$W_{6,3} =$	$W_{7,3} =$	$W_{8,3} =$	$W_{9,3} =$
$E_{tot} = 3$	$E_{tot} = 4$	$E_{tot} = 5$	$E_{tot} = 6$	$E_{tot} = 7$	$E_{tot} = 8$	$E_{tot} = 9$
$W_{3,tot} =$	$W_{4,tot} =$	$W_{5,tot} =$	$W_{6,tot} =$	$W_{7,tot} =$	$W_{8,tot} =$	$W_{9,tot} =$

(a) For each total energy, fill in the distinguishable occupancies consistent with that energy value. Note that not all of the ladders will be needed. For example, there is only one distinguishable arrangement with $E_{tot} = 3$ (shown).

(b) For each distinguishable arrangement above, give the number of indistinguishable configurations (i.e., the multiplicities). Next, give the total number of configurations (distinguishable and indistinguishable) for each E_{tot} value. You can use the chart above for your answers.

(c) For each value of E_{tot}, calculate the entropy (or if you like, S/k). Plot E_{tot} on the y-axis versus the entropy (or S/k) on the x-axis. Taking the value of E_{tot} to be equivalent to the internal energy, interpret the shape of your plot given the thermodynamic identity for temperature above.

5.3 Come up with a function for A (the Helmholtz free energy), and derive the differential form that reveals A as a potential (Equation 5.20).

5.4 Under what conditions does the enthalpy act as a potential. Under what conditions does the Helmholtz free energy (A) act as a potential? Which one of these would be easier to achieve experimentally?

5.5 Calculate the tangent lines depicted in Figure A5.1.2 for the exponential decay function, and give the $\{D_i, \Lambda_i\}$ pairs, which are plotted as points in Figure 5.3. The x_i points used to make these tangent lines are $\{-1.25, -1.0, -0.75, -0.5, -0.25, 0.0, 0.25, 0.5, 0.75, 1.0, 1.25\}$.

5.6 Calculate the Legendre transform of a parabola $y = x^2$. Also calculate the back transform.

5.7 Use the derivative energy identities for S and V to demonstrate that the p–T Maxwell relation resulting from G (Equation 5.35) is equivalent to the statement that the cross-derivative of a state function is independent of its order.

$$\left(\frac{\partial \left(-\frac{\partial G}{\partial T} \right)}{\partial p} \right)_T = \left(\frac{\partial \left(\frac{\partial G}{\partial p} \right)}{\partial T} \right)_p$$

$$\left(\frac{-\partial^2 G}{\partial p \partial T} \right)_T = \left(\frac{\partial^2 G}{\partial T \partial p} \right)_p$$

5.8 Approximating water and ethanol as having ideal mixing volumes (Figure 5.9A), calculate the total volume of a mixture of 2.2 moles of water and 4.3 moles of ethanol.

5.9 Show the following relationships (Equations 5.6 and 5.7) to be true:

$$\left(\frac{\partial S}{\partial T}\right)_V = \frac{C_V}{T}$$

$$\left(\frac{\partial S}{\partial V}\right)_T = \left(\frac{\partial p}{\partial T}\right)_V$$

5.10 Integrate Equation 5.9, that is,

$$dS = \frac{C_V}{T} dT + \frac{nR}{V} dV \qquad (5.9)$$

to come up with an expression for the entropy of an ideal gas as a function of temperature and volume. As described in the text, you can do this in two steps where one variable is fixed, giving a single integration variable.

5.11 Starting with Equation 5.11 for $S(V,T)$, derive Equation 5.12, expressing $U(S,V)$. Equation 5.11 gives entropy as a function of T and V relative to a reference $S_r = S(V_r,T_r)$:

$$S = S_r + C_V \ln \frac{T}{T_r} + nR \ln \frac{V}{V_r}$$

To get $U(S,V)$, we need to substitute T for U (using the monatomic gas relation $U = 3nRT/2$) and solve for U:

$$S = S_r + C_V \ln \frac{2U}{3nRT_r} + nR \ln \frac{V}{V_r}$$

Exponentiating gives

$$e^{S-S_r} = e^{C_V \ln \frac{2U}{3nRT_r} + nR \ln \frac{V}{V_r}}$$

$$= e^{\ln\left(\frac{2U}{3nRT_r}\right)^{C_V}} e^{\ln\left(\frac{V}{V_r}\right)^{nR}}$$

$$= \left(\frac{2U}{3nRT_r}\right)^{C_V} \left(\frac{V}{V_r}\right)^{nR}$$

Rearranging gives

$$\left(\frac{2U}{3nRT_r}\right)^{C_V} = e^{(S-S_r)} \left(\frac{V_r}{V}\right)^{nR}$$

$$\frac{2U}{3nRT_r} = e^{(S-S_r)/C_V} \left(\frac{V_r}{V}\right)^{nR/C_V}$$

$$U = \frac{3}{2} nRT_r \left(\frac{V_r}{V}\right)^{2/3} e^{(S-S_r)/C_V}$$

The volume exponent in the last line comes from the relationship $C_V = 3nR/2$.

5.12 In Figure 5.3, interpret the slope of the internal energy with entropy at constant volume. Where, on this surface, is the temperature highest? Where is it lowest? Make a temperature contour map of this surface (contouring temperature values as a function of S and V, as in Figure 5.3B).

5.13 Explain, in words, the observation that for a monatomic ideal gas, the enthalpy goes up with pressure at constant enthalpy. Also, explain why it has a soft (i.e., $p^{2/5}$, less than linear) dependence at constant entropy (Figure 5.5B).

5.14 For a closed system in equilibrium, the differential of Gibbs energy is

$$dG = -SdT + Vdp$$

(the equality part of Equation 5.18) Assume entropy and the volume have the following dependences on temperature and pressure:

$$S = \sigma T - \alpha p \qquad (5P.1)$$

$$V = V_0 - \beta p + \alpha T \qquad (5P.2)$$

where σ, β, and α are positive constants. Demonstrate that Equations 5P.1 and 5P.2 are acceptable functional forms, that is, when combined with expression for dG, they give an exact differential.

5.15 Using the results in Problem 5.14 above, give an expression for G, relative to $G_0 = G(T = 0, p = 0) = 0$, as a function of T, P, σ, β, and α. (Hint: one way to do this is by direct integration along a defined path, but another is simply to try to identify a function (G) that differentiates to the expression for S and V). If you want to check your answer, you can compare it to the following two plots, which show $G(p,T)$:

5.16 Determine the function $G(T,p)$ for a monatomic ideal gas, by Legendre transform of $H(S,p)$ (Equation 5.27).

5.17 (a) Starting with the equation $U = G + TS - pV$ (you should convince yourself that this is true), show that Equation 5.79 holds, that is, that

$$\mu_A = \left(\frac{\partial U}{\partial n_A} \right)_{S,V,n_{j \neq A}}, \quad \mu_B = \left(\frac{\partial U}{\partial n_B} \right)_{S,V,n_{j \neq B}}, \dots$$

5.17 (b) Starting with the equation $A = G - pV$ (convince yourself this is true), show that Equation 5.80 holds, that is, that

$$\mu_A = \left(\frac{\partial A}{\partial n_A} \right)_{T,V,n_{j \neq A}}, \quad \mu_B = \left(\frac{\partial A}{\partial n_B} \right)_{T,V,n_{j \neq B}}, \dots$$

5.18 Starting with the fact that $\mu_i = \bar{H}_i - T\bar{S}_i$, and the relationships that

$$H = \sum_{i=1}^{n\,species} n_i \bar{H}_i \quad \text{and} \quad S = \sum_{i=1}^{n\,species} n_i \bar{S}_i$$

demonstrate that

$$H = \sum_{i=1}^{n\,species} \mu_i n_i + TS$$

5.19 Derive the relations connecting chemical potential to partial molar volume and entropy in Figure 5.10, starting with the Maxwell relations.

5.20 For a system of C independent non-interconverting species (more precisely, C components) at constant p and T, how many Maxwell relations are there relating chemical potential to volume and entropy? How many Maxwell relations link the chemical potentials and composition of multiple species? (See Figure 5.12 and the surrounding text for more on the distinction between these two types of Maxwell relations. Note, for this problem, invoke the Gibbs free energy as a potential.)

5.21 Show that if Equation 5.88 is true (partial pressure is the mole fraction of the total pressure), then Equation 5.89 is true (i.e., that the total pressure of a gas mixture is given by the sum of the partial pressures).

Appendix 5.1: Legendre Transforms of a Single Variable

The internal energy differential (Equation 5.3) is a fundamental equation, because it incorporates the inequality provided by the second law, but it focuses on the system without requiring isolation from the surroundings. However, the fundamental variables of the internal energy potential (S and V) are not convenient. S is neither easy to control (and hold constant) nor easy to measure. We are interested in systems with variable S (and to some extent, V), but constant T and p. The transforms we made to the internal energy inequality introduced new energy functions (H, G, and A) which, in differential form, have one or both of these conjugate variables swapped. In general, this swapping procedure comes from a mathematical procedure called the Legendre transform.

Here, we will introduce the general approach for the Legendre transform using a function of a single variable, and apply it to a simple exponential decay function,

$$f(x) = e^{-2x} \qquad (A5.1.1)$$

(for a similar analysis with a parabola, see Problem 5.6). The derivative of this function is

$$D = \frac{df(x)}{dx} = -2e^{-2x} \qquad (A5.1.2)$$

where we use D as a shorthand to represent the derivative (analogous to, e.g., $-S$ as a shorthand for $(\partial G/\partial T)_p$). Rearranging to differential form,

$$df(x) = Ddx = -2e^{-2x}dx \qquad (A5.1.3)$$

The Legendre transform is equivalent to finding new function, Λ, where the differential $d\Lambda$ is

$$d\Lambda = xdD \qquad (A5.1.4)$$

Note that the differential and nondifferential role of D and x are swapped in going from Equations A5.1.3 (left equality) to A5.1.4.

A problem with specifying if Λ with Equation A5.1.4 is that, we have lost some of the information from our original function in differentiation. Remember from Chapter 2, because there is an entire family of different functions can have the same value at a given value of D (**Figure A5.1.1**). However, if we also include the y-intercept of the tangent line to f(x), we specify a single member of these shifted curves (see the family of red tangent lines in **Figure A5.1.2**). The Legendre transform takes a starting curve f(x) and extracts from it a new function using both the slopes D and intercepts of the tangent lines to the original function as variables. The quantity Λ in Equation A5.1.4 is the intercept value. Since it is a function of the slope D of the original function, we can write it as $\Lambda(D)$, just as we write our original function as f(x). Although $\Lambda(D)$ is a distinctly different curve than f(x), the two functions contain all the same information, and thus, $\Lambda(D)$ can be back-transformed to give f(x).

To find $\Lambda(D)$, we need to calculate the tangent lines T to f(x) and the intercepts at different values of x, to evaluate their slopes. We will refer to specific values of x and the corresponding tangent line as x_i and $T_{y,i}$, respectively. These tangent lines can be written as

$$T_{y,i} = D_i x_i + \Lambda_i \qquad (A5.1.5)$$

The i subscripts on the slope (D_i) and intercept (Λ_i) are a reminder that these values are calculated at all values of x_i. Keep in mind, Λ_i is not the y-intercept of the function $y = y(0)$, it is the intercept for a specific tangent line $T_{y,i}$.

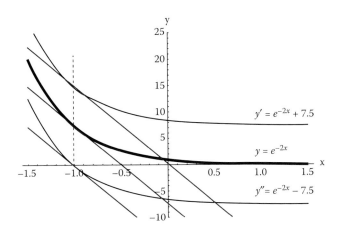

Figure A5.1.1 The derivative alone is not enough to uniquely specify a function. The thick black curve shows the exponential decay corresponding to Equation A5.1.1. The thin black curves are displaced by $+/-$ 7.5 y-units. Although these three curves are clearly different, they all have identical slopes, as is shown by the tangent lines at $x = -1$. Thus, a function based solely on the slope at each x value cannot resolve these curves. One way to specify the starting function is to provide the y-intercepts of the tangent lines along with the slopes.

We can relate Λ_i to its slope D_i by simple geometry. The slope of $T_{y,i}$ is simply the rise over the run, which we can determine from two points: $\{x_i, y(x_i)\}$ and $\{0, \Lambda_i\}$:

$$D_i = \frac{rise}{run} = \frac{y_i - \Lambda_i}{x_i - 0} \tag{A5.1.6}$$

Thus,

$$\Lambda_i = y(x_i) - D_i x_i \tag{A5.1.7}$$

Although this function gives Λ as a function of D, as intended, it still has the old variable x_i in it. The first term in Equation A5.1.7 can typically be substituted as some rearrangement of D_i. The x_i can typically be eliminated from the second term by solving the expression D_i for x_i. As a result, the desired form

$$\Lambda = \Lambda(D) \tag{A5.1.8}$$

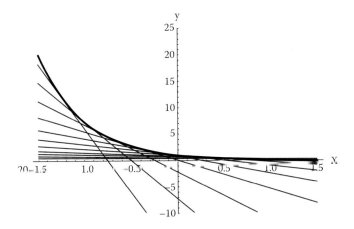

Figure A5.1.2 A family of tangent lines for an exponential decay curve. Equation A5.1.1 is shown in black, with a family of equally spaced (in x) tangent lines in red. At negative x_i values, the tangent slopes D_i are large and negative, and have very negative y-intercepts (Λ_i values). As xi values increase, tangent slopes become less negative (but always remain less than zero, and y-intercepts actually increase to positive values, before decreasing back to zero. Thus, we would expect the Legendre transform $\Lambda = \Lambda(D)$ to be defined exclusively on the right half of the D versus Λ graph, and to go through a maximum at a moderately negative value of D.

is obtained. The elimination in this last step is difficult to illustrate in abstract form, but can be shown more clearly using our example. For the exponential function given in Equation A5.1.1, the family of tangent lines (calculated using Equation A5.1.5) is shown in Figure A5.1.2. To obtain an initial expression for Λ_i (involving x_i), we can substitute our expression for $y(x_i)$ into Equation A5.1.7:

$$\Lambda_i = y(x_i) - D_i x_i$$
$$= e^{-2x_i} - D_i x_i \qquad \text{(A5.1. 9)}$$

Our goal is to replace x_i values in Equation A5.1.9 with expressions involving D_i. The x_i in the first term in Equation A5.1.9 can be replaced by slightly rearranging the expression for D_i (Equation A5.1.2):

$$D_i = -2e^{-2x_i}, \quad \text{thus,}$$
$$-\frac{D_i}{2} = e^{-2x_i} \qquad \text{(A5.1.10)}$$

Likewise, the x_i in the second term of Equation A5.1.9 can be substituted by further rearrangement of D from Equation A5.1.10:

$$-2x_i = \ln\left(\frac{-D_i}{2}\right)$$
$$x_i = -\frac{1}{2}\ln\left(\frac{-D_i}{2}\right) \qquad \text{(A5.1.11)}$$

Substitution of Equations A5.1.10 through A5.1.11 into Equation A5.1.9 gives

$$\Lambda_i = -\frac{D_i}{2} + \frac{D_i}{2}\ln\left(-\frac{D_i}{2}\right)$$
$$= \frac{D_i}{2}\left[\ln\left(-\frac{D_i}{2}\right) - 1\right] \qquad \text{(A5.1.12)}$$

This is true for all $\{D_i, \Lambda_i\}$ pairs; thus, the i subscripts can be omitted in expression the general transformed function $\Lambda = \Lambda(D)$, as shown in **Figure A5.1.3**.

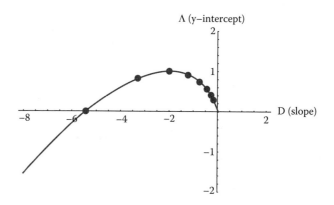

Figure A5.1.3 The Legendre transform of an exponential decay (Equation A5.1.1). The points shown are slope, intercept pairs $\{D_i, \Lambda_i\}$ from the tangent lines in Figure A5.1.2 (except the point from the steepest tangent line, which is off-scale). The solid line shows the Legendre transform, ($\Lambda(D)$, Equation A5.1.12).

CHAPTER 6

Using Chemical Potentials to Describe Phase Transitions

Goals and Summary

The main goal of this chapter is to build on the concept of chemical potential in Chapter 5 to develop free-energy expressions for systems undergoing physical transformations. The simplest form of physical transformation is one in which a single component converts from one distinct phase to another. The condition required for coexistence of multiple phases is that their chemical potentials must be identical. If the chemical potentials differ, the phase with the lowest potential is formed at the exclusion of the other phases. Chemical potential and its derivative molar quantities ($\bar{H}$, $\bar{S}$, $\bar{V}$, and $\bar{C}$) provide a means to quantitatively analyze changes of phase, and to predict phase properties under various conditions. A phase diagram for water is developed using measured heat capacities, molar volumes, and enthalpies of melting and boiling, along with classical thermodynamic relationships developed in previous chapters. From this simple diagram, and from the dependence of chemical potentials on temperature and pressure, considerable insight can be obtained relating to the underlying states of matter, their coexistence, and the cooperativity of transition between these very different phases. The effects of having multiple components on phase equilibria will be discussed, and Gibbs' phase rule will be developed as a means to connect the number of components with the number of phases.

As developed in the previous chapter, the Gibbs free energy plays a central role in thermodynamic analysis for systems at constant temperature and pressure. The chemical potential allows the total Gibbs energy to be broken down into contributions from individual species and components. In Chapter 7, will use this dissection to analyze chemical reactions and chemical equilibrium. Here, we analyze simple systems where a single chemical component can partition reversibly into different types of bulk materials with distinct properties.

PHASES AND THEIR TRANSFORMATIONS

When materials have large-scale differences in their bonding patterns, they are considered to be different phases. These atomic-level differences in bonding produce discernably different mechanical and thermal properties on the macroscopic scale. These include differences in densities, compressibilities, viscosities, and heat capacities, as well as electromagnetic and optical properties such as polarizabilities and refractive indices.

Perhaps the most dramatic (and most familiar) phase transitions are among the three common forms of matter: solid, liquid, and gas. For pure, single-component systems, changes among these three phases involve change in noncovalent interactions, but not covalent bonding. Differences in physical properties among the phases give rise to obvious differences in how different phases distribute in a container: gas phases expand to fill the entire container, liquid phases collect in the

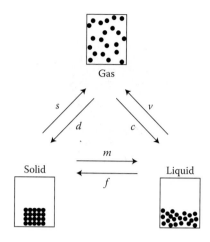

Figure 6.1 Phase transitions between solid, liquid, and gas for a single-component system. Transitions are labeled *s* for *sublimation*, *d* for *deposition*, *c* for *condensation*, *v* for *vaporization*, *m* for *melting* (sometimes referred to as *fusion*), and *f* for *freezing*. The gas phase has a much larger molar volume (inverse density) than the other two phases. Typically, solid has the lowest molar volume (highest density), although for water, the solid phase has a higher molar volume.[¶] Also, the gas phase has a much larger entropy (and enthalpy) than the other two phases, whereas the solid is the lowest entropy (and enthalpy).

bottom but flow to conform to the container's shape, and solid phases (also at the bottom) are rigid, retaining their shape rather than that of the container. The six transitions among these three basic phases are given in **Figure 6.1**.

The most familiar examples of phase transitions are those of water between ice, liquid water, and steam.[†] Though the covalent structure of these phases is the same, the interactions among the water molecules differ significantly in the three phases, and as a result, the bulk properties differ as well.

In addition to the transitions shown in Figure 6.1, single-component materials can have phase transitions among different solid states. Common ice is referred to as ice-I_h (where h denotes a hexagonal lattice structure), but there are many other phases of ice that form at lower temperatures and higher pressures (these other forms are rare outside the laboratory). In principle, single-component liquids can also form multiple phases, although this is quite rare. An example is the low temperature (at about $T = 2\,K$) phase transition of helium-4 from an "ordinary" liquid to a "superfluid."[‡] Because there are very few interactions between molecules in the gas phase (none in the ideal gas), separate phases gas phases are neither expected nor observed.

In contrast, mixtures of multiple components can (and do) show multiple liquid and solid phases. Examples include immiscible (and partly miscible) liquids such as cyclohexane and water, which form two separate liquid phases in nearly all proportions. The greater number of accessible phases that can coexist for multicomponent systems will be discussed in the context of the "Gibbs phase rule" at the end of the chapter.

Finally, materials with different covalent bonding can also be regarded as different phases. For example, elemental carbon can exist as graphite, diamond, graphene, and closed form structures such as fullerenes and carbon nanotubes. Because covalent bonds are much harder to break than noncovalent interactions, the kinetics of these covalent phase transitions is quite slow. Thus, there is often a kinetic component to phase transitions (especially covalent ones) such that the phase that is present differs from the predictions of equilibrium thermodynamics (because equilibration is very slow). For example, at room temperature and pressure, graphite is a more stable form of carbon than diamond, although the kinetic barrier for phase transition is so high that our jewelry is not at risk. In addition, noncovalent phase transitions often have a kinetic component that influences which phase is present at a given time. For example, very pure liquid water can be supercooled to below its freezing point, and can remain in a liquid "metastable" state over a considerable period of time.

THE CONDITION FOR EQUILIBRIUM BETWEEN TWO PHASES

Here we will use equilibrium thermodynamics to analyze phase transitions and phase stability. We will ignore kinetic effects through patience, giving enough time for the system to display equilibrium properties. Phase stability will be analyzed by assigning a chemical potential to each phase, μ^{π}, where the superscript indicates a particular phase (here using a generic π).[§] Because we will analyze the phase transitions of a single component in this section, we don't need the i subscript introduced in Chapter 5 for partial molar Gibbs energy. We will also leave off the

---

[¶] An exception to the generalization that the gas phase has a higher molar volume than the liquid phase occurs at the critical point, where the density (and entropy) of the gas and liquid are equal. Because of this, these two phases lose their separate identities, and become a single supercritical fluid.

[†] Note that steam is commonly misidentified as the white wispy stuff that appears over hot water, especially in cold air. Although this material is often physically associated with water vapor, the white stuff is really tiny water droplets that have condensed in the air. Unlike these little liquid droplets, which scatter light, pure steam is transparent.

[‡] This *⁴He* superfluid phase is quite exotic, as is the transition to it. The transition is significantly broader than other transitions. The resulting superfluid lacks viscosity, showing quantum mechanical properties at the bulk level.

[§] The Greek lowercase *p*, for phase.

* superscript designating purity for each phase, since all phases analyzed in this section will be pure (i.e., single component).

These phase-specific chemical potentials (μ^π) can then be connected to the total Gibbs free energy of each phase and to the multi-phase system as a whole, through equations developed in Chapter 5. For example, consider a pure liquid and a pure solid phase in contact at constant temperature and pressure (**Figure 6.2**). Labeling one phase as α and the other as β, we can express the variation in the Gibbs free energy in response to changes in the number of moles of each phase[†]:

$$dG = \mu^\alpha dn^\alpha + \mu^\beta dn^\beta \tag{6.1}$$

Although Equation 6.1 permits material to be added to (or taken away from) both phases, we are interested in changes where the total number of moles of material is fixed (i.e., a closed system). With this constraint, material that transfers into the α phase (dn^α) is matched by transfer of an equal amount from the β phase (dn^β), that is,

$$dn^\alpha = -dn^\beta \tag{6.2}$$

This conservation rule simplifies Equation 6.1, allowing variation in Gibbs energy to be given in terms of variation in just one mole number:

$$dG = (\mu^\alpha - \mu^\beta)dn^\alpha \tag{6.3}$$

As discussed in Chapter 5, the direction of spontaneous change at constant temperature and pressure is that which decreases G (i.e., $dG < 0$). If the chemical potential of the α phase happens to be larger than that of the β phase, then G will decrease when dn^α is negative, that is, when material moves from the α phase to the β phase. In Figure 6.2, this corresponds to melting of the solid to produce more liquid. At constant pressure and temperature, the α phase (having the higher chemical potential of the two phases) will completely transform to the β phase. As a result, the volume will change in proportion to the number of moles transformed and the difference in the molar volumes of the two phases $[(\bar{V}^\beta - \bar{V}^\alpha)dn^\beta]$ and the system will contract or expand accordingly (Problems 6.1, 6.2). Likewise, heat will be given off or taken up depending on the molar enthalpies of the two phases $[(\bar{H}^\beta - \bar{H}^\alpha)dn^\beta]$.

However, if the chemical potentials for the two phases are equal, both phases will be present at equilibrium. This can be seen by setting μ^α equal to μ^β in Equation 6.3:

$$dG = (\mu^\alpha - \mu^\alpha)dn^\alpha \tag{6.4}$$

With equal chemical potentials, the equilibrium condition ($dG = 0$) is maintained regardless of whether dn^α is positive or negative, since the zero results from the chemical potential difference. Thus, when the chemical potentials of the phases are equal, materials can go reversibly from one phase to the other without disrupting equilibrium, a condition we will refer as "phase coexistence" (although it is often referred to as "phase equilibrium").[‡]

Summarizing, if the chemical potential of one phase is greater than that of the other, material spontaneously converts (ignoring possible kinetic barriers) from high to low chemical potential by changing phases. If, part way through the conversion, the chemical potentials become equal (this might happen if pressure and temperature are not held constant, Problems 6.4 and 6.5), both phases will remain in coexistence (though they will be in different relative amounts than when they started). If not, the material will all be converted to a single phase (the one with the lower chemical

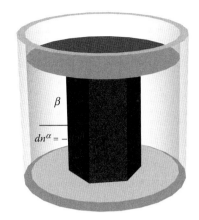

Figure 6.2 Material equilibrium between two pure phases. Here, a liquid (labeled β) and a crystalline solid (α) are depicted. The system is covered by a thin impermeable membrane (top) that can move to maintain constant pressure but prevents material exchange between system and surroundings. Although not depicted, the system can exchange heat with a reversible thermal reservoir, maintaining constant temperature. Because the membrane closes the system to material exchange, any material that adds to the α-phase will be removed from the β-phase, and vice versa (Equation 6.2).

[†] Equation 6.1 neglects surface effects between the two phases.

[‡] One problem with the expression "phase equilibrium" is that under conditions where a single phase has the lowest chemical potential, the system is still in equilibrium (populating only one phase). Under such circumstances, coexistence is clearly *not* achieved, though equilibrium is.

potential). This framework generalizes to more than two phases, although there is an important limitation on phase number that will be discussed below.

HOW CHEMICAL POTENTIALS OF DIFFERENT PHASES DEPEND ON TEMPERATURE AND PRESSURE: DERIVING A *T–P* PHASE DIAGRAM FOR WATER

The analysis above provides a rule for determining whether phases coexist under a specific set of conditions, and if not, which phase is formed. To predict the details of phase coexistence for a given component, we need actual values of chemical potentials of each phase as a function of temperature and pressure. Here, we will derive $\mu^{\pi}(T,p)$ expressions for the common phases of water—steam, liquid water, and ice. These expressions will be developed using easily measured values, including heat capacities, molar volumes, and the temperatures and heats (enthalpies) of melting and vaporization. With these expressions, we will be able to generate a phase diagram, simply by evaluating which chemical potential is lowest at a given T and p value. Water is chosen because of its importance in life at all scales, from solvation of macromolecules, to hydration of cells and tissues, all the way up to determining local ecology and global climatology. In Problems 6.22 through 6.29, an analogous approach is taken to map the phase diagram of carbon dioxide.

How do we get the chemical potentials of phases as a function of T and p? Remembering that chemical potentials are molar free energies, we can express $\mu^{\pi}(T,p)$ with the same relations used for G. In integrated form,

$$\mu^{\pi} = \bar{H}^{\pi} - T\bar{S}^{\pi} \tag{6.5}$$

in analogy to Equation 5.16 (see Problem 6.6). We will first use this equation at a single pressure (1 atm), taking enthalpies and entropies of each phase from measurements of heats of melting and vaporization (boiling). Temperature dependences of $\bar{H}^{\pi}$ and $\bar{S}^{\pi}$ for each phase will be determined from measurements of the heat capacity of each phase.[†] With the temperature dependence in hand, we will build in the pressure dependences using molar volumes of each phase, and for the gas phase, compressibility, assuming an ideal gas. This will give us chemical potential surfaces for each phase on the p–T plane. The lowest chemical potential surface gives the stable phase at a given p, T locus. Intersections between surfaces mark phase coexistence curves.

Temperature dependence of heat capacities for ice, water, and steam

For the first step in our approach, we need expressions for $\bar{H}^{\pi}$ and $\bar{S}^{\pi}$. A critical quantity for both of these expressions is the partial molar heat capacity $(\bar{C}_p^{\pi})$. Fortunately, $\bar{C}_p^{\pi}$ is directly measurable over the temperature range where each phase is stable. **Figure 6.3** shows the heat capacity of water at $p = 1$ atm, in the solid ice (the I_h phase, below $T_m = 273.15$ K), in the liquid phase (273.15–373.15 K), and in the gas phase (above $T_v = 373.15$ K).[‡]

The heat capacity of ice increases significantly with temperature. The heat capacity of liquid water is highest among the three phases, but is rather insensitive to temperature, showing a weak positive concavity. The heat capacity of steam is about the same as that of solid ice at the melting point, and appears to be insensitive to

---

[†] Though we cannot measure the absolute values of $\bar{H}^{\pi}$ and $\bar{S}^{\pi}$ for reasons discussed in Chapter 2, the analysis here depends only on the relative values for each phase.

[‡] T_m and T_v refer to melting and vaporization temperatures. Although the values of T_m and T_v depend on pressure, in our analysis of water, we will generally use these abbreviations to indicate the 1 atm values, to avoid having to write the pressures out explicitly.

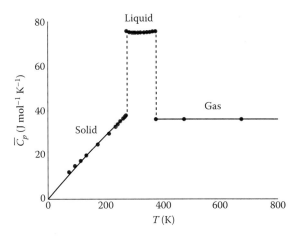

Figure 6.3 Molar heat capacity of pure water in ice, liquid, and gas phases at *p* = 1 atm. Of the three phases, liquid water has the highest heat capacity, suggesting an internal structure that is gradually disrupted through this region. At the phase transition boundaries (dotted lines), the heat capacity appears to become infinite, due to the heat of fusion and melting. Data (red circles) are from the *CRC Handbook of Chemistry and Physics*. Lines are from fitting of Equations 6.6 through 6.8 to the heat capacity data for each phase.

temperature. Note that the heat capacity of steam at one atmosphere ($\sim$36.3 J mol^{-1} K^{-1} at 373.15 K) is close to that expected of a classical bent triatomic ideal gas at constant pressure (Problem 6.8).

To analyze chemical potentials, enthalpies, and entropies of the three phases, we need analytical expressions for heat capacities. These can be obtained by fitting. Because the heat capacity of the vapor phase shows no discernable temperature dependence, we can "fit" it with a single constant (Equation 6.6). The liquid phase can be fitted with a quadratic that includes a constant and linear offset terms (Figure 6.8). Although the solid (ice) phase could be well-fitted by a linear equation with a positive offset, it is important that heat capacities go to zero at absolute zero, and that it plateau at high temperature. The equation for $\bar{C}_p^{ice}$ in Figure 6.3 meets both these criteria.

$$\bar{C}_p^{steam} = a^{steam}$$
$$= 36.3 \tag{6.6}$$

$$\bar{C}_p^{liq} = a^{liq} + b^{liq}T + c^{liq}T^2$$
$$= 101 - 0.163T + 2.56\times10^{-4}T^2 \tag{6.7}$$

$$\bar{C}_p^{ice} = \frac{a^{ice}T}{b^{ice}+T} = \frac{228.3T}{1387+T} \tag{6.8}$$

In Equations 6.6 through 6.8, units are in *J mol^{-1} K^{-1}*; numerical values for coefficients are fitted values determined by least-squares fitting.

Temperature dependence of enthalpies and entropies of ice, liquid water, and steam

In addition to the $\bar{C}_p^\pi(T)$ expressions determined from the data in Figure 6.3, several pieces of information are required to express the enthalpies and entropies of each of phase as a function of temperature. First, we need to choose molar enthalpy and entropy values at a reference temperature for one of the three phases. Because enthalpy should increase with temperature (Problem 6.10), and because ice should have the lowest enthalpy of the three phases (we expect the solid phase to have the highest degree of bonding), we will simply set the enthalpy of ice to zero at T_m (again, this is arbitrary). The reference point for entropy can be made in a slightly less arbitrary way, by assuming the entropy of ice to zero at absolute zero.[†] This is

---

[†] Since we are only considering the three major phases, our model does not consider the phase transitions of ice at low temperature and high pressure. A more detailed analysis would include entropies of all phase transitions encountered between 0 K and T_m.

consistent with the high degree of order of crystalline solids, and the absence of thermal disorder at absolute zero. Equation 6.8 can then be used to calculate the entropy and enthalpy of ice at other temperatures.

Second, to calculate enthalpies and entropies of the water and vapor phases, we need experimentally determined T_m and T_v values, and enthalpies for melting and vaporization ($\Delta \bar{H}_m$ and $\Delta \bar{H}_v$). These quantities can be determined directly from calorimetry (Problem 6.11). Because the free energy of phase transition is zero at the coexistence temperature, the entropies of the melting and vaporization are

$$\Delta \bar{S}_m = \frac{\Delta \bar{H}_m}{T_m} \tag{6.9}$$

$$\Delta \bar{S}_v = \frac{\Delta \bar{H}_v}{T_v} \tag{6.10}$$

The enthalpy and entropy of liquid water at T_f is the sum of the ice reference values and the enthalpy and entropy of melting (**Table 6.1**). Again, the enthalpy and entropy of liquid water at other temperatures can be determined by integrating Equation 6.7 from T_m. By integrating to T_v using and combining with $\Delta \bar{H}_v$ and $\Delta \bar{S}_v$, the molar enthalpy and entropy of steam can be calculated at T_v (Table 6.1). Note that even though our choice of $\bar{H}^{ice}$ and $\bar{S}^{ice}$ reference values was arbitrary, $\Delta \bar{H}_m$ and $\Delta \bar{H}_v$ are experimentally measured quantities (and $\Delta \bar{S}_m$ and $\Delta \bar{S}_v$ are defined by these values, see Equations 6.9 and 6.10), the *relative* stabilities of the three phases are not affected by arbitrary selection of the reference values (although the appearance of the $\mu_i^\pi(T)p_r$, curves are).

By integrating the temperature dependence of heat capacity (Equations 6.6 through 6.8) and combining with enthalpies and entropies at reference points for melting and vaporization, we can obtain expressions for $\bar{H}^\pi(T)$ and $\bar{S}^\pi(T)$, and thus, $\mu^\pi(T)$ (Equation 6.5). For each phase, the enthalpy variation with temperature is[†]

$$d\bar{H}^\pi = \bar{C}_p^\pi dT \tag{6.11}$$

Table 6.1 Enthalpy and entropy values for the three major phases of water at the melting and boiling points at 1 atmosphere

T	Ice	Liquid	Steam
T_m	$\bar{H}^{ice}(T_m) \equiv 0$	$\bar{H}^{liq}(T_m) = \Delta \bar{H}_m = 6.01$	
T_m	$\bar{S}^{ice}(T_m) = \int_0^{T_m} \frac{\bar{C}_p^{\,ice}}{T} dT = 0.041$	$\bar{S}^{liq}(T_m) = S^{ice} + \Delta \bar{S}_m = 0.063$	
T_v		$\bar{H}^{liq}(T_v) = \bar{H}^{liq}(T_m) + \int_{T_m}^{T_v} \bar{C}_p^{\,liq} dT = 13.5$	$\bar{H}^{steam}(T_v) = \bar{H}^{liq}(T_v) + \Delta \bar{H}_v = 54.2$
T_v		$\bar{S}^{liq}(T_v) = \bar{S}^{liq}(T_m) + \int_{T_m}^{T_v} \frac{\bar{C}_p^{\,liq}}{T} dT = 0.087$	$\bar{S}^{steam}(T_v) = \bar{S}^{liq}(T_v) + \Delta \bar{S}_v = 0.195$

Note: T_m and T_v are melting and vaporization temperatures at 1 atm (237.15 and 373.15 K, respectively). Enthalpies and entropies are in kJ mol^{-1} and kJ mol^{-1} K^{-1}, respectively. The molar enthalpy of ice at the melting point, $\bar{H}^{ice}(T_m)$, is arbitrarily set to zero. The molar entropy of ice at the melting point, $\bar{S}^{ice}(T_m)$, is set to the *standard* molar entropy of ice, $\bar{S}^{ice,\circ}$, which assumes that the entropy of ice is zero at absolute zero. The molar enthalpies of fusion and vaporization are $\Delta \bar{H}_m = 6.01$ and $\Delta \bar{H}_v = 40.68$ kJ mol^{-1}, respectively. Combining these enthalpies with T_m and T_v (Equations 6.9 and 6.10) gives molar entropies of melting and vaporization of $\Delta \bar{S}_m = 0.022$ and $\Delta \bar{S}_v = 0.109$ kJ mol^{-1} K^{-1}, respectively. Values are from the CRC Handbook of Chemistry and Physics.

[†] Unlike Chapter 3, we are using partial molar quantities rather than total extensive analogs, but the same relations apply.

Integration from a reference temperature T_r (here, T_m or T_v) gives

$$\int_{\bar{H}^\pi(T_r)} d\bar{H}^\pi = \int_{T_r} \bar{C}_p^\pi dT \qquad (6.12)$$

or

$$\bar{H}^\pi(T) = \bar{H}^\pi(T_r) + \int_{T_r} \bar{C}_p^\pi dT \qquad (6.13)$$

The integrated form of the right-hand side will depend on the temperature dependence of the heat capacity function (Equations 6.6 through 6.8). For example, for ice

$$\bar{H}^{ice}(T) = \bar{H}^{ice}(T_m) + \int_{T_m} \left(\frac{a^{ice}T}{b^{ice} + T} \right) dT$$
$$= a^{ice}(T - T_m) - a^{ice}b^{ice}\ln\left(\frac{b^{ice} + T}{b^{ice} + T_m} \right) \qquad (6.14)$$

In the second line in Equation 6.14, the reference enthalpy is dropped because we set it to zero above (see Problem 6.12). Derivation of analogous molar enthalpies for liquid water and steam are left to the reader (Problem 6.14).

Likewise, from Chapter 5, we can express molar entropy variation with temperature as

$$d\bar{S}^\pi = \frac{\bar{C}_p^\pi}{T} dT \qquad (6.15)$$

As with the enthalpy analysis above, integration from a reference temperature T_r gives

$$\bar{S}^\pi(T) = \bar{S}^\pi(T_r) + \int_{T_r} \frac{\bar{C}_p^\pi}{T} dT \qquad (6.16)$$

Again, using the ice phase as an example, the temperature-dependent heat capacity in Equation 6.6 integrates to

$$\bar{S}^{ice}(T) = \bar{S}^{ice}(T = 0) + a^{ice} \int_{T=0} \left(\frac{1}{b^{ice} + T} \right) dT$$
$$= a^{ice}\ln\left(\frac{b^{ice} + T}{b^{ice}} \right) \qquad (6.17)$$

(Problem 6.13). Derivation of analogous molar entropies for liquid water and steam are left to the reader (Problem 6.15). By combining the temperature dependences with the measured melting and vaporization data, enthalpy and entropy of the three phases can be obtained for their coexistence points.

Temperature dependences of chemical potentials for ice, water, and steam

With temperature-dependent expressions for molar enthalpy and entropy of ice, chemical potentials for each phase are obtained using the expression

Figure 6.4 The chemical potential, the molar enthalpy, and the molar entropy of ice as a function of temperature, at $p = 1$ atm. Curves are generated using the fitted heat capacity function from Figure 6.3 (Equation 6.6). The reference enthalpy is set to zero at $T_m = 273.15$ K (Table 6.1). To simplify the plot, the entropy is set to zero at zero Kelvin, which leads to a slight offset from the molar entropy in Table 6.1. Red lines show $\bar{S}^{ice}$ (dashed) and $T\bar{S}^{ice}$ (solid). Black lines show $\bar{H}^{ice}$ (dashed) and μ^{ice} (solid). Although ice is not the dominant phase above T_m because its chemical potential is above that of liquid water (and perhaps steam), Equations 6.14 and 6.17 remain valid throughout. Note that the entropy curve is on a different vertical scale (J mol^{-1} K^{-1}) than the other three curves (kJ mol^{-1}).

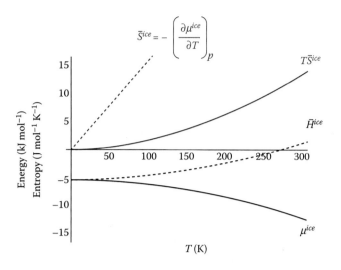

$\mu^{ice} = \bar{H}^{ice} - T\bar{S}^{ice}$ (Equation 6.5). **Figure 6.4** shows the temperature dependence of the chemical potential of ice, along with molar enthalpy and entropy values (Equations 6.14 and 6.17).

The curves in Figure 6.4 illustrate five key points:

1. In the low temperature limit, the chemical potential is equal to the molar enthalpy. This results from of the fact that T multiplies the molar entropy in the expression $T\bar{S}^\pi$ (and from the fact that we set entropy to zero at zero Kelvin).[†]

2. As the temperature increases, the molar enthalpy increases. Recall that the heat capacity and the enthalpy are related by the formula

$$\bar{C}_p^\pi = \left(\frac{\partial \bar{H}^\pi}{\partial T} \right)_p \qquad (6.18)$$

Since the heat capacity is always positive (temperature always rises when a system is heated), the slope of $\bar{H}^\pi$ with T must be positive.

3. As the temperature increases, the molar entropy increases. Recall that the heat capacity and the entropy are related by the formula

$$\left(\frac{\partial \bar{S}}{\partial T} \right)_p = \frac{\bar{C}_p}{T} \qquad (6.19)$$

As with molar enthalpy, the restriction that the heat capacity is always positive (along with absolute temperature) means that the molar entropy rises with temperature.

4. As temperature increases, the chemical potential decreases. This results from yet another derivative,

$$\left(\frac{\partial \mu^\pi}{\partial T} \right)_p = -\bar{S}^\pi \qquad (6.20)$$

and also from a reasonable expectation that entropy is positive everywhere (point 3 supports this expectation).

[†] For materials with multiple ground states, entropy will remain above zero. Also, see point 3.

5. As temperature increases, the (negative) slope of chemical potential increases (i.e., μ^π has negative curvature). This can be seen by combining Equations 6.19 and 6.20, which gives

$$\left(\frac{\partial^2 \mu^\pi}{\partial T^2}\right) = -\left(\frac{\partial \bar{S}^\pi}{\partial T}\right) = -\frac{\bar{C}_p{}^\pi}{T} \tag{6.21}$$

As in points 2 and 3 above, this restriction in the curvature of the molar free energy is a result of positive heat capacity. The implications of this last point are more profound than might be expected from first glance, providing a fundamental restriction on thermodynamic stability.

Temperature-driven phase transitions

The chemical potentials of the liquid and steam phases of water can be determined in an analogous manner as was done for ice above (that is, combining temperature-dependent expressions for $\bar{H}^\pi$ and $\bar{S}^\pi$ using Equation 6.5). Plotting the chemical potentials of ice, liquid water, and steam as a function of temperature at 1 atm (**Figure 6.5**) shows the relationship between chemical potentials, phase stability, and coexistence.

As developed at the beginning of this chapter, if the chemical potentials of two phases are equal, the two phases coexist (Equation 6.4), whereas if chemical potentials differ, the phase with the lowest potential forms. Starting at low temperature in Figure 6.5, the ice phase is formed, because it has lowest chemical potential (at 1 atm). This makes sense, because at low temperature, the solid phase is expected to have the lowest enthalpy (best bonding), and according to Point 1 above, the chemical potential is expected to be dominated by enthalpy at low temperature. However, as temperature increases, μ^{liq} decreases more than μ^{ice}, because liquid water has a larger entropy (less order), and therefore a greater negative slope. As a result, μ^{liq} and μ^{ice} intersect at T_m (the melting point, left vertical line, Figure 6.5), and the ice and liquid phases coexist.

Above the melting temperature, the chemical potential of liquid is lower than that of ice, so liquid water is formed exclusively. However, because steam has a very large entropy compared to both condensed phases (because a gas has **many** configurations that are energetically accessible), μ^{steam} decreases very sharply with temperature. The steam becomes more stable than ice around 350 K,[†] but more importantly, at 373 K (T_v, the boiling point), the chemical potential of steam matches that of the dominant phase, liquid water (right vertical line, Figure 6.5);

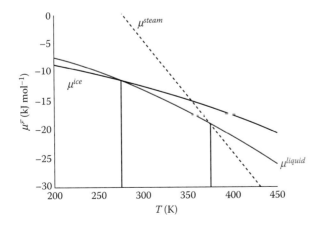

Figure 6.5 Chemical potentials of water in the ice, liquid, and gas phases at $p = 1$ atm. Chemical potentials were determined by combining molar enthalpies and entropies (Equations 6.14, 6.17 respectively) using Equation 6.5. Phase transitions occur where chemical potentials intersect. At these intersection points, the differences in slopes (molar entropies, and thus the molar enthalpies, since chemical potentials are the same) give the molar enthalpies of melting and vaporization.

[†] Since ice is not stable at 350 K, we do not observe the phase transition from ice to steam ("sublimation") at 1 atm pressure and ambient temperature. For carbon dioxide, this is the major phase transition seen under ambient conditions.

thus, at 373 K, liquid water and steam coexist. Above this temperature, steam has a lower chemical potential than liquid water, so the steam phase is formed exclusively.

The chemical potentials drawn in Figure 6.5 give information on the temperature dependence of each phase, whether it is stable or not. However, in practical terms, the total thermodynamic functions of the system reflect only the stable phase (or at coexistence points, phases). These total thermodynamic functions can be expressed in terms of equilibrium mole fractions for each phase, x^{π}:

$$\bar{G}_T = x^{ice}\mu^{ice} + x^{liq}\mu^{liq} + x^{steam}\mu^{steam} = \sum_{\pi=1}^{n\,phases} x^{\pi}\mu^{\pi} \tag{6.22}$$

$$\bar{H}_T = x^{ice}\bar{H}^{ice} + x^{liq}\bar{H}^{liq} + x^{steam}\bar{H}^{steam} = \sum_{\pi=1}^{n\,phases} x^{\pi}\bar{H}^{\pi} \tag{6.23}$$

$$\bar{S}_T = x^{ice}\bar{S}^{ice} + x^{liq}\bar{S}^{liq} + x^{steam}\bar{S}^{steam} = \sum_{\pi=1}^{n\,phases} x^{\pi}\bar{S}^{\pi} \tag{6.24}$$

The summations on the right-hand side generalize the equations to n phases.[†]

These "population-weighted average" functions are shown in **Figure 6.6** for water at 1 atm. Though the total molar free energy (Figure 6.6B) is continuous, its slope changes abruptly at the melting and boiling points, corresponding to the ice–liquid and liquid–steam phase transitions. On either side of a given coexistence point, the total molar free energy is equal to the chemical potential of the phase that is formed (i.e., to the phase with lowest chemical potential). The abrupt change in the slope of the molar free energy at the transition temperature reflects the fact that the different phases have different total entropies (Figure 6.6C).[‡] These entropy differences, which appear as discontinuities in a plot of $\bar{S}_T$ versus T, are the entropies of fusion and vaporization. Similar discontinuities appear in the temperature profiles of the total molar enthalpy (Figure 6.6D).

One fairly unique feature of the phase properties of water is that the molar enthalpy and entropy of fusion (melting) is significantly smaller than for vaporization. This is consistent with a model in which water has a significant amount of structure in the liquid phase, resulting in a low enthalpy (strong bonding) and a low entropy (high ordering). These thermodynamic properties result from the strong and stereospecific hydrogen bonding properties of water. A high amount of structure is also consistent with the high heat capacity of liquid water compared to ice (Figure 6.3), which requires not only that an enthalpically stable structure can be formed, but also that the structure is labile when heated (i.e., the structure can be melted).

Note that since the heat capacity is the temperature derivative of the enthalpy (and is also closely related to the temperature derivative of the entropy, Equation 6.21), the total heat capacity becomes infinite at the melting and boiling temperatures (Figure 6.6E). These two "singularities" occur because the total enthalpy (and entropy) is discontinuous at the melting and boiling temperatures, reflecting the heat required to break bonds in the solid and liquid, respectively. The slope (and thus, the heat capacity) at a disontinuity is infinite. Another way to understand these singularities in heat capacity is that the heat capacity is equal to the variance in the enthalpy (see Chapter 10). When phases with very different enthalpies coexist, the overall heat capacity of the system is high because there is an enormous variance in enthalpy.[§] Interpreted structurally, heat put into the system under

---

[†] Although for single-component systems, only three phases are needed; see below.
[‡] Except at the critical point (see below).
[§] This is why the heat capacity in Figure 2.15 is high at the reaction midpoint.

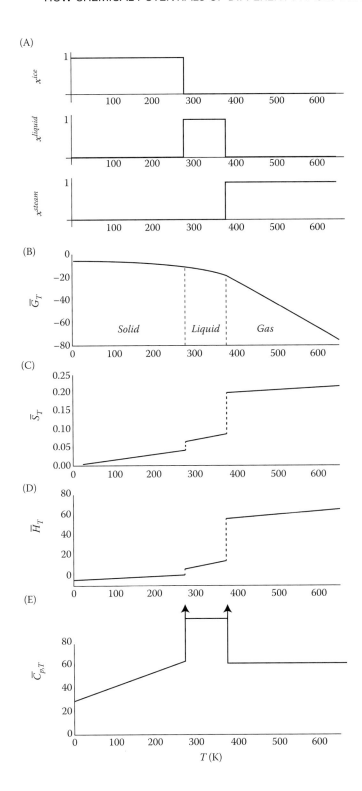

Figure 6.6 The thermodynamic properties of water in phase equilibrium at 1 atm. (A) Mole fractions of the ice, liquid, and steam phases of water. Below $T_m = 273.15$ K, the ice phase predominates. At T_m, the ice melts, the enthalpy and entropy increase modestly (C and D) but abruptly (reflecting the enthalpy and entropy of vaporization), and the heat capacity doubles (E). The liquid phase remains stable until $T_v = 373.15$ K where the liquid water boils. Upon vaporization, the heat capacity decreases by a factor of two, and the enthalpy and entropy (of vaporization) increase substantially. At T_m and T_v, heat capacity abruptly goes to infinity (indicated by upward arrows). Qualitatively, these changes to the thermodynamic parameters (discontinuities in the heat molar capacity, enthalpy, and entropy) are seen for many materials undergoing phase transitions. In contrast, the molar free energy (μ) is continuous (B), although its slope (the negative entropy) is not. G and H have units of kJ mol^{-1}; S has units of kJ mol^{-1} K^{-1}, C_p has units of J mol^{-1} K^{-1}.

conditions of phase coexistence simply shifts material from the low enthalpy phase to the high enthalpy phase, breaking lattice or liquid bonds (for solid and liquid, respectively), without changing the temperature.

Adding the effects of pressure to chemical potential relationships

So far we have examined the effects of temperature on phase equilibrium at constant pressure. What about the effects of pressure on phase equilibrium? The effects of pressure result from the $\bar{V}^{\pi}dp$ term in the differential expression for molar free energy variation:

$$d\bar{G}^{\pi} = d\mu^{\pi} = -\bar{S}^{\pi}dT + \bar{V}^{\pi}dp \qquad (6.25)$$

(see Chapter 5; a formal derivation for molar quantities is given below, Equations 6.32 through 6.37). For phases with large molar volumes, chemical potential will vary strongly with pressure. At moderate pressures, the volume of the gas phase is typically much larger than the volumes of the condensed phases. For example, for water at the boiling point at 1 atm of pressure, one mole of water vapor occupies about 30 liters, whereas ice and liquid water occupy 0.0186 and 0.0187 liters, respectively; that is, the gas phase has more than 1000 times greater molar volume than the condensed phases. Moreover, the compressibility and the thermal expansivity coefficients (κ_T and α) are much larger for gas phases than for condensed phases.

Thus, as a crude approximation, pressure effects can be ascribed exclusively to the chemical potential of the gas, ignoring effects on the condensed phases. A slightly more sophisticated (first order) approximation is to include pressure effects on liquid and solid phases by including the molar volumes (18.012 and 19.64 mL mol^{-1} for water at 273.15 K), but assuming these values to be independent of temperature and pressure (i.e., $\alpha = \kappa_T = 0$). Though imperfect (see Problems 6.16 through 6.18 for a more accurate representation), this approximation captures several key features of the experimentally determined p–T phase diagram.

Applying this crude approximation to our analysis of the phase equilibrium of water, we can derive the pressure dependence of the chemical potential of steam by introducing the ideal gas law into Equation 6.6 at constant temperature:

$$\begin{aligned} d\mu^{steam} &= \bar{V}^{steam}dp \\ &= \frac{RT}{p}dp \end{aligned} \qquad (6.26)$$

Integrating Equation 6.26 from a reference pressure of 1 atm leads to

$$\int_{\mu_{p^{\bullet}}}^{2} d\mu^{steam} = RT\int_{p^{\bullet}}^{\,} \frac{dp}{p}$$

$$\mu^{steam}(p) - \mu_{p^{\bullet}}^{steam} = RT\ln\left(\frac{p}{p^{\bullet}}\right) \qquad (6.27)$$

$$\mu^{steam} = \mu_{p_r}^{steam} + RT\ln p$$

In Equation 6.27, we used the used a special notation for the reference pressure of 1 atm, that is, $p_r = p^{\bullet}$. In Chapter 7, we will refer to this special reference as a "standard state pressure." Because the reference pressure $p^{\bullet}$ has unit value, it is omitted for simplicity.[†]

The expression we generated for the temperature dependence of chemical potential of steam at 1atm (dotted black curve, Figure 6.5; see Problems 6.14 and 6.15) can be substituted into Equation 6.27 for $\mu_{p_r}^{steam}$ to give a chemical potential surface for steam as a function of temperature and pressure:

[†] Though p_r is not explicitly written in the final term in Equation 6.27, its presence is implied by the requirement of logarithms that their arguments be dimensionless. We will return to this issue when we define "standard states" for reaction thermodynamics in Chapter 7.

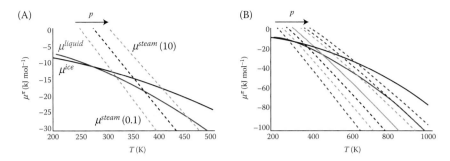

Figure 6.7 The chemical potential of steam as a function of temperature at different pressures. Chemical potentials for ice and liquid water (1 atm values) are plotted as in Figure 6.5 (black and red lines). The effect of pressure on the steam phase (Equation 6.28) is shown using dashed lines. Making the approximation that the condensed phases do not change their potentials with pressure, the shift in intersection of μ^{liq} and μ^{steam} with increasing pressure corresponds to an increase in the boiling temperature. For clarity, panel (A) only shows three pressures, each differing by a factor of ten. The black dashed curve shows μ^{steam} at $p^\bullet = 1$ atm. Panel (B) shows a broader range of pressures (and temperatures), and indicates a low pressure triple-point (~0.005 atm, red dashed curve) and a high temperature critical point (~330 atm, second dashed curve from right dashed curve). These approximate values differ from experimental triple-and critical-point values for water (0.006 atm and 220 atm, respectively), due to the approximations in the analysis (no volume dependences for the condensed phases, ideal gas dependence for steam).

$$
\begin{aligned}
\mu^{steam}(p,T) &= \bar{H}_{p^\bullet}^{steam} - T\bar{S}_{p^\bullet}^{steam} + RT\ln p \\
&= \bar{H}_{p^\bullet,T_v}^{steam} + a^{steam}(T-T_v) - T\left(\bar{S}_{p^\bullet,T_v}^{steam} + a^{steam}\ln(T/T_v)\right) + RT\ln p \\
&= \bar{H}_{p^\bullet,T_v}^{steam} - a^{steam}T_v - T\left\{\bar{S}_{p^\bullet,T_v}^{steam} + a^{steam} + a^{steam}\ln(T/T_v) - R\ln p\right\} \\
&= \mu_{p^\bullet,T_v}^{steam} - a^{steam}T_v - T[a^{steam}\{1+\ln(T/T_v)\} - R\ln p]
\end{aligned}
\tag{6.28}
$$

In principle, Equation 6.28 should have the same form as Equation 5.28, since both describe the free energy of an ideal gas, although superficially at least, the two equations look rather different. You should convince yourself that they are more similar than they look (Problem 6.19). The advantage of Equation 6.28 is that it we constructed it to directly relate to the chemical potentials of liquid water and ice.

Treating the chemical potentials of the condensed phases as pressure independent, we can get a rough sense of how pressure affects phase behavior of water by perturbing μ_{steam} (**Figure 6.7**). As pressure is increased, the μ^{steam} curve is shifted to the right, increasing the boiling point significantly (at 10 atm, the crossing point in Figure 6.7A appears to be 420 K, a 50 K upshift). Conversely, as pressure decreases, the boiling point decreases. Looking at a broader range of pressure variation (Figure 6.7B), two notable features can be seen. First, at very low pressure (about 0.005 atm), μ^{steam} intersects the crossing point (melting point) for μ^{ice} and μ^{liquid}. This point of crossing (which, as we will show below, is indeed confined to a single point on the T-p plane), is referred to as a "triple point," since all three phases coexist.

Second, at very high pressure (about 330 atm) μ^{steam} intersects μ^{liquid} as a tangent, that is, with the same slope (second dashed curve from right and solid red curve, Figure 6.7B). Because μ^{steam} and μ^{liquid} have the same slopes with temperature at this intersection, they have the same molar entropy, thus, $\Delta \bar{S}_{vap} = 0$. Since μ^{steam} and μ^{liquid} are identical at this point (and are thus in phase equilibrium), $\Delta \bar{H}_{vap} = 0$ as well. In addition, it is found that at this intersection point, the molar volumes and compressibilities of the two phases become identical. The fact that all of these thermodynamic parameters (e.g., enthalpy, entropy, and density) become identical at this intersection point is consistent with the two phases (liquid and vapor) becoming indistinguishable from one another. At this point, which is referred to as the "critical point," the two phases become one, and above this point, a single "supercritical" fluid phase exists.

Combining pressure and temperature into a single phase diagram

As given in Equation 6.28, the chemical potential of steam is given by a surface that depends on both T and p. The same can be done for the condensed phases, though the ideal gas law is no longer appropriate as an equation of state. Instead, empirical data can be used to approximate equations of state. Under the zero-order approximation described above, the chemical potential surfaces of the condensed phases would be a flat projection of the temperature dependence in the direction of pressure. Using the more realistic first-order approximation (that is, including the volume contributions of the condensed phases but assuming $\bar{V}^{ice}$ and $\bar{V}^{liq}$ are independent of p and T), the pressure dependence of μ^{ice} and μ^{liq} is obtained by integration of an equation analogous to the first line of Equation 6.26:

$$\int_{\mu_{p^\bullet}} d\mu^\pi = \bar{V}^\pi \int_{p^\bullet} dp$$

$$\mu^\pi = \mu_{p^\bullet}^\pi + \bar{V}^\pi(p - p^\bullet)$$

$$(6.29)$$

where π represents the liquid or solid phase. Combining this pressure dependence with the temperature-dependent enthalpy and entropy functions (Problem 6.14) gives the chemical potential surfaces for the condensed phases. For ice,

$$\mu^{ice}(p,T) = \bar{H}_{p^\bullet}^{ice}(T) - T\bar{S}_{p^\bullet}^{ice}(T) + V^{ice}(p - p^\bullet)$$

$$= a^{ice}(T - T_m) - a^{ice}b^{ice}\ln\left(\frac{b^{ice} + T}{b^{ice} + T_m}\right)$$

$$(6.30)$$

$$- Ta^{ice}\ln\left(\frac{b^{ice} + T}{b^{ice}}\right) + V^{ice}(p - p^\bullet)$$

The subscript $p^\bullet$ has been dropped from the first two terms of the second equality, since $\bar{H}_{T_f}^{ice}$ and $\bar{S}_{T_f}^{ice}$ are assumed to be pressure independent (based on a Maxwell relation, Problem 6.20). Likewise, for liquid water,

$$\mu^{liq}(p,T) = \bar{H}_{p_r}^{liq} - T\bar{S}_{p_r}^{liq} + \bar{V}^{liq}(p - p^\bullet)$$

$$= \bar{H}^{liq}(T_m) + a^{liq}(T - T_m) + \frac{b^{liq}}{2}(T^2 - T_m^2) + \frac{c^{liq}}{3}(T^3 - T_m^3)$$

$$- T\left\{\bar{S}^{liq}(T_m) + a^{liq}\ln(T/T_m) + b^{liq}(T - T_m) + \frac{c^{liq}}{2}(T^2 - T_m^2)\right\} + \bar{V}^{liq}(p - p^\bullet)$$

$$(6.31)$$

Plotting these three chemical potentials surfaces as a function of T and p provides a detailed map of phase behavior. Unlike the single variable picture in Figure 6.5, where phase transitions are represented by single points of chemical potential crossing, the intersections of p–T surfaces define *line curves* of coexistence (**Figure 6.8A**). Away from these surface crossings, the phase with the lowest chemical potential predominates. For example, at high temperatures and low pressures, the steam phase has the lowest chemical potential surface (Figure 6.8A), and is formed at the expense of the condensed phases. However, as temperature decreases

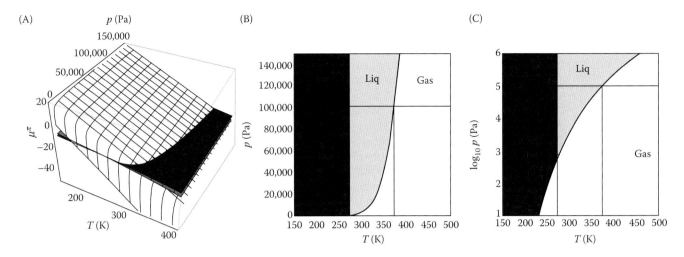

Figure 6.8 Phase diagram for a simple model of water. (A) Three dimensional plot of the chemical potential surfaces for ice (red), liquid water (gray), and steam (white). The intersections of these surfaces give two-dimensional curves where the chemical potentials of different phases are equal. As long as these curves of intersection are at lower potential than that of the third phase, the intersecting phases coexist simultaneously. (B) Projection of coexistence curves onto the p–T plane. In between the coexistence curves, the phase with the lowest chemical potential is populated exclusively. Solid lines separating these regions represent lines of phase coexistence. (C) Transformation to a $log_{10}p$ versus T plane (solid lines), which better shows the triple point, and the direct sublimation from ice to vapor at low temperature (the liquid–vapor critical point is *not* shown in this model). Note that although the T–p phase diagram (B) can be approximated by viewing the chemical potential surfaces from below, the approximations used to render such surfaces graphically can lead to fairly substantial distortions of coexistence lines. Thus, panels (B) and (C) were used with the Mathematica command "RegionPlot," which gives a much more accurate representation of the boundaries of the most stable phase.

and pressure increases, the potential surfaces of the condensed phases drop below that of steam, with liquid dominating at intermediate temperatures, and ice dominating at low temperature.

One nice way to visualize which phases are populated is to project the intersections between chemical potential surfaces onto the T–p plane (**Figure 6.8B**). In regions between the coexistence lines, only a single phase is stable. Along a coexistence line, the two bordering phases are in equilibrium. In the plot in Figure 6.8C, the same phase stability information is plotted as a function of the log of the pressure, to better display both low- and high-pressure behavior. This log plot reveals a coexistence line between solid and vapor (referred to as "sublimation"), and also the triple point, where all three phases coexist. The triple point identified from this approximate phase diagram ($p = 0.006$ atm, $T = 273.19$) compares surprisingly well with the experimental value ($p_3 = 0.006$ atm, $T_3 = 273.16$ K), given the simplicity of the model used to generate the coexistence curves.[†]

ADDITIONAL RESTRICTIONS FROM THE PHASE DIAGRAM: THE CLAUSIUS–CLAPEYRON EQUATION AND GIBBS' PHASE RULE

The details of phase diagrams, such as the specific temperatures of phase transitions, triple-points, and critical points, depends on the molecular details of the material under study (e.g., water versus carbon dioxide). However, there are important general principles that can be extracted from geometric constructs such as Figures 6.6 and 6.7. We will derive some of these principles here. One of these principles is the "phase rule," a relationship discovered by J.W. Gibbs (Gibbs, 1878)[‡] that

---

[†] This is a significant improvement over the zero-order approximation (Figure 6.7), where the predicted triple-point was at 0.005 atm, 273.03 K (although even that prediction was not that bad).

[‡] This (rather dense) paper, along with Gibbs' other works, has been reprinted and is available in "The Scientific Papers of J. Willard Gibbs," Volume 1: Thermodynamics. Oxbow Press, Woodbridge, CT (1993).

restricts the maximum number of phases that can coexist, and how many free variables (i.e., macroscopic degrees of freedom) can be accommodated given particular coexistence. The phase rule also takes into account the number of different chemical components and the number of reactions among components, although in this section we will remain focused on single component systems. Another principle is the "Clausius–Clapeyron" equation, which relates coexistence lines (and surfaces and points) to the thermodynamic properties of the materials in each phase. Though the phase rule and Clausius-Clapeyron equation are often derived separately (Adkins) by evaluating Gibbs–Duhem relationships for each phase geometrically on the phase diagram, here we will present a derivation using differentials that generates both principles at once in a logical and intuitively accessible way.

A three-dimensional representation of the Gibbs–Duhem equation

From Chapter 5, the Gibbs–Duhem relation for a single component system can be written as

$$n\,d\mu + S\,dT - V\,dp = 0 \qquad (6.32)$$

where n, S, and V are extensive quantities. As with the phase relationships above, the i subscript is omitted since we are treating a single-component system. Thus, we can divide by the number of moles and rearrange Equation 6.32 to get

$$d\mu = -\bar{S}\,dT + \bar{V}\,dp \qquad (6.33)$$

For a multiphase single-component system at equilibrium, the system variables T and p will be the same throughout each phase, although the molar entropy and molar volume will generally differ from one phase to the next. Thus, there is a Gibbs–Duhem equation associated with each phase[†]:

$$d\mu^{ice} = -\bar{S}^{ice}\,dT + \bar{V}^{ice}\,dp \qquad (6.34)$$

$$d\mu^{liq} = -\bar{S}^{liq}\,dT + \bar{V}^{liq}\,dp \qquad (6.35)$$

$$d\mu^{steam} = -\bar{S}^{steam}\,dT + \bar{V}^{steam}\,dp \qquad (6.36)$$

Now consider a set of conditions (i.e., specific p and T values) where two phases coexist. For example, at $p = 1$ atm, $T = 373.15$ K, steam and liquid water are in stable coexistence, that is, $\mu^{steam} = \mu^{liq}$. A key question is "what restrictions are there on the system variables (p and T) if coexistence is to be maintained?" Changing p and T will change chemical potentials in both phases through each associated Gibbs–Duhem Equation 6.33. In general, phase potentials will change by different amounts (i.e., $d\mu^{steam} \neq d\mu^{liq}$). This can be shown graphically in **Figure 6.9**, where most arbitrary variations dp and dT will move the system off the coexistence curve to a single-phase region. It is only a limited subset of simultaneous changes to both p and T that maintain the relationship $\mu^{steam} = \mu^{liq}$ will remain on the coexistence curve. For this to occur, the chemical potentials of the two phases must not only start out equal ($\mu^{steam}_{initial} = \mu^{liq}_{initial}$), *they must end up equal* ($\mu^{steam}_{final} = \mu^{steam}_{initial} + d\mu^{steam}_{initial} = \mu^{liq}_{initial} + d\mu^{liq}_{initial} = \mu^{liq}_{final}$). This means that $d\mu^{steam} = d\mu^{liq}$. For example starting at the steam-water coexistence starting at $p = 1$ atm, $T = 373.15$ K (Figure 6.9), if an arbitrary change is made to pressure (by dp), then a compensating change must be made to T (by dT) that is fixed by

---

[†] Here we have continued with the ice, liquid, steam phase markers to connect with the previous discussion of phase equilibrium in water, although in general, any number of phases could be labeled α, β, γ, etc.

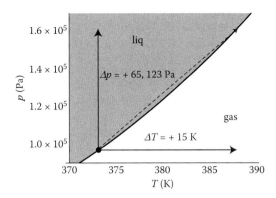

Figure 6.9 Changing state variables to maintain phase coexistence. An expanded version of the water phase diagram from Figure 6.8. For a system starting from the 1 atm boiling point, where $\mu^{liq} = \mu^{steam}$, arbitrary changes to either p or T (by Δp or by ΔT) without a compensating change in the other variable would change μ^{steam} and μ^{liq} by different amounts. As a result, the system would move from the coexistence line to a single phase. region of the phase diagram. To maintain phase coexistence, the change $\Delta T = +15$ K shown above would need to be accompanied by a specific change in pressure ($dp = +65{,}123$ Pa), so that the two chemical potentials of the two phases change by the same amount. Such changes keep the system on the coexistence curve (dashed arrow).

the requirement that the final chemical potentials must be equal. If this condition is met, the system "travels" diagonally up the coexistence line, producing the same change to chemical potential ($d\mu$) in both phases.

To see how the magnitude of $d\mu$ is controlled by the variations dT and dp, and to explore the restrictions imposed by phase coexistence, it is useful to plot the Gibbs–Duhem equation for each phase in terms of its *differential* variables. For liquid water and steam at the boiling point, the Gibbs–Duhem equations are planes given by

$$d\mu^{liq} = -\bar{S}^{liq}dT + \bar{V}^{liq}dp$$
$$= -68dT + 18.02 \times 10^{-6}dp \tag{6.37}$$

and

$$d\mu^{steam} = -\bar{S}^{steam}dT + \bar{V}^{steam}dp$$
$$= -195dT + 30.6 \times 10^{-3}dp \tag{6.38}$$

where molar entropy units are J mol^{-1} K^{-1}, and molar volume units are m^3 mol^{-1}. Equations 6.37 and 6.38 describe planes for each phase, Γ^{π}, in a space with axes dp, dT, and $d\mu$ (**Figure 6.10**). A moment of thought should convince you that no matter what their slopes, these Gibbs–Duhem planes intersect the axes at their origins.[†]

Geometrically, the planes Γ^{steam} and Γ^{liq} have different slopes in arbitrary directions, due to the differences in molar volumes and molar entropies of the two phases. Thus, arbitrary variations to T and p from the origin will typically cause the two planes to separate (i.e., $d\mu^{liq} \neq d\mu^{steam}$), consistent with the general conversion to a single phase described above. Only along the line of dp, dT values for which the two planes intersect is the relation $d\mu^{liq} = d\mu^{steam}$ satisfied (red line, Figure 6.9C), maintaining phase coexistence. Note that along this line, since the chemical potentials of both phases vary by the same amount, we can substitute a generic $d\mu$ for $d\mu^{liq}$ and $d\mu^{steam}$.

[†] The fact that the origin is intersected makes intuitive sense: if $dT = dp = 0$, then $d\mu = 0$. It also makes algebraic sense, since there is no offset term in Equations 6.37 and 6.38.

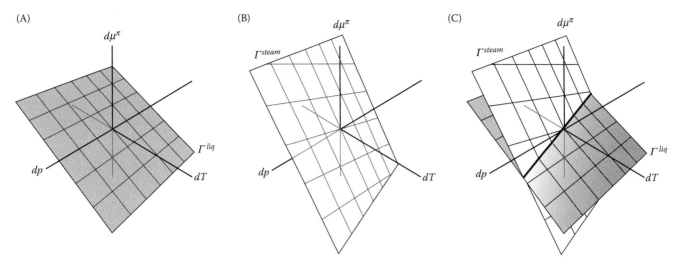

Figure 6.10 Gibbs–Duhem planes for different phases. For each phase π, the Gibbs–Duhem equation defines a plane Γ^π that gives the differential variation in μ with increments in p and T. (A) The Gibbs–Duhem plane for liquid water (Equation 6.37) at 1 atm, 373.15 K. (B) The Gibbs–Duhem plane for steam (Equation 6.38) at 1 atm, 373.15 K. (C) The intersection of the liquid and steam planes (red line) corresponds to simultaneous changes in p and T that maintain phase equilibrium. Although the chemical potentials of both phases change along this line, they change by the same amount ($d\mu^{liq} = d\mu^{steam} \neq 0$).

Intersection of Gibbs–Duhem planes and the Clausius–Clapeyron equation

So how do we find the coexistence line from the two planes Γ^{steam} and Γ^{liq}? The coexistence line corresponds to the intersection of the two planes; from analytical geometry, the intersection of two planes is given by the cross product of the normal vectors to the each plane.[†] The normal vector for the Gibbs–Duhem plane for phase π is given as

$$\vec{n}^\pi = \left\langle -\bar{S}^\pi, \bar{V}^\pi, 1 \right\rangle \tag{6.39}$$

where the angled brackets denote the three components of the vector. Thus, the intersection line is

$$
\begin{aligned}
\vec{n}^{liq} \times \vec{n}^{steam} &= \left\langle -\bar{S}^{liq}, \bar{V}^{liq}, 1 \right\rangle \times \left\langle -\bar{S}^{steam}, \bar{V}^{steam}, 1 \right\rangle \\[4pt]
&= \begin{vmatrix} i & j & k \\ -\bar{S}^{liq} & \bar{V}^{liq} & 1 \\ -\bar{S}^{steam} & \bar{V}^{steam} & 1 \end{vmatrix} \\[4pt]
&= \left(\bar{V}^{liq} - \bar{V}^{steam} \right) i - \left(-\bar{S}^{liq} + \bar{S}^{steam} \right) j + \left(-\bar{S}^{liq}\bar{V}^{steam} + \bar{S}^{steam}\bar{V}^{liq} \right) k \\[4pt]
&= \Delta\bar{V}^{s \to l} i + \Delta\bar{S}^{s \to l} j + \left(\bar{S}^{steam}\bar{V}^{liq} - \bar{S}^{liq}\bar{V}^{steam} \right) k
\end{aligned}
\tag{6.40}
$$

where i, j, and k are unit vectors in the direction dT, dp, and $d\mu$, respectively, and the $\Delta\bar{V}^{s \to l}$ and $\Delta\bar{S}^{s \to l}$ are molar changes in going from the steam to liquid phase. The equation for this differential coexistence line can be written as three parametric equations using the free parameter[‡] x:

$$\Gamma^{s \to l} = x \left\langle \Delta\bar{V}^{s \to l}, \Delta\bar{S}^{s \to l}, \left(\bar{S}^{steam}\bar{V}^{liq} - \bar{S}^{liq}\bar{V}^{steam} \right) \right\rangle \tag{6.41}$$

[†] This is a general result from vector calculus, and can be found most introductory texts on the subject.
[‡] Often the free parameter t is used when writing parametric equations. Here x is used instead to avoid confusion with temperature T.

The terms in the angled brackets in $\Lambda^{s \to l}$ give the magnitude of change in the directions dT, dp, and $d\mu$. For a one unit increment of the parameter x, dT increments by $\Delta \bar{V}^{s \to l}$, and dp increments by $\Delta \bar{S}^{s \to l}$. Accompanying this change in the chemical potentials of both phases by $\bar{S}^{steam}\bar{V}^{liq} - \bar{S}^{liq}\bar{V}^{steam}$. In general (for an arbitrary value t of the parameter x), the variation in dp with dT in the direction of phase coexistence is given by

$$\left(\frac{\partial p}{\partial T}\right)_{\mu^{steam}=\mu^{liq}} = \frac{t\Delta\bar{S}^{s \to l}}{t\Delta\bar{V}^{s \to l}} = \frac{\Delta\bar{S}^{s \to l}}{\Delta\bar{V}^{s \to l}} \tag{6.42}$$

Equation 6.42 is known as the Clausius–Clapeyron equation. For most phase transitions, the higher entropy (more disordered) phase has greater volume, so the slope of the coexistence line is positive ($\Delta\bar{V}^{\alpha \to \beta}$ and $\Delta\bar{S}^{\alpha \to \beta}$ have the same sign). For a few materials, the volume change is negative, resulting in a negative coexistence slope. One important example is the ice to liquid transition of water at 1 atm, for which entropy increases but volume decreases.

The number of coexisting phases and the phase rule

Finally, we will use Gibbs–Duhem planes to examine the restrictions on simultaneous coexistence of three phases. Is there a way to vary T and p and maintain coexistence of all three phases? For this to be the case, the differential coexistence line of the first and second phase would need to be *in the Gibbs–Duhem plane* of the third phase (**Figure 6.11A**). While this condition can be satisfied geometrically, it puts restrictions on the thermodynamic properties of the third phase. If the coexistence line $\Lambda^{\alpha \to \beta}$ of the first two phases (α and β) is in the plane of the third phase (γ), the dot product of the line $\Lambda^{\alpha \to \beta}$ with the normal of the third phase must be zero (because they must be perpendicular). Algebraically, this results in the following restriction (Problem 6.21)

$$\bar{S}^{\gamma} = \frac{\bar{S}^{\beta}\bar{V}^{\alpha} - \bar{S}^{\alpha}\bar{V}^{\beta}}{\Delta\bar{V}^{\beta \to \alpha}} + \left(\frac{\Delta\bar{S}^{\beta \to \alpha}}{\Delta\bar{V}^{\beta \to \alpha}}\right)\bar{V}^{\gamma} \tag{6.43}$$

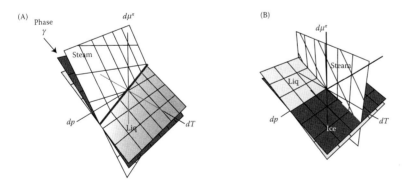

Figure 6.11 Differential coexistence planes for three phases in a single component system. (A) A hypothetical intersection of three Gibbs–Duhem planes along a single coexistence line. The water steam and liquid phases are shown at the triple point (0.006 atm, 273.16 K) along with a hypothetical phase (γ) that has an intermediate molar volume ($\bar{V}^{\gamma} = 0.04\,m^3\,mol^{-1}$). According to Equation 6.43, the molar entropy of the third phase is dependent on the steam and liquid phases. Thus, although this hypothetical third phase has the same coexistence line as the water and ice phases, its thermodynamic properties are dependent on the other two phases; thus it is not an independent phase. (B) The actual intersection of the three phases of water at the triple point. Unlike the two-phase coexistence (Figure 6.10C), which permits variation of p and T along a line (Equation 6.43), the three-phase coexistence (ice, liquid, and steam) is restricted to a single point, for which $dT = dp = d\mu^{\pi} = 0$. Thus, there are no free variables for the triple coexistence point, as is given by Gibbs' phase rule (Equation 6.44).

Equation 6.43 says that the entropy and volume of the third phase are not independent; rather, these two quantities depend on the properties of the other two phases. For example, if a hypothetical "third phase" that coexistent with liquid water and steam, the value of the molar entropy of the third phase is fixed by the properties of the steam and liquid water phases. In short, there is no third independent coexistent phase.

The only way that a *unique* third phase can coexist with two others is if the plane of the third phase *does not* contain the coexistence line for the first and second phases (**Figure 6.11B**). In this case, there is only a *point* of intersection of the three differential planes (at $dT = dp = d\mu^\pi = 0$). This means that T and p cannot be varied without disrupting coexistence among three phases of a single component. Such a coexistence can be described as having "no degrees of freedom" ($D = 0$), as opposed to a two-phase coexistence, which has one degree of freedom (e.g., coexistence of liquid water and steam is not disrupted by variation in T, as long as p is also varied by an amount given by the Clausius–Clapeyron equation). For the single component system ($C = 1$), this can be summarized as

$$C - \#\pi + 2 = D \tag{6.44}$$

where $\#\pi$ is the number of coexistent phases. This expression is known as "Gibbs' phase rule." You should convince yourself (perhaps using a diagram like **Figure 6.9**) that for a single-phase system, the number of degrees of freedom given by the phase rule is consistent with the interpretations of the phase diagram above.

Problems

6.1 Starting with one mole of liquid water and one mole of ice at the freezing point (1 atm), calculate the volume change if all the ice gets converted to liquid water at constant T and p.

6.2 Calculate the enthalpy change in Problem 6.1 above.

6.3 One of the features that distinguishes intensive variables from extensive variables is that in multiphase systems, many of the intensive variables (but not all) are the same across all phases, whereas the extensive ones are not (except, perhaps, by coincidence). Consider a multiphase system at equilibrium. Identify which of the quantities are intensive versus extensive, and identify which of the intensive quantities are identical across all phases in equilibrium with one another.

Variable	Intensive/Extensive?	The same in all phases?
E		
C_p		
G		
V		
p		
T		
$\bar{V}$		
X_i		
$\bar{S}_i$		
μ_i		
$\bar{G} = G/n_T$ (n_T is total mole number)		

6.4 If a system has two phases and one is higher in chemical potential than the other, phase transitions can sometimes go all the way to the lowest phase, but sometimes they can convert some of the high chemical potential phase but come to an equilibrium in which the two phases coexist. As mentioned in the text, the first type of behavior (complete phase conversion) occurs at constant pressure and temperature, whereas the second does not. Why should temperature and pressure be the variables that determine which type of transition results?

6.5 To help illustrate the concept explored in Problem 6.4, consider a volume of liquid water that is superheated to 400 K and is put into contact with its vapor phase at 1 atm atmosphere in two different circumstances. In the first, the superheated liquid is placed in contact with a heat reservoir, and a pressure reservoir of 1 atmosphere, allowing expansion to any final volume. In the second, the liquid and vapor phase contact a heat reservoir, but are sealed in fixed volume vessel. Describe possible equilibrium states in these two circumstances.

6.6 Demonstrate the validity of Equation 6.5,

$$\mu^\pi = \bar{H}^\pi - T\bar{S}^\pi$$
$$= \bar{E}^\pi + p\bar{V}^\pi - T\bar{S}^\pi$$

starting with the assumption that it is valid in extensive form for the system as a whole, that is,

$$G = H - TS = E + pV - TS$$

6.7 For purposes of integrating $d\mu(p,T)$, show that if molar volume is independent of temperature, the molar entropy is independent of pressure.

6.8 Calculate the expected heat capacity at constant pressure of steam assuming ideal behavior and compare it with the experimentally determined value at 373.15 K (37.5 J mol^{-1} K^{-1}). Provide a possible molecular explanation for any deviation you find between the measured and calculated values.

6.9 Although enthalpies of phase transitions can be measured directly by calorimetry (as a heat of transition at constant pressure), entropies cannot. Given the enthalpies of fusion and vaporization given in Table 6.1, how can entropies be determined for the two phase transitions? Show that the numbers determined by this simple method match the molar entropies of fusion and vaporization given in Table 6.1.

6.10 In the text, it was stated that in general, molar enthalpy should increase with temperature. Come up with an explanation that supports this statement.

6.11 Pure ethanol has a melting temperature of $T_m = 159$ K at $p = 1$ atm, and a melting enthalpy of $\Delta \bar{H}_m° = 31$ J mol^{-1} K^{-1}. Determine the entropy of melting of ethanol.

6.12 Derive the expression for the enthalpy of ice as a function of temperature (Equation 6.14).

6.13 Derive the expression for the entropy of ice as a function of temperature (Equation 6.17).

6.14 Derive expressions for the molar enthalpy of liquid water and steam as a function of temperature, at $p_r = 1$ atm.

6.15 Derive expressions for the molar entropy of liquid water and steam as a function of temperature, at $p_r = 1$ atm.

6.16 Using the data below for the molar volume of ice as a function of temperature and pressure, fit an expression (in three dimensions) to come up with an expression for $\bar{V}^*(p,T)$. I would suggest using a polynomial with as many terms as you think necessary to avoid nonrandom residuals, but no more terms than that. Use this expression to derive an expression for the chemical potential of ice as a function of T and p.

6.17 Using the data below for the molar volume of liquid water as a function of temperature and pressure, derive $\bar{V}^{liq}(p,T)$ and $\mu^{liq}(p,T)$ as in Problem 6.16.

6.18 Along with the expression $\mu^{steam}(p,T)$ from Equation 6.28, plot the three chemical potentials in μ, p, T space, highlighting their intersections. Project these intersection lines onto the p, T plane. Do they appear different than the phase diagram approximated in Figure 6.7?

6.19 Rearrange Equation 6.28 to see how close you can make it look to Equation 5.28 (or vice versa). When you have done your best, what differences remain, and what do they represent? Plot the two functions to visualize their similarities and differences.

6.20 If the molar volume of ice is temperature independent (i.e., $\alpha^{ice} = 0$), then what can you say about the entropy constant temperature (e.g., T_m)?

6.21 For three hypothetical phases that all share the same coexistence line (Figure 6.11B), derive Equation 6.43, which stipulates that the three phases are not independent.

6.22 You are probably familiar with the fact that solid carbon dioxide ("dry ice") at atmospheric pressure does not melt to a liquid, but goes directly to gaseous form (this is why they call it "dry"). At 1 atm pressure, the temperature of this phase transition (called "sublimation") is $T_{sub} = 194\,K$, and the enthalpy of sublimation ($\Delta\bar{H}_{sub}$) is 25,200 J mol^{-1}. What is the entropy of sublimation ($\Delta\bar{S}_{sub}$) at T_{sub}?

6.23 The molar heat capacities of dry ice and CO_2 gas, as a function of temperature, at 1 atm (constant) pressure, are given below. Plot these, come up with a reasonable function that describes each of them, and fit them using nonlinear least squares. Note, avoid log functions and for the solid, a one-parameter model (with a zero intercept) would be best for extrapolation to 0 K.

Phase	T (K)	C_p (J mol^{-1} K^{-1})
Solid	15.52	2.53
	146.48	47.11
	198.78	54.5
Gas	198.15	33.89
	273.15	36.33
	288.15	36.61
	373.15	38.01
	673.15	43.81
	1273.15	50.87
	2273.15	56.91

6.24 For solid CO_2, generate an equation for the temperature dependence of the molar enthalpy [i.e., $\Delta\bar{H}^{solid}(T)$], by setting the enthalpy equal to zero at T_{sub}.

6.25 Next, generate an equation for the temperature dependence of the molar entropy of dry ice [i.e., $\Delta\bar{S}^{solid}(T)$], by assigning the entropy to be zero at $T = 0$.

6.26 Using your answers from Problems 6.24 and 6.25, derive an equation for the chemical potential of dry ice as a function of temperature, at 1 atm.

6.27 Come up with an expression for the chemical potential of the gas at 1 atm, using the enthalpy and entropy of sublimation (Problem 6.22).

6.28 Using the molar volume of dry ice (2.7×10^{-5} m³/mol^{-1}) and assuming the gas is ideal, give expressions for the pressure dependence of the chemical potentials of the two phases at T_{sub}.

6.29 Combine the pressure and temperature parts of the chemical potentials of dry ice and CO_2 gas to get chemical potential surfaces. Plot these surfaces over the T–p plane.

Further Reading

Gibbs, J.W. 1878. On the equilibrium of heterogeneous substances. *Trans. Connecticut Academy*. III, 108–248.

Gill, S.J. 1962. The chemical potential. *J. Chem. Edu.* 39, 506–510.

CHAPTER 7

The Concentration Dependence of Chemical Potential, Mixing, and Reactions

Goals and Summary

The goals of this chapter are to show how the free energies of mixtures depend on the concentrations of the various species, and how these concentration dependences determine the thermodynamics of mixing and of chemical reactions. As with single-component phase transitions described in Chapter 6, the chemical potential plays a central role in analyzing equilibrium in mixing and in reaction. However, unlike single-component phase transitions, where each phase is pure, here we will need to calculate chemical potentials of species that are combined in various proportions. As we will see, the chemical potential varies logarithmically with concentration. In the process of deriving this logarithmic dependence, we will introduce the concept of "standard state" concentrations. Together, the standard state and logarithmic concentration terms will be used to develop expressions for mixing and reaction thermodynamics. A key concept that emerges from the reaction free energy equation is the equilibrium constant. We will discuss the relationship between the equilibrium constant and the standard state, and how the equilibrium constant depends on temperature and pressure.

THE DEPENDENCE OF CHEMICAL POTENTIAL ON CONCENTRATION

The analysis of *single-component* phase transition in Chapter 6 shows how chemical potentials can be used to identify equilibrium conditions when different forms of *pure material* are present. When multiple species are present, Equations 5.64 and 5.65 can be used to give free energies of mixtures in terms of chemical potentials. We can combine these ideas to identify equilibrium conditions for mixtures, but first need to answer a central question: *"how does chemical potential (and thus Gibbs free energy) depend on concentration?"*

Concentration did not come into our discussion of single-component phase equilibrium, because phase transitions simply move molecules from one pure phase to another, in what can be regarded as physically distinct locations (**Figure 7.1**). During this process, the chemical potentials of the pure phases do not change. If the two chemical potentials differ, the total Gibbs energy decreases linearly as material moves from one phase to the other.

Figure 7.1 Single-component phase transitions involve transfer of material between pure, spatially separate phases. Two distinct phases containing a single component (black dots) are depicted as red and white lattices. From top to bottom, material is transferred from the α-phase (red) to the β-phase (white), which are spatially distinct. Thus, the chemical potential of each phase ($\mu^{*,\alpha}$ and $\mu^{*,\beta}$) is independent of the extent of material transferred between the two phases, and the total Gibbs free energy G is simply a mole-weighted linear combination of $\mu^{*,\alpha}$ and $\mu^{*,\beta}$. Note that the number of molecules in each phase must be divided by Avogadro's number (L_0) to give the total Gibbs free energy from the chemical potentials.

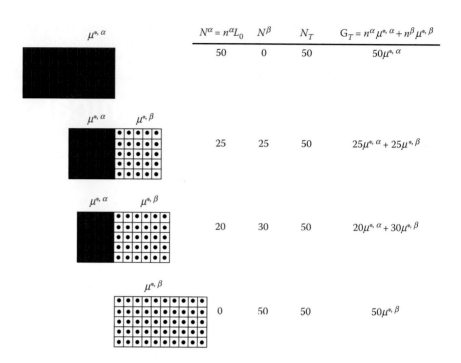

The situation is different for mixtures of different species (**Figure 7.2**). In this case we have a single physical phase[†] that is composed of multiple species. We have seen from statistical analysis of lattice models in Chapter 4 the mixing process produces a nonlinear variation in entropy with concentration (e.g., Figure 4.15D). Since the total entropy can be written in terms of molar entropies (Chapter 5),

$$S_T = \sum_{i=1}^{s} n_i \bar{S}_i \ , \quad \text{or dividing by } n_T,$$

$$\frac{S_T}{n_T} = \sum_{i=1}^{s} x_i \bar{S}_i$$

(7.1)

Figure 7.2 Lattice chemical reaction resulting in a mixture of distinct species. Molecules that can interconvert between distinct species (*A* and *B*) are depicted as black and white dots. From top to bottom, reaction progresses from *A* to *B*. Although the endpoints (top and bottom) consist of a pure phase of a single species (with chemical potential μ_A^* and μ_B^*), all intermediate extents of reaction involve a mixture; thus, chemical potentials differ from pure values. As a result, the total Gibbs free energy depends on concentration implicitly through chemical potentials μ_A and μ_B, as well as through the explicit n_A and n_B multipliers.

	$N_A = n_A L_0$	N_B	N_T	$G = n_A \mu(x)_A + n_B \mu(x)_B$
	50	0	50	$50\mu_A^*$
	25	25	50	$25\mu(x)_A + 25\mu(x)_B$
	20	30	50	$25\mu(x)_A + 30\mu(x)_B$
	0	50	50	$50\mu_B^*$

[†] Actually, this is only partly true for mixing, where the different species start out in distinct physical locations (like an artificial phase equilibrium), though the end state is a single, homogeneous phase. Also note that mixtures can also partition into multiple phases. Such behavior forms the basis for important processes like distillation, formation of solid alloys, and in culinary preparations such as salad dressing.

it follows that the partial molar entropies must also be nonlinear in concentration, because the explicit concentration dependence, represented using mole fraction in Equation 7.1, is linear. This nonlinear dependence of $\bar{S}_i$ on concentration results in a nonlinear concentration dependence in chemical potential through the relationship:

$$\mu_i = \bar{H}_i - T\bar{S}_i \tag{7.2}$$

Thus, unlike phase equilibrium, the total Gibbs free energies of mixing and of reaction are nonlinear with respect to composition.

Concentration scales

Before tackling the concentration dependence of chemical potentials and the total Gibbs free energy, we need to introduce the different concentration scales that are used in thermodynamics (**Table 7.1**). Though "molarity" is most familiar to chemistry and biology students, other scales often lead to simpler expressions for chemical potential, and as a result, end up being more intuitive. Selecting the best concentration scale depends on the type of mixture being analyzed (gas, mixed cosolvent, or dilute solution). An important consequence of selecting a particular concentration scale is that the scale directly determines the reference, or "standard" state, which in turn, impacts calculated quantities such as reaction free energies and equilibrium constants.

Regardless of the scale, concentration refers to the amount of material (either by moles or by mass—in either case, an extensive quantity) normalized to some

Table 7.1 Some commonly used concentration scales, and their relation to the mole fraction

Volume-normalized scales		
Molarity[a]	Moles species i per liter	$M_i = [i] = \dfrac{n_i}{V_T} = \dfrac{x_i}{\sum_{j=1}^{s} x_j \bar{V}_j}$
Partial density	Grams species i per mL	$\rho_i = \dfrac{n_i \times MW_i}{V_T} = \dfrac{MW_i x_i}{\sum_{j=1}^{s} x_j \bar{V}_j}$
Partial pressure[b]	Force exerted by species i per unit of surface area	$p_i = p - \sum_{j \neq i}^{s-1} p_j = \dfrac{RTx_i}{\sum_{j=1}^{s} x_j \bar{V}_j}$
Species-normalized scales		
Mole fraction	Moles species i per total moles of all species	$x_i = \dfrac{n_i}{n_T}$
Molality[c]	Moles species i per kilogram solvent	$m_i = \dfrac{n_i}{1000 n_s MW_s} = \dfrac{x_i}{1000 x_s MW_s}$

Note: Here, units are based on convention. Care should be used in calculations involving these non-SI concentration scales, to make sure that results are at least internally consistent, especially when expressing volumes in different scales. The T subscript on the volume (V_T) is included as a reminder that the volume is a total (extensive) system volume, rather than a molar volume.

[a] As in Chapter 3, M_i is used to indicate molarity generically; when discriminating among multiple species in a mixture, square brackets are used.

[b] The left-most equality for partial pressure is a rearrangement of Dalton's law (Equation 5.90); the right-most equality applies only to ideal gas mixtures.

[c] For molality, the subscript s refers to the component designated as the solvent. Though designation of a species as solvent can be somewhat arbitrary, when describing the thermodynamics of dilute solutions, the choice is usually obvious.

extensive variable that represents the size of the system.[†] This normalization can be to *total volume*, as with molarity and density, or to the number of moles of all (mole fraction) or just one (as with molality) of the components. As with the molar quantities developed in Chapter 5, concentration is an intensive quantity.

It is often necessary to convert between concentration scales. Each of the concentration scales in Table 7.1 can be related to mole fraction. Thus, mole fraction provides an easy means to convert between different concentration scales (Problems 7.1 and 7.2). Note that at dilute concentrations, the different scales all tend to be linearly proportional to each other (though at higher concentrations they are not, Problem 7.3). As a result, analysis of dilute solutions is comparatively simple, allowing substitution of different concentration scales with little more effect with changing slope and units. For quantities that depend logarithmically on concentration (like chemical potential), this change in slope becomes an additive offset.

The difference between concentrations and mole amounts

From the expressions for concentration scales in Table 7.1, it is clear that concentrations and mole amounts are closely related quantities. Owing to this close relationship, it is easy to mix up the two quantities. To avoid potential confusion, we will highlight the differences, and show how these differences matter. Nowhere is making the distinction between mole amounts and concentrations more important than in expressions that connect total Gibbs free energy to composition and chemical potential. For a system with species A through S,

$$G_T = n_A \mu_A + n_B \mu_B + \ldots + n_S \mu_S \tag{7.3}$$

(Equation 5.73). The leading n_i terms show that mole amounts have an explicit linear influence on the total Gibbs energy. Depending on the sign of the chemical potential,[‡] increasing the mole amounts at fixed concentration (increasing the extent of the system—e.g., doubling, tripling it) can either increase or decrease G_T (**Figure 7.3A**). In contrast, concentrations have an implicit, nonlinear influence on G_T, through the chemical potential terms (shown in the following section). When concentrations go up without changing total mole amounts (e.g., by

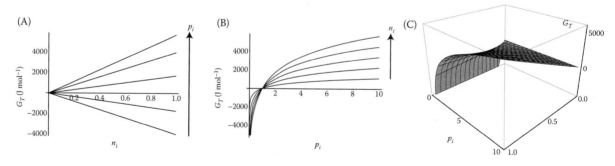

Figure 7.3 Dependence of free energy on concentration and mole amount for a single species system.
(A) Linear dependence on the number of moles of species *i* at different (fixed) concentrations. (B) Logarithmic dependence of total Gibbs energy on concentration, for different (fixed) mole numbers (on the partial pressure scale from 0.2 to 10 atm). (C) Dependence of Gibbs free energy on both concentration and mole number, according to Equations 7.3 and 7.11. For all three panels, the temperature is fixed to 298.15 K, and ideal gas behavior is assumed.

[†] An apparent exception to this is partial pressure, which can be thought of as a decomposition of total pressure (an intensive quantity) using Dalton's law. However, for an ideal gas, partial pressure is also equal to the number of moles divided by the total volume.

[‡] As with other variables in classical thermodynamics (e.g., internal energy), chemical potentials are given relative to a reference state. Relative to a reference state, chemical potential will be negative at concentrations lower than the reference state, and also tend to be negative at high temperatures.

squeezing fixed mole amounts into a smaller volume), G_T always goes up, as a result of increasing the corresponding chemical potentials (**Figure 7.3B**).[†] The combined effect of concentration and mole amounts is shown in **Figure 7.3C**.

Concentration dependence of chemical potential for an ideal gas

Since pressure is a direct measure of concentration (Table 7.1), we can evaluate the concentration dependence of chemical potential using the thermodynamics of gases we developed in Chapters 3 and 4, along with the partial molar quantities developed in Chapter 5. A good starting point for this from the Gibbs–Duhem equation:

$$\sum_{i=1}^{s} n_i d\mu_i = Vdp - SdT$$

$$= \left(\sum_{i=1}^{s} n_i \bar{V}_i\right) dp - \left(\sum_{i=1}^{s} n_i \bar{S}_i\right) dT \tag{7.4}$$

Focusing on from the second line, we can regard Equation 7.4 as the sum of s equations (one for each species) of the following form:

$$n_i d\mu_i = -n_i \bar{S}_i dT + n_i \bar{V}_i dp \tag{7.5}$$

Since each of these equations is independent of the others, each can be evaluated on its own. Dividing the *ith* equation by n_i leads to the equation

$$d\mu_i = -\bar{S}_i dT + \bar{V}_i dp \tag{7.6}$$

Equation 7.7 can also be obtained by carrying out Legendre transforms from the partial molar internal energy for a given species (Problem 7.5).

If we hold temperature constant and vary the pressure we get

$$d\mu_i = \bar{V}_i dp \tag{7.7}$$

If the gas behaves as an ideal mixture, the variation in the total presssure dp is equal to the sum of the variations in the pressure contributed by each species (the partial pressures):

$$dp = d\left(\sum_{i=1}^{s} p_i\right) = dp_1 + dp_2 + \ldots + dp_s \tag{7.8}$$

As suggested by the middle term in Equation 7.8, this expression is obtained by taking the differential of Dalton's law. If we only vary the concentration (pressure) of species i, Equation 7.8 simplifies to $dp = dp_i$. Substituting this result into Equation 7.7 and integrating leads to

$$\int_{\mu_i^\bullet} d\mu_i = \mu_i - \mu_i^\bullet = \int_{p_i^\bullet} \bar{V}_i dp_i \tag{7.9}$$

[†] Because the multiplying n_i values are always positive.

In the next section, more will be said about the quantities $p^{\bullet}$ and $\mu^{\bullet}$, and how they define a "standard state." For now, just consider them to be integration limits. As long as each species behaves like an ideal gas (i.e., each species interacts neither with itself nor the other species), the molar volume of each species is given a version of the ideal gas law for that species, that is, $\bar{V}_i = RT/p_i$. Using this expression to substitute for $\bar{V}_i$ in Equation 7.9 leads to

$$\mu_i - \mu_i^{\bullet} = \int_{p_i^{\bullet}}^{p} \frac{RT}{p_i} dp_i = RT \ln\left(\frac{p_i}{p_i^{\bullet}}\right) \tag{7.10}$$

or rearranging,

$$\mu_i = \mu_i^{\bullet} + RT \ln\left(\frac{p_i}{p_i^{\bullet}}\right) \tag{7.11}$$

Equation 7.11 gives an important result that we will encounter in many different contexts, and use over and over again:

Chemical potential, that is, partial molar free energy, increases **logarithmically** with concentration.

Although we have achieved our goal of expressing chemical potential in terms of concentration (at least for ideal gas mixtures), Equation 7.11 still looks as much like the gas phase pressure thermodynamics of previous chapters as it looks like the thermoynamics of solutions, which involve molarities. At this point, we could substitute the pressure terms in Equation 7.11 with molarities (Problems 7.6 and 7.7):

$$\begin{aligned}\mu_i &= \mu_i^{\bullet} + RT \ln \frac{\bar{V}_i^{\bullet}}{\bar{V}} \\ &= \mu_i^{\bullet} + RT \ln \frac{[i]}{[i]^{\bullet}}\end{aligned} \tag{7.12}$$

where $\bar{V}^{\bullet}$ and $[i]^{\bullet}$ are molar volume and molarity at the reference state that is connected with the "$\bullet$" integration limit. Although Equation 7.12 (especially the second equality) may have a more familiar look to solution chemists and biologists, the pressure version (Equation 7.11) is better for gas-phase thermodynamics, where pressure is easily measurable.

Choosing standard states

Integration from the Gibbs–Duhem equation to obtain a concentration dependence for chemical potential requires us to make two choices. First, we must choose the scale for representing concentration. This choice is often guided by the type of system being analyzed. Second, as discussed in Chapter 3, thermodynamic quantities such as chemical potential are not determined in absolute terms in classical thermodynamics. Instead, these quantities are to a reference. Though formally introduced as constants of integration in Equation 7.11, $\mu^{\bullet}$ and $p^{\bullet}$ can be viewed as reference state values. By choosing the same reference state value for all gases,[†] we will make the chemical potentials of different gases easy to compare. We will refer to our chosen value as a "standard state."

In principle, we are free to choose any arbitrary reference point for the standard state. However, it is best to select a point that is similar to the states studied by experiment. For example, studies of gases are often performed near pressures of 1 atm, because such conditions can be reached without specialized pumps or

---

† We will also adopt common reference states for mixtures of liquids and solutions, although these reference states will be different from the gas-phase reference.

Table 7.2 Common standard state definitions for different systems.

Type of System	Concentration scale	Standard state
Gas phase	Pressure	$p_i^{\bullet} \equiv 1\,atm$
Liquid cosolvents	Mole fraction	$x_i^{*} \equiv 1$
Dilute solution	Molarity	$[i]^{\circ} \equiv 1\,molar$

vessels. Thus, in describing gas thermodynamics, where we use pressure to represent concentration, we will define the standard state to be 1 atm (**Table 7.2**). We will use the filled circle (•) to represent this standard state; for component i, we define $p_i^{\bullet} \equiv 1\,atm$. Likewise, the chemical potential of component i at this standard state will be denoted as $\mu_i^{\bullet}$. For gas mixtures, we will invoke the same standard state for all components, that is, $p_1^{\bullet} = p_2^{\bullet} = \ldots = p_s^{\bullet} = 1\,atm$. However, it will generally *not* be the case that $\mu_1^{\bullet} = \mu_2^{\bullet} = \ldots = \mu_s^{\bullet}$, due in large part to differences in the structural and energetic features of each component.

For other types of systems, different concentration scales are more convenient, as are different standard states. For example, in liquid mixtures where multiple components are at high concentrations (e.g., the water/ethanol mixture in Figure 5.7), mole fraction is a convenient concentration scale. For such systems, we will reference thermodynamic quantities for each component to its pure liquid. That is, the standard state is a mole fraction of one. Because this corresponds to pure cosolvent, we will continue to use an asterisk (i.e., x_i^{*}) to represent this standard state. Note that although the $x_i^{*} = 1$ standard state satisfies the goal of being experimentally accessible, it is clearly impossible to achieve the standard state simultaneously for multiple components. In other words, a good standard state should be experimentally accessible, but it can have aspects that are not achievable.

For dilute solutions, where the solvent component makes up the bulk of the system, we will use a molarity scale to represent the concentration scale of the dilute components, and will choose a one molar standard state. We will use an open circle (°) to represent this standard state. If thermodynamic calculations require the solvent to be included explicitly, we can also put it on the molar scale with a 1 M standard state, or we can retain the mole fraction standard state ($x_s^{*} = 1$) for that component alone. The one molar standard state is typically used for free energy calculations in solutions of biological macromolecules.[†] These standard state choices are compared in Table 7.2. Although any mixture could, in principle, be represented by any standard state, the molecular environments in gasses, cosolvent liquids, and dilute solutions differ significantly (**Figure 7.4**). Matching the standard states to these three types of systems as described above allows us to quantify the thermodynamics of different species in a relevant context.[‡]

The choice of unity for each of the different standard states simplifies calculations of chemical potential at a given concentration. For example, in Equation 7.11, the argument in the log simplifies to the pressure of the system of interest:

[†] The one molar standard state is often accessible in chemical studies of small molecules. However, in reality, one molar concentrations are not accessible for studies of biological macromolecules, which are often limited to the millimolar or micromolar range. An alternative standard state at one of these lower molar concentrations would seem more appropriate, but it is not the convention. For this and other reasons, it is worth regarding the one molar standard state as a "hypothetical" state for macromolecules in solution.

[‡] Note that it would, in principle, be possible for two (or even three) of these standard states to be the same, but this would be quite a coincidence. For example, a pure solvent (at the $x^* = 1$ standard state) can have a vapor pressure of 1 atm (the $p^{\bullet} = 1$ standard state) but it would require the liquid to be at its boiling temperature at 1 atm pressure. Likewise, a pure solvent can have a molarity of 1 molar (the $[i] = 1$ standard state) but the the molecules would need a molar volume roughly 55 times greater than liquid water.

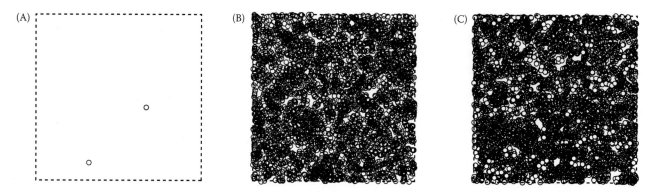

Figure 7.4 Mixtures in the gas, liquid cosolvent, and dilute liquid phases. Two distinct species (*A* and *B*) are shown as red and black circles, each of radius 0.14 nm (the approximate radius of water). A slab of molecules 1 nm thick is projected onto a 10 × 10 nm plane. In these images, excluded volume is *not* accounted for. (A) A snapshot of a gas mixture at a pressure of about 1 atm, where $x_A = x_B = 0.5$. (B) A liquid consisting of cosolvents at $x_A = x_B = 0.5$. The density of the liquid depicted is about that of liquid water at room temperature and pressure. (C) A dilute solution consisting of a concentrated solvent ($x_B = x_S = 0.999$) and a dilute solute ($x_A = 0.001$). At the density shown, this corresponds to molarities of around 55 molar and 55 millimolar for solvent and solute, respectively (again, the density of liquid water).

$$\mu_i - \mu_i^\bullet = RT \ln p_i \tag{7.13}$$

Expressions using the mole fraction and molar standard states have the same form [i.e., $\ln(x_i/x_i^*)$ and $\ln([i]/[i]^\circ)$], and thus, simplify in the same way. While this convenient shorthand is quite common, but it should be remembered that there is a standard state term hidden in the argument of the logarithm (making the argument of the log dimensionless, as good log arguments should be).

Standard states are like reference points on maps

To help make standard state selection more intuitive, we will develop an analogy between selecting standard states and making maps. When building a map, as with defining a standard state, different dimensions can be used and different reference points are included, depending on the type of information to be conveyed.

Dimensions in maps are selected based on the size of the object to be mapped. To represent cities and local geography, Cartesian distances work nicely, with coordinate axes running north, south, east, and west. To represent global features (such as continents and oceans), spherical coordinates based on angles (as in latitude and longitude) work nicely. The best scale for a map is also determined by the size of the object. For a map of a house or a treasure map, footsteps work nicely (e.g., feet, and meters). For long hikes or drives, something larger (miles and kilometers) is more appropriate. For global navigation maps on the sphere of the earth, latitude and longitude are often preferable to distances. And for maps of space (the solar system, galaxy, and universe), astronomical units (average earth–sun distance), light years, Parsecs, and even redshift can be used.

Like the standard state, a good reference point on a map should be directly accessible. For a treasure map, the west edge of the island is probably a better point than the middle of the lagoon, if the scale of measure is paces. For latitudes and longitudes, a nearby spot on the earth's surface is an appropriate reference. Since many early mapmakers lived in England, a spot in the city of Greenwich, England, was chosen as the reference (zero) for longitude; the equator was chosen as a reference for latitude, given its simple symmetric relationship to the polar earth.

For elevation, represented in distance, a particular spot on the surface of the earth (like Greenwich) would be a really bad choice for a reference. The tallest mountains in the Himalayas would have a smaller "elevation" value than the ocean floor off

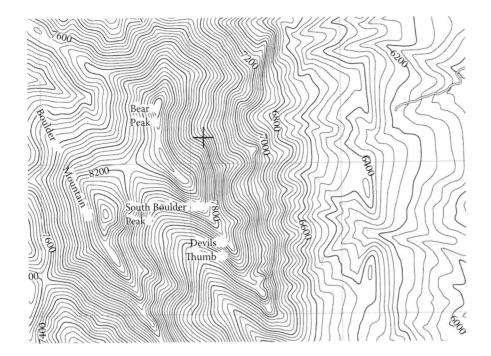

Figure 7.5 Elevation from sea level as a convenient standard state for local geography. Compared to a lot of bad reference points we could use to measure the height of mountains (e.g., Greenwich England, the center of the earth, the surface of the sun), the elevation from sea level is convenient both because it matches our ideas about "high" and "low," and because it is "experimentally accessible." In the map above, the contour lines indicate the height of mountains in the Rockies (in feet) relative to the sea. From the Eldorado Springs 7.5" quadrangle, United States Geological Survey.

the Antipodes Islands.[†] A better choice would be the center of the earth. However, the center of the earth is not accessible to people who use maps to climb mountains—thus, although it dimensionally matched to the problem, it is very inconvenient location (variation from the highest mountain to the lowest spot in the ocean would only be about 0.5%). Instead, sea level is chosen as a reference point (**Figure 7.5**). The convenience of this choice is like that in choosing 1 atm instead of a perfect vacuum as a reference pressure for gas experiments.

The concentration dependence of chemical potential for "ideal" liquid mixtures

As indicated above, for mixtures of liquids where more than one component is at high concentration, it is convenient to use mole fractions to represent the concentration of components. The high concentrations of multiple components mean that mole fractions will have large numerical values. Thus, a mole fraction of one is a reasonably close reference point to use as a standard state.

To express chemical potentials of liquid cosolvents in terms mole fraction, we will make use of phase equilibrium between the components in solution and in vapor phases. As long as each component behaves as an ideal gas in the vapor phase, we can use the expression for the concentration dependence of chemical potential in the gas phase derived above (relative to a 1 atm standard state pressure; Equation 7.11) to give the chemical potential of the liquid phase:

$$\mu_i^{liq} - \mu_i^{gas}$$
$$= \mu_i^{\bullet} + RT\ln\left(\frac{p_i}{p_i^{\bullet}}\right) \tag{7.14}$$

Though Equation 7.14 appears to be just another gas phase equation (p_i is the partial pressure of species i in the gas phase), since we have constructed a phase

---

[†] These islands, which are off the coast of New Zealand, are named because they are opposite to England.

equilibrium between the gas phase and the liquid, the p_i values are also properties of the liquid phase. In this context, p_i is often referred to as the "vapor pressure" of a liquid. The vapor pressure is the pressure of a gas phase were it to coexist with the vapor phase. A way to measure such a pressure would be to partly fill a vessel with a liquid at constant temperature, pull a vacuum on the "head space" above the liquid (at which point the liquid would boil), seal the liquid and its head space off. As the liquid boils, the pressure in the head space would rise as it gets filled with boiled-off liquid vapor. Once the concentration of gas in the head space is high enough that the rate of material flow from the liquid phase to the gas phase is equal to the rate of condensation from the gas phase back to the liquid phase, equilibrium is established. The pressure that results is the vapor pressure of the liquid.

To express μ_i^{liq} in terms of mole fractions in the liquid phase, we need an expression that relates the liquid phase vapor pressures p_i with x_i values. The simplest expression that makes this connection is "Raoult's law," which states that the vapor pressure of a species i in a liquid is linearly proportional to its mole fraction:

$$p_i = p_i^* x_i \tag{7.15}$$

In Equation 7.15, the constant p_i^* is the vapor pressure of pure liquid i.[†] Mixtures that conform to Raoult's law are referred to as "ideal solutions."

An example of two species that closely conform to Raoult's law is shown in **Figure 7.6A**. Ethylene and propylene dibromide (*EDB* and *PDB*) are chemically quite similar (with formulas CH_2BrCH_2Br and $CH_3CHBrCH_2Br$, respectively). Thus their interactions with each other are quite similar to their interactions with themselves. Because of this similarity, mixing *EDB* and *PDB* together in various proportions in solution neither enhances nor diminishes their volatility. As can be seen in Figure 7.6A, the vapor pressure of PDB increases linearly with the mole fraction of PDB (as does that of EDB), demonstrating the ideal solution behavior described by Raoult's law. The slightly higher vapor for *EDB*

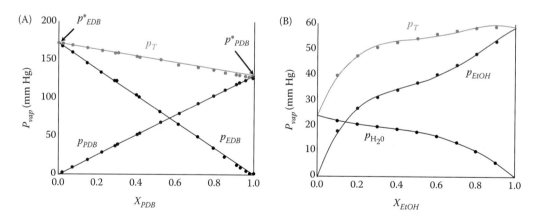

Figure 7.6 Partial pressures of two different two-component liquids. (A) The partial and total vapor pressures of a binary (meaning two-component) mixture of ethylene dibromide (EDB) and propylene dibromide (PDB), two similar components that closely approximate an ideal solution. Vapor pressures were measured at 358 K (Von Zawidzki, *Z. Phys. Chem.* 35, 129 [1900]). This mixture conforms quite closely to Raoult's law (solid lines) with zero intercepts at $x_{PDB} = 0$ and 1 for *PDB* and *EDB*, respectively. (B) The partial and total pressures of water and ethanol, two very different components that show significant solution nonideality. Solid lines result from fitting polynomials of order 3 or 4 mixture with zero intercepts at $x = 0$ and $x = 1$. In both mixtures, the total vapor pressure (p_T) is given as the sum of the partial vapor pressures, according to Dalton's law. Vapor pressures were measured at 298 K (Guggenheim and Adam, *Proc. Roy Soc. A.* 139, 231 [1933]).

[†] Which, conveniently, is at the standard that we wish to use, that is, $x_i^* = 1$.

at standard state may result the smaller size of *EDB*, which should make fewer interactions to keep it in the liquid phase than *PDB*.

It should be kept in mind that unlike ideal gases, the kind of liquid phase ideality represented in Raoult's law does not mean that there are no interactions in the liquid phase. To the contrary, formation of a liquid in preference to a gas requires favorable interactions between molecules to offset the unfavorable entropy of condensation. Rather, Raoult's law requires that the strength of interactions that bind the components into solution are the same between components (i.e., nonself *A:B* interactions) as those within components (self *A:A* and *B:B* interactions). In the example in Figure 7.6, EDB and PDB have quite similar structures and functional groups, the two components are likely to have interactions with each other (nonself interactions) that are similar to those they have with themselves, consistent with the observed Raoult's law behavior.

For ideal solutions, Equations 7.14 and 7.15 can be combined to give

$$\mu_i^{liq} = \mu_i^{gas}$$
$$= \mu_i^{\bullet} + RT \ln\left(\frac{p_i^* x_i}{p_i^{\bullet}}\right) \tag{7.16}$$
$$= \mu_i^{\bullet} + RT \ln\left(\frac{p_i^*}{p_i^{\bullet}}\right) + RT \ln(x_i)$$

In the third line, the first two terms combine to give the chemical potential of the pure liquid *i*, in terms of its vapor pressure and the 1 atm standard state (using Equation 7.11). We can take this combination of terms as defining a new standard state for pure component *i* (that is, at a mole fraction of one in the liquid phase):

$$\mu_i^{\bullet} + RT \ln\left(\frac{p_i^*}{p_i^{\bullet}}\right) \equiv \mu_i^* \tag{7.17}$$

Introducing this shorthand into Equation 7.16 gives

$$\mu_i^{liq} = \mu_i^* + RT \ln(x_i) \tag{7.18}$$

Equation 7.19 effectively switches from the $p_i^{\bullet} = 1$ standard state to the $x_i^* = 1$ standard state. Although this is a different (and more convenient) way to express the chemical potential, it should be stressed that changing the concentration variable and the reference point does not change the chemical potential on the left-hand side. There is only one chemical potential, and it is independent of how we choose our standard state.

Using Equation 7.18, chemical potentials of binary (i.e., two component) solutions are plotted in **Figure 7.7** as a function of mole fraction concentration. In the left panel, the standard state chemical potentials are identical ($\mu_A^* = \mu_B^*$), whereas in the right panel they differ by a factor of ten. These differences in μ_i^* shift the chemical potential curves vertically; aside from this shift, each component has the same concentration dependence (i.e., $RT \ln x_i$).[†]

For a fixed total mole amount ($n_T = n_A + n_B$), the total Gibbs free energy can be calculated from chemical potentials as a function of composition, which plays a central role in the thermodynamics of mixing and of reaction (Problems 7.8 and 7.9), and can be extended to more than two ideal species (Problem 7.11).

[†] Except for the trivial difference that when the potential of one species is plotted versus the mole fraction of the other species, it goes as $1 - x_i$ rather than x_i.

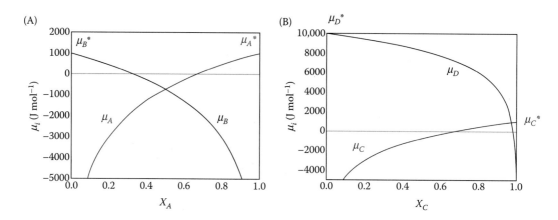

Figure 7.7 Chemical potentials of components in two different binary ideal solutions. Chemical potentials are plotted against the mole fraction of species A (left) and species C (right), using Equation 7.18, which assumes Raoult's law (Equation 7.15). In (A), standard state chemical potentials are identical ($\mu_A^* = \mu_B^* = 1000 \, \text{J mol}^{-1}$). In (B), standard state potentials differ ($\mu_C^* = 1000$, $\mu_D^* = 10,000 \, \text{J mol}^{-1}$).

The Gibbs free energy of mixing of ideal solutions

In Chapter 4, we calculated the multiplicity of mixing two distinct components on a lattice (Figure 4.2), which can be converted to an entropy of mixing using Boltzmann's formula (Equation 4.46, see Problem 4.25). With the mole fraction dependence of chemical potential developed above, it is a simple matter to calculate the free energy and entropy of mixing of an ideal solution using classical thermodynamic formulas. The agreement between the statistical and classical expressions (Equation 7.22; see Problem 7.10) is a satisfying one, since the two methods approach the mixing problem from very different perspectives.

The free energy of mixing of ideal components can be calculated from the chemical potentials (Equation 7.18) by applying Equation 5.65 to the unmixed and mixed states, and taking the difference:

$$\Delta G_{mixing} = G_{mixed} - G_{unmixed}$$
$$= (n_A \mu_A + n_B \mu_B)_{mixed} - (n_A \mu_A + n_B \mu_B)_{unmixed} \tag{7.19}$$

Using the mole fraction concentration scale, the chemical potentials of the two unmixed liquids are simply the standard state values, since the two liquids are pure before mixing. After mixing, chemical potentials are given by Equation 7.18, giving

$$\Delta G_{mixing} = n_A\left(\mu_A^* + RT\ln x_A\right) + n_B\left(\mu_B^* + RT\ln x_B\right) - \left(n_A\mu_A^* + n_B\mu_B^*\right) \tag{7.20}$$

The unmixed terms cancel the standard states of the mixed terms, leaving

$$\Delta G_{mixing} = n_A RT\ln x_A + n_B RT\ln x_B \tag{7.21}$$

Dividing by the total moles mixed together, that is, $n_T = n_A + n_B$, gives

$$\Delta\bar{G}_{mixing} = \left(\frac{n_A}{n_T}\right)RT\ln x_A + \left(\frac{n_B}{n_T}\right)RT\ln x_B$$
$$= RT\left(x_A\ln x_A + x_B\ln x_B\right) \tag{7.22}$$

The molar free energy of mixing can be plotted as a function of mole fraction of either of the two components (by the substitution $x_B = 1 - x_A$). For a two component ideal solution, the free energy of mixing has a symmetrical shape about $x_A = x_B = 0.5$ (**Figure 7.8**). As an example of this symmetry, the free energy of mixing 0.75 moles of A with 0.25 moles of B is the same as that of mixing 0.25 moles of

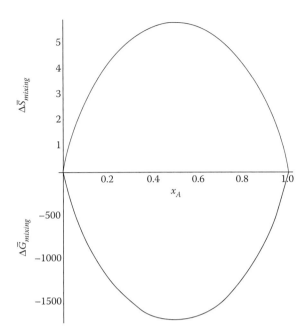

Figure 7.8 Free energy and entropy of mixing of a two-component ideal solution.
Free energy (black) from Equation 7.22, in $J\ mol^{-1}$; entropy (red) from Equation 7.23 in J mol^{-1} K^{-1}. Temperature is 298.15 K.

A with 0.75 moles of B. This symmetry relates to the fact that although we can distinguish the A and B molecules, there is nothing "special" about either one. Since our chemical potential equations assume solution ideality, meaning that the A and B molecules interact identically with themselves and with each other, our choice of labels (A versus B) does not affect the thermodynamics.

Another related point about our derivation of the free energy of mixing (Equations 7.20 through 7.22) is that the cancellation of standard state terms means that any difference between the standard state chemical potentials of components does not affect the mixing energy. For example, even though the two pairs of chemical potentials in Figure 7.7A and B cross at very different mole fractions, the same mixing energy and entropy is obtained in both cases, with a symmetric minimum at a mole fraction of 0.5. It is only when the mixed components can interconvert via chemical reaction that the standard state chemical potential values influences the process.

By taking the temperature derivative of Equation 7.22, we can obtain the molar entropy of mixing:

$$\Delta \bar{S}_{mixing} = -\left(\frac{\partial \Delta \bar{G}_{mixing}}{\partial T}\right)_p$$

$$= -\left(\frac{\partial RT\{x_A \ln x_A + x_B \ln x_B\}}{\partial T}\right)_{p,n_A,n_B} \quad (7.23)$$

$$= -R\{x_A \ln x_A + x_B \ln x_B\}$$

As can be seen in Figure 7.8, the entropy of mixing (red curve) has the same shape as the free energy of mixing, though it is inverted about the x-axis due to the negative sign in the thermodynamic identity relating S and G.

Finally, the enthalpy of mixing can be calculated from the entropy and free energy of mixing from the relationship $\Delta G = \Delta H - T\Delta S$. Inserting Equations 7.22 and 7.23 and rearranging leads to the result that $\Delta H_{mixing} = 0$. Thus, for mixing of ideal solutions the driving force is entirely entropic. This makes sense, because there is no preferential bonding between ideal components (by assertion). Thus, any enthalpic interactions that maintain the solution in its condensed (liquid) phase should be the same for both self and nonself interactions, rendering enthalpy independent of composition. That the same result was obtained for lattice mixing (Chapter 4, and see Problem 7.10) when we analyzed a model lacking any specific interactions supports the concept of an entropic driving force.

Though the ideal solution approach is quite useful, most solvent mixtures display preferential interactions, and as a result, show some degree of nonideal behavior. The water–ethanol mixture in Figure 7.6B is one example, where partial pressures show significant deviations from Raoult's law. In a more extreme example, mixing liquid water with nonpolar liquids such as those found in oil have such strong nonidealities that they fail to mix, despite the statistical driving force. In the next few sections, we will deal with nonideal mixtures. First we will use a classical approach by modifying Raoult's law (and defining a new standard state in the process). Second, we will return to the lattice model of mixtures, to introduce preferential interactions among components. Though this model makes a number of approximations, it can mimic liquid–liquid phase separation like that seen for mixtures of water and oil.

Concentration dependence of chemical potentials for nonideal solutions

Here we will modify the concentration dependence of chemical potential to account for deviations from Raoult's law. One approach for doing this is to obtain vapor pressure measurements as in Figure 7.6B, and fit with functions capable of capturing the nonlinearity of the p_i versus x_i values. While this *can* produce an excellent analytical expression that relates p_i to x_i, (see the equation in the legend to Figure 7.6B, e.g., and also Problem 7.12), and can be used to give an adequate representation of chemical potential across a broad concentration range (Problem 7.13), this approach has several limitations. First, the equations that result are not transferrable across mixtures of different components: they work on the mixture they were derived from, but that is all. Second, it is often difficult to obtain vapor pressure data across a broad range of mole fractions; for large macromolecules with very low volatility, it is not possible to measure vapor pressures at all.[†] Even when vapor pressure measurements can be made, they are likely to be restricted to a range of small mole fractions. For macromolecules, even getting to 1% weight fraction[‡] is often a formidable challenge.

While this second restriction is a limitation, it also suggests a way forward. We can restrict analysis of nonideal solutes to dilute solution conditions. Species with significant solution nonideality at intermediate mole fractions often show a linear concentration–vapor pressure relationship when dilute. The relation that describes this limiting behavior, which has the same functional form as Raoult's law, is "Henry's law":

$$p_i = K_{H,i}x_i \tag{7.24}$$

$K_{H,i}$ is a constant (Henry's law constant) that substitutes for p_i^* in Raoult's law, but reflects the dilute solvated environment of the solute.[§] This environment is made

---

[†] In principle, the Gibbs–Duhem equation can be used to determine the chemical potential of a component that cannot be measured experimentally, as long as all the chemical potential of all the other species can be measured. However, if chemical potentials of two components cannot be measured, neither can be determined by this approach.

[‡] Note that a macromolecule weight fraction of 0.01 corresponds to mole fraction that is significantly lower (likely lower than 10^{-4}), due to the large size of macromolecules.

[§] Until now we have been using the term "cosolvent" to describe species in solution, because our goal has been to analyze mixtures where components are present at high concentrations. When we restrict analysis to low concentrations, we will refer to dilute species as "solutes." Since for a solution, not all species can be highly dilute, we will still have (at least) one primary "solvent" species. For studies of biomolecules, the solvent is almost always water.

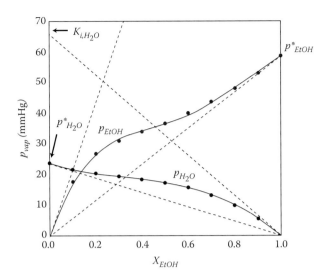

Figure 7.9 Limiting vapor pressures for water and ethanol at high dilution. The data from Figure 7.6 are fitted, and are compared with ideal solution lines (Raoult's law, dashed), and ideal dilute solution lines (Henry's law, dot dashed). For this binary mixture, the Henry's law constants ($K_{H,i}$, corresponding to the slope at $x_i = 0$), are significantly higher than the vapor pressure of the pure species (p_i^*; note the value $K_{H,EtOH}$ lies well off the scale). The Henry's law lines are obtained by taking the slopes (first derivatives in x_i) from evaluated at $x_i = 0$. The Raoult's law lines are drawn to connect p_i^* at $x_i = 1$ to zero pressure at $x_i = 0$.

up almost entirely of solvent molecules; thus, $K_{H,i}$ can be thought of as the *hypothetical* vapor pressure that species *i* would have as a pure liquid, if it experienced the same solvation environment as in dilute solution. Note that in a dilute solution, the solvent molecules are also in a solvated environment; for the most part, solvent molecules just interact with other solvent molecules. Thus, the vapor pressure of the solvent *s* is approximately p_s^*, and over the limited solute concentration range, the solvent vapor pressure follows Raoult's law (Equation 7.15). Dilute solutions where the solutes follow Henry's law and the solvent follows Raoult's law are referred to as "ideally dilute solutions."

More generally, for component *i* in a nonideal solution, Raoult's and Henry's law can be thought of as limiting behaviors at $x_i \to 0$ and $x_i \to 1$. These limiting behaviors are shown in **Figure 7.9** for water and ethanol. Graphically, the Henry's law constants $K_{H,i}$ are the slope of the Henry's law line (and to the p_i at low x_i), and also to the vapor pressure value at $x_i \to 1$. For both water and ethanol, the slopes at high dilution are significantly greater than given by Raoult's law (thus, $K_{H,i} > p_i^*$). This suggests that both water and ethanol make considerably better solvents for themselves than they do for each other. This is likely to be due to the strong hydrogen bonding potential of water, relative to ethanol, and to the somewhat hydrophobic character of ethanol.

The chemical potential of an ideally dilute solution is obtained by substitute the Henry's law concentration dependence into Equation 7.14, leading to

$$
\begin{aligned}
\mu_i^{liq} &= \mu_i^\bullet + RT \ln\left(\frac{K_{H,i} x_i}{p_i^\bullet}\right) \\
&= \mu_i^\bullet + RT \ln\left(\frac{K_{H,i}}{p_i^\bullet}\right) + RT \ln(x_i)
\end{aligned}
\tag{7.25}
$$

As with the ideal solution, the first two terms on the right-hand side of Equation 7.25 can be thought of as a chemical potential at a new reference pressure. In this case, the reference pressure is the hypothetical pure vapor pressure that solute *i* would have if its interactions were the same as at infinite dilution. We will mark this reference with the symbol $\otimes$:

$$
\mu_i^\otimes \equiv \mu_i^\bullet + RT \ln\left(\frac{K_{H,i}}{p_i^\bullet}\right)
\tag{7.26}
$$

Combining Equations 7.25 and 7.26 gives a chemical potential expression (Equation 7.27) where concentration remains on the mole fraction scale, but the reference point is changed:

$$\mu_i^{liq} = \mu_i^{\otimes} + RT\ln(x_i) \tag{7.27}$$

A SIMPLE LATTICE MODEL FOR NONIDEAL SOLUTION BEHAVIOR

As discussed above, a lattice model with no interactions gives the same form of mixing entropy, and thus, free energy of mixing, as the classical treatment using Raoult's law. Here we will extend the lattice model to include interaction energies between molecules at adjacent lattice sites. Specifically, we will allow the energies of self- and nonself interaction to differ (**Figure 7.10**), leading to attractions (or repulsions) depending on chemical identity.

Once interaction energies are included, models can get quite complex, even when structural features are simplified using a lattice. Here we will develop the simplest possible model that includes interactions. When developing such a model, it is important to clearly state all the simplifications and approximations at the outset. In the present model, the following simplifications (or "rules") are made:

1. All lattice sites are identical, and they are all filled, containing one and only one molecule (of distinguishable A or B type).
2. Interactions can only occur between nearest-neighbors on the lattice that share a common edge (Figure 7.10).
3. Interactions are pairwise, and are independent of their context. The type of interaction energy is determined only by the pair of molecules sharing the interaction site. There is no influence of neighboring molecules not sharing the site.
4. There are only three interaction energies: two self-interactions ε_{AA}, ε_{BB}, and one nonself-interaction ε_{AB}.
5. For simplicity, the two types of self-interaction energies will be treated as the same, that is, $\varepsilon_{AA} = \varepsilon_{BB} = \varepsilon_{self}$. With this simplification, the critical energy variable is the difference between the self- and nonself energies.
6. The lattice is assumed to be large enough that we can ignore edge effects—nearest-neighbor interactions will be approximated using the bonding patterns of internal sites.

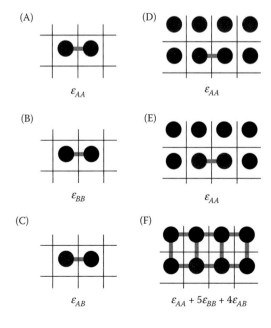

Figure 7.10 A lattice model for interactions in a two-component solution. The two components, in red and black, interact only with neighbors that share a common lattice cell edge (gray line). (A) and (B), pairwise self-interactions between nearest-neighbors, and (C) nonself interaction between nearest-neighbors. The energies interactions are unaffected by the surrounding environment; for example, the *AB* interactions shown in (D) and (E) are the same. (F) The total interaction energy is assumed to be additive.

7. The distribution of molecules over lattice sites will be taken to be random. That is, the probability that a molecule is found at site i is not influenced by the identities of molecules at adjacent sites j, k, etc. Although large differences in interaction energies should favor one type of interaction over the other, we will ignore such clustering effects.

Solution models that invoke rule 7, which results in a nonzero enthalpy of mixing but retains ideal mixing entropy (Equation 7.23), are referred to as "regular solutions." Models that invoke approximation (3) are common in statistical thermodynamics, and are referred to as "nearest-neighbor models." We will use them in Chapter 12, where they will be applied helix-coil conformational transitions in polypeptides.

With these approximations, we can tabulate the total interaction energy, which we will equate with the enthalpy,[†] as a function of mole fraction. To do this, we need to calculate the probability of the three different types of interactions, p_{AA}, p_{AB}, and p_{BB}. If we take the interaction energies in units of Joules per mole of interactions, the average energy per interaction will be given by the population-weighted average over the different interaction energies:

$$\bar{H}/\text{mole interactions} = p_{AA}\varepsilon_{AA} + p_{BB}\varepsilon_{BB} + p_{AB}\varepsilon_{AB} \qquad (7.28)$$

Because there are multiple interaction sites per lattice cell, we need to multiply Equation 7.28 by the number of interactions per cell (the coordination number, Z):

$$\bar{H} = Z(p_{AA}\varepsilon_{AA} + p_{BB}\varepsilon_{BB} + p_{AB}\varepsilon_{AB}) \qquad (7.29)$$

For a two-dimensional square lattice, $Z = 4$; for a three-dimensional cubic lattice, $Z = 6$ (**Figure 7.11**).[‡] The task remaining is to connect the probabilities of different types of interaction, p_{XX}, to the mole fractions, both in the mixed state and in the unmixed state.

In the mixed state, this can be done through a simple conditional probability statement relating the three p_{XX} terms to the probability that a particular pair of adjacent lattice cells are occupied by a pair of molecules. For example, the probability that an interaction between two lattice sites i and j is of type A:A is given by the probability that site i is occupied by a A molecule times the probability that site j is also occupied by an A, given that site i is occupied by an A molecule:

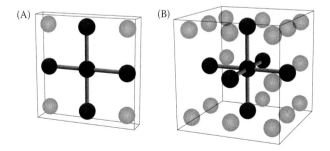

Figure 7.11 Number of nearest-neighbor interactions in a two- and three-dimensional lattice. (A) The two-dimensional square lattice, and (B) the three-dimensional cubic lattice. For a central molecule (black), there are four and six nearest-neighbors (red) in the 2D and 3D lattices. If contacts are made with only these nearest neighbors (dark gray connectors) and not to diagonal neighbors (light gray), the resulting coordination numbers are $Z = 4$ and $Z = 6$, respectively.

[†] Though the enthalpy will differ by the internal energy by pV, we will assume the lattice volume remains constant on mixing, and as a result, $\Delta(pV) = 0$.

[‡] Again, as per rule (6), we are treating all sites as internal, and are ignoring edge effects.

$$p_{AA}^{ij} = p_A^i \times p_{A|A^i}^j \qquad (7.30)$$

However, by rule above, (molecules don't influence their neighbors), $p_A^i = p_{A|A^i}^j$. And by rule 1 above (all sites are identical), Equation 7.30 applies equally to all sites on the lattice, so we can drop the i and j superscripts. These two simplifications lead to the expressions

$$p_{AA} = p_A^2$$
$$p_{BB} = p_B^2 \qquad (7.31)$$

(the expression for the probability of BB interaction follows exactly as for AA interaction). To express the probability of a nonself interaction AB, the argument goes the same as for AA (and BB).

$$p_{AB}^{ij} = p_A^i \times p_{B|A^i}^j$$
$$= p_A^i \times p_B^j \qquad (7.32)$$
$$= p_A \times p_B$$

However, it must be remembered that each pair of lattice sites i and j, there are two ways to get a nonself interaction depending on whether we put the A in site i or site j:

$$p_{AB} = 2p_A \times p_B \qquad (7.33)$$

As a last, trivial step connecting enthalpy to mole fraction, we can recognize that the probability of occupancies are the same as the mole fractions, that is, $p_A = x_A$, $p_B = x_B$. Each of these interaction probabilities p_{XX} is plotted in **Figure 7.12**.

This plot gives the intuitive result that the nonself interactions are maximized at a mole fraction of $x_A = x_B = 0.5$. Figure 7.12 also confirms that Equations 7.31 and 7.33 correctly account for probabilities of interaction; that is, the sum $p_{AA} + p_{BB} + 2p_{AB} = 1$. If we had not doubled the probability for one of the two nonself interactions, we would get a probability deficit (of 0.75) at 0.5 mole fraction.

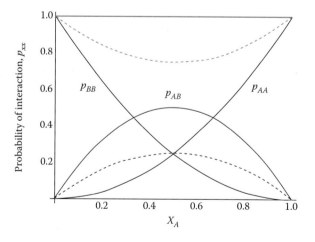

Figure 7.12 Probabilities of different types of interactions in a binary lattice fluid. Black curves show the probability of self interaction (*AA and BB*) between two adjacent sites (Equation 7.31, substituting mole fraction for probability). The solid red curve shows the probability of nonself interaction (*AB*; Equation 7.33). The sum of the self- and nonself interactions (gray flat line) has a value of one at all compositions. However, if the statistical factor of 2 is omitted from the nonself interactions (red dashed curve), not all interactions are accounted for (gray dashed curve) in the sum.

These results can also be shown algebraically (Problem 7.14). Substituting these interaction probabilities into the enthalpy Equation 7.29 gives

$$\overline{H}_{mixed} = Z\left(x_A^2\varepsilon_{AA} + x_B^2\varepsilon_{BB} + 2x_Ax_B\varepsilon_{AB}\right) \tag{7.34}$$

Rearranging and using rule 5 leads to a form highlighting different aspects of the model (see Problem 7.15):

$$\overline{H}_{mixed} = Z\left(2\left\{x_A - x_A^2\right\}\left\{\varepsilon_{AB} - \varepsilon_{self}\right\} + \varepsilon_{self}\right) \tag{7.35}$$

This form is simple, and highlights just two energy parameters: the difference in energy between the AB interaction and the self-interactions, and the absolute value of the self-interaction. The term multiplying the interaction energy difference is simply $2p_Ap_B = p_{AB}$, the probability an interaction across a lattice cell is nonself.

In the unmixed state, all the A molecules are in maximal contact with other A molecules, and likewise for B molecules. Thus the enthalpy should simply be the *total* mole fraction-weighted average[†] of the interaction energies, times the coordination number:

$$\overline{H}_{unmixed} = Z(x_A\varepsilon_{AA} + x_B\varepsilon_{BB})$$
$$= Z\varepsilon_{self} \tag{7.36}$$

Again, the second line comes from rule 5.

The enthalpy of mixing is simply the difference of Equations 7.35 and 7.36:

$$\Delta\overline{H}_{mixing} = \overline{H}_{mixed} - \overline{H}_{unmixed}$$
$$= Z\left(2\left\{x_A - x_A^2\right\}\left\{\varepsilon_{AB} - \varepsilon_{self}\right\} + \varepsilon_{self}\right) - Z\varepsilon_{self} \tag{7.37}$$
$$= 2Z\Delta\varepsilon\left\{x_A - x_A^2\right\}$$

Thus, as long as the energy of the two self-interactions are equal, they mostly drop out; one ε_{self} term is retained, but can be thought of as defining a difference energy between the self and nonself energies $\varepsilon_{AB} - \varepsilon_{self} \equiv \Delta\varepsilon$. This term can be thought of as a single parameter that controls the behavior of the mixing process. This enthalpy of mixing is plotted in the figure as a function of x_A for different values of $\Delta\varepsilon$. Equation 7.37 can be combined with the ideal entropy of mixing Equation 7.24 to yield the Gibbs free energy of mixing of a regular solution:

$$\Delta\overline{G}_{mixing} = \Delta\overline{H}_{mixing} - T\Delta\overline{S}_{mixing}$$
$$= 2Z\Delta\varepsilon\left\{x_A - x_A^2\right\} + RT\{x_A\ln x_A + x_B\ln x_B\} \tag{7.38}$$

The free energy of mixing is plotted as a function of mole fraction and the energy difference (nonself minus self interaction) in **Figure 7.13**. As expected, the energy minimum at $x_A = 0.5$ becomes deeper than for ideal mixing when $\Delta\varepsilon$ becomes negative (i.e, when nonself interactions are more stable than self-interactions; compare the red ideal curve in Figure 7.13B with that below it). However, for positive values of $\Delta\varepsilon$, the free energy minimum at $x_A = 0.5$ flattens, and then becomes a *maximum* at large $\Delta\varepsilon$ values, separating two symmetrical minima at low and high mole fraction (compare the red ideal curve in Figure 7.13 with the curves above it).

[†] Here the total more fraction refers not to the mole fraction of A and B within each unmixed compartment, but as the composite system as a whole. It simply provides a count of the size of the two chambers in which all AA and BB interactions are formed.

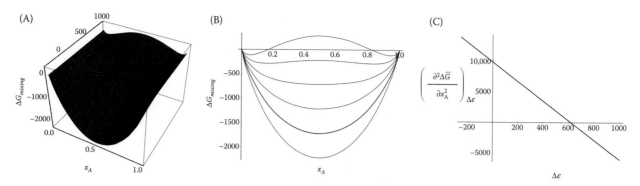

Figure 7.13 Gibbs free energy of mixing for a two-component interacting lattice mixture. (A) Free energy of mixing on a square lattice versus mole fraction and the energy difference $\Delta\varepsilon = \varepsilon_{AB} - \varepsilon_{self}$. The two self-interaction energies are assumed to be the same ($\varepsilon_{AA} = \varepsilon_{BB} = \varepsilon_{self}$). Other details of the model are illustrated in Figure 7.10 and in rules 1–7 in the text. (B) Mixing free energies at various self energies, ranging from –250 (bottom) to +1000 J mol interactions^{-1}. The red curve corresponds to a $\Delta\varepsilon$ value of zero (ideal mixing). At large positive $\Delta\varepsilon$ values, two separate free energy minima are seen, corresponding to liquid–liquid phase separation. (C) The second derivative of mixing free energy with respect to x_A, as a function of $\Delta\varepsilon$, evaluated at a mole fraction of 0.5. The second derivative becomes zero at $\Delta\varepsilon = +620$ J mol interactions^{-1}, the point at which phase separation sets in. Temperature is fixed to 298.15 K, free energy units are J mol^{-1}.

A mixing curve with two minima is shown in expanded scale in **Figure 7.14A** ($\Delta\varepsilon = +750$ J mol interactions^{-1}). These two minima and the intervening maximum at $x_A = 0.5$, which can be identified from taking the first derivative of the mixing free energy with respect to mole fraction (**Figure 7.14B**; Problem 7.16), have rather profound consequences on mixing behavior. Though at all mole fractions the free energy of mixing is negative, at total mole fractions in between the two minima, free energy can be further decreased by producing two separate regions, or phases, of low and high mole fractions (corresponding to the x_A values for the two minima, Figure 7.14B). Because the free energy of the phase-separated mixture is lower than for the randomly dispersed mixture, such separation is spontaneous.

For example, if equal proportions of A and B molecules totaling one mole were mixed (i.e., 0.5 moles each), under the conditions given in Figure 7.14

Figure 7.14 Phase separation in the two-component interacting lattice fluid. (A) Free energy of mixing for a $\Delta\varepsilon$ value of 750 J mol interaction^{-1} at 298.15 K, on a 2D square lattice. (B) First derivative, showing three critical points at $x_A = 0.165, 0.5$, and 0.835. The middle critical point corresponds to the free energy maximum. The two symmetrically displaced critical points correspond to mixing free energy minima (red dotted lines). In the mole fraction interval $0.165 \leq x_A \leq 0.835$, the binary lattice fluid can minimize its free energy by separating to two different phases, one at each of these mole fractions, in a proportion given by their total relative mole amounts. Allowing for this phase separation, the free energy of mixing would follow the flat dotted line in panel (A) through this region.

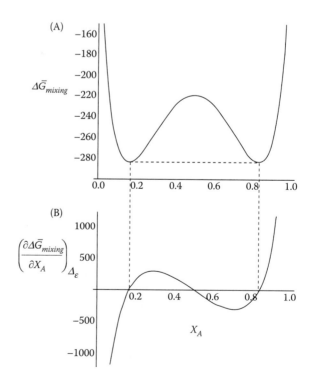

($\Delta\varepsilon = +750$ J mol interactions^{-1}, $T = 298.15$ K), the randomly dispersed single-phase mixture, would have a mixing free energy of -218 J mol^{-1}. In contrast, if the same material was split into equal separate solution phases at mole fractions of $x_A = 0.165$ ($x_B = 0.835$) and $x_A = 0.835$ ($x_B = 0.165$), the molar free energy of the each phase would be -283 J mol^{-1} (Problem 7.19). Thus, the system spontaneously separates into two phases with mole fractions corresponding to the two minima. Starting with an equimolar mixture, equal proportions of the two phases would be formed at equilibrium (Problem 7.20). Starting at other molar mixtures within the two-phase region ($0.165 \leq x_A \leq 0.835$), the two phases form in unequal proportions, based on the equilibrium mole fractions of the two phases (Problem 7.20)

In terms of the nonself:self energy difference, $\Delta\varepsilon$, the point at which the minimum at $x_A = 0.5$ becomes a maximum (i.e., the onset of phase separation) can be determined analytically from the second derivative of the mixing free energy (Figure 7.13C, Problem 7.22). This transition from one to two liquid phases occurs when the second derivative (evaluating at $x_A = 0.5$) is zero, corresponding to a change in curvature from positive to negative. The relationship between free energy curvature and phase stability is important in a variety of contexts (see Callen, 1985).

Chemical potential on the molar scale

As Equations 7.11, 7.18, and 7.27 suggest, our standard state is determined by the concentration scale we select. Though the mole fraction scale is useful for cosolvent mixtures, it is less convenient for dilute solutions, where most of the volume is taken up by the solvent (compare Figures 7.4B and 7.4C). As our discussion of Henry's law indicates (Figure 7.9), the greater the nonideality of the solute, the greater the difference between the hypothetical ($K_{H,i}$) and actual standard state vapor pressures (approaching p_i^*). This difference can be minimized by selecting a lower concentration for the standard state of the solute than a mole fraction of one. In biology and chemistry, the one molar standard state is commonly used for dilute solutes, and will be adopted here.

To convert to the molar concentration scale, we will take advantage of two facts related to dilute solutions with a single dominant solvent (such as water). First, the mole fraction of a solute i is adequately approximated by its mole ratio to the solvent:

$$x_i = \frac{n_i}{n_T} \approx \frac{n_i}{n_s} \tag{7.39}$$

where the s subscript designates solvent. Second, the total volume of the solution is adequately approximated by the solvent volume:

$$V = \sum_{j\ species} n_j \bar{V}_j \approx n_s \bar{V}_s^* \tag{7.40}$$

Rearranging Equations 7.39 and 7.40 and combining gives an expression for molarity in terms of solute mole fraction and solvent molar volume:

$$[i] = \frac{n_i}{V} = \frac{n_s x_i}{n_s \bar{V}_s^*} = \frac{x_i}{\bar{V}_s^*} \tag{7.41}$$

Substituting a rearranged form of (7.41) for x_i in Equation 7.27 leads to

$$\begin{aligned}
\mu_i^{liq} &= \mu_i^{\otimes} + RT \ln\left(\bar{V}_s^*[i]\right) \\
&= \mu_i^{\otimes} + RT \ln \bar{V}_s^* + RT \ln[i] \\
&= \mu_i^{\otimes} - RT \ln[s]^* + RT \ln[i]
\end{aligned} \tag{7.42}$$

where $[s]$ is the molarity of the solvent (the reciprocal of the solvent molar volume; for water, $[s] = 55$ moles per liter). As with the mole fraction relationships for chemical potential (Equations 7.18 and 7.27), the two terms on the right-hand side of Equation 7.42 (bottom line) can be combined to give a new standard state chemical potential, in this case, at a solute concentration of one molar:

$$\mu_i^\circ \equiv \mu_i^{liq}(1\,M) = \mu_i^\otimes - RT\ln[s]^* \tag{7.43}$$

We will mark the "one molar solute" standard state with an open circle (∘). With this new standard state definition,[†] we can write Equation 7.42 as

$$\mu_i^{liq} = \mu_i^\circ + RT\ln[i] \tag{7.44}$$

Admittedly, the combination of the terms that define the one molar standard state (right-hand side of Equation 7.43) is not as intuitive as the combination of partial pressure terms in Equations 7.17 and 7.26. Subtraction of the solvent molarity term can be though of as an approximate correction to the molarity of the solute in pure form (the standard state corresponding to $\mu_i^\otimes$) down to the new one molar reference. However, one would intuitively expect this correction to involve the molarity of the *solute* in pure solution, not the *solvent*. This apparent mismatch results from the approximation we made that the volume is contributed only by the solvent (Equation 7.41). Although this approximation affects the absolute chemical potential of dilute solutes and the total Gibbs free energy, these variations cancel out in calculations that compare molar free energies, such as free energies associated with chemical reactions, which is our next topic.

CHEMICAL REACTIONS

In the analysis of mixtures above, we treated the components as unreactive. Though unreactive mixtures can display fairly exotic behaviors such as phase transitions, the complexity of chemical and biological systems cannot be described without chemical reaction. Here we will develop a framework to analyze the thermodynamics of chemical reactions. Much of the reaction thermodynamics used here was developed almost a century ago by Lewis and Randall (1961) and Guggenheim (1933), using ideas about chemical potential developed by Gibbs a half century before (Gibbs, 1876). Though we will go beyond mixtures in this analysis, it will become clear that mixing of reactants and products plays a fundamental role in reaction thermodynamics.

In keeping with the nomenclature introduced in Chapter 5, we will refer to components that convert through chemical reaction as "species." Species can convert either covalent bond rearrangement (e.g., in ATP hydrolysis) or rearrangement of noncovalent interactions (e.g., binding reactions and conformational changes). Although the state of a system is specified by giving the component amounts, analysis of chemical equilibrium requires the mole amounts of all reactive species.

A formalism for chemical reactions

To start, we will describe a generic chemical reaction with reactants (A, B, etc.) on the left side and products (G, H, etc.) on the right:

$$aA + bB \rightleftharpoons gG + hH \tag{7.45}$$

Though single arrows are often used to emphasize the "direction" of the reaction, we will use the double arrow because it emphasizes reversibility: the reaction can proceed in either direction, allowing all species to convert, and as a result, achieve equilibrium. The coefficients a, b, g, h are positive, rational, and dimensionless

---

[†] Unlike $\mu_i^\otimes$, we will use μ_i° again and again.

values; we will refer to these values as "reaction numbers."[†] A more compact (though perhaps less intuitive) way to write the reaction looks more like an equation:

$$gG + hH - aA - bB = 0 \qquad (7.46)$$

where reactants are subtracted from products (by convention) and set to zero. This type of expression for chemical reaction can be generalized using a summation, replacing the reaction numbers a, b, g, and h with "stoichiometric coefficients" ν_i:

$$\sum_{i \; species} \nu_i I = 0 \qquad (7.47)$$

For products, the ν_i symbols are equal to the reaction numbers (g, h in Equation 7.45). For reactants, the ν_i symbols are equal to the negative of the reaction numbers (a, b in Equation 7.45). For example, in the reaction

$$2H_2 + O_2 \rightleftharpoons 2H_2O \qquad (7.48)$$

the stoichiometric coefficients are $\nu_{H_2O} = 2$, $\nu_{H_2} = -2$, and $\nu_{O_2} = -1$. These coefficients provide a key constraint, allowing us to describe the extent of reaction of each species using a single variable.[‡]

We will develop reaction free energies in two ways, first by taking the difference between the free energies of reactant and product chemical potentials, and second, by evaluating the differential of free energy with respect to the moles of reaction. In both treatments, the chemical potential and its concentration dependence plays a central role. The first approach leads to the concept of equilibrium constant, and its relationship with reaction free energy. The second approach illustrates how the reaction free energy serves as a potential for predicting equilibrium and spontaneous change as a function of reaction coordinate, as we described in Chapter 5.

Reaction free energy from a finite difference approach

Here, we take a very direct approach of representing the free energy of reaction as the difference between the free energy of products and the reactants. If a (finite) number of moles η_{prod} of products are formed,[§] the free energy of reaction is simply the Gibbs energy increment (or decrement) that results from consumption of reactants and formation of products. In terms of extensive energies,

$$\Delta G_{rxn} = \eta_{react}\overline{G}_{react} + \eta_{prod}\overline{G}_{prod} \qquad (7.49)$$

where η_{react} is negative if reactants convert to products. It should be kept in mind that the Δ in Equation 7.49 describes transformation from reactants to products. Though, as will be shown below, ΔG_{rxn} depends on the concentrations of reactants and products, in the derivation that follows, these concentrations are assumed to be fixed. The concept of running a reaction while keeping reactant and product concentrations fixed can be confusing,[¶] we will return to it after our derivation.

[†] Fixed and simple numbers for stoichiometric coefficients were first recognized by John Dalton (of partial pressure fame) in the early 1800s, and stated as the "Law of Multiple Proportions."

[‡] This constraint is the reason we only need to give the amount of a single component in specifying the state of the system.

[§] We will represent the number of moles of reaction using the Greek letter eta (η). The use of η to give the number of moles of reaction avoids confusion with the total number of moles of each species in the system, n.

[¶] The conceptual challenge is similar to the challenge we encountered when defining partial molar volumes in Chapter 5. If we add some number of moles of a component to a mixture we are interested in to measure the volume change, we change the concentrations of the mixture.

The quantities η_{react} and η_{prod} in Equation 7.49 are coupled through their stoichiometric coefficients. If the number of moles of one component (e.g., A in Equation 7.45) changes by some amount (e.g., η_A), all of the other reactants and products will be changed in proportion to their stoichiometric coefficients (e.g., $\eta_G = (\nu_G/\nu_A)\eta_A$). We can standardize these relationships by defining a number of moles of reaction as a whole, η_{rxn}, and combining with the stoiciometric coefficients. For the ith reacting species,

$$\eta_i = \nu_i \eta_{rxn} \tag{7.50}$$

We will take η_{rxn} to be positive when reactants convert to products. For the reaction in Equation 7.48, $\eta_{H_2O} = 2\eta_{rxn}$, $\eta_{H_2} = -2\eta_{rxn}$, and $\eta_{O_2} = -1\eta_{rxn}$.

Using these mole numbers, we can express the free energy of reaction in terms of η_{rxn}. For reaction (7.45),

$$\Delta G_{rxn} = \nu_A \eta_{rxn} \bar{G}_A + \nu_B \eta_{rxn} \bar{G}_B + \nu_G \eta_{rxn} \bar{G}_G + \nu_H \eta_{rxn} \bar{G}_H$$
$$= \eta_{rxn}(\nu_A \mu_A + \nu_B \mu_B + \nu_G \mu_G + \nu_H \mu_H) \tag{7.51}$$

Dividing both sides by the number of moles of reaction gives a molar free energy of reaction:

$$\Delta \bar{G}_{rxn} = \nu_A \mu_A + \nu_B \mu_B + \nu_G \mu_G + \nu_H \mu_H \tag{7.52}$$

To express the reaction free energy as a function of reaction and product concentration, we can substitute Equation 7.44, which gives the chemical potentials of the reactants and products relative to a one molar standard state:

$$\Delta \bar{G}_{rxn} = \nu_A\left(\mu^\circ_A + RT\ln[A]\right) + \nu_B\left(\mu^\circ_B + RT\ln[B]\right)$$
$$+ \nu_G\left(\mu^\circ_G + RT\ln[G]\right) + \nu_H\left(\mu^\circ_H + RT\ln[H]\right) \tag{7.53}$$

Equation 7.53 has two types of terms: standard state potentials, and concentration-dependent terms. Combining like terms gives

$$\Delta \bar{G}_{rxn} = \nu_A\mu^\circ_A + \nu_B\mu^\circ_B + \nu_G\mu^\circ_G + \nu_H\mu^\circ_H + RT\ln([A]^{\nu_A}[B]^{\nu_B}[G]^{\nu_G}[H]^{\nu_H}) \tag{7.54}$$

If we evaluate the free energy of reaction under standard state conditions (i.e., one molar concentrations of all reactants and products), then the right-most term in Equation 7.54 has a value of zero. This demonstrates that the sum of coefficient-weighted standard state potentials in Equation 7.54 is equivalent to the reaction free energy at standard state, which we will represent as $\Delta \bar{G}^\circ_{rxn}$.

Note that for a given set of conditions (p, T, and solvent conditions), $\Delta \bar{G}^\circ_{rxn}$ is a constant; it is *not* a function of concentrations of reactants or products. The concentration dependence of $\Delta \bar{G}_{rxn}$ comes from second term, which is sometimes referred to as the "reaction quotient" (Q). Although Q does not look much like a quotient, it should be remembered that for the reactants, the stoichiometric coefficients are negative. Thus, their concentrations (and powers thereof) end up in the denominator. This is easy to see by converting back to specific reaction numbers in Equation 7.45:

$$Q = [A]^{\nu_A}[B]^{\nu_B}[G]^{\nu_G}[H]^{\nu_H} = [A]^{-a}[B]^{-b}[G]^g[H]^h = \frac{[G]^g[H]^h}{[A]^a[B]^b} \tag{7.55}$$

Substituting $\Delta \bar{G}_{rxn}$ and Q into Equation 7.54 gives

$$\Delta \bar{G}_{rxn} = \Delta \bar{G}^\circ_{rxn} + RT\ln Q \tag{7.56}$$

One important quantity that derives from Equation 7.56 is the equilibrium constant, K_{eq}. When a chemical reaction is at equilibrium, the free energy of reaction is zero. Thus, Equation 7.56 becomes

$$0 = \Delta \overline{G}^{\circ}{}_{rxn} + RT \ln Q_{eq} \tag{7.57}$$

or

$$\begin{aligned}
\Delta \overline{G}^{\circ}{}_{rxn} &= -RT \ln Q_{eq} \\
&= -RT \ln \left([A]^{\nu_A}_{eq}[B]^{\nu_B}_{eq}[G]^{\nu_G}_{eq}[H]^{\nu_H}_{eq} \right) \\
&= -RT \ln \left(\frac{[G]^g [H]^h}{[A]^a [B]^b} \right)_{eq}
\end{aligned} \tag{7.58}$$

The *eq* subscripts indicate equilibrium concentrations. Since there is only one set of equilibrium concentrations (for a fixed total number of moles of reactant plus product), the molarities that make up Q_{eq} are fixed. Thus, Q_{eq} is a constant, which we refer to as an equilibrium constant, K_{eq}:

$$K_{eq} \equiv \left(\frac{[G]^g [H]^h}{[A]^a [B]^b} \right)_{eq} \tag{7.59}$$

Combining Equations 7.58 and 7.59 gives

$$\Delta \overline{G}^{\circ}{}_{rxn} = -RT \ln K_{eq} \tag{7.60}$$

Equation 7.60 suggests yet another way to describe the standard state free energy of reaction: $\Delta \overline{G}^{\circ}{}_{rxn}$ is the ratio of equilibrium reactant and product concentrations, raised to the power of their respective reaction numbers. An important consequence is that by simply measuring equilibrium concentrations, the standard state reaction free energy can be determined, providing access (at least as a difference) to standard state chemical potentials for the reactants and products.

The two terms in Equation 7.56 play very different roles in reaction thermodynamics, and given the importance of this equation, it is essential to understand these differences. The first term that imparts uniqueness to a given reaction, and determines specificity and cooperativity in complex systems. If a drug is to be developed with a high affinity for a macromolecular target, it is $\Delta \overline{G}^{\circ}{}_{rxn}$ that needs to be optimized. Though the second term leads to variation to the free energy of reaction in response to reactant and product concentrations (through Q), it does so in a predetermined way. In fact, all reactions of a given stoichiometry have identical Q terms. In such cases, it is $\Delta \overline{G}^{\circ}{}_{rxn}$ that reflects specific differences among reactions. In much of the remaining text, we will focus on $\Delta \overline{G}^{\circ}{}_{rxn}$ values through their relation to the equilibrium constant (Equation 7.60).

Now that our derivation is complete, let's return to the somewhat confusing point of running a finite amount of reaction without changing reactant and product concentrations. Although Q allows the free energy of reaction to be calculated at any concentration of reactant and product, the reaction free energy represented by Equation 7.56 is evaluated at *fixed* reactant and product concentrations. This important point can be difficult to reconcile with the concept that $\Delta \overline{G}_{rxn}$ is the free energy generated when a mole of reactant converts to product. One might expect that mole of reaction to significantly lower reactant concentration and increase product concentration. Though changes in concentration can occur, especially for reaction in a closed system, they are not amenable to direct analysis by Equation 7.56.

Given this fixed-concentration requirement, one might question the usefulness of Equation 7.56 to analyze real reactions. For a closed system, Equation 7.56 tells us the free energy of reaction at a given set of conditions at the start of the reaction. To maintain constant reactant and product concentrations for a finite amount of reaction, we would need to add fresh reactant as it is consumed and remove product when it is produced. Although this "chemostatic" approach takes a lot of effort in the lab,[†] it is quite common in cellular metabolism. In cells, tissues, and organisms, reactants (and products) are typically produced (and consumed) by a network of controlled biochemical reactions, and are balanced by nutrient uptake and waste removal. In such cases, "steady state" levels can be faithfully maintained, and Equation 7.56 allows a relevant reaction free energy to be calculated at cellular concentrations. For detailed and elegant treatises of the thermodynamic and kinetic properties of such networks, see Beard and Qian (2008) and Hill (2005).

Reaction free energy from differentials

One way to avoid the conceptual difficulties of the finite approach is to allow only a small mole amount of reactant to convert to product. For a small amount of reaction, the chemical potentials of reactants and products don't change *much*. In the limiting case of infinitesimal reaction amount, concentrations remain constant (as do chemical potentials), so we don't need to worry about adding reactant or removing product. This differential approach to reactions is analogous to the differential approach to molar volumes in Chapter 5 (Equations 4.36 and 4.37).

Mathematically, we can represent this infinitesimal approach using the differential variation in Gibbs energy in response to changes in mole numbers of the reactant and product species. For the reaction in Equation 7.45, the variation in mole numbers brings a change in the total (extensive) Gibbs energy of

$$dG = \mu_{prod}dn_{prod} + \mu_{react}dn_{react}$$
$$= \mu_G dn_G + \mu_H dn_H + \mu_A dn_A + \mu_B dn_B \tag{7.61}$$

(Equation 5.76) where the n_i terms represent in the total number of moles of each species present. Though Equation 7.61 is general, we can use it to analyze a chemical reaction by restricting the variations dn_i in proportion to the reaction stoichiometry (Equation 7.45).

One advantage of the differential approach is that it allows us to focus on the total amount of reactant and product present (η_i in Equation 7.61, rather than η in Equation 7.49). As before, we will describe the "extent of reaction" using a single variable, which we will represent with the symbol ξ.[‡] The coordinate ξ has units of moles, and its value ranges from zero (when no product is formed) to n_T (the total number of moles, when no reactant remains, **Figure 7.15**).[§] The position along the

---

[†]   A chemostatic reaction chamber would require sensors for reactant and product concentrations, semipermeable membranes, stirrers, and pumps, all of which that all respond faster than the reaction itself. These types of reactors are common in biotechnology (for microbial growth) and industry, but they are not the subject of the present analysis.

[‡] ξ is also referred to as a "reaction coordinate." The fact that we are using the same variable as was used to develop the concept of potentials in Chapters 4 and 5 is no coincidence. Chemical reactions are exactly the type of variations that can be analyzed using potential equations $dS \geq 0$ and $dG \leq 0$, for isolated and for constant T, p, systems.

[§]   Note that for reactions with complicated stoichiometries, there are some complexities associated with mapping the extent of reaction ξ to the total number of moles, n_T. First off, because reactants and products have different reaction coefficients (a, b, g, h in Equation 7.45), we need to decide which species to use to define the total number of moles, n_T. A simple convention is to use a species with a reaction number of one. For some reactions, *none* of the participants have reaction numbers of one. For example, consider the reaction $4NH_3 + 7O_2 \rightarrow 4NO_2 + 6H_2O$. In this case, rather than retaining integer reaction numbers, we can divide by the smallest reaction number to give $NH_3 + 7/4\ O_2 \rightarrow NO_2 + 3/2\ H_2O$. An extent of reaction $\xi = 1$ mole means that there is one mole of NO_2 and 1.5 moles of H_2O. If a reaction were prepared by mixing one mole of NH_3 1.75 mole of O_2, 0.5 mole of NO_2, and 0.75 mole of H_2O, the limiting (maximum) value of ξ would be $n_{NH_3} + n_{NO_2} = 1 + 0.5 = 1.5$ moles.

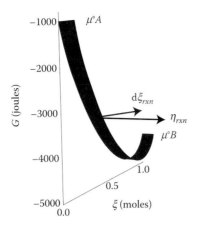

Figure 7.15 Difference between the variation in mole number in the finite approach to reaction thermodynamics and the extent of reaction in the differential approach. The free energy as a function of the extent of reaction, ξ, is depicted as a red ribbon for a simple reaction $A \rightleftharpoons B$ (Equation 7.66). In the finite difference approach, the number of moles of reaction (η_{rxn}, black arrow) leaves composition unchanged, and thus can be considered to be perpendicular to ξ. In contrast, in the differential approach, the variation in the extent of reaction (red arrow) is along the ξ coordinate. Standard state chemical potentials for A and B are set to -1000 and -4000 J mol^{-1}. For this figure, the molar volumes of both A and B are set to 1L per mole, and are assumed to be independent of composition. The total number of moles ($n_A + n_B$) is set to 1, and the temperature is set to 298.15 K.

coordinate ξ describes how much the reactant has converted to product. Because the finite reaction formalism above involves no change in reactant or product concentration, it occurs at a fixed value of ξ. In other words, the number of moles η_{rxn} in the finite approach can be thought of as perpendicular to the reaction coordinate (black arrow, Figure 7.15).

In the differential approach to reaction thermodynamics, we can consider variation in the direction the extent of reaction, because our variations ($d\xi$) are infinitesimal, and leave chemical potential essentially unchanged. As in the finite approach (Equation 7.50), we will use stoichiometric coefficients ν_i to express the moles of reaction of each species dn_i using a single quantity, in this case $d\xi$:

$$dn_i = \nu_i d\xi \tag{7.62}$$

Inserting Equation 7.62 into Equation 7.61 gives

$$dG = \nu_G \mu_G d\xi + \nu_H \mu_H d\xi + \nu_A \mu_A d\xi + \nu_B \mu_B d\xi \tag{7.63}$$

Dividing the extensive quantity dG in Equation 7.63 by $d\xi$ gives

$$\left(\frac{\partial G}{\partial \xi}\right)_{T,p} = \nu_G \mu_G + \nu_H \mu_H + \nu_A \mu_A + \nu_B \mu_B \tag{7.64}$$

The right-hand side of Equation 7.64 is simply the molar free energy of reaction (Equation 7.52); i.e.,

$$\left(\frac{\partial G}{\partial \xi}\right)_{T,p} = \Delta \overline{G}_{rxn} \tag{7.65}$$

An additional complication arises when the relative amounts of the reactants (and products) are out of proportion with their reaction numbers. In such cases, one or more reagents (and procuts) are "limiting," and the extent of reaction x may be limited to a subset of values between 0 and n_T.

Figure 7.16 Gibbs energy as a potential that determines the value of reaction free energy. (A) A unimolecular reaction of the type $A \rightleftharpoons B$, with parameters given in Figure 7.15. The Gibbs energy, given by Equation 7.67, has a minimum at an extent of reaction of $\xi = 0.77$. This minimum is the point of reaction equilibrium. For $\xi < 0.77$, the slope of G with respect to ξ is negative; thus, the direction of spontaneous change is toward product. For $\xi < 0.77$, the reverse is true. (B) The free energy of reaction, calculated as the derivative of extensive Gibbs energy with respect to ξ. The free energy of reaction is zero at $\xi = 0.77$, corresponding to the minimum in (A).

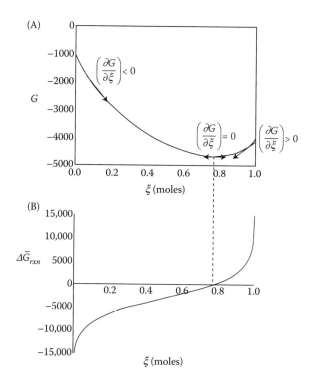

The significance of Equation 7.65 can be appreciated by recalling that at constant temperature and pressure, Gibbs free energy acts as a potential. If G decreases as ξ increases, spontaneous change will go in the direction that increases ξ, converting reactants to products. If instead, G increases as ξ increases, spontaneous change converts products back to reactants. Thus, when plotted against the extent of reaction, total Gibbs free energy serves as a reaction potential. Accordingly, Equation 7.65 states that the free energy of reaction is the slope of G versus ξ plot; this slope can be regarded as a driving force for reaction.

These ideas are illustrated in **Figure 7.16**, assuming a simple unimolecular reaction,

$$A \rightleftharpoons B \tag{7.66}$$

where $n_T = n_A + n_B = 1$ mole, with standard state chemical potentials arbitrarily set to $\mu_A^{\circ} = -1000 \, \mathrm{J\,mol^{-1}}$ and $\mu_B^{\circ} = -4000 \, \mathrm{J\,mol^{-1}}$. To simplify things, we will assume that $\overline{V}_A$ and $\overline{V}_B$ are identical (and thus can each be represented as $\overline{V}$) and independent of composition. This assumption allows us to express the extensive Gibbs free energy as

$$G = (1-\xi)\left\{\mu_A^{\circ} + RT\ln(1-\xi)\right\} + \xi\left\{\mu_B^{\circ} + RT\ln\xi\right\} - RT\ln V \tag{7.67}$$

This equation derives from the expression for the total Gibbs energy (Equation 5.65), along with a separation of molarity in Equation 7.44 into moles and liters (see Problem 7.24). For a reaction in dilute solution, the volume term in Equation 7.67 is approximately that of the solvent. This equation can be simplified further by setting V equal to 1 L. In addition to eliminating the volume term in Equation 7.67, the lower and upper limits of ξ (0 and $n_T = 1$ mole) become the standard states of A and B.

Plotting G versus ξ (Figure 7.16A) gives the potential for this reaction (Equation 7.66). The slope of the curve in Figure 7.16A gives the reaction free energy (Equation 7.65). On the left side of Figure 7.16A, the total Gibbs energy decreases with ξ, reflecting spontaneous conversion of reactant to product. Consistent with this, the slope is negative and thus the free energy of reaction is negative. On the right side, Gibbs energy increases with ξ, reflecting spontaneous conversion of product to reactant; the positive slope means a positive free energy of reaction. In between (more precisely, at $\xi = 0.77$; Problem 7.26), G assumes a minimum value, corresponding to chemical equilibrium. From Equation 7.65, this corresponds to $\Delta \bar{G}_{rxn} = 0$.

An explicit expression for $\Delta \bar{G}_{rxn}$ can be calculated by differentiating Equation 7.67 with respect to ξ (Problem 7.25):

$$\Delta \bar{G}_{rxn} = \left(\frac{\partial G}{\partial \xi} \right)_{T,p}$$

$$= \mu^{\circ}_B - \mu^{\circ}_A + RT \ln \frac{\{\xi\}}{\{1 - \xi\}} \tag{7.68}$$

$$= \Delta \bar{G}^{\circ}_{rxn} + RT \ln \frac{\{\xi\}}{\{1 - \xi\}}$$

Note that even when $V \neq 1$, it does not contribute to $\Delta \bar{G}_{rxn}$, as long as it is independent of composition. With a little bit of comparison, you should be able to convince yourself that Equation 7.68 is essentially the same as the finite reaction thermodynamics Equation 7.57 for the simple A to B reaction (Equation 7.66).

Equation 7.68 is plotted in Figure 7.16B. As ξ goes to zero (all reactant, no product), the free energy of reaction goes negatively infinite, that is, there is a strong driving force to form product. The opposite is true when ξ goes to one.[†] As anticipated, $\Delta \bar{G}_{rxn}$ crosses zero (the equilibrium point for the reaction) at $\xi = 0.77$. Though the finite and differential approaches gives identical reaction free energy equations (as they must), the differential approach shows clearly how the Gibbs free energy acts as a potential for chemical reaction.

SIMILARITIES (AND DIFFERENCES) BETWEEN FREE ENERGIES OF REACTION AND MIXING

The total Gibbs energy plot for reacting systems (Figure 7.16A) is similar in shape to that of the free energy of mixing (Figure 7.8). This similarity comes from the logarithmic terms (Equations 7.67 and 7.23, respectively). To develop this similarity further, we will consider the Gibbs energy of reaction on a mole fraction scale (rather than our molar scale) for our two-component system (Equation 7.66):

$$G_T = n_A \mu_A + n_B \mu_B$$

$$= n_A \left(\mu_A{}^* + RT \ln x_A \right) + n_B \left(\mu_B{}^* + RT \ln x_B \right) \tag{7.69}$$

$$= n_T x_A \left(\mu_A{}^* + RT \ln x_A \right) + n_T x_B \left(\mu_B{}^* + RT \ln x_B \right)$$

Collecting like terms gives

$$G_T = n_T \left\{ \left(x_A \mu_A{}^* + x_B \mu_B{}^* \right) + \left(x_A RT \ln x_A + x_B RT \ln x_B \right) \right\}$$

$$= n_T \left\{ \left\langle \bar{G}^* \right\rangle + \Delta \bar{G}_{mixing} \right\} \tag{7.70}[‡]$$

Equation 7.70 reveals the mixing term as part of the total free energy of a reacting system. However, the total free energy contains an additional population-weighted average standard state chemical potential, abbreviated as $\left\langle \bar{G}^* \right\rangle$. Rearrangement of Equation 7.70 shows that

---

[†]   Note that these infinities are not obvious from the plot, because the plot is truncated to a reasonable range of free energy of reaction (and because Mathematica samples a finite number of points in making line plots).

[‡]   At this point, you should resist the urge to divide Equation 7.70 through by n_T. Soon we will obtain a molar free energy of reaction, but we will do it by differentiating with respect to the reaction mole number (Equation 7.65).

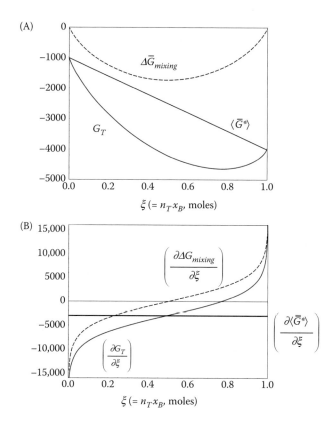

Figure 7.17 Relationship between the free energy of mixing to the free energy of reaction. (A) The total Gibbs energy (red) versus the extent of reaction, which is equal to the total number of moles (taken here to be $n_T = 1$), for the simple A to B reaction (Equation 7.66). The Gibbs energy can be decomposed into a linear standard-state term (black solid line, Equation 7.71) and a mixing term (black dashed line, see Equation 7.70). The minimum in the G_T curve results from the mixing term. However, the position of the minimum (the point of reaction equilibrium) is shifted from $x_A = x_B = 0.5$ as a result of the standard state term. (B) The free energy of reaction (red), obtained by differentiating the total Gibbs energy (red curve, A) with respect to the extent of reaction. The reaction free energy contains two parts. The derivative of the standard state term is a flat offset at the value $\mu_B{}^* - \mu_A{}^*$ (here, -3000, thick black line). The derivative of the mixing term (dashed line) is a nonlinear function that increases with extent of reaction, crossing zero at $x_A = x_B = 0.5$. Because the standard state derivative is added to the mixing derivative to get the free energy of reaction, the position where $\Delta \overline{G}_{rxn}$ crosses zero is shifted from $x_A = x_B = 0.5$ to the minimum in (A).

$$\left\langle \overline{G}^* \right\rangle = x_B \left[\mu_B{}^* - \mu_A{}^* \right] + \mu_A{}^* \tag{7.71}$$

Because $\left\langle \overline{G}^* \right\rangle$ is linear in mole fraction, it has no minimum (**Figure 7.17A**), other than at the limit $x_B = 1$ or 0 (depending on whether the standard state chemical potential of A is higher or lower than B).

To see how the mixing and standard state terms in the total Gibbs free energy influence the free energy of reaction, Equation 7.70 is differentiated with respect to η_{rxn}. Recognizing that $\eta_{rxn} = n_T x_B$, and thus, $d\eta_{rxn} = n_T dx_B$ (n_T is a constant),

$$
\begin{aligned}
\Delta \overline{G}_{rxn} &= \left(\frac{\partial G_T}{\partial \eta_{rxn}} \right)_{T,p} \\
&= n_T \left(\frac{\partial \left\langle \overline{G}^* \right\rangle}{n_T \partial x_B} \right)_{T,p} + n_T \left(\frac{\partial \Delta \overline{G}_{mixing}}{n_T \partial x_B} \right)_{T,p} \\
&= \mu_B{}^* - \mu_A{}^* + RT \ln x_B - RT \ln(1 - x_B)
\end{aligned}
\tag{7.72}
$$

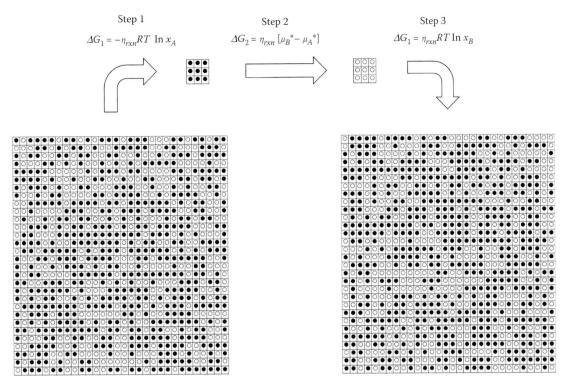

Step 1
$$\Delta G_1 = -\eta_{rxn} RT \ln x_A$$

Step 2
$$\Delta G_2 = \eta_{rxn} [\mu_B^* - \mu_A^*]$$

Step 3
$$\Delta G_1 = \eta_{rxn} RT \ln x_B$$

Figure 7.18 A three-step model for chemical reaction. An A to B reaction scheme (Equation 7.66) is depicted on a lattice for simplicity, with an x_B value of 0.4. In step 1, nine reactant A molecules (black) are removed from the mixture and put in a separate container. In step 2, the isolated A molecules are allowed to react to B molecules. In step 3., these newly formed B molecules are reintroduced to the mixture. The sum of these three free energy terms is equal to the (extensive) reaction free energy. Note that no change in concentration is depicted, because the mixture is so much larger than the transformed molecules. This is consistent with the definition of $\Delta \overline{G}_{rxn}$, which is calculated at fixed reactant and product concentrations.

The derivative of the average standard state term shifts $\Delta \overline{G}_{rxn}$ vertically by a constant amount (which is equal to the difference in standard state chemical potentials). The derivative of the mixing term results in the nonlinear dependence of $\Delta \overline{G}_{rxn}$ on extent of reaction, generates extreme low and high values at limiting values of ξ, and ensures that $\Delta \overline{G}_{rxn}$ crosses zero, corresponding to the chemical equilibrium point (**Figure 7.17B**).

Although the mixing term in the expression for G_T influences the free energy of reaction, it does not remain intact in the equation for $\Delta \overline{G}_{rxn}$. Instead, differentiation propagates pieces of the free energy of mixing. To understand this connection, we will analyze the A to B reaction using a hypothetical three-step finite process (**Figure 7.18**).[†] In this analysis, we will assume the reaction to be occurring in a very large reaction mixture ($\eta_{rxn} \ll n_T$), so that the overall concentrations remain constant.

Step 1: Remove η_{rxn} moles of A from a very large mixture ($n_T \gg \eta_{rxn}$) of reactant and product and place the A's in separate container, giving η_{rxn} moles of pure A. The (extensive) free energy change from this step is

$$\Delta G_1 = \eta_{rxn}\mu_A^* - \eta_{rxn}\mu_A$$
$$= \eta_{rxn}\mu_A^* - \eta_{rxn}\left(\mu_A^* + RT \ln x_A\right) \qquad (7.73)$$
$$= -\eta_{rxn}RT \ln x_A$$

---

† Since free energy is a state function, the hypothetical three-state path will give the same result as an actual chemical reaction.

The reason we can get away with representing the loss of free energy from removing the A's from the mixture as $\mu_A{}^* + RT \ln x_A$ is that the mole fractions do not change appreciably as long as $n_T \gg \eta_{rxn}$.

Step 2: Convert all of separated A's (by reaction) to B's. For this process, the starting and ending points are both pure, so the free energy change for this step is

$$
\begin{aligned}
\Delta G_2 &= \eta_{rxn}\mu_B{}^* - \eta_{rxn}\mu_A{}^* \\
&= \eta_{rxn}\left(\mu_B{}^* - \mu_A{}^*\right)
\end{aligned}
\tag{7.74}
$$

Step 3: Put the η_{rxn} moles of converted B back into the original mix. The free energy change for this step is

$$
\begin{aligned}
\Delta G_3 &= \eta_{rxn}\mu_B - \eta_{rxn}\mu_B{}^* \\
&= \eta_{rxn}\left(\mu_B{}^* + RT \ln x_B\right) - \eta_{rxn}\mu_B{}^* \\
&= \eta_{rxn} RT \ln x_B
\end{aligned}
\tag{7.75}
$$

Summing these three terms gives an extensive version of the free energy of reaction:

$$
\begin{aligned}
\Delta G_{rxn} &= \Delta G_1 + \Delta G_2 + \Delta G_3 \\
&= \eta_{rxn}\left(\mu_B{}^* - \mu_A{}^*\right) + \eta_{rxn} RT \ln x_B - \eta_{rxn} RT \ln x_A
\end{aligned}
\tag{7.76}
$$

Dividing by η_{rxn} gives the molar reaction free energy:

$$
\Delta \overline{G}_{rxn} = \mu_B{}^* - \mu_A{}^* + RT \ln x_B - RT \ln x_A
\tag{7.77}
$$

The mole fraction terms in Equation 7.77 are the "pieces" of the mixing free energy produced from differentiation. In the stepwise process, the $RT \ln x_B$ term results from mixing the pure product into the bulk (stepwise), and the $-RT \ln x_A$ term results from separating (unmixing) the reactant from the bulk into a pure phase. Thus, the nonlinear dependence of $\Delta \overline{G}_{rxn}$ on the extent of reaction can be thought of as a result of unmixing of reactant and a mixing of product. At reaction mixtures that are highly enriched in reactant, it is not difficult to separate out the reactant— almost any molecules you grab will be reactant. In contrast, adding product to a system that is mostly reactant is very favorable—there are very many distinguishable ways the product can be configured. This gives a strong driving force for reaction, that is, a highly negative $\Delta \overline{G}_{rxn}$ value. In contrast, in mixtures highly enriched in product, it is very difficult to find and remove reactant molecules, creating a large positive $\Delta \overline{G}_{rxn}$ value.

HOW CHEMICAL EQUILIBRIUM DEPENDS ON TEMPERATURE

Measuring the position of chemical equilibrium provides direct determination of $\Delta \overline{G}{}^\circ$. By studying how the position of equilibrium depends on intensive variables such as temperature, pressure, and concentrations of nonreacting species (cosolvents, ligands, denaturants, and pH), we can obtain a more detailed (and more complete) thermodynamic description of reactions. Here we will examine how temperature affects chemical equilibrium, and what we can learn by varying the temperature.

A key relationship in the analysis of the temperature dependence of reaction comes from the definition of the Gibbs free energy ($G = H - TS$; Equation 5.16). By using this definition to take the difference between the molar Gibbs energies of reactant and product,

$$\bar{G}^{\circ}_{prod} - \bar{G}^{\circ}_{react} = \bar{H}^{\circ}_{prod} - \bar{H}^{\circ}_{react} - T\left(\bar{S}^{\circ}_{prod} - \bar{S}^{\circ}_{react}\right) \tag{7.78}$$

or

$$\Delta\bar{G}^{\circ} = \Delta\bar{H}^{\circ} - T\Delta\bar{S}^{\circ} \tag{7.79}$$

where the difference in Equation 7.78 is taken at the same temperature for the reactant and product, and is taken at the standard state concentrations.[†]

Equation 7.79 gives an explicit temperature dependence to the free energy of reaction. $\Delta\bar{G}^{\circ}$ varies in direct proportion (but with opposite sign) to $\Delta\bar{S}^{\circ}$. This dependence is consistent with the thermodynamic identity relating free energy, enthalpy, and temperature (Chapter 5, see Equation 5.32). If $\Delta\bar{H}^{\circ}$ and $\Delta\bar{S}^{\circ}$ are independent of temperature, $\Delta\bar{G}^{\circ}$ varies linearly with temperature (**Figure 7.19A**). In addition, the values of $\Delta\bar{H}^{\circ}$ and $\Delta\bar{S}^{\circ}$ may on temperature, sometimes producing a nonlinear dependence of $\Delta\bar{G}^{\circ}$ on temperature (as is seen for protein folding; see Chapter 8).

By combining Equations 7.60 and 7.79, we can obtain a fundamental relationship for the temperature dependence of the equilibrium constant:

$$-RT\ln K_{eq} = \Delta\bar{H}^{\circ} - T\Delta\bar{S}^{\circ} \tag{7.80}$$

Separating the equilibrium constant from the temperature gives

$$\ln K_{eq} = \frac{\Delta\bar{S}^{\circ}}{R} - \frac{\Delta\bar{H}^{\circ}}{RT} \tag{7.81}$$

Equation 7.81 is often referred to as the "Van't Hoff equation." Plotting $\ln K_{eq}$ as a function of T shows an inverse temperature dependence (**Figure 7.19B**). As temperature approaches zero, $\ln K_{eq}$ becomes infinite (with either a positive or a negative sign, depending on the sign of $\Delta\bar{H}^{\circ}$). This corresponds to an equilibrium constant of infinity or zero. At high temperature, the enthalpy term in Equation 7.81

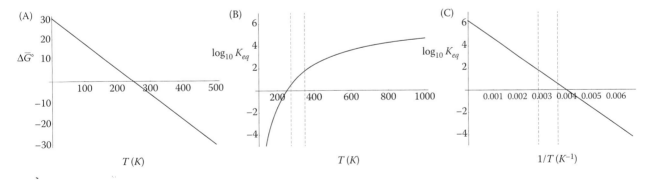

Figure 7.19 Temperature dependence of the free energy and equilibrium constant for chemical reaction. (A) The temperature dependence of free energy (in kJ mol⁻¹) for a reaction that has temperature-independent $\Delta\bar{H}^{\circ}$ and $\Delta\bar{S}^{\circ}$ values of 30 kJ mol⁻¹ and 0.12 J mol⁻¹ K⁻¹, respectively. The slope of this plot gives the negative of the reaction entropy, and the intercept gives the reaction enthalpy. (B) The log of the equilibrium constant as a function of temperature. At high temperature, the equilibrium constant reaches a plateau that is determined by the entropy of reaction. However, because reactions in aqueous solutions are limited to a relatively narrow range of temperature (e.g., vertical dashed lines), such a plateau is not likely to be seen. Note that a base-10 log is used to represent tenfold changes in K_{eq}. (C) A Van't Hoff plot, with a slope of $0.434 \times \Delta\bar{H}^{\circ}/R$, and an intercept of $0.434 \times \Delta\bar{S}^{\circ}/R$ (the value $0.434 \approx \log_{10} e$ results from using a base-10 log scale). Reciprocal temperature limits (vertical bars) correspond to limits shown in (B). Values of $\Delta\bar{H}^{\circ}$ and $\Delta\bar{S}^{\circ}$ from panel (A) are used in panels (B) and (C).

[†] Equation 7.79 holds at any set of concentrations; here we use standard states (one molar reactant and product) to connect to chemical reactions, and to the equilibrium constant in particular.

decays to zero, and $\ln K_{eq}$ converges to a value determined by the entropy. In other words, the position of equilibrium is determined by enthalpy at low temperature, and entropy at high temperature.[†]

One historically popular way to represent Equation 7.81 is to plot $\ln K_{eq}$ versus inverse temperature (**Figure 7.19C**). The slope of this so-called Van't Hoff plot gives the enthalpy of reaction, and the y-intercept[‡] gives the entropy of reaction. If $\Delta \bar{H}°$ is independent of temperature, the Van't Hoff plot is linear. Deviation from linearity indicates that $\Delta \bar{H}°$ is temperature dependent.[§] Though testing for linearity provides has conceptual appeal, the Van't Hoff plot requires data (in the form of K_{eq}) to be transformed logarithmically. As described in Chapter 2, such transforms negatively impact least-squares fitting, and can distort the values of fitted parameters (in this case, $\Delta \bar{H}°$ and $\Delta \bar{S}°$).

HOW CHEMICAL EQUILIBRIUM DEPENDS ON PRESSURE

As with the temperature analysis above, we can use pressure as an intensive variable to shift populations between chemical reactants and products. Since populations are related to the free energy of reaction (through the equilibrium constant), the pressure dependence of chemical equilibrium provides additional thermodynamic information about the reaction. Whereas temperature variation reveals the molar entropy difference between the reactants and products, pressure variation reveals the difference in molar volumes between the reactants and products.

To demonstrate this, we will start with a differential expression for free energy (Equation 5.18):

$$dG = -SdT + Vdp \tag{7.82}$$

We can use this expression to give differences in molar quantities:

$$d\Delta \bar{G} = -\Delta \bar{S}dT + \Delta \bar{V}dp \tag{7.83}$$

At constant temperature ($dT = 0$), Equation 7.83 simplifies to

$$d\Delta \bar{G} = \Delta \bar{V}dp \tag{7.84}$$

Rearranging gives

$$\Delta \bar{V}° = \left(\frac{\partial \Delta \bar{G}°}{\partial p} \right)_T \tag{7.85}$$

which is analogous to one of the thermodynamic identities in Chapter 5 (Table 5.1). Including standard state symbols in Equation 7.85 is justified because this equation is valid at all concentrations of reactants and products, including the standard state.[¶] Equation 7.85 states that the pressure dependence (at fixed temperature) of

---

[†] This temperature dependence will become clear when we consider populations from a statistical thermodynamic perspective in Chapters 10–12.

[‡] Infinite temperature on the $1/T$ scale.

[§] In addition, a nonlinear temperature dependence of the Van't Hoff plot indicates that $\Delta \bar{S}°$ depends on temperature. The temperature dependence of $\Delta \bar{H}°$ and $\Delta \bar{S}°$, which is a universal feature of protein folding, will be described in Chapter 8.

[¶] The standard state marker is often omitted from the molar reaction volume. Omitting this marker is only a problem if $\Delta \bar{V}$ depends on reactant and product concentrations.

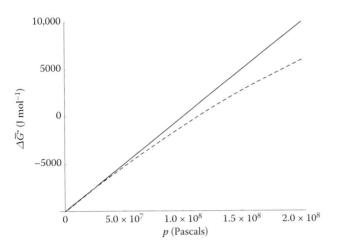

Figure 7.20 Pressure dependence of the free energy of reaction. A constant $\Delta \bar{V}$ (here, -100 mL mol^{-1}) leads to a linear increase in free energy with pressure (black line). The slope of the line is equal to $\Delta \bar{V}$. A constant compressibility difference (here 10^{-7} mL/Pa) leads to a quadratic free energy–pressure relationship (red dashed line). The slope of the line (from Equation 7.89, which assumes a reference pressure of zero) decreases with pressure; this reflects a decrease in magnitude of $\Delta \bar{V}$ due to the positive reaction compressibility.

the reaction free energy is the difference in molar volumes between the reactant and product, just as the temperature dependence (at fixed pressure) is the entropy difference.

To obtain an expression for the reaction free energy as a function of pressure, we need to integrate Equation 7.84:

$$\int_{\Delta G^\circ_{ref}} d\Delta \bar{G}^\circ = \int_{p_{ref}} \Delta \bar{V}^\circ dp \qquad (7.86)$$

If $\Delta \bar{V}^\circ$ is independent of pressure (i.e., if the compressibilities of the reactant and product are equal; see Problem 7.29) Equation 7.86 integrates to

$$\Delta \bar{G}^\circ(p) - \Delta \bar{G}^\circ_{ref} = (p - p_{ref})\Delta \bar{V}^\circ \qquad (7.87)$$

or

$$\Delta \bar{G}^\circ(p) = \Delta \bar{G}^\circ_{ref} + (p - p_{ref})\Delta \bar{V}^\circ \qquad (7.88)$$

In other words, the free energy varies linearly with pressure, with a slope of $\Delta \bar{V}^\circ$ (**Figure 7.20**). If the products have a larger molar volume than the reactants, increasing pressure will increase $\Delta \bar{G}^\circ$, favoring reactant. This is an intuitive manifestation of "Le Chatelier's principle," which states that a reaction will shift to accommodate stress.

If the compressibilities of the reactants and products differ, the free energy of reaction will have a nonlinear pressure dependence (Figure 7.20). For example, if the difference in compressibility is a constant (other than zero), a quadratic pressure dependence is obtained (Problem 7.31). If we choose a reference pressure of zero,[†] this quadratic dependence becomes particularly simple:

$$\Delta \bar{G}^\circ(p) = \Delta \bar{G}^\circ_{ref} + p\Delta \bar{V}_{ref} + \frac{p^2}{2}\left(\frac{\partial \Delta V}{\partial p}\right)_T \qquad (7.89)$$

In Equation 7.89 we need to specify a reference pressure for the molar volume change (i.e., $\Delta \bar{V}_{ref}$), because $\Delta \bar{V}$ depends on pressure as a result of the compressibility term.

[†] Although a reference temperature of zero might seem inconvenient since it is experimentally unattainable (except at absolute zero), there is often no difference between reference states of zero and 1 atm since the pressure dependences of chemical reactions are so modest.

Finally, it should be noted that for typical reactions in solution, very high pressures are required to significantly shift chemical equilibrium. This is because reaction volume changes are quite small. In Figure 7.20, where a $\Delta \bar{V}_{ref}$ of 100 mL mol^{-1} was assumed,[†] a pressure of 2×10^8 Pascals (2000 atm) is required to shift the free energy by 20 kJ mol^{-1}. This corresponds to a shift in equilibrium constant of about 3000-fold at room temperature; although this shift is within the measurable range, the pressure range demands specialized high-pressure equipment.

Problems

7.1 Demonstrate that molarity, partial density, and partial pressure (in the ideal gas limit) can be expressed using mole fractions as given in Table 7.1.

7.2 Derive an expression for partial density and for partial pressure (in the ideal gas limit) in terms of molarity.

7.3 For a two component A, B system with a fixed total mole number $n_A + n_B = 1$ mole, make a three-dimensional plot of molarity (z-axis) versus mole fraction (x-axis, 0 to 1) and total volume (y-axis, from 0 to 1 l). At a fixed total volume (e.g., 0.5 L), is molarity linear with mole fraction? At a fixed mole fraction (e.g., $x_A = 0.5$), is molarity linear with volume?

7.4 For the same two-component system as in the previous problem (binary mixture, fixed $n_T = 1$ mole), derive an expression for molarity as a function of mole-fraction, assuming the molar volumes of A and B are independent of composition (i.e., mole fraction), and have values $\bar{V}_A = 20$ mL mol^{-1}. Plot molarity (y-axis) as a function of mole-fraction of A (x-axis).

7.5 Derive Equation 7.6 by converting the equation $dU_i = TdS_i - pdV_i$ to an expression for dG_i using Legendre transforms.

7.6 Derive the first equality in Equation 7.12, giving the chemical potential of the ith species of an ideal gas mixture in terms molar volume.

$$\mu_i - \mu_i^\bullet = RT \ln \frac{\bar{V}_i^\bullet}{\bar{V}_i}$$

7.7 Derive the second equality in Equation 7.12, giving the chemical potential of the ith species of an ideal gas mixture in terms of molarities.

$$\mu_i - \mu_i^\bullet = RT \ln \frac{[i]}{[i]^\bullet}$$

7.8 Derive and plot expressions for the chemical potentials of two nonreactive species A and B in an ideal solution, using a mole fraction concentration scale, and assuming standard state values $\mu_A^* = 1000$ and $\mu_B^* = 4000$ J per mole. Identify these standard state chemical potentials on the plot.

7.9 Using the chemical potential expressions from Problem 7.8, calculate the total Gibbs energy for a fixed total number of moles, $n_T = n_A + n_B = 1$. Plot this total Gibbs energy versus x_A.

7.10 Starting with the Boltzmann entropy formula from Chapter 4 (Equation 4.46) for the statistical entropy of mixing of a binary ideal solution (from Problem 4.25),

---

[†] Though a value of 100 mL mol^{-1} is small on the absolute scale, it is quite large for simple chemical reactions. Keep in mind that atoms are neither being made nor destroyed in a chemical reaction; rather, they are rearranging as bonds form and break. Because the volume of atoms (so-called "van der Waals volumes") are essentially constant, changes in volume between reactants and products must result from packing defects among atoms and changes in solvation. These changes most likely to contribute to reactions involving large (macro) molecules, and in particular, wholesale folding–unfolding transitions (see Chapter 8).

rearrange to express entropy in terms of mole fractions, instead of N_A and N_B. Get it as close as you can to the classical expression for mixing entropy (Equation 7.23).

7.11 Assuming an ideal solution of three nonreacting species A, B, and C, derive the chemical potentials for all three species in terms of two mole fractions (x_A and x_B). Plot all three chemical potentials as a function of these two mole fractions. Over what region of the x_A-x_B plane are these plots defined?

7.12 Using the following vapor pressure data for water in a binary mixture with ethanol, use nonlinear least squares fit an "appropriate" equation to the data to give $p_{H_2O}(x)$. You should cast this equation so that the vapor pressure at zero water concentration is zero, as this clearly must be the case. This will be easiest if you fit your data as a function of the mole fraction of water, rather than the mole fraction of ethanol.

7.13 Use the expression you generated in Problem 7.12, that is, $p_{H_2O}(x)$, to generate an expression giving the chemical potential of water as a function of mole fraction.

7.14 For the lattice mixing problem with interactions, show algebraically (i.e., without graphs) that the probabilities of interactions account for all of the interactions, that is, they sum to one.

7.15 Show that for the regular solution lattice model with interactions, the molar enthalpy given by Equation 7.34 can be arranged to the form given by Equation 7.35.

7.16 Make a phase diagram of the binary two-dimensional interacting lattice fluid model as a function of mole fraction and $\Delta\varepsilon$ at 298.15 K. Do this by generating a three-dimensional plot of chemical potential versus the variables x_A and $\Delta\varepsilon$ as variables, and projecting the minima of the ΔG surface onto a 2-D$\times_A$ versus $\Delta\varepsilon$ plane.

7.17 For a given value of $\Delta\varepsilon$, does the point of transition between single- and two-phase behavior depend on temperature? Why or why not?

7.18 Calculate the temperature dependence of the phase diagram for the binary interacting lattice fluid model (assume 2d cubic structure) with a $\Delta\varepsilon$ value of $+750$ J mol interactions^{-1}. Do this as in Problem 7.16, by generating a three-dimensional plot of chemical potential versus the variables x_A and T as variables, and projecting the minima onto a 2Dx_A versus T plane. What is the temperature of transition between single- and two-phase behavior?

7.19 Calculate the numerical values of the Gibbs energy of mixing of 0.5 moles of molecule A with 0.5 moles of molecule B to a single phase and to a phase separated mixture, using the 2d cubic lattice interaction model and the parameters in Figure 7.14. For the single-phase calculation, assume a random distribution of A's and B's on the lattice. For the phase separated mixture, use mole fractions corresponding to the minima in the mixing surfaces, that is, $x_A = 0.165$ ($x_B = 0.835$) and $x_A = 0.835$ ($x_B = 0.165$). Assume that $T = 298.15$ K.

7.20 For the mixture in Problem 7.19, calculate the amount of each phase in the phase-separated mixture. Can the system produce any other ratio of these two phases? Why or why not?

7.21 For an arbitrary total mole fraction in the two-phase region ($0.165 \leq x_A \leq 0.835$) in Figure 7.14, calculate the relative amounts of each solution phase formed at equilibrium, the mole fraction of each phase, and the molar free energy.

7.22 For the binary two-dimensional square interacting lattice model, calculate the second derivative of ΔG_{mixing} with respect to x_A, and use it to identify the $\Delta\varepsilon$ value at the transition point from one-to-two phases (see Figure 7.13C).

7.23 Demonstrate that $\Delta \bar{G}^{\circ}{}_{rxn}$ is the free energy of reaction at standard state concentrations by directly evaluating Equation 7.52 for the reaction

$$A + 2B \rightleftharpoons C$$

at the standard state.

7.24 For the reaction described in Figure 7.16, derive the expression for the extensive Gibbs free energy as a function of mole extent of reaction ξ, as given in Equation 7.67. Assume the mechanism in Equation 7.66, the μ_i° values given in Figure 7.16 and 7.25, the total mole number $n_T = n_A + n_B = 1$ mole, and that molar volumes are identical and independent of composition.

7.25 Derive an explicit expression for $\Delta \bar{G}_{rxn}$ in terms of ξ, based on the assumptions in Figure 7.16, by differentiating the total Gibbs energy with respect to ξ.

7.26 For the reaction described in Figure 7.16, determine the equilibrium point of the reaction by setting the derivative of G, with respect to ξ (Problem 7.25), equal to zero. Show that the same value is obtained by analyzing Equation 7.56.

7.27 Derive an expression for G and calculate its derivative with respect to ξ, assuming for the reaction in Figure 7.16, assuming the molar volumes molar volumes of $\bar{V}_A^* = 0.02$ and $\bar{V}_A^* = 0.05$ L mol^{-1}, and assuming $\bar{V}_A$ and $\bar{V}_B$ do not depend on concentration (i.e., the molar volumes behave ideally).

7.28 For the hydrolysis reaction of ATP, that is,

$$ATP + H_2O \rightleftharpoons ADP + PO_4$$

the standard state reaction free energy is

$$\Delta \bar{G}^{\prime \circ} = -37 \text{ kJ mol}^{-1}$$

The odd standard state symbol (with the prime) is a "biochemist's standard state" where all concentrations are one molar except water, which is at a standard state of 55 M (and a hydrogen ion concentration of 10^{-7} M, corresponding to a pH of 7). Calculate the free energy of hydrolysis under cellular concentrations, where [ATP] = 10 mM, [ADP] = 0.6 mM, and $[PO_4^{2-}] = 20$ mM.

7.29 For the simple reaction

$$A \rightleftharpoons B$$

show that the molar volume of reaction (at fixed temperature) is given by the expression

$$\Delta \bar{V} = \Delta \bar{V}_{ref} + \int_{P_{ref}}^{P} \left(\frac{\partial \Delta \bar{V}}{\partial p} \right)_T dp$$

7.30 For the simple reaction in Problem 7.29, show that if the reactant and product have the same pressure dependence of their molar volumes at all pressures, that is,

$$\left(\frac{\partial \bar{V}_A}{\partial p} \right)_T = \left(\frac{\partial \bar{V}_B}{\partial p} \right)_T$$

then the molar free energy of reaction is linear in pressure. You can use the expression given in Problem 7.29 as a starting point.

7.31 For the simple reaction

$$A \rightleftharpoons B$$

derive an expression for the pressure dependence of the free energy of reaction (at constant temperature), assuming that the difference in compressibilities of the reactant and product is independent of pressure but is nonzero. Use an arbitrary reference temperature for your integration.

7.32 Setting the reference pressure to zero atmospheres, use your answer from Problem 7.31 to derive the simplified version of the reaction free energy as a function of pressure given in Equation 7.89.

Further Reading

Historical Development

Gibbs, J.W. 1876–1878. On the equilibrium of heterogeneous Substances. *Trans. Conn. Acad. Arts Sci.* III. 108–244 and 343–524.

Gill, S.J. 1962. The chemical potential. *J. Chem. Edu*. 39, 506–510.

Guggenheim, E.A. 1933. *Modern Thermodynamics by the Methods of Willard Gibbs*. Methuen & Co., London.

Guggenheim, E.A. 1967. *Thermodynamics: An Advanced Treatment for Chemists and Physicists*. North Holland Publishing, Amsterdam.

Lewis, G.N. and Randall, M. 1961. *Thermodynamics and the Free Energy of Substances*, 2nd edition. McGraw-Hill, New York.

Curvature, Convexity, and Phase Stability

Callen, H.B. 1985. *Thermodynamics and Thermostatistics*, Second edition. John Wiley & Sons, New York.

Reaction Thermodynamics

Beard, D.A. and Qian, H. 2008. *Chemical Biophysics: Quantitative Analysis of Chemical Systems*. Cambridge University Press, UK.

Hill, T. 2005. *Free Energy Transduction and Biochemical Cycle Analysis*. Dover Books, Mineola, NY.

CHAPTER 8

Conformational Equilibrium

Goals and Summary

The goal of this chapter is to apply reaction thermodynamics to equilibrium conformational transitions of proteins. We will begin with a description of the structural variables associated with both proteins and nucleic acids, to underscore the structural complexity of macromolecular conformational transitions. Despite this high level of complexity, a surprisingly good (and useful) representation of conformational equilibrium can be obtained for many proteins using a model with just two thermodynamic states, corresponding to fully folded ("native") and unfolded ("denatured") ensembles. We will analyze conformational equilibrium of unimolecular protein folding reactions using this "two-state" model. To experimentally quantify conformational transitions, and to connect to energies, entropies, and the like, equilibrium conformational transitions (informally referred to as "melts") are required. The general properties of such melts will be discussed, along with some subtleties that are important for data analysis.

Using the concept of two-state melts, several types of protein unfolding transitions will be described, including thermal unfolding transitions, denaturant unfolding transitions, and pressure unfolding transitions. In addition to providing access to free energies and equilibrium constants, these various methods allow measurement of different "derivative" quantities, including enthalpy, entropy, heat capacity, denaturant and cosolute interactions, charge interactions, and volume changes.

In principle, the classical thermodynamic equations developed in Chapter 7 can be used to describe any type of reaction, regardless of reactant and product structures, as long as a mechanism can be written in terms of species and their stoichiometric coefficients. In practice, there are two further requirements for measuring and analyzing reaction thermodynamics. First, the reaction must be a reversible one—that is, the system can move from product back to reactant as well as from reactant to product, as required for chemical equilibrium. Second, the experimental method used to monitor the position of equilibrium must be able to distinguish the contributions of each species to whatever is being measured. If some (or all) of the species in a proposed mechanism look alike, then reaction among these species will not be detected. For complex reactions, experimental resolution of each species can be challenging.

For simple chemical reactions, where one or a few bonds are made, broken, or rearranged, mechanisms are easy to write, since there are a limited number of species, and they are easy to keep track of. Such reactions fit nicely into the framework of reaction thermodynamics developed in Chapter 7. Surprisingly, this framework can be extended to reactions that would seem to be much more complicated: order-disorder transitions in macromolecules. In this chapter we will discuss a beautiful example of one such a reaction: the "folding–unfolding" reactions of proteins. Though proteins have many hundreds or thousands of atoms that are organized into precise structures, by applying the reaction thermodynamic framework from Chapter 7, we will determine, their folding–unfolding transitions can often be treated with very simple mechanisms.

MACROMOLECULAR STRUCTURE

Describing the structure of complex macromolecules requires a different viewpoint than that used to describe the simpler systems discussed so far. Simple nonbonded systems (e.g., ideal gases) can be described by giving the position of each atom, either as a set of continuous coordinates, or as occupancies in a lattice. In such cases, there is little internal structure to worry about. In contrast, for large macromolecular systems where many atoms are joined together through covalent bonding, the description of structure shifts from external to internal coordinates. These internal coordinates (mainly bond "torsion" angles) define the positions of bonded atoms relative to each other within a molecule.

Though the lengths of covalent bonds and angles between connected triples of bonded atoms are, to a good approximation, fixed by local chemistry, rotation about single bonds is quite facile, and is the main variable that determines macromolecular structure. These internal rotation coordinates are referred to as "dihedral" (or sometimes "torsion") angles, because the angle is given by two planes that both include the bond on which the dihedral centers, as well as adjacent bonded atoms.

For proteins, each residue has two backbone dihedral angles that can vary substantially.[†] These two angles, referred to as ϕ and ψ, are centered on the bond between the amide N and adjacent Cα atoms, and on the bond between the Cα and carbonyl C atoms, respectively (**Figure 8.1A**). Note that although ϕ and ψ can vary, they are restricted by neighboring atoms. The allowed configurations adopted by the ϕ and ψ dihedral angles determine the type of secondary structure (α-helix, β-sheet, tight turn) adopted, and collectively, these angles determine the overall fold architecture. In addition, for all but the smallest side chains, there are side-chain dihedrals that determine internal packing in the folded state.

Nucleic acids have even more backbone conformational degrees of system per residue (**Figure 8.1**), reflecting the larger number of bonds per residue, although the side chain nucleobases are less variable due to their rigid planarity.

Because each residue contributes conformational degrees of freedom, the long strings of residues that make up proteins and nucleic acids have an astronomical

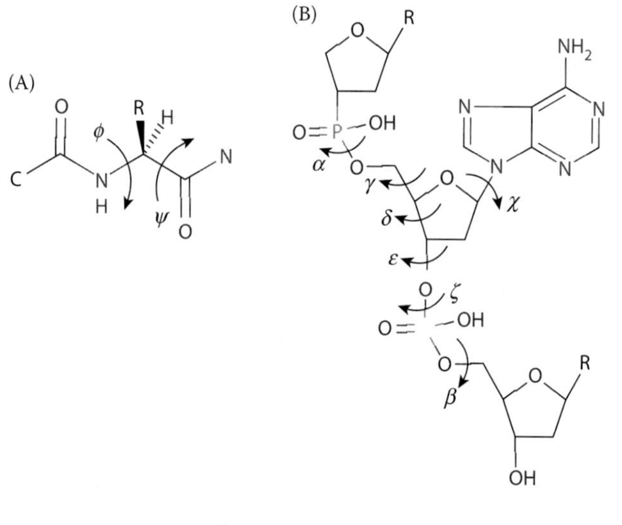

Figure 8.1 Backbone dihedral angles in polypeptides and in DNA. (A) In polypeptides, there are just two backbone dihedral angles (ϕ and ψ, see Box 8.1) for each residue, and a variable number of side-chain dihedrals depending on the residue. (B) In both DNA and RNA, there are six unconstrained backbone dihedrals for the phosphoanhydride backbone (note that β is displaced by one residue for clarity). In addition, there are four constrained dihedrals for each (deoxy) ribose ring, leading to several ribose conformations (the most common being 2′ endo and 3′ endo). Further, there is one dihedral for the glycosidic bond attaching each nucleobase; for purines, there are two allowed values, leading to the *syn* and *anti* conformations, but pyrimidines are restricted to the *anti* conformation.

___

[†] The dihedral centered on the peptide bond does not vary significantly due to double-bond character.

number of conformations (where each conformation can be thought of as a unique set of dihedral values). Despite these numbers, many proteins and some nucleic acids fold into well-ordered structured (or "native") states that have little conformational variation. Native states of proteins show hydrogen-bonded secondary structures (α-helices and β-sheets, connected by tight turns), burial of hydrophobic groups within the interior to form well-packed hydrophobic cores. As a result of this intramolecular packing and hydrogen bonding, native proteins have compact, globular shape. Native states of nucleic acids include the well-known Watson–Crick hydrogen bonded base pairs and double helical structures. Though these structures are dominant in chromosomal duplex DNA; RNA structures also include loops, unconventional base pairs, triples, and quadruplexes, long-range tertiary interactions, and can be quite compact.

Given the very large number of conformations available, one might expect the analysis of conformational equilibrium in proteins and nucleic acids to be quite complex. Models with separate energy terms corresponding to each conceivable conformation would be cumbersome, especially using the classical reaction thermodynamic approach described above. Although the statistical thermodynamics described in later chapters is much better at representing complexity, the relative energy values of each conformation are not known in advance. Even with the organizing principles of statistical thermodynamics, little prediction can be made with a model that includes of all possible states, but not their relative energies. In principle, such energies can determined through experimental measurement, although in practice, measurable quantities that reflect macromolecular structure are too low in resolution to resolve large numbers of structures at once. Fortunately, many macromolecular folding transitions, especially those of single-domain proteins, are much simpler than their structural complexity might suggest, facilitating simple analysis using reaction thermodynamics.

A SIMPLE TWO-STATE MODEL FOR CONFORMATIONAL TRANSITIONS

To good approximation, a number of conformational transitions of single-domain proteins[†] can be described using just two thermodynamic states. These states include a folded (or "native," N) state, and a broad collection of unfolded configurations, which together make up a "denatured" (D) state[‡]. With these two thermodynamic states, unimolecular folding reactions can be represented with the scheme

$$D \underset{}{\overset{K_{fold}}{\rightleftharpoons}} N \qquad\qquad (8.1)$$

where

$$K_{fold} = \frac{[N]}{[D]} \qquad\qquad (8.2)$$

The beauty of the simple two-state mechanism is that it allows deterimination the thermodynamic quantities K and ΔG° by simply measuring the relative concentrations of the two macromolecular species N and D simultaneously.[§] However, as

---

[†] Multidomain proteins and many structured nucleic acids tend to unfold via more complex mechanisms.

[‡] The terns "native" and "denatured" have historical origins from the early days of macromolecular biochemistry. In these early studies, materials were extracted from biological samples and tissues, being careful not to disrupt their "native" structures and activities. Studies of conformational equilibria were then carried out by disrupting (i.e., "denaturing") these native structures.

[§] In this context, "simultaneously" means "under the same conditions," rather than "at the same time" Because our goal here is to describe equilibrium thermodynamics, there should be no time dependence.

described in the next section, the *simultaneous* requirement can pose an experimental challenge; this challenge is resolved by measuring a "melt," in which the native state is converted to the denatured state.

SIMULTANEOUS VISUALIZATION OF *N* AND *D*

In principle, measuring just two concentrations (*N* and *D*) should be an easy task. All that is needed is a technique that can "see" the difference between native and denatured macromolecules in solution. There are a number of spectroscopic[†] techniques to detect macromolecules and can tell the difference between structure and lack of structure. The simplest spectroscopic method for studying macromolecules in solution is absorbance spectroscopy. For proteins, the side chains of tyrosine and tryptophan strongly absorb UV light in the range of 270–290 nm, and for some proteins this absorbance changes on unfolding, thereby reporting on the unfolding transition. Nucleic acids almost always show absorbance changes upon unfolding (due to the nucleobase chromophores). Though many tryptophan residues in proteins don't change their absorbance on unfolding, almost all buried tryptophans change their fluorescence on folding, providing a sensitive probe of the folding reaction.

Another spectroscopic method that is good at detecting the folding of proteins is circular dichroism (CD) spectroscopy, which results from the difference in absorbance of left and right circularly polarized light. CD in the wavelength range of 200–230 nm detects secondary structure. In particular, α-helix display strong negative "ellipticity"[‡] at these wavelengths, due to the regular, repetitive geometry of neighboring backbone amide chromophores (see Cantor and Schimmel, 1980). Denatured proteins display little ellipticity in this region (**Figure 8.2**). Proteins with β-sheet structure also have significant negative ellipticity at these wavelengths.

Although these methods can tell the difference between *N* and *D*, measuring the amounts of both species in solution in a single sample can be difficult. This

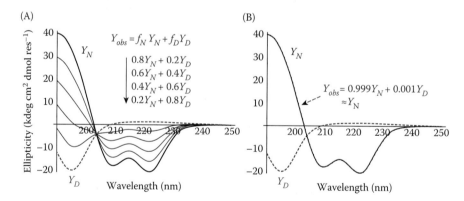

Figure 8.2 CD spectra of a native α-helical and denatured protein (black and gray) and linear combinations of those two spectra (red). The CD spectrum of α-helical secondary structure shows two negative peaks at 222 and 208 nm, and a positive peak at 190 nm. Denatured polypeptide shows little CD (ellipticity) above 200 nm, but has a negative peak around 195 nm. Panel A shows spectra of mixtures from the measurable equilibrium range (red; equilibrium constants from 4 to 0.25, corresponding to 80% to 20% *N*, respectively). Panel B shows a spectrum from a mostly folded mixture (red; $K_{fold} = 10^3$, which is not atypical for a small globular protein); the contribution of the 0.1% denatured protein to the spectrum is not easily resolved from that of the *N* state (black). Spectra were generated by summing one positive and two negative Gaussian functions (means of 222, 208, and 190 nm; standard deviations of 5, 5, and 8 nm; amplitudes of −20, −20, and +40 ellipticity units, respectively).

[†] Such methods use light to probe molecular structure.

[‡] Ellipticity is a measure of CD strength, and is proportional to the amount of structured molecules present.

simultaneous detection is easy when the equilibrium constant is around one (Figure 8.2A), where the sample consists of equal concentrations of folded and proteins. Such a method can also likely simultaneously determine N and D in samples that are 80% folded protein (20% unfolded). The sensitivities of CD spectra (ellipticity vs. wavelength) to the relative concentration of N and D is shown in Figure 8.2. In this range, spectra of mixtures that contain significant spectral features of both N and D. These spectra fall in between spectra of the pure native and denatured states (solid black and gray dashed lines, Figure 8.2). If, in contrast, the equilibrium constant for folding is significantly greater (or less) than one, the spectrum of the corresponding equilibrium "mixture" will be indistinguishable (within error) from the relevant limiting spectrum (either N, as in Figure 8.2B, or D).

We will refer to the range over which the two populations can simultaneously be detected as the "measurable equilibrium" range.[†] The limits of this range will depend on the method, the instrument, and the sample, but a reasonable rule of thumb is

$$0.1 \leq K_{fold} \leq 10 \tag{8.3}$$

At body temperature (310 K), this corresponds to a reaction free energy range (in kcal mol^{-1}) of

$$1.42 \geq \Delta \bar{G}°_{fold} \geq -1.42 \tag{8.4}$$

The red CD spectra in Figure 8.2A are within this range ($0.25 \leq K_{fold} \leq 4$), whereas the red spectrum in Figure 8.2B is not ($K_{fold} = 10^3$).

Since most proteins with *structured* native states are not biologically active in their unfolded states,[‡] folding equilibrium constants are biased toward the native state. Experimentally determined folding free energies are often in the range[§] of -5 to -10 kcal mol^{-1}, giving equilibrium constants of 10^3 or greater ($f_N \geq 0.999$; Problem 8.4). For such proteins, the folding equilibrium constant must be decreased to bring it into the measurable equilibrium range. Nearly all measurements of macromolecular folding energetics are obtained by generating unfolding transitions (or "melts") in which conditions are varied to shift the K_{fold} into the measurable range.

THE THERMAL UNFOLDING TRANSITION AS A WAY TO DETERMINE K_{FOLD} AND $\Delta G°$

As described above, to analyze the thermodynamics of unfolding, we need a way to vary the equilibrium constant. At first glance, "variation" and "constant" seem somewhat at odds. Certainly under a specified set of conditions, the equilibrium constant is fixed to a specific value.[¶] However, the value of this constant depends on the intrinsic variables of the system (T, p, and solution composition), which we can adjust externally. If our goal is to bring the equilibrium constant into a measurable range, we can usually find at least one way to bring an appropriate population shift

---

[†] "Measurable" is the key word here; equilibrium is still achieved outside this range, but we can't accurately measure it.

[‡] In fact, when unfolded states are significantly populated, they can aggregate. Cellular protein aggregation is associated with a number of diseases, often with neurodegenerative symptoms. Other proteins never fold, but function in their "disordered" states. These "intrinsically disordered" proteins (IDPs) tend to be more soluble, avoiding aggregation.

[§] Though this range may be biased by the data set, which requires small, well-behaved proteins, with well-resolved baselines.

[¶] This must be true, because the equilibrium constant is uniquely determined by the free energy. Like the free energy, the equilibrium constant is a state function.

using one of these three variables.† Moreover, determining the sensitivity of the equilibrium constant (and the free energy) to T, p, and solution composition reveals "derivative" quantities such as enthalpy, heat capacity, and volume change, and can provide mechanistic insight into structural features like cavity formation, surface area exposure, and the state of ionization. Thus, by measuring an equilibrium unfolding transition, we gain access to this type of information.

Here, we will introduce the simplest (experimentally, if not conceptually) type of unfolding transition, the *thermal* unfolding transition. By increasing the temperature above physiological values, the equilibrium constants of most proteins can be shifted toward the denatured state. An example of such a transition is shown in **Figure 8.3A**. At low temperature, the protein is nearly fully folded, and a spectroscopic probe of folding (referred to generically as Y_{obs}) appears insensitive‡ to temperature changes. In this "native baseline region," the observed signal is that of the native state, Y_N. However, once the equilibrium constant shifts to the measurable equilibrium range, the spectroscopic signal changes sharply with temperature, transitioning rather abruptly toward the denatured state. Above this transition region, Y_{obs} converges to the baseline signal Y_D and becomes insensitive§ to temperature. The center of this transition (where the sample is half N and half D) is referred to as a midpoint temperature, T_m.

In short, the thermal melt brings K_{fold} into a measurable range near T_m. The question is, how do we extract K_{fold} from a set of Y_{obs} values? The key idea, which was introduced in Figure 8.2A, is that the signal Y_{obs} is linearly proportional to the amount of each state present. This is most easily expressed as fractional populations f_N and f_D:¶

$$Y_{obs} = f_N Y_N + f_D Y_D \tag{8.5}$$

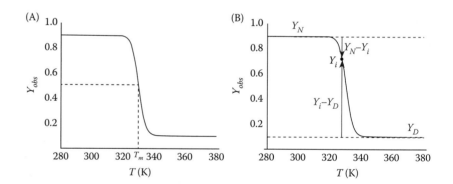

Figure 8.3 A thermal protein folding transition with constant reaction enthalpy and entropy. (A) The observed signal (solid black line) shows a sharp transition between native and denatured baselines. Here, the folding enthalpy is set to −100 kcal mol⁻¹, and the folding entropy is 0.303 kcal mol⁻¹ K⁻¹. This fixes the T_m value to 330 K (black dashed line, see Equation 8.20). Y_N and Y_D are set to 0.9 and 0.1, respectively. (B) The baselines play a key role in determining the value of fractional populations and equilibrium constants. For an arbitrary point Y_i within the transition, the fractional populations f_N and f_D and K_{fold} can be described by the red arrowheads connecting the point to the extended native and denatured baselines Y_N and Y_D (red dotted lines).

---

† Really, more than three, since there are many ways to change solution composition, changing salt, pH, denaturant, or osmolyte concentrations.

‡ Relatively insensitive, but often not completely insensitive. Baseline spectroscopic signals often show modest, roughly linear dependence on external variables such as temperature and denaturant concentration.

§ Again, modest linear dependence is often seen for Y_D.

¶ At equilibrium, these fractional populations, which range from zero to one, can be thought of as probabilities, as long as the sample is large enough for good averaging. To me, fractional population seems more natural and concrete when discussing a finite sample than a probability of being in a certain state.

In Equation 8.5, Y_{obs} is a population-weighted average of signals that result from the native and denatured states (Y_N and Y_D). This is equivalent to invoking a type of Beer's law for the N and D states (Y_{obs} proportional to concentration of each), and keeping the total concentration ($[N] + [D]$) constant.

A simple geometric way to connect Y_{obs} to K_{fold}

One simple way to relate the position of the unfolding transition to the equilibrium constant involves a point-by-point comparison of the observed signal to the baselines (Problem 8.3). Though the point-by-point method is not the best approach to obtain thermodynamic parameters for unfolding (the best way is to fit a modified version of Equation 8.5, to the data; see below), it is a convenient way to think about what is going on in the transition.

In the point-by-point method, we use f_N (and f_D) to connect Y_{obs} to K_{fold}. First, we can arrange Equation 8.5 to get an expression for f_N in terms of the spectroscopic variables (Y_{obs}, Y_N, and Y_D) by invoking a conservation of mass constraint:

$$f_N + f_D = 1 \tag{8.6}$$

Equation 8.6 can be regarded as another way to state the reaction mechanism. Solving for f_D and substituting into Equation 8.5 gives

$$\begin{aligned} Y_{obs} &= f_N Y_N + (1 - f_N) Y_D \\ &= Y_D + f_N Y_N - f_N Y_D \\ &= Y_D + f_N (Y_N - Y_D) \end{aligned} \tag{8.7}$$

thus,

$$f_N = \frac{Y_{obs} - Y_D}{Y_N - Y_D} \tag{8.8}$$

By a similar manipulation (Problem 8.1), we can express f_D as

$$f_D = \frac{Y_N - Y_{obs}}{Y_N - Y_D} \tag{8.9}$$

Equations 8.8 and 8.9 have a similar form. The denominator of each is simply the difference between the native and denatured signals (the baselines). In Figure 8.7B, this is the distance between the two dashed red lines (numerically, 0.9–0.1). The numerator of each is the distance between the observed signal and one or the other baselines. For f_N, this difference ($Y_{obs} - Y_D$) is the line segment that connects the denatured baseline to the observed signal (red arrow pointing up) and for f_D, this difference ($Y_N - Y_{obs}$) is the line segment that connects the native baseline to the observed signal (red arrow pointing down). In other words, the fractions folded simply correspond to the distance of the measured point from the opposite baseline.

We can use these geometric descriptions of the fractional populations to give a similar description of the equilibrium constant. In terms of concentrations, the fractional populations are

$$f_N = \frac{[N]}{[M]_T} = \frac{[N]}{[N] + [D]} \tag{8.10}$$

and

$$f_D = \frac{[D]}{[M]_T} = \frac{[D]}{[N] + [D]} \tag{8.11}$$

where $[M]_T$ is the sum of all of the protein species.[†] If we form the ratio of these two quantities, we get the equilibrium constant:

$$\frac{f_N}{f_D} = \frac{[N]/([N]+[D])}{[D]/([N]+[D])}$$
$$= \frac{[N]}{[D]} \quad (8.12)$$
$$= K_{fold}$$

Substituting the geometric descriptions for the fractional populations (Equations 8.8 and 8.9) into Equation 8.12 gives

$$K_{fold} = \frac{f_N}{f_D} = \frac{(Y_{obs}-Y_D)/(Y_N-Y_D)}{(Y_N-Y_{obs})/(Y_N-Y_D)} = \frac{Y_{obs}-Y_D}{Y_N-Y_{obs}} \quad (8.13)$$

Equation 8.13 states that the equilibrium constant is the ratio of the red arrows in **Figure 8.3B**, that is, the extent to which the transition has advanced from from one baseline to the other. For this reason we need to sample enough points in the baseline to know where to start these arrows to define each baseline. As such, the folding transition should be thought of as containing three critical regions: the native baseline, the transition region, and the denatured baseline. Though the direct fitting method does not involve calculating the "arrows" in Figure 8.3C, the baselines are still essential for the same reason: forming an equilibrium constant from progress through the transition requires knowledge of where the transition started, and where it is going.

An equation to fit unfolding transitions

Though Equation 8.12 provides a simple relation between K_{fold} and the fractional populations, our main goal is to introduce K_{fold} into Y_{obs} (Equation 8.5), which directly describes unfolding transitions (and can be used to fit folding transition data). Thermodynamics can be introduced through the fractional populations. By dividing the numerator and denominator of f_N and f_D by $[D]$, we replace molar concentrations with K_{fold}:

$$f_N = \frac{[N]/[D]}{[N]/[D]+[D]/[D]} = \frac{K_{fold}}{K_{fold}+1} \quad (8.14)$$

$$f_D = \frac{[D]/[D]}{[N]/[D]+[D]/[D]} = \frac{1}{K_{fold}+1} \quad (8.15)$$

Substituting these expressions into Equation 8.5 gives Y_{obs} in terms of K_{fold}:

$$Y_{obs} = \frac{K_{fold}}{1+K_{fold}}Y_N + \frac{1}{1+K_{fold}}Y_D$$
$$= \frac{Y_D+Y_N K_{fold}}{1+K_{fold}} \quad (8.16)$$

Equation 8.16 connects the observed signal with a thermodynamic quantity, the equilibrium constant. However, Equation 8.16 cannot be fitted directly to thermal folding data because it does not have an explicit temperature dependence. This equation providing a means both to represent unfolding transitions and to fit experimental data.

[†] If we had a more complicated model with additional states, Equations 8.10 and 8.11 would be modified by including additional terms to the denominator.

An explicit temperature dependence can be obtained from K_{fold} using the formula $\Delta \bar{G}° = -RT \ln K_{eq}$ (Chapter 7). Rearranging and exponentiating gives

$$K_{fold} = e^{-\Delta \bar{G}°/RT} \tag{8.17}$$

Though it is possible that the inverse temperature term in the exponent of Equation 8.17 accounts for the entire temperature variation of K_{fold} (see Problem 8.2), this simplistic temperature model is unlikely.[†] Rather, $\Delta \bar{G}°$ is likely to depend on temperature. This temperature dependence is given through the equation $\Delta \bar{G}° = \Delta \bar{H}° - T \Delta \bar{S}°$. Inserting this expression into Equation 8.17 gives

$$K_{fold} = e^{-(\Delta \bar{H}° - T \Delta \bar{S}°)/RT}$$
$$= e^{-\Delta \bar{H}°/RT} e^{\Delta \bar{S}°/R} \tag{8.18}$$

Substituting Equation 8.18 into Equation 8.16 provides a useful connection between the signal change during an unfolding transition and the underlying thermodynamics of the transition:

$$Y_{obs} = \frac{Y_N e^{\Delta \bar{S}°/R} e^{-\Delta \bar{H}°/RT} + Y_D}{1 + e^{\Delta \bar{S}°/R} e^{-\Delta \bar{H}°/RT}} \tag{8.19}$$

Equation 8.19 expresses the observed spectroscopic signal in terms of the enthalpy and entropy of the folding transition. As with any reaction that is driven by heating, a conformational transition that shifts from native to denatured upon temperature increase is endothermic; that is, $\Delta \bar{H}°$ is positive for *unfolding*. Thus, in the *folding* direction, $\Delta \bar{H}°$ is negative. Because the free energy near the transition is small (zero at T_m), $\Delta \bar{S}°$ (in the folding direction) must also be negative in the folding direction to cancel $\Delta \bar{H}°$ in Equation 8.4. These values are consistent with expectation for a disorder–order transition, where bond formation in the native state decreases enthalpy, and conformational restriction decreases entropy (**Figure 8.4**). Two important questions are, what is the value of $\Delta \bar{S}°$ and $\Delta \bar{H}°$, and do *they* depend on temperature?

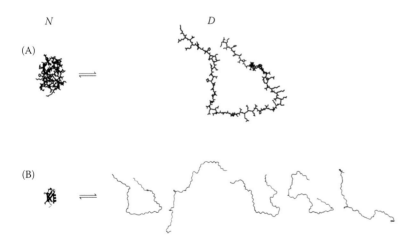

Figure 8.4 Illustration of an unfolding transition for ubiquitin, a small (76 residue) protein. Native (left) and denatured conformation (right) are shown in (A) all-atom representation (hydrogen atoms are omitted for clarity) and in (B) ribbon representation. In (B), multiple unfolded conformations are shown as a reminder of the large number of conformational states available (and high entropy) to the denatured state. Note that for each equilibrium, the same scale is used for the native and denatured conformations, emphasizing the compactness of the native state, and the expanded structures of denatured conformations. Unfolded states were generated by computer simulation by Dr. Rohit Pappu (Washington University).

[†] This simplistic scenario would require the reaction entropy to be zero. A reaction as complex as protein folding is unlikely to proceed without entropy change.

A SIMPLE MODEL FOR THERMAL TRANSITIONS: CONSTANT $\Delta \bar{H}^\circ$ AND $\Delta \bar{S}^\circ$

In this section we will evaluate thermal unfolding transitions based on Equation 8.19, assuming that ΔS° and ΔH° are nonzero but are independent of temperature (resulting in a linear temperature dependence on $\Delta \bar{G}^\circ$). The thermal transition in Figure 8.3 was generated using this model. Though this treatment of ΔS° and ΔH° as temperature independent misses an important aspect of the protein folding reaction,[†] it clearly illustrates the relationship among the thermodynamic quantities Y_{obs}, K_{fold}, $\Delta \bar{G}^\circ$, $\Delta \bar{H}^\circ$, and $\Delta \bar{S}^\circ$.

Within this model, the values of $\Delta \bar{H}^\circ$ and $\Delta \bar{S}^\circ$ influence two distinct features of the folding transition. First, the two terms together determine the midpoint temperature. Since $\Delta \bar{G}^\circ = 0$ at T_m, we can express T_m as

$$T_m = \frac{\Delta \bar{H}^\circ}{\Delta \bar{S}^\circ} \tag{8.20}$$

The second feature of the folding transition influenced by ΔH° and ΔS° is its slope. To understand this connection, we will take the temperature derivative of Y_{obs}. Assuming the two baselines are independent of temperature for simplicity, this slope is

$$\frac{dY_{obs}}{dT} = (Y_N - Y_D)\frac{K}{(1+K)^2}\frac{\Delta \bar{H}^\circ}{RT^2} \tag{8.21}$$

This slope is composed of three terms. The first term is just the "amplitude" of the unfolding transition, that is, how much the signal changes (i.e., ΔY_{obs}); this term is more related to spectroscopy than to thermodynamics. The second term on the right-hand side of Equation 8.21 is closely related to populations. By separating the square in the denominator, the second term can be recognized as a product of the fractional populations:

$$\frac{K}{(1+K)^2} = \frac{K}{1+K} \times \frac{1}{1+K}$$
$$= f_N f_D \tag{8.22}$$

This second term peaks at T_m at a value of 0.25 (because $f_N = f_D = 0.5$ at T_m), and decays to zero in the two baseline regions (**Figure 8.5A**). We will refer to this term as a "spread over states."[‡] The spread over states term appears in a variety of conformational equilibria, such as ligand-binding reactions, and will show up again in our discussion of scanning calorimetry (below). For the two-state conformational transition Equation 8.22 represents the extent to which the population is spread out over N and D.

Though the maximum value of the spread term is always the same, its width depends on $\Delta \bar{H}^\circ$ (Figure 8.5B). High enthalpy values narrow the peak, whereas low enthalpy values broaden it. This results from a temperature dependence of the equilibrium constant (and thus the populations) on $\Delta \bar{H}^\circ$ (see the Van 't Hoff equation, Chapter 7). In addition, $\Delta \bar{H}^\circ$ affects the overall slope of the transition directly through the third term in Equation 8.21 (Figure 8.5C): a large $\Delta \bar{H}^\circ$ value

---

[†] Specifically, in the simple analysis in this section, the heat capacities of N and D are implicitly assumed to be identical. In practice, the D always has a higher heat capacity than N. The consequences of this will be developed in subsequent sections.

[‡] Note that our spread over states is often referred by others as a "fluctuation" term. We avoid this term, because fluctuation can be misinterpreted as expressing dynamics. Though all states described by equilibrium thermodynamics must be kinetically accessible on the timescale of measurement, the values of the rates of interconversion are otherwise irrelevant for the spread term.

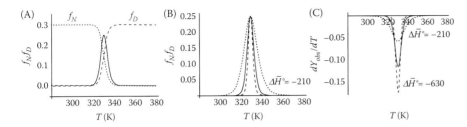

Figure 8.5 Population spreading through a thermal two-state transition. (A) The product of f_N and f_D (black line, Equation 8.22) for the two-state unimolecular transition in Figure 8.3 (T_m of 330 K, $\Delta \bar{H}°$ of −420 kJ mol⁻¹). Red dotted and dashed lines show individual f_N and f_D profiles, but are scaled to overlay on the same plot. (B) Comparison of population spreads for different enthalpies, but the same T_m values (dotted line: $\Delta \bar{H}° = -210$ kJ mol⁻¹; solid line, $\Delta \bar{H}° = -420$ kJ mol⁻¹; dashed line, $\Delta \bar{H}° = -630$ kJ mol⁻¹). (C) Resulting slopes (Equation 8.21) for the transitions described in (B), with $Y_N − Y_D$ set to one. Around T_m, the magnitude of the slope (here negative) is proportional to the reaction enthalpy: the larger the $\Delta \bar{H}°$, the greater the slope, and thus the sharper the transition.

increase the height of derivative peak, whereas small enthalpy values decrease the height.

When the three terms in Equation 8.21 are combined, the overall result is that a large enthalpy value results in a sharp transition (high slope over a narrow region), whereas a small enthalpy value results in a broad transition (low slope, spread out over a wide region). This is shown in **Figure 8.6A**, where three conformational transitions are shown with the same T_m values but variable $\Delta \bar{H}°$. The transition with an enthalpy of 210 kJ mol⁻¹ is spread over about 30 degrees, whereas the transition with an enthalpy of 630 kJ mol⁻¹ is narrowed to about 10 degrees.

Although the argument in the previous paragraph makes it sound like the enthalpy determines the steepness of the transition, a similar argument can be made for the entropy. The three transitions in Figure 8.6A also have significant differences in $\Delta \bar{S}°$ values. A rearranged form of Equation 8.20 makes this clear:

$$\Delta \bar{H}° = T_m \Delta \bar{S}° \qquad (8.23)$$

Since all three transitions in Figure 8.6A have the same T_m values, the variation in $\Delta \bar{H}°$ must be compensated by variation in $\Delta \bar{S}°$. In other words, if T_m is specified,[†]

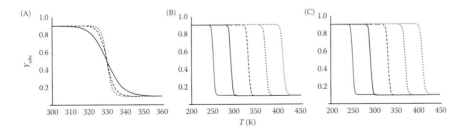

Figure 8.6 The effects of $\Delta H°$ and $\Delta S°$ on a simple thermal folding transition. (A) Three transitions with the same T_m (330 K), but different $\Delta H°$ values (210, 420, and 630 kJ mol⁻¹ in solid, dashed, and dotted lines) and $\Delta S°$ values (630, 1270, and 1815 J mol⁻¹ K⁻¹). (B) Transitions with a common $\Delta S°$ (1270 J mol⁻¹ K⁻¹) but different $\Delta H°$ (314–523 kJ mol⁻¹), and therefore different T_m values. (C) Transitions with a common $\Delta H°$ (420 kJ mol⁻¹) but different values of $\Delta S°$ (1670–1000 J mol⁻¹ K⁻¹ from left to right) and therefore different T_m values. In all three panels, the dashed black curves are the same transitions. In panels (B) and (C), T_m values match for each pair of curves. Though there are subtle differences in shapes of the extreme curves (red), these differences are barely discernable by the eye.

[†] This is equivalent to specifying the free energy of reaction at a particular temperature (thus, specifying the equilibrium constant).

$\Delta \bar{S}°$ and $\Delta \bar{H}°$ are not both free parameters—if one is doubled, the other must also double in order for T_m to remain fixed.

The only way to examine how $\Delta \bar{H}°$ affects the unfolding transition without simultaneous variations in $\Delta \bar{S}°$ (and vice versa) is to allow T_m to vary as well. This is shown in **Figure 8.6B**, where $\Delta \bar{S}°$ is held constant and $\Delta \bar{H}°$ is varied (along with T_m), and in **Figure 8.6C**, where $\Delta \bar{H}°$ is held constant and $\Delta \bar{S}°$ is varied (along with T_m). In contrast to Figure 8.6A, there is little variation the slope of the transition from low to high $\Delta \bar{H}°$ (Figure 8.6B) and $\Delta \bar{S}°$ values (Figure 8.6C), and little difference between these two approaches to changing the free energy.

FITTING CONFORMATIONAL TRANSITIONS TO ANALYZE THE THERMODYNAMICS OF UNFOLDING

Just as with simple chemical systems (e.g., gases, liquids, and solids), thermodynamic data on macromolecular conformational transitions are obtained as discrete points, not as smooth curves and functions. For thermal transitions, such data are in the form of temperature and Y_{obs}. Given a data set of this type, Equation 8.19 can be directly fitted to estimate thermodynamic unfolding parameters, and to test the applicability of the two-state model. In addition to estimating thermodynamic parameters, fitting Equation 8.19 requires estimation of parameters describing the native and denatured baselines (Y_N, Y_D, and any associated temperature dependence). As long as both baseline regions are adequately defined, these baseline parameters should be well-determined by the data, and should not compromise estimates of thermodynamic parameters, which are of primary interest.

The explicit thermodynamic variables of Equation 8.19 are $\Delta \bar{H}°$ and $\Delta \bar{S}°$. Though these two parameters both provide complementary insights into thermodynamic mechanism, in the context of fitting, it is easier to use the thermal unfolding midpoint, T_m and *one* of the two variables $\Delta \bar{H}°$ or $\Delta \bar{S}°$ (Figure 8.3) as adjustable parameters. As described above, there are only two independent variables among these three parameters. Fitting the T_m has the advantage of being the most conspicuous feature of a conformational transition,[†] and thus, a good initial guess for fitting T_m can be made without much effort. Likewise, a fitted T_m value is easy to compare against the primary data as a check on the success of a fitting. The choice of $\Delta \bar{H}°$ versus $\Delta \bar{S}°$ as a second fitting parameter is less clear-cut. Because enthalpy changes can directly be measured by calorimetry, to fitting $\Delta \bar{H}°$ rather than $\Delta \bar{S}°$ may be preferable, though Figure 8.6 shows that they both have very similar effects on the slope of a transition.

To introduce T_m as one of the two fitted thermodynamic variables, the identity introduced in Equation 8.20 can be rearranged to isolate $\Delta \bar{S}°$:

$$\Delta \bar{S}° = \frac{\Delta \bar{H}°}{T_m} \qquad (8.24)$$

Substituting into the Y_{obs} Equation 8.19 and rearranging gives

$$Y_{obs} = Y_D + (Y_N - Y_D) \frac{e^{\Delta \bar{H}°(T-T_m)/RTT_m}}{1 + e^{\Delta \bar{H}°(T-T_m)/RTT_m}} \qquad (8.25)$$

Figure 8.7 shows a thermal unfolding transition of ubiquitin, measured by CD spectroscopy. The model fits reasonably well to the data (even though, as described below, the model leaves out an important temperature dependence), and gives to fitted parameters that are quite consistent with the data. The fitted T_m value is 341 K (68°C), which is indeed squarely in the middle of the unfolding

---

[†] Loosely, T_m is the closest parameter to the measured data.

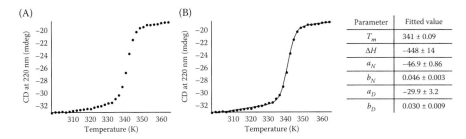

Parameter	Fitted value
T_m	341 ± 0.09
ΔH	-448 ± 14
a_N	-46.9 ± 0.86
b_N	0.046 ± 0.003
a_D	-29.9 ± 3.2
b_D	0.030 ± 0.009

Figure 8.7 Thermal unfolding of a small globular protein fitted with a simple two-state model. Panel (A) shows the temperature dependence of the CD signal for hen lysozyme, a 129 residue β+α protein, in 20 mM glycine (a buffer) at pH 2.5. Panel (B) shows the data fitted by a two-state model (Equation 8.25) in which the enthalpy (and thus entropy) is treated as independent of temperature. The baselines are fitted with separate linear temperature dependences (two parameters per baseline: a_N and a_D are intercepts (at zero Kelvins!) and b_N and b_D are slopes). The table to the right shows best-fitted parameter values and standard errors. Units for T_m and $\Delta H°$ are kJ mol^{-1}, respectively.

transition. The fitted $\Delta \bar{H}°$ value is -448 kJ mol^{-1}; although your intuition about this value is not likely to be as strong as for T_m, we can say that at the very least, the fitted $\Delta \bar{H}°$ value comes out less than zero, which is consistent with an exothermic folding reaction (i.e., an endothermic unfolding reaction). Since lysozyme unfolds as the temperature is increased, the *unfolding* reaction must be endothermic.

The entropy associated with the folding transition, calculated from Equation 8.24, is -1.31 kJ mol^{-1} K^{-1}. Interpreted statistically, this value is consistent with a significant decrease in disorder upon folding, as expected (Figure 8.4).

Extrapolation of conformational transitions

Without fitting a transition, all that can be said with much certainty about the thermodynamics of unfolding is that at the free energy of unfolding is zero at the midpoint of the curve, negative below the midpoint, and positive above it. In addition (and with some experience), a qualitative statement of the size of $\Delta \bar{H}°$ and $\Delta \bar{S}°$ can be made from the steepness of the transition. Fitting replaces this quantitative description with a quantitative one, providing numerical estimates of T_m and $\Delta \bar{H}°$ (and through they, $\Delta \bar{G}°$ and $\Delta \bar{S}°$). Moreover, although these parameters are determined from equilibrium data in a fairly narrow transition region, they can be combined with the original fitting model to evaluate thermodynamic parameters (for this model, $\Delta \bar{G}°$ and K_{fold}) outside the transition region. This is referred to as "extrapolation."

Extrapolation is particularly useful because, for reasons stated above, unfolding transitions often require harsh conditions (here, high temperature), but stability parameters are usually of greatest interest under physiological (or laboratory) conditions (where they cannot be directly measured). For example, hen egg lysozyme functions at the temperature of a hen egg[†] (about 315 K), but a significantly higher temperature is needed to observe the unfolding transition (about 336–346 K; see **Figure 8.8A**). Extrapolation allows stability to be estimated outside the transition, at physiological temperature.[‡] The black line in Figure 8.8B shows extrapolation of the folding free energy of lysozyme unfolding away from the transition, using the fitted parameters in Figure 8.7.

However, there are *potential* dangers associated with extrapolation. First, like any other experimentally determined quantity, fitted parameter values have errors

---

[†] With a hen on top of the egg.

[‡] Note that at this point we should not get too carried away with "physiology," since the data in Figure 8.7 were collected at pH 2.5 and at very low salt. These unusual conditions were chosen because they maintain reversibility. As discussed below, reversibility is essential for equilibrium thermodynamic measurements.

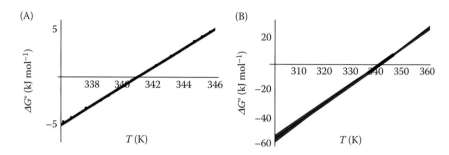

Figure 8.8 Extrapolation of free energy of unfolding of hen lysozyme away from the thermal transition region. Circles are data points from Figure 8.7 that fall within the unfolding transition region, transformed to free energies using fitted baseline parameters (a and b values in Figure 8B). (A) Data in the transition region, and (B) extrapolation outside of the transition region. The solid black line is the best-fit line through the experimental free energy estimates. The red band, which shows uncertainties in extrapolated $\Delta G°$ values (at the 95% confidence level; see Problem 8.5), expand significantly outside the transition region, showing that large extrapolations magnify uncertainties in parameter estimates.

associated with them (see Chapter 2). Thus, when the fitted parameters T_m and $\Delta \bar{H}°$ are used to calculate free energies, these errors will be propagated to the free energy values. Though these free energy errors may be quite small at temperatures within the transition, errors are amplified at temperatures far from the transition (i.e., for long extrapolations). In large part, this compounding of error away from the transition comes from uncertainty in $\Delta \bar{H}°$, which is proportional to the slope of the transition. The red band in **Figure 8.8B** shows the uncertainties in $\Delta \bar{G}°$ away from the transition, using a simple formula for error propagation from uncertainties from $\Delta \bar{H}°$ and T_m (see Problem 8.5). This uncertainty band is narrow in the transition region, but gets large both at low and at high extrapolated temperatures. Another type of transition that is subject to extrapolation error is chemical denaturation, which will be described later in this chapter.

A second type of potential danger that is compounded by long extrapolation has to do with the model itself more than the parameters. If the model is imperfect, the quantity being extrapolated (e.g., K_{fold}, $\Delta \bar{G}°$) can diverge significantly from the true value. The model can be thought of as having two components: a reaction mechanism (here, two-state), and its dependence on temperature (or other external factors). If the mechanism is correct but its temperature dependence is wrong, then near the transition (where equilibrium is actually measured), thermodynamic values obtained from the fit will not be too far off. But as the extrapolation goes away from the transition region, the extrapolated values will become progressively worse approximations of the true values. This effect will be shown in the section below, where a more realistic temperature dependence is developed for the protein folding reaction. Note that, if the mechanism itself (here, the two-state mechanism) is incorrect, then the meaning of the fitted parameters is questionable, even in the transition region, because the fundamental assumption that Y_{obs} results from a population-weighted average of two states is wrong. Though the data may still be interpreted in terms of how resistant protein structure is to temperature, interpretation of fitted values of equilibrium constant and related quantities is inappropriate.

A MORE REALISTIC MODEL FOR THERMAL UNFOLDING OF PROTEINS: THE CONSTANT HEAT CAPACITY MODEL

Although the simple model above appears to fit the thermal unfolding transitions of many proteins quite well, in data of very high resolution, deviation from the simple model can be seen. In some cases, this can be seen as a systematic deviation from fits to thermal transitions, but deviations are more obvious when comparing extrapolated parameters from thermal and chemical denaturation, and by direct measurement of unfolding using calorimetry (see below). The problem with the simple model is its treatment of the enthalpy and entropy as temperature-independent. For protein folding, both $\Delta \bar{H}$ and $\Delta \bar{S}$ depend on temperature.[†] Here we will derive a model for the reaction free energy (through the equation $\Delta \bar{G} = \Delta \bar{H} - T \Delta \bar{S}$) that takes this temperature dependence into account, and will consider its rather profound consequences on protein stability.

As discussed in Chapter 3, the temperature variation of the enthalpy is proportional to the heat capacity. In differential form, we can write

$$dH_N = \bar{C}_{p,N} dT \tag{8.26}$$

Here, the N subscripts indicate that the molar enthalpy and heat capacity values are for the native state. An analogous expression holds for the denatured state. Integrating Equation 8.26 gives the temperature dependence of the native state enthalpy:

$$\bar{H}_N - \bar{H}_{N,T_{ref}} = \int_{H_{N,T_{ref}}} d\bar{H}_N = \int_{T_{ref}} \bar{C}_{p,N} dT \tag{8.27}$$

Single-limit integration from T_{ref} makes $\bar{H}_{N,T_{ref}}$ a constant of integration. To evaluate the right-hand side, we need either an explicit temperature dependence for $\bar{C}_{p,N}$, or an assumption that it is temperature independent over the temperature range under consideration. We will do the latter, which results in the expression

$$\bar{H}_N = \bar{H}_{N,T_{ref}} + \bar{C}_{p,N}(T - T_{ref}) \tag{8.28}$$

In Equation 8.28, the reference enthalpy has been moved to the right side of the equation to isolate the temperature-dependent enthalpy, $\bar{H}_N$. Repeating this expression for denatured state gives

$$\bar{H}_D = \bar{H}_{D,T_{ref}} + \bar{C}_{p,D}(T - T_{ref}) \tag{8.29}$$

Subtracting Equation 8.29 from Equation 8.28 gives the temperature-dependent reaction enthalpy:

$$\begin{aligned} \Delta \bar{H} = \bar{H}_N - \bar{H}_D &= \bar{H}_{N,T_{ref}} - \bar{H}_{D,T_{ref}} + \bar{C}_{p,N}(T - T_{ref}) - \bar{C}_{p,D}(T - T_{ref}) \\ &= \Delta \bar{H}_{T_{ref}} + \Delta \bar{C}_p(T - T_{ref}) \end{aligned} \tag{8.30}$$

The term $\Delta \bar{C}_p$ is the difference between the heat capacities of the N and D states. As a result of the assumption that $\bar{C}_{p,N}$ and $\bar{C}_{p,D}$ are independent of temperature (e.g., Equations 8.27 and 8.28), $\Delta \bar{C}_p$ is independent of temperature.[‡] As a result, the reaction enthalpy is *linear* in temperature, with a slope equal to $\Delta \bar{C}_p$. The connection to slope reflects the definition of C_p as the temperature derivative of H (Chapter 3).

The temperature dependence in the entropy is determined in an analogous way, starting from

$$d\bar{S}_N = \frac{\bar{C}_{p,N}}{T} dT \tag{8.31}$$

[†] In fact, if the enthalpy is temperature dependent, the entropy must also be temperature dependent. See Problem 8.5.

[‡] In fact, ΔC_p could be independent of temperature even if $C_{p,N}$ and $C_{p,D}$ depend on temperature, as long as their temperature dependence is *the same* (so it cancels in the difference).

(from the equality part of the Clausius relation). Integrating from a reference temperature gives

$$\bar{S}_N = \bar{S}_{N,Tref'} + \bar{C}_{p,N} \ln\left(\frac{T}{T_{ref}'}\right) \tag{8.32}$$

Again, the heat capacity is assumed to be temperature independent. Note that since this is a separate integration from that for enthalpy (Equation 8.27), the two reference temperatures need not be the same (this is why the reference temperature for entropy is marked with a prime in Equation 8.33). Repeating this integration for the D state and subtracting from Equation 8.32 gives

$$\Delta\bar{S} = \Delta\bar{S}_{Tref'} + \Delta\bar{C}_p \ln\left(\frac{T}{T_{ref}'}\right) \tag{8.33}$$

As with the enthalpy Equation 8.30, the temperature dependence of the entropy is determined by ΔC_p, although for the entropy this dependence is nonlinear.

Equations 8.30 and 8.33 can be combined to give the temperature dependence of the free energy of folding:

$$\Delta\bar{G}° = \Delta\bar{H} - T\Delta\bar{S}$$

$$= \Delta\bar{H}_{Tref} + \Delta\bar{C}_p(T - T_{ref}) - T\Delta\bar{S}_{Tref'} - T\Delta\bar{C}_p \ln\left(\frac{T}{T_{ref}'}\right) \tag{8.34}$$

This expression is often referred to as the "Gibbs–Helmholtz" equation.[†] Here, we will refer to the model that leads to Equation 8.34 as the "**c**onstant **h**eat **c**apacity" (or "CHC") model. The CHC model has three unknown parameters. Like the linear free energy model, it has entropy and enthalpy terms (here, $\Delta\bar{H}_{Tref}$ and $\Delta\bar{S}_{Tref'}$). The additional parameter is $\Delta\bar{C}_p$, which gives curvature of the folding free energy with temperature. Note that although T_{ref} and $T_{ref'}$ also appear in Equation 8.34, they are arbitrarily chosen as integration limits. Their chosen values determine the values of $\Delta\bar{H}_{Tref}$ and $\Delta\bar{S}_{Tref'}$, but they do not act as additional unknown parameters.

A form of the constant heat capacity model that is good for fitting data

Because the integration limits T_{ref} and $T_{ref'}$ are arbitrarily chosen, we are free to pick values that are most convenient for analysis. Since all of the thermodynamic parameters are directly connected to the transition region, it makes sense to pick a reference temperature in the transition. The most obvious candidate is the midpoint temperature, T_m. Using T_m for both the entropy and enthalpy integration limits gives a slightly modified form of the CHC equation,

$$\Delta G° = \Delta\bar{H}_{T_m} + \Delta\bar{C}_p(T - T_m) - T\Delta\bar{S}_{T_m} - T\Delta\bar{C}_p \ln\left(\frac{T}{T_m}\right) \tag{8.35}$$

Since T_m is not known prior to data analysis, and depends both on the solution conditions and the protein under study, it must be regarded as an unknown (i.e., a fitted) parameter. Introducing an additional parameter into an equation that we intend to use to fit to data would seem like a bad idea. However, the actual free energy curve described is not changed by our change in parameters, and if three

---

† Although Equation 8.34 is often called the Gibbs–Helmholtz equation, this name has also been associated with a differential equation that resembles the Van 't Hoff equation (J.A. Schellman, personal communication).

parameters were sufficient to describe one representation of the free energy curve, only three parameters should be required to describe another. If we choose to promote our integration temperature to an unknown parameter, one of the unknowns in the original CHC Equation 8.34 must become redundant. We can identify (and eliminate) this redundant parameter by recalling that at T_m, $\Delta \bar{H}_{Tm}$, and $\Delta \bar{S}_{Tm}$ are not independent; we can substitute either $\Delta \bar{H}_{Tm}$ and $\Delta \bar{S}_{Tm}$ with the other variable (and T_m). Substituting for entropy gives

$$\Delta \bar{G}^{\circ} = \Delta \bar{H}_{Tm} + \Delta \bar{C}_p (T - T_m) - \frac{T \Delta \bar{H}_{Tm}}{T_m} - T \Delta \bar{C}_p \ln \left(\frac{T}{T_m} \right) \qquad (8.36)$$

Equation 8.36 has three unknown parameters: $\Delta \bar{C}_p$, T_m, and $\Delta \bar{H}_{Tm}$. This representation makes it easy to come up with initial guesses in fitting. T_m is the easiest parameter to eyeball from raw data. Though $\Delta \bar{H}$ is bit harder to eyeball, it is easiest to do at T_m, where the slope of the transition is most apparent. Away from the transition (i.e., in the baseline regions), the enthalpy is modulated by $\Delta \bar{C}_p$, which is also an unknown parameter. Thus, in generating initial guesses for fitting, and as a description of the shape of a thermal transition, $\Delta \bar{H}_{Tm}$ is closest to the data.

A form of the constant heat capacity model that is suited for entropy and enthalpy analysis

The T_m-based form of the CHC equation emphasizes the unfolding transition itself. As described above, this region is the experimental access point for folding thermodynamics populations. A second form of the equation can be used to focus on the enthalpy and entropy functions, and how they combine to shape the free energy function. This approach provides insight into common features of protein folding thermodynamics among different proteins, and it reduces the free energy equation to a fairly simple form.

In this approach, separate temperatures are used for the entropy and enthalpy integrations. As with the T_m parameter above, these temperatures will be used as unknown fitting parameters at which some thermodynamic condition is achieved. The condition for T_m was $\Delta \bar{G}^{\circ} = 0$, and thus $\Delta \bar{H}_{Tm} = T_m \Delta \bar{S}_{Tm}$. Here, we will define two new temperatures T_h and T_s, where the enthalpy and entropy change are zero. Using T_h and T_s, the temperature dependences of the enthalpy and entropy (Equations 8.30 and 8.33) simplify to

$$\Delta \bar{H} = \Delta \bar{H}_{T_h} + \Delta \bar{C}_p (T - T_h)$$
$$= \Delta \bar{C}_p (T - T_h) \qquad (8.37)$$

and

$$\Delta \bar{S} = \Delta \bar{S}_{T_s} + \Delta \bar{C}_p \ln \left(\frac{T}{T_s} \right)$$
$$= \Delta \bar{C}_p \ln \left(\frac{T}{T_s} \right) \qquad (8.38)$$

The simplifications in the second lines of Equations 8.37 and 8.38 result from the fact that $\Delta \bar{H}_{T_h}$ and $\Delta \bar{S}_{T_s}$ are zero by definition of T_h and T_s. Combining Equations 8.37 and 8.38 gives

$$\Delta \bar{G} = \Delta \bar{C}_p \left(T - T_h - T \ln \left(\frac{T}{T_s} \right) \right) \qquad (8.39)$$

As with the other two versions of the CHC free energy equation (Equations 8.34 and 8.35), there are three unknown parameters in Equation 8.39. In this case, they are $\Delta \bar{C}_p$, T_h, and T_s.

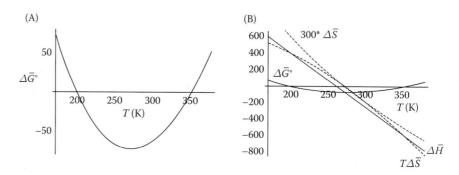

Figure 8.9 Free energy, enthalpy, and entropy of protein folding as a function of temperature. A heat capacity change on folding of -6.4 kJ mol^{-1} K^{-1} is assumed, along with T_h and T_s values of 261 and 272 K. These values are approximately those for the folding of hen lysozyme. (A) The folding free energy profile, showing a minimum at T_s and two zero crossings corresponding to heat and cold denaturation. (B) A wider energy scale to show the relationship between the free energy (black solid line) and the derivative thermodynamic quantities. The enthalpy (red solid line) is linear, with a slope of ΔC_p. The entropy (red dashed line, multiplied by 300 to provide similar scale to the other curves), is nonlinear, and has positive concavity. Multiplying the entropy by T (black dashed line) gives negative concavity, producing two intersection points with the enthalpy. These two intersection points correspond to the heat and cold denaturation temperatures. The enthalpy and entropy have x-intercepts at T_h and T_s.

Cold denaturation of proteins

Figure 8.9 shows the CHC free energy curve, along with enthalpy and entropy curves, for the folding of a typical globular protein, using Equations 8.37 through 8.39.[†] The free energy of folding as a function of temperature looks like a shifted parabola[‡] with a minimum and two zero crossings (Figure 8.8A). The zero crossings (where $\Delta \bar{G}^\circ = 0$) correspond to temperatures where $\Delta \bar{H}^\circ$ and $T\Delta \bar{S}^\circ$ have the same values (red solid line and black dashed curve, Figure 8.9B). Though the high-temperature zero crossing is expected (this is where the native protein unfolds as the temperature is increased; Figure 8.7), the low-temperature zero crossing is somewhat surprising. At this temperature, the native protein unfolds as the temperature *decreases*, a process referred to as "cold denaturation." Taking free energy as a measure of stability, these two denaturation points (sometimes referred to as T_g and $T_g{}'$, in analogy with T_s and T_h) define a region of stability for the native protein, akin to the boundaries on a phase diagram. Proteins unfold outside this region, both at high and at low temperatures.[§]

One of the reasons the cold denaturation of proteins seems unexpected is that it is difficult to observe experimentally. The cold denaturation transition is typically predicted to occur at temperatures significantly below the freezing point of water. Using parameters for hen lysozyme (Figure 8.9), the predicted temperature for cold denaturation is around 200 K (Problem 8.12). For lysozyme and many other proteins, the temperature of minimum free energy (maximum stability, corresponding to T_s) is around freezing, so that the observed temperature dependence of protein stability is as expected all the way down to the lowest accessible solution temperatures.

Another surprising feature of cold denaturation is the change of sign of the enthalpy and entropy functions from the high-temperature denaturation. Unfolding at high

---

[†] Recognize though that the same curves would be obtained using the equations including T_m (Equation 8.36). The free energy curve describes a property of the thermodynamic system, in this case a protein in solution, not how we parameterize it.

[‡] It *looks like* a parabola, but is not a parabola. The nonlinearity comes from the logarithmic term in Equation 8.39.

[§] Unlike a phase transition, the protein unfolding transitions centered at T_g and $T_{g'}$ are broad. Rather than saying "proteins unfold outside this region," it would be more appropriate to say that outside this region, proteins are more unfolded than folded.

temperatures is opposed by enthalpy, (that is, it is endothermic), but it is driven by entropy. This matches our expectation (Figure 8.4) that favorable noncovalent bonds must be broken to unfold the protein (an enthalpy increase), but conformational freedom increases (an entropy increase). However, the opposite is true for low-temperature unfolding: the enthalpy and entropy decrease upon unfolding. This is likely a result of changes in the structure of the water that solvates the hydrophobic groups that become exposed in the unfolded state. In addition, there may be changes in the unfolded state that favor more compact and organized conformations at low temperature (see Babu et al., 2004).

If we cannot measure cold denaturation, how do we know it is relevant to protein folding thermodynamics? One answer to this question is that if we can confirm the nonlinear shape of the free energy under conditions that are experimentally accessible, then we can be confident that the experimentally inaccessible part of the free energy curve will arc back to zero. All we need to do to establish nonlinearity is to demonstrate that there is a negative heat capacity change associated with folding, and the equations do the rest. The thermodynamic relations developed in Chapter 5 can be used to show that the curvature in the free energy profile is directly proportional to $\Delta \bar{C}_p$ (Problem 8.11). If $\Delta \bar{C}_p$ is equal to zero, the free energy is linear in temperature (with slope equal to a temperature *independent* $\Delta \bar{S}$), and can only cross the temperature axis once. Conversely, if $\Delta \bar{C}_p$ is nonzero, the free energy is nonlinear, and can cross the temperature axis multiple times. One of the best ways to quantify ΔC_p is to measure it directly using calorimetric methods, as described in the next section.

MEASUREMENT OF THERMAL DENATURATION BY DIFFERENTIAL SCANNING CALORIMETRY

In the methods described above, spectroscopic measurements are used to monitor the position of folding equilibrium as a function of temperature, and these data are fitted to determine the free energy and enthalpy of the folding reaction. The heat capacity of the folding reaction provides additional insights into folding, and should be included in any attempt to extrapolate free energy away from the transition region. Unfortunately, the influence of the heat capacity change on the shape of a single folding transition is subtle, and is difficult to accurately determine by fitting. Even the enthalpy can be a challenge to fit, since it is related to the slope of the transition.[†]

An alternative and much more direct approach to measuring thermal folding transitions in proteins is a type of high-precision calorimetry called "differential scanning calorimetry," or DSC[‡] for short. In the DSC experiment, temperature is increased, and the difference in heat capacity (also referred to a "differential heat capacity," δC_p) of the protein solution relative to a buffer solution is measured as a function of temperature (Appendix 8.1). Because DSC measures heat capacity directly, and folding enthalpy is determined from the area under the heat capacity curve (as detailed below), DSC can give a much more accurate determination of $\Delta \bar{C}_p$ and $\Delta \bar{H}°$ than can spectroscopic methods.

There is a fundamental difference between folding transitions monitored by DSC, and those monitored by spectroscopic methods. Whereas spectroscopic transitions vary as a simple population-weighted average (Equation 8.5), the heat capacity monitored by DSC shows a large peak centered on the folding transition (**Figure 8.10A**). This peak results from a shift in population from the low-enthalpy

---

[†] Inasmuch as the enthalpy is related to the slope of the thermal folding transition, the heat capacity is related to the slope of the slope (a second derivative, remember, $C_p = \partial \Delta H / \partial T$), an even harder quantity to fit from experimental data than the slope.

[‡] In addition to representing the method, the acronym "DSC" also refers to the instrument itself, that is, the differential scanning calorimeter.

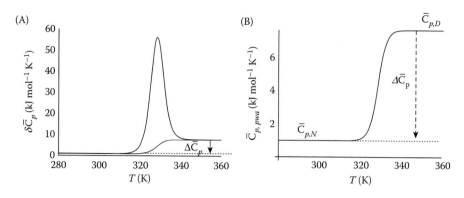

Figure 8.10 DSC-monitored unfolding of hen egg lysozyme. (A) DSC folding transition (black line, Equation 8.48), showing a peak ($\bar{C}_{p,\,ex}$) superimposed on the population-weighted average heat capacity ($\bar{C}_{p,\,pwa}$, Equation 8.40). The area between the peak and the population-weighted average (red) is the enthalpy of the unfolding reaction (positive in the *unfolding* direction). The thermodynamic parameters ($\Delta\bar{H}_{T_m} = -429\,\text{kJ mol}^{-1}$, $T_m = 328K$, $\Delta\bar{C}_p = -6.7\,\text{kJ mol}^{-1}\,K^{-1}$) are from the calorimetric study of Pfiel and Privalov (*Biophys. Chem.* 4, 23[1976]). The heat capacity of the native state ($\bar{C}_{p,N}$) is arbitrarily set to 1000 J mol^{-1}. (B) An expanded view of $\bar{C}_{p,\,pwa}$; note that the difference between the two baselines is equal to $\Delta\bar{C}_p$.

native state to the high-enthalpy denatured state as the temperature is increased through the transition. Heat is consumed in driving the reaction, rather than simply increasing the thermal energy of the system, resulting in a high heat capacity. The reaction is driven most strongly at T_m (that is, populations are shifted the most), resulting in a heat capacity peak.

Another way to understand this peak involves a fundamental relationship between heat capacity and the energy (in this case, enthalpy) distribution. As described in Chapter 10, the heat capacity is proportional to the spread of the enthalpy values a system can access (Equation 10.65). If both the native and denatured states are populated, there is a large spread in enthalpy among these states. This spread is maximized at T_m (as illustrated in Figure 8.5).

In the resulting heat capacity profile, this "excess" heat capacity peak ($\bar{C}_{p,\,ex}$) is added to a smaller population-weighted average heat capacity to yield the measured differential heat capacity:

$$\delta\bar{C}_p = \bar{C}_{p,\,pwa} + \bar{C}_{p,\,ex} \tag{8.39A}$$

The population-weighted average (*pwa*) is analogous to the signal changes observed in a spectroscopically monitored folding transition (Equation 8.5), that is,

$$\begin{aligned}\bar{C}_{p,\,pwa} &= f_N\bar{C}_{p,N} + f_D\bar{C}_{p,D} \\ &= f_N\bar{C}_{p,N} + (1-f_N)\bar{C}_{p,D} \\ &= \bar{C}_{p,D} + f_N\Delta\bar{C}_p\end{aligned} \tag{8.40}$$

An expanded view of $\bar{C}_{p,\,pwa}$ is shown in **Figure 8.10B**. Since the heat capacity of the denatured state of proteins is larger than that of the native state, $\bar{C}_{p,\,pwa}$ increases sigmoidally through the transition from $\bar{C}_p^N$ to $\bar{C}_p^D$, with a midpoint of T_m, and a steepness given by $\Delta\bar{H}°$.

To better understand the shape of the DSC profile, and to directly analyze DSC data, we need an analytical expression for the heat capacity peak, that is, $\bar{C}_{p,\,ex}$. This can be obtained by analyzing the differential enthalpy (δH). Unlike the heat capacity term, δH is a simple weighted average of the native and denatured states. In terms of mole numbers n_N and n_D,

$$\delta H = n_N\bar{H}_N + n_D\bar{H}_D \tag{8.41}$$

Dividing by the total number of moles of protein gives

$$\frac{1}{n_N + n_D} \delta H = \frac{n_N}{n_N + n_D} \bar{H}_N + \frac{n_D}{n_N + n_D} \bar{H}_D \tag{8.42}$$

which can be rewritten as a simple population-weighted average:

$$\begin{aligned} \delta \bar{H} &= f_N \bar{H}_N + f_D \bar{H}_D \\ &= \bar{H}_D + f_N \Delta \bar{H} \end{aligned} \tag{8.43}$$

Because the DSC measures heat capacity, we need to convert $\delta \bar{H}$ into a heat capacity. This is done by differentiation with respect to temperature:

$$\begin{aligned} \delta \bar{C}_p &= \left(\frac{\partial \delta \bar{H}}{\partial T} \right)_p \\ &= \left(\frac{\partial \bar{H}_D}{\partial T} \right)_p + f_N \left(\frac{\partial \Delta \bar{H}}{\partial T} \right)_p + \Delta \bar{H} \left(\frac{\partial f_N}{\partial T} \right)_p \end{aligned} \tag{8.44}$$

The first two derivatives on the right-hand side are simple heat-capacity terms, and together combine to give $\bar{C}_{p,pwa}$ (Equation 8.40):

$$\begin{aligned} \delta \bar{C}_p &= \bar{C}_{p,D} + f_N \Delta \bar{C}_p + \Delta \bar{H} \left(\frac{\partial f_N}{\partial T} \right)_p \\ &= \bar{C}_{p,pwa} + \Delta \bar{H} \left(\frac{\partial f_N}{\partial T} \right)_p \end{aligned} \tag{8.45}$$

Comparing Equations 8.39A and 8.45 shows that the heat capacity peak ($\bar{C}_{p,ex}$) is equal to the second term in Equation 8.45, that is,

$$\bar{C}_{p,ex} = \Delta \bar{H} \left(\frac{\partial f_N}{\partial T} \right) \tag{8.45A}$$

Because $\Delta \bar{H}$ is linear in temperature, the peak in $\bar{C}_{p,ex}$ must result from the $\partial f_N / \partial T$ term. We can see this by considering the slope of f_N with temperature. Away from the transition (both below and above T_m), f_N is flat (no slope). In the high temperature transition region, f_N decreases sharply with T (high negative slope). Accordingly, $\partial f_N / \partial T$ should have a sharp minimum at T_m, that is, an inverted peak centered at T_m. It is this peak that gives rise to the heat capacity peak in the DSC transition.

Because the steepness of folding transition is affected by $\Delta \bar{H}$, features of the $\partial f_N / \partial T$ peak (and the resulting $\bar{C}_{p,ex}$ peak) are influenced by the value of $\Delta \bar{H}$. Figure 10.11A shows two thermal folding transitions (represented by f_N) with identical T_m values but different folding enthalpies. Since the f_N curves differ in steepness, the widths of the resulting $\partial f_N / \partial T$ peaks differ. The larger enthalpy value produces a narrower peak (red curve, **Figure 8.11A**). The two $\partial f_N / \partial T$ peaks also differ in their heights (Problem 8.14), because the slope of the sharper transition (larger negative $\Delta \bar{H}$) is greater at T_m. Another way to justify inverse relation between the width and height is that the temperature integral of $\partial f_N / \partial T$ from baseline to baseline (i.e., the area under the $\partial f_N / \partial T$ peak) is the same for all peaks; thus, a narrower peak must be taller.

This difference in height between the two $\partial f_N / \partial T$ peaks is further amplified (**Figure 8.11B**) by multiplication by $\Delta \bar{H}$. Because the areas under (over) the $\partial f_N / \partial T$ peaks are the same for both curves (Problem 8.13), the areas under the $\Delta \bar{H} \times \partial f_N / \partial T$ peaks are proportional to $\Delta \bar{H}$ (Problem 8.15), and the heights are proportional to $\Delta \bar{H}^2$.

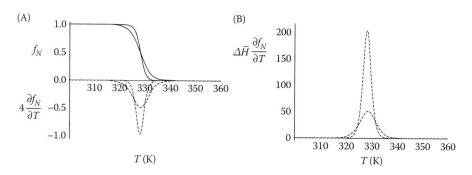

Figure 8.11 Heat capacity peaks in DSC folding transitions result from temperature-induced population shifts. (A) Thermal unfolding transitions, represented by the fraction native (f_N; solid lines, top), and the temperature derivative of the thermal unfolding transition ($\partial f_N/\partial T$, Equation 8.47; dashed lines, bottom), for lysozyme (black, parameters in Figure 10.8) and for a protein with the same T_m and $\Delta \bar{C}_p$ but twice the folding enthalpy ($\Delta \bar{H}° = -858\,\text{kJ}\,\text{mol}^{-1}$, red). Note that both dashed peaks are multiplied by a factor of four to make them easier to compare on this scale. The protein with the larger enthalpy (and steeper transition) has a $\partial f_N/\partial T$ peak that is both narrow and tall. (B) The product of $\Delta \bar{H}°$ and $\partial f_N/\partial T$, which gives the heat capacity peak in the DSC folding transition, for the two proteins from (A). Because both $\Delta \bar{H}°$ and the height of the $\partial f_N/\partial T$ peak are larger for the "red" protein by a factor of two, the heat peak is taller by a factor of four.

To use Equation 8.45 to fit DSC folding transitions, we need to express $\partial f_N/\partial T$ in terms of protein folding parameters. Substituting the folding equilibrium constant into f_N (Problem 8.15),

$$\left(\frac{\partial f_N}{\partial T}\right) = \frac{K}{(1+K)^2} \frac{\partial \ln K}{\partial T} \tag{8.46}$$

Using the Van 't Hoff equation from Chapter 7 (Equation 7.81), Equation 8.46 can be written in terms of enthalpy:

$$\left(\frac{\partial f_N}{\partial T}\right) = \frac{K}{(1+K)^2} \frac{\Delta \bar{H}}{RT^2} \tag{8.47}$$

Note that the first term on the right-hand side is simply $f_N f_D$. This is the "spread over states" term discussed above (see Equation 8.22), which peaks at T_m (see Figure 8.5).

Combining Equation 8.45, Equation 8.45, and Equation 8.47 gives

$$\delta \bar{C}_p = \bar{C}_{p,D} + f_N \Delta \bar{C}_p + \frac{\Delta \bar{H}^2}{RT^2} \frac{K}{(1+K)^2} \tag{8.48}$$

Equation 8.48 can be directly fitted to DSC folding transitions, with parameters $\bar{C}_{p,D}$, $\Delta \bar{C}_p$, $\Delta \bar{H}_{T_m}$, and T_m. In addition to providing direct experimental access to $\Delta \bar{C}_p$, only four parameters need to be fitted when analyzing DSC data with the two-state CHC model,[†] compared to seven for spectroscopic data. In essence, DSC replaces optical baseline parameters (which are of marginal interest in the analysis of macromolecular stability) with thermodynamic parameters, which are of primary interest.

Although a decreased parameter number is a good thing in terms of fitting statistics, the quality of fits to DSC data using four (or five) parameter can be low, showing large non-random residuals. One potential source for poor fitting is the determination of sample concentration. Since the raw DSC data must be divided by macromolecular concentration to obtain a molar heat capacity, concentration errors will affect the vertical scale, and thus the area under the DSC peak. Because

---

[†] Five parameters if the heat capacities of the native and denatured states are dependent on temperature.

$\Delta\bar{H}$ is proportional to the area under the curve, concentration errors will also affect the fitted value of $\Delta\bar{H}$. However, $\Delta\bar{H}$ is also sensitive to the shape of the DSC transition, which is *not* affected by concentration errors. Thus, a single value of $\Delta\bar{H}$ will not be unable to capture both the area under the curve and the shape of the curve.

One solution to this problem is to introduce two enthalpy parameters, one for terms derived from the equilibrium constant and its temperature sensitivity (Equation 8.47), and the other associated with the height of the heat capacity peak (Equation 8.45A). Expressing $\Delta\bar{C}_p$ in terms of these "two enthalpies" (termed $\Delta\bar{H}°_{\text{van't Hoff}}$ and $\Delta\bar{H}°_{\text{cal}}$) gives a variant of Equation 8.48:

$$\delta\bar{C}_p = \bar{C}_{p,D} + f_N\Delta\bar{C}_p + \frac{\Delta\bar{H}°_{\text{van't Hoff}}\,\Delta\bar{H}°_{\text{cal}}}{RT^2}\frac{K}{(1+K)^2} \tag{8.49}$$

where $\Delta\bar{H}°_{\text{van't Hoff}}$ is used in all occurrences of K and f_N. A downside of this approach is that it introduces another parameter. Another downside is that if the two-state model is correct, there should only be one enthalpy value. If a second enthalpy value is required (i.e., if Equation 8.48 fits poorly to a DSC transition), it may be an indication that the two-state model is incorrect—in such a case, the equilibrium constant and free energy should not be taken too seriously.[†] Alternatively, it may be an indication that the concentration used to convert to molar heat capacity is in error. In this case, it may be more appropriate to analyze the data using a parameter that captures this concentration error (σ_c):

$$\delta\bar{C}_p = \sigma_c\left(\bar{C}_{p,D} + f_N\Delta\bar{C}_p + \frac{\Delta\bar{H}^2}{RT^2}\frac{K}{(1+K)^2}\right) \tag{8.50}$$

This function is equivalent to Equation 8.49 (differing only in parameterization), but it is internally consistent as a thermodynamic model.

CHEMICAL DENATURATION OF PROTEINS

As described above, the measurement of protein folding thermodynamics requires equilibrium to be perturbed so that both the native and denatured state are significantly populated. Thermal unfolding transitions provide this measurable equilibrium, and also reveal baselines, which are essential for quantifying these populations. Another common way to perturb the protein folding equilibrium is to add destabilizing additives, referred to as "denaturants." This approach, which is referred to as "chemical denaturation," has a few advantages over thermal denaturation. First, chemical denaturation tends to be reversible, whereas thermal denaturation often shows limited reversibility. This difference in reversibility likely results from the fact that denatured proteins have low solubility at high temperatures, but have high solubility in high concentrations of denaturant. If the unfolding transition is not reversible, the system is not in equilibrium; thus, extracting equilibrium thermodynamic parameters is questionable (at best). Second, the denaturant dependence of folding free energy is simpler than the temperature dependence.

The two most commonly used protein denaturants are urea and guanidine hydrochloride (**Figure 8.12**). Urea is a neutral but polar diamide in aqueous solution, whereas guanidine hydrochloride dissociates to a chloride salt. Urea and guanidine hydrochloride can be used in analogous ways to temperature to measure denaturant-induced unfolding transitions. **Figure 8.13A** shows a urea-induced unfolding transition of hen lysozyme at constant temperature (308.15 K). As with

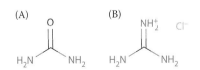

Figure 8.12 Urea (A) and Guanidine Hydrochloride (B). These two denaturants shift the protein folding equilibrium toward the denatured state. Whereas urea remains neutral, guanidine hydrochloride dissociates into a salt in aqueous solution (often referred to as guanidinium chloride).

[†] Although the $\Delta\bar{H}°_{\text{cal}}$ parameter is still likely to reflect the thermodynamic enthalpy of reaction from the native to denatured state, since it is determined directly as a heat flow, and does not depend on the path from N to D.

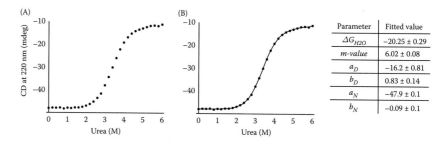

Parameter	Fitted value
ΔG_{H2O}	-20.25 ± 0.29
$m\text{-}value$	6.02 ± 0.08
a_D	-16.2 ± 0.81
b_D	0.83 ± 0.14
a_N	-47.9 ± 0.1
b_N	-0.09 ± 0.1

Figure 8.13 A urea-induced unfolding transition of hen lysozyme. Panel (A) shows raw CD signal in 20 mM glycine (a buffer), pH 2.5°C, 35°C. Panel (B) shows the data fitted by a two-state model (Equation 8.61) in which the free energy depends linearly on urea concentration (Equation 8.62). The baselines are fitted with separate linear urea dependences (two parameters per baseline: a_N and a_D are intercept values at zero molar denaturant, and b_N and b_D are slopes). The table to the right shows best-fitted parameter values and standard errors. Units for $\Delta G°_{H2O}$ and m-value are kJ mol^{-1} and kJ mol^{-1} M^{-1}, respectively.

thermal denaturation transitions, chemical denaturation shows a single sigmoidal transition between two baseline regions that have modest, linear denaturant dependences. The transition defines a midpoint concentration (often referred to as a C_m) where the equilibrium constant for folding has a value of unity, and the associated free energy for folding is zero.

We can again analyze the thermodynamics of unfolding by again assuming the CD signal is a population-weighted averaged of signals from the native and denatured states (Equation 8.5). By substituting populations with equilibrium constants (Equations 8.14 through 8.16), and equilibrium constants with free energies (Equation 8.17), we can express the observed signal in terms of free energy:

$$Y_{obs} = \frac{Y_D + Y_N e^{-\Delta \bar{G}°/RT}}{1 + e^{-\Delta \bar{G}°/RT}} \tag{8.51}$$

To fit Equation 8.61 to denaturant-induced unfolding data, we need to express the folding free energy as a function of denaturant concentration. Based on studies of a large number of proteins, it has been shown that folding free energy varies linearly with denaturant concentration, that is,

$$\begin{aligned} \Delta \bar{G}° &= \Delta \bar{G}_{H2O} + m[denaturant] \\ &= \Delta \bar{G}_{H2O} + m[x] \end{aligned} \tag{8.52}$$

where $[x]$ in the second line is shorthand for molar denaturant concentration.

Figure 8.13B shows a fit of Equation 8.51 to the lysozyme urea unfolding transition, using the linear denaturant dependence of Equation 8.52. Unlike the CHC model for thermal unfolding, the fit to the urea unfolding transition only involves two thermodynamic parameters, $\Delta \bar{G}°_{H2O}$ and the m-value. The $\Delta \bar{G}°_{H2O}$ parameter can be considered to be the intercept in a plot of $\Delta \bar{G}°$ versus denaturant concentration. We use the subscript "H_2O" because the intercept occurs when the denaturant concentration is zero, that is, the solvent is pure water. This quantity is of primary interest; although we need to use denaturants to bring about an unfolding transition, we are not typically interested in the free energy of unfolding in high concentrations of denaturant. It should be kept in mind that the same extrapolation errors described in Figure 8.8 apply to $\Delta \bar{G}°_{H2O}$, which is often determined by extrapolation from equilibration at high denaturant concentrations.[†]

---

† Note that we don't actually *do* the extrapolation, instead we fit the entire melt. However, due to the correlation between the m-value and ΔG_{H2O}, extrapolation errors apply.

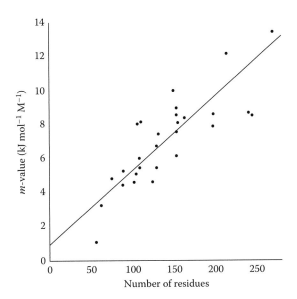

Figure 8.14 The dependence of urea *m*-values on protein size. Each dot shows the urea *m*-value measured for a protein of a given length (in residues). Values were compiled from the literature and analyzed in Myers et al., *Protein Science 4*, 2138–2148, 1994. Although *m-value* measurements are made different conditions (pH, salt concentration, and temperature), a strong linear correlation can be seen. The solid line is a linear least-squares fit to the data ($m = 0.044 \times n_{res} + 0.89$).

The other parameter determined in the fit is referred to as an "*m*-value" (m for slope in the equation for a line). The *m*-value is the denaturant sensitivity coefficient for the folding free energy,

$$m = \left(\frac{\partial \Delta \bar{G}}{\partial [x]} \right)_{p,T} \tag{8.53}$$

Because the free energy model (Equation 8.52) is linear in denaturant concentration, the m-value is constant, depending only on the type of denaturant, and the protein being studied. For a given protein, urea *m*-values are about half of guanidine hydrochloride *m*-values. For different proteins, *m*-values tend to scale in proportion to protein size (**Figure 8.14**).

The *m*-value directly reflects the steepness of the unfolding transition (Problem 8.16). A large *m*-value reflects a very steep sigmoidal folding transition. Loosely speaking, this is analogous to cooperativity in ligand binding reactions (see Chapters 12–14), where steep ligand saturation curves correspond to high cooperativity in binding. For large proteins, which have large *m*-values, a large number of groups (residues) are coupled into a single folding transition. However, the analogy is a bit misplaced, because the all-or-none feature of the transition is built into the model, which asserts that there are only two thermodynamic states: fully folded, and fully unfolded. As long as the folding reactions are two-state, large proteins are no more all-or-none than small proteins.[†]

Problems

8.1   Verify the expression for f_D in Equation 8.9, namely,

$$f_D = \frac{Y_N - Y_{obs}}{Y_N - Y_D}$$

8.2   If the only temperature dependence in the unfolding transition comes from the denominator of Equation 8.17, what are the implications for obtaining an unfolding transition?

[†]   In reality, the two-state approximation is likely to break down as for large proteins, which tend to be structured into separate domains. In such cases, fitted *m*-values are smaller than expected based on chain-length (Figure 8.14).

8.3 Derive an expression (Equation 8.21) for the *slope* of a two-state thermal unfolding transition with the assumption that $\Delta\bar{H}°$ and $\Delta\bar{S}°$ are *independent* of temperature. For simplicity, you can assume the native and denatured baselines are independent of temperature.

8.4 This problem explores the direct point-by-point conversion of a folding transition into an equilibrium constant (and free energy values), and the pitfalls of this approach.

The below data from a simulated T-melt are presented for a protein folding transition. The Y_{obs} values are equal to the fraction of native protein (f_N), but they contain 2% Gaussian error (normally a transition would be offset by unknown baseline parameters, but here we will keep it simple). Calculate equilibrium constants and folding free energies at each temperature value from the Y_{obs} values, assuming that they are a direct measure of f_N. Plot the free energy values as a function of temperature (using ListPlot if you are using Mathematica). How would you describe the distribution of the points over the temperature range? If you fitted these free energies directly to get enthalpy and entropy, what problems might you run into?

T (K)	Y_{obs}	T (K)	Y_{obs}	T (K)	Y_{obs}
310	1.0091	324	0.8818	338	0.0834
312	1.0227	326	0.8028	340	0.0559
314	0.9903	328	0.6733	342	0.0294
316	0.9969	330	0.4964	344	0.0145
318	0.9793	332	0.3436	346	0.0229
320	0.9639	334	0.1972	348	−0.0236
322	0.9369	336	0.1075	350	0.0221

Note: You are likely to run into trouble with the free energy calculations for a few of the points. If so, just eliminate those from your calculations.

8.5 Using the simple temperature-independent $\Delta\bar{H}°$ model for thermal unfolding, determine how the uncertainty in $\Delta\bar{H}°$ and T_m ($\sigma_{\Delta\bar{H}°}$ and σ_{T_m}) and propagate into $\Delta\bar{G}°$? To do this, use the error propagation formula for the multivariable function $f(x,y)$, where the uncertainty in f (σ_f) is given by

$$\sigma_f = \sqrt{\left(\frac{\partial f}{\partial x}\right)_y^2 \sigma_x^2 + \left(\frac{\partial f}{\partial y}\right)_x^2 \sigma_y^2}$$

where σ_x and σ_y are uncertainties in x and y. Use numerical values for $\Delta\bar{H}°$, T_m, $\sigma_{\Delta\bar{H}°}$, and σ_{T_m} from the table in Figure 8.7B.

8.6 Show that if the reaction enthalpy for protein folding (i.e., $\Delta\bar{H}° = \bar{H}^N - \bar{H}^D$) is temperature dependent, then the entropy ($\Delta\bar{S}°$) must also be temperature dependent.

8.7 Consider three proteins that fold by a two-state mechanism, and have the following values of T_h and T_s:

$$T_h = 265\text{K}$$
$$T_s = 275\text{ K}$$

Plot stability curves ($\Delta\bar{G}°$ for folding vs. *T*) over a broad temperature range, for the following $\Delta\bar{C}_p$ values:

Protein A $\Delta C_p = -1$ kcal / mol $*$ K
Protein B $\Delta C_p = -2$ kcal / mol $*$ K
Protein B $\Delta C_p = -4$ kcal / mol $*$ K

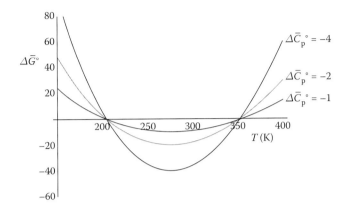

8.8 For proteins A, B, and C in Problem 8.5, find T_g and T_g' (the high and low temperature midpoints for folding). Hint: there is an easy way to do this, which involves numerical approximation, and a hard way, which involves transcendental functions that you probably have not encountered yet.

Protein	T_m (K)	ΔH_{Tm} (kcal mol^{-1})	ΔC_p (kcal mol$^{-1} \cdot$ K)
D	330	-50	-2
E	330	-100	-2
F	330	-200	-2

8.9 Proteins D, E, and F have the following folding parameters:

Make a plot of fraction native as a function of temperature, choosing a temperature range that clearly shows both the native and denatured baselines and the high-temperature unfolding transition. Which protein has the steepest transition, and why?

8.10 For proteins D, E, and F in Problem 8.8, calculate the free energy of folding at 310 K (body temperature).

8.11 Assuming a negative value for ΔC_p for protein folding, what can be said of the free energy curve (Figure 8.9) if the value of T_s is less than T_h?

8.12 To demonstrate the relationship between the curvature of the folding free energy in the CHC model and ΔC_p, calculate the second derivative of the free energy with respect to temperature.

8.13 Based on the thermodynamic parameters for unfolding of hen lysozyme (see legend to Figure 8.9), determine the temperature midpoint for cold denaturation.

8.14 Determine the area under (or over) the $\partial f_N / \partial T$ peak for a two-state protein folding reaction (Figure 8.11A), from a temperature where the protein is fully folded ($T_N \gg T_m$) to a temperature where the protein is fully unfolded () and show that the area it is independent of the steepness of the transition (i.e., independent of $\Delta \bar{H}$).

8.15 Show that for a two-state folding transition, the height of the $\partial f_N / \partial T$ curve is proportional to $\Delta \bar{H}$.

8.16 Based on your answer to Problem 8.13, determine the area under the $\Delta\bar{H} \times \partial f_N/\partial T$ peak, assuming $\Delta\bar{H}$ is independent of temperature. Show that the derivative $\partial f_N/\partial T$ is given by Equation 8.46, that is,

$$\left(\frac{\partial f_N}{\partial T}\right) = \frac{K}{(1+K)^2}\frac{\partial \ln K}{\partial T}$$

8.17 Generate folding transitions for three proteins (proteins A, B, and C) where the urea midpoints (C_m values) are fixed at 2 Molar, and the m-values are 1, 2, and 4 kcal mol^{-1} M^{-1}, respectively. Determine the $DG^o_{H_2O}$ values for each of these proteins.

8.18 Assuming the CHC model (i.e., $\Delta\bar{H}$ depends linearly on temperature), solve an expression for the integral of the excess heat peak and determine its relationship to $\Delta\bar{H}_{T_m}$.

8.19 Integrate the black DSC transition for lysozyme in Figure 8.10A, where the $T_m = 328$ K, 429 kJ mol^{-1}, 6.7 kJ mol^{-1} K^{-1}. Use a lower integration temperature limit of 280 K. Plot this integral along with the original heat peak as a function of temperature. What is the numeric value of this integral when it is evaluated at an upper limit of $T = 360$ K? Plot this integral along with the heat peak, and pay careful attention to the shape of the integral.

8.20 Continuing from Problem 8.19, calculate the temperature integral of the excess heat capacity function (Equations 8.45A combined with Equation 8.47). Plot this integral along with the excess heat capacity function itself, as you did for the DSC curve in Figure 8.18. Calculate the integral over the same limits (280 K to 360 K), and compare it to your answer from Problem 8.19.

Appendix 8.1: Differential Scanning Calorimetry

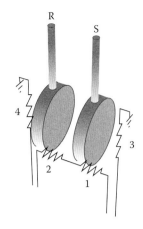

Although the measurement of the heat capacity of a sample (the heat required to bring about a change in temperature) is rather straightforward, there are a number of challenges in applying this method to protein solutions in a buffered aqueous solution. The biggest challenge is that the contribution of the protein to the overall heat capacity of a sample is quite small. Typically, protein concentrations are limited to 1–2 mg mL^{-1} (about 0.1% of the total mass of the sample). The bulk of the sample is water, which has a very large heat capacity. In a direct measurement of the heat capacity of a dilute protein–water solution, the contribution from the protein would be obscured by the large signal from the water.

What is needed is a method to subtract the water contribution away from the total heat capacity of the sample. To achieve this, the difference in heat capacity measured between a protein sample and a carefully matched[†] buffer solution is determined. The DSC instrument has two carefully matched cells that have the same volume and thermal properties. The sample (protein) is loaded into one cell, and the matched buffer is loaded into the other (termed the "reference" cell). Starting at low temperature, both the sample and buffer cells are slowly heated up, by delivering a constant amount of heat per unit time to each through a pair of matched resistors (**Figure A8.1.1**).

Figure A8.1.1 Schematic of a differential scanning calorimeter. Sample containing analyte (here, folded protein) is loaded into the sample cell (S, right), and a precisely matched buffer is loaded into the reference cell (R, left). Both cells are slowly heated (at scan rate $\gamma = dT/dt$) using matched resistors 1 and 2, and any difference in temperature is balanced by either adding heat through auxiliary resistor 3 (sample) or 4 (reference). Since resistances 3 and 4 are both known, the difference in power (δP, Equation A8.1.1) is determined. Adapted from Plotnikov et al. (1997).

If the heat capacity of the sample and reference cells differ, this matched heating protocol would produce a temperature difference (δT, where we will use the lower case delta to indicate a difference between the sample and reference cells) that is proportional to the difference in heat capacity.

Rather than allowing a significant temperature difference to accumulate between the sample and reference cells, the DSC monitors the temperature of both, and adds a small amount of heat (δq) through auxiliary resistors to whichever cell is lagging in temperature. This auxiliary heat is recorded as a "differential power" (energy per unit time; Equation A8.1.1) during the temperature scan:

$$\delta P = \frac{d(\delta q)}{dt} \tag{A8.1.1}$$

where the lowercase t indicates time (not temperature!). By dividing the differential power by the scan rate ($\gamma = dT/dt$), a differential heat capacity is obtained:

$$\delta C_p = \frac{\delta P}{\gamma} = \frac{d(\delta q)/dt}{dT/dt} = \frac{d(\delta q)}{dT} \tag{A8.1.2}$$

In the DSC experiment, temperature is scanned at a constant rate (typically $\gamma = 1$ Kelvin per second), and δC_p is recorded as a function of time. Using the scan rate and starting temperature, time is converted to temperature:

$$T = T_0 + \gamma t \tag{A8.1.3}$$

where T_0 is the temperature at the start of the experiment (where $t = 0$). Dividing δC_p by the number of moles of protein in the sample cell gives the differential molar heat capacity, $\delta \overline{C}_p$, which, when plotted versus temperature, gives a heat capacity profile like that shown in Figure 8.10A.

[†] This match ensures that the sample and the reference have the same pH, buffer components, and salt concentration. This match is usually achieved by dialysis.

References

Differential Scanning Calorimetry:

Plotnikov, V.V., Brandts, J.M., Lin, L-N., and Brandts, J.A. 1997. A new ultrasensitive scanning calorimeter. *Anal. Biochem*. 250, 237–244.

Privalov, P. L. 2012. *Microcalorimetry of Macromolecules: The Physical Basis of Biological Structures*. John Wiley & Sons, Hoboken, NJ.

CHAPTER 9

Statistical Thermodynamics and the Ensemble Method

Goals and Summary

The goals of this chapter are to give students an understanding of the ensemble method and how the ensemble method is used to go from molecular models to bulk thermodynamic averages. Molecular models provide a comprehensive list of allowed of microstates with different properties (such as energy, position, and molecular configuration), but on their own, they do not define the equilibrium distribution. Such an equilibrium distribution is needed to calculate average thermodynamic properties. By providing equilibrium distributions for molecular models, statistical thermodynamics provides a framework to connect molecular models to bulk thermodynamic properties.

Here we will describe the ideas behind the statistical thermodynamic approach by considering the properties of an isolated system, the eight-particle heat exchange ladder discussed in Chapter 4. Using this model, we will show that one way to get thermodynamic averages is to build a dynamic (i.e., time-dependent) model and average over time. However, for large systems, constructing dynamic models to obtain time averages becomes impractical. Instead, in statistical thermodynamics, a much simpler averaging method is used where a large number of random, uncorrelated snapshots are collected into an "ensemble," and averages are taken over the ensemble. One way this ensemble can be constructed (mentally) is to replicate the thermodynamic system of interest a large number of times, allow equilibration among the replicates, and then "freeze" time. This process generates the desired collection of random snapshots, and effectively takes time out of the process. Using the ensemble approach, we will find the most probable distribution (using the method of Lagrange multipliers) for an isolated ensemble, also referred to as a "microcanonical ensemble." We will define an important function, the "partition function," which encodes the probability of each microstate. For the microcanonical partition function, each microstate has the same probability, consistent with the "principle of equal a priori probabilities." The microcanonical partition function provides a direct route to calculate the ensemble entropy, closely connecting the partition function to the thermodynamic potential of the isolated system.

The classical thermodynamic approach in previous chapters provides a formalism to analyze processes such as mixing, phase transitions, and chemical reactions. Though such reactions can be quite complex (e.g., the folding transitions in Chapter 8 can involve thousands of atoms), the free energy changes we have discussed so far apply to conversion about a single equilibrium arrow. Most biochemical processes involve multiple equilibrium reactions that are linked to one another in a complex network. Examples include ligand binding to polyvalent macromolecules, and covalent transformations in metabolic networks. Even for unimolecular conformational transitions, if partly folded states are significantly populated,[†] they must be taken into account. The helix–coil transition in short polypeptides is an example of a unimolecular conformational transition that involves many populated intermediates. Keeping track of this kind of complexity one reaction at a time using classical thermodynamics would be cumbersome. The formalism of statistical thermodynamics deals with complex equilibria in a compact and powerful way.

A major difference between the statistical and classical approach to thermodynamics is the role that models play in the two approaches. In statistical thermodynamics, analysis begins with a molecular model, often including molecular energies, motions, configurations, and interactions. In classical thermodynamics, molecular models are generally absent, apart from descriptions of chemical reactions.[‡] And even when simple models do appear in classical thermodynamics, the models themselves don't provide information about relative energies, entropies, or equilibrium positions; this information has to be supplemented from experiment. In contrast, in statistical thermodynamics, such properties can be predicted directly from the models. Another significant difference between the statistical and classical approaches is that whereas the classical approach only provides changes in thermodynamic quantities (such as ΔU, ΔS, and ΔG), statistical thermodynamics can provide absolute values (U, S, and G).

Despite these differences, both approaches provide a framework for thermodynamic analysis of real systems. Thus, they must be consistent with each other, and they must give congruent descriptions. In the next three chapters, we will develop and refine an approach to go from a molecular model to appropriately averaged thermodynamic properties.

THE RELATIONSHIP BETWEEN THE MICROSTATES OF MOLECULAR MODELS AND BULK THERMODYNAMIC PROPERTIES

A key tenant of classical thermodynamics is that the bulk state of an equilibrium system is determined uniquely by specifying a small number of variables (for instance, energy, volume, and mole numbers). Likewise, when a system is in an equilibrium state, the values of these variables are determined by the state of equilibrium. Given the complexity of equilibrium systems at the molecular level (see Figure 4.1), the simplicity of the bulk description is somewhat surprising. Certainly at the molecular level, a precise description of the state of the molecules at any given instance in time would require many more variables than, for example, U, V, and N.[§] And depending on how the system and surroundings interact, we might expect the system to be able to adopt molecular arrangements that differ in U, V, and even N. Somehow, this molecular complexity must average out, giving simple macroscopic behavior.

To illustrate how this complexity averages out, we will return to the heat exchange model introduced in Chapter 4 (Figures 4.3 and 4.15). If we look at the energy difference between the left and right subsystems, $E_L - E_R$, on a time-scale fine enough to capture individual microstates, we would see a jerky variation in $E_L - E_R$, which might look something like the time sequence shown in Figure 9.1 (see Problem 9.1 and its descendants for the construction and analysis of such time sequences). This sequence is jerky because the molecules can, by definition, only exist in the allowed microstates. As the system passes from one microstate to the next, the energy difference $(E_L - E_R)$ varies.[¶]

[†] That is, when the two-state mechanism for conformational transitions breaks down.

[‡] In some sense, chemical identities in mixtures, and phase properties are also models that are part of classical thermodynamics.

[§] For a full description, we would expect to need to specify the positions and velocities of each atom to give the molecular state of the system.

[¶] Although the total energy remains constant at $E_L + E_R = 14$. We constructed the model so that the combined subsystem pair is isolated.

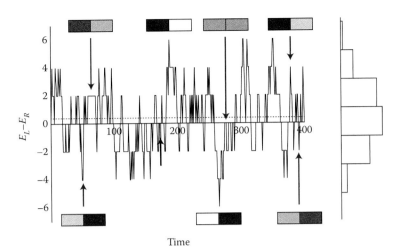

Figure 9.1 Time variation of the energy difference between the two sides of the heat exchange model. A dynamic model was constructed (Problem 9.1) that allows conversion of nearby configurations for the system shown in Figure 4.3 (four particles in an even energy ladder in each subsystem), and a simulation was performed for 400 time steps. On the scale of individual time steps, the system jumps from one discrete microstate to the next, producing a jerky variation in the energy distribution (indicated by red shading). Because each microstate has the same total energy ($E_L + E_R = 14$), each microstate is equiprobable. However, since there are many more ways to have the same energy on the left and right subsystems (Figure 4.15), the system spends most of its time in microstates where energy is shared between the two subsystems (the pink pair of boxes). The horizontal histogram on the right shows the resulting energy distribution over the 400 time-step trajectory. Because experiments that measure bulk properties typically take a long-time relative to interconversion times, they give long-time averages. A measurement of $E_L - E_R$ that took up the 400 time steps shown would yield an average of 0.40 energy units (dotted horizontal line). For longer measurement times, this average would converge to zero.

If a measurement of the energy distribution could be made instantaneously, it would be sensitive to this variation, leading to the conclusion that $E_L - E_R$ is not uniquely determined by the system. However, real measurements are not made instantaneously, but require averaging. Over the period of time it would likely take to measure $E_L - E_R$, the system will travel through a very large number of microstates, yielding an average internal energy difference of zero[†] (i.e., with each subsystem having seven units of energy).

We could, in principle, determine average properties from a molecular system by building a time-dependent model like that used to generate **Figure 9.1**, and computing the average from a long-time simulation. We would need to be careful to run the simulation long enough to get good averaging of different states. Otherwise, we might be biased by the starting point for our simulation. For the 400 step trajectory in Figure 9.1, the energy appears to swing back and forth between extremes several times, suggesting that the system extensively reconfigures on the order of 100 time steps.

A more rigorous measure of the reconfiguration time can be obtained by calculating an "autocorrelation" function. For a quantity Y that varies in time, the autocorrelation function is given by

$$C_Y(\tau) = \frac{\sum\limits_{i=0}^{n-\tau}(Y_i - \langle Y \rangle)(Y_{i+\tau} - \langle Y \rangle)}{\sum\limits_{i=0}^{n}(Y_i - \langle Y \rangle)^2} \tag{9.1}$$

---

† Note that the relatively short time average in Figure 9.1 leads to an average value of 0.4 energy units.

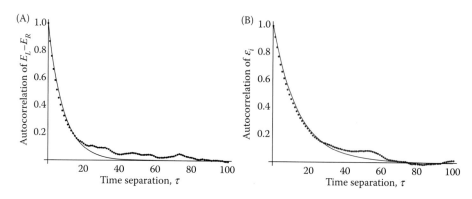

Figure 9.2 The loss of correlation in the heat exchange model over time. The dynamic model in Figure 9.1 for the heat-exchange model (Figure 4.3) was run for 10,000 time steps (much longer than in Figure 9.1), and the autocorrelation function (Equation 9.1) was calculated on the subsystem energy difference ($E_L - E_R$, panel A), and on the energy level of individual particles (panel B). Solid lines show fitted exponential decays (with a single decay constant, an amplitude of one, and a decay to zero). By both measures, the system is uncorrelated after 100 time steps (although the particle energy level remains correlated slightly longer than the subsystem energy difference).

where $\langle Y \rangle$ is the average value of Y over the time series, and τ is the time interval separating microstates. **Figure 9.2** shows different autocorrelations (the energy difference between subsystems (A), and the energy of a single particle (B)) from the simulation in Figure 9.1. Both autocorrelations start out at a value of one at $\tau = 0$ (meaning that each point in the trajectory is completely correlated with itself) and decay to zero at long τ values (complete lack of correlation). The time it takes to lose correlation, 50–100 time steps in our example, is consistent with the variations we described Figure 9.1.

With a quantitative measure of correlation time, we should be able to get an unbiased picture of bulk properties from a model by running the dynamic model for a very long time, and sampling the distribution at time points that are separated by the correlation time. This will give us a collection of uncorrelated configurations from which we can compute average properties. For the heat exchange model, the correlation analysis above indicates that separating snapshots by 100 time steps should give us a collection of uncorrelated microstates. By collecting enough of these, we could get good averages for the all aspects of the distribution, including the energies of the individual particles, the energies of the individual subsystems, and the spread in their energies. The histogram of $E_L - E_R$ values in **Figure 9.3A** was constructed using this method, where the simulation shown in Figure 9.1 was run for one million time steps, and the distribution was determined from 10,000 points that are each 100 steps apart.

Though the type of simulation described above provides is legitimate way to obtain average properties, and it has the advantage that (within the limits of the model) it is a real representation of what a system would do under measurement, this approach suffers several practical limitations. First, even a modest increase in system size greatly increases the amount of computation required for a statistically significant simulation. Simulations of samples containing on the order of thousands of particles are feasible (but not trivial), but those with Avogadro-sized numbers of particles are not. Moreover, the dynamic model itself is likely to be a rather imprecise approximation of actual dynamics. For example, the model used in Figure 9.1 assumes that each time step involves an attempt at a single unit of energy exchange between just two particles. Real dynamics would be more complicated, especially systems of large size. In the next section, we will describe an alternative approach, the "ensemble method," which builds model-based distributions of large systems without the need for dynamic models or time simulations. This approach is the basis of statistical thermodynamics.

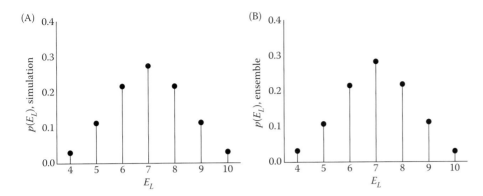

Figure 9.3 The energy distribution in the heat exchange model from long-time simulation (A) and from the assumption of equal *a priori* probabilities (B). Histograms of the energy of one of the subsystems from Figure 4.3, from (A) a single simulation of one million time steps, sampling every 100 steps (10,000 snapshots), and (B) a random sampling of 10,000 microstates from a distribution in which each allowed microstate is equally probable. As described below, this collection is referred to as a "microcanonical ensemble." Within error, populations from these two very different sampling procedures are the same, consistent with the ergodic hypothesis.

THE ENSEMBLE METHOD AND THE ERGODIC HYPOTHESIS

In the approach above, we collected a large number (10,000) of uncorrelated snapshots from our dynamic model, and used those snapshots to compute average properties that we would measure in bulk. But if the snapshots are really uncorrelated, the details and application of our dynamic model—how the system gets from one state to the subsequent one—don't matter all that much. Instead, we will take a short cut, by building a large collection of microstates that are equivalent to the uncorrelated snapshots from the time simulation. We will refer to this collection of microstates as an "ensemble."

The ensemble approach is the central method in statistical thermodynamics. Obviously, the challenge is to get the right number of each microstate so that the ensemble we build matches the bulk average. In the next two sections we will formalize the concepts and variables of the ensemble method, and we will derive the rules for how the available microstates are populated, that is, how to get the right number of each microstate. These rules give us everything we need to know, as they define the thermodynamic distribution of the ensemble, and the bulk properties of the system represented by the ensemble, including the entropy and free energy.

Before we describe the details of the ensemble approach, we will compare the distribution from our dynamics calculation on the heat exchange model (Figure 9.3A) with a distribution from the ensemble approach, to demonstrate that they give the same results. The rule we will use to populate the ensemble is to assume that all microstates of the combined pair of subsystems are equally likely, since they have the same total energy (14 units). This rule, which is often referred to as the "principle of equal *a priori* probabilities," applies when each microstate has the same internal energy. The distribution in **Figure 9.3B** comes from randomly sampling equally (and without replacement) from all heat exchange microstates. This simple method produces a distribution that looks very much like that obtained from our dynamic model (Figure 9.3A).

The advantage of the simple ensemble-based method for computing bulk properties of systems is that we do not need to build a dynamic model of our system, nor do we need to carry out lengthy time simulations (remember, we threw out 99% of our dynamics to generate the distribution in Figure 9.3A). In short, we eliminate time from the problem.[†] We have replaced an average of microstates taken at different times from a single system with an average of the microstates of many systems at an arbitrary point in time. The idea that the ensemble average is equivalent to the average of a time series is referred to as the "***ergodic hypothesis***,"[‡] and is a central to connecting statistical thermodynamic analysis to experiment.

Simple as it is, the example above depends on the assumption of equal *a priori* probabilities to generate the ensemble, which we introduced without justification. Figure 9.3 demonstrates that, for this system, the equal *a priori* assumption works quite well, but it would be good to generalize it. Moreover, this equal *a priori* probability distribution does not apply to microstates that differ in energy, which we need to deal with if we are to apply statistical thermodynamics to systems that exchange energy with their surroundings (Chapter 10). Fortunately, the ensemble construct can be used to calculate the underlying microstate probability distributions, in addition to calculating thermodynamic distributions and averages from specific models. We will use the ensemble to calculate various probability distributions in the next section, but first, we need to define more precisely the features of the ensemble.

THE ENSEMBLE METHOD OF BUILDING PARTITION FUNCTIONS

The ensemble is a mental (or computer-based) construct in which we generate a very large number of copies of the microstates that are compatible with a real system. We will represent the total number of copies, or "replicas" in the ensemble using the symbol $\mathbb{A}$.[§] Each replica in our ensemble is consistent with the real system, that is, it conforms to any restrictions on energy, volume, particle number, etc. If there are m microstates available to the real system, we will expect each to be represented in our ensemble many times (as long as the ensemble is large enough, i.e., $\mathbb{A} \gg m$). The number of replicas in a given microstate i will be represented with the symbol $\mathbb{N}_i$. The sum of all of various $\mathbb{N}_i$ values over the m different microstates gives an important conservation relation:

$$\sum_{i=1}^{m} \mathbb{N}_i = \mathbb{A} \tag{9.2}$$

Our goal in statistical thermodynamics is to build our ensemble in such a way that the distribution of the number of replicas in each microstate, which we will represent as the set

$$\{\mathbb{N}_i\} \equiv \{\mathbb{N}_1, \mathbb{N}_2, \mathbb{N}_3 \dots, \mathbb{N}_m\} \tag{9.3}$$

is proportional to the probability of each microstate in the real system. If we achieve this goal, we can calculate these probabilities from the statistics of the ensemble

$$\{p_1, p_2, p_3, \dots, p_m\} \equiv \frac{\{\mathbb{N}_1, \mathbb{N}_2, \mathbb{N}_3 \dots, \mathbb{N}_m\}}{\mathbb{A}} \tag{9.4}$$

Thus, if we have a mechanism to predict this set of "occupancy numbers," we have the equilibrium probability distribution of the real system we are interested in.

To make the idea of an ensemble more concrete, it is helpful to draw some pictures. Though these pictures are too small (that is, they have too few replicas) to give

---

[†] Which seems reasonable when we are concerned with equilibrium thermodynamics.

[‡] The name "ergodic," which is attributed to Boltzmann, derives from the Greek words "ergos"" (work) and "odos" (path, or way). The connection of these roots to the ergodic hypothesis seems a bit obscure.

[§] We have chosen "$\mathbb{A}$" to indicate that the number of replicas is very large, on the order of Avogadro's number. This is not to say that we must have 6.03×10^{23} replicas in our ensemble, but simply that a number the size of Avogadro's would be an appropriately large choice. This type of double-lined font (Euclid Math Two, sometimes called "blackboard boldface" in the classroom) will be used to identify quantities related to ensemble size.

5	4	5	2	5
5	1	2	5	4
5	3	3	4	2
3	3	2	5	3
5	5	5	3	3
4	1	3	4	5
3	2	4	5	4
2	4	1	4	5
4	2	2	5	3
3	4	3	4	5

$\{\mathbb{N}_1,\mathbb{N}_2,\mathbb{N}_3,\mathbb{N}_4,\mathbb{N}_5\} = \{3,8,12,12,15\}$

4	2	5	4	5
1	5	1	1	2
4	4	4	2	3
2	2	4	3	2
4	2	3	2	1
3	5	1	3	3
1	2	4	5	5
2	5	3	5	5
3	3	3	4	3
3	2	2	5	3

$\{\mathbb{N}_1,\mathbb{N}_2,\mathbb{N}_3,\mathbb{N}_4,\mathbb{N}_5\} = \{6,12,13,9,10\}$

1	3	2	2	4
1	4	3	2	1
2	1	5	5	5
3	4	1	1	1
4	4	2	1	4
3	5	3	1	3
1	4	4	3	3
1	1	1	5	5
4	4	3	2	4
3	2	5	3	5

$\{\mathbb{N}_1,\mathbb{N}_2,\mathbb{N}_3,\mathbb{N}_4,\mathbb{N}_5\} = \{13,7,11,11,8\}$

Figure 9.4 Some examples of an ensemble. Three different randomly generated ensembles are shown, each containing $\mathbb{A} = 50$ replicas of a system with $m = 5$ equiprobable microstates. The number of replicas in each of the microstates ($\mathbb{N}_i$) is given below the ensemble. Values for each $\mathbb{N}_i$ differ significantly among the three ensembles, because $\mathbb{A} = 50$ replicates is too small to get good statistics.

accurate populations and thermodynamic averages, they provide a conceptual connection between the system of interest and the corresponding statistical thermodynamics. In general, we will draw an ensemble using a two-dimensional grid,[†] where each cell houses a microstate that is compatible with the system of interest. The small ensembles in **Figure 9.4** are based on a system with $m = 5$ allowed microstates, and we have built ensembles with $\mathbb{A} = 50$ replicas. The microstates for the three ensembles in Figure 9.4 were all chosen by randomly sampling from an even distribution (equal probability of 0.2 for each microstate), but the occupancy numbers vary significantly among the three ensembles.

In the description above, we stipulated that the number of replicas in each microstate matches the distribution in the real system, but we did not say how we achieve this match. We could build in the correct distribution if we knew the population distribution in advance, but if we knew the distribution in advance, we would not need to build an ensemble in the first place. One way to imagine creating an ensemble with the correct distribution is to include an intermediate step in ensemble building, where rather than populating with individual microstates, we replicate the thermodynamic system of interest ($\mathbb{A}$ times), and collect these thermodynamic systems in our grid.[‡] Each of these thermodynamic replicas will explore different microstates over time, and will equilibrate to whatever degree they are allowed.[§] Once they are equilibrated, we freeze time, leaving each one in an individual microstate. I will refer to this as the "musical chairs" approach to building an ensemble.[¶] Because these microstates result from thermodynamic equilibrium of the system (now systems) of interest, they represent the underlying thermodynamic equilibrium. In this method, we are letting nature pick how many replicates of each microstate will be in the ensemble.

Isolated thermodynamic systems and the microcanonical ensemble

The simplest and most intuitive ensemble is the one that models the isolated system. Recall that the isolated system has fixed energy, volume, and particle number. Thus, in building an ensemble to mimic an isolated system, each of the replicas has the same internal energy (in addition to volume and particle number). We refer to

---

[†] Though the geometry of the grid does not matter.

[‡] Though admittedly, this seems like a step in the *wrong* direction: we are building the ensemble to try and simplify the behavior of the original system, and here we have made a bunch of systems of the same complexity.

[§] Each replica will certainly equilibrate internally, and if we construct the ensemble so that replicas can exchange energy, volume, or material, they will equilibrate with the other replicas as well.

[¶] In the game of musical chairs, a group of people (usually children) run around, exchanging chairs while music plays. When the music stops, everybody (except the loser) picks a chair and sits still. Unlike the game, there is no loser in statistical thermodynamics. Everybody has a chair.

this type of ensemble as a "microcanonical" ensemble. If the small ensembles in Figure 9.4 are microcanonical ensembles, each of the $\mathbb{A} = 50$ replicas would have the same internal energy.

The question we need to answer is "How many times should each microstate occur?"[†] In general, answering this question requires us to find m unknowns (each of the $\mathbb{N}_i$ values defining the distribution). We will do this by maximizing the number of configurations available to the ensemble. If, for example, the ensemble only includes a single microstate repeated over and over (out of m allowed microstates), there is only one configuration. Ensembles with more than one microstate have more configurations available, and are more likely than the one with a single microstate. Because the ensembles under consideration are very large, this is a very strong trend. So strong, it turns out, that the ensemble with the maximum number of configurations dominates over other configurations, and adequately represents the distribution (and the thermodynamics) of the real system.

With m different microstates, the number of configurations an ensemble can adopt is given by the multinomial coefficient (Chapter 1). Using the ensemble variables defined above, the multinomial coefficient is

$$W(\{\mathbb{N}_i\}; \mathbb{A}) = \frac{\mathbb{A}!}{\mathbb{N}_1! \mathbb{N}_2! \dots \mathbb{N}_m!} \tag{9.5}$$

In principle, we can find the maximum in W and the corresponding occupancy numbers (we will refer to these as W^* and $\{\mathbb{N}_i\}^*$) by differentiating with respect to $\mathbb{N}_i$ and setting to zero (see Chapter 2). It turns out that, due to the factorials in Equation 9.5, it is easier to maximize $\ln W$ (and importantly, the maxima W^* and $\ln W^*$ occur at the same place).[‡] Taking the log of Equation 9.5 and applying Stirling's approximation gives (Problem 9.5):

$$\ln W = \mathbb{A} \ln \mathbb{A} - \sum_{i=1}^{m} \mathbb{N}_i \ln \mathbb{N}_i \tag{9.6}$$

Since $\ln W$ is a multivariable function of its m configuration variables, its maximum $\ln W^*$ is obtained by differentiating with respect to each $\mathbb{N}_i$ and setting each derivative to zero:

$$\left(\frac{\partial \ln W}{\partial \mathbb{N}_1} \right)_{\mathbb{N}_{j \neq 1}} = 0$$

$$\left(\frac{\partial \ln W}{\partial \mathbb{N}_2} \right)_{\mathbb{N}_{j \neq 2}} = 0$$

$$\vdots \tag{9.7}$$

$$\left(\frac{\partial \ln W}{\partial \mathbb{N}_m} \right)_{\mathbb{N}_{j \neq m}} = 0$$

The maximum in $\ln W$ occurs at the configuration $\{\mathbb{N}_i\}^*$ where all the equations in Equation 9.7 are simultaneously satisfied. Note that this system of m equations can be combined into a single vector equation[§]:

[†] We have already stated the answer to this question earlier in this chapter, in the principle of equal *a priori* probabilities. Here we will show that this principle is general for all microcanonical ensembles.

[‡] This is true for any function $f(x)$, as long as $f(x) > 0$. For configurations we are interested in, multiplicity W is always one or larger. This correspondence between maxima is explored further in Problems 9.6 and 9.7.

[§] As described in Chapter 2, the "gradient" operator (∇) produces a vector function from W, in which the m partial derivatives form coefficients in each of the m directions. For this reason, a gradient of zero requires that each coefficient (Equation 9.7) be zero independently (equal and opposite partial derivatives give a nonzero gradient). The gradient form has advantages of being compact (one equation vs. m), and of having a useful geometrical meaning, as described below.

$$\nabla \ln W = 0 \tag{9.8}$$

The derivatives are evaluated from by differentiating Equation 9.6 with respect to each occupancy number:

$$
\begin{aligned}
\left(\frac{\partial \ln W}{\partial \mathbb{N}_i}\right)_{\mathbb{N}_j} &= \left(\frac{\partial \mathbb{A} \ln \mathbb{A}}{\partial \mathbb{N}_i}\right) - \left(\frac{\partial \mathbb{N}_i \ln \mathbb{N}_i}{\partial \mathbb{N}_i}\right) \\
&= \mathbb{A}\left(\frac{\partial \ln \mathbb{A}}{\partial \mathbb{N}_i}\right) + \ln \mathbb{A}\left(\frac{\partial \mathbb{A}}{\partial \mathbb{N}_i}\right) - \mathbb{N}_i\left(\frac{\partial \ln \mathbb{N}_i}{\partial \mathbb{N}_i}\right) - \ln \mathbb{N}_i\left(\frac{\partial \mathbb{N}_i}{\partial \mathbb{N}_i}\right) \\
&= \frac{\mathbb{A}}{\mathbb{A}}\left(\frac{\partial \mathbb{A}}{\partial \mathbb{N}_i}\right) + \ln \mathbb{A}\left(\frac{\partial \mathbb{A}}{\partial \mathbb{N}_i}\right) - \frac{\mathbb{N}_i}{\mathbb{N}_i}\left(\frac{\partial \mathbb{N}_i}{\partial \mathbb{N}_i}\right) - \ln \mathbb{N}_i \\
&= (1 + \ln \mathbb{A})\left(\frac{\partial \mathbb{A}}{\partial \mathbb{N}_i}\right) - 1 - \ln \mathbb{N}_i
\end{aligned}
\tag{9.9}
$$

Equation 9.9 can be simplified by recognizing that $\mathbb{A}$ is the sum of all the occupancy numbers $(\mathbb{N}_1, \mathbb{N}_2, \ldots, \mathbb{N}_m)$ but only the $\mathbb{N}_i$ term contributes the derivative, that is,

$$\left(\frac{\partial \mathbb{A}}{\partial \mathbb{N}_i}\right)_{\mathbb{N}_j} = \left(\frac{\partial \sum_{i=1}^{m} \mathbb{N}_i}{\partial \mathbb{N}_i}\right)_{\mathbb{N}_j} = \left(\frac{\partial \mathbb{N}_i}{\partial \mathbb{N}_i}\right)_{\mathbb{N}_j} = 1 \tag{9.10}$$

Combining Equations 9.9 and 9.10 gives

$$
\begin{aligned}
\left(\frac{\partial \ln W}{\partial \mathbb{N}_i}\right)_{\mathbb{N}_j} &= 1 + \ln \mathbb{A} - 1 - \ln \mathbb{N}_i \\
&= \ln \mathbb{A} - \ln \mathbb{N}_i
\end{aligned}
\tag{9.11}
$$

Figure 9.5 shows $\ln W$, a partial derivative with respect to $\mathbb{N}_i$ (Equation 9.11), and some vectors from $\nabla \ln W$ for an ensemble with just two $m = 2$ microstates (here we choose $m = 2$ so that $\ln W$ can be plotted as a function of both variables). As Figure 9.5 shows, $\ln W$ (and thus W) is maximized by increasing the number of replicas in the

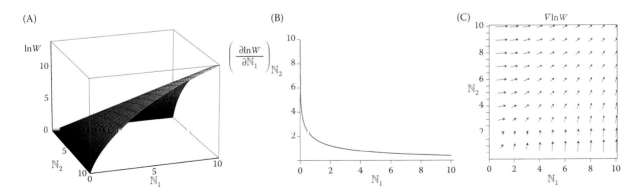

Figure 9.5 Unconstrained maximization of the number of configurations in an ensemble. (A) The natural logarithm of the number of configurations, using Stirling's approximation, for an ensemble with just two microstates. ln$W$ increases with the number of replicates in each microstate; this increase is roughly logarithmic with either occupancy number alone ($\mathbb{N}_1$ or $\mathbb{N}_2$) and is linear when both are increased in fixed proportion ($\mathbb{N}_1 + \mathbb{N}_2$ along the diagonal $\mathbb{N}_1 + \mathbb{N}_2$, see Problem 9.10). (B) The partial derivative of ln$W$ with respect to the occupancy number for either microstate, with the other occupancy number fixed to five. The derivative is positive for all values of $\mathbb{N}_i$. (C) The gradient, ∇ln$W$. There are no points where ∇ln$W$ is zero (no minima or maxima). Rather, the gradient points in the direction of increasing ensemble size ($\mathbb{N}_1 + \mathbb{N}_2$).

ensemble. The derivative of $\ln W$ with respect to each occupation number is positive for all ensemble configurations (Figure 9.5B), reflecting a monotonic increase in $\ln W$ with its variables. Consistent with this, the gradient of $\ln W$ points in the direction of increasing both $\mathbb{N}_1$ and $\mathbb{N}_2$.

Though increasing ensemble size increases $\ln W$ (this also holds for ensembles with more than two microstates), it does not provide a solution to the problem we are trying to solve. We want to find the relative distribution of microstates, given an ensemble of a *fixed size*. The size ($\mathbb{A}$) needs to be suitably large so that the distribution is well-determined, but it should be held constant during the maximization. This puts a constraint on the maximization variables $\{\mathbb{N}_1, \mathbb{N}_2, \ldots, \mathbb{N}_m\}$; namely, that the sum of $\mathbb{N}_i$ is constant (Equation 9.2 with constant $\mathbb{A}$).

There are several approaches to perform this type of constrained maximization. One approach is direct substitution. In the example in Figure 9.5 with just two maximization variables, one variable (e.g., $\mathbb{N}_2$) could be substituted using the other variable (as $\mathbb{A} - \mathbb{N}_1$, see Problem 9.9). However, when there are many microstates, the simplicity of this substitution method is lost. A more general method, which we will use here, is that of Lagrange (or "undetermined") multipliers. This approach works well for systems with many microstates (and thus, a large number of maximization variables). Application of the Lagrange method to the microcanonical ensemble is a good prequel to a related problem where we will maximize $\ln W$ subject to constraints in both ensemble size ($\mathbb{A}$) and total ensemble energy ($\mathbb{E}$).[†]

We will introduce the Lagrange constrained maximization approach using the two-microstate ensemble (Figure 9.5), because the maximization problem is easy to visualize with just two variables. **Figure 9.6A** reproduces the $\ln W$ surface as a function of $\mathbb{N}_1$ and $\mathbb{N}_2$, along with a constraint of ten total replicas (i.e, $\mathbb{A} = 10$). This constraint is expressed by the equation:

$$\mathbb{N}_1 + \mathbb{N}_2 = 10 \tag{9.12}$$

This equation describes a line in the $\mathbb{N}_1$, $\mathbb{N}_2$ plane. Our goal is to maximize $\ln W$ along this line. Though there will be points off the constraint line with higher $\ln W$ values (corresponding to larger values of $\mathbb{N}_1$ and/or $\mathbb{N}_2$), they must be excluded from the maximization, since they violate the constraint. To help identify the part of the $\ln W$ surface that is "along" the constraint line, we can represent the constraint as a plane parallel to the $\ln W$ axis.[‡] The constrained maximum is located on the intersection between $\ln W$ and the constraint plane (the gray rectangle, Figure 9.6A).

In addition to providing a nice conceptual picture of constrained maximization, Figure 9.6A suggests a general approach to constrained maximization. This approach relies on two mathematical results. The first is that the constrained maximum occurs at the point where the contour line of $\ln W$ (also known as the "level curve") and the constraint plane are parallel (**Figure 9.6B**). If the level curve and constraint are parallel, moving along the constraint line neither increases nor decreases the function being maximized (i.e., $\ln W$ is level in the direction of the constraint). The second result is that the gradient of a function at any point is perpendicular to the level curve of the function at that point (see Chapter 2). Since the level curve of $\ln W$ and the constraint are parallel at the constrained maximum, the gradient of $\ln W$ must be parallel to the gradient of the constraint equation. This can be expressed as

$$\nabla \ln W = \alpha \, \nabla(\mathbb{N}_1 + \mathbb{N}_2) \tag{9.13}$$

[†] This maximization is applied to the "canonical" ensemble where the energy of each replica can vary, but the total energy of the ensemble ($\mathbb{E}$) is fixed. This ensemble reproduces a system in thermal equilibrium with its surroundings.

[‡] This projection is okay because the constraint is independent of $\ln W$, and is equally valid at all points on the $\ln W$ axis.

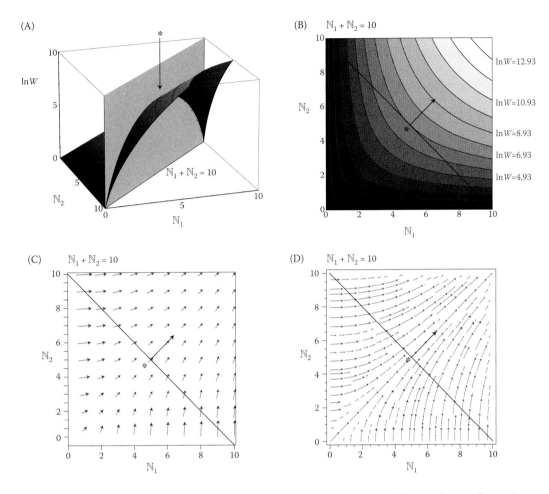

Figure 9.6 Maximizing the number of ways a two-microstate ensemble can be configured, subject to the constraint $\mathbb{N}_1 + \mathbb{N}_2 = 10$. (A) The red surface shows $\ln W$ (using Stirling's approximation) as a function of the number of replicas in microstates 1 and 2. The gray plane shows the constraint on the ensemble size. Geometrically, constrained maximization can be viewed as maximization of $\ln W$ along the intersection with the constraint plane (asterisk). (B) Contour plot of $\ln W$, with level curves in dark red, and the constraint line in black. At the constrained maximum (asterisk), the level curve of $\ln W$ is tangent to the constraint. (C–D) The relationship between the gradient of $\ln W$ (red arrows) and the constraint line (black). In (C) the gradient is calculated at intervals of one replica. In (D), gradient vectors are lined up to provide a better sense of connectivity. The constrained maximum is the point where $\nabla \ln W$ is perpendicular to the constraint plane. Since the gradient of the constraint is perpendicular to the constraint plane, the two gradients are parallel at the constrained maximum.

Here, α is a constant of proportionality. This constant is required because the analysis above only states that the two gradients are parallel, not that they are the same.[†] Explicitly writing each partial derivative in the vector Equation 9.13 gives

$$\left(\frac{\partial \ln W}{\partial \ln \mathbb{N}_1}\right)_{\mathbb{N}_2} = \alpha \left(\frac{\partial \{\mathbb{N}_1 + \mathbb{N}_2\}}{\partial \mathbb{N}_1}\right)_{\mathbb{N}_2} = \alpha \left(\frac{\partial \mathbb{N}_1}{\partial \mathbb{N}_1}\right)_{\mathbb{N}_2} = \alpha$$

$$\left(\frac{\partial \ln W}{\partial \ln \mathbb{N}_2}\right)_{\mathbb{N}_1} = \alpha \left(\frac{\partial \{\mathbb{N}_1 + \mathbb{N}_2\}}{\partial \mathbb{N}_2}\right)_{\mathbb{N}_1} = \alpha \left(\frac{\partial \mathbb{N}_2}{\partial \mathbb{N}_2}\right)_{\mathbb{N}_1} = \alpha$$

(9.14)

[†] In general, the two gradients don't even have the same dimensions. In Equation 9.13, the dimensions of $\nabla \ln w$ is inverse replicas ($\mathbb{N}_i^{-1}$), whereas $\nabla(\mathbb{N}_1 + \mathbb{N}_2)$ is dimensionless. Thus, the α multiplier provides dimensional conversion as well as length scaling.

Combining Equations 10.9 and 10.12 gives

$$\ln \mathbb{A} - \ln \mathbb{N}_1^* = \alpha$$
$$\ln \mathbb{A} - \ln \mathbb{N}_2^* = \alpha$$

(9.15)

Rearranging Equation 10.13 gives

$$\frac{\mathbb{N}_1^*}{\mathbb{A}} = \frac{1}{e^\alpha}$$
$$\frac{\mathbb{N}_2^*}{\mathbb{A}} = \frac{1}{e^\alpha}$$

(9.16)

The system (10.14) has three unknowns ($\mathbb{N}_1^*$ and $\mathbb{N}_2^*$, which we are trying to solve for, and α, which results from the Lagrange method), but comprises only two equations. Fortunately, we can supplement this system with the constraint Equation 9.2 to solve for $\mathbb{N}_1^*$, $\mathbb{N}_2^*$, and α. By substituting for $\mathbb{A}$ in Equation 9.16 using the constraint equation, the system becomes

$$\frac{\mathbb{N}_1{}^*}{\mathbb{N}_1{}^* + \mathbb{N}_2{}^*} = \frac{1}{e^\alpha}$$
$$\frac{\mathbb{N}_2{}^*}{\mathbb{N}_1{}^* + \mathbb{N}_2{}^*} = \frac{1}{e^\alpha}$$

(9.17)

Note that the expressions on the left-hand side are populations of microstates 1 and 2 in the ensemble. Adding Equation 10.15 gives

$$\frac{\mathbb{N}_1{}^* + \mathbb{N}_2{}^*}{\mathbb{N}_1{}^* + \mathbb{N}_2{}^*} = 2e^{-\alpha}$$

(9.18)

or

$$e^\alpha = 2$$

(9.19)

In turn, Equation 9.19 can be substituted back into Equation 10.15 to give

$$\frac{\mathbb{N}_1{}^*}{\mathbb{N}_1{}^* + \mathbb{N}_2{}^*} = p_1 = \frac{1}{2}$$
$$\frac{\mathbb{N}_2{}^*}{\mathbb{N}_1{}^* + \mathbb{N}_2{}^*} = p_2 = \frac{1}{2}$$

(9.20)

Before discussing the significance of Equations 9.19 and 9.20, we will extend these results to a general microcanonical ensemble with m microstates. The method of Lagrange multipliers (Equation 9.13) applies to general systems with m-microstates in the exact same way as for the two-state system above (Problem 9.11). Constrained maximization produces an m-dimensional Lagrange multiplier equation,

$$\nabla \ln W = \alpha \nabla \left(\sum_{i=1}^{m} \mathbb{N}_i \right)$$

(9.21)

which can be viewed as a system of m equations with $m + 1$ unknowns ($\mathbb{N}_1$, $\mathbb{N}_2$, ... , $\mathbb{N}_m$, and α). By combining with the constraint Equation 9.2, this system of equations can be used to solve for α:

$$e^\alpha = m$$

(9.22)

In turn, this result can be combined with the system of m equations to give

$$\frac{\overset{*}{\mathbb{N}}_1}{\overset{*}{\mathbb{N}}_1 + \overset{*}{\mathbb{N}}_2 + \ldots + \overset{*}{\mathbb{N}}_m} = p_1 = \frac{1}{m}$$

$$\frac{\overset{*}{\mathbb{N}}_2}{\overset{*}{\mathbb{N}}_1 + \overset{*}{\mathbb{N}}_2 + \ldots + \overset{*}{\mathbb{N}}_m} = p_2 = \frac{1}{m}$$

$$\vdots$$

$$\frac{\overset{*}{\mathbb{N}}_m}{\overset{*}{\mathbb{N}}_1 + \overset{*}{\mathbb{N}}_2 + \ldots + \overset{*}{\mathbb{N}}_m} = p_m = \frac{1}{m}$$

$$(9.23)$$

Equation 9.23 says that each microstate in the microcanonical ensemble has the same probability, consistent with the principle of equal *a priori* probability. Note that if the ensemble included microstates of different energy, these microstates would have different probabilities.[†] For example, if replicas exchange energy through heat flow during the equilibration phase (the "music" period of musical chairs), the resulting microstate probabilities decay exponentially with energy, as will be shown in the next chapter.[‡]

Though the Lagrange method provides a rigorous analytical derivation of the distribution of states in the microcanonical ensemble, and is essential for other types of ensembles, it should be pointed out that we could have skipped some of the steps above and still arrived at the equal *a priori* distribution, by recognizing the symmetry in the equations for the microcanonical distribution. The key is that each microstate influences the multiplicity in the same way (Equation 9.5). Though the details of their molecular rearrangements may differ, there is nothing special about any of the microstates. For the two-microstate example, the two equations (Equation 9.14 and all those that follow) that result from the Lagrange method are the same, except for an exchange of labels. The same is true with an arbitrary number m of microstates (Problem 9.11). Thus, these equations *must* result in the same probability for each microstate. This is similar to our analysis of mixing thermodynamics (Chapter 6), where we argued that the entropy was maximized when all species are at equal concentrations (and by mole fraction, equal probabilities).

Before going on to apply the results of our constraint maximum to analyze thermodynamics, it is worth considering the Lagrange method a little deeper. Compared to maximization without constraint (where $\nabla \ln W = 0$, Equation 9.8), the condition for constrained maximization is $\nabla \ln W = \alpha \nabla(\sum \mathbb{N}_i)$. Thus, at the constrained maximum, there *is* variation in $\ln W$ with the variables $\mathbb{N}_1, \mathbb{N}_2, \ldots \mathbb{N}_m$, but it is *all* in the direction *perpendicular* to the constraint (see Figure 9.6). This means there is *no* variation in $\ln W$ *along* the constraint—that is, $\ln W$ is maximized in the direction that variation is permitted.

We can also say more about the value of α. Notice that the gradient on the right-hand side of Equation 9.21 (also 9.12) is a vector with length components in each of the $\mathbb{N}_i$ directions (Problem 9.12), that is,

$$\nabla \left(\sum_{i=1}^{m} \mathbb{N}_i \right) = 1\hat{x}_1 + 1\hat{x}_2 + \ldots + 1\hat{x}_m \qquad (9.24)$$

where $\hat{x}_i$ is the unit vector in the $\mathbb{N}_i$ direction.[§] In the Lagrange multiplier equations (Equations 9.13 and 9.21), α amplifies this unit variation (for each of the m directions) to match the variation in $\ln W$ in the same direction. Since α is proportional to the number of microstates (Equation 9.22), the Lagrange multiplier equation states that for the most probable distribution, the variation in $\ln W$ produced by increasing

---

[†] Though microstates with the same energy would have still have the same probabilities.
[‡] This type of ensemble mimics a diathermal system, and is referred to as the "canonical ensemble."
[§] To see this, just take the derivative of the sum with respect to each $\mathbb{N}_i$.

the number of replicates in any microstate ($\mathbb{N}_i$) is proportional[†] to the total number of microstates available.

The Microcanonical partition function and entropy

As indicated by Equation 9.17, the Lagrange multiplier α plays an important role in the microcanonical ensemble, determining the population of each of the microstates. In its exponential form (e^α), the multiplier can be viewed as a quantity that describes the partitioning of the ensemble into the different microstates. Specifically, the population of each microstate is given by

$$p_i = \frac{1}{e^\alpha} \tag{9.25}$$

(by combining Equations 9.22 and 9.23). Given its importance, the denominator of Equation 9.25 is referred to as a "partition function" for the microcanonical ensemble (the "microcanonical partition function" for short[‡]), abbreviated using a capital omega:

$$\Omega = e^\alpha = m \tag{9.26}$$

Thus, the population of the ith microstate in the ensemble is

$$p_i = \frac{1}{\Omega} \tag{9.27}$$

By the ergodic hypothesis, this is also the time-averaged population of the ith microstate in the real system on which the ensemble is based.

The partition function is a recurring quantity in statistical thermodynamics. Though there are differences between the partition functions of different types of ensembles, there are many similarities. First, all partition functions can be regarded as sums[§] over the available ensemble microstates. Each term in the sum gives the relative likelihood of the corresponding microstate, and will be referred to as a "Boltzmann Factor."[¶] For the microcanonical ensemble, each microstate has the same probability, and thus the same Boltzmann factor (abbreviated BF). Combining these ideas with Equation 9.26, it is clear that each Boltzmann factor in the microcanonical partition function Ω has a value of one:

$$\Omega = \sum_{i=1}^{m} BF_i = \sum_{i=1}^{m} 1 = m \tag{9.28}$$

The partition functions corresponding to other types of ensembles will include Boltzmann factors with a range of values, corresponding to the unequal probabilities of different microstates.

Before going on to apply the results of our constraint maximum to analyze thermodynamics, it is worth considering the Lagrange method a little deeper. Second, partition functions can all be used to give the relative probabilities of different microstates. The probability of microstate i is calculated by dividing the Boltzmann factor for that state by the partition function. This can be written as

$$p_i = \frac{BF_i}{\sum\limits_{i=1}^{m} BF_i} \tag{9.29}$$

[†] Logarithmically proportional; see Equation 9.22.
[‡] A little shorter.
[§] Or integrals over states, if there are a lot of states and they closely resemble one another.
[¶] Sometimes Boltzmann factors are referred to as "statistical weights," though this term is easy to confuse with statistical factors that are sometimes combined with microstate Boltzmann factors, when partition functions are summed over indices with multiple microstates.

Equation 9.29 is consistent with Equation 9.27 for the microcanonical ensemble, where each of the Boltzmann factors is one. Equation 9.29 also applies to other types of ensembles, where the Boltzmann factors differ from one.

A third feature of partition functions is that they all give an indication of the spread of the ensemble over different microstates. For partition functions where the lowest energy state (the "ground state") has a statistical weight of one, the partition function can be interpreted directly as the average number of microstates in the ensemble. As described below, a Boltzmann factor of one corresponds to an energy value of zero, which we will often refer to as the "ground state" for obvious reasons.[†] For the microcanonical ensemble, all the microstates can be thought of as ground states, since they all have the same energy value.

A final important feature of partition functions is that they are all related to thermodynamic functions. In particular, for each type of partition function, the logarithm of the function gives the thermodynamic potential that matches the real system described by the ensemble. For example, as described in Chapter 4, the thermodynamic potential[‡] for the isolated system is the entropy. That is, an isolated system is at equilibrium when its entropy is maximized. Based on the assertion above, the logarithm of the microcanonical partition function should give the ensemble entropy:

$$S = k \ln \Omega \qquad (9.30)$$

This is the "Boltzmann entropy formula" that we described in Chapter 4. As with the partition function, this entropy is independent of the size of the ensemble ($\mathbb{A}$). Just as Ω is a measure of the number of microstates available to the system on which the ensemble is based, the entropy in the Boltzmann formula is a measure of the entropy of the system being modeled. This is not to say that the Boltzmann entropy is intensive. If the real system of interest is increased in size, Ω increases (sharply, because m will increase sharply), and S increases (linearly, as long as Stirling's approximation holds).

Combining Boltzmann's entropy formula with the relationship between microstate population and Ω (Equation 9.27) gives

$$S = k \ln \frac{1}{p_j} = -k \ln p_j \qquad (9.31)$$

That is, for the microcanonical ensemble, the entropy is obtained from the logarithm of any of the microstate populations.[§] If there are many microstates available (large m), each population term will be small, and the entropy will be large (owing to the minus sign in Equation 9.31; the log of a small population is a large negative number). By multiplying Equation 9.31 by the sum of the populations (the sum is equal to one), the Gibbs entropy formula (Chapter 4) can be obtained:

$$S = -k \ln p_j \left(\sum_{i=1}^{m} p_i \right)$$
$$= -k \sum_{i=1}^{m} p_i \ln p_j \qquad (9.32)$$

Though there appears to be a bit of a mismatch in the population indices (i vs. j), since each p_i is the same, we can change the j subscript to track the summation index,

---

[†] For partition functions we will encounter in subsequent chapters (e.g., partition functions for ligand binding and conformational equilibria), it will sometimes be convenient to assign a high Boltzmann factor to the low energy state. Although in such cases the value of the partition function loses its connection to the spread over states, connection can be restored by shifting microstate energies so that the lowest state has an energy of zero.

[‡] Remember, thermodynamic potentials determine the position of equilibrium.

[§] Any of the m p_i values, because they are all the same in the microcanonical ensemble. Here we arbitrarily pick the jth value to facilitate the derivation that follows.

giving the "Gibbs entropy formula." Though it is not obvious from our derivation of Equation 9.32, the Gibbs entropy formula also applies to ensembles where equilibrium microstate populations differ (i.e., where microstate energies differ), though this remains to be shown.

A MICROCANONICAL ENSEMBLE FROM THE HEAT EXCHANGE MODEL

To help illustrate the ensemble method, and to provide an example of how to calculate (and use) the partition function, we will use the heat exchange model introduced in Chapter 4 to build a few (small) microcanonical ensembles. As a reminder, the model consists of two subsystems, each with four particles, and each particle can adopt one of four energy levels (with values of 1, 2, 3, and 4; Figure 4.3). As in Chapter 4 and Figure 9.1, we will combine these two subsystems to form a single composite isolated system with a fixed total energy of 14 units. Before tackling the combined system, we will build a microcanonical ensemble from just one of the subsystems (four particles, **Figure 9.7**), to help illustrate the equal roles that individual microstates play (both distinguishable and indistinguishable) in populating the ensemble, and to explore how the size of the system being modeled affects the statistics.

A four-particle isolated system

Let's start with the four particle isolated system with a total energy of seven units.[†] How do we "build" an ensemble for this simple system? For our purposes, we can do

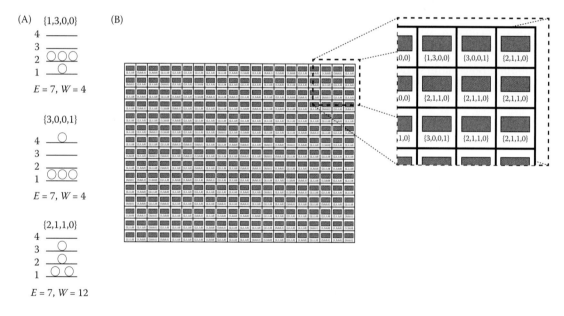

Figure 9.7 A microcanonical ensemble for a four particle energy ladder. The ensemble represents an isolated system with a total energy of seven units. (A) There are three distinguishable arrangements ({3,0,0,1}, {1,3,0,0}, and {2,1,1,0}) consistent with this energy constraint. Each of these arrangements comprises multiple indistinguishable microstates. (B) A microcanonical ensemble with $\mathbb{A} = 300$ replicas (very small for a statistical thermodynamic ensemble). Each replica has the same total energy (seven units, indicated by the same pink shading). A small section of the ensemble is expanded on the right, to help show the different arrangements among replicas. For a large ensemble, the number of replicas representing each distinguishable arrangement would be expected to occur in proportion to the number of indistinguishable microstates (4 each for {3,0,0,1} and {1,3,0,0}, 12 for {2,1,1,0}). For the relatively small ensemble shown here, the precise number in each of the three states would be expected to deviate noticeably from the expected values of $\mathbb{N}_{\{2,1,1,0\}} = 180$ and $\mathbb{N}_{\{3,0,0,1\}} = \mathbb{N}_{\{1,3,0,0\}} = 60$ (see Figure 9.8).

[†] We use a total energy value of seven, because this matches the average subsystem energy when the combined subsystems have 14 units of energy.

this in four steps. We choose the ensemble size, identify all the configurations to be included, determine the probability of each, and fill the ensemble based on these probabilities.

Though we are free to choose any ensemble size, good thermodynamic averaging (the goal of the ensemble method) requires a large ensemble. "Large" means that there should be many more replicas[†] in our ensemble than microstates available to the system. Even for our little four-particle isolated system, which has 3 distinguishable and 20 total microstates, it would be hard to draw a picture of an ensemble of adequate size on a single page. We picked an ensemble size that is too small ($\mathbb{A} = 300$) for good averaging in Figure 9.7), in order to generate a manageable picture. This point will be illustrated below in Figure 9.8.

The configurations to be included are determined by the fact that the system is isolated. Thus, every replica must have the same energy ($E_T = 7$). There are three different distinguishable particle arrangements consistent with this total energy (configurations {2,1,1,0}, {3,0,0,1}, and {1,3,0,0}; Figure 9.7A). Thus, each of these arrangements will be included in the ensemble. The probabilities for these configurations are determined by the principle of equal *a priori* probability (because we are building a microcanonical ensemble). Although it is tempting to apply this principle directly to the three distinguishable arrangements, each of these comprises multiple indistinguishable microstates. *It is these microstates that have equal probabilities in the ensemble. Even though they cannot be distinguished, they contribute to the statistics, just the same as if they could be distinguished.* Since there are a greater number of microstates associated with the {2,1,1,0} configuration, it has a higher probability ($p_{\{2,1,1,0\}} = 12/20 = 0.6$, Problem 9.14) than the other two configurations.

To complete our picture, we will fill the ensemble by randomly selecting from our allowed microstates (there are 20 of them in this case) using our predetermined probabilities (1/20th in this case). Alternatively, we could fill it by randomly selecting configurations using configuration probabilities (e.g., 0.6 for the {2,1,1,0} configuration). The ensemble in Figure 9.7B was generated using this approach. Each replica is the same pink color, because the total energy does not vary. In the expanded region, the occupation numbers are consistent with the highest population for the {2,1,1,0} configuration.

Note that when we populate the ensemble using our random selection procedure, we are *not* simply filling the ensemble with 5% of each microstate (nor with 60% of {2,1,1,0} and 20% each of {3,0,0,1} and {1,3,0,0}). Rather, we are randomly sampling based on these probabilities. We can expect the number of microstates in the ensemble to deviate from the most probable values. This deviation is particularly conspicuous for small ensembles (Figure 9.8; Problem 9.16).

Using the four-particle subsystem, **Figure 9.8** shows how this expected deviation relates to ensemble size. For the smallest ensemble ($\mathbb{A}$, Figure 9.8A), although the probabilities for the three distinguishable arrangements have their maxima at the expected value (at a ratio of 1:1:3), the peaks are quite broad. That is, there is significant probability that an ensemble of this size would have a different composition from that expected. For $\mathbb{A} = 30$, the ensemble average would be *unlikely* to match the long-time average for the real system on which the ensemble is based. However, as the ensemble gets larger, the peaks get significantly narrower, when plotted on a scale from 0 to $\mathbb{A}$. For the $\mathbb{A} = 300$ ensemble, the probability peaks are quite narrow (Figure 9.8C). For very large ensembles (for $\mathbb{A}$ on the order of Avogadro's number), the probability distributions become infinitely sharp.[‡] In such cases, the ensemble average matches the time average of the real system, consistent with the ergodic hypothesis.

[†] Many *orders of magnitude* more replicas! This becomes particularly important if energy is not held constant. In such cases, we need good statistical sampling of rare, high-energy states.

[‡] Note that the "narrowing" of the probability distributions with ensemble size is seen when plotted as a function of $\mathbb{N}$, from 0 to $\mathbb{A}$. In this type of plot, the position along the x-axis represents the fraction of the replicates that are in a particular configuration. In terms of raw values of $\mathbb{N}$, the distribution actually gets *wider* as $\mathbb{A}$ increases, but since $\mathbb{N}$ grows slower than $\mathbb{A}$ (by $\mathbb{A}^{1/2}$), the distribution narrows on a fractional scale.

Figure 9.8 Probability distributions for distinguishable arrangements in microcanonical ensembles of different sizes. In each plot, the three curves show probabilities for the three different distinguishable arrangements for the four-particle isolated system with $E_T = 7$ (Figure 9.7). $\mathbb{N}_{\{N_1,N_2,N_3,N_4\}}$ is the number of replicates in each of these three states, for ensemble sizes ranging from $\mathbb{A} = 30$, 300, and 3000 in panels A–C, respectively. Probabilities are calculated from the binomial distribution (see Problem 9.16). For all three ensemble sizes, the most probable value is the same, but the distribution becomes narrower when plotted from $\mathbb{N} = 0$ to $\mathbb{N} = \mathbb{A}$. This reflects the fact that the average over a large ensemble matches well to thermodynamic averages of real systems, but averages of small systems do not.

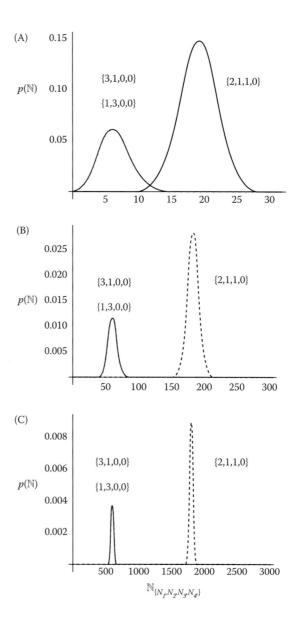

An isolated system with the two subsystems combined

Now consider an isolated system made of two combined four-particle subsystems. The two subsystems are allowed to exchange energy with each other, but not with the surroundings, so the total energy is fixed at 14 units. We will build the ensemble using the same steps as for the single four-particle. That is, we will pick an ensemble size ($\mathbb{A}$), decide what microstates are permitted, determine their probabilities, and populate the ensemble. Importantly, we will be able to "see" the energy distribution, and compare it to that described in Chapter 4: equal partitioning between the two subsystems.

For illustration, we will again choose a very small ensemble size ($\mathbb{A} = 300$). For the two-subsystem model, there will be many more allowed arrangements than for the single subsystem, because each subsystem in the pair can vary in energy from 4 to 10 units (as long as the other subsystem varies oppositely, from 10 to 4).[†] **Figure 9.9**

[†]   Neither system can go below these four units that is the lowest energy the model permits. Though the subsystem model permits energies higher than 10 units, when combined with constraints the other subsystem must contribute at least 4 units of energy and the total energy is 14 units, the maximum energy per subsystem is 10 units.

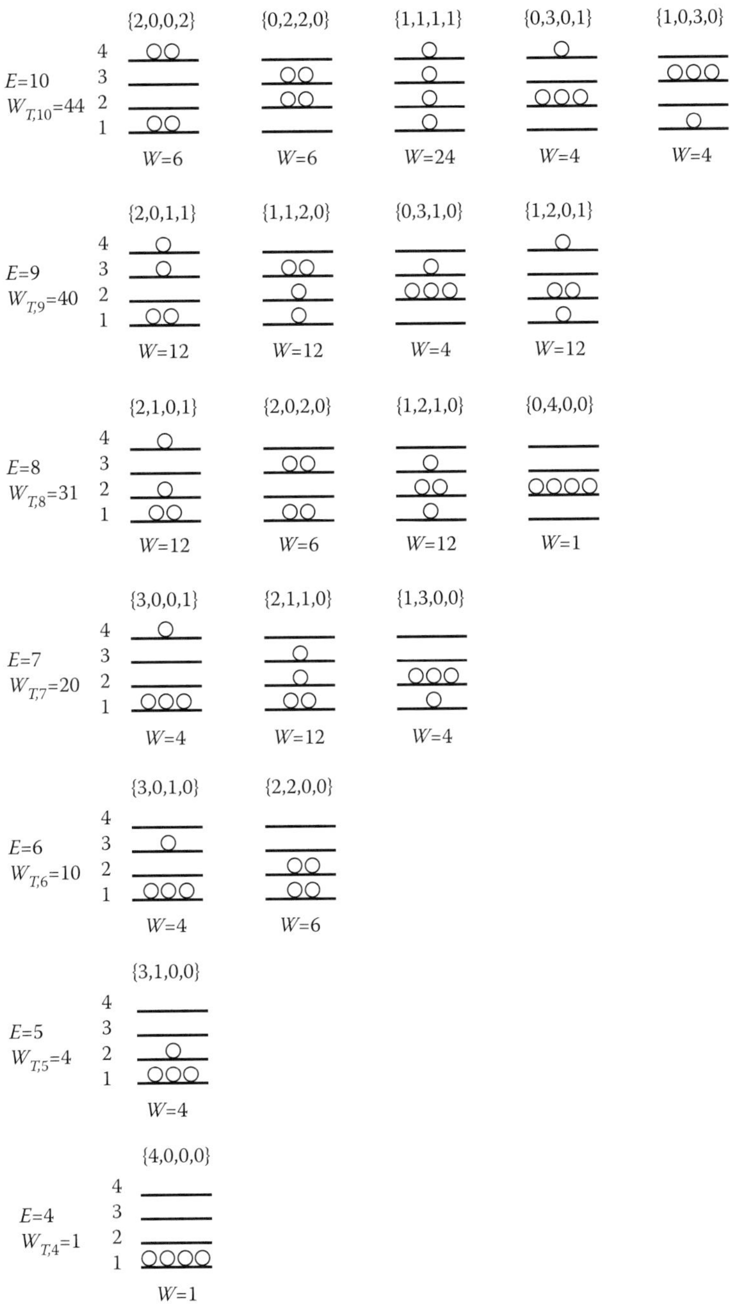

Figure 9.9 Distinguishable energy distributions for a single four-particle subsystem between the limit $4 \leq E \leq 10$. Each distinguishable configuration is shown for the allowed energy levels, along with the number of microstates, (W), consistent with each arrangement. Since each arrangement is mutually exclusive, the total number of microstates at each energy value W_T, is the sum of the W's for all the microstates of the same energy.

shows all of the distinguishable arrangements for a single subsystem, with the energy $4 \leq E \leq 10$.

As long as the energies of the two combined subsystems sum to 14 units, we can consider the two subsystems to be independent. This simplifies calculation of the number of distinguishable arrangements and of the number of microstates that compose these arrangements. We simply combine all of the microstates for two subsystems of appropriate energy by multiplication. These combinations are shown in **Figure 9.10A**.

As an example, consider the situation where the left subsystem has six units of energy (and thus, the right subsystem must have eight units of energy). There are two distinguishable arrangements with $E_L = 6$, and four with $E_R = 8$ (labeled A_L and A_R in Figure 9.10A), giving a total of $A_L \times A_R = 8$ distinguishable arrangements for the pair. The total number of microstates (regardless of arrangement) for energies six and eight are $W_6 = 10$ and $W_8 = 31$ (Figure 9.9). This means that among these eight

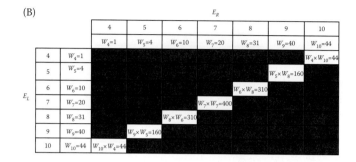

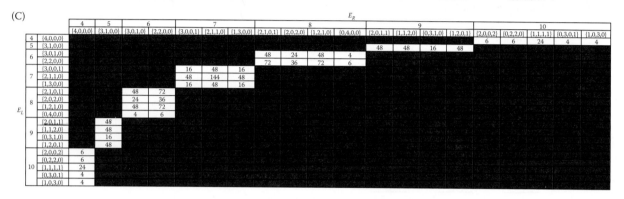

Figure 9.10 The numbers of distinguishable arrangements and microstates in the isolated two-subsystem heat exchange model. Each subsystem can adopt the energies, arrangements, and microstate numbers given in Figure 9.9, provided the two subsystems have a total energy of 14 units. (A) The number of distinguishable arrangements, and (B) the number of microstates for each energy combination. (C) The number of microstates for each distinguishable arrangement. Combinations of subsystems that are disallowed by the energy constraint $E_L + E_R = 14$ are shaded red. In total, there are 43 distinguishable arrangements (sum of white cells in panel A) and 1436 microstates (sum of white cells in panel B).

distinguishable arrangements, there are 310 microstates (**Figure 9.10B**). The distribution of these microstates overall the distinguishable 6, 8 arrangements, which is obtained by multiplying the multiplicities for each arrangement (Figure 9.9), is shown in **Figure 9.10C**. This approach also gives the number of microstates for the other allowed[†] energy combinations. From this collection of microstates (both distinguishable and indistinguishable, there are 1436 total), we can fill our ensemble by randomly selecting $\mathbb{A} = 300$ microstates (each with a probability of 1/1436). Equivalently, we could fill it by randomly selecting distinguishable arrangements using probabilities from Figure 9.10C (e.g., 48/1436 ≈ 0.0336 for the {2,1,0,1}, {3,0,1,0} configuration).

The ensemble in **Figure 9.11** was generated by randomly selecting from the 1436 heat-exchange microstates. The resulting distribution shows general trends in the energy distribution. There are a lot of snapshots that have the same shade of pink on both sides ($E_L = E_R = 7$). In addition, there are a number of snapshots that are more pink on one side than the other (a bit uneven in energy distribution), and a few snapshots that are white on one side and red on the other. These trends are qualitatively consistent with the expected energy distribution (Figure 9.3). But because the ensemble in Figure 9.11 is small, and was built by **_random_** selection, it does not exactly match the expected distribution due to statistical fluctuation (e.g., the average value of E_L is 7.05 units, not the expected 7.00).

A quantitative comparison of the energy distributions of the expected and randomly sampled $\mathbb{A} = 300$ ensemble is shown in **Figure 9.12**. The number of snapshots with different E_L values[‡] (red crosses) and the expected values from the

---

[†] $E_L + E_R = 14$.

[‡] The energy distribution of the right subsystem would look like the mirror image of Figure 9.12A, because of the energy constraint $E_L + E_R = 14$.

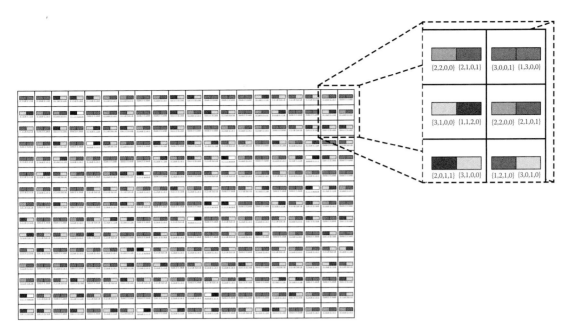

Figure 9.11 A depiction of a microcanonical ensemble for the two-subsystem model. Random microstates ($\mathbb{A} = 300$ of them) are shown for the isolated heat-flow model (Figure 4.3), with a total energy of 14 units per replica. Colors correspond to the total energies of the left and right subsystems. For each color, there are multiple distinguishable molecular configurations (given in curly brackets, see expanded inset, right), and for each distinguishable configuration, there are multiple indistinguishable microstates (Figure 9.9). This ensemble was generated by randomly selecting from each of the 1436 allowed microstates, according to the microcanonical partition function.

microcanonical partition function show peaks at $E_L = 7 \ (=E_R)$, with decreases in the number of states for energies both above and below this peak value. However, there are significant deviations at each expected energy E_L value (of $+/- 5$ snapshots). Generating different $\mathbb{A} = 300$ ensembles would produce similar distributions, with deviations of similar magnitude.

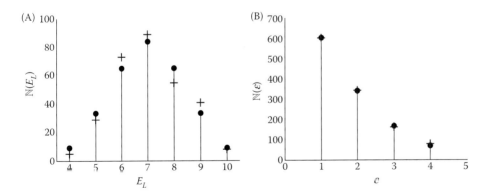

Figure 9.12 The energy distribution in the two-subsystem microcanonical ensemble. Red plus signs show the value for the $\mathbb{A} = 300$ ensemble in Figure 9.11. Black circles show the expected values from the microcanonical partition function, with the 1436 microstates given in Figure 9.10C. (A) The number of subsystems (here the left one was chosen) in the $\mathbb{A} = 300$ ensemble (red) with a given total energy (the sum of the energy of the four particles), and the expected value (black). The average value for the randomly generated $\mathbb{A} = 300$ ensemble is $E_L = 7.05$ units (i.e., the energy is nearly equally shared between the two subsystems). (B) The number of particles at each energy value (ε). Although the most probable total subsystem energy exceeds the lowest possible value (seven vs. four), the most common particle energy is one unit, the lowest energy level. The number of particles in higher levels decays roughly exponentially with increase in energy.

As shown in Figure 9.8 (for the single four-particle isolated system), we would expect better agreement between the ensemble and the expected distribution by increasing the ensemble size. Though increasing the ensemble size is easy in principle, actually drawing a really large ensemble in the style of Figure 9.11 would go beyond the limits of normal human patience. Moreover, the resulting figure would be too intricate to look at in any comprehensive way. Fortunately, we don't need to make these kinds of pictures to use statistical thermodynamics. The statistical thermodynamic approach rests on the premise that if we were to make an appropriately large ensemble, we would get a precise match to the underlying probability distribution, which in this case is the microcanonical partition function. With this premise, we can simply use the partition function to calculate ensemble properties, without actually drawing the implied picture.

Another way to look at the energy distribution

The ensembles we have calculated in this chapter are all for isolated systems. For the ensemble based on the single four-particle subsystem (Figure 9.7), the snapshot energy "distribution" is a single peak at $E = 7$, with no spread. For the two-subsystem heat exchange model, the total energy would be similarly peaked at $E_L + E_R = 14$. Though energy exchange between the two subsystems spreads the energy of each subsystem compared to the single subsystem, the total energy is still peaked at $E = 7$.

However, a different picture is obtained if we calculate the energy distribution among individual particles. We can compute this simply by going one snapshot at a time through our ensemble. For example, a snapshot in the {2,1,1,0}, {3,0,1,0} configuration has five particles with $\varepsilon = 1$, one particle with $\varepsilon = 2$, one particle with $\varepsilon = 3$, and none with $\varepsilon = 4$. Summing over the $\mathbb{A} = 300$ snapshots in Figure 9.11 gives the distribution in Figure 9.12B (red crosses). The distribution of *particle* energies decays with increasing energy, even though the total system and subsystem energies do not (Figure 9.12A). This kind of decay is consistent with the "Boltzmann distribution," which we will derive in the next chapter when we apply the ensemble method to analyze a diathermic (heat-exchanging) system. In this derivation, the temperature of the bath will emerge as the parameter that describes the sharpness of the energy decay of snapshots in the ensemble. For the isolated systems in this chapter, the particle energy decay can be thought of as resulting from the particles acting as heat baths for each other (see Problems 9.21 and 9.22).

Problems

9.1 For the two-subsystem heat exchange model (Figure 4.3), come up with a dynamic (i.e., time-based) model that allows the system to convert among allowable microstates as a function of time. Use the following rules to build your model:

(a) The total energy $(E_L + E_R)$ must be conserved at 14 units.

(b) Each set of four particles must remain on the left and right sides—they cannot cross from one system to the other, and they cannot leave (or enter) the system.

(c) For each time step, select a pair of particles randomly (they can be in the same subsystem or opposite subsystems.

(d) In a single time step, particles selected can only increase or decrease energy by one unit.

(e) If the energy of one of the selected particles goes up, the other has to go down to conserve energy (rule i. above).

(f) If the two particles selected cannot exchange energy (if both particles are in the highest or the lowest energy levels), they are left unchanged, but time is incremented by one step.

(g) If the two particles can exchange energy in only one of two ways (e.g., if one particle is in the highest or lowest level), energy is exchanged (as per rule iv.), and time is incremented by one step.

(h) If energy exchange can occur in both directions (e.g., both particles are in intermediate energy levels), pick a direction randomly, exchange energy, and increment the time.

Run your model for 100,000 time steps (you can start at whatever microstate you like), and plot time trajectories for the following time intervals: 1 to 100 steps, 1 to 1000 steps, and 1 to 10,000 steps.

9.2 Calculate and plot the autocorrelation function for the total energy in the "left" subsystem in your time series. How long does it take for the model to become uncorrelated with its former self?

9.3 Plot a histogram of the distribution of energy values on the "left" subsystem. How would this distribution compare with the distribution on the right side, and the difference $E_L - E_R$?

9.4 Compare your distribution from Problem 9.3 to that expected taking probabilities from the microcanonical partition function for the heat exchange model by plotting, and by comparing the mean and standard deviation of each.

9.5 Demonstrate the following equality (Equation 9.6):

$$\ln W = \mathbb{A}\ln\mathbb{A} - \sum_{i=1}^{m}\mathbb{N}_i\ln\mathbb{N}_i$$

9.6 In statistical thermodynamics, in order to maximize W, we instead choose to maximize $\ln W$. For the following functions,

$$(1)\quad f(x) = -x^2 + 3x \qquad (2)\quad f(x) = e^{-(x-3)^2}$$

demonstrate that the maximum of each function $f(x)$ occurs at the same value of x as the maximum of the log of each function ($\log f(x)$).

9.7 For the following functions, determine whether the critical points (min or max) of $f(x)$ and $\ln f(x)$ occur at the same value of x.

$$(1)\ f(x) = 3x^2 \qquad\qquad (2)\ f(x) = e^{-4x}$$

9.8 Demonstrate the following equality:

$$\left(\frac{\partial\ln W}{\partial\mathbb{N}_i}\right)_{\mathbb{N}_j} = \ln\mathbb{A} - \ln\mathbb{N}_i$$

9.9 Maximize $\ln W$ for the two microstate system subject to the constraint $\mathbb{N}_1 + \mathbb{N}_2 = \mathbb{A}$, by direct substitution of one maximization variable by the other.

9.10 For the Stirling's approximation of $\ln W$ for a system with two microstates, show that $\ln W$ varies linearly with $\mathbb{N}_1 + \mathbb{N}_2$ along the main diagonal defined by $\mathbb{N}_1 = \mathbb{N}_2$.

9.11 Derive an expression for the Lagrange multiplier α and populations (Equations 10.18 and 10.20) for the general microcanonical ensemble with m microstates.

9.12   Demonstrate that the gradient of the constraint on ensemble size is equal to the vector given in Equation 9.24. Plot this vector for an ensemble with two microstates (using VectorPlot in Mathematica) and an ensemble with three microstates (using VectorPlot3D).

9.13   Show that for the microcanonical ensemble, the Lagrange multiplier is linearly proportional to the entropy of the system being modeled.

9.14   Calculate the number of microstates (i.e., the multiplicities) for the three distinguishable arrangements of the four-particle system with seven total energy units (Figure 9.7).

9.15   For a microcanonical ensemble function such as that in Figure 9.7, give an expression for the maximal probability of being in the distinguishable arrangement $\{3,0,1,0\}$. For an ensemble with $\mathbb{A}$ replicas, what composition $\{\mathbb{N}_{\{3,0,1,0\}}, \mathbb{N}_{\{1,3,0,0\}}, \mathbb{N}_{\{2,1,1,0\}}\}$ does this correspond to?

9.16   For the microcanonical ensemble in Figure 9.7 ($\mathbb{A} = 300$, $E_T = 7$ units per replica), derive an expression for the probability of the ensemble having $\mathbb{N}_{\{2,1,1,0\}}$ microstates with distinguishable arrangement $\{2,1,1,0\}$. Your expression should give the black dashed curve plotted in Figure 9.8B.

9.17   Plot the distribution ($p_{\{3,0,1,0\}}$ vs. $\mathbb{N}_{\{3,0,1,0\}}$) for ensembles with $\mathbb{A} = 10$, 100, and 10,000 replicas (as in Figure 9.3). How does the root-mean-square deviation of the energy scale with the size of the ensemble?

9.18   For the microcanonical ensemble in Figure 9.7, give an expression for the "average occupancy numbers" of the particles in the four energy levels, that is, the average number in level 1, 2, 3, and 4.

9.19   Combine what you have learned in Problems 9.18 and 9.17) by generating *random* microcanonical ensembles containing 10, 100, and 10,000 snapshots, and calculating the number of $\{3,0,1,0\}$ states and the occupancy numbers for each. What happens to these values, compared to the expected averages calculated in Problems 9.17 and 9.18 as the ensemble gets large?

9.20   Using the isolated system introduced in Chapter 4 for heat flow between two subsystems (four particles per subsystem, an energy ladder of equal spacing, and a total energy of 14 units, see Figures 4.3 and 4.15), generate microcanonical ensembles with 10, 100, and 1000 replicas (you don't need to draw them). Calculate the ensemble averaged energy distribution for one of the two subsystems (E_L or E_R) for your three different ensembles.

9.21   Calculate the distribution of individual molecules in different energy levels for the single four particle subsystem, with total energy $E = 7$ (Figure 9.7).

9.22   For the isolated following systems (each with two units of energy per particle, calculate the expected molecular energy distributions (such as that in Figure 9.12B) from the corresponding microcanonical partition functions. Which ones show decaying energy distributions?

$N=1, E=2$	$N=2, E=4$	$N=3, E=6$
4 ――――	4 ――――	4 ――――
3 ――――	3 ――――	3 ――――
2 ――○――	2 ―○○―	2 ―○○○―
1 ――――	1 ――――	1 ――――
	Plus other arrangements	Plus other arrangements

CHAPTER 10

Ensembles That Interact with Their Surroundings

Goals and Summary

The goal of this chapter is to extend the ensemble method to systems that exchange energy with their surroundings. A "canonical" ensemble will be constructed in which thermodynamic replicas are in contact with a large thermal reservoir at a defined temperature. After equilibration with the reservoir through heat flow, the ensemble is isolated. The equilibrium distribution of microstates is found by maximizing the multiplicity using the Lagrange method, this time allowing all microstates to be populated regardless of energy, but holding the total energy constant. In the resulting distribution, which is described by the "canonical partition function", microstate probabilities decay exponentially with energy (the "Boltzmann distribution"), with the steepness of the decay being determined by the reservoir temperature. One important extension of this distribution, which is demonstrated with a simple energy ladder, is that when multiplicities combine with this exponential energy decay, the resulting probabilities appears to be nonexponential in energy. From the Boltzmann distribution, which is represented through the canonical partition function Q, we can derive expressions for thermodynamic quantities. The most straight forward derivation produces the Helmholtz free energy, A, which is the potential for systems at constant temperature and volume.

To connect to systems at constant pressure, we will introduce a third type of ensemble that allows volume change as well as heat flow by coupling to a mechanical (constant pressure) reservoir. Maximizing this "isothermal–isobaric ensemble" using the Lagrange method produces an exponential decay with enthalpy, and is directly related to the Gibbs energy G.

In the previous chapter, we introduced the ensemble method as a way to calculate average thermodynamic properties from molecular models. We did this by creating many replicas of our system of interest (in the previous chapter, an isolated system), freezing time, and taking ensemble averages from the resulting collection of snapshots. For the ensemble based on the isolated system (the microcanonical ensemble), the distribution was obtained by finding the maximum number of ways the ensemble could be configured, given a fixed ensemble size (A replicates). In that distribution, each allowed microstate is populated with equal probability.

Here we will consider an ensemble based on a system that can exchange energy with its surroundings in the form of heat flow (a diathermic system). The resulting "canonical" ensemble is built by replicating a thermodynamic system that is in equilibrium with a thermal reservoir. As a result, replicas can vary in energy. We will again find the equilibrium distribution by maximizing the multiplicity, but this time with the additional constraint that the total energy of the ensemble, $\mathbb{E}_T$, is constant. The equilibrium distribution is represented by another type of partition function, the "canonical partition function," in which the probabilities of microstates decay exponentially with energy.

HEAT EXCHANGE AND THE CANONICAL ENSEMBLE

To build an ensemble that describes a diathermic system, we will replicate a system enclosed by walls that are rigid and impermeable, but permit heat flow. Thus, each of the $\mathbb{A}$ replicas has a fixed number of particles (N) and a fixed volume (V), but the energy of each replica (E_i) can vary. Each replica will be put in contact with a large thermal reservoir at temperature T and with each other (**Figure 10.1**). As a result, the replicas come into thermal equilibrium with each other, and with the reservoir.[†] Freezing time gives a collection of microstates with different energies. Our goal is to find this distribution of microstates.

Before we proceed, it is worth discussing the relationship of the temperature to the snapshots in the canonical ensemble. In Figure 10.1, the temperature is defined by the thermal reservoir. Clearly, the reservoir temperature affects the ensemble distribution as a whole. However, we cannot associate the temperature with individual microstates. Though each microstate has a definite energy (E_i), the temperature is a property of the ensemble. This leads to an important but subtle difference between the microcanonical and canonical ensembles. Although each microstate of the microcanonical ensemble has the same value of N, V, and E, we cannot simply swap the roles of T and E to go from the microcanonical ensemble to the canonical ensemble.[‡]

The fact that we cannot associate temperature with individual ensemble members might seem to pose a problem, since our goal is to mimic a constant temperature system with the properties of the individual microstates. The temperature is defined by the thermal reservoir, but we don't want to include the details of the surroundings in our analysis of the ensemble. The good news is that once thermal equilibrium is achieved between the replicas and the thermal reservoir, we can forget about the reservoir—the information about the temperature is imparted on

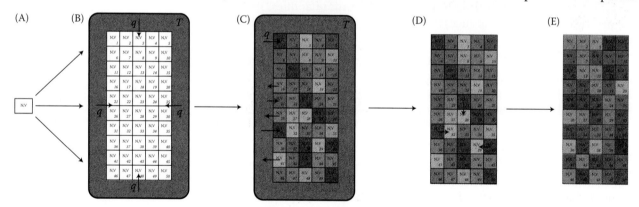

Figure 10.1 Generating a canonical ensemble to describe a diathermic system. A system that has fixed particle number and volume, but can exchange heat with its surroundings (A) is replicated $\mathbb{A}$ times (50 in this example) and is put in contact with a large heat reservoir at temperature T (approximated by the pink surroundings, B). In this configuration, heat can flow (black arrows) between the reservoir and the collection of replicas, and between replicas. As a result of heat flow, the collection of replicas comes to thermal equilibrium with the reservoir (C). At any given instant, the energy of each replica can vary, as shown by the variation in the pink shading. Once thermal equilibrium has been established, the collection of replicas can be isolated from the reservoir (D) without changing its statistical properties. These collectively isolated replicas can exchange energy with each other (compare D with E), but the total energy ($\mathbb{E}_T$, Equations 10.1 and 10.2) is fixed. By freezing time on the isolated replicas, a set of snapshots are obtained that have fixed N and V, but variable energies, E_i.

[†] The reservoirs don't need to be in contact with each other to come to thermal equilibrium, so long as they are in contact with a common reservoir. This idea is sometimes referred to as the "zeroth law of thermodynamics".

[‡] The fairly common familiar-sounding "*NVE* ensemble": and "*NVT* ensemble" in place of the somewhat baroque terms "microcanonical" and "canonical" ensembles is likely to contribute to confusion on this point.

the energy distribution among the replicas. At this point, we can isolate the entire collection of (thermally equilibrated) replicas from the reservoir (Figure 10.1D).

Isolating the collection of replicas fixes the *total* energy of the ensemble, which we will refer to as $\mathbb{E}_T$. Because energy is extensive, $\mathbb{E}_T$ is simply the sum of the instantaneous energy of each replica (E_j):

$$\mathbb{E}_T = \sum_{j=1}^{\mathbb{A}} E_j \tag{10.1}$$

This sum also applies once we freeze time, allowing $\mathbb{E}_T$ to be expressed as a sum of over microstates:

$$\mathbb{E}_T = \sum_{i=1}^{m} \mathbb{N}_i E_i \tag{10.2}$$

where E_i is the energy of a particular microstate. By introducing Equation 10.2 as a second constraint, we can use the Lagrange method to find the most probable distribution for the canonical ensemble (i.e., the canonical partition function) in terms of the microstate energy values.

THE CANONICAL PARTITION FUNCTION

As with our derivation of the microcanonical partition function in the previous chapter, we will derive the canonical partition function for a few simple systems and a general one. For comparison, we will begin with an ensemble with two microstates, to see how the total energy condition further constrains the system. Because this added constraint leads to a rather trivial maximum for the two-microstate system, we will introduce a third microstate to help illustrate how the Lagrange method works with two constraints. Finally, we will generalize to an ensemble of m microstates.

A canonical ensemble with just two microstates

We will begin with a very simple canonical ensemble with just two microstates, as with our treatment of the microcanonical ensemble in Chapter 9. However, unlike the microcanonical ensemble treatment, our microstates will have different energies, with values E_1 and E_2. We will build an ensemble of $\mathbb{A}$ replicas (of our own choosing). Again, $\mathbb{A}$ provides a constraint on $\mathbb{N}_1$ and $\mathbb{N}_2$:

$$\mathbb{N}_1 + \mathbb{N}_2 = \mathbb{A} \tag{10.3}$$

Once thermal equilibrium is achieved between our replicas and the thermal reservoir, we can use the total ensemble energy $\mathbb{E}_T$ as a second constraint on the numbers of replicates in each microstate and their energies (E_1 and E_2):

$$\mathbb{N}_1 E_1 + \mathbb{N}_2 E_2 = \mathbb{E}_T \tag{10.4}$$

In both constraint Equations 10.3 and 10.4, the unknowns are $\mathbb{N}_1$ and $\mathbb{N}_2$, which define the distribution. The values E_1 and E_2 are defined by our starting model, and $\mathbb{E}_T$ is a constant (though, unlike $\mathbb{A}$, the actual value of $\mathbb{E}_T$ may not be known to us until we complete our analysis[†]). The energy constraint Equation 10.4 is a line in

---

[†] Since we built the ensemble, we know its size $\mathbb{A}$. Though $\mathbb{E}_T$ is uniquely defined by the temperature of the thermal reservoir (a high value of T will result in a high ensemble energy) and by the energy model, we often find ourselves in a situation where we know T (and the energy model), but not $\mathbb{E}_T$, because we do not know the distribution of microstates at the outset. The partition function that we derive by applying the Lagrange method to the canonical ensemble (the canonical partition function) will give us this distribution, allowing us to calculate $\mathbb{E}_T$ (and other quantities).

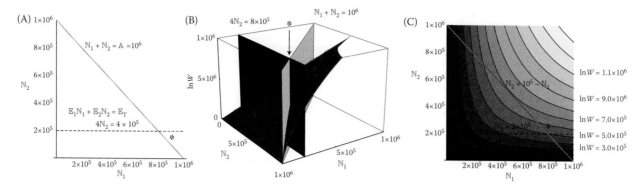

Figure 10.2 A canonical ensemble with just two microstates. In this example, we have are analyzing an ensemble with $\mathbb{A} = 10^6$ replicas, and just two different microstates with energies $E_1 = 0$ and $E_2 = 4$ units. Setting the total ensemble energy $\mathbb{E}_T$ to a fixed value (equivalent to isolating the ensemble) leads to a constraint on allowed values of $\mathbb{N}_1$ and $\mathbb{N}_2$ in the ensemble. In this example, $\mathbb{E}_T$ is 8×10^5 units, leading to the black dashed constraint line in panel A. This energy constraint intersects the ensemble size constraint (gray diagonal line, panel A) at a single point (*, where $\mathbb{N}_1 = 8 \times 10^5$, $\mathbb{N}_2 = 2 \times 10^5$), uniquely specifying the equilibrium distribution. (B) As in Figure 10.7, these constraint lines can be extended as vertical planes perpendicular to the $\mathbb{N}_1$, $\mathbb{N}_2$ plane, defining the equilibrium value of lnW (approximately 5.0×10^5, seen more clearly in panel C).

the $\mathbb{N}_1$, $\mathbb{N}_2$ plane, with an intercept proportional to $\mathbb{E}_T$, and a slope proportional to the ratio E_1/E_2 (**Figure 10.2**). And as in Chapter 10, the ensemble size constraint is a line in this plane with an intercept of $\mathbb{A}$, and a slope of minus one. The requirement that both constraints be satisfied means that these two lines must intersect. Here we are interested in the situation where the constraint lines intersect at a single point in the $\mathbb{N}_1$, $\mathbb{N}_2$ plane (because the lines have different slopes, see Figure 10.2A).[†] The equilibrium distribution of the canononical ensemble is at this point of intersection.

In Figure 10.2, we have taken E_1 and E_2 values of 0 and 4 units, a total energy of 8×10^5 units, and an ensemble size of $\mathbb{A} = 10^6$.[‡] With these values, the two constraint equations become

$$\mathbb{N}_1 + \mathbb{N}_2 = 10^6 \tag{10.5}$$

$$4\mathbb{N}_2 = 8 \times 10^5 \tag{10.6}$$

The $\mathbb{N}_1$ term is omitted from the energy constraint Equation 10.6 because here E_1 (the coefficient of $\mathbb{N}_1$) is zero. These two constraint Equations 10.5 and 10.6 have two unknowns, which can be solved by substitution. Equation 10.6 gives $\mathbb{N}_2 = 2 \times 10^5$; substituting this value into the first equation gives $\mathbb{N}_1 = 8 \times 10^5$, corresponding to the intersection point in Figure 10.2A.

Because the model in Figure 10.2 has only two microstates, we do not have to maximize $\ln W$ to find the equilibrium distribution. Our two constraints (Equations 10.5 and 10.6) are enough to specify a single equilibrium point. Though our interests generally lie in systems with more than two microstates, which will require maximization, we can gain further insight with our simple two-microstate system into how total energy relates to the ensemble distribution and to temperature, and the role that Lagrange multipliers will play the canonical ensemble.

First, let's see what happens when we choose different values for the total ensemble energy, $\mathbb{E}_T$, keeping the microstate energy values the same (here, $E_1 = 0$, $E_2 = 4$). This variation shifts the y-intercept (in this case, $\mathbb{N}_2$) of the energy constraint equation (corresponding to $\mathbb{N}_2$ values; horizontal lines, **Figure 10.3**). As long as the

---

[†] They can also intersect at all points, if they are the same line (see Problem 10.1).

[‡] This ensemble size is considerably larger than the two-microstate microcanonical ensemble in Chapter 10, because the Lagrange analysis to follow requires that $\ln W$ be accurately approximated by Stirling's equation. A good approximation is achieved at high values of $\mathbb{A}$ (and $\mathbb{N}_1$ and $\mathbb{N}_2$).

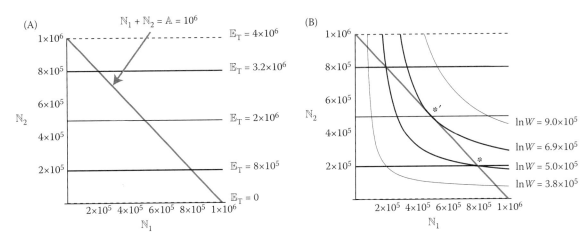

Figure 10.3 How the total energy of the two-microstate canonical ensemble shifts the populations. For the two microstate model in Figure 10.2, with energies $E_1 = 0$ and $E_2 = 4$, different energy constraint Equations 10.4 are shown, with total energy values ranging from zero to 4×10^6 units (A). Here, these constraints are horizontal lines, because E_1 (the slope of the line in the $\mathbb{N}_1$–$\mathbb{N}_2$ plane) is zero. The anti-diagonal gray line is the ensemble size constraint ($\mathbb{A} = 10$, Equation 10.5). The points of intersection of this size constraint with the different energy constraints give different equilibrium values for $\mathbb{N}_1$ and $\mathbb{N}_2$. From $\mathbb{E}_T = 0$ to $\mathbb{E}_T = 2 \times 10^6$, these points correspond to increasing reservoir (and ensemble) temperatures from absolute zero to infinite temperature. Higher total energies correspond to negative absolute temperatures (red lines, see Problem 5.3). (B) A set of contour lines for the *lnW* surface are drawn in dark red, along with approximate values of *lnW*. The thick contours pass through constraint intersections at $\mathbb{E}_T = 8 \times 10^5$ (at point ★, where $\mathbb{N}_1 = 9 \times 10^5$, and $\mathbb{N}_2 = 1 \times 10^5$) and $\mathbb{E}_T = 2 \times 10^6$ (at point ★′, where $\mathbb{N}_1 = \mathbb{N}_2 = 5 \times 10^5$).

ensemble size is kept fixed (here, $\mathbb{A} = 10^6$), changing $\mathbb{E}_T$ shifts the intersection point with the ensemble size constraint (Equation 10.5), and thus changes the equilibrium distribution.

Comparing these intersection points shows that the lower the total energy, the greater the number of low-energy snapshots ($\mathbb{N}_1$). At the lowest possible energy ($\mathbb{E}_T = 0$), all the replicas are in the low-energy microstate ($\mathbb{N}_1 = \mathbb{A}$, $\mathbb{N}_2 = 0$; lower right corner of Figure 10.3A). This would correspond to a temperature of absolute zero. As $\mathbb{E}_T$ increases, the number of snapshots in microstate 2 increases. Perhaps surprisingly, the high-temperature limit corresponds to an intersection point midway between the $\mathbb{N}_1$ and $\mathbb{N}_2$ axes (where $\mathbb{N}_1 = \mathbb{N}_2$; the point marked ★′). Intersections above this point correspond to negative absolute temperatures. As discussed in Chapter 5 (see Problem 5.3), an intersection on the $\mathbb{N}_2$ axis ($\mathbb{N}_1 = 0$, $\mathbb{N}_2 = \mathbb{A}$; upper left corner of Figure 10.3A) would correspond to negative absolute zero.

Though we don't need to use Lagrange's method for the two-microstate system, this simple system can provide a clear illustration the Lagrange method, and what the multipliers do. In **Figure 10.4**, the two equilibrium points ★ and ★′ are shown, along with the constraints that determine them and the ln$W$ contour that passes through the constraint intersection.

The Lagrange method says that at the equilibrium point, the gradients of the two constraints (shown as the arrows labeled $\nabla \mathbb{E}_T$ and $\nabla \mathbb{A}$) scaled by their multipliers (α and β) sum to the gradient of ln$W$ (arrows labeled $\nabla \ln W$):

$$\nabla \ln W = \alpha \nabla \mathbb{A} + \beta \nabla \mathbb{E}_T \qquad (10.7)$$

A closer look at the vector components and multipliers in the Lagrange Equation 10.7 is shown in **Figure 10.5**. When the Lagrange equation is evaluated at the point marked ★ from Figure 10.4 ($\mathbb{N}_1 = 8 \times 10^5$, $\mathbb{N}_2 = 2 \times 10^5$), a pair of equations results:

$$\alpha = 0.2231 \qquad (10.8)$$

$$\alpha + 4\beta = 1.6094 \qquad (10.9)$$

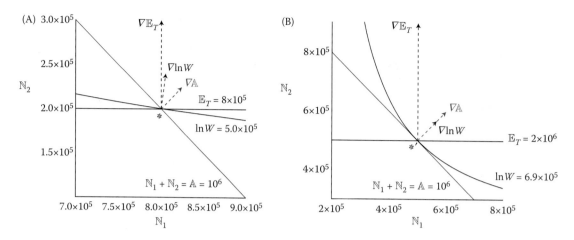

Figure 10.4 A Lagrange-type analysis of the two-microstate canonical ensemble. For the two equilibrium points from Figure 10.3B ($N_1 = 8 \times 10^5$, $N_2 = 2 \times 10^5$, panel A; $N_1 = N_2 = 5 \times 10^5$, panel B), the constraint equations and corresponding lnW contours are shown along with their gradients. To allow gradients to be compared to the curves from which they are derived, each gradient arrow is scaled (by 2.5×10^4 in panel A, and by 1×10^5 in panel B; note, the *relative* lengths of the arrows are maintained within each panel).

Figure 10.5 A geometric picture of the Lagrange multipliers for the two-microstate canonical ensemble. (A) For the equilibrium distribution $N_1 = 8 \times 10^5$, $N_2 = 2 \times 10^5$, the gradient vectors ∇A and $\nabla \mathbb{E}_T$ (dashed gray and black arrows) are scaled by $\alpha = 0.2231$ and $\beta = 0.3466$ (solid arrows). (B) When these scaled vectors are added (by placing the tail of one at the head of the other), the sum is equal to ∇lnW, consistent with the Lagrange equation.

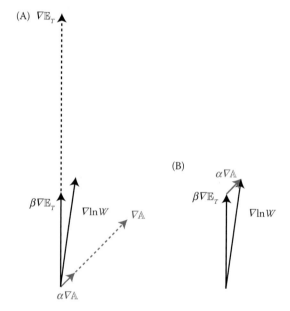

These two equations come from partial differentiation of lnW with respect to N_1, (10.8) and N_2; (10.9; see Problem 10.4). The first of these equations gives the value for α at point*, and substituting this value of α (Equation 10.8) into Equation 10.9 gives the value for β (0.3466). When these values of α and β are used to scale the ∇A and $\nabla \mathbb{E}_T$ vectors and added, the resulting sum is equal to the the ∇lnW vector (Figure 10.5B).

In this way, these three vectors determine the multipliers α and β.

To consider how α and β change for different equilibrium points, consider the point $N_1 = N_2 = 5 \times 10^5$, (marked *' in Figure 10.4B, which is the equilibrium distribution for a total ensemble energy of 2×10^6 units). At this locus, the Lagrange Equation 10.7 gives the pair of equations

$$\alpha = 0.6931 \tag{10.10}$$

$$\alpha + 4\beta = 0.6931 \tag{10.11}$$

Substituting Equation 10.8 into 10.9 gives $\beta = 0$. This can be seen from the vector diagram in Figure 10.4B (Problem 10.5): the $\nabla\mathbb{A}$ and $\nabla\ln W$ vectors are parallel, whereas the $\nabla\mathbb{E}_T$ vector is not. No multiple of $\nabla\mathbb{E}_T$ (other than zero) would combine with $\nabla\mathbb{A}$ to point in the same direction as $\nabla\ln W$. The only way to maintain the equality in the Lagrange equation is to shrink $\nabla\mathbb{A}$ (by the factor $\alpha = 0.693$), and combine it with zero multiples of $\nabla\mathbb{E}_T$. Going back to our interpretation of Figure 10.3 in terms of temperature, this says that the $\nabla\mathbb{E}_T$ constraint plays a large role at low temperature (where the Lagrange multiplier β is large), but a small role at high temperature (where β goes to zero).

A canonical ensemble with three microstates, and Lagrange maximization

For canonical ensembles with three or more microstates, the ensemble size and total energy constraints are insufficient to provide a unique equilibrium distribution of microstates. In such cases, constrained maximization of $\ln W$ is required.[†] Here we will include a third microstate in our analysis, with the goal of illustrating the constrained maximization procedure while retaining a graphical picture of the procedure. Although we cannot plot the three composition variables and a fourth $\ln W$ variable, we can represent different values of $\ln W$ in a three-dimensional space of microstate numbers $\mathbb{N}_1$, $\mathbb{N}_2$, and $\mathbb{N}_3$. (**Figure 10.6A**). This is the three-dimensional equivalent of the two-microstate $\mathbb{N}_1$, $\mathbb{N}_2$ contour plot (Figure 10.2C). In this representation, the different $\ln W$ values appear as nonintersecting surfaces, or leaflets.

Applying the approach we used for the two-state canonical ensemble, we invoke two constraint equations, one on the ensemble size $\mathbb{A}$, and one for the total ensemble energy $\mathbb{E}_T$:

$$\mathbb{N}_1 + \mathbb{N}_2 + \mathbb{N}_3 = \mathbb{A} \tag{10.12}$$

$$E_1\mathbb{N}_1 + E_2\mathbb{N}_2 + E_3\mathbb{N}_3 = \mathbb{E}_T \tag{10.13}$$

In the example in Figure 10.6, we set the ensemble size to $\mathbb{A} = 10^6$, the microstate energy levels to $E_1 = 0$, $E_2 = 1$, and $E_3 = 3$ units, and the total ensemble energy to $\mathbb{E}_T = 10^6$. Thus, the constraint equations become

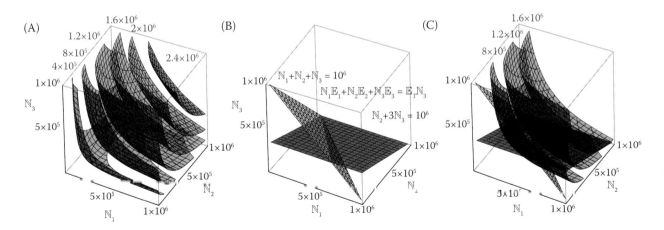

Figure 10.6 A canonical ensemble with three microstates. (A) Contour surfaces showing constant values of *lnW* (red numbering) as a function of the number of replicas in microstate 1, 2, and 3. (B) Constraint equations setting both the ensemble size $\mathbb{A}$ and total ensemble energy $\mathbb{E}_T$ to 10^6, assuming microstate energy values of $E_1 = 0$, $E_2 = 1$, and $E_3 = 3$. The intersection of these two constraint planes (a line) gives allowable distributions of the ensemble. (C) The most probable (equilibrium) distribution is that which maximizes *lnW* along this intersection. For clarity, only three *lnW* surfaces are shown.

[†] As in Chapter 10, we will assume that the most probable distribution of $\ln W$ is so much more probable than the other values that the (constrained) maximum gives an accurate representation of the equilibrium distribution (see Figure 10.9).

$$\mathbb{N}_1 + \mathbb{N}_2 + \mathbb{N}_3 = 10^6 \tag{10.14}$$

$$\mathbb{N}_2 + 3\mathbb{N}_3 = 10^6 \tag{10.15}$$

Applying the Lagrange method (Equation 10.7) gives a set of three equations:

$$\ln 10^6 - \ln \mathbb{N}_1 = \alpha \tag{10.16}$$

$$\ln 10^6 - \ln \mathbb{N}_2 = \alpha + \beta \tag{10.17}$$

$$\ln 10^6 - \ln \mathbb{N}_3 = \alpha + 3\beta \tag{10.18}$$

These equations can be rearranged to give populations of each of the three microstates:

$$p_1 = 10^{-6} \mathbb{N}_1^* = e^{-\alpha} \tag{10.19}$$

$$p_2 = 10^{-6} \mathbb{N}_2^* = e^{-\alpha} e^{-\beta} \tag{10.20}$$

$$p_3 = 10^{-6} \mathbb{N}_3^* = e^{-\alpha} e^{-3\beta} \tag{10.21}$$

As in Chapter 10, the asterisks indicate the distribution that maximizes $\ln W$. Though these three equations contain five unknowns ($\mathbb{N}_1^*$, $\mathbb{N}_2^*$, $\mathbb{N}_3^*$, α, and β), which sounds like a bad situation, we can supplement this trio with our two constraint Equations 10.14 and 10.15, giving us enough equations to solve these unknowns. We can solve for α by summing Equations 10.19 through 10.21,

$$10^{-6}(\mathbb{N}_1^* + \mathbb{N}_2^* + \mathbb{N}_3^*) = e^{-\alpha} + e^{-\alpha} e^{-\beta} + e^{-\alpha} e^{-3\beta} \tag{10.22}$$

and then substituting the constraint Equation 10.14 into the left-hand parenthesis,

$$10^{-6}(10^6) = e^{-\alpha} + e^{-\alpha} e^{-\beta} + e^{-\alpha} e^{-3\beta} \tag{10.23}$$

Rearranging gives

$$e^{\alpha} = 1 + e^{-\beta} + e^{-3\beta} \tag{10.24}$$

Equation 10.24 provides a means to eliminate α from Equations 10.19 through 10.21, and express our distribution in terms of the microstate energies[†] and the remaining undetermined multiplier β. As with the microcanonical ensemble, e^{α} plays a central role in defining the equilibrium distribution of the canonical ensemble. We will refer to e^{α} as the "canonical partition function," Q. For this three-microstate system, we can write

$$Q = e^{\alpha} = 1 + e^{-\beta} + e^{-3\beta}$$

$$= \sum_{i=1}^{m=3} e^{-E_i \beta} \tag{10.25}$$

The sum is taken over each microstate (three of them in this case), not over energy values.[‡] As with the microcanonical partition function (Ω), we refer to the terms in the sum (Equation 10.25) as Boltzmann factors. Whereas the Boltzmann factors in Ω

[†] At this point, the microstate energies are the coefficients (0, 1, and 3) that multiply β.
[‡] This distinction becomes important for models where different microstates have the same energy values.

all have values of unity, the Boltzmann factors in the canonical ensemble decay exponentially from unity[†] as microstate energy increases (assuming β to be positive).

By combining Equations 10.19 through 10.21 with Equation 10.25, we arrive at set of key equations for equilibrium populations:

$$p_1 = \frac{1}{Q} \tag{10.26}$$

$$p_2 = \frac{e^{-\beta}}{Q} \tag{10.27}$$

$$p_3 = \frac{e^{-3\beta}}{Q} \tag{10.28}$$

Though Equations 10.26 through 10.28 provide important expressions for the equilibrium distribution of the canonical ensemble, they include the multiplier β, which remains undetermined. Fortunately, we have one constraint equation left (the energy constraint) that we can use to eliminate β. For the three microstate model, dividing Equation 10.13 by $\mathbb{A}$ gives

$$E_1 \frac{\mathbb{N}_1}{\mathbb{A}} + E_2 \frac{\mathbb{N}_2}{\mathbb{A}} + E_3 \frac{\mathbb{N}_3}{\mathbb{A}} = \frac{\mathbb{E}_T}{\mathbb{A}} \tag{10.29}$$

or in terms of populations,

$$E_1 p_1 + E_2 p_2 + E_3 p_3 = \frac{\mathbb{E}_T}{\mathbb{A}} \tag{10.30}$$

Including specific microstate energy levels from our model, along with constraint values $\mathbb{A} = 10^6$ and $\mathbb{E}_T = 10^6$ gives

$$p_2 + 3p_3 = 1 \tag{10.31}$$

Substituting our population Equations 10.27 and 10.28 into Equation 10.31 to give

$$\frac{e^{-\beta}}{1 + e^{-\beta} + e^{-3\beta}} + 3 \frac{e^{-3\beta}}{1 + e^{-\beta} + e^{-3\beta}} = 1 \tag{10.32}$$

Equation 10.32 can be rearranged to give

$$e^{-\beta} + 3e^{-3\beta} = 1 + e^{-\beta} + e^{-3\beta} \tag{10.33}$$

or

$$2e^{-3\beta} = 1 \tag{10.34}$$

Taking the log of Equation 10.34 and rearranging gives

$$\ln 2 - 3\beta = 0 \tag{10.35}$$

Solving gives $\beta = \ln(2)/3 \approx 0.231$. Populations of the three microstates (using Equations 10.26 through 10.28) are shown for this β-value and two other positive values in **Figure 10.7**.

[†] The decay is from unity so long as the energy value of the lowest microstate is set to zero, as we have done for our three-microstate model.

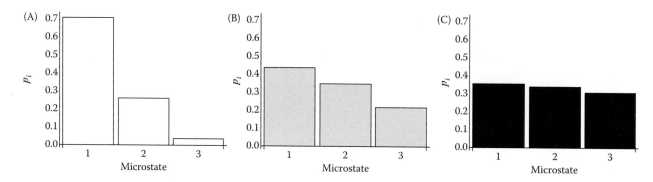

Figure 10.7 Equilibrium populations for the three-microstate canonical ensemble. Microstates have energies of 0, 1, and 3 units (our model has no microstate with an energy of two, hence the gaps). Populations are calculated using Equations 10.26 through 10.28, for β-values of 1.0 (panel A), 0.231 (panel B), and 0.05 (panel C). This corresponds to increasing temperatures from panel A to panel C, and to increasing ensemble-averaged energies (from 0.36 to 1.00 to 1.26 units in A, B, and C, respectively).

For all three values of β, microstate population decreases exponentially with increasing microstate energy. This decay is sharpest when β is large (Figure 10.7A). When β approaches zero, the populations are nearly equal, regardless of microstate energy (Figure 10.7C). Because the sum of the microstate populations is equal to one, this shift toward equal populations corresponds to an increasing total ensemble energy ($\mathbb{E}_T$, and thus to an average microstate energy, $<E_i> = \mathbb{E}_T/\mathbb{A}$; see Problems 10.7–10.9).

The connection between β and the average energy illustrated in Figure 10.7 (i.e., a high value of β corresponds to a low average energy) is worth a bit more discussion. A key feature of the canonical ensemble, shown in Figure 10.1, is its equilibration with a thermal reservoir at temperature T, but so far, temperature has not made an explicit appearance in our equations for the canonical partition function Q, or in populations derived from Q. The shift in the ensemble-averaged energy with β is like that associated with a change in temperature, although the dependences are opposite: energy values increase with temperature, but decrease with β. This suggests an inverse relation between β and T, that is,

$$\beta \propto \frac{1}{T} \tag{10.36}$$

Though we will justify this inverse dependence below by appealing to classical thermodynamic equations, it should be pointed out that Equation 10.36 is not the functional form that would lead to an increase in energy with temperature. For example, an inverse square dependence would also produce an increase.

We can also go a bit further in anticipating the relationship between β and T by recognizing that β must have units of inverse energy.[†] Thus, we can expect a proportionality constant in Equation 10.36 to have units of kelvins per joule. The constant $1/k$ has these units, suggesting $\beta = 1/kT$.

A canonical ensemble with an arbitrary number m of microstates

In this section, we will generalize the results above to $m > 3$ microstates. Though we cannot easily make plots of distributions and constraints, the equations generalize in a very straightforward way, retaining the same form as for the three-microstate example above. With m microstates, the constraints on ensemble size and total ensemble energy are

---

[†] Since the product βE_i appears in the exponents of our Boltzmann factors, the product must be dimensionless. Thus, β must have dimensions of inverse energy.

$$\sum_{i=1}^{m} \mathbb{N}_m = \mathbb{A} \tag{10.37}$$

$$\sum_{i=1}^{m} E_m \mathbb{N}_m = \mathbb{E}_T \tag{10.38}$$

Applying Lagrange's method gives

$$\nabla \ln W = \alpha \nabla \mathbb{A} + \beta \nabla \mathbb{E}_T \tag{10.39}$$

which is a system of m equations with $m + 2$ unknowns ($\mathbb{N}_1{}^*, \mathbb{N}_2{}^*, ..., \mathbb{N}_m{}^*, \alpha,$ and β):

$$\begin{aligned}
\ln \mathbb{A} - \ln \mathbb{N}_1{}^* &= \alpha + \beta E_1 \\
\ln \mathbb{A} - \ln \mathbb{N}_2{}^* &= \alpha + \beta E_2 \\
&\vdots \\
\ln \mathbb{A} - \ln \mathbb{N}_m{}^* &= \alpha + \beta E_m
\end{aligned} \tag{10.40}$$

Exponentiating and rearranging gives

$$\begin{aligned}
\frac{\mathbb{N}_1{}^*}{\mathbb{A}} &= e^{-\alpha} e^{-\beta E_1} \\
\frac{\mathbb{N}_2{}^*}{\mathbb{A}} &= e^{-\alpha} e^{-\beta E_2} \\
&\vdots \\
\frac{\mathbb{N}_m{}^*}{\mathbb{A}} &= e^{-\alpha} e^{-\beta E_m}
\end{aligned} \tag{10.41}$$

Summing this set of equations gives

$$\frac{\left(\mathbb{N}_1{}^* + \mathbb{N}_2{}^* + \cdots + \mathbb{N}_m{}^* \right)}{\mathbb{A}} = e^{-\alpha} \left(e^{-\beta E_1} + e^{-\beta E_2} + \cdots + e^{-\beta E_m} \right) \tag{10.42}$$

Substituting the constraint Equation 10.37, we can solve for α in terms of β, and in the process, obtain a general expression for the canonical partition function Q:

$$\frac{\mathbb{A}}{\mathbb{A}} = e^{-\alpha} \left(e^{-\beta E_1} + e^{-\beta E_2} + \cdots + e^{-\beta E_m} \right) \tag{10.43}$$

or

$$e^{\alpha} = \sum_{i=i}^{m} e^{-\beta E_i} \equiv Q \tag{10.44}$$

Note that if we give the lowest energy microstate to zero (i.e., $E_1 = 0$),[†] the term for the $i = 1$ microstate in the partition function is equal to one. Although this sort of shift changes the value of Q (and of each Boltzmann factor that Q comprises), it does not change the population values we derive from Q. As will be demonstrated below, applying this kind of shift makes Q into a measure of the average number of microstates that are populated in the canonical ensemble.[‡]

[†] And we shift all of the other energies accordingly.
[‡] In the same way that W provides a measure of the average number of microstates populated in the microcanonical ensemble.

As with the two- and three-state models, we can use Q (shifted or not) to express populations of different microstates in the ensemble, by taking a ratio of the Boltzmann factor to the partition function. For any give microstate j, this is written as

$$p_i = \frac{e^{-\beta E_j}}{Q} = \frac{e^{-\beta E_j}}{\sum_{i=1}^{m} e^{-\beta E_i}} \tag{10.45}$$

At this point in our three-microstate analysis, we solved for β by invoking our energy constraint equation, setting $\mathbb{E}_T$ to a known value. As long as we know the total energy, this would work here as well. However, in most cases we don't know the total energy, we simply know the temperature of the reservoir used to equilibrate the ensemble. It would be very nice, at this point, to replace β with a function of the thermodynamic temperature. Here we will justify the substitution we anticipated above, that is, $\beta = 1/kT$, by connecting Q with classical thermodynamic quantities (E and S).

Thermodynamic variables and the canonical partition function

As with the microcanonical partition function (Ω), the canonical partition function is directly connected to thermodynamic quantities. The log of Ω gives the entropy (Equation 9.30), which acts as the thermodynamic potential for the isolated system (which the microcanonical ensemble models). As we will show below, the log of Q gives the Helmholtz energy A, which acts as the thermodynamic potential for the isothermal system at fixed volume (which the canonical ensemble models). It is gratifying that in both cases, the position of equilibrium identified by the ensemble method is so closely connected to the thermodynamic potential for the type of system modeled by the ensemble.

The internal energy

We will first derive the relationship between Q and the internal energy U. The internal energy can be calculated as the population-weighted average of the microstate energies[†]:

$$U = \langle E_i \rangle = \sum_{i=1}^{m} p_i E_i \tag{10.46}$$

In some treatments, an additional offset term (U_0) is added ensemble-averaged energy, which accounts for modes of energy that are not included in the partition function. For clarity, we will omit this extra term.

We can substitute for the p_i values in Equation 10.46 using Equation 10.45:

$$U = \sum_{i=1}^{m} \frac{e^{-\beta E_i}}{Q} E_i = \frac{1}{Q} \sum_{i=1}^{m} E_i e^{-\beta E_i} \tag{10.47}$$

The rearrangement on the right-hand side of Equation 10.47 takes advantage of the fact that Q is independent of the index of summation. The expression inside the sum is equal to the derivative of $e^{-\beta E_i}$ with respect to β:

$$U = \frac{1}{Q} \sum_{i=1}^{m} -\frac{de^{-\beta E_i}}{d\beta} = -\frac{1}{Q} \frac{d}{d\beta} \sum_{i=1}^{m} e^{-\beta E_i} \tag{10.48}$$

[†] We could also calculate U as the average of E_i over each of the A replicas in the ensemble; however, it is best to get away from this arbitrary number.

The rearrangement on the right-hand side simply moves the derivative outside the sum.[†] The sum that remains is simply the canonical partition function:

$$U = -\frac{1}{Q}\frac{dQ}{d\beta} = -\frac{d\ln Q}{d\beta} \tag{10.49}$$

This expression certainly has a simple form, though we will hold off discussing it until we have established the relationship between β and T.

The entropy

The entropy can be obtained from Gibbs' formula (which is derived for the canonical ensemble in Problem 10.13):

$$S = -k\sum_{i=1}^{m} p_i \ln p_i \tag{10.50}$$

We will substitute the population formula (Equation 10.45) inside of the log term to give

$$S = -k\sum_{i=1}^{m} p_i \ln\left(\frac{e^{-\beta E_i}}{Q}\right) = -k\sum_{i=1}^{m} p_i\left(\ln e^{-\beta E_i} - \ln Q\right) \tag{10.51}$$

Again, Q can come outside the sum, giving

$$S = k\sum_{i=1}^{m} \beta E_i p_i + k\ln Q\sum_{i=1}^{m} p_i \tag{10.52}$$

For the first sum in Equation 10.52, β comes out, leaving the average internal energy. The second sum is equal to one. Thus,

$$S = k\beta U + k\ln Q \tag{10.53}$$

The thermodynamic value of β

With expressions for the energy and the entropy (Equations 10.49 and 10.53), we can determine the value of the Lagrange multiplier β and replace it an expression involving the reservoir temperature. From classical thermodynamics, we can express T from U and S through the thermodynamic identity we developed in Chapter 5:

$$\left(\frac{\partial U}{\partial S}\right)_V = T \tag{10.54}$$

The constant volume part of the identity is appropriate to the canonical ensemble, where each replica has the same value of V. At this point, neither U (Equation 10.49) nor S (Equation 10.53) is in great form to evaluate the derivative in Equation 10.54, but they are both functions of β, so we will evaluate Equation 10.54 by taking the ratio of a pair of derivatives with respect to β[‡]:

$$\left(\frac{\partial U}{\partial S}\right)_V = \frac{\left(\frac{\partial U}{\partial \beta}\right)_V}{\left(\frac{\partial S}{\partial \beta}\right)_V} = T \tag{10.55}$$

[†] This stunt (setting the sum derivatives to the derivative of the sum) is fairly common in statistical thermodynamics.

[‡] This is equivalent to using the chain-rule to evaluate, for example, $dU/d\beta$, taking U to be a function of S, and S to be a function of β.

Differentiating S from Equation 10.53 simplifies the denominator of Equation 10.55:

$$\left(\frac{\partial S}{\partial \beta}\right) = \left(\frac{\partial}{\partial \beta}\{k\beta U + k\ln Q\}\right)_V$$
$$= k\left[U + \beta\left(\frac{\partial U}{\partial \beta}\right)_V + \left(\frac{\partial \ln Q}{\partial \beta}\right)_V\right] \tag{10.56}$$

By substituting Equation 10.49 for U, the first and third terms in Equation 10.56 cancel, leaving

$$\left(\frac{\partial S}{\partial \beta}\right)_V = k\beta\left(\frac{\partial U}{\partial \beta}\right)_V \tag{10.57}$$

Inserting this into the denominator of Equation 10.55 gives

$$\left(\frac{\partial U}{\partial S}\right)_V = \frac{\left(\dfrac{\partial U}{\partial \beta}\right)_V}{k\beta\left(\dfrac{\partial U}{\partial \beta}\right)_V} = T \tag{10.58}$$

The derivatives in the numerator and denominator of the middle term cancel, rearranging to give the relationship we seek:

$$\beta = \frac{1}{kT} \tag{10.59}$$

With this fundamental expression for β, it is worth rewriting both the canonical partition function (Equation 10.44) and the results we have derived so far with this substitution. The canonical partition function can be written as

$$Q = \sum_{i=1}^{m} e^{-E_i/kT} \tag{10.60}$$

Microstate populations (Equation 10.45) are given as

$$p_j = \frac{e^{-E_j/kT}}{Q} = \frac{e^{-E_j/kT}}{\displaystyle\sum_{i=1}^{m} e^{-E_i/kT}} \tag{10.61}$$

Here the j is used to avoid any confusion with the index of the sum.

Thermodynamic relations from the canonical partition function

Now that we have an expression for Q as a function of the reservoir temperature (and by thermal equilibrium, the ensemble temperature), we can derive some important expressions for the thermodynamic functions such as U, A, and C_v. Note that as long as we know the absolute energy levels, we can calculate absolute values for these quantities, rather than the difference quantities we were restricted to in classical thermodynamics.

The internal energy (Equation 10.49) is

$$U = -\frac{d\ln Q}{d(1/kT)} = -\frac{kd\ln Q}{d(1/T)} \tag{10.62}$$

By recognizing that $d(1/T) = -dT/T^2$,[†] Equation 10.62 can be simplified to

$$U = kT^2 \frac{d\ln Q}{dT} \tag{10.63}$$

By substituting $\beta = 1/kT$ in the equation (Equation 10.53), we can derive a formula for the entropy of the canonical ensemble:

$$S = \frac{U}{T} + k\ln Q \tag{10.64}$$

As with the microcanonical ensemble, the entropy is proportional to the log of the canonical partition function through the second term in Equation 10.64. However, there is an additional contribution to the entropy from the (always positive-valued) term U/T. This term is similar in form to our classical definition of entropy from the Clausius equation (Chapter 4), namely, $dS = dq_{rev}/T$. There we argued that a injecting heat into a low-temperature system produces a large amount of entropy, because the cold system is not likely to convert the associated energy to another system.

Using this expression for the entropy, we can express the free energy[‡] in term of Q, using our definition from classical thermodynamics:

$$\begin{aligned} A &= U - TS \\ &= U - T\left(\frac{U}{T} + k\ln Q\right) \\ &= -kT\ln Q \end{aligned} \tag{10.65}$$

As we saw with the microcanonical ensemble, the logarithm of the canonical partition function corresponds to the thermodynamic potential for the system described by the ensemble (constant temperature and constant volume).

Another important quantity that can be calculated from Q is the heat capacity, C_V.[§] Recall from Chapter 4 that C_V is the temperature derivative of the internal energy. Using equation (Equation 10.63), we obtain the following:

$$C_V = \left(\frac{\partial U}{\partial T}\right)_V = 2kT\frac{\partial \ln Q}{\partial T} + kT^2 \frac{\partial^2 \ln Q}{\partial T^2} \tag{10.66}$$

Although this expression is rather cumbersome to apply by hand, especially when Q has many terms, it is easily computed in packages such as Mathematica.

There is another important expression for the heat capacity that we are now in position to derive. As alluded to in previous chapters, the heat capacity is closely related to the variance of the energy. Specifically,

$$C_V = \frac{\left\langle \left(E_i - \langle E_i \rangle\right)^2 \right\rangle}{kT^2} = \frac{\langle E_i^2 \rangle - \langle E_i \rangle^2}{kT^2} = \frac{\langle E_i^2 \rangle - U^2}{kT^2} \tag{10.67}$$

where E_i are microstate energies, and the angled-brackets indicate averages over the ensemble. An analogous relation holds between C_p and the variance of $E_i - pV_i$.

[†] This simple mathematical identity can be remembered by taking the derivative an inverse and rearranging. For example, the derivative of x^{-1} with respect to x is

$$\frac{d}{dx}\left(\frac{1}{x}\right) = -\frac{1}{x^2}$$

Moving the dx to the right gives $dx^{-1} = -x^{-2}dx$.

[‡] Again, because the canonical ensemble is at constant volume, this is the Helmholtz energy, A.

[§] C_v in this case, since the volume of the replicas in Q are fixed.

The second equality follows from the expansion developed in Chapter 1 (Equation 1.34). Here we will derive Equation 10.67 starting with the population-weighted average of energy (Equation 10.46):

$$U = \langle E_i \rangle = \sum_{i=1}^{m} E_i p_i = \frac{1}{Q} \sum_{i=1}^{m} E_i e^{-E_i/kT} \tag{10.68}$$

Differentiating this Equation 10.65 with respect to T gives the C_V:

$$\frac{\partial \langle E_i \rangle}{\partial T} = \frac{1}{Q} \sum_{i=1}^{m} \frac{E_i^2}{kT^2} e^{-\varepsilon_i/kT} - \frac{1}{Q^2} \left(\frac{\partial Q}{\partial T} \right) \sum_{i=1}^{m} E_i e^{-E_i/kT} \tag{10.69}$$

Equation 10.69 makes use of the product rule for differentiation, since both Q and the exponential are temperature dependent. Rearranging the two terms in Equation 10.65 gives

$$\begin{aligned}
C_V = \frac{\partial U}{\partial T} &= \frac{1}{kT^2} \sum_{i=1}^{m} E_i^2 \frac{e^{-E_i/kT}}{Q} - \frac{\partial \ln Q}{\partial T} \sum_{i=1}^{m} E_i \frac{e^{-E_i/kT}}{Q} \\
&= \frac{1}{kT^2} \sum_{i=1}^{m} E_i^2 p_i - \frac{1}{kT^2} \langle E_i \rangle \sum_{i=1}^{m} E_i p_i \\
&= \frac{1}{kT^2} \langle E_i^2 \rangle - \frac{1}{kT^2} \langle E_i \rangle \langle E_i \rangle \\
&= \frac{\langle E_i^2 \rangle - \langle E_i \rangle^2}{kT^2}
\end{aligned} \tag{10.70}$$

Equation 10.70 provides a unique and very fundamental view of heat capacity. If, for a canonical ensemble, we generate a histogram of microstate energies, the spread of the energies is proportional to the heat capacity. At a given temperature, a narrow distribution of microstate energies (**Figure 10.8A**) corresponds to a low heat capacity, whereas one with a broad distribution corresponds to a high heat capacity (see Figure 10.8).

Note that when the two distributions shown in Figure 10.8A and B are combined (Figure 10.8C), an even larger energy variance (by two- and five-fold, respectively; see Problem 10.18), and thus a larger heat capacity. The heat capacity peak seen in the thermal denaturation of proteins by DSC is a direct result of the large energy variation, in this case, between the native and denatured states.

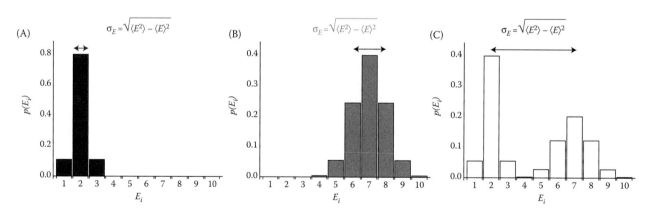

Figure 10.8 Energy distributions and their relation to heat capacity. Populations of integer energy levels are given for (A) a narrow energy distribution, (B) a broad distribution, and (C) a distribution that is a combination of (A) and (B). Double-headed arrows show the RMSDs of energy for the three distributions, σ_E. The square of σ_E is proportional to the heat capacity through Equation 10.70. At a given temperature, the narrow distribution (A) has the smallest heat capacity; though the broader distribution (B) has a higher heat capacity, it is smaller than that of (C), which has a high-energy variance as a result of the two peaks at energy values of two and seven.

A CANONICAL ENSEMBLE REPRESENTING A THREE PARTICLE ISOTHERMAL SYSTEM

To illustrate the behavior of the canonical ensemble and the thermodynamic functions we have derived above, we will return to our energy ladder model. We will consider a three-particle system to help keep things manageable. Unlike the microcanonical ensemble, where we only included replicas with a specific energy (see Figure 10.7), the canonical ensemble includes all microstates regardless of their energy. The allowed arrangements for this model are shown in **Figure 10.9**, along with multiplicities for each. Note that in this example, energies in molar units, to keep the numbers simple (see **Box 10.1**).

The canonical partition function allows us to determine *how many* replicas are in each allowed microstate (and thus, the probability of each microstate). However, since microstate with the same energies have the same probability (you can think of replicas with the same energy as constituting a microcanonical sub-ensemble within the canonical ensemble), each Boltzmann factor with the same microstate energy will be the same. This allows us to sum over energy levels rather than microstate identities, multiplying Boltzmann factors by the number of microstates at that energy. Continuing with the microcanonical sub-ensemble picture, we will use the number Ω_i (the microcanonical partition function, Chapter 9) to represent the number of microstates with energy i. For the three-particle ladder, the canonical partition function can be written as

$$Q = \sum_{i=1}^{7} \Omega_i e^{-\bar{E}_i/RT}$$

(10.78)

$$= 1 + 3e^{-1000/RT} + 6e^{-2000/RT} + 7e^{-3000/RT} + 6e^{-4000/RT} + 3e^{-5000/RT} + e^{-6000/RT}$$

As described in Box 10.1, $\hat{E}_i$ is the energy per mole of replicas. In addition to being shorter than a partition function over microstates, the summation over energy values is easier to index than a sum over indistinguishable microstates. We will often

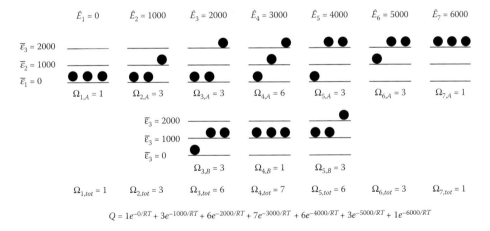

$$Q = 1e^{-0/RT} + 3e^{-1000/RT} + 6e^{-2000/RT} + 7e^{-3000/RT} + 6e^{-4000/RT} + 3e^{-5000/RT} + 1e^{-6000/RT}$$

Figure 10.9 The energy levels, multiplicities, and the canonical partition function for a three-particle energy ladder. Here we are using *molar* energies for particle energy levels ($\bar{\varepsilon}_i$). For a give replica, the sum of $\bar{\varepsilon}_i$ values over the particles can be considered energies per mole of *replica* ($\hat{E}_i$, see Box 10.1). There are two distinguishable configurations arrangements with $\hat{E}_i$ values of 2000, 3000, and 4000 J mol⁻¹. For most distinguishable arrangements there are multiple indistinguishable microstates (with multiplicities $\Omega_{i,A}$ and $\Omega_{i,B}$). Because each microstate is mutually exclusive to the others, multiplicities add. Thus, the total multiplicity of a given $\hat{E}_i$ level is the sum of the multiplicities from the two distinguishable states ($\Omega_{i,tot} = \Omega_{i,A} + \Omega_{i,B}$). Using these total multiplicities, a canonical partition function can be constructed (Equation 10.78) with seven terms, one for each $\hat{E}_i$ value. This is equivalent to summing over energy levels (instead of microstates), where the prefactors for each term are the multiplicities at each $\hat{E}_i$ value.

Box 10.1: Molar energies in statistical thermodynamics

In the example in Figure 10.9, we are using *molar* energies (i.e., energies per mole of particles) instead of absolute, extensive energies. Molar energies have two advantages over molecular energies. First, molar energies are ordinary-sized numbers, whereas molecular energies (i.e., amount of energy per molecule) are tiny quantities, and working with them in SI units can be tedious. For example, the energy of a mole of particles in energy level 2 of Figure 10.9 has a value of $\bar{\varepsilon}_2 = 1\,\mathrm{kJ\,mol^{-1}}$ (as in Chapter 5, we will use an overbar to indicate molar quantities); per molecule this corresponds to an energy of $\varepsilon_2 \approx 1.66 \times 10^{-21}$ J. Partition functions full of numbers like this are awkward to read and write, and mistakes are easy to make. Second, chemists and biologists typically express thermodynamic quantities (such as energy) per mole, not per molecule. Thus, using molar quantities puts statistical thermodynamics into a more familiar representation.

In statistical thermodynamic formulas, there are two places where the issue of energy units comes into play: in the exponents of Boltzmann factors, and in calculations of thermodynamic quantities based on logarithms of partition functions. In Boltzmann factors, molecular energies are always divided by k (and T) to produce a dimensionless quantity:

$$BF = e^{-E_i/kT} = e^{-\left(\sum_{j=1}^{N} \varepsilon_j\right)_i \Big/ kT} \tag{10.71}$$

The second equality uses the fact that the energy E_i of a microstate of particles is the sum of the molecular energies of each of the N molecules:

$$E_i = \sum_{j=1}^{N} \varepsilon_j \tag{10.72}$$

To express the energy of each molecule on a molar basis, the molecular energy is multiplied by Avogadro's number, $L_0 = 6.02 \times 10^{23}$, that is,

$$\bar{\varepsilon}_j = L_0 \varepsilon_j \tag{10.73}$$

In terms of molecules, I like to think of L_0 having units of "molecules per mole." Thus, the units for the product $L_0 \varepsilon_j$ are

$$Dim(L_0 \varepsilon_i) = \left(\frac{\mathrm{molecules}}{\mathrm{mole}}\right) \times \left(\frac{\mathrm{joules}}{\mathrm{molecule}}\right) = \left(\frac{\mathrm{joules}}{\mathrm{mole}}\right)$$

Rearranging Equation 10.73 and substituting for molecular energy in Equation 10.71 gives

$$BF_i = e^{-\left(\sum_{j=1}^{N} \frac{\bar{\varepsilon}_j}{N_A}\right)_i \Big/ kT} = e^{-\left(\sum_{j=1}^{N} \bar{\varepsilon}_j\right)_i \Big/ RT} = e^{-\tilde{E}_i/RT} \tag{10.74}$$

The middle equality comes from the fact that the product $L_0 k = R$ (you can think of k as a gas constant for a single molecule, and R as the gas constant for a mole of molecules). The last equality gives the energy *of a mole of replicas*. Since it is not quite a molar energy (i.e., it is not the energy of a mole of molecules), we use a bent over-bar instead of a flat one.

Making these substitutions in the canonical partition function gives

$$Q = \sum_{i=1}^{m} e^{\sum_{j=1}^{N} -\bar{\varepsilon}_j/RT} = \sum_{i=1}^{m} e^{-\tilde{E}_i/kT} \tag{10.75}$$

(Continued)

Box 10.1 (*Continued*): Molar energies in statistical thermodynamics

Similarly, in calculating thermodynamics quantities from partition functions, we can use Avogadro's number to express thermodynamic quantities in molar units. For example, the Helmholtz energy A can be expressed as

$$\hat{A} = L_0 \times A = -L_0 kT \ln Q = -RT \ln Q \tag{10.76}$$

This expression gives the Helmholtz energy per mole of replicas ($\hat{A}$), and each replica contains N molecules (e.g., $N = 3$ in Figure 10.9). To obtain the Helmholtz energy per mole of molecules, Equation 10.76 must be divided by N_A:

$$\bar{A} = \frac{L_0}{N} \times A = -\frac{RT}{N} \ln Q \tag{10.77}$$

index partition functions this way, summing over a shared or collective parameter for simplicity (such as the total energy, the number of ligands bound over multiple binding sites, and the extent of structure formation), being careful to account for each microstate through the multiplicity term Ω_i.

The temperature dependence of the canonical partition function for the three-particle ladder is given in **Figure 10.10A**.

As the temperature approaches zero, Q approaches a value of one:

$$\lim_{T \to 0} Q = \Omega_1 e^{-\bar{E}_1/R \times 0} + \sum_{i=2}^{7} \Omega_i e^{-\bar{E}_i/R \times 0}$$

$$= 1 e^{-0/0} + \sum_{i=2}^{7} \Omega_i e^{-\infty} \tag{10.79}^{\dagger}$$

$$= e^{-0/0} = 1$$

(the "low-temperature limit" of the partition function). As the temperature increases, Q increases monotonically. This increase is modest at low temperature, but picks up

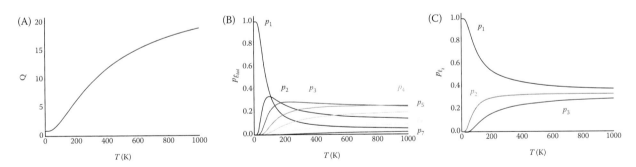

Figure 10.10 The dependence of the canonical partition function and energy-level populations on temperature for the three-particle ladder (A) The partition function Q (Figure 10.9, bottom) increases monotonically with temperature from a low-temperature limit of unity. At 300 K, Q has a value of about 10, indicating that the ensemble is spread out over about 10 microstates. (B) Temperature dependence of populations over the seven allowed energy levels (Figure 10.9). For the lowest energy level (with just a single microstate), the fractional population decreases sharply from unity to 1/27 at high temperature. For higher energy levels, populations increase from zero to limiting values of $\Omega_i/27$. These increases occur at different temperatures, and for some intermediate levels (e.g., level 2), this increase is followed by a decrease to the limiting value. (C) Temperature dependence of the population of individual particles over the three allowed energy levels. Unlike the distribution in panel (B), the lowest energy state ($\bar{\varepsilon}_1 = 0$) always has the highest population, reflecting the simple exponential decay of the Boltzmann distribution.

$\dagger$ The last line in the equality can be obtained by applying l'Hopital's rule to the quotient in the exponent, which is $\bar{E}_1/RT = 0/RT$, and differentiating numerator and denominator with respect to T. The result is equal to zero, giving $e^0 = 1$.

as temperature is increased. Finally, at high temperature, Q reaches a plateau.[†]
For the three-particle ladder, this "high-temperature limit" has a value of 27:

$$\lim_{T \to \infty} Q = \sum_{i=1}^{7} \Omega_i e^{-\bar{E}_i / R \times \infty}$$

$$= \sum_{i=1}^{7} \Omega_i e^0 \tag{10.80}$$

$$= \sum_{i=1}^{7} \Omega_i = 27$$

Values for Ω_i (given in Figure 10.9) sum up to 27 for the three-particle model.

The temperature limits of Q have special significance in terms of the population of different states. In general, partition functions are proportional to the number of microstates that are populated. Large values of Q mean that many microstates are populated, small values mean few are populated. In this example, because we set the ground-state energy to zero, this relationship becomes more than a proportionality: *Q is equal to the average number of states populated.*[‡]

Population distributions as a function of temperature

The populations of different microstates can be calculated as a function of temperature using an equation analogous to (Equation 10.61). As with the partition function above (Equation 10.78), we will calculate populations at different energy values (rather than populations of indistinguishable microstates) by multiplying microstate probabilities and energy level multiplicities:

$$p_{\bar{E}_i} = \Omega_{i,tot} \times p_i = \Omega_{i,tot} \times \frac{e^{-\bar{E}_i / RT}}{Q} \tag{10.81}$$

At absolute zero, only the ground-state energy level is populated (a single microstate in this model, **Figure 10.10B**). This can be derived from Equation 10.61:

$$\lim_{T \to 0} p_{\bar{E}_1} = \Omega_{1,tot} \frac{e^{0/0}}{Q} = 1 \tag{10.82}$$

Physically, this is because there is there is not enough thermal energy to populate higher energy states.

As the temperature increases, populations increase in the higher energy levels (Figure 10.10B), reflecting the increase in available thermal energy. At very high temperatures, the energy differences between microstates become negligible compared to the very large thermal energies. In this situation, all microstates (27 of them in this model) become equally populated regardless of their energy values. This means that the population at each energy level is equal to the number of microstates at that level:

$$\lim_{T \to \infty} p_{\bar{E}_i} = \Omega_{i,tot} \frac{e^{-\bar{E}_i / R\infty}}{Q} = \frac{\Omega_{i,tot}}{27} \tag{10.83}$$

[†] The high temperature plateau occurs for models where there are a finite number of microstates with finite energies. For models with an infinite number of microstates of increasing energy, the increase of Q with temperature is unbounded.

[‡] Obtaining this one-to-one correspondence is the main reason to set the ground state energy to zero.

In this regard, the high-temperature canonical ensemble is similar to the microcanonical ensemble, in which all states are uniformly populated. Note that if the energy model does not have an upper bound (i.e., if microstates are allowed with infinitely large energies), then Q continues to increase with temperature (Problem 10.11).

Another way to represent the equilibrium distribution is to plot total populations versus microstate energy level at different temperatures (**Figure 10.11**). This is equivalent to taking cuts at different temperatures in Figure 10.10B. At low temperature (Figure 10.11A), the three-microstate ladder shows what appears to be an exponential decay in population as a function of energy, consistent with the exponential form of the Boltzmann factors in the canonical ensemble (see, e.g., Equation 10.61).

However, a markedly different distribution is seen at higher temperature (Figure 10.11B, C). Rather than simple exponential decay, the distribution peaks at intermediate energy levels. This is because the probabilities in Figure 10.11 are for having a given *energy value* (Equation 10.83), not for being in a given microstate (Equation 10.61). Though the probabilities of individual microstates decay exponentially with energy (**Figure 10.12A**), the number of microstates with a given energy ($\Omega_{i,tot}$) are peaked at intermediate energy (**Figure 10.12B**). Unless the exponential decay of microstates is very sharp, the product of the microstate probabilities and $W_{i,tot}$ (Equation 10.83) will resemble $\Omega_{i,tot}$, peaking at intermediate energy values.

This type of nonexponential decay in population verses energy level is seen in many types of systems, including the kinetic energy distribution of an ideal gas (Problem 10.12).

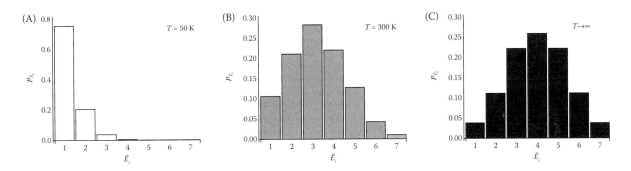

Figure 10.11 Distribution of replicas among different energy microstate energies. Using the three-particle ladder with energies in Figure 10.10, the population at each three-particle energy value (E_i, here in kJ mol⁻¹) is plotted at three different temperatures. At low temperature (A), the distribution approximates an exponential decay. However, at intermediate temperature (B), the distribution is peaked at $\hat{E}_i = 3$ energy units. In the high-temperature limit (C), this peak becomes symmetric, reflecting only the number of microstates at each energy value ($\Omega_{i,tot}$).

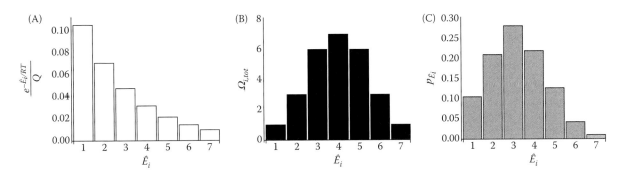

Figure 10.12 Individual probabilities decay exponentially with energy, but populations at different energies do not. Microstate probabilities (A, Equation 10.61) are evaluated for the three-particle ladder at a fixed temperature ($T = 300$ K). Since Q is a constant at fixed temperature, the microstate probability (A) decays exponentially with increasing microstate energy. However, owing to the increased number of microstates at intermediate energy levels (with a maximum at $\hat{E}_i = 4$ energy units, panel B), the population of replicas at different energy levels peaks at 3 energy units (panel C).

Bulk thermodynamic properties for the three-particle ladder

With probabilities in hand, we can describe the bulk ensemble properties based on the properties of the individual microstates. Combining these populations with relationships from classical thermodynamics gives ensemble energies, entropies, and the like.

The internal energy can be calculated from the temperature derivative of the partition function (Equation 10.63). For the three-particle model, the molar internal energy (per mole of molecules) is

$$\bar{U} = \frac{\hat{U}}{N} = \frac{RT^2}{N}\frac{dQ}{dT}$$

$$= \frac{1000}{3} \times \frac{3e^{-1000/RT} + 12e^{-2000/RT} + 21e^{-3000/RT} + 24e^{-4000/RT} + 15e^{-5000/RT} + 6e^{-6000/RT}}{1 + 3e^{-1000/RT} + 6e^{-2000/RT} + 7e^{-3000/RT} + 6e^{-4000/RT} + 3e^{-5000/RT} + e^{-6000/RT}}$$

$$(10.84)$$

(see Problem 10.12). The temperature dependence of the internal energy for the three-particle ladder is shown in **Figure 10.13A**. The energy starts at zero (where only the lowest energy microstate is populated), and it increases monotonically to an average energy value of 1000 units at high temperature (where each microstate becomes equally populated). Like Q, the increase in $\bar{U}$ lags at low temperature, because increases in temperature are too small to promote higher energy microstates in appreciable numbers. It is only when the temperature difference (more precisely, the product RT) approaches the energy of the excited states that a population shift occurs (Figure 10.10B), increasing the energy.

An important but subtle point is that the high-temperature limit of the ensemble-averaged internal energy remains well below the energy of highest allowed microstate (2000 kJ mol⁻¹ per particle). Achieving such an internal energy would require each replica to adopt the highest energy microstate, effectively inverting the population from the low temperature limit. This would result in a very low ensemble entropy, and a negative absolute temperature (approaching zero). Rather, in the high-temperature limit (Figure 10.10B), the ensemble maximizes entropy, spreading population evenly over all microstates regardless of their energy value.

As described in the previous section, the entropy derived from the canonical ensemble is the sum of two terms, the average energy (Equation 10.84 for the three-particle ladder) divided by the temperature, and the log of the canonical partition

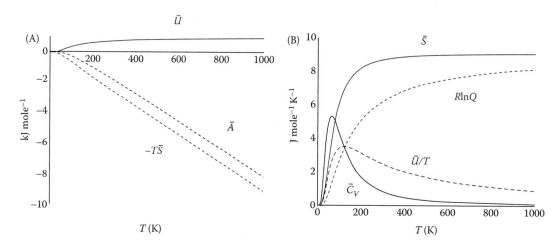

Figure 10.13 Ensemble-averaged thermodynamic functions for the three-particle ladder. (A) The ensemble-averaged molar internal energy (red), minus T times the entropy (red dashes) and Helmholtz energy (black). (B) The entropy and its parts (red), and the heat capacity (black). Note the difference in energy scale in (A) versus (B). All quantities are molar quantities, that is, per mole of particles.

function, Q (Equation 10.78). The log of Q increases monotonically from zero with temperature, eventually reaching a plateau at high temperature (**Figure 10.13B**). In contrast, $\bar{U}/T$ displays a maximum at intermediate temperatures (Problem 10.14). This more nonmonotonic temperature dependence results from an increase at low temperature as a result of the increase in $\bar{U}$ with temperature, and a decay to zero at high temperature resulting from the factor $1/T$.

It is worth a bit more consideration of what these two terms contribute to the temperature dependence of the entropy. Since $\bar{U}/T = 0$ both at low and at high temperatures, $\bar{S} = (R/N)\ln Q$. At low temperatures where $Q = 1$, $\bar{S} = 0$. At high temperature, $\bar{S} = (R/N)\ln m$ where m is the number of allowed microstates. In this regard, at high temperature, the canonical ensemble behaves like a microcanonical ensemble, where microstates have the same energies. Although the microstate energies in the canonical ensemble are not the same, what matters is the energy values relative to the temperature. At high temperature, all the $\bar{E}_i/T$ values converge to zero.

The $\bar{U}/T$ term impacts the entropy at intermediate temperatures. As seen for the three-particle ladder from 50 to 400 K, the entropy is significantly larger than the $R\ln Q$ term. This reflects an entropy contribution in the canonical ensemble that is not accounted for in the spread over microstates.

The molar replica free energy ($\bar{A}$, since each replica has a constant volume) can be calculated from Equation 10.65:

$$\bar{A} = -\frac{RT}{N}\ln Q$$

$$= -\frac{1{,}000\,RT}{3}\ln(1 + 3e^{-1{,}000/RT} + 6e^{-2{,}000/RT} \qquad (10.85)$$

$$+ 7e^{-3{,}000/RT} + 6e^{-4{,}000/RT} + 3e^{-5{,}000/RT} + e^{-6{,}000/RT})$$

Because of the negative sign, $\bar{A}$ *decreases* monotonically with T (Figure 10.13A). And because of the premultiplier T in Equation 10.85, the decrease in $\bar{A}$ with T is linear at high temperature, where Q plateaus. Thus, the high-temperature limit of $\bar{A}$ is negative infinity. This can be understood from the increasing contribution that entropy makes to free energy at high temperature (through $T\bar{S}$, Figure 10.13A).

The heat capacity can be determined by differentiating the internal energy (Equation 10.84) with respect to temperature. As with $\bar{U}/T$, the heat capacity peaks at intermediate temperature, and has a value of zero both at both low- and high-temperature limit (Figure 10.13B). The peak in the heat capacity function can be understood in a couple of different ways. Recall that the heat capacity can be viewed as the ability of the system to store heat. To store heat, replicas need to shift to excited energy states. At very low temperature, all the replicas of the canonical ensemble are in the ground state. To bring about a significant population shift to the first excited state, the temperature must be increased to a value around $(\bar{E}_2 - \bar{E}_1)/R$ (for the present model, this is about 50 K). Below this temperature, the ensemble stores energy[†] poorly, so the heat capacity is low.[‡] In contrast, at temperatures significantly higher than $(\bar{E}_7 - \bar{E}_1)/R$, the replicas are evenly spread out among all seven energy states. At this point, the ensemble has very little capacity to adsorb additional heat, and what little heat can be taken up drives a large temperature increase. Another way to understand the peak in the heat capacity is through as a heat capacity variance (see Problem 10.15).

[†] Heat flow in this case, because the volume of each replicate is fixed.

[‡] By this reasoning, it should be quite difficult to get to absolute zero temperature. Even the smallest amount of heat that leaks into a very cold system is going to raise the temperature significantly.

THE ISOTHERMAL–ISOBARIC ENSEMBLE AND GIBBS FREE ENERGY

The two ensembles we have discussed so far (the microcanonical and canonical ensembles) parallel our development of classical thermodynamic potentials in Chapters 4 and 5. In Chapter 4, we developed the entropy as a potential for isolated systems. Building an ensemble from an isolated system leads to the microcanonical partition function (Ω), which is proportional (logarithmically) to the entropy. In Chapter 5, we developed potentials for other types of systems by making Legendre transforms. In one transform, we swapped the roles of T and S, which led to the Helmholtz energy A as a potential for a diathermic (heat-exchanging) system at constant volume. Building an ensemble from the diathermic system leads to the canonical ensemble (Q), which is proportional (logarithmically) to A.

Given the emphasis we placed on the Gibbs energy in our development of classical thermodynamics, it seems fitting that we come up with an ensemble that is proportional (logarithmically) to G. In Chapter 5, we obtained G by exchanging the roles of p and V (using a Legendre transform from A), resulting in a potential at constant pressure as well as temperature. Following this approach, we will modify the roles of p and V from that of the canonical ensemble. Instead of fixing each replica at the same volume, we will allow volumes of our replicas to vary. Though this ensemble can't be represented on a square lattice with shared walls (e.g., Figure 10.1), we can make a decent picture by lining up the replicas in a single row, and letting one of the unshared walls move (**Figure 10.14**).

Since we are allowing V to vary from replica to replica, it seems we should somehow fix p. However, p is not a quantity we can easily associate with individual microstates. Rather, p is defined as an average thermodynamic quantity, in the same way that T is defined as an average quantity. Thus, we will fix the pressure of a large mechanical reservoir that can equilibrate with our ensemble, just as we fixed the temperature when we developed the canonical ensemble. Here, we will put the entire ensemble in contact with a large mechanical reservoir at fixed pressure (in addition to a thermal reservoir at fixed temperature). In this way, the reservoir pressure defines the equilibrium volume distribution of the ensemble. In fact, once the ensemble is mechanically equilibrated with the reservoir, we can isolate the system from the reservoir, thereby fixing its total volume $\mathbb{V}_T$ (and its energy, $\mathbb{E}_T$).

Figure 10.14 An isothermal–isobaric ensemble. A very small ensemble is shown ($\mathbb{A} = 12$ replicas). (A) Each of these replicas can expand and contract against a very large reservoir at pressure p, through a moving wall, thereby changing replica volume. In addition, each replica can exchange energy with the reservoir (at temperature T) through heat flow. (B) A frozen collection of three-particle microstates. For simplicity, just three different volumes (V_1, V_2, and V_3) are shown. For many molecular systems (such as particles moving inside a box), microstate energy values change with volume. Here, the energy ladder becomes more closely spaced as volume increases. The equilibrium distribution is one that has the greatest multiplicity, subject to constraints on replica number $\mathbb{A}$, total energy $\mathbb{E}$, and total volume $\mathbb{V}$, and is given by the isothermal–isobaric partition function Θ (Equation 10.95).

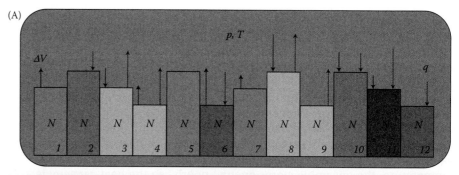

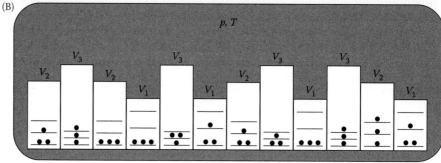

In the musical chairs analogy above, when we freeze time, each system will have a definite volume. Though the volume of individual snapshots can differ, the total volume will be well-determined by the pressure of the surroundings.

The isothermal–isobaric partition function

Following the procedure used above, we will find the equilibrium distribution of the isothermal–isobaric ensemble by maximizing the number of ways the ensemble can be obtained, subject to constraints. In addition to constraints on the total number of replicas ($\mathbb{A}$) and the total energy of the ensemble ($\mathbb{E}_T$), we will subject our ensemble to a constraint on the total volume ($\mathbb{V}_T$).[†] The three resulting constraint equations should have the form of summations over the different allowed energy values and the different volumes. We will represent the number of replicas with a given energy and volume value with the symbol $\mathbb{N}_{i,j}$, where i corresponds to the energy value, and j corresponds to the volume. Assuming there are m energy microstates, and n volumes, there will be a total of $n \times m$ $\mathbb{N}_{i,j}$ terms.

With this notation, the constraint on ensemble size is

$$\sum_{j=1}^{n}\sum_{i=1}^{m}\mathbb{N}_{i,j}=\mathbb{A} \tag{10.86}$$

where the inside (right) sum is over m allowed energy microstates at a given volume, and the outside (left) sum is over n allowed volume values.[‡]

The constraint on the total ensemble volume is

$$\sum_{j=1}^{n}\sum_{i=1}^{m}\mathbb{N}_{i,j}V_j=\sum_{j=1}^{n}\left(V_j\sum_{i=1}^{m}\mathbb{N}_{i,j}\right)=\mathbb{V} \tag{10.87}$$

Because the volume of each replica is determined only by the position of the movable wall (and not on the microstate of the particles inside), the replica volume on the left-hand side of Equation 10.87 is indexed only to the volume sum, allowing it to come outside the energy sum in the middle equality).

The constraint on the total ensemble energy is

$$\sum_{j=1}^{n}\sum_{i=1}^{m}\mathbb{N}_{i,j}E_{i,j}=\mathbb{E} \tag{10.88}$$

Generally the microstate energies depend on volume, so there will be a different energy value for each volume and energy microstate.

The number of ways an ensemble can be configured, given a distribution of microstates $\{\mathbb{N}_{i,j}\}$, is given by the multinomial coefficient

$$W=\frac{\mathbb{A}!}{\mathbb{N}_{1,1}!\mathbb{N}_{2,1}!\mathbb{N}_{3,1}!...\mathbb{N}_{m,n}!}=\frac{\mathbb{A}!}{\displaystyle\prod_{j=1}^{n}\prod_{i=1}^{m}\mathbb{N}_{i,j}!} \tag{10.89}$$

[†] It may seem strange to say that the total volume is fixed, since we let the volume of each replica vary. However, once we "freeze time" (the musical chairs analogy above), the volume becomes fixed. Recall that we made the same argument for the total energy $\mathbb{E}_T$ when we developed the canonical ensemble. As long as the ensemble is in equilibrium with the reservoir, we can isolate it without changing its average properties (and its total energy).

[‡] The fact that we are treating volume as a discrete variable in this analysis may seem a bit contrived. Note, though, that by making the steps in the outside sum small (making n large), we could smooth these steps out. Another way to say this is that we could replace the outside sum with an integral over volumes.

The expression on the right-hand side organizes the $n \times m$ factorials using product notation. Taking the log of W (Problem 10.19) and maximizing subject to our three constraints using the Lagrange method gives the equation

$$\nabla \ln W = \alpha \nabla \mathbb{A} + \beta \nabla \mathbb{E} + \delta \nabla \mathbb{V} \tag{10.90}$$

where δ is a new Lagrange multiplier for the volume constraint. Equation 10.93 is a set of $n \times m$ equations with $n \times m + 3$ unknowns (each $\mathbb{N}_{i,j}$, α, β, and δ), each of the form

$$\ln \mathbb{A} = \ln \mathbb{N}_{i,j} = \alpha + \beta E_{i,j} + \delta V_j \tag{10.91}$$

As before, we can rearrange and exponentiate to give

$$\frac{\mathbb{N}_{i,j}}{\mathbb{A}} = e^{-(\alpha + \beta E_{i,j} + \delta V)} \tag{10.92}$$

or

$$p_{i,j} = \frac{e^{-\beta E_{i,j}} e^{-\delta V_j}}{e^\alpha} \tag{10.93}$$

Summing all $n \times m$ of these equations gives

$$\begin{aligned}
1 &= e^{-\alpha} \left(e^{-\beta E_{1,1}} e^{-\delta V_1} + e^{-\beta E_{1,2}} e^{-\delta V_1} + \ldots + e^{-\beta E_{m,n}} e^{-\delta V_n} \right) \\
&= e^{-\alpha} \sum_{j=1}^{n} \sum_{i=1}^{m} e^{-\beta E_{i,j}} e^{-\delta V_j}
\end{aligned} \tag{10.94}$$

As can be seen in Equation 10.93, e^α serves as the partition function for the isothermal–isobaric ensemble, and the $e^{-\beta E_{i,j}} e^{-\delta V_j}$ terms are the Boltzmann factors for each microstate. Rearranging Equation 10.94 gives an expression for the isothermal–isobaric partition function, which we will represent with using the symbol Θ (Equation 10.95).[†]

$$\Theta \equiv e^\alpha = \sum_{j=1}^{n} \sum_{i=1}^{m} e^{-\beta E_{i,j}} e^{-\delta V_j} \tag{10.95}$$

Again, the V_j term can come outside the right-hand sum, since it is independent of i, giving

$$\begin{aligned}
\Theta &= \sum_{j=1}^{n} \left(e^{-\delta V_j} \sum_{i=1}^{m} e^{-\beta E_{i,j}} \right) \\
&= \sum_{j=1}^{n} e^{-\delta V_j} Q(V_j)
\end{aligned} \tag{10.96}$$

As indicated in the second line, the summation over the energy levels is simply the canonical partition function, this time, as a function of volume.

The Lagrange multipliers of the isothermal–isobaric partition function, and relationship to thermodynamic quantities

Although Equation 10.96 gives the functional form of Θ, we still need to determine the Lagrange multipliers in terms of physical parameters. We already know from

---

† I have chosen the symbol Θ (a capital theta) because it is the Greek letter for Q. This highlights the connection between the canonical and isothermal–isobaric partition functions. Some texts use the capital delta (Δ) instead.

our derivation of the canonical ensemble that $\beta = 1/kT$. Thus, we simply need to find the value of δ.

As before, we appeal to classical thermodynamics and to the Gibbs entropy formula. By substituting the log population term with Equation 10.93, we can obtain an expression for the entropy of the isothermal–isobaric ensemble (Problem 10.20):

$$S = \frac{\langle E_{i,j} \rangle}{T} + k\delta\langle V_j \rangle + k\ln\Theta \tag{10.97}$$

Note that in the present derivation, we will stick with molecular (rather than molar quantities) to save work. You should spend some time developing analogous expressions based on molar quantities.

Next, from definition of the Gibbs free energy from classical thermodynamics (Chapter 5), we can calculate the average Gibbs free energy per replica in the isothermal–isobaric ensemble:

$$\begin{aligned} G &= U + pV - TS \\ &= \langle E_{i,j} \rangle + p\langle V_j \rangle - TS \end{aligned} \tag{10.98}$$

The angle brackets (ensemble averages) go on the quantities that are directly associated with individual replica microstates. G and S are themselves ensemble-averaged properties, as are T and p (although we have also defined these two intensive quantities from the reservoir). Next, substituting the entropy average (Equation 10.97) gives

$$\begin{aligned} G &= \langle E_{i,j} \rangle + p\langle V_j \rangle - T\left(\frac{\langle E_{i,j} \rangle}{T} + k\delta\langle V_j \rangle + k\ln\Theta \right) \\ &= p\langle V_j \rangle - k\delta T\langle V_j \rangle - kT\ln\Theta \\ &= \langle V_j \rangle(p - k\delta T) - kT\ln\Theta \end{aligned} \tag{10.99}$$

For the microcanonical and canonical ensembles, the thermodynamic potentials (S and A, respectively) are equal to the log of the partition functions. For the isothermal–isobaric ensemble, where the classical potential is the Gibbs free energy, we can expect that $\langle G \rangle \propto \ln\Theta$. Equation 10.99 gives this equality when the first term, $\langle V_j \rangle(p - k\delta T)$, is zero. One way to achieve this is when $\langle V_j \rangle$ is zero, although we are not interested in systems with no volume. Thus, the zero must come from the equality

$$p - k\delta T = 0 \tag{10.100}$$

Rearranging Equation 10.100 provides a value for the Lagrange multiplier associated with volume variation:

$$\delta = \frac{p}{kT} \tag{10.101}$$

Thus, we can write the isothermal–isobaric partition function as

$$\Theta = \sum_{j=1}^{n} e^{-pV_j/kT} Q(V_j) \tag{10.102}$$

Equation 10.102 can be regarded as a weighted sum over canonical partition functions where the replicas of different partition functions have different volumes. The weighting term $(e^{-pV_i/kT})$ decays sharply with volume, so that the $Q(V_i)$ terms at smaller volumes make a large contribution to Θ, whereas those at larger volumes make a small contribution. This weighting is most pronounced at high pressures, consistent with the idea that increasing pressure drives a system to smaller volume. It is worth noting that the volume dependence of $Q(V)$ tends to counterbalance the pressure dependence to some degree. As shown in Figure 10.13B, as volume increases, microstate energy levels become closer, resulting in a greater multiplicity and a higher entropy at higher volume.

Further insight into the meaning of Θ as the volume sum over canonical ensembles can be obtained by expressing $Q(V)$ in terms of A. By exponentiating Equation 10.65, we obtain

$$Q(V) = e^{-A(V)/kT} \tag{10.103}$$

where the parentheses indicate that both Q and A are functions are volume. Combining with Equation 10.102 gives

$$\Theta = \sum_{j=1}^{n} e^{-pV_j/kT} e^{-A_j/kT} \tag{10.104}$$
$$= \sum_{j=1}^{n} e^{-(A_j + pV_j)/kT}$$

where the index on A_j refers to different volumes. Using the definitions of the Helmholtz and Gibbs free energies in Chapter 5, we can express Equation 10.104 as

$$\Theta = \sum_{j=1}^{n} e^{-G_j/kT} \tag{10.105}$$

Equation 10.105 gives the isothermal–isobaric partition function as the sum of Boltzmann factors corresponding to each allowed volume, where the Boltzmann weighting is given by the Gibbs energy at each volume. We will return to this relationship below when we calculate thermodynamic quantities from the isothermal–isobaric ensemble.

Though Equation 10.105 is nice and simple, it should be recognized that the terms in the summation do not refer to individual microstates. The microstates are grouped in each of the $Q(V_i)$'s. In terms of individual microstates, we will return to the double sum of Equation 10.95, replacing β with $1/kT$, and δ with p/kT:

$$\Theta = \sum_{j=1}^{n} \sum_{i=1}^{m} e^{-E_{i,j}/kT} e^{-pV_i/kT} \tag{10.106}$$
$$= \sum_{j=1}^{n} \sum_{i=1}^{m} e^{-(E_{i,j} + pV_i)/kT}$$

Recall that in Chapter 5 that the sum of U and pV is the enthalpy. The quantity $E_i + pV_i$ is similar to the enthalpy, but applies to single microstates, rather than ensemble averages. In some contexts it will be helpful to emphasize this similarity by defining the quantity

$$h_i \equiv E_i + pV_i \tag{10.107}$$

We use a lower-case h_i as a reminder that this quantity is not the thermodynamic average enthalpy value, but instead applies to microstate i with energy E_i and volume V_i, given the reservoir pressure p.

As with the canonical partition function, we can obtain microstate populations from the isothermal–isobaric ensemble by the ratio of corresponding Boltzmann factor to the isothermal–isobaric partition function:

$$p_{i,j} = \frac{e^{-(E_{i,j}+pV_j)/kT}}{\Theta} \qquad (10.108)$$

Finally, we can obtain ensemble-averaged values for enthalpy and volume using population weighted averages from Equation 10.108:

$$H = U + pV$$
$$= \langle E_{i,j} \rangle + p \langle V_j \rangle = kT^2 \left(\frac{\partial \ln\Theta}{\partial T} \right)_p \qquad (10.109)$$

$$\hat{V} = \langle \hat{V}_j \rangle = -kT \left(\frac{\partial \ln\Theta}{\partial p} \right)_T \qquad (10.110)$$

The derivations of these quantities follow the derivation of $U = \langle E_i \rangle$ from the canonical ensemble (Equations 10.46 through 10.49), and are left to the reader (Problems 10.22 and 10.23). Because both enthalpy and volume are both positive quantities, these two equations require that the isothermal–isobaric partition function increases with temperature (as replicas access microstates with high values of the quantity $E_{i,j} + pV_i$), but decreases with pressure (as replicas are pushed to small volume microstates). Note the similarity between Equation 10.106 and the van't Hoff relation (Chapter 7), where $\ln K$ plays an analogous role to $\ln\Theta$.

Finally, by combining Equations 10.97 and 10.109 we obtain an expression for the Gibbs free energy of the ensemble:

$$G = -kT\ln\Theta \qquad (10.111)$$

(see Problem 10.24).

Problems

10.1 In our derivation of the two-microstate canonical ensemble, we stated that the ensemble size and total energy constraints (Equations 10.3 and 10.4) could intersect on a line. What does that mean in terms of the microstate energy levels E_1 and E_2? In this case, how do we find the equilibrium distribution, and what does the ensemble become equivalent to? In this case, if $\mathbb{A} = 10^6$, what are the values of $\mathbb{N}_1$ and $\mathbb{N}_2$?

10.2 For the two-microstate canonical ensemble in Figure 10.2, calculate $\ln W$ using both forms of Stirling's approximation (the two-term and the more accurate three-term approximation) and compare to the exact value.

10.3 From the Stirling's approximations in Problem 10.2, calculate the gradient of $\ln W$ using the two approximations. Compare the differences in the two approximations of $\ln W$ to the differences in their gradients.

10.4 Derive the pair of Equations 10.8 and 10.9 from the Lagrange Equation 10.7, and solve for α and β, for the equilibrium distribution $N_1 = 8 \times 10^5$, $N_2 = 2 \times 10^5$, (i.e., where $E_T = 8 \times 10^5$).

10.5 Show that for the two-microstate model (Figure 10.2, $E_1 = 0$, $E_2 = 4$) $N_1 = N_2 = 5 \times 10^5$, the Lagrange multiplier Equation 10.7 yields the multipliers $\alpha = 0.693$, $\beta = 0$. What temperature does this correspond to? Give the value of the canonical partition function at this temperature.

10.6 Determine the ensemble energy E_T for the three-microstate model where the multiplier β has a value of zero. What are the populations of each of the three microstates in this case?

10.7 Using Equation 10.25, determine the values of the canonical partition function Q for the three-microstate model in Figure 10.7, for β-values of 0.05, 0.231, and 1.0.

10.8 Using Equations 10.26 through 10.28, determine the values of the populations p_1, p_2, and p_3, for the three-microstate model in Figure 10.7, for β-values of 0.05, 0.231, and 1.0.

10.9 By taking a population-weighted average (Equation 10.30), determine the values of the average energy per replica (E_T/A) for the three-microstate model in Figure 10.7, for β-values of 0.05, 0.231, and 1.0.

10.10 Determine the value of β for the three-microstate model ($E_1 = 0$, $E_2 = 1$, and $E_3 = 3$) for an $A = 10^6$ ensemble with a total ensemble energy $E_T = 2 \times 10^6$. Using this value of β, give the populations for each of the three microstates. In a few words, what does this correspond to?

10.11 Derive an expression for a canonical partition function for an energy ladder with an infinite number of levels with equal spacing. Assume the spacing to be $\Delta\hat{E} = 1\,\mathrm{Jmol}^{-1}$, and take the lowest energy state to be $\hat{E}_1 = 0$. Note that you should be able to write your answer in closed form using the closed-form formula for the sum below:

$$\sum_{i=1}^{\infty} x^i = \frac{x}{1-x}$$

Using your result, plot Q as a function of temperature.

10.12 Derive an expression for the average energy, U, from the canonical ensemble of the three particle energy ladder (Figure 10.9), as a function of temperature (i.e., derive Equation 10.84).

10.13 Demonstrate that the Gibbs formula (Equation 10.50) correctly provides a measure of the average entropy per replica (and thus the thermodynamic entropy of the original system) by dividing up the ensemble into "sub-ensembles" that have the same energy, E_j.

10.14 For the three-particle ladder model (Figure 10.9, energy values in Figure 10.10), find the temperature corresponding to the largest deviation between the entropy and $k\ln Q$. That is, find the maximum of the U/T curve in Figure 10.13B.

10.15 Using the three-parameter model, calculate the heat capacity CV as a function of temperature by calculating the internal energy variance (Equation 10.70) as a function of temperature. Plot and compare to the heat capacity determined by differentiation (black curve, Figure 10.13B). Variance in energy and the heat capacity for the three-particle system.

10.16 Calculate the allowed energy levels, the number of distinguishable configurations, and the total number of microstates for the four particle, four microstate system from Chapter 10 (Figure 10.10).

10.17 Give the canonical partition function and populations for the four-particle, four-microstate system from Problem 10.16, setting the energy of the lowest energy microstate to zero, and assuming an energy increment of 1000 J mol^{-1} between steps in the ladder.

10.18 The populations for each of the energy distributions in Figure 10.8 are given below.

$E/(1000*k)$	Figure 10.8A	Figure 10.8B	Figure 10.8C
1	0.108	0.000	0.054
2	0.7980	0.000	0.399
3	0.108	0.000	0.054
4	0.000	0.004	0.002
5	0.000	0.054	0.027
6	0.000	0.242	0.121
7	0.000	0.399	0.199
8	0.000	0.241	0.121
9	0.000	0.054	0.026
10	0.000	0.004	0.002

Calculate the energy variances and the heat capacities for each distribution, assuming a temperature of 300 K.

10.19 Starting with the expression for W in Equation 10.89, derive an expression for $\ln W$ using Stirling's approximation. Use summations in your answer.

10.20 Using the Gibbs' entropy formula, derive an expression for the total entropy using the isothermal–isobaric ensemble, and substitute Θ to obtain Equation 10.97 (hint: see the derivation of Equation 10.53).

10.21 Using our definition of the Gibbs free energy from classical thermodynamics, that is, $G = U + PV - TS$, calculate the Gibbs free energy of the isothermal–isobaric ensemble.

10.22 Derive the expression for enthalpy of the isothermal–isobaric ensemble (Equation 10.109), using the probability Equation 10.108.

10.23 Derive the expression for the average volume per replica of the isothermal–isobaric partition ensemble (Equation 10.110), starting with the probability Equation 10.108.

10.24 Derive the expression for the Gibbs free energy of the isothermal–isobaric ensemble (Equation 10.104) from the average enthalpy and entropy.

10.25 How is temperature defined for the microcanonical ensemble?

CHAPTER 11

Partition Functions for Single Molecules and Chemical Reactions

Goals and Summary

The goal of this chapter is to develop partition functions for ensembles of single molecules and for chemical reactions between molecules. This greatly simplifies the ensemble method by avoiding enumeration of multiparticle microstates and their multiplicities, focusing directly on molecular configurations. We will develop the so-called "molecular partition functions" for the canonical ensemble and the isothermal–isobaric ensemble, and show how these simpler partition functions can be used to build multiparticle partition functions. A requirement for this second step is that particles must not interact with one another. Although this requirement would seem to limit the use of molecular partition functions to describe complex chemical and biological systems, we will show how the molecular partition function can be combined with reaction mechanisms, to build partition functions for reactions of arbitrary complexity.

The canonical and isothermal–isobaric partition functions developed in the previous chapter allow us to calculate thermodynamic properties of collections of N molecules by applying molecular models of the collection. These models determine energy levels, which can derive from motions, vibrations, and interactions. In addition, the models contain statistical terms describing the number of ways that each collection of molecules can be configured. The problem with this approach is that for systems with a large number of molecules, both the energy levels and (in particular) the statistics gets unwieldy. In practice, direct ensemble analysis of replicas the size of experimental systems ($\sim 10^{20}$ molecules) using the multiparticle approach in the previous chapter would be impossible.

One way to solve this problem is to go the other way–that is, to build ensembles with just a single molecule in each replica. Since the number of molecules in a replica (N) is a parameter of our own choosing, there is nothing keeping us from setting it to whatever value we want, including one. Writing a partition function over the states of a single molecule eliminates the statistics associated with combining the states of different molecules. And as long as molecules don't interact in solution, the molecular partition function can be used to give the distribution of systems of N molecules (i.e., Q), and the thermodynamic quantities associated with Q.

A CANONICAL PARTITION FUNCTION FOR A SYSTEM WITH ONE MOLECULE

We will begin by building a canonical ensemble where each replica contains just one molecule. To illustrate such an ensemble, we will return to the three-level

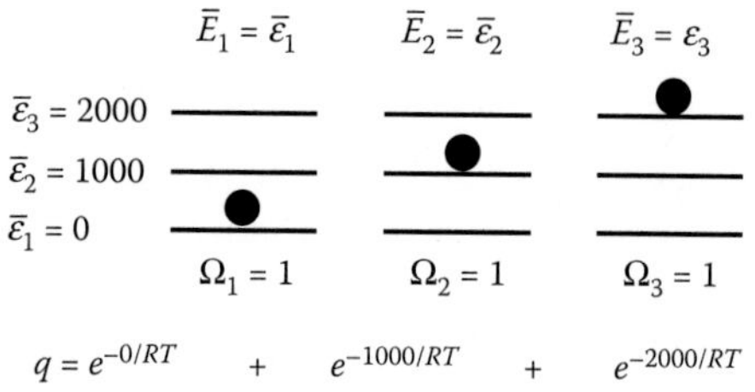

$$\bar{E}_1 = \bar{\varepsilon}_1 \qquad \bar{E}_2 = \bar{\varepsilon}_2 \qquad \bar{E}_3 = \varepsilon_3$$

$$\bar{\varepsilon}_3 = 2000 \quad \underline{\quad\quad} \qquad \underline{\quad\quad} \qquad \underline{\;\bullet\;}$$

$$\bar{\varepsilon}_2 = 1000 \quad \underline{\quad\quad} \qquad \underline{\;\bullet\;} \qquad \underline{\quad\quad}$$

$$\bar{\varepsilon}_1 = 0 \quad \underline{\;\bullet\;} \qquad \underline{\quad\quad} \qquad \underline{\quad\quad}$$

$$\Omega_1 = 1 \qquad \Omega_2 = 1 \qquad \Omega_3 = 1$$

$$q = e^{-0/RT} \quad + \quad e^{-1000/RT} \quad + \quad e^{-2000/RT}$$

Figure 11.1 The energy levels and multiplicities for a one-particle energy ladder. With a single molecule, there are just three microstates, each with a different (distinguishable) energy (in joules per mole; note that since there is one particle per replica, the molar replica energy is equal to the molar particle energy, i.e., $\bar{E}_i = \bar{E}_i = \bar{\varepsilon}_i$). This greatly simplifies the expression for the partition function for the canonical ensemble. This single-molecule partition function is referred to as the "molecular partition function," q.

energy ladder described above (Figure 10.9), but with a single molecule per replica instead of three (**Figure 11.1**):

The development of the single-molecule canonical ensemble is exactly the same as for the multiparticle canonical ensemble in Chapter 10. We equilibrate a collection of replicas (now $N = 1$ molecule each) against a thermal reservoir at temperature T. Once the ensemble is equilibrated with the bath, we can isolate the ensemble from the bath, giving it a fixed total energy $\mathbb{E}_T$. By freezing time, we obtain a collection of snapshots, each with a single molecule. A small single-particle ensemble is shown in **Figure 11.2**.

The partition function associated with the single-molecule ensemble is often referred to as the "***molecular partition function***," and is represented by a lower case q.[†] The distribution can be found the using the same approach as in Chapter 10, by maximizing the log multiplicity of the distribution subject to constraints on total ensemble size and energy ($\mathbb{A}_T$ and $\mathbb{E}_T$). However, we already solved this problem with the canonical ensemble. As described above, the partition function Q applies no matter how many molecules each replica contains. We need only identify all the m molecular microstates for our single molecule, and sum them up:

$$q = \sum_{i=1}^{m} e^{-\bar{\varepsilon}_i/RT} \tag{11.1}$$

Here the energies of the molecular states are given per mole (hence the over-bar and division by R rather than k). The molecular partition function for the three-particle ladder (Figure 11.1) has only three terms, one for each microstate:

$$q = 1 + e^{-1000/RT} + e^{-2000/RT} \tag{11.2}$$

This is much simpler than the three-particle canonical partition function, which has 7 energy terms (Equation 10.78) that together comprise 27 microstates.[‡]

Temperature dependence of the molecular partition function

With Equation 11.2 we can compare the temperature dependence of the molecular partition function (**Figure 11.3**) with that of the three-particle partition function with the same model (Figure 10.9). At a qualitative level, the temperature dependences of the partition functions are similar. Both functions have low-temperature values of one,[§] and both increase monotonically (after low temperature lag), plateauing at high

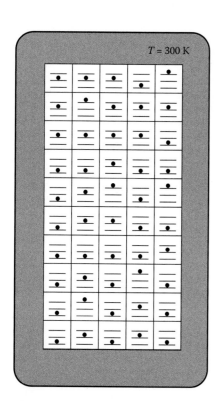

$T = 300$ K

Figure 11.2 A single-molecule ensemble for the three-level energy ladder. Fifty replicas, each with a single molecule, are shown equilibrating with thermal reservoir at $T = 300$ K. Snapshots were selected randomly, using probabilities taken from the Boltzmann distribution for energy levels of zero, 1, and 2 kJ mol⁻¹. The ensemble configuration shown {24, 21, 5} is a rough approximation of these probabilities ($p_1 = 0.472$, $p_2 = 0.316$, $p_3 = 0.212$); again, the approximation would be much better for a larger ensemble.

[†] As implied by the choice of the letter q, the molecular partition function represents a canonical ensemble of single molecule replicas. There is no reason we cannot build other types of single-molecule (i.e., molecular) ensembles.

[‡] The four-particle is even worse, with 13 energy terms and 256 microstates.

[§] Again, this is because we set the ground state energy to zero.

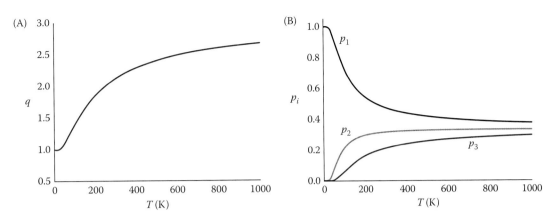

Figure 11.3 The molecular partition function and populations for a three-level ladder. Energy levels are $\hat{E}_1 = 0$, $\hat{E}_2 = 1000$, and $\hat{E}_3 = 2000$ kJ/mol. (A) The molecular partition function increases with temperature with the same overall profile as the three-particle canonical partition function (Figure 11.9), with an initial lag at low temperature at a value of one, and a plateau of three at high temperature (the number of microstates). (B) The populations of the three microstates shifts from all ground state at low temperature to equal microstate population at high temperature ($p_1 = p_2 = p_3 = 1/3$).

temperature. However, there are quantitative differences in the two partition functions as well. Most notably, because the number of available microstates is larger for the three-particle ensemble, Q is larger than q at all temperatures above absolute zero.[†] This is consistent with the proportionality of log of the canonical ensemble to the free energy (Equation 10.65): the three-particle system should have a larger free energy[‡] than the one particle system. The quantitative relationship between molecular and multi-particle partition functions will be developed in the next section.

The probabilities of each of the three microstates from the single-molecule ensemble is obtained in the same way as for the multiparticle canonical ensemble, dividing the Boltzmann factor for the microstate of interest by the partition function (Equation 10.61). The temperature dependences of the three microstate probabilities are shown in Figure 11.3B. Significantly, this distribution is *identical* to the probabilities we calculated in the three-particle canonical ensemble (Figure 10.10C). This makes sense, since we have assumed that the particles in the three-particle system don't interact with each other. Since we get the same results for populations with the molecular and canonical partition functions, we should calculate quantities that depend on to relative probabilities (equilibrium constants and extent of structure formation) using the simpler molecular partition function q.

Thermodynamic quantities from q

We can also use the molecular partition function to calculate energies and entropies, using equations analogous to Equations 10.63 through 10.65. In this case, since the replicas only have a single molecule, the energies and entropy calculated from q are "per molecule" (rather than "per replicas"). This simplifies the substitution to molar quantities, (for example, in Equation 10.77, $N = 1$ molecule). In other words, thermodynamic calculations from molecular partition functions give molar quantities simply by replacing k with R:

$$\bar{U} = RT^2 \frac{d\ln q}{dT} \tag{11.3}$$

$$\bar{S} = \frac{\bar{U}}{T} + R\ln q \tag{11.4}$$

$$\bar{A} = -RT\ln q \tag{11.5}$$

[†] This is most clearly seen in comparing the high temperature limits (3 and 27 for q and Q, respectively).

[‡] A larger *negative* free energy, since $\bar{A} = -RT\ln Q$.

THE RELATIONSHIP BETWEEN THE MOLECULAR AND CANONICAL PARTITION FUNCTIONS

As we saw above, for systems where molecules do not interact with each other, the molecular (single-molecule) partition function gives the same populations as those calculated from the canonical (*N-particle*) partition function. And although the energies and entropies calculated from q (Equations 11.3 through 11.5) differ from those calculated from Q, these differences are simply a result of the number of particles in the ensembles. For example, for the free (Helmholtz) energy

$$\frac{\widehat{A}}{N} = \bar{A} \tag{11.6}$$

(see Equation 10.77). Substituting expressions using the canonical and molecular ensembles into the left and right sides of Equation 11.6 (from Equations 10.65 and 11.5, respectively) gives

$$-\frac{RT}{N}\ln Q = -RT\ln q \tag{11.7}$$

Rearranging Equation 11.7 gives

$$\ln Q = N\ln q \tag{11.8}$$

or

$$Q = q^{N} \tag{11.9}$$

Equation 11.9 provides a fundamental relationship between q and Q, but taking it at face value we will miss some important underlying statistics. We can obtain a deeper understanding by thinking about the partition function in terms of combinations and probabilities, as described in Chapter 1. Recall that each term in the partition function is proportional to a probability of a particular microstate (or group of microstates, such as microstates of the same energy level). Thus, the same combination rules for probabilities also apply to the terms in the partition function. To obtain a relationship between q and Q, we need to combine all of the states of one molecule with the states of a second, and combine all of those with the states of a third, and so on up to N molecules. Since we are assuming each molecule to be independent, we *multiply* the statistical weights of each state of each molecule. Because q gives a probability term for each molecular state, we combine the states of two (independent) molecules by multiplying their molecular partition functions. For N identical molecules, this multiplication is done N times, giving the product q^{N} (Equation 11.9).[†]

As an example, we will build a three-particle canonical partition function out of our three-level molecular partition function (**Figure 11.4**). Doing this in a stepwise manner, we combine the molecular states of two independent molecules (multiplying their molecular partition functions), and then combine all these states with the states of a third molecule. Because the molecular partition function has three terms (corresponding to the three microstates), the first multiplication gives $3^2 = 9$ terms, and the second multiplication gives $9 \times 3 = 3^3 = 27$ terms (Figure 11.4). However, since the q's we are multiplying are identical (they represent the microstate probabilities of identical molecules), many of the terms in the multiplication

---

[†]   Note that if we had two different types of molecules N_A and N_B, we would multiply corresponding values of their partition functions q_A and q_B to calculate Q (see Problems 11.3 and 11.4).

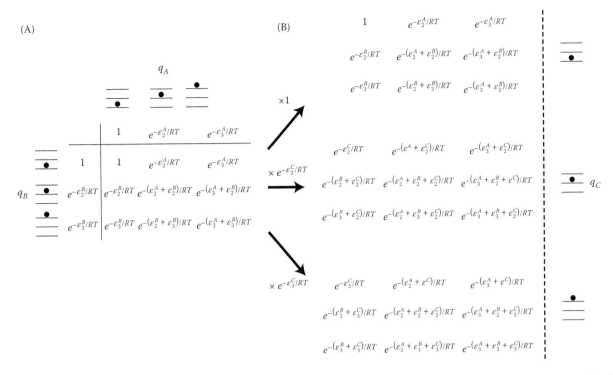

Figure 11.4 The product of three molecular partition functions leads to the canonical partition function for three identical independent molecules. For two noninteracting molecules (labeled *A* and *B*), the product of the two molecular partition functions (q_A and q_B, left) gives nine terms, one for each combination of microstates. For a third noninteracting molecule (*C*), additional multiplication by q_C gives 27 terms. So that the contribution of each molecule can be identified, energies are labeled based on which molecule they are associated with, as well as the energy level itself. If the three molecules have the same energy levels, we can drop these labels, but we need to keep track of their statistical contribution.

can be combined. For a three-term partition function, the product q^N is trinomial, and can be expanded in much the same way as with the binomial theorem (Problems 11.1 through 11.3). For the trinomial expansion with $N = 3$ molecules, there will be 10 terms.[†]

$$
\begin{aligned}
Q = q^3 &= \left(1 + e^{-\bar{\varepsilon}_2/RT} + e^{-\bar{\varepsilon}_3/RT}\right)^3 \\
&= 1 + e^{-3\bar{\varepsilon}_2/RT} + e^{-3\bar{\varepsilon}_3/RT} \\
&\quad + 3\left(e^{-2\bar{\varepsilon}_2/RT} + e^{-2\bar{\varepsilon}_3/RT} + e^{-\bar{\varepsilon}_2/RT} + e^{-(\bar{\varepsilon}_2+2\bar{\varepsilon}_3)/RT} + e^{-\bar{\varepsilon}_3/RT} + e^{-(2\bar{\varepsilon}_2+\bar{\varepsilon}_3)/RT}\right) \\
&\quad + 6e^{-(\bar{\varepsilon}_2+\bar{\varepsilon}_3)/RT}
\end{aligned} \tag{11.10}
$$

Further simplification resulting from the even spacing of the ladder reduces Q to seven terms, the same as for the three-particle canonical partition function in Equation 10.71 (Problem 11.8).

Indistinguishable particles

There is one situation[‡] where taking the product of N molecular partition functions (Equation 11.9) does not give the correct expression for the canonical partition function, and that is when particles are genuinely indistinguishable from each other. Although this is not typically the situation in analysis of the thermodynamics of solutions, it is good to be aware of this complication when thinking about simple systems.

---

† In general, a trinomial $(a+b+c)^N$ will have $(N+2)(N+1)/2$ terms, giving 10 for $N=3$.
‡ In addition to when molecules interact with one another.

When particles are genuinely indistinguishable, we cannot treat the configuration where molecule A is in state i and B is in state j as different from where A is in j and B is in i. These are the same microstate. Thus when we multiply q_A and q_B, we over-count by a factor of two.[†] Combining with a third molecule C overcounts further, by a factor of three ($[AB]_{ij}C_k$, $[AB]_{ik}C_j$, and $[AB]_{kj}C_i$), giving a total overcount of 2×3. For N indistinguishable molecules, the overcount is $N!$. Thus, for indistinguishable molecules this case,

$$Q = \frac{q^N}{N!} \tag{11.11}$$

To understand when the correction in Equation 11.11 should be applied, we need to more carefully define what we mean by "indistinguishable." In this context, we mean that molecules cannot be distinguished from one another *even in principle*. Although chemically identical molecules in solution of molecules in the lab may seem like they should be indistinguishable (e.g., we add only one type of water, one type of buffer, and a defined set of macromolecules—the "components" described in Chapter 5), each molecule will occupy a different position at any given time, and each is very likely to be in a subtly different structural state (including subtle structural and vibrational variations, and differences in translational and rotational energies). Though we will almost certainly be ignorant of these subtle differences, their existence means that we *could* use them as a basis for distinction, in principle.

This type of distinguishability is equivalent to tossing multiple coins at once. Regardless of whether we have bothered to distinguish the coins,[‡] we are much more likely to get a mix of heads and tails than to get all heads or all tails. This is because the coins are different; heads, tails is a different outcome than tails, heads, and each contributes to the statistics. If coin-tosses were to follow the statistics of indistinguishability, there would be a single outcome corresponding to each *composition* of heads and tails. For three coins there would be four outcomes: H_3T_0 (three heads and one tail), H_2T_1, H_1T_2, and H_0T_3. Each of these four outcomes would occur with equal probability, at least in a statistical sense. In terms of familiar statistical trials, this would be more like a toss of a four-sided die than a toss of three coins.

AN ISOTHERMAL–ISOBARIC MOLECULAR PARTITION FUNCTION

As described above, most reactions in chemical and biological studies occur at constant pressure and temperature, and are governed by the Gibbs free energy potential. The corresponding isothermal–isobaric ensemble described in Chapter 10 is equilibrated against a mechanical (constant pressure) reservoir as well as a thermal (constant temperature) reservoir, and leads to the partition function Θ. The Boltzmann factors in Θ are exponentially weighted over microstate h_i levels (where $h_i = E_i + pV_i$; Equations 10.97 and 10.98), and the log of Θ is proportional to G (Equation 10.111).

Following the same rationale as in the previous section, we can develop a single molecule version of Θ, which we will designate as θ (a lower case theta). Each replica in the ensemble is allowed to heat with the reservoir, and can vary in volume (analogous to Figure 10.14), but contains only a single molecule in each replica (**Figure 11.5**).

[†] Or nearly a factor of two. For m different microstates, straight multiplication of q_A and q_B gives m^2 terms. Of these, m involve the same microstate (e.g., A in i, B in i) and don't need to be corrected (though these terms can lead to unique statistics based on quantum mechanical exclusions). It is the remaining $m^2 - m$ terms that have been double-counted; thus, there are $m^2 - (m^2 - m)/2 = (m^2 + m)/2$ unique microstates. However, for molecules with a large number of thermally accessible microstates, this difference is negligible.

[‡] Either by applying markings to the coins, using different years, or throwing them in order.

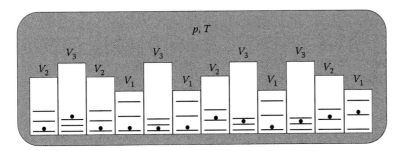

Figure 11.5 A single molecule isothermal–isobaric ensemble. Replicas can vary in volume against a constant pressure mechanical reservoir at pressure *p*, and can exchange thermal energy through heat flow with a thermal reservoir at temperature *T*. Each replica contains a single molecule, which, in this example, partitions among three volume-dependent energy levels.

The partition function for the single-molecule isothermal–isobaric ensemble is obtained simply by evaluating Θ with $N = 1$ molecule. Again, with only one molecule, there are no combinatorial factors for different microstates. Each term in the summation is simply an exponential in molecular enthalpy:

$$\theta = \sum_{i=1}^{m} e^{-(\bar{\varepsilon}_i - pV_i)/RT} \tag{11.12}$$

The molecular isothermal–isobaric partition function θ has the same relationships to the thermodynamic functions as does Θ, that is,

$$\bar{H} = \langle \bar{\varepsilon}_i + p\bar{V}_i \rangle = RT^2 \left(\frac{\partial \ln\theta}{\partial T} \right) \tag{11.13}$$

$$\bar{V} = RT \left(\frac{\partial \ln\theta}{\partial p} \right) \tag{11.14}$$

$$\bar{G} = -RT\ln\theta \tag{11.15}$$

Likewise, for replicas with N identical but distinguishable[†] particles,

$$\Theta = \theta^N \tag{11.16}$$

A STATISTICAL THERMODYNAMIC APPROACH TO CHEMICAL REACTION

When we initiated the need to develop statistical thermodynamics in Chapter 9, a major goal was the analysis of complex chemical reactions with many species. In this section, we will show how partition functions can be used to analyze systems in chemical equilibrium. This will allow us to determine the equilibrium distribution of reactants and products, and how these distributions depend on temperature, pressure, and solution conditions. One important application of this analysis is testing different models for interactions within and between reactants and products, and extracting interaction parameters from experimental data.

Here we will consider two types of reactions. The first type of reaction is unimolecular, that is, all the stoichiometric numbers have values of one:

$$A \rightleftharpoons B \rightleftharpoons C \rightleftharpoons \ldots \tag{11.17}$$

This type of reaction scheme is useful for describing complex conformational transitions such as the helix–coil transition (Chapter 12). Note that for this type of

---

[†]   As a reminder, we only need distinguishability *in principle*. For systems in chemistry and biology, this kind of distinguishability is satisfied.

reaction (as well as the one described below), although we have specified the reaction scheme as a specific sequence, the present analysis works equally for any other sequence.[†] Moreover, we could connect each of the species by multiple equilibrium arrows, forming a reaction network.[‡]

The second type of reaction includes multimolecular steps. The formation and depletion of different species occurs in different proportions (with different stoichiometric coefficients) and can involve couplings in which different species transform together. A specific example of such a scheme is

$$aA + bB \rightleftharpoons cC \rightleftharpoons dD + eE + fF \qquad (11.18)$$

The lower case a, b, c, ... , are positive-valued reaction numbers (see chapter 7). One specific example of such a scheme is a reaction in which multiple ligands (x) bind to a macromolecule (M):

$$nx + M \rightleftharpoons (n-1)x + Mx \rightleftharpoons \; ... \; \rightleftharpoons x + Mx_{n-1} \rightleftharpoons Mx_n \qquad (11.19)$$

This type of binding reaction is detailed in Chapters 13 and 14. Another specific example that involves multimolecular association is the polymerization reaction,

$$nM \rightleftharpoons (n-2)M + M_2 \rightleftharpoons (n-3)M + M_3 \rightleftharpoons \cdots \rightleftharpoons M_n \qquad (11.20)$$

In such reaction, a single molecular species (often a macromolecule) polymerizes into a linear fiber, a two-dimensional membrane, or a three-dimensional aggregate. Note that the first steps of these reactions are often quite unfavorable compared to subsequent steps; as a result, such "nucleated" reactions are often kinetically (rather than thermodynamically) limited.

SINGLE MOLECULE ENSEMBLES FOR CHEMICAL REACTIONS

The picture developed above for the molecular partition function can accommodate chemical reaction by allowing chemical transformation within each replica. For the unimolecular transformation, each replica will contain a single molecule (**Figure 11.6A**), and chemical transformation among the reactants and products will be allowed during the equilibration phase. For the multimolecular scheme, each replica will contain a minimal group of molecules that corresponds to a single step in the chemical reaction into each replica. For example, in **Figure 11.6B**, we could put in one A and one B molecule, or two C molecules, or alternatively one D and one E and one F molecule. We will refer to these three groupings as "reaction cohorts."

Using the method we developed in Chapter 9 to build ensembles (the musical chairs analogy), the contents of each replica would be free to transform among all the species in the reaction scheme.[§] When we freeze time, each replica in the unimolecular reaction scheme will contain a single molecule, but the chemical identity will differ from replica to replica (Figure 11.6A). For the multimolecular reaction scheme, each replica will contain a single set of molecules (a cohort, Figure 11.6B).[¶]

[†] Though many reactions involve a simple sequence of events that specify a logical order (e.g., in progressive binding of ligands as described in Chapters 13 and 14).

[‡] This is because we are only considering equilibrium concentrations. For kinetics analysis such systems, the placement of arrows between directly interconverting species is essential.

[§] Note, we will not concern ourselves with the possibility that the *mechanism* of a specific reaction step may be promoted (catalyzed) by other reactants or products—this would affect the dynamics but not the equilibrium distribution, which is the subject of our analysis.

[¶] The unimolecular scheme turns out to be consistent with the multimolecular framework, where for each step, one species has a stoichiometric coefficient of one, and the others have coefficients of zero. Since the unimolecular and true multimolecular schemes have conceptual and well as analytical differences, we will treat them separately.

(A) $A \rightleftharpoons B \rightleftharpoons C$

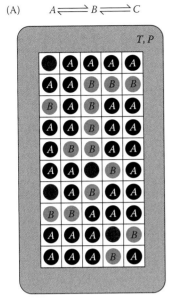

(B) $A + B \rightleftharpoons 2C \rightleftharpoons D + E + F$

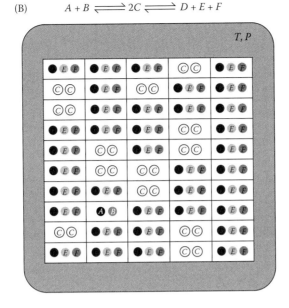

Figure 11.6 Ensembles of single molecules undergoing chemical reaction. Ensembles for unimolecular (A) and multimolecular (B) reactions (mechanisms given above). The replicas in the unimolecular ensemble each contain a single molecule. Replicas in the multimolecular ensemble contain multiple molecules ("reaction cohorts") corresponding to a single unit of reaction. To connect with the Gibbs free energy of reaction, we will use isothermal–isobaric reaction ensembles (though for simplicity, the volume variations among replicas are not depicted here).

BUILDING A SINGLE MOLECULE REACTION PARTITION FUNCTION

Because the chemical species in each replica are mutually exclusive, we can build an overall reaction partition function (which we represent with the symbol $\rho^{\dagger}$) by adding partition functions for each reaction step. We *add* partition functions because each replica can contain the contents of the first reaction step *or* the second reaction step, *or* ... the last step. The reaction cohorts from different steps are mutually exclusive.

For unimolecular reaction, this summation involves individual molecular partition functions for individual species. For i species,

$$\rho = \sum_{i \, species} \theta_i \tag{11.21}$$

For the three species in Figure 11.6A,

$$\rho = \theta_A + \theta_B + \theta_C \tag{11.22}$$

We are using the isothermal–isobaric partition function θ_i since reactions are typically described using Gibbs free energy (Chapter 7), and since θ_i is related (exponentially) to the Gibbs energy:

$$\theta_i = e^{-\bar{G}_i / RT} = e^{-\mu_i / RT} \tag{11.23}$$

(here we use molar Gibbs energies) The second equality in Equation 11.23 is a reminder that the molar free energy of a species is equal to its chemical potential, which plays a central role in expressions for free energies of reaction.

In our development of the molecular partition function above, we defined θ in as a sum over energy levels and volumes for each molecular microstate.[‡] Using this

---

[†] ρ is the lower-case Greek letter rho (**r**, for **r**eaction). We use lowercase rho because we are constructing the unimolecular reaction partition function as a molecular (i.e., "single-molecule") partition function.

[‡] Recall that these molecular enthalpy levels are analogous to molecular energy levels in q, but they include variations in replica volume.

approach for the reaction partition function requires a sum of these sums, one for each the θ_i terms in Equation 11.21 for each chemical species:

$$\rho = \sum_{\substack{i\ species}} \sum_{\substack{j\ enthalpy \\ levels}} e^{-\bar{h}_j^i/RT} \tag{11.24}$$

Figure 11.7 shows a series of enthalpy levels for three species (A, B, and C). Though correct, this double summation often provides more molecular detail than we need (or want) for analysis of reaction thermodynamics. Instead, we will simply combine Equations 11.21 and 11.23 to give the reaction partition function in terms of the molar free energies of the reactive species:

$$\rho = \sum_{\substack{i\ species}} \theta_i = \sum_{\substack{i\ species}} e^{-\bar{G}_i/RT} \tag{11.25}$$

Equation 11.25 can be thought of as resulting from a ladder of Gibbs free energies of each species, the same way that energy and enthalpy ladders defined q and θ. For the example in Figure 11.6A, the reaction partition function can be viewed as a Boltzmann weighting of the Gibbs free energy levels of species A, B, and C (Figure 11.7).

If we set the free energy of a reference state species to zero ,(for example, species A in Figure 11.7), then the free energies of the other species become free energy differences relative to this reference state. This is equivalent to expressing all of the exponents as *differences* in free energies:

$$\rho' = \sum_{\substack{i\ species}} e^{-(\bar{G}_i - \bar{G}_A)/RT}$$
$$= \sum_{\substack{i\ species}} e^{-\Delta\bar{G}_i/RT} \tag{11.26}$$

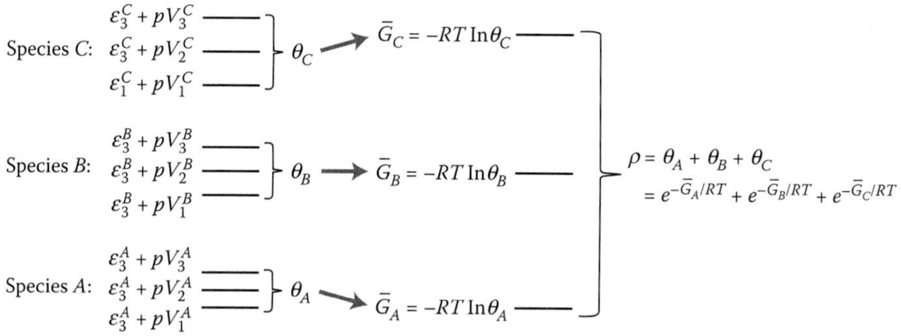

Figure 11.7 Representing a reaction partition function with the molecular partition functions of reactive species. On the left, microstates for three different species (A, B, C) are shown, along with h_j ladders, (that is, $\varepsilon_j + pV_j$) for each species. Each set of h_j levels defines the molecular partition function (in this case, the isothermal–isobaric θ) for the corresponding species. The logarithm of each θ_i function determines the molar Gibbs free energy of each species. In turn, these free energies can be viewed as an energy ladder. Because the enthalpy levels of the three species shown in Figure 11.7 differ both in position and in spacing, the three free energies are shifted from the enthalpies, especially at intermediate temperatures (see Problems 11.9 and 11.10). This free energy ladder can be used to generate a "reaction partition function," ρ. Because the three species are mutually exclusive, the three molecular partitions are added to generate ρ. Adding the partition functions is equivalent to exponential weighting of the free energies of each species (Figure 11.31 below).

The prime indicates that we have changed the numerical value of the reaction partition function by shifting the Gibbs energies relative to the reference.[†] As with the "shifted" canonical ensemble, the reference-state Boltzmann factor in ρ has a value of one.

Recall from Chapter 7 that exponentials of the form in Equation 11.26 are related to corresponding equilibrium constants (see Equation 7.59). In this case, the relevant equilibrium constants relate species i to the reference species A.[‡] Thus, we can write the reaction partition function as

$$\rho = \sum_{i \; species} K_i \tag{11.27}$$

Equation 11.27 provides a *very* simple way to represent a complex chemical reaction in statistical thermodynamic terms, as will be demonstrated in Chapter 12 for the helix–coil transition in peptides.

BUILDING A MULTIMOLECULAR REACTION PARTITION FUNCTION

As with unimolecular reactions, the reaction cohorts of multimolecular reactions are mutually exclusive. Thus, as in Equation 11.21, we can build a reaction partition function by summing partition functions for each cohort. However, unlike the unimolecular scheme, these cohort partition functions involve multiple molecules (e.g., A and B; two C; D, E, and F in Figure 11.6B). Since they are independent molecules,[§] their molecular partition functions multiply to give a cohort partition function, θ_i. For example, for the first step in Figure 11.6B, the cohort partition function is

$$\theta_i = \theta_A \theta_B \tag{11.28}$$

Summing these cohort products gives the reaction partition function. For the example in Figure 11.6B,

$$\rho = \theta_A \theta_B + \theta_C{}^2 + \theta_D \theta_E \theta_F \tag{11.29}$$

In terms of partial molar free energies, we can express Equation 11.29 as

$$\rho = e^{-\bar{G}_A/RT} e^{-\bar{G}_B/RT} + e^{-2\bar{G}_A/RT} + e^{-\bar{G}_D/RT} e^{-\bar{G}_E/RT} e^{-\bar{G}_F/RT} \tag{11.30}$$
$$= e^{-(\bar{G}_A+\bar{G}_B)/RT} + e^{-2\bar{G}_A/RT} + e^{-(\bar{G}_D+\bar{G}_E+\bar{G}_F)/RT}$$

As with the unimolecular partition function, we can modify the reaction partition function so that one of the cohorts serves as a reference. Referencing to A and B gives

$$\rho' = e^0 + e^{-(2\bar{G}_C-\bar{G}_A-\bar{G}_B)/RT} + e^{-(\bar{G}_D+\bar{G}_E+\bar{G}_F-\bar{G}_A-\bar{G}_B)/RT} \tag{11.31}$$
$$= 1 + e^{-\Delta\bar{G}_I/RT} + e^{-\Delta\bar{G}_{II}/RT}$$

Again, the prime indicates the shifted reaction partition function (Equation 11.31) has a different value than the original (Equation 11.30).

[†] We will usually omit such primes, but it is included here to emphasize that Equations 11.25 and 11.26 are different (by a multiple of $e^{G_A/RT}$).

[‡] Chapter 13 and 14 we will refer to this kind of equilibrium constant as an "overall" (rather than a "stepwise") equilibrium constant.

[§] Obviously, the molecules within a cohort are not independent in the sense of reaction, but they are independent in terms of their configurations and energies, as described in θ.

where the subscripts *I* and *II* refer to the reaction of *A* and *B* to 2*C* and to *D*, *E*, and *F*, respectively.

Using reaction partition functions

As with partition functions described in Chapters 9 and 10, the reaction partition function can be used to calculate populations. For the unimolecular mechanism, populations of individual species are given by

$$p_i = \frac{\theta_i}{\rho} = \frac{K_i}{\sum_{i\,species} K_i} \tag{11.32}$$

These populations provide the equilibrium concentrations of all of the reactive species. A similar equation (in which a cohort term θ_i replaces a species partition function) is obtained for the multimolecular mechanism, but the probabilities become cohort probabilities. Obtaining species populations (and concentrations) requires correction using the stoichiometric coefficients (Problem 11.14).

Because each of the θ_i terms depends on temperature, populations can be expressed as a function of temperature. For conformational transitions involving more than two-states, Equation 11.32 provides a detailed description of thermal unfolding. Likewise, the denaturant dependence of the θ_i terms (using a linear free energy dependence, e.g., Equation 8.52) can be used to represent multistate chemical denaturation (Problems 11.15 through 11.18). Moreover, since the isothermal–isobaric partition function allows for volume variation, multistate populations can be expressed as a function of pressure.

In addition to calculating species populations, Equation 11.32 provides a route to calculate population-weighted averaged quantities. Examples include directly observable quantities such as spectroscopic properties (Problem 11.18), and thermodynamic quantities such as average enthalpy, entropy, and heat capacity.

Finally, it should be emphasized that in order to use molecular partition functions to describe chemical reaction (regardless of molecularity), it must be the case that the individual species (either within or across cohorts) do not interact with each other. Of course, this does not mean they don't transform through chemical reaction. It means that the energy molecular energy levels of each species are unaffected by the other species.

Problems

11.1 For the following trinomial, demonstrate that the product is equal to the

$$(a + b + c)^3 = a^3 + b^3 + c^3$$
$$+ 3(ab^2 + ac^2 + a^2b + bc^2 + a^2c + b^2c)$$
$$+ 6abc$$

11.2 Demonstrate that the expansion above corresponds to the "trinomial" distribution, namely

$$(a + b + c)^N = \sum_{\substack{i,j,k; \\ i+j+k=N}}^{N} \left(\frac{N!}{i! \times j! \times k!} \right) a^i b^j c^k$$

11.3 Based on the binomial (Chapter 1) and trinomial distributions (Problem 11.2), give a compact series that expands the following multinomial:

$$(a_1 + a_2 + \cdots + a_m)^N$$

11.4 Calculate a molecular partition function q_B for a molecule of type "B" with a six-level energy ladder with energy values of 0, 500, 900, 1300, 1500, 1600 J mol^{-1}. This ladder has uneven spacing, as would be expected for energy levels in the vibration of a chemical bond.

11.5 Calculate and plot the probabilities of the six microstates of molecule B from in problem 11.4 as a function of temperature from $T = 0$ to a temperature approaching the high-temperature limit.

11.6 Now consider a canonical ensemble with two molecules, one of type "B" with six energy levels given in Problem 11.4, and one molecule with three levels as given in Figure 11.1 (call it "type A"). Using a drawing like Figure 11.4A, identify all the available microstates. Write the canonical partition function for the two-particle ensemble, and confirm that it corresponds to the microstates you identified.

11.7 Now consider a canonical ensemble where each replica has N_A molecules of type "A," and N_B of type "B" (see Problem 11.6). Give an expression for the canonical partition function Q. For your answer, you should assume A's are identical (but in principle distinguishable), B's are identical (also distinguishable in principle), A's and B's are nonidentical.

11.8 Demonstrate that the ten-term expression for Q (Equation 11.10) simplifies to the seven-term expression (Equation 10.66) when combined with values $\bar{\varepsilon}_1 = 0$, $\bar{\varepsilon}_2 = 1000$, and $\bar{\varepsilon}_3 = 2000$.

11.9 In Figure 11.7, the Gibbs free energy values for species A, B, and C are shifted compared to the average enthalpy values for each species. Rationalize this shift at intermediate temperatures.

11.10 Would you expect the Gibbs energy shift in Figure 11.7 (see Problem 11.9) at really high temperatures (and why)? Would you expect the shift at really low temperatures (and why)?

11.11 Calculate the free (Helmholtz) energies of the three-particle system (Figure 10.8) and the one-particle system (Figure 11.1), and plot them as a function of temperature. What is the ratio of these two free energies?

11.12 Show that the cube of the molecular partition function for the three-particle ladder (Equation 11.10) is equal to the canonical partition function for a three-particle system (Equation 10.71), using the "trinomial theorem," which is equivalent to the binomial theorem but for trinomials:

$$(a + b + c)^N = \sum_{i,j,k}^{N} \left(\frac{N!}{i!\,j!\,k!} \right) a^i b^k c^j$$

11.13 For combinations of two indistinguishable particles, show that the correction factor for over-counting of indistinguishable combinations (e.g., $A_i R_j$ and $A_j R_i$) converges to 2 as the number of accessible microstates becomes large.

11.14 For the reaction in Figure 11.6B, give an expression for the equilibrium concentration of [C] as a function of chemical potentials and equilibrium concentrations of the other species, starting with the probability Equation 11.32.

11.15 Consider the unfolding reaction of a protein that unfolds through an intermediate, partly folded conformation:

$$N \underset{}{\overset{K_{NI}}{\rightleftharpoons}} I \underset{}{\overset{K_{ID}}{\rightleftharpoons}} D$$

where the equilibrium constants are

$$K_{NI} = \frac{[I]}{[N]} \text{ and } K_{ID} = \frac{[D]}{[I]}$$

Write the reaction partition function, and the populations of the native, intermediate, and unfolded proteins using the native state as the reference.

11.16 From the three-state unfolding reaction described in Problem 11.15, give an expression for the temperature dependence of the native, intermediate, and unfolded states. For simplicity, assume that all three species have the same heat capacities (i.e., $\Delta C_p = 0$ for both reactions; see Chapter 8).

11.17 Plot the fraction of native, intermediate, and unfolded states as a function of temperature assuming the following parameters:

	N → I	I → D
T_m	320 K	345 K
$\Delta \bar{H}$	−400 kJ mol⁻¹	−300 kJ mol⁻¹

11.18 Using the parameters above, plot the temperature dependence of the CD signal at 222 nm (monitoring α-helix formation), assuming the following CD signals:

Species	Ellipticity (kdeg cm² dmol⁻¹)
N	−14,000
I	−7000
D	−1000

CHAPTER 12

The Helix–Coil Transition

Goals and Summary

The goal of this chapter is to apply the partition function formalism to analyze the multistate conformational transitions of linear polymers in general, and the polypeptide α-helix-to-coil transition in particular. This α-helix–coil transition shows positive cooperativity, where structure formation within each repeating unit (individual residues) promotes structure formation in neighboring subunits. Owing to the repetitive symmetry of the α-helix, simple "nearest-neighbor" models can be used to capture this coupling and to build partition functions to analyze structure formation and resolve local energies from interaction energies between subunits. Specific models to be described are (1) a noninteracting model, which serves as a simple reference for noncooperative conformational transition, (2) a "one-helix" approximation that simplifies the partition function by leaving out rare chimeric structures, and (3) a matrix formalism that includes all partly structured states. The matrix formalism is particularly useful for modeling "heteropolymers," where subunits have different sequences and energies.

In Chapter 8, we discussed conformational transitions that can be treated using simple two-state models. These models do a good job capturing the equilibrium unfolding transitions of single-domain globular proteins. Larger, multi-domain proteins often unfold by complex mechanisms involving partly folded intermediate states, which may not be a surprise, given their larger, more complex structures. It may come as a surprise that *smaller* polypeptides also tend to unfold by more complex mechanisms as well. This may be due to the decreased number of stabilizing contacts between distant segments of a chain, a situation that is particularly pronounced for proteins that form elongated structures.

One type of protein conformational transition involving an extended structure is α-helix melting in peptides. The α-helix is a common element of protein structure, and was proposed from simple bonding considerations by Pauling before high-resolution protein structures had been determined. The α-helix has a regular architecture in which each residue has the same backbone conformation, with both ϕ and ψ dihedral angles around $-60°$ (see Chapter 8). If this backbone orientation is repeated in three adjacent residues ($i + 1$ to $i + 3$), a hydrogen bond can form between the previous backbone C=O group (residue i) and the next backbone NH group four residues toward the C-terminus (residue $i + 4$, **Figure 12.1**). When multiple adjacent residues adopt this conformation, an extended α-helix is formed.

The α-helix is a common secondary structure element in folded, globular proteins. In most cases, these helices are stabilized by contacts with other parts of the proteins, since when these sequences are excised, they typically lose much of their helical structure. The resulting equilibrium between helix and unstructured

(A)

(B)

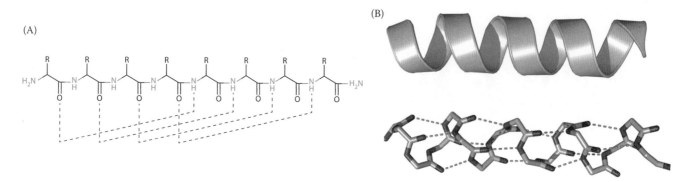

Figure 12.1 The structure of the α-helix. (A) Schematic showing the regular hydrogen bonding pattern (dashed lines) between main-chain carbonyl CO and NH groups with four-residue spacing. To form each hydrogen bond, the backbone dihedral (ϕ, ψ) angles for the three intervening residues must be fixed to around $-60°$. Side-chains are designated as "R" groups. (B) A three-dimensional model of the α-helix, as a ribbon (top), and as sticks. Again, hydrogen bonds are shown as dashed lines.

polypeptide, which we will refer to as the "helix–coil transition"[†] is the subject of this chapter. Using spectroscopy (often circular dichroism, see Chapter 8) to monitor α-helicity in short monomeric peptides of well-defined sequence, the energetics of helix formation can be determined.

As Figure 12.1 shows, in the absence of a surrounding protein environment, the α-helix lacks long-range stabilizing interactions. As a result distant segments can unfold and fold independently; thus, the simple two-state model we used in Chapter 8 is too simple to model the helix formation. Instead, in this chapter we will build models that treat the structural transition of *individual residues*. Though it might be argued that this is going a bit overboard in terms of complexity, the selection of the individual residue as the structural unit nicely matches the repeating structure of the polypeptide chain. Taking advantage of such symmetry often simplifies models, and helps connect fitted thermodynamic parameters to underlying structure.

If we are to treat each residue explicitly, then for a peptide that is N residues in length, if each residue has x conformations available, our model needs to account for x^N different microstates. For large peptides, this is *a lot* of microstates. To help keep numbers down, we will build models that only have two different states, helix (h) and coil (c).[‡] If we build a molecular partition function from microstates representing every different combination of h's and c's, we will have 2^N terms. For a short pentapeptide ($N = 5$), these $2^5 = 32$ terms are shown in **Figure 12.2**. Not surprisingly, the list of microstates in Figure 12.2 looks the same as the list of permutations of heads and tails in five tosses. That is, the number of microstates is governed by the binomial coefficients. Though as we will show, the distribution over these microstates only follows the binomial distribution if helix formation is noncooperative.

As a final step in building a helix–coil model, we need to describe how each residue behaves. There are two key questions that need to be addressed. First, are the residues all *identical*, as we would expect from a peptide made up of a single type of residue? Such sequences, which we refer to as a "homopolymers," greatly simplify the model. Second, is helix formation at different positions *independent* of the other positions? For an n residue peptide, when we evaluate helix thermodynamics at a given position i, do we have to worry about the conformations at the $n - 1$ other positions? If so, how far away to we have to worry? Models where sites are all independent (noncooperative

---

[†] The "coil" part of this name can be a little bit confusing, since α-helices look like coils as well. In this context, you can think of the "coil" part of the equilibrium as a disordered chain, lacking the helical hydrogen bond and characteristic dihedral angles.

[‡] Thus, the model is two-state at the level of individual residues, but multistate at the level of entire peptides.

C_5	C_4H	C_3H_2	C_2H_3	CH_4	H_5

$$ccccc \rightleftharpoons \begin{matrix} hcccc \\ chccc \\ cchcc \\ ccchc \\ cccch \end{matrix} \rightleftharpoons \begin{matrix} hhccc \\ hchcc \\ hcchc \\ hccch \\ chhcc \\ chchc \\ chcch \\ cchhc \\ cchch \\ ccchh \end{matrix} \rightleftharpoons \begin{matrix} hhhcc \\ hhchc \\ hchhc \\ chhhc \\ hhcch \\ hchch \\ chhch \\ hcchh \\ chchh \\ cchhh \end{matrix} \rightleftharpoons \begin{matrix} hhhhc \\ hhhch \\ hhchh \\ hchhh \\ chhhh \end{matrix} \rightleftharpoons hhhhh$$

Figure 12.2 The microstates for the helix–coil transition in a pentapeptide. Each of the $2^N = 32$ microstates corresponds to a single permutation of i h's and $5-i$ c's. The capital letters above indicate combinations of microstates with i helical (and $5-i$ coil) residues. For the noncooperative heteropolymer model, the distribution over these microstates is given by binomial (e.g., coin-toss) statistics. For more complicated models, the distribution depends on sequence variation and interaction among neighboring sites.

models) are the simplest types of models. The simplest type of cooperative models are "nearest-neighbor" models, where adjacent residues are coupled, but nonadjacent residues are independent. Before going into specific models, we will develop some general statistical thermodynamic relationships for the helix–coil transition.

GENERAL REACTION PARTITION FUNCTIONS FOR THE HELIX–COIL TRANSITION

Regardless of the model, we will calculate thermodynamics and populations using the unimolecular reaction partition function. The most general way to carry out the summation is over each microstate. With a two-conformation model (h or c at each residue), the partition function will contain 2^N terms:

$$\rho_H = \sum_{i=1}^{m=2^N} e^{-(\bar{G}_i^\circ - \bar{G}_0^\circ)/RT} \tag{12.1}$$

The subscript H is to remind us that the partition function is for the helix–coil transition. Throughout this chapter, we will take the all-coil state ($ccccc$ in Figure 12.2) to be the ground or "reference" state, with energy $\bar{G}_0^\circ = \bar{G}_{coil}^\circ$. As described in Chapter 11, the reaction partition function can be written as

$$\rho_H = \sum_{i=1}^{m=2^N} e^{-\Delta\bar{G}_i^\circ/RT} \tag{12.2}$$

where

$$\Delta\bar{G}_i^\circ = \bar{G}_i^\circ - \bar{G}_{coil}^\circ \tag{12.3}$$

With this set of equations, we can give the population of microstates in the usual way:

$$p_i = \frac{e^{-\Delta\bar{G}_i^\circ/RT}}{\rho_H} \tag{12.4}$$

From these populations we can calculate average molecular properties. For the helix–coil reaction, the most important (and directly observable) average property is the fraction helix, $\langle f_H \rangle$. In terms of the 2^N microstates, this is given by

$$\langle f_H \rangle = \sum_{i=1}^{m=2^N} p_i f_{H,i} = \sum_{i=1}^{m=2^N} f_{H,i} \frac{e^{-\Delta \bar{G}_i^\circ / RT}}{\rho_H} \tag{12.5}$$

where $f_{H,i}$ is the fraction helix of the ith microstate.

Though the general scheme above (Equations 12.1 through 12.5) is correct, it is somewhat limited in its application. The problem is that we must specify 2^N microstate free energy values in order to evaluate populations and calculate helicity.[†] We are not likely to know values for the 2^N free energies *a priori*, and are even less likely to extract these values from experimental data. In addition, as described in Chapter 10 for the three-particle ladder, it would be cumbersome to come up with a unique labelling system for each of the 2^N microstates to explicitly use them in the summations.

Instead, we will simplify these sums to get more practical expressions. Most of the simplification comes from specific models discussed below (e.g., homopolymer and nearest neighbor models). However, some simplification can be obtained by grouping microstates with the same number of residues in helical conformation[‡]:

$$\rho_H = \sum_{j=0}^{N} e^{-\Delta \bar{G}_j^\circ / RT} \tag{12.6}$$

where $\Delta \bar{G}_j^\circ$ represents the free energy of the collection of microstates with j helical residues, relative to the ground state (Problem 12.1). For the pentapeptide example in Figure 12.2, Equation 12.6 represents a sum over the six columns (from C_5 to H_5); in general, summation over helical levels shortens the partition function from 2^N terms to $N + 1$.[§] Note that $\Delta \bar{G}_j^\circ$ in Equation 12.6 is *not* a sum of microstate free energies. Rather, the *exponential* of $\Delta \bar{G}_j^\circ$ (the Boltzmann factor in Equation 12.6) is a sum of the *exponentials* of each microstate free energy. Interpreting the free energy exponentials as Boltzmann factors, this means that the statistical weight of the jth helix level is equal to the sum of the Boltzmann factors of the microstates with j helical residues.

In addition to simplifying the helix–coil partition function, summation over the number of helical residues simplifies the expression for the fraction of helical residues, f_H. This is because f_H is equal to the average number of helical residues divided by length of the peptide, that is,

$$\langle f_H \rangle = \frac{\langle j \rangle}{N} = \frac{1}{N} \sum_{j=0}^{N} j \times p_j \tag{12.7}$$

Thus, f_H can be expressed using the same sum as is used for the helix–coil partition function ρ_H (Equation 12.6), indexed to the number of helix residues:

$$\langle f_H \rangle = \frac{1}{N} \sum_{j=0}^{N} j \times \frac{e^{-\Delta \bar{G}_j^\circ / RT}}{\rho_H} \tag{12.8}$$

Although the summation over the number of helical residues provides a significant simplification of the partition function, a lot of terms remain (N; still to many to determine from simple experiments).[¶] Below, we will develop some models that

---

[†] Actually, we only need $2^n - 1$, since we know that the constant representing the all-coil reference state has a value of one. Still, it is a lot of free energies.

[‡] This is analogous to our sum over energy levels rather than microstates in the canonical partition function (Chapter 11).

[§] There are $N + 1$ (as opposed to N) helical levels because we must count the level where there are no helical residues (our all-coil reference state) as well as the levels with helical residues. In the sum in Equation 12.7, this is accounted for by starting the index at zero.

[¶] Generally, the helix–coil analysis is applied to peptides longer than five residues, since five residue peptides do not form a measureable helix. This further increases the number of $\Delta \bar{G}_j$ required.

can be used to give these N terms using just a few parameters. These models significantly reduce the complexity of the partition function, and in some cases can be expressed in "closed form," that is, as a simple algebraic expression instead of an explicit summation. Though simple, these models can give considerable insight into key aspects of helix–coil energetics.

THE NONCOOPERATIVE HOMOPOLYMER MODEL

The simplest model for helix formation is one where each residue is identical and is independent of its neighbor. Although α-helix formation in peptides does not adhere to the "independent" part, this simple model is a useful starting point to comparing with models where adjacent positions are coupled.

The assumption that all sites are identical to one another means that we can use a single free energy to describe a large number of reactions that are distinct at the microstate (and helix) level. For example, if all five sites in the pentapeptide in Figure 12.2 are identical, each of the five microstate reactions in the first step in Figure 12.2 can be given the same free energy (which we will refer to as $\Delta \bar{G}^{\circ}_{c \rightarrow h}$):

$$ccccc \xrightleftharpoons{\Delta \bar{G}^{\circ}_{c \rightarrow h}} hcccc \qquad (12.9)$$

$$ccccc \xrightleftharpoons{\Delta \bar{G}^{\circ}_{c \rightarrow h}} chccc \qquad (12.10)$$

$$\vdots$$

$$ccccc \xrightleftharpoons{\Delta \bar{G}^{\circ}_{c \rightarrow h}} cccch \qquad (12.11)$$

This is because each reaction involves formation of a single residue of helix. Thus, we can express the term in the helix–coil partition function (Equation 12.6) corresponding to the single-residue helix ($N = 1$) as

$$e^{-\Delta \bar{G}^{\circ}_1 / RT} = 5e^{-\Delta \bar{G}^{\circ}_{c \rightarrow h} / RT} \qquad (12.12)$$

As a simplification, we will define a variable for the free energy of conversion of a single residue of coil to helix as

$$\kappa \equiv e^{-\Delta \bar{G}^{\circ}_{c \rightarrow h} / RT} \qquad (12.13)$$

The quantity κ can be thought of as equilibrium constant for single-residue helix formation (convince yourself by taking the log of Equation 12.13).[†] With this definition, the single-residue ($N = 1$) term in the helix–coil partition function (Equation 12.12) can be written as 5κ.

We can use $\Delta \bar{G}^{\circ}_{c \rightarrow h}$ (and κ) to simplify terms corresponding to higher helical levels. For example, each of the microstates with two helical residues (e.g., the third column in Figure 12.2) involves two $c \rightarrow h$ reactions from the all-coil reference. Thus, the free energy difference of each microstate is *twice* the value of $\Delta \bar{G}^{\circ}_{c \rightarrow h}$. For the pentapeptide, two examples (out of ten) are

$$ccccc \xrightleftharpoons{\Delta \bar{G}^{\circ}_{c \rightarrow h}} hcccc \xrightleftharpoons{\Delta \bar{G}^{\circ}_{c \rightarrow h}} hchcc \qquad (12.14)$$

$$ccccc \xrightleftharpoons{\Delta \bar{G}^{\circ}_{c \rightarrow h}} hcccc \xrightleftharpoons{\Delta \bar{G}^{\circ}_{c \rightarrow h}} hhccc \qquad (12.15)$$

[†] It is important to understand that κ is *not* the equilibrium constant for the equilibrium constant that combines the N conformations that make up the $j = 1$ residue helix. It is for a single microstate (just one of the N conformations).

These two reactions differ in that Equation 12.14 describes helix formation in nonadjacent helical residues, whereas Equation 12.15 describes helix formation in adjacent residues. If there were an interaction between adjacent helix residues, Equation 12.15 would need to be modified to include the free energy of interaction. However, here we are assuming that all sites are independent. Thus, $\Delta \bar{G}^{\circ}_{c \rightarrow h}$ applies to each single-residue helix formation step, regardless of the conformation of neighboring (and nonneighboring residues).

The equilibrium constant connecting the end-points for each of these two-step reactions is obtained by taking the exponential of the summed single-step free energies. For example, for reaction (12.14),

$$
\begin{aligned}
e^{-\Delta \bar{G}^{\circ}_{hchcc}/RT} &= e^{-\left(\Delta \bar{G}^{\circ}_{c \rightarrow h} + \Delta \bar{G}^{\circ}_{c \rightarrow h}\right)/RT} \\
&= e^{-2\Delta \bar{G}^{\circ}_{c \rightarrow h}/RT} \\
&= \kappa^2
\end{aligned} \tag{12.16}
$$

Equation 12.16 states that the equilibrium constant for the two-step reaction is the product of the single-step equilibrium constants. This can be confirmed by expressing κ as a ratio of microstate concentrations. For example,

$$
\begin{aligned}
\kappa^2 &= \kappa \times \kappa \\
&= \frac{[hcccc]}{[ccccc]} \times \frac{[hchcc]}{[hcccc]} \\
&= \frac{[hchcc]}{[ccccc]}
\end{aligned} \tag{12.17}
$$

In the second line of Equation 12.17, we can choose any pair of microstates we want, as long as they differ by a single residue switching from coil (in the denominator) to helix (in the numerator). The two partly helical conformations cancel to give the third line, giving an equilibrium constant for the *hchcc* microstate, relative to the all-coil reference (Equation 12.14). Because there are ten microstates with two helical residues, the $n = 2$ term in the helix–coil partition function is $10\kappa^2$.

More generally, for any microstate with j helical residues, as long as sites are identical and independent, the free energy relative to the coil state is $j \times \Delta \bar{G}^{\circ}_{c \rightarrow h}$, and the equilibrium constant is κ^j. The term in the helix–coil partition function at helix level j is the product of κ^j and the number of ways j helical residues can be arranged over N sites. This statistical factor is the same as the number of ways to get N_H heads in N coin tosses, and is given by the binomial coefficient (Chapter 1). Thus, the helix–coil partition function (Equation 12.6) becomes

$$
\begin{aligned}
\rho_H &= \sum_{j=0}^{N} e^{-\Delta \bar{G}^{\circ}_j / RT} \\
&= \sum_{j=0}^{N} \frac{N!}{(N-j)! \times j!} e^{-j \Delta \bar{G}^{\circ}_{c \rightarrow h}/RT} \\
&= \sum_{j=0}^{N} \binom{N}{j} \kappa^j
\end{aligned} \tag{12.18}
$$

where the free energy in the first line represents the free energy of a single microstate with j helical residues. Using the binomial theorem, Equation 12.18 simplifies to

$$
\rho_H = (1+\kappa)^N \tag{12.19}
$$

For the pentapeptide example ($N = 5$), Equation 12.19 becomes

$$\rho_H = (1+\kappa)^5$$
$$= 1 + 5\kappa + 10\kappa^2 + 10\kappa^3 + 5\kappa^4 + \kappa^5 \qquad (12.20)$$

Each of the $N+1$ terms in Equation 12.20 is a Boltzmann factor, or statistical weight, of the set of conformations with j helical residues.

Another simple way to obtain the closed-form expression for ρ_H is to think about the conformations available at each position in the peptide, rather than the conformations at each helical level. We can write a helix–coil partition function at a single position ($\rho_{H,i}$ at position i).[†] Since there are only two conformations at each position (coil and helix), this partition function has only two terms:

$$\rho_{H,i} = 1 + \kappa \qquad (12.21)$$

Since we have assumed each residue is identical, each of the $N\rho_{H,i}$ single-residue partition functions is identical. To get the overall helix–coil partition function, we need to combine each of the position partition functions in a way that combines all states. Remember, partition functions and the terms they comprise represent the relative probabilities of different mutually exclusive states. Because each position is *independent*, the product of the position partition functions combines each of the states at each position with each of the states at the other positions:

$$\rho_H = \rho_{H,1} \times \rho_{H,2} \times \rho_{H,3} \times \cdots \times \rho_{H,N-1} \times \rho_{H,N}$$
$$= (1+\kappa_1)(1+\kappa_2)(1+\kappa_3) \times \cdots \times (1+\kappa_{N-1})(1+\kappa_N) \qquad (12.22)$$
$$= (1+\kappa)^N$$

The "closed form" of the partition function (Equation 12.19) is particularly easy to use in calculations of populations and population-based averages. The population of peptides with j residues in helical conformation is

$$p_j = \binom{N}{j} \frac{\kappa^j}{(1+\kappa)^N} \qquad (12.23)$$

The probability for any microstate is simply the second part of the right-hand side of Equation 12.23 (the same expression but without the binomial coefficient). Pentapeptide populations for each of the six helix-levels for different values of κ are shown in **Figure 12.3**.

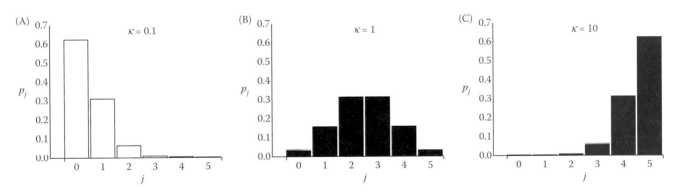

Figure 12.3 Populations for pentapeptide helix formation from the identical independent site model. The population of peptides that have j residues in helical conformations (the six columns in Figure 12.2) is calculated from Equation 12.23 for three different values of κ. (A) With a low value of κ, most peptides populate the all-coil microstate, or the five microstates with a single helical residue. (B) With a κ of 1.0, most peptides populate partly helical configurations, with equal numbers of low- and high-helix configurations. This distribution matches the fair coin toss distribution (see Problem 12.2). (C) With a high value of κ, most peptides populate all- or mostly-helical microstates.

[†] Here, i is used as a position marker along the peptide, from residue 1 to residue N. This is different from the j we have been using to count the number of helical residues.

Overall, helix formation tracks the value of κ (and thus $\Delta \bar{G}^\circ_{c \to h}$), shifting from mostly coil to mostly helix as κ increases. However, the spread over the different helix levels is considerable; even when κ differs by a factor of 10 (e.g., compare Figures 12.3A and 12.3B), there is considerable overlap. This is due in large part to the increased number of microstates in the center of the distribution (Figure 12.2), which shifts the low κ distribution up (Figure 12.3A), and the high κ distribution down (Figure 12.3C). Note that for long peptides, this overlap diminishes (Problem 12.3).

The populations of the different helix levels can be used to calculate fractional helix formation $\langle f_H \rangle$. For short peptides (such as the pentapeptide), it is fairly simple to calculate populations and take a population-weighted average, using Equation 12.7. However, for long peptides, the use of this summation becomes cumbersome. Instead, we will use Equation 12.8, which results in considerable simplification. Substituting ρ_H and $\Delta \bar{G}^\circ_j$ with expressions based on κ (Equations 12.18 and 12.19) gives

$$\langle f_H \rangle = \frac{1}{N \rho_H} \sum_{j=0}^{N} j \times e^{-\Delta \bar{G}^\circ_j / RT}$$

$$= \frac{1}{N \rho_H} \sum_{j=0}^{N} \binom{N}{j} j \times \kappa^j \tag{12.24}$$

The summation in Equation 12.24 looks *almost* like the helix–coil partition function (Equation 12.18). Expression of $\langle f_H \rangle$ in terms of ρ_H would be nice, because ρ_H can be expressed in closed form (Equation 12.19). The difference between $\langle f_H \rangle$ (Equation 12.18) and ρ_H is that the summand in Equation 12.24 includes multiplication by the index j. We can eliminate the factor j by substituting $j \times \kappa^j$ with the following derivative:

$$j \times \kappa^j = \kappa (j \times \kappa^{j-1}) = \kappa \left(\frac{d}{d\kappa} \kappa^j \right) \tag{12.25}$$

Substituting into Equation 12.24 gives

$$\langle f_H \rangle = \frac{1}{N \rho_H} \sum_{j=0}^{N} \binom{N}{j} \kappa \left(\frac{d\kappa^j}{d\kappa} \right)$$

$$= \frac{\kappa}{N \rho_H} \frac{d}{d\kappa} \sum_{j=0}^{N} \binom{N}{j} \kappa^j \tag{12.26}$$

This set of manipulations (differentiation and moving outside the sum) is similar to the manipulation we used to calculate the average energy of the canonical ensemble (see Equations 10.47 through 10.49), and will be used again in analysis of ligand binding. As a result of this manipulation, the factor of j is eliminated, leaving the helix–coil partition function:

$$\langle f_H \rangle = \frac{\kappa}{N \rho_H} \frac{d\rho_H}{d\kappa} \tag{12.27}$$

Substitution of the closed-form expression for ρ_H (Equation 12.19) gives a very simple expression for $\langle f_H \rangle$ (Problem 12.4):

$$\langle f_H \rangle = \frac{\kappa}{1 + \kappa} \tag{12.28}$$

That is, the fraction helix has a simple rectangular hyperbolic dependence on the helix equilibrium constant κ (**Figure 12.4**).

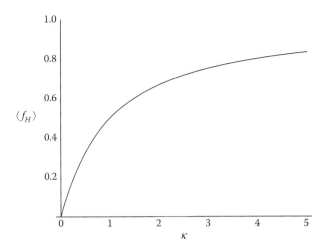

Figure 12.4 Fraction helix as a function of the residue equilibrium constant for the independent identical site model. Helix formation increases with κ in a simple rectangular hyperbola (Equation 12.28) regardless of peptide length.

The simple hyperbolic shape shown in Figure 12.4 is characteristic of a noncooperative reaction with a single type of subunit. However, it is difficult to do experiments to directly verify (or falsify) that a given peptide conforms to this curve. As experimentalists, we can measure $\langle f_H \rangle$, but we don't have a *direct* means to vary κ to generate points on a graph such as Figure 12.4. Under a fixed set of conditions, κ is (by definition) a constant. The best we can do is vary κ indirectly by changing conditions. The easiest way to do this is by changing the temperature. Assuming a simple van't Hoff temperature dependence, we might expect obtain a melting curve such as one of the three in **Figure 12.5**. These curves treat the helix formation reaction as exothermic (consistent with hydrogen bond formation between peptide groups), but have different enthalpy changes. Calorimetrically measured $\Delta \bar{H}^\circ$ changes for helix formation match the black dashed curve (Scholtz et al., 1991),[†] which would suggest that the helix–coil transition would be much too broad to see by temperature variation, if it conformed to the identical independent (i.e., noncooperative) model. Although experimental helix–coil transitions are broad, they are not *that* broad (they are more like the gray curve in Figure 12.5). This suggests that adjacent residues are not independent.

One of the most important features of Equation 12.28 is that $\langle f_H \rangle$ does not depend on N for the identical independent model. That is, the same helix content is predicted

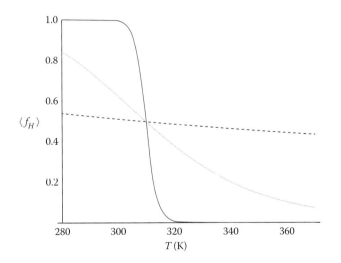

Figure 12.5 A temperature melt of helical structure for the independent identical site model. A simple van't Hoff model (with ΔC_p set to zero) is used to impart a temperature dependence on κ in Equation 12.28. The three curves have a T_m of 310 K, and ΔH values of −4 (black dashed), −40 (gray), and −400 (red) kJ mol⁻¹. The calorimetrically measured enthalpy of helix formation is closest to −4 kJ mol⁻¹; the noncooperative model would predict a *very* broad transition (dashed curve, much broader than observed), suggesting that the there is significant cooperativity in helix formation.

[†] The fact that the enthalpy change is quite small, compared to enthalpy changes of protein unfolding (Chapter 8) is consistent with the small size of the peptide group compared to a large protein (∼100 residues). The red curve in Figure 12.5 is the right order of magnitude for a protein, and has an enthalpy 100 times that of the black dashed curve.

for short and long peptides.[†] Taking this to the extreme, Equation 12.28 should be applicable to a "single-residue peptide." Although the concept of a single-residue helix creates some conceptual problems,[‡] the functional dependence makes sense in the context of the noncooperative model. The single-residue peptide is modeled as a two-state reaction, which should follow a rectangular hyperbola (Problem 12.6). The fact that the same expression applies to long peptides[§] is a result of the model treating each residue as independent. Though multiple residues are on the same peptide, according to the independent model they don't "see" each other even if they are nearest neighbors, so they might as well be on separate peptides.

Our justification for the form of $\langle f_H \rangle$ in the previous paragraph is simpler (and more elegant) than our lengthy derivation of Equation 12.28, as was our derivation of ρ_H using Equation 12.22. For this model, we could have skipped much of the math (in particular, Equations 12.24 through 12.27) and substituted some careful thinking. However, for models that involve coupling between neighbors, we will not be able to arrive at expressions for through words and logic, no matter how clever. In such cases, the analytical approach above (writing the partition function in as simple a form as possible, and using this form to calculate populations and average properties) is the only way to go. Before we analyze models with coupling, we will treat a more complex model where careful thinking can give us expressions for ρ_H and $\langle f_H \rangle$ without the math.

THE NONCOOPERATIVE HETEROPOLYMER MODEL

Here we will derive the helix–coil partition function and fraction helix for a model in which all residues are independent, but are not identical. We will consider two different types of residues, which we will call A residues and K residues.[¶] All the A residues will be identical to each other, as will the K residues. For a peptide that has N_T residues, we will have some number n_A of A residues, and some number $N_K = N_T - N_A$ of K residues.

Because we now have different residues, we need to specify the sequence of our peptide, that is, the order of the A's and K's from the N-terminus to the C-terminus. We will consider a peptide where the A and K residues are in some specific sequence, for example, *AAKAAKAAKAAK*, which is just one of many sequences (see Problem 12.7) with $N_A = 8$, $N_K = 4$. For the independent heteropolymer model, some quantities don't depend on sequence (the partition function and fraction helix), whereas some do (the amount of helix at each site).

Since all the sites are independent, we can construct the helix–coil partition function as we did in Equation 12.22, by making a sub-partition functions for each position (which has a specific residue since we defined the sequence) and multiply the sub-partition functions (again, multiplication is the right way to combine probabilities, since the sites are independent). Since we have two different residue types, we will have two different position-specific partition functions:

$$\rho_{H,A} = 1 + \kappa_A \tag{12.29}$$

$$\rho_{H,K} = 1 + \kappa_K \tag{12.30}$$

The overall partition helix–coil partition function for the sequence above is the twelve-term product

---

[†] This would mean that the thermal transitions shown in Figure 12.5 would also be independent of peptide length, which contradicts experiment.

[‡] Helix formation involves interaction of neighboring residues (Figure 12.1), which don't exist in a single-residue peptide. In fact, it is questionable to call a single residue a peptide (rather than an amino acid). But the model does not know such things.

[§] Despite the fact that the partition function increases (exponentially) with the length of the peptide.

[¶] Here, the A and K residues could be alanine and lysine, which are both pretty good helix formers.

$$\rho_H = \rho_{H,A} \times \rho_{H,A} \times \rho_{H,K} \times \cdots \times \rho_{H,A} \times \rho_{H,K}$$

$$= \rho_{H,A}{}^{N_A} \times \rho_{H,A}{}^{N_K} \tag{12.31}$$

$$= (1 + \kappa_A)^{N_A} (1 + \kappa_K)^{N_A}$$

Although the terms in the first line of Equation 12.31 is written in the order of the peptide sequence for clarity, it does not matter what order the twelve terms are multiplied. Obviously, the simpler forms are preferred for manipulations.

As above, we could calculate the fraction helix by differentiation, as we did in Equation 12.27 (see Problem 12.10). Instead, we will apply some simple ideas, rather than detailed derivations, to get an expression for $\langle f_H \rangle$. Because each site is independent, the fraction helix formation in a heteropolymer would be the same as if each type of residue were separated single-residue peptides. The fraction helix of each of these single-residue peptides would be given by Equation 12.28, that is,

$$\langle f_H \rangle_A = \frac{\kappa_A}{1 + \kappa_A} \tag{12.32}$$

$$\langle f_H \rangle_K = \frac{\kappa_K}{1 + \kappa_K} \tag{12.33}$$

However, the two types of single-residues peptides would not contribute equally to the total fraction helix, because they will be present in different proportions of residues in solution, given by N_A and N_K.[†] Thus, we need a residue-weighted average of the fraction helix from each residue type, that is,

$$\langle f_H \rangle = \frac{N_A}{N_A + N_K} \langle f_H \rangle_A + \frac{N_K}{N_A + N_K} \langle f_H \rangle_K$$

$$= \frac{N_A}{N_T} \frac{\kappa_A}{1 + \kappa_A} + \frac{N_A}{N_T} \frac{\kappa_A}{1 + \kappa_A} \tag{12.34}$$

For the *AAKAAKAAKAAK* sequence, Equation 12.34 becomes

$$\langle f_H \rangle = \frac{2}{3} \left(\frac{\kappa_A}{1 + \kappa_A} \right) + \frac{1}{3} \left(\frac{\kappa_K}{1 + \kappa_K} \right) \tag{12.35}$$

This expression is plotted in **Figure 12.6** as a function of κ_A and κ_K. As with the homopolymer curve, the fraction helix shows a rectangular hyperbolic dependence on each equilibrium constant, although cuts where one of the constants is fixed typically have limiting values partway between zero and one (Problem 12.11).

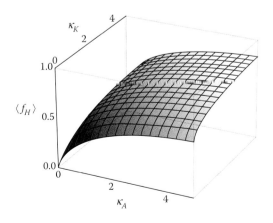

Figure 12.6 Fractional helix formation in a peptide with two types of independent residues. Here, $\langle f_H \rangle$ is shown as a function of the single site equilibrium constants for the two types of residues (κ_A and κ_K). The composition of the peptide is 2/3 *A* and 1/3 *K* residues.

[†] Unless the heteropolymer had equal composition of the two residue types, that is, $N_A = N_K$.

Note that surface plotted in Figure 12.6 applies to all peptides of composition $N_A = 2N_K$, regardless of length or the specific sequence. This is at odds with experimental data for helix formation of peptides of different lengths and sequences, again suggesting the assumption of independent sites is incorrect. In the next section, we will develop a model that allows for coupling between adjacent sites.

COUPLING BETWEEN THE SITES AND THE BASIS FOR COOPERATIVITY

Although the models above that treat adjacent sites as independent have the advantage of simplicity, they do not account for several aspects of the helix–coil transition in peptides. Most notably, helix formation depends on chain length. This can be seen in **Figure 12.7**, where thermal unfolding transitions of peptides of different lengths are monitored by circular dichroism spectroscopy (see Chapter 8). At a given temperature, long peptides of a given composition form more helices than short peptides, suggesting that helices are hard to initiate, but are easy to propagate. An energetic barrier to nucleation, combined with an energetic recompense for propagation, would rarify short helices and favor long ones, making helix formation cooperative.

To capture this aspect of the reaction, and to reconcile the steepness of the thermal transition with the measured enthalpy of helix formation, we need a model in which helix formation of neighboring residues are coupled (i.e., sites are *not* independent).

COUPLING BETWEEN RESIDUES THROUGH "NEAREST-NEIGHBOR" MODELS

There are many models that we could propose that couple helix formation among the N residues of a polypeptide. Not knowing which model to favor, it is generally best to consider simple models over complicated ones, both for ease of manipulation and analysis, and on philosophical grounds.[†] One of the simplest models for coupling is a "nearest neighbor" model, in which the helix–coil equilibrium at a given site is directly influenced by its adjacent neighbor(s), but not by nonadjacent residues.[‡]

Figure 12.7 The length dependence of helix formation in peptides with the repeating sequence block AEAAKA. Peptides with the sequence Ac-Y(AEAAKA)$_x$F-NH$_2$, where Ac- and -NH$_2$ groups substitute the terminal charges with peptides bonds. Peptide lengths range from 14 residues ($x = 2$, open circles), 20 residues ($x = 3$ filled circles), 26 residues ($x = 4$, open triangles), 32 residues ($x = 5$, filled triangles), 38 residues ($x = 6$, open squares), and 50 residues ($x = 8$, filled squares). Data are from Scholtz et al. (1991). At a given temperature, longer peptides show more helix formation (larger negative CD values), an indication that helix formation is positively coupled between neighboring residues.

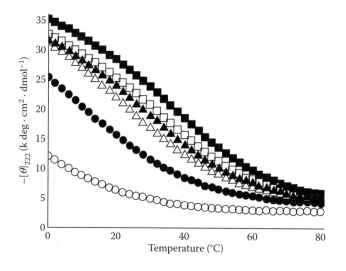

[†] There is a viewpoint in philosophy, sometimes referred to as "Occam's razor," those simple models should be favored over more complicated ones (as long as the simple models work).

[‡] Nonadjacent residues do have an indirect influence on helix formation at a given position, but this effect is mediated through the adjacent residue.

We can express nearest-neighbor coupling by modifying the equilibrium constant for helix formation when the residue converting from coil to helix is next to a residue that is already helical. We will represent helix formation in a coil background with the equilibrium constant $\kappa^\dagger$, and will modify this constant to account for interaction with a helical neighbor by the factor "τ." For example, the three single-residue reactions below would have the following equilibrium constants:

$$ccccc \underset{}{\overset{\kappa}{\rightleftharpoons}} chccc \underset{}{\overset{\kappa\tau}{\rightleftharpoons}} chhcc \underset{}{\overset{\kappa\tau}{\rightleftharpoons}} chhhc \tag{12.36}$$

Based on the reactions in Equation 12.36, we can associate κ with "helix initiation," and the product $\kappa\tau$ with "helix propagation." Including κ in both initiation and propagation recognizes features common to both processes.

Using κ and τ, we can write equilibrium constants for helix formation over multiple residues, relative to the coil reference state:

$$ccccc \underset{}{\overset{\kappa^2\tau}{\rightleftharpoons}} chhcc \tag{12.37}$$

$$ccccc \underset{}{\overset{\kappa^3\tau^2}{\rightleftharpoons}} chhhc \tag{12.38}$$

These equilibrium constants‡ have a power of κ for every helical residue and a power of τ for junction between helical residues.

One feature of the nearest-neighbor coupling model is that, for a given number of helix and coil residues, different arrangements can have different equilibrium constants relative to the all-coil reference, and thus, different Boltzmann factors. Consider the following three reactions:

$$ccccc \underset{}{\overset{\kappa^3\tau^2}{\rightleftharpoons}} chhhc \tag{12.39}$$

$$ccccc \underset{}{\overset{\kappa^3\tau}{\rightleftharpoons}} chhch \tag{12.40}$$

$$ccccc \underset{}{\overset{\kappa^3}{\rightleftharpoons}} hchch \tag{12.41}$$

Each conformation on the right has $j = 3$ helical residues, but each differs in the number of hh junctions, and thus in the power of τ. Thus, unlike the noncooperative model, we cannot generate a partition function by summing over helix levels and substitute a microstate Boltzmann factor along with a multiplicity.

THE ZIPPER MODEL—A "ONE-HELIX" APPROXIMATION FOR COOPERATIVE HOMOPOLYMERS

One approach to writing a partition function, given the complexity demonstrated in Equation 12.39, would be to attempt a double summation over both helical residues and hh pairs. This is a lot harder than it sounds.§ Instead we will simplify the partition function by ignoring microstates with multiple stretches of helical residues separated by one or more coil residues. This simplification is often referred to as the "one-helix approximation" or the "zipper model" (because once helices form, they zip up in continuous block).

With this simplification, the τ term appears in the partition function in a very simple way: each microstate Boltzmann factor is of the form $\kappa^j\tau^{j-1}$. Leaving out

---

† This is the same symbol we used for the noncooperative model—in both cases, κ represents intrinsic helix formation without the influence of neighbors.
‡ Which are Boltzmann factors in the reaction partition function summed over microstates.
§ A bit frightening, since it *sounds* hard to begin with!

microstates with multiple separate helices is a good approximation as long as the numerical value of τ is large. For the helix–coil transition, the value of τ is around 10^3. Thus, the Boltzmann factors of sequences with two separate helices will be 1–10,000 times smaller than those with a single helix (Problem 12.12).

The simplification provided with the one-helix approximation allows us to write the partition function as a sum over helical levels (j from 1 to N) using the Boltzmann factor $\kappa^j \tau^{j-1}$. All we need is the number of ways that a j-residue helix can be placed on an N residue peptide. Probably the easiest way to count the ways is to start with a $j = N$ residue helix, there is only one way to place it: with the h residue at the N-terminus, and the last residue at the C-terminus. If we take away one residue (an $N - 1$ residue helix), there are two ways to place it: with the first h at the N-terminus (residue 1 of the peptide), or with the first h at residue 2 of the peptide. Moving the $N - 1$ residue helix any farther toward the C-terminus would force the last h residue to "hang off the end," which is not a physically meaningful state. Following this logic, an $N - 2$ residue helix could be placed in three distinct positions, an $N - 3$ residue helix could be placed in four positions, and so-on. This trend is shown in **Figure 12.8**.

This relationship can be generalized by the following equation:

$$W = N - j + 1 \tag{12.42}$$

Note that this relation does not apply to a helical stretch of zero residues (the coil state), which can be arranged in only one way (not $N + 1$ ways, as would be predicted from Equation 12.42).

By combining the multiplicities from Equation 12.42 with the factor $\kappa^j \tau^{j-1}$, we can write a one-helix partition function with nearest-neighbor interactions:

$$\rho_H = 1 + \sum_{j=1}^{N} W \kappa^j \tau^{j-1}$$
$$= 1 + \sum_{j=1}^{N} (N - j + 1) \kappa^j \tau^{j-1} \tag{12.43}$$

The leading term of 1 is the Boltzmann factor for the reference (all coil) state, which cannot be included in the sum due to the limited range of W discussed above.

As with the noncooperative partition function, it would be preferable to express the one-helix partition function in closed form. This can be achieved by some mathematical substitutions. First, we will modify the sum in Equation 12.43 so that κ and τ are raised to the same power:

Figure 12.8 The number of ways (*W*) a helical stretch of length *j* can be accommodated on a peptide of *N* = 10 residues. Left, an example of the six ways that a five-residue helical stretch can be positioned on the 10 residue peptide. Right, a plot of *W* for various helix lengths from 1 to 10 residues. This relationship is straight line with a slope of minus one, and an intercept of 11 (*N* + 1). Note, although the plot is extended to the *y*-intercept, the linear relationship only applies over the range $1 \leq j \leq N$.

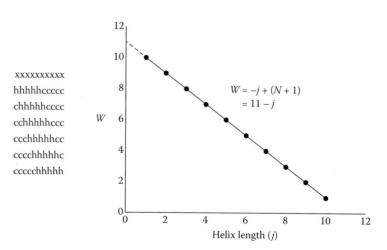

$$\rho_H = 1 + \frac{1}{\tau} \sum_{j=1}^{N} (N - j + 1)(\kappa\tau)^j \qquad (12.44)$$

The sum in Equation 12.44 can be separated into two sums:

$$\rho_H = 1 + \frac{1}{\tau} \left[\sum_{j=1}^{N} (N + 1)(\kappa\tau)^j - \sum_{j=1}^{N} j(\kappa\tau)^j \right]$$

$$= 1 + \frac{1}{\tau} \left[\sum_{j=1}^{N} (N + 1)(\kappa\tau)^j - \kappa \frac{\partial}{\partial\kappa} \sum_{j=1}^{N} (\kappa\tau)^j \right] \qquad (12.45)$$

In the derivative in the second line, τ is held constant. The closed form solution is obtained by recognizing that summations in the second line of Equation 12.45 are power series, and can be written as

$$\sum_{j=1}^{N} x^j = \frac{x(x^N - 1)}{x - 1} \qquad (12.46)$$

(Problem 12.13). Letting $\kappa\tau$ take the place of x in Equation 12.46 and substituting (twice) into Equation 12.45 gives

$$\rho_H = 1 + \frac{1}{\tau} \left[(N + 1) \frac{\kappa\tau[(\kappa\tau)^N - 1]}{\kappa\tau - 1} - \kappa \frac{\partial}{\partial\kappa} \left\{ \frac{\kappa\tau[(\kappa\tau)^N - 1]}{\kappa\tau - 1} \right\} \right] \qquad (12.47)$$

Evaluating the derivative in Equation 12.47 (Problem 12.14) and substituting back into Equation 12.47 (Problem 12.15) gives

$$\rho_H = 1 + \kappa \left[\frac{(\kappa\tau)^{N+1} - N(\kappa\tau) - (\kappa\tau) + N}{(\kappa\tau - 1)^2} \right] \qquad (12.48)$$

This closed form solution is certainly more complicated than the noncooperative partition function (Equation 12.19), but is much more compact than the open sum (Equation 12.43) for long peptides.

As before, we can calculate the fraction helix for the zipper model using probabilities for each helix level:

$$\langle f_H \rangle = \frac{1}{N} \sum_{j=0}^{N} j \times p_j = \frac{1}{N\rho_H} \sum_{j=1}^{N} j \times (N - j + 1)\kappa^j \tau^{j-1} \qquad (12.49)$$

Because the $j = 0$ term in the middle sum has a value of zero, the sum can be started at $j = 1$, making it look more like the partition function. Again, we can express the sum in Equation 12.49 as a derivative with respect to κ:

$$\langle f_H \rangle = \frac{\kappa}{N\rho_H} \sum_{j=1}^{N} (N - j + 1)\tau^{j-1} \frac{d\kappa^j}{d\kappa}$$

$$= \frac{\kappa}{N\rho_H} \frac{d}{d\kappa} \left(\sum_{j=1}^{N} (N - j + 1)\kappa^j \tau^{j-1} \right) \qquad (12.50)$$

Note that in Equation 12.50 we can add the value 1 inside the derivative without affecting the equality (since the derivative of 1 is zero):

$$\langle f_H \rangle = \frac{\kappa}{N\rho_H} \frac{d}{d\kappa} \left(1 + \sum_{j=1}^{N} (N-j+1)\kappa^j \tau^{j-1} \right)$$

$$= \frac{\kappa}{N\rho_H} \frac{d\rho_H}{d\kappa} \qquad (12.51)$$

The second equality in Equation 12.51 is the same result we obtained for the non-cooperative homopolymer model (Equation 12.27).[†] Differentiating the closed-form expression for ρ_H (Equation 12.48) gives the rather cumbersome expression[‡]:

$$\langle f_H \rangle = \frac{2\kappa^2 \tau [1 - (\kappa\tau)^N] + N\kappa[1 - \kappa\tau][1 + (\kappa\tau)^{1+N}]}{N(\kappa\tau - 1)\{1 + \kappa(N-2\tau) + \kappa^2\tau[\tau - 1 - N + (\kappa\tau)^N]\}} \qquad (12.52)$$

Fractional helix formation as a function of κ is calculated for different peptide lengths (**Figure 12.9A**) and coupling constants τ (**Figure 12.9B**). One obvious feature of these curves is that helix formation as a function of κ appears to be cooperative. This is in contrast to our previous model with independent sites. Whereas at low values of κ depends only modestly on κ, as κ increases, this sensitivity increases. Once κ is high enough to promote helix formation somewhere along the peptide, other residues join in due to the favorable nearest-neighbor interaction.

Another clear trend is that longer peptides show more helix formation at a given value of κ, and show sharper fraction helix transitions as a function of κ (Figure 12.9A). This is because longer peptides have more sites for initiation, and once a helix has initiated, longer stretches of helix can propagate. Whether or not residues add to an existing helix is given by the product $\kappa\tau$, and specifically, whether this product is greater than one. Whether a helix can be formed from coil requires

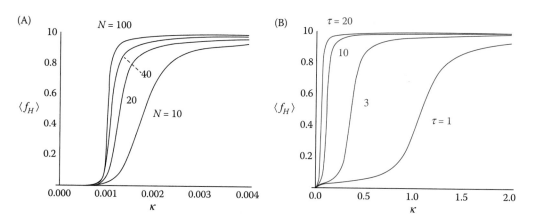

Figure 12.9 Helix formation modeled with the one-helix approximation. Fractional helix formation is calculated using Equation 12.52, which assumes only a single helical stretch with no gaps (partition function from Equation 12.43). (A) Helix formation assuming a $\tau = 1000$ (strong coupling between adjacent residues), for peptides of different length. (B) Helix formation for a peptide of length $N = 30$ residues, with various values of τ. Note that although $\tau = 1$ should be equivalent to the noncooperative homopolymer model, it has a complex, nonhyperbolic shape (see text).

[†] The reason this trick works is that the number of helical residues is given by the power of κ (that is, j) in the partition function. Taking the derivative with respect to κ brings down this power, and multiplies it by the Boltzmann factor divided by the partition function (i.e., the probability).

[‡] Though I would normally be inclined to ask you to derive this as a problem at the end of the chapter, this is an expression that is best evaluated by inserting Equation 12.47 into 12.54 in Mathematica, and evaluating the result with the command "Simplify[]."

products of $\kappa\tau$ to offset the unfavorable initiation penalty. For long helices, many residues are available to contribute to this offset.

Not surprisingly, as τ decreases, helix formation becomes less cooperative (and less favorable; Figure 12.9B). One rather surprising (though perhaps subtle) aspect of this trend is that at $\tau = 1$, the curve remains largely sigmoidal. At this value of τ, there should be no favorable (or unfavorable) interaction between neighboring residues, so we would expect a rectangular hyperbolic curve (Figure 12.4). This discrepancy is caused by the approximation we made in the one-helix model—the partition function used to calculate $\langle f_H \rangle$ leaves out all the partly helical states that have gaps. This is okay if τ is large (so that gapped states have negligible populations) but not as τ approaches one. For low τ values, the zipper model leaves out most of the significantly populated states. This is the reason the $\tau = 1$ curve in Figure 12.9B looks like it has been "scooped out" at intermediate κ values.

An exact nearest-neighbor description using matrices

Although the zipper model is useful for analyzing the helix–coil transition of homopolymers with high cooperativity, it not applicable to heteropolymers,[†] and gives a poor approximation when cooperativity is modest. Although it may be possible to modify the microstate sum (Equation 12.2) to fix these shortcomings,[‡] such modifications are arduous at best. Here we will introduce a method for nearest-neighbor coupling that represents the partition function as a product of N small (two by two) matrices. Each matrix represents the conformational state of a residue (helix vs. coil), correlated with the state of the neighboring residue. This method is exact in that it includes every state, and it can be easily modified to accommodate sequence variation. Although multiplying these N matrices by hand would be a challenge for peptide sequences larger than a few residues, such operations are trivial on a computer, even for peptides of 100 residues or more. And for homopolymers, additional matrix manipulations[§] can be applied to simplify calculations.

The elements of the correlation matrix are given in **Table 12.1**. The columns give the Boltzmann factor contributions for residue i in the helix and coil states, contingent on the conformation (helix and coil, rows) of the $i - 1$ position.

When residue i is in the helical state, it contributes a power of κ the corresponding Boltzmann factors. If its neighbor on the N-terminal side is also helical, it contributes a power of τ as well. In contrast, when residue i is in the coil state, it leaves the corresponding Boltzmann factors unchanged, relative to the coil reference. In matrix notation, Table 12.1 can be represented as

$$w = \begin{pmatrix} \kappa\tau & 1 \\ \kappa & 1 \end{pmatrix} \qquad (12.53)$$

Matrix w is sometimes referred to as a "weight" or "transfer" matrix.

Table 12.1 Correlation between residues i and $i - 1$

$i-1$ \ i	h_i	c_i
h_{i-1}	$\kappa\tau$	1
c_{i-1}	κ	1

[†] More specifically, it is not good for heteropolymers where different residues have significantly different values of κ and/or τ.

[‡] To keep track of the number of microstates with different numbers of gaps, or the effects of sequence variation on microstate energies.

[§] Diagonalization and eigenvalue analysis; see Cantor and Schimmel (1980) and; Aksel and Barrick (2009).

Figure 12.10 The product of *N* correlation matrices generates each of the terms in the helix coil partition function. Matrix products are compared to helix–coil partition functions ($\rho_{H,N}$) for (A) $N = 1$, (B), $N = 2$, and (C) $N = 3$ residues. The terms in each partition function are represented in the lower row of the corresponding matrix product, highlighted in red. When multiplied by the appropriate row and column vector (Equation 12.54), these terms are extracted from the matrix product as a single sum.

(A)
$$w = \begin{pmatrix} \kappa\tau & 1 \\ \kappa & 1 \end{pmatrix}$$
$$h \quad c$$
$$\rho_1 = 1 + \kappa$$

(B)
$$w^2 = \begin{pmatrix} \kappa\tau & 1 \\ \kappa & 1 \end{pmatrix}\begin{pmatrix} \kappa\tau & 1 \\ \kappa & 1 \end{pmatrix}$$

$$= \begin{pmatrix} \kappa^2\tau^2 + \kappa & \kappa\tau + 1 \\ \kappa^2\tau + \kappa & \kappa + 1 \end{pmatrix}$$
$$hh \quad ch \quad hc \ cc$$
$$\rho_2 = \kappa^2\tau + \kappa + \kappa + 1$$

(C)
$$w^3 = \begin{pmatrix} \kappa^2\tau^2 + \kappa & \kappa\tau + 1 \\ \kappa^2\tau + \kappa & \kappa + 1 \end{pmatrix}\begin{pmatrix} \kappa\tau & 1 \\ \kappa & 1 \end{pmatrix} = \begin{pmatrix} (\kappa^2\tau^2 + \kappa)\,\kappa\tau + (\kappa\tau + 1)\kappa & \kappa^2\tau^2 + \kappa + \kappa\tau + 1 \\ (\kappa^2\tau + \kappa)\,\kappa\tau + (\kappa + 1)\kappa & \kappa^2\tau + \kappa + \kappa + 1 \end{pmatrix}$$

$$= \begin{pmatrix} \kappa^3\tau^3 + \kappa^2\tau + \kappa^2\tau + \kappa\kappa^2\tau^2 + \kappa + \kappa\tau + \kappa & \kappa^2\tau^2 + \kappa + \kappa\tau + 1 \\ \kappa^3\tau^2 + \kappa^2\tau + \kappa^2 + \kappa & \kappa^2\tau + \kappa + \kappa + 1 \end{pmatrix}$$
$$hhh \quad chh \quad hch \ cch \quad hhc \ chc \ hcc \ ccc$$
$$\rho_3 = \kappa^3\tau^2 + \kappa^2\tau + \kappa^2 + \kappa + \kappa^2\tau + \kappa + \kappa + 1$$

To build the partition function, we need to generate 2^N different terms corresponding to each microstate (12.2). This is done by multiplying matrix w together N times. A few examples are shown in **Figure 12.10**, where w^1, w^2, and w^3 are compared to corresponding helix–coil partition functions.

The bottom row of each matrix product in Figure 12.10 contains the terms corresponding to the partition function, in two separate columns. We can combine these terms (and leave behind the first row, which has extra powers of τ)[†] by left-multiplying by a row vector (n in Equation 12.54), and by right-multiplying by a column matrix (c in Equation 12.54). This produces a scalar (the summation of the bottom row), which is the correct dimension for the partition function.

$$\rho_H = \begin{pmatrix} 0 & 1 \end{pmatrix}\begin{pmatrix} \kappa\tau & 1 \\ \kappa & 1 \end{pmatrix}^N\begin{pmatrix} 1 \\ 1 \end{pmatrix}$$
$$= nw^N c \tag{12.54}$$

The zero in row vector n eliminates the top row of w^N but retains the bottom row, which the column vector c combines into a single sum.

For illustration, we will apply the matrix method to an $N = 10$ residue homopolymer. It is hoped that the reader will gain three things from this exercise: (1) a better understanding of how to apply the deceptively simple Equation 12.54, (2) an appreciation for the complexity of the resulting partition function, and (3) a view of some underlying principles and familiar ideas that emerge despite this complexity (Problems 12.24 and 12.25). Evaluating Equation 12.54 for $N = 10$ residues gives

$$\rho_H = nw^{10}c$$
$$= \begin{pmatrix} 0 & 1 \end{pmatrix}\begin{pmatrix} \kappa\tau & 1 \\ \kappa & 1 \end{pmatrix}\begin{pmatrix} \kappa\tau & 1 \\ \kappa & 1 \end{pmatrix}\begin{pmatrix} \kappa\tau & 1 \\ \kappa & 1 \end{pmatrix}\cdots\begin{pmatrix} \kappa\tau & 1 \\ \kappa & 1 \end{pmatrix}\begin{pmatrix} 1 \\ 1 \end{pmatrix} \tag{12.55}$$

[†] The extra powers of τ in the top row of w^N come from the first term ($\kappa\tau$) of the first w matrix in the product. This term anticipates that the previous ($i - 1$) residue is helical, but for the first residue, there is no $i - 1$ residue.

Multiplying this matrix product out gives a rather large expression not surprisingly, with $2^{10} = 1024$ terms. However, *many* of these terms are identical, and can be combined into a simpler expression:

$$
\begin{aligned}
\rho_H = {} & 1 \\
& + 10\kappa \\
& + 9\kappa^2(4 + \tau) \\
& + 8\kappa^3(7 + 7\tau + \tau^2) \\
& + 7\kappa^4(5 + 15\tau + 9\tau^2 + \tau^3) \\
& + 6\kappa^5(1 + 10\tau + 20\tau^2 + 10\tau^3 + \tau^4) \\
& + 5\kappa^6(\tau + 10\tau^2 + 20\tau^3 + 10\tau^4 + \tau^5) \\
& + 4\kappa^7(5\tau^3 + 15\tau^4 + 9\tau^5 + \tau^6) \\
& + 3\kappa^8(7\tau^5 + 7\tau^6 + \tau^7) \\
& + 2\kappa^9(4\tau^7 + \tau^8) \\
& + \kappa^{10}\tau^9
\end{aligned}
\tag{12.56}
$$

Expressions such as Equation 12.56 can be obtained from matrix products (Equation 12.55) generated in Mathematica using the command "`Simplify[…]`." From top to bottom, each term in Equation 12.56 has an additional power of κ, corresponding to an increased number of helical residues. The terms in parenthesis increase in powers of τ, corresponding to increased numbers of adjacent helical residues. The number of ways of each of these $\kappa^j \tau^{k<j}$ states can be arranged is given by the product of integers outside and inside the parentheses. For example, the fourth line of Equation 12.56 gives the Boltzmann factors for all microstates in the $N = 10$ residue peptide with three helical residues. These can be in a single block ($\kappa^3\tau^2$; there are 8 such arrangements[†]), or can have one or two gaps ($\kappa^3\tau$ and κ^3; there are $8 \times 7 = 56$ arrangements for each).

Fraction helix from the matrix partition function

As shown in the $N = 10$ example above, although the number of multiples of τ for each microstate depends on the arrangement of helical residues, the multiples of κ simply scale with the number of helical residues (as κ^j, Equation 12.56). Thus, expressing fraction helix in terms of the helix level j, that is,

$$
\begin{aligned}
\langle f_H \rangle &= \frac{1}{N} \sum_{j=1}^{N} j \times p^j \\
&= \frac{1}{N\rho} \sum_{j=1}^{N} f(\tau, j, N) \times j \times \kappa^j
\end{aligned}
\tag{12.57}
$$

gives the same $j\kappa^j$ term as in the noncooperative and zipper models, and allows us to express $\langle f_H \rangle$ by differentation with respect to κ:

$$
\begin{aligned}
\langle f_H \rangle &= \frac{\kappa}{N\rho} \sum_{j=1}^{N} f(\tau, j, N) \frac{\partial \kappa^j}{\partial \kappa} \\
&= \frac{\kappa}{N\rho} \frac{\partial}{\partial \kappa} \left[\sum_{j=1}^{N} \kappa^j f(\tau, j, N) \right] \\
&= \frac{\kappa}{N\rho} \frac{\partial \rho_H}{\partial \kappa}
\end{aligned}
\tag{12.58}
$$

[†] This is just the $N - 3 + 1$ factor from the zipper model.

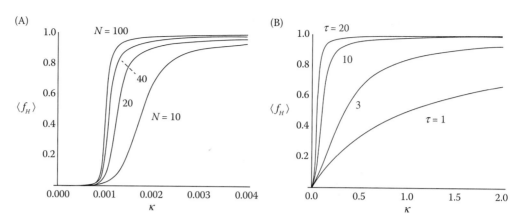

Figure 12.11 Homopolymer helix formation modeled with the nearest-neighbor matrix method. Fractional helix formation is calculated by differentiating the matrix-based partition function (Equation 12.54) with respect to κ. (A) Helix formation for peptides of different length, assuming $\tau = 1000$ (strong coupling between adjacent residues). (B) Helix formation for a peptide of length $N = 30$ residues, with various values of τ. Here the $\tau = 1$ curve is equivalent to the noncooperative homopolymer model.

In the second line of Equation 12.58, we can take the derivative outside the sum because $f(\tau, j, N)$ does not depend on κ. The third line is obtained by inserting the Boltzmann factor for the reference coil state (unity) into the derivative in the second line, which differentiates to zero.

Because the matrix model produces a polynomial in κ, evaluation of the fraction helix by differentiation (Equation 12.58) is straightforward.[†] **Figure 12.11** shows fraction helix calculated by differentiating $\langle f_H \rangle$ expressions generated using the matrix method, for the same parameters as with the zipper model in Figure 12.9. As long as τ is large, nearly identical $\langle f_H \rangle$ curves are obtained with the exact and zipper models (compare Figures 12.9A and 12.11A, where $\tau = 1000$). However, for low values of τ, $\langle f_H \rangle$ curves differ significantly. For the $\tau = 1$ curve, the matrix model shows the (correct) noncooperative transition, not the "scooped out" curve of the zipper model (compare Figures 12.11B and 12.9B). Even when $\tau = 20$, there are clear differences between the exact model and zipper approximation, with the exact transition showing less helix formation and a slightly lower slope with κ. The slightly more advanced math associated with the exact model pays off significantly in accuracy.

Extending the matrix approach to accommodate sequence variation

In addition to providing an exact partition function for homopolymers, the matrix method can easily be adapted to accommodate site-specific variations in helix formation. This type of variation is important in helix formation in real peptides and proteins, where sequence variation is the rule, not the exception. To account for the effects of sequence on helix formation, we will have different w matrices for each position, depending on the sequence.

In principle, differences in w matrices could result from sequence variation in either κ or τ values.[‡] In practice, it is far simpler to restrict sequence effects to the κ parameter, using a single value of τ for all residues. This is because κ is a local parameter, whereas τ describes coupling of adjacent residues. Representing the 20 naturally occurring amino acids in proteins using variation in κ parameters requires 20

---

† Albeit a little tedious for long peptides, and best left to a computer.
‡ Or both κ and τ!

parameters; using τ would require as many as 400 parameters, one for each pair of residues. Here, we will adopt the simpler approach, which gives weight matrices of the form

$$w_i = \begin{pmatrix} \tau \kappa_i & 1 \\ \kappa_i & 1 \end{pmatrix} \qquad (12.59)$$

As an example, we can write the helix–coil partition function (ρ_H) for the sequence *ADAQGAMNKALELFRKDIAAKYKE* (which corresponds to the H-helix of sperm whale myoglobin) as

$$\rho_H = n w_A w_D w_A w_Q w_G w_A w_M w_N w_K w_A w_L w_E w_L w_F w_R w_K w_D w_I w_A w_A w_K w_Y w_K w_E c \qquad (12.60)$$

Equation 12.60 is certainly unwieldy in the extreme, and unlike the homopolymer matrix partition function, it does not easily factor into a matrix product. Nonetheless, it can be calculated easily on a computer and can be used to determine populations and fractional helix formation. Although we will not provide a justification here (see Problem 12.38), fractional helix formation can be calculated using an equation analogous to Equation 12.58, taking partial derivatives with respect to each type of residue:

$$
\begin{aligned}
\langle f_H \rangle &= \frac{1}{\rho_H N}\left[\kappa_A \frac{\partial \rho_H}{\partial \kappa_A} + \kappa_C \frac{\partial \rho_H}{\partial \kappa_C} + \cdots + \kappa_Y \frac{\partial \rho_H}{\partial \kappa_Y} \right] \\
&= \frac{1}{\rho_H N}\left[\sum_{j \in \{20\ aa's\}} \kappa_j \frac{\partial \rho_H}{\partial \kappa_j} \right]
\end{aligned}
\qquad (12.61)
$$

For generality, the summation in Equation 12.61 is taken over the 20 naturally occurring amino acids in proteins. For a peptide with fewer types of residues, fewer partial derivatives are required (e.g., the sequence above has only 13).

A number of peptide systems have been used to measure the energetics of helix formation for the 20 naturally occurring residues in proteins. These systems often consist of short "host" peptides of defined sequence and length. These host peptides often contain a number of alanine residues, which promotes helix formation, and often contain charged residues to promote solubility in water. By varying the identity of "guest" residues and measuring fractional helix content using CD spectroscopy, nearest-neighbor parameters for helix formation can be determined for each residue (**Table 12.2**).

The two columns in Table 12.2 can be viewed as equilibrium constants for helix initiation (κ) and for propagation ($\kappa\tau$). Because each residue was analyzed using the same value of τ (434), the two columns are directly proportional. For all residues, the equilibrium constant for initiation is quite small, indicating that the coil configuration is strongly favored over a single isolated helical residue. Propagation constants are significantly larger, though most are below one, slightly disfavoring propagation. The only residue that significantly favors helix propagation is alanine. Arginine promotes helix propagation, although not to as great an extent as alanine. On the other side of the scale, glycine strongly disfavors propagation, which is likely to result from the the fact that can adopt many alternative conformations, since it lacks a side chain. Side chains that are β-branched (Ile, Val, and Thr) also disfavor helix propagation, possibly a result of steric clash between the extra bulk attached to the β-carbon.

Table 12.2 Helix–coil equilibrium constants for the 20 naturally occurring residues of proteins

Residue	κ	$\kappa\tau$	$\Delta\bar{G}_p^\circ = -RT\text{Ln}(\kappa\tau)$
Ala	3.55×10^{-3}	1.54	−0.98
Arg+	2.53×10^{-3}	1.1	−0.22
Leu	2.12×10^{-3}	0.92	0.19
Lys+	1.80×10^{-3}	0.78	0.56
Glu^0	1.45×10^{-3}	0.63	1.05
Met	1.38×10^{-3}	0.60	1.16
Gln	1.22×10^{-3}	0.53	1.44
Glu−	9.91×10^{-4}	0.43	1.92
Ile	9.68×10^{-4}	0.42	1.97
Tyr	$8.52–11.5 \times 10^{-4}$	0.37–0.50	2.26–1.57
His^0	8.29×10^{-4}	0.36	2.32
Ser	8.29×10^{-4}	0.36	2.32
Cys	7.60×10^{-4}	0.33	2.52
Asn	6.68×10^{-4}	0.29	2.81
Asp-	6.68×10^{-4}	0.29	2.81
Asp^0	6.68×10^{-4}	0.29	2.81
Trp	$6.68–8.29 \times 10^{-4}$	0.29–0.36	2.81–2.32
Phe	6.45×10^{-4}	0.28	2.89
Val	5.07×10^{-4}	0.22	3.44
Thr	2.99×10^{-4}	0.13	4.63
His+	1.38×10^{-4}	0.06	6.39
Gly	1.16×10^{-4}	0.05	6.80
Pro	2.30×10^{-6}	0.001	15.7

Note: Values are from Chakrabartty et al. (1994). Values for κ were determined by setting the Lifson–Roig v' parameter (see Appendix 12.1) to 0.048, which is equivalent to a τ value of 434. $\Delta\bar{G}_p^\circ$ are helix propagation free energies (kJ mol^{-1}), evaluated at 273.15 K. For ionizable residues, parameters for both the neutral and charged forms are given. For the aromatic residues, the range of parameters results from uncertainties based on contributions of the aromatic chromophores to the CD spectra.

Problems

12.1 This problem explores the relationship between the free energy difference between individual *microstates* with j helical residues ($\Delta\bar{G}_{m\in j}^\circ$, where the delta compares to all-coil reference state), and the free energy of *the collection* with j helical residues ($\Delta\bar{G}_j$). These are the two types of free energies in the partition functions that sum over individual microstates (Equation 12.2) and helical levels (Equation 12.6).

(a) Using the molecular partition function for microstate m with j helical residues ($\theta_{m\in j}$), give the free energy difference $\Delta\bar{G}_{m\in j}^\circ$.

(b) Using molecular partition functions, give the molar Gibbs energy for the collection of microstates with j helical residues, and the free energy difference $\Delta\bar{G}_j^\circ$.

(c) Using your results from Problems 12.1a and b, what is the relationship between the $m\ \Delta\bar{G}^{\circ}_{m\in j}$ terms and the $\Delta\bar{G}^{\circ}_{j}$ term?

(d) If each of the m microstates in helix level j has the same Gibbs free energy (i.e., each of the $\Delta\bar{G}^{\circ}_{m\in j}$ terms is the same), what is the relationship between $\Delta\bar{G}^{\circ}_{j}$ and $\Delta\bar{G}^{\circ}_{m\in j}$?

12.2 For the independent identical pentapeptide distributions in Figure 12.3, imagine that the populations result from coin-tosses, where the faces of the coin are labeled h and c (rather than heads and tails). What would be the probability corresponding to throwing a single coin and getting heads and tails for the three distributions (i.e., what are the values of p_h and p_c)?

12.3 For the independent identical model for helix formation, plot the distributions for 20- and for 100-residue peptides over their j helix levels (21 and 101 levels, respectively), for $\kappa = 0.1, 1.0,$ and 10, such as in Figure 12.3. How do the distributions for different lengths (including the pentapeptide distribution, Figure 12.3) compare to each in terms of overlap and sharpness?

12.4 Starting with Equation 12.27, derive Equation 12.28, namely,

$$\langle f_H \rangle = \frac{\kappa}{1+\kappa}$$

12.5 Starting with Equation 12.27, demonstrate the following relationship:

$$\langle f_H \rangle = \frac{1}{N}\frac{d\ln\rho_H}{d\ln\kappa}$$

12.6 Writing the equilibrium constant for the independent identical model using a "single residue peptide," that is,

$$\kappa = \frac{[h]}{[c]}$$

show that the fraction helix for a single-residue peptide should follow Equation 12.28.

12.7 For a heteropolymer with N_A A residues and N_K K residues, how many different peptide sequences can be made?

12.8 For a heteropolymer model with independent positions, how does the fraction helix, $\langle f_H \rangle$, differ for the two peptides?

Sequence 1:   AAAAAAKKKKKK
Sequence 2:   AKAKAKAKAKAK

Assume that $\kappa_A = 2$ and $\kappa_K = 0.5$.

12.9 For the two peptides and κ values in Problem 12.8, how do the distributions of helix differ along the peptide, assuming independent positions? Plot the fraction helix at each residue for the two peptides.

12.10 For a heteropolymer with independent sites containing N_A A residues and N_K K residues, calculate the total fraction helix by differentiation of the partition function (Equation 12.31) with respect to N_A and N_K, and show that your result is equivalent to Equation 12.34. Hint: this requires summing two partial derivatives.

12.11 For a heteropolymer with independent sites containing $8A$ residues and $4K$ residues, plot the fraction helix as a function of κ_K, assuming a κ_A value of 10.

Next, plot fraction helix as a function of κ_A, assuming a κ_K value of 0.1. What are the lower and upper limits for the two plots?

12.12 Assuming a homopolymer nearest-neighbor model, give Boltzmann factors for the following sequences:

$$\text{chhhc} \quad \text{chhch} \quad \text{hchch}$$

Use the coil state as a reference, and the values $\kappa = 0.001$, $\tau = 0.001$.

12.13 Demonstrate that the power series

$$\sum_{j=1}^{N} x^j$$

can be written in the closed-form given by Equation 12.46

12.14 Show that the derivative in Equation 12.47 is equal to

$$\frac{d}{d\kappa}\left\{\frac{\kappa\tau[(\kappa\tau)^N - 1]}{\kappa\tau - 1}\right\} = \tau\left(\frac{N(\kappa\tau)^N[\kappa\tau - 1 - 1/N] + 1}{(\kappa\tau - 1)^2}\right)$$

12.15 Substitute the derivative from Problem 12.14 into Equation 12.47 and show that ρ_H (the one-helix approximation) given by Equation 12.48, that is,

$$\rho_H = 1 + \kappa\left[\frac{(\kappa\tau)^{N+1} - N(\kappa\tau) - (\kappa\tau) + N}{(\kappa\tau - 1)^2}\right]$$

12.16 Write the partition function for a tetrapeptide where each residue can adopt either helix or coil conformation using the noncooperative model.

12.17 What are the terms that are left out of the partition function for the one-helix zipper model?

12.18 For a value of $\kappa = 1$, give the value of the partition function for the noncooperative model.

12.19 For values of $\kappa = 0.00164$, $\tau = 5 \times 10^3$, give the value of the partition function for the one-helix zipper model.

12.20 Using the κ and τ values in Problem 8.1.E, give the value of the sum of the terms left out of the one-helix zipper model (part 8.1.C).

12.21 Using κ and τ values from Problems 8.1.D and 8.1.E (for noncooperative and zipper, respectively), give populations of each of the five macrostates (collections of configurations with the same number of helical residues, i) and organize your results in the following table:

	c4	c3h	c2h2	ch3	h4
Noncooperative					
Zipper					

12.22 Using κ and τ values from Problems 8.1.D and 8.1.E (for noncooperative and zipper, respectively), make a graph of the population of each of the macrostates for the noncooperative and zipper models as a function of i, the number of helical residues in each macrostate.

12.23 Using κ and τ values from Problems 8.1.D and 8.1.E (for noncooperative and zipper, respectively), calculate the fraction helix for the two models.

12.24 Using the matrix method, write a partition function for the helix–coil transition of a homopolymer helix a length of $N = 20$. Use Mathematica to build this partition function with two-by-two transfer matrices. Here is some helpful Mathematica syntax.

If you want to generate a matrix

$$w = \begin{pmatrix} a & b \\ c & d \end{pmatrix}$$

It is a list of lists, generated with the following input:

```
w = {{a,b},{c,d}}
```

To multiply the same matrix 20 times, use the command `MatrixPower[w,20]`. And to multiply two matrices, or a matrix by a vector (row or column) use a "dot" (.):

```
{0,1}.w.{1,1}
```

12.25 Rearrange your answer for ρ_{20} from Problem 12.42 by grouping like powers of κ. (See Equation 12.56; you can use the Mathematica "Simplify[...]" command to help with this.) Interpret the first two (κ^0 and κ^1) and last two terms (κ^{19} and κ^{20}) in terms of microstates and arrangements.

12.26 What is the sum of the numerical coefficients in Equation 12.56 ($N = 10$ residues) equal to?

12.27 Assuming a τ of 10^3, plot the fraction helix by differentiating with respect to κ (Equation 12.58), with the partition function from Problem 12.42 ($N = 20$). Use an appropriate range of κ to help visualize the curve.

12.28 Generalize your answer from Problem 12.45 over a range of τ values, by making a 3D plot of fraction helix from $\kappa = 0$ to $\kappa = 3$, and from $\tau = 1$ to $\tau = 5$. Does this plot suffer from the problem the zipper model has at $\tau = 1$? Why or why not?

The next set of problems (Problems 12.28 through 12.33) explores the extent of helix formations at different positions in an N residue peptide. To make things simple, assume a homopolymer. To make things interesting, assume a nearest-neighbor coupling model. To make things precise, use the matrix model for the exact partition function (Equation 12.54).

12.29 For a four-residue peptide, write the conformations (in terms of strings of h's and c's) that are helical at position 2 (e.g., *chcc* is one of them).

12.30 Write a "sub-partition function" that sums up the position-2 h conformations from Problem 12.28 (I refer to this sub-partition function as ρ_{2h}).

$$\rho_{2h} = \kappa + \kappa^2 + 2\kappa^2\tau + 2\kappa^3\tau^2 + \kappa^3\tau + \kappa^4\tau^3$$

12.31 Use the matrix approach to generate the sub-partition function in Problem 12.29. Do this by modifying the entries in the w matrix corresponding to position 2 to eliminate the coil-containing states. Multiply the matrix product out (you can do this by hand if you want to, but it is okay to use Mathematica) to demonstrate that your expression is the same as in Problem 12.29.

12.32 Come up with an expression that sums up all the microstates with a coil conformation at position 2 (hint—you can either modify the approach outlined in Problem 12.30, or you can combine your result from Problem 12.31 with Equation 12.55).

12.33 Use you answers from Problems 12.22 and 12.23 to give expressions for (a) the fraction of helix at position 2 (i.e., the fraction of the conformations in which position 2 is in a helical state) and an equilibrium constant giving the ratio of helical conformations at position 2 to coil conformations at position 2.

12.34 Using the concepts and general relations you developed in Problems 12.20 through 12.25, generate a distribution of fraction helix at each position in a ten residue homopolymer peptide, using a nearest-neighbor model with $\kappa = 0.02$ and $\tau = 80$. Plot these ten values as a function of position. What does the plot tell you about how helix distributes in short peptides?

12.35 For a nearest-neighbor heteropolymer, justify that the fraction helix is given by the sum of partial derivatives in Equation 12.61, that is,

$$\langle f_H \rangle = \frac{1}{\rho_H N} \left[\sum_{j \in \{20 \text{ aa's}\}} \kappa_j \frac{\partial \rho_H}{\partial \kappa_j} \right]$$

12.36 Modify the Zimm–Bragg partition function by building a correlation matrix for four residues, $i - 1$, i, and $i + 1$, and $i+2$, where the conformations in which one and two helical residues flanked by coil residues are omitted (with equilibrium constant zero).

Appendix 12.1: Other Representations of the Helix–Coil Transition

The κ, τ representation of the nearest-neighbor model used here is conceptually simple, and it arises as a straightforward extension of the noncooperative model. However, it is not the only formalism for modeling cooperativity in the helix–coil transition. Here we describe two alternative statistical thermodynamic models that have played a central role in the development and analysis the helix–coil transition: the Zimm–Bragg ("ZB") model, and the Lifson–Roig ("LR") model.

The ZB model is formally equivalent to the nearest-neighbor model developed above, but uses a different parameterization for helix initiation and propagation. The initiation step is described by the product of two parameters, σ and s, and the propagation step is described by the parameter s. In this representation, the equilibria from Equation 12.36 are

$$ccccc \xrightleftharpoons{\sigma s} chccc \xrightleftharpoons{s} chhcc \xrightleftharpoons{s} chhhc \tag{A12.1.1}$$

As with κ, s captures features common to both initiation and propagation. However, σ can be thought of as a penalty to initiation, whereas τ acts as a reward for propagation. By comparing Equations 12.36 and A12.1.1, these four constants can be related as follows:

$$\sigma = \frac{1}{\tau} \tag{A12.1.2}$$

$$s = \kappa\tau \tag{A12.1.3}$$

$$\kappa = \sigma s \tag{A12.1.4}$$

In terms of the ZB parameters, the correlation matrix becomes

$$w_{ZB} = \begin{pmatrix} s & 1 \\ \sigma s & 1 \end{pmatrix} \tag{A12.1.5}$$

Note, this is the same matrix as w in Equation 12.53 but with different symbols.

The partition function for the helix–coil transition (ρ_H) is generated using the same matrix multiplication formula as with the κ, τ representation (Equation 12.54):

$$\rho_H = \begin{pmatrix} 0 & 1 \end{pmatrix} \begin{pmatrix} s & 1 \\ \sigma s & 1 \end{pmatrix}^N \begin{pmatrix} 1 \\ 1 \end{pmatrix} \tag{A12.1.6}$$

$$= nw_{ZB}^N c$$

The fractional helix formation can be expressed as a derivative taken with respect to s (because s counts the number of helical residues):

$$\langle f_H \rangle = \frac{s}{N\rho} \frac{\partial \rho_H}{\partial s} \tag{A12.1.7}$$

The LR model differs from the ZB (and κ, τ) model in that it captures the conformational state of three adjacent residues. Since there are eight conformations ($2^{n=3}$), more matrix elements are required than can be accommodated by a two-by-two matrix. A three-by-three matrix has enough elements (and can be arranged to build the helix–coil partition; see Poland and Scheraga 1970), but the four-by-four has the advantage that the conformational states at each position can be arrayed quite simply. In terms of the eight tripeptide conformations, the LR matrix is arranged according to the following table

i−1	i	h (h)	c (h)	h (c)	c (c)
		(i+1) h	c	h	c
h	h	hhh	hhc		
h	c			hch	hcc
c	h	chh	chc		
c	c			cch	ccc

(A12.1.8)

For each of the eight entries in the table, the conformation of the $i-1$ residue is given by the left-most column, and the conformation of the $i+1$ residue is given in the top row. The conformation of the ith residue is given in both the second column and row. Note that eight table cells are left empty because they correspond to "impossible conformations" in which residue i would need to be in both helix and coil conformations.

The question is, what equilibrium constants should we use to describe residue i in the table above, in its $i-1$, $i+1$ context? The original LR formalism used three "statistical weights": u (the energetic contribution of a coil at position i), v (the energetic contribution of a helix at position i when one or both adjacent residues are coil), and w (the energetic contribution of a helix at position i when both its neighbors are also helix). With these three constants, we can write the correlation matrix as

$$W_{LR} = \begin{pmatrix} w & v & 0 & 0 \\ 0 & 0 & u & u \\ v & v & 0 & 0 \\ 0 & 0 & u & u \end{pmatrix} = \begin{pmatrix} w' & v' & 0 & 0 \\ 0 & 0 & 1 & 1 \\ v' & v' & 0 & 0 \\ 0 & 0 & 1 & 1 \end{pmatrix}$$

(A12.1.9)

The matrix elements of impossible conformations are set to zero—if they are impossible, they should not contribute to the partition function. In the second matrix, the each element is divided by u, which converts the statistical weights to equilibrium constants relative to a coil reference state (distinguished by primes). Although this was not part of the original LR formalism, it is both conceptually simpler and is a better reflection of the experimentally accessible parameters (two, not three).

As with the two-by-two representations above, the partition function for the helix–coil transition is given by the product of N w_{LR} matrices, multiplied from the left and right by four element row and column vectors:

$$\rho_H = (0 \quad 0 \quad 1 \quad 1) \begin{pmatrix} w' & v' & 0 & 0 \\ 0 & 0 & 1 & 1 \\ v' & v' & 0 & 0 \\ 0 & 0 & 1 & 1 \end{pmatrix}^N \begin{pmatrix} 0 \\ 1 \\ 0 \\ 1 \end{pmatrix}$$

(A12.1.10)

$$= n_{LR} \, w_{LR}^N \, c_{LR}$$

A fractional helix value can be obtained by differentiating with respect to w':

$$\langle f_H \rangle^* = \frac{w'}{N\rho} \frac{\partial \rho_H}{\partial w}$$

(A12.1.11)

This expression for fraction helix differs slightly from those above (hence the asterisk): Equation A12.1.10 only counts helical residues that are flanked by helical residues on both sides (since w' counts such residues). For large and cooperative peptides, the difference between these two expressions is quite small.

Given the increased size of the w_{LR} matrix, one might wonder what (if any) advantage the LR model provides. The strength of the LR model is that it is better matched to the structural details of the protein α-helix than the more compact *ZB* and κ, τ models. Recall from Figure 12.1 that formation of a helical hydrogen bond requires the three intervening residues to adopt a helical configuration. In the *LR* model, a power of *w′* is contributed to the Boltzmann factor only when three adjacent residues are helical. Thus, *w′* can be regarded as a measure of hydrogen bond formation. Likewise, the *v′* parameter corresponds to residues that have adopted a helical conformation but don't contribute a helical hydrogen bond, because they don't have helical neighbors (either because they are in a one- or two-*h* sequence, or because they are on the end of a long helix). Assuming the helical hydrogen bond to be stabilizing, the parameter *w′* should have a larger value than *v′*. In this regard, w′ and v′ are similar to s and σs, and to $\kappa \tau$ and τ (Problem 12.35).

CHAPTER 13

Ligand Binding Equilibria from a Macroscopic Perspective

Goals and Summary

The goals of this chapter are to develop an analytical formalism to describe how macromolecules interact with ligands using statistical thermodynamics, and to explore how different types of functional behavior arise. Different binding schemes will be considered, from simple one-to-one binding to multistep binding of several ligands to a macromolecule. Ligand titration curves will be developed as a fundamental graphical representation of macromolecule–ligand interaction. Different representations of the binding curve will be developed (linear and logarithmic ligand scales, as well as derivative and the Hill transformation) to highlight different features of binding, such as binding capacity, positive and negative cooperativity, and linkage among multiple ligands. In addition, some important features (and limitations) of experimental binding data will be presented.

In this chapter, binding schemes will be presented at the macroscopic level, using macroscopic binding constants (both stepwise and overall) to represent average affinities at different ligation states. The "binding polynomial," a partition function for ligand binding, will be developed, both for analytical simplicity and to provide analytical insight. The binding polynomial can be easily generalized to include binding of multiple different types of ligands, and will be used to develop general linkage relationships. This formalism will be carried forward in the next chapter, where binding will be represented with specific, microscopic models.

Over the next two chapters, we will analyze binding reactions of ligands, which we will refer to generically as "x,"[†] to biological macromolecules referred to as "M." Typically, macromolecules analyzed using this formalism are proteins, nucleic acids, and/or assemblies of both. Ligands are often small inorganic molecules (e.g., O_2, H^+, Mg^{2+}, SO_4^{2-}), small organic molecules (e.g., sugars, organic phosphates, amino acids, and nucleotides), large organic molecules (fatty acids, steroid hormones, oligosaccharides, and peptides), or other macromolecules. Binding reactions are central to nearly every biochemical processes, regulating enzyme activity, controlling gene regulation through binding of proteins to specific sites on DNA, controlling ion channel activity, and regulating cellular response to external stimuli and foreign pathogens. Binding reactions involving macromolecules are particularly important in medicine and the pharmaceutical industry, as a large number of the drugs used to treat diseases are small molecules that bind and inhibit (or in some cases activate) receptor macromolecules.

To describe binding, we will use a representation where the total macromolecule concentration is kept constant, and ligand concentration is varied. By invoking a specific binding model, populations of macromolecules with different numbers of

---

[†] And sometimes as y, z, etc., when more than one type of ligand can bind.

ligands are calculated, and used to predict the extent of binding (the primary experimental observable) as a function of ligand concentration. This formalism (macromolecule M titrated with ligand x) matches how biophysical binding experiments are typically done. This experimental set-up is used because macromolecules are often hard to get, especially at high concentrations, whereas small-molecule ligands are easy to get and can be prepared at high concentrations. Thus, it is easier to titrate low concentrations of macromolecules with high concentrations of macromolecules than the reverse. However, as long as the thermodynamics are described by mass action relationships (Chapter 4), the mechanisms, methods, and equations described here are completely general, and can be equally applied to macromolecules titrated with macromolecules, small molecules titrated with small molecules, and small molecules titrated with macromolecules.

In this and the next chapter we present complementary descriptions of binding. Here, we consider binding from a macroscopic perspective, in which bound conformations of macromolecules are grouped in terms of the number of ligands bound, but not their arrangement in sites. In the next chapter, we will consider binding from a microscopic perspective, where different arrangements of ligands bound to macromolecules will be treated. Each approach has its own set of strengths; we will develop these strengths, and also how these two perspectives relate to each other.

LIGAND BINDING TO A SINGLE SITE

We begin with a simple bimolecular binding reaction of a single ligand x to a single site on macromolecule M. In addition to introducing nomenclature, we will use this example to show some features of different types of binding curves, as well as some practical considerations and complications associated with measuring ligand binding reactions. We can represent this simple binding reaction with the following equilibrium equation:

$$M + x \rightleftharpoons Mx \tag{13.1}$$

The equilibrium constant for binding is given by mass-action:

$$K = \frac{[Mx]}{[M][x]} \tag{13.2}$$

This association constant, with units of reciprocal molar $(L \cdot mol^{-1})$ is often inverted to give the dissociation constant:

$$K_d = K^{-1} = \frac{[M][x]}{[Mx]} \tag{13.3}$$

The dissociation constant is convenient because it has units of concentration (perhaps more intuitive than inverse concentration), and represents the free ligand concentration at which the macromolecule is half-saturated ($[Mx] = [M]$; see below). The standard state free energy of binding[†] is obtained as

$$\Delta \bar{G}^{\circ} = -RT \ln K \tag{13.4}$$

Fractional saturation and average ligation number for the single-site model

Equations 13.1 through 13.4 provide a simple thermodynamic description of the single-site binding mechanism. However, to be useful to experimentalists, these

---

[†] As described in Chapter 7, the standard state for chemical reaction in solution is typically one molar reactants (M and x) and one molar product (Mx). Equation 13.4 gives the free energy of forming one mole of Mx from M and x at fixed standard-state (one molar) concentrations of all species.

equations must be connected to actual binding data. Two related quantities that connect thermodynamics to binding data are the average "fractional saturation" $\langle f_b \rangle$ and the "average ligation number" $\langle x_b \rangle$. These quantities are analogous to the fractional quantities we used in Chapters 8 and 12 to represent conformational transitions.

Fractional saturation is the fraction of macromolecular binding sites that have ligand bound. For the single-site model (Equation 13.1), this is simply

$$\langle f_b \rangle = \frac{[Mx]}{[M] + [Mx]} \qquad (13.5)$$

As with the fractional quantities in previous chapters, $\langle f_b \rangle$ ranges from zero to one, depending both on the affinity and the free ligand concentration.

The average ligation number is the average number of ligands bound per macromolecule. $\langle x_b \rangle$ is closely related to $\langle f_b \rangle$, and indeed, if there is only one binding site, as in the present example, the two quantities are equal. However, when there is more than one binding site on the macromolecule (we will use s to indicate the number of sites), then $\langle x_b \rangle$ is larger than $\langle f_b \rangle$ (the former ranges from 0 to s), and can be related to $\langle f_b \rangle$ as $\langle f_b \rangle = \langle x_b \rangle / s$.

By solving for the concentration of bound macromolecule in the mass–action relationship (i.e., by rearranging Equation 13.2 to $[Mx] = K[M][x]$) and substituting into Equation 13.5, $\langle f_b \rangle$ can be expressed using the equilibrium binding constant:

$$\begin{aligned} \langle f_b \rangle &= \frac{K[M][x]}{[M] + K[M][x]} \\ &= \frac{K[x]}{1 + K[x]} \end{aligned} \qquad (13.6)$$

Equation 13.6 connects the thermodynamics of ligand binding (through the binding constant) to the experimentally accessible fractional saturation, and to the free ligand concentration, which will be treated as the independent variable in ligand binding experiments.

Graphical representation of the fractional saturation

As with conformational transitions, we will evaluate expressions for fractional saturation as a function of ligand concentration to gain insight into how molecular mechanisms and binding energy translate into observable behavior, and conversely, how these quantities can be extracted from experimental data. For conformational transitions, we plotted the fraction of structured macromolecules as a function of folding propensity (equilibrium constant or free energy of folding), either directly as for the helix–coil transition (Figure 12.4), or by modulating folding propensity using an external variable such as temperature or denaturant concentration. Although we could construct the same type of plot for ligand binding, that is, fractional saturation as a function of K, the ability to modulate saturation simply by changing ligand concentration allows us to characterize the binding equilibrium without significantly changing physical conditions. The resulting binding curves (either $\langle f_b \rangle$ or $\langle x_b \rangle$ vs. $[x]$) provide a means to estimate binding constants and binding energies from data using fitting methods.

For single-site ligand binding, the binding curve described by Equation 13.6 is a rectangular hyperbola (**Figure 13.1**).

When ligand concentration is low, saturation is low, but it rises steeply, and then plateaus as saturation is approached. As with conformational transitions, the midpoint of the binding curve is an important parameter, with direct connection to thermodynamics. This can be seen by evaluating Equation 13.6 at the midpoint ligand concentration, $[x]_m$:

$$\langle f_b \rangle_m = 0.5 = \frac{K[x]_m}{1 + K[x]_m} \qquad (13.7)$$

Figure 13.1 A ligand binding curve for a simple single-site binding reaction (Equation 13.1). Here, $[x]$ represents the free ligand molarity. The fraction bound ($\langle f_b \rangle$), which in this case is equal to $\langle x_b \rangle$, ranges from zero to one. At half-saturation ($\langle f_b \rangle = 0.5$), $[x] = K^{-1} = K_d$. In this example, $K = 1$ M^{-1}.

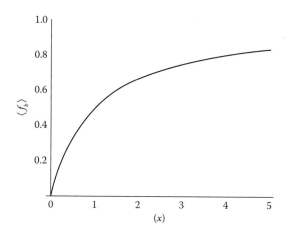

Solving for $[x]_m$ gives

$$[x]_m = \frac{1}{K} = K_d \tag{13.8}$$

Thus, for single-site binding, K_d represents the concentration of free ligand required to achieve half-saturation; likewise, the half-saturation point can be used to eyeball K_d from experimental data.

As Equation 13.8 demonstrates, the numerical value of K strongly influences the binding curve. For larger values of K, the position of equilibrium in Equation 13.1 lies further to the right (toward the Mx complex). As a result, at a given free ligand concentration $[x]$, the concentration of Mx is increased and the concentration of M is decreased. Thus, increasing the value of K shifts the $\langle f_b \rangle$ curve to the left, and results in a sharper response to $[x]$. These shifts can be seen in **Figure 13.2**, where $\langle f_b \rangle$ is plotted for different values of K, both as a function of ligand concentration (Figure 13.2A) and of the logarithm of ligand concentration (Figure 13.2B). In both the linear plot ($\langle f_b \rangle$ vs. $[x]$) and the log plot ($\langle f_b \rangle$ vs. log$[x]$), higher binding constants shift the curve to the left.

Although both plots in Figure 13.2 depict increases in saturation with ligand concentration, these two representations of the binding curve each highlight different aspects of binding. The linear concentration scale reflects the experimental "space" in which the ligand binding data were collected, whereas the logarithmic concentration scale reflects the chemical potential of the ligand, and is thus a representation of the molar free energy of the ligand. For the linear concentration scale, the value of K (i.e., the affinity) affects the steepness of saturation, as well as the midpoint, whereas for the logarithmic scale, the shape (or steepness) of the curves is not changed.

As we see below, one advantage of the linear plot is that key features associated with cooperative ligand binding influence the shape of the binding curve in a way that is both easily visualized and is physiologically important. One advantage of the logarithmic plot is that binding curves that have the same mechanism but different affinities can be closely compared.[†] A second advantage of the logarithmic binding plot is that the extent to which experimental binding data define the unbound and fully bound portions of the curve is readily apparent. As with conformational transitions, these limits are often unknown; thus, obtaining precise estimates of unbound and fully bound signals directly from the data are as important for accurately determining thermodynamic parameters as obtaining data around the midpoint (see below). It should be kept in mind that although the curves are sigmoidal on the logarithmic scale, *the curves result from a simple, noncooperative mechanism.*

[†] For the range of K's depicted in Figure 13.2, the linear plot only shows partial saturation for the lowest affinity curve, and very little detail can be seen in the highest affinity curves. In contrast, the curves in the logarithmic plot are nicely spread out.

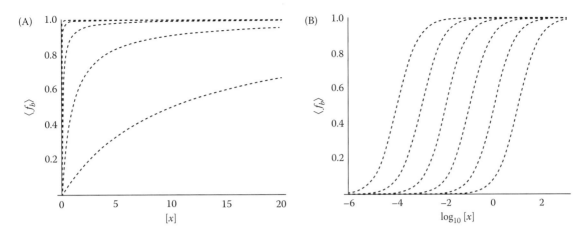

Figure 13.2 The effect of different values of *K* on binding curves for a single-site reaction. *K* values range from 10^4 M^{-1} to 10^{-1} M^{-1} by factors of ten. Plotted as a linear function of [*x*] (left), increasing *K* A-shifts curves and increases the steepness at the midpoint. Plotted as a function of $\log_{10}$[*x*] (B), the curves shift by a constant amount (one log unit, because *K* is incremented by factors of ten), but all the curves have the same shape, reflecting simple noncooperative binding. Note that the logarithmic curve better allows curves with very different affinities to be compared.

The binding capacity

When plotted as a function of the $\log_{10}$[*x*], binding curves for a single-site reaction have maximal slope around their midpoint (Figure 13.2B). This high slope reflects the high sensitivity of binding to increases in ligand free energy (i.e., chemical potential, which is proportional to $\log_{10}$[*x*]). In general, the slope of a binding curve is an important quantity, representing the extent to which a unit change in ligand concentration (or free energy on the log scale) increases the saturation of the macromolecule. A high value, such as is seen at the midpoint of a single-site binding curve, means the macromolecule has a high capacity to binding ligand. For this reason, we will refer to the slope of the binding curve as the "binding capacity," C_x. The binding capacity can be calculated directly by differentiating the fractional saturation with respect to $\log_{10}$[*x*]:

$$C_x = \frac{df_{bound}}{d\log_{10}[x]} = 2.303 \times \frac{df_{bound}}{d\ln[x]} \qquad (13.9)$$

Because the binding capacity is evaluated with respect to the log of the ligand concentration, it is a dimensionless quantity. The value 2.303 converts the base ten log to the natural log.[†] For the single-site mechanism, differentiating the expression for $\langle f_b \rangle$ (Equation 13.6) gives

$$
\begin{aligned}
C_x &= 2.303 \times \frac{df_{bound}}{d\ln[x]} = 2.303[x] \times \frac{d}{d[x]}\left(\frac{K[x]}{1+K[x]}\right) \\
&= 2.303[x]\left(\frac{K}{1+K[x]} - \frac{K^?[x]}{(1+K[x])^2}\right) \\
&= 2.303K[x]\left(\frac{1+K[x]}{(1+K[x])^2} - \frac{K[x]}{(1+K[x])^2}\right) \\
&= 2.303\left(\frac{K[x]}{(1+K[x])^2}\right)
\end{aligned}
\qquad (13.10)
$$

[†] This value comes from the relationship $\ln(x)/\log(x) = 2.303....$

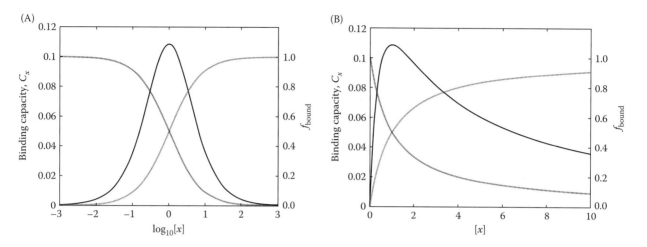

Figure 13.3 Binding capacity for a single-site binding reaction. The binding capacity, C_x, is shown as a solid black line, along with $\langle f_b \rangle$ and $\langle f_u \rangle$ (dotted gray and red lines, respectively), as a function of both $\log[x]$ (A) and $[x]$ (B). The two curves assume a binding constant of $K = 1\ \text{M}^{-1}$.

The right-hand side of Equation 13.10 can be rearranged into a particularly useful form:

$$C_x = 2.303 \left(\frac{K[x]}{(1 + K[x])^2} \right) = 2.303 \left(\frac{K[x]}{1 + K[x]} \right) \left(\frac{1}{1 + K[x]} \right)$$
$$= 2.303 \times f_{bound} \times f_{free}$$

(13.11)

The last line of Equation 13.11 has the same form as the excess heat capacity term in our analysis of two-state protein unfolding by DSC (Equations 8.47 and 8.48). Away from the midpoint of the simple binding mechanism, either $\langle f_b \rangle$ or $\langle f_u \rangle$ goes to zero, and thus, so does the product $f_{bound} \times f_{free}$. Only at the midpoint, where both $\langle f_b \rangle$ and $\langle f_u \rangle$ are nonzero, is this product nonzero. As a result, C_x has a maximum at the midpoint of the simple binding curve, which is particularly conspicuous when plotted as $\log[x]$ (**Figure 13.3A**).

The maximum in C_x is also apparent when plotted as a linear function of ligand concentration (**Figure 13.3B**). Although this maximum may seem surprising, given the slope of $\langle f_b \rangle$ monotonically decreases with $[x]$, it should be kept in mind that C_x was calculated as the slope as a function of $\log[x]$ (Equation 13.9). In terms of slope versus linear ligand concentration, C_x takes the form

$$C_x = 2.303 \times \frac{df_{bound}}{dln[x]} = 2.303[x] \times \frac{df_{bound}}{d[x]}$$

(13.12)

Although the slope of the linear binding curve monotonically decreases (i.e., it increases as $[x]$ goes to zero), multiplication by $[x]$ decreases the value of C_x as $[x]$ goes to zero.

The maximum in C_x is analogous to the maximum in the heat capacity function at the temperature midpoint of a conformational transition. The maximum in the heat capacity results both from a difference in enthalpy between conformational states, and a significant population of both conformational states. These two factors combine to give a large enthalpy variance. Here the variance is that of the extent of ligation, which results from high populations that differ in the number of ligands bound.

PRACTICAL ISSUES IN MEASURING AND ANALYZING BINDING CURVES

The goal of measuring binding curves is to see whether ligands bind, to determine the reaction mechanism, and to quantify affinities and binding energies. For example, can the interaction of a specific macromolecule with ligand be described using a rectangular hyperbola, such as that predicted for single-site binding, and if so, what is the binding constant? Although these answers should be easy to get from fractional saturation curves such as those in Figure 13.2, real experimentally determined binding data differ from idealized curves in four fundamental ways:

1. Saturation curves are collected a collection of discrete points at selected ligand concentrations. A reasonable selection of points must be made.
2. Measurement of $\langle f_b \rangle$ is not typically made directly. Rather, some sort of signal that varies with ligand concentration is measured, and the limits must be estimated from the resulting curve.
3. In most ligand binding experiments, free ligand concentration is not known. Instead, what is known is total ligand concentration ($[x]_{tot}$; i.e., $[Mx] + [x]$ in a simple single-site mechanism).
4. Measurements of binding are subject to experimental error.

These four features of experimental binding data dictate how ligand binding data should be collected and analyzed, and set some limitations in what can be obtained with ligand titrations.[†] These issues will be discussed in turn below.

Discrete data and data selection

In contrast with smooth curves generated from analytical functions, there are limits (amount of material, solubility of the ligand, patience of the experimentalist) in the number of data points that can be practically collected. And like conformational transitions, there are several important regions of the binding curve that must be defined experimentally. Obviously, ligand concentrations around the midpoint ($[x] = K^{-1}$) are important for measuring the equilibrium of free and bound macromolecules. But because binding measurements are indirect (see point 2 below), it is important to get enough data points at very low and very high degrees of saturation. As described above, the logarithmic representation of the binding curve clearly represents these three regions, much as a well-defined conformational transition shows the baselines as well as the midpoint region.

Although in favorable cases these three different ligand concentration ranges can be accessed experimentally, the spread of these three regions into very different ligand concentration ranges makes the linear sampling of ligand concentrations impractical. For example, If we wanted to get data in all three regions of the plot, covering for example from 5% saturation to 95% saturation, and we collected data at a constant ligand step-size Δx (i.e., at constant increments of ligand concentration such as 0.05 M additions in the plot above), it would take a large number of steps (400 data points for the plot above), as illustrated in **Figure 13.4**.

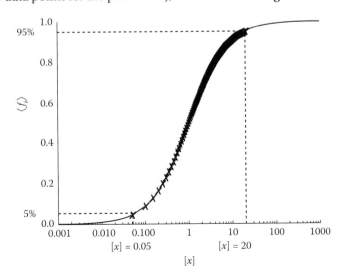

Figure 13.4 The number of equal steps required to fully define a ligand binding curve. For a single-site mechanism with a binding constant $K = 1\,M^{-1}$, the ligand concentration range shown on the plot (20-fold below to 20-fold above the K_d) corresponds to 5%–95% fractional saturation. For equally spaced increments of ligand concentration defined by the starting ligand concentration (0.05 M), 400 points (red x's) are required to span this range.

[†] Features 1, 2, and 4 are also limitations in measurements of conformational transitions; feature 3 is unique to ligand binding.

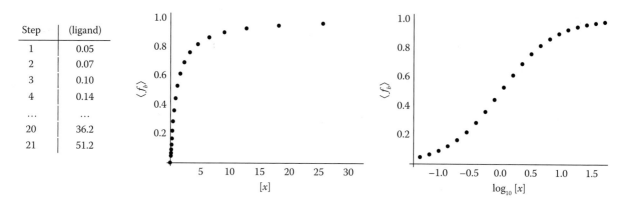

Figure 13.5 A simple ligand titration using increments of ligand concentration of $\sqrt{2}$. The table shows a list of successive ligand concentrations in which each is increased by 1.41. The middle titration is for a simple single-site mechanism with $K = 1$ M^{-1}. The black circles show an increment in ligand concentration by this factor; an additional point at zero ligand concentration (red circle) is included. This point is usually easy to measure, but it is not part of the multiplicative series. Plotted on a log scale (right), ligand concentrations are evenly spaced. Note the red circle at $[x] = 0$ cannot be depicted in this plot, since $\log_{10}(0) = -\infty$.

If a larger Δx is used, not enough points will be sampled in the unbound and transition regions (low and intermediate $[x]$ values). If fewer points are taken, the saturated baseline will not be determined. A solution is to increment ligand concentration by a constant multiplicative factor, rather than a constant additive factor. Handsome, well-sampled curves are generated when that factor is the square root of two:

$$[x]_{i+1} = \sqrt{2} \times [x]_i$$
$$\approx 1.4 \times [x]_i, \quad \text{or} \tag{13.13}$$
$$\Delta[x] = [x]_{i+1} - [x]_i \approx 1.4 \times [x]_{i+1} - [x]_i = 0.4 \times [x]_i$$

Figure 13.5 shows a partial list of ligand concentrations that would be generated using $\sqrt{2}$ multiplication, and how a simple binding curve (with no error) would look using this ligand concentration spacing:

Indirect measurement of $\langle f_b \rangle$

The fact that in in most types of ligand binding experiments we do not directly measure $\langle f_b \rangle$, but rather measure a changing signal Y_{obs}, is typically not difficult to deal with. If the binding-responsive signal originates from the macromolecule, we just assume that our signal comes from a population-weighted average from unbound and bound macromolecule as below:

$$Y_{obs} = Y_{free} \times f_{free} + Y_{bound} \times f_{bound}$$
$$= Y_{free} \times (1 - f_{bound}) + Y_{bound} \times f_{bound}$$
$$= Y_{free} - Y_{free} \times f_{bound} + Y_{bound} \times f_{bound}$$
$$= Y_{free} + (Y_{bound} - Y_{free}) \times f_{bound} \tag{13.14}$$
$$= Y_{free} + \Delta Y \times \frac{K[x]}{1 + K[x]}$$

ΔY is the "amplitude" or total deflection swept out by the binding curve. Binding curves described by Equation 13.14 (**Figure 13.6**) look like the $\langle f_b \rangle$ versus $[x]$ curves described above (Figures 13.1 and 13.2), but they range from Y_{free} to Y_{bound}, rather than from zero to one.

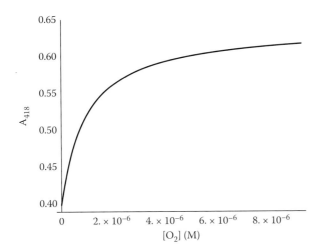

Figure 13.6 Binding of O_2 to sperm whale myoglobin, monitored by absorbance change. Dioxygen (O_2) binding to the heme group of myoglobin leads to an absorbance change, which is especially pronounced at wavelengths around 418 nm. At a myoglobin concentration of 5 μM, the absorbance changes from $A_{free} = 0.4$ by $\Delta A = 0.24$ units upon saturation. In this example, $K = 1.06 \times 10^6$ M^{-1} (pH 7.0°C, 20°C; values from Antonini, E., and Brunori, M. 1971. North Holland Publishing Company, Amsterdam).

Determining K from titration data then amounts to fitting the three parameters in the equation above using nonlinear least-squares: Y_{free}, ΔY, and K.[†] This approach is analogous to our treatment of experimental conformational transitions, where observable signals such as absorbance, fluorescence, or circular dichroism are detected instead of fraction folded. Note that if the ligand being titrated also contributes Y_{obs}, Equation 13.14 must be modified to take into account the change in ligand concentration through the titration.

Total (rather than free) ligand as an independent variable

In most ligand binding experiments, a solution of the macromolecule is prepared at a known concentration and volume, and then a specific amount of ligand is added. After adding ligand, we know the total ligand concentration ($[x]_{tot}$), but only a fraction remains free in solution (as $[x]$), since some is bound by the macromolecule (here as $[Mx]$). The amount bound is predicted by the equilibrium constant, but we don't know the equilibrium constant *a priori* (that is a main reason to measure ligand binding in the first place). The same is true for total macromolecule concentration: we knew the concentration at the start of the experiment ($[M]_{tot}$), but only a fraction remains in solution. By invoking the constraint of mass conservation, we can express these ideas as

$$[x]_{tot} = [x] + [Mx]$$
$$[M]_{tot} = [M] + [Mx] \qquad (13.15)$$

These two expressions can be rearranged to give expressions for free $[x]$ and $[M]$.

$$[x] = [x]_{tot} - [Mx]$$
$$[M] = [M]_{tot} - [Mx] \qquad (13.16)$$

These two equations allow $[x]$ and $[M]$ to be expressed in terms of two known quantities and only a single unknown. By introducing these substitutions into the mass action expression for K, that is,

$$K = \frac{[Mx]}{[M][x]} = \frac{[Mx]}{([M]_{tot} - [Mx])([x]_{tot} - [Mx])} \qquad (13.17)$$

we can express K in terms of one unknown (Equation 13.17, right-hand side) rather than three (Equation 13.17, middle). Equation 13.17 can be solved for $[Mx]$ in terms

---

[†] Alternatively, Y_{bound} could be fitted instead of ΔY.

of the known parameters $[M]_{tot}$ and $[x]_{tot}$, and for the unknown K, which in turn can be used to express $\langle f_b \rangle$ and $\langle f_u \rangle = 1 - \langle f_b \rangle$ in Equation 13.14. When the denominator of Equation 13.17 is multiplied out, a quadratic in $[Mx]$ is obtained:

$$K([M]_{tot} - [Mx])([x]_{tot} - [Mx]) - [Mx] = 0$$

$$[Mx]^2 + \left(-[M]_{tot} - [x]_{tot} - \frac{1}{K}\right)[Mx] + [M]_{tot}[x]_{tot} = 0 \qquad (13.18)$$

This can be solved using the quadratic formula to give

$$[Mx] = \frac{[M]_{tot} + [x]_{tot} + (1/K) \pm \sqrt{\left(-[M]_{tot} - [x]_{tot} - (1/K)\right)^2 - 4[M]_{tot}[x]_{tot}}}{2} \qquad (13.19)$$

By evaluating the $[Mx]$ expression above at $[x]_{tot} = 0$, it can easily be demonstrated that the negative root is the one that describes the binding curve (Problem 13.6). This expression can then be inserted into $\langle f_b \rangle$, by first writing $\langle f_b \rangle$ in terms of what is known, namely $[M]_{tot}$ and $[x]_{tot}$:

$$Y_{obs} = Y_{unbound} + \Delta Y \times f_{bound}$$

$$= Y_{unbound} + \Delta Y \times \frac{[Mx]}{[M] + [Mx]}$$

$$= Y_{unbound} + \Delta Y \times \frac{[Mx]}{[M]_{tot}}$$

$$= Y_{unbound} + \Delta Y \times \frac{[M]_{tot} + [x]_{tot} + \frac{1}{K} - \sqrt{\left(-[M]_{tot} - [x]_{tot} - \frac{1}{K}\right)^2 - 4[M]_{tot}[x]_{tot}}}{2[M]_{tot}}$$

$$(13.20)$$

Although Equation 13.20 looks considerably more complicated than Equation 13.14, both equations have the same three unknowns ($Y_{unbound}$, ΔY, and K). Thus, both equations might be expected to be equally suited to fit to a set of binding data to extract K. Indeed, as long as $[M]_{tot} \ll K^{-1}$ (or perhaps more intuitively, $[M]_{tot} \ll K_d$), Equations 13.14 and 13.20 are identical, and thus binding curves described by these two expressions look the same. However, when $K_d \ll [M]_{tot}$, the shape of the curve described by Equation 13.20 changes (**Figure 13.7**). Rather than showing

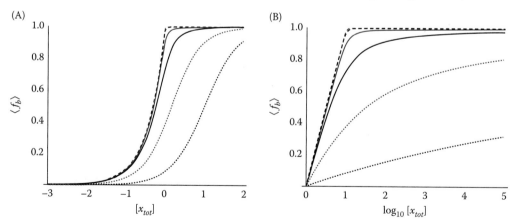

Figure 13.7 The saturation binding regime for a single-site mechanism. Fractional binding is plotted as a function of total ligand concentration (free plus bound) for a titration of 1 M total macromolecule ($[M]_{tot}$). As K changes from 0.1 M^{-1} to 10^4 M^{-1} (the same range as in Figure 13.2), the shape of the single-site binding curve changes from a rectangular hyperbola to an intersection of two lines at $[x]_{tot} = [M]_{tot}$ (A). Although the intersection point can be useful for analyzing the stoichiometry of the binding reaction (in this case 1:1), it prevents determination of K: the two highest affinity curves are nearly identical, although they differ in affinity by a factor of 10. The saturation limit also distorts the shape when fractional binding is plotted versus $[x]_{tot}$ on a log scale (B).

gradual curvature through the binding transition, the curve approaches two intersecting straight lines. The first line climbs from $Y_{obs} = Y_{unbound}$ at $[x]_{tot} = 0$ ($\langle f_b \rangle = 0$) up to $Y_{obs} = Y_{bound}$ at $[x]_{tot} = [M]_{tot}$ ($\langle f_b \rangle = 1$). Within this region, all added ligand binds to macromolecule to form Mx. At $[x]_{tot} \geq [M]_{tot}$, the curve plateaus at $Y_{obs} = Y_{bound}$. Thus, there is virtually no free macromolecule in solution: the macromolecule is saturated. This type of binding is referred to as "saturation binding."

One important but subtle feature of saturation binding is that in the saturation limit, binding curves become insensitive to the value of K. Whereas there is a big difference between the curves corresponding to $K = 0.1\ \text{M}^{-1}$ and $K = 1\ \text{M}^{-1}$ ($K_d \geq [M]_{tot}$) in Figure 13.7, there is very little difference between the curves corresponding to $K = 10^3\ \text{M}^{-1}$ and $K = 10^4\ \text{M}^{-1}$. Although stoichiometry is well-determined, K cannot be accurately fitted from binding data in the saturation limit (Problems 13.4 and 13.5)

Measurements of binding are subject to experimental error

Because experimental measurements of binding have associated error, fitted values of binding constants determined real data sets will also have errors. This is particularly true when, as is typically the case, $\langle f_b \rangle$ is estimated indirectly from a binding-sensitive signal Y_{obs}.

To illustrate the effects of measurement error on the values of fitted binding constants, ligand titrations were simulated with different amounts of Gaussian error on Y_{obs}, and binding constants were fitted (**Figure 13.8A through C**). By repeating this procedure (1,000 times), the effects of random measurement error on fitted binding constant value can be determined. The resulting binding constant distributions (**Figure 13.8D through F**) show considerably greater spread at higher error levels: at 2.5% error in the absorbance amplitude (ΔY in Equation 13.14), the standard deviation in the binding constant values is around 10%. For more complicated binding schemes, uncertainties in fitted binding constants can become significantly larger, partly as a result of correlations among binding constants for different steps.

BINDING OF MULTIPLE LIGANDS

Equation 13.20 notwithstanding, the single-site binding mechanism is delightfully simple, and it applies to many important biological (and nonbiological) equilibrium reactions (as well as "steady-state" kinetic reactions such as enzyme catalysis of a single-step reaction at a single site). However, many macromolecules have multiple binding sites, and the single-site binding scheme is insufficient to describe and analyze ligand binding curves, in addition to being mechanistically incorrect. Common examples of multisite macromolecules include multisubunit proteins with one site per subunit for the same ligand, linear repetitive macromolecules such as DNA, which can have multiple nonoverlapping (and overlapping) sites, and ribonucleoprotein particles that have multiple sites for different proteins and RNAs. When just a single-type ligand (x) binds to multiple sites on a macromolecule (we represent the number of sites on the macromolecule with the letter $s > 1$), we refer to binding as "homotropic," as represented by the following scheme:

$$M_0 + sx \rightleftharpoons Mx + (s-1)x \rightleftharpoons Mx_2 + (s-2)x \rightleftharpoons \cdots \rightleftharpoons Mx_s \quad (13.21)$$

or simply

$$M_0 \rightleftharpoons Mx \rightleftharpoons Mx_2 \rightleftharpoons \cdots \rightleftharpoons Mx_s \quad (13.22)$$

where we have omitted free ligand concentration. Each reaction is stated explicitly in **Figure 13.9**. When multiple distinct ligands (x, y, ...) bind to sites on a macromolecule, we refer to binding as "heterotropic" (see below).

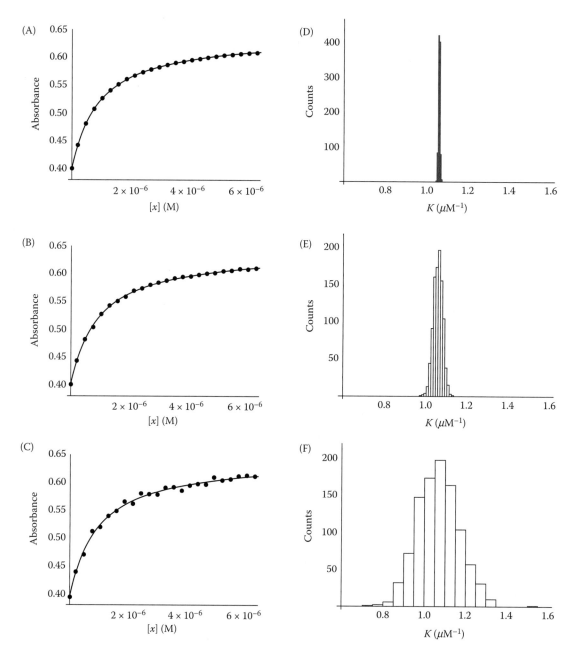

Figure 13.8 Effects of error on fitted binding constants for the single-site mechanism. (A–C) Simulated titrations (absorbance vs. free ligand [x]) with different amounts of error. Twenty-four values for were calculated using with $\Delta[x]$ increments of $\sqrt{2}$ (see Figure 13.5), using parameters for the single-site model in Figure 13.6 (O_2 binding to myoglobin; $K = 1.06 \times 10^6$ M^{-1}, $A_{free} = 0.4$, $\Delta Y = 0.24$ units). Random errors were added to each absorbance value using Gaussian distributions with standard deviations of 0.1% (A), 0.5% (B), and 2.5% (C) of the total amplitude of the titration, Δy. Solid lines show fits to model from which data were generated (solid lines). (D–F) Distributions of fitted binding constants for 1,000 fits to data with error levels from panels (A–C). The spread in fitted values for K increases significantly with error in the measured absorbance value: an error of 2.5% (0.006 absorbance units) a standard deviation of ~10% in the fitted binding constant distribution.

Sometimes ligands bind to the different sites on a macromolecule independently. In other cases, binding sites are coupled energetically, that is, binding of ligand to one site influences the affinity of the other sites (either by strengthening or weakening the interaction). This coupling is referred to as **cooperative** ligand binding, and is essential for many of the switch-like and graded properties that form the basis of biological regulation. Understanding how distant sites are coupled remains an ongoing challenge in structural biology.

In the remainder of this chapter, we will develop the framework for modeling multisite equilibria and for determining affinities and interaction energies. As with

(A)

Stepwise reaction	Stepwise binding constant	Units
$M_0 + x \rightleftharpoons Mx$	$K_1 = \dfrac{[Mx]}{[M_0][x]}$	M^{-1}
$Mx + x \rightleftharpoons Mx_2$	$K_2 = \dfrac{[Mx_2]}{[Mx][x]}$	M^{-1}
...	...	...
$Mx_{s-1} + x \rightleftharpoons Mx_s$	$K_s = \dfrac{[Mx_s]}{[Mx_{s-1}][x]}$	M^{-1}

(B)

Overall reaction	Overall binding constant	Units
$M_0 + x \rightleftharpoons Mx$	$\beta_1 = \dfrac{[Mx]}{[M_0][x]}$	M^{-1}
$M_0 + 2x \rightleftharpoons Mx_2$	$\beta_2 = \dfrac{[Mx_2]}{[M_0][x]^2} = K_1 K_2$	M^{-2}
...	...	...
$M_0 + sx \rightleftharpoons Mx_s$	$\beta_t = \dfrac{[Mx_s]}{[M_0][x]^s} = \prod\limits_{i=1}^{s} K_i$	M^{-s}

Figure 13.9 Stepwise and overall macroscopic binding formalisms. For a macromolecule with s binding sites, ligand binding can be described in a series of single ligand steps (A), in which reactants and products differ by the binding of one ligand. Alternatively, binding can be described as a collection overall binding reactions (B), in which reactants are the unligated macromolecule (M_0) and i ligand molecules (ix) and the product is contains i bound ligands (Mx_i). Both formalisms can be described by as set of s binding constants (referred to as K_i's and β_i's for stepwise and overall schemes). Although each set of constants determines the other, they generally differ in their numerical values and dimensions.

our treatment of single-site binding, we will connect binding mechanisms and thermodynamic parameters to experimental observables using fractional saturation (i.e., probability) expressions. As will become apparent, these expressions have a form very similar to partition functions (and their derivatives) developed in previous chapters, with the important difference that the terms in these sums have ligand concentrations raised to various powers.[†] Thus we will refer to such expressions as "binding polynomials." As with partition functions in other systems, the binding polynomial will lead to a very compact, convenient, and intuitive way to represent complex binding schemes. In fact, by comparing the forms that binding polynomials take for different mechanisms, very fundamental connections between the partition function and basic probability concepts (Chapter 1) can be seen.

We will develop two complementary descriptions of multisite binding: a macroscopic formalism (in the remainder of this chapter) and a microscopic formalism (in the chapter that follows). The macroscopic formalism represents the total number of ligands bound to the macromolecule, but ignores their distribution among sites. The microscopic formalism represents the occupancy of each individual site, and depending on the model, the coupling between sites. The macroscopic approach has the advantage that it is analytically simple, and can be applied only with knowledge of molecular details. The microscopic approach has the advantage that it is more fundamental, and explicitly represents site-specific affinities and coupling terms. However, the mechanistic detail of the microscopic formalism comes at a cost of more complex expressions. For this reason, microscopic models often invoke simplifications (e.g., equivalence of sites, symmetry of subunits) to minimize the complexity (and the number of parameters). These simplified microscopic treatments can be quite elegant and can provide significant insight, but their underlying approximations should be kept in mind.

A MACROSCOPIC REPRESENTATION OF MULTIPLE LIGAND BINDING

The macroscopic formalism considers equilibrium of states of a macromolecule with different total numbers of ligands bound, but does not discriminate between the reactivity of the s different sites or different ways that i ligands can be bound to the s sites. Rather, all of the different microscopic arrangements on i ligands are lumped into a single macrostate concentration $[Mx_i]$. Within the macroscopic formalism, there are two ways to write binding reactions and equilibrium constants to connect the different ligation states (Figure 13.9). "Stepwise" binding constants

---

[†] The ligand concentration terms can be viewed as canceling the dimensions of the binding constants in the sum, which is required to keep all of the terms dimensionless.

represent each consecutive step in which a single ligand is added to the macromolecular species. Thus, the set of stepwise binding reactions describes the formation of Mx_i one ligand at a time. "Overall" binding constants describe the formation of different ligation states Mx_i with multiple ligands (ix) from the unligated state (M_0) in a single "concerted"[†] reaction.

The equilibrium constants for the stepwise and overall treatments will be represented as K_i and β_i, respectively, where i indicates the constant describes formation of the Mx_i complex, (i.e., the complex with i ligands bound). You should convince yourself that the overall binding constants can be related to the stepwise constants as

$$\beta_i = K_1 K_2 \ldots K_i = \prod_{j=1}^{i} K_j \qquad (13.23)$$

The stepwise binding constants are easier to think about because they all have the same units (inverse molarity, M^{-1}); if a step in ligand binding is favored, this is reflected in a larger value for the corresponding stepwise binding constant (although there is a hidden statistical factor that must be accounted for; see Chapter 8). In contrast, because the units of overall binding constants are increasing powers of inverse molarity (a result of the $[x]^i$ term in the denominator in the mass action expression), the numerical values of the overall equilibrium constants for different steps tend to become extremely large (or small) at for large values of i. As a result, the relative affinities for different steps are not so easily deduced from the relative values of overall binding constants. Although this is a liability, overall binding constants lead to more compact equations for ligand binding.

These numerical features are illustrated in **Table 13.1**, which compares the four stepwise and overall constants for binding of molecular oxygen (O_2) to hemoglobin, a four-subunit protein with $s = 4$ sites.

Within a statistical correction (Chapter 14), the increase in the stepwise binding constants for the third and fourth ligands (K_3 and K_4) over the first two (K_1 and K_2) reflect an increase in affinity as saturation proceeds, and as will be illustrated below, positive cooperativity. Although the same saturation behavior must also be represented in the overall binding constants (see Problem 13.8), this modest (~ 8-fold) increase is dwarfed by the large changes in the β_i due to differences in their dimensions (Table 13.1).

Table 13.1 Stepwise and overall constants for O_2 binding to hemoglobin (Hb). At pH 7.0

Formalism (units)	HbO$_2$	Hb(O$_2$)$_2$	Hb(O$_2$)$_3$	Hb(O$_2$)$_4$
Stepwise (M^{-1})	$K_1 = 2.7 \times 10^4$	$K_2 = 2.3 \times 10^4$	$K_3 = 1.2 \times 10^5$	$K_4 = 1.75 \times 10^5$
Stepwise (mm Hg^{-1})	$K_1 = 0.049$	$K_2 = 0.043$	$K_3 = 0.22$	$K_4 = 0.32$
Overall (M^{-1})	$\beta_1 = 2.7 \times 10^4$	$\beta_2 = 6.2 \times 10^8$	$\beta_3 = 7.4 \times 10^{13}$	$\beta_4 = 1.3 \times 10^{19}$
Overall (mm Hg^{-1})	$\beta_1 = 0.05$	$\beta_2 = 2.1 \times 10^{-3}$	$\beta_3 = 4.7 \times 10^{-4}$	$\beta_4 = 4.9 \times 10^{-8}$

Source: Antonini, E., and Brunori, M. 1971. North Holland Publishing Company, Amsterdam.
Note: For both the stepwise and overall formalism, binding constants are given both in molarities and in O_2 partial pressures. The two quantities are converted using the relationship 1 mm Hg = 1.82×10^{-6} M for oxygen in aqueous solution at 20°C (see Problem 13.7).

[†] See the discussion on the next page for a clarification of what is and what is *not* meant by "concerted."

One final comment about the overall binding constants. Although the β_i's describe overall equilibrium reactions involving many reactant molecules, this by no means implies that these reactants all associate simultaneously in a single mechanistic step. Indeed, such a mechanism would be highly improbable. Rather, the stoichiometric equations associated describe an overall equilibrium connecting M_0 with the Mx_i. Because we will describe these reactions in terms of state functions (free energies and their related equilibrium constants), it does not matter what path, (i.e., through what partly ligated intermediates) the reaction follows.

Average ligation number and fractional saturation

As for single-site binding, we will develop saturation expressions in terms of species concentrations (macroscopic species in this treatment) and then substitute in equilibrium constants ($K's$ and $\beta's$). When multiple ligands bind, the fractional saturation and average ligation number ($\langle f_b \rangle$ and $\langle x_b \rangle$) differ. $\langle f_b \rangle$ ranges from 0 to 1, whereas $\langle x_b \rangle$ ranges from 0 to s. Fractional saturation is a bit more intuitive, because, you need not remember how many sites s the macromolecule has. An $\langle f_b \rangle$ value of 0.9 means that ninety of the sites on the macromolecule are 90% occupied. For this reason, we will develop saturation expressions for $\langle f_b \rangle$, but you should remember that the two expressions can be converted with the relation $\langle x_b \rangle = s \times \langle f_b \rangle$.

Analytically, we can represent $\langle f_b \rangle$ as the sum of the concentrations of all forms of macromolecule with ligand bound, times the fractional saturation of the macromolecule *in each corresponding ligation state*, divided by the concentrations of all the forms of macromolecule:

$$\langle f_b \rangle = \frac{\frac{1}{s}[Mx] + \frac{2}{s}[Mx_2] + \cdots + \frac{s}{s}[Mx_s]}{[M_0] + [Mx] + [Mx_2] + \cdots + [Mx_s]}$$

$$= \frac{1}{s} \times \frac{[Mx] + 2[Mx_2] + \cdots + s[Mx_s]}{[M_0] + [Mx] + [Mx_2] + \cdots + [Mx_s]}$$

(13.24)

Since the unligated form M_0 carries zero ligands, it is left out of the sum in the numerator. The $1/s$ term in the second expression will be in all expressions we derive for fractional saturation when $s > 1$. Given the relationship between $\langle f_b \rangle$ and $\langle x_b \rangle$, it should be clear that the second term in Equation 13.24 is equal to $\langle x_b \rangle$.

To represent $\langle f_b \rangle$ in more thermodynamic terms, we will substitute the $[Mx_i]$ terms in Equation 13.24 with rearranged versions of the overall binding constants (i.e., $[Mx_i] = \beta_i [M_0][x]^i$) to yield

$$\langle f_b \rangle = \frac{1}{s} \times \frac{\beta_1 [M_0][x] + 2\beta_2 [M_0][x]^2 + \cdots + s\beta_s [M_0][x]^s}{[M_0] + \beta_1 [M_0][x] + \beta_2 [M_0][x]^2 + \cdots + \beta_s [M_0][x]^s}$$

$$= \frac{1}{s} \times \frac{\beta_1 [x] + 2\beta_2 [x]^2 + \cdots + s\beta_s [x]^s}{1 + \beta_1 [x] + \beta_2 [x]^2 + \cdots + \beta_s [x]^s}$$

(13.25)

This can be expressed as in terms of stepwise binding constants by substituting Equation 7.17:

$$\langle f_b \rangle = \frac{1}{s} \times \frac{K_1 [x] + 2K_1 K_2 [x]^2 + \cdots + s[x]^s \prod_{i=1}^{s} K_i}{1 + K_1 [x] + K_1 K_2 [x]^2 + \cdots + [x]^s \prod_{i=1}^{s} K_i}$$

(13.26)

As advertised, the stepwise representation (Equation 13.26) is a little more cumbersome than the overall representation (Equation 13.25).

THE BINDING POLYNOMIAL P: A PARTITION FUNCTION FOR LIGAND BINDING

Notice that the denominator of Equations 13.25 and 13.26 are both simple polynomials in ligand concentration ($[x]$, of order s). For a given macromolecule under a given set of conditions, the $\beta_i's$ and the $K_i's$ are constants. We will refer to these sums "binding polynomials," abbreviated with the letter P:

$$P = 1 + \beta_1[x] + \beta_2[x]^2 + \cdots + \beta_s[x]^s = 1 + \sum_{i=1}^{s} \beta_i[x]^i$$

$$= 1 + K_1[x] + K_1K_2[x]^2 + \cdots + \prod_{i=1}^{s} K_i[x] = 1 + \sum_{i=1}^{s}\left([x]^i \prod_{j=1}^{i} K_j\right)$$

(13.27)

P has many of the same properties as partition functions we have seen in previous chapters (Chapters 10 through 12). P is a sum of different terms, one for each macromolecular species. Each term contains one (or more, for the stepwise formalism) equilibrium constants, which are exponentially related to free energies of reaction. The leading term in P represents the reference state, in this case unligated species M_0. Each of the i subsequent terms is a thermodynamic weight for the corresponding Mx_i macrostate, relative to the reference species M_0.

However, there is one notable difference in the functional form of P compared to other partition functions we have seen. Being a polynomial, P contains $[x]^i$ terms. Such terms have not appeared explicitly in the molecular partition functions discussed in Chapters 9 through 11. However, by recalling that concentration terms can be expressed in terms of chemical potentials (Chapter 7), each concentration term in P can be written as an exponential in energy:

$$[x]^i = \left\{e^{-(\mu_x{}^\circ - \mu_x)/RT}\right\}^i$$
$$= e^{-i(\mu_x{}^\circ - \mu_x)/RT}$$

(13.28)

By substituting Equation 13.28 into 13.27, P takes the form of a partition function that allows the chemical potential of one of the reactants ($[x]$) to be varied, which is appropriate for the ligand binding experiment (Problem 13.11). This is analogous to a type of partition function in statistical thermodynamics called the "grand canonical partition function" (see Hill, 1987, for more details on this).

The presence of the ith power of the ligand concentration means that the more ligands a species binds, the more its weight will come to dominate at high ligand concentration. Thus, the macromolecules will be driven to the fully ligated macrostate, Mx_s, at high ligand concentration. Another consequence of this is that at high $[x]$, P becomes infinite, although there are only a finite number of states. This is quite different from the effect of temperature on the molecular partition function q, which converges to the number of available states when temperature becomes large, reflecting the fact that each state becomes equally populated. These differences can be expressed formally as limits

$$\lim_{[x]\to\infty} P = \infty$$

(13.29)

$$\lim_{T\to\infty} q = i$$

(13.30)

where i represents the total number of states described by q. Obviously, these differences result from the different functional forms of $P([x])$ versus $q(T)$ with respect to the variables $[x]$ and T (polynomial vs. negative inverse exponential). But a more fundamental way to think about this difference (and to retain the status of P as a partition function) is to recognize that changing the ligand concentration effectively changes the relative free energy levels of the Mx_i macrostates. As ligand concentration increases, the Mx_i forms decrease their free energies (in proportion to the number of ligands bound), and as a result, the M_0 reference species becomes the highest energy state. For q, the energy levels are fixed; the reference state remains the lowest level at all temperatures. This concept is explored in Problem 13.13.

A concise relationship between the fractional saturation and *P*

The approach we used above to generate the fractional saturation (writing out macrostate concentrations and ligation number, and substituting equilibrium constants, Equations 13.24 and 13.25) is conceptually straightforward, but can be a cumbersome process, especially for macromolecules with large numbers of sites. A simpler approach is to begin with the binding polynomial, and differentiating it. The numerator of the fraction saturation expressions (Equations 13.25 and 13.26) has the form ix^i, which can be written as x times the derivative of x^i with respect to i as follows:

$$
\begin{aligned}
\langle f_b \rangle &= \frac{1}{s} \times \frac{\displaystyle\sum_{i=1}^{s} i\beta_i[x]^i}{P} = \frac{1}{s} \times \frac{[x]\displaystyle\sum_{i=1}^{s} i\beta_i[x]^{i-1}}{P} = \frac{[x]\displaystyle\sum_{i=1}^{s} \beta_i(d[x]^i/d[x])}{P} \\[2ex]
&= \frac{1}{s} \times \frac{[x](d/d[x])\left(\displaystyle\sum_{i=1}^{s} \beta_i[x]^i\right)}{P} \\[2ex]
&= \frac{[x](d/d[x])\left(1+\displaystyle\sum_{i=1}^{s} \beta_i[x]^i\right)}{sP} \\[2ex]
&= \frac{[x]}{sP}\frac{dP}{d[x]} \\[2ex]
&= \frac{1}{s} \times \frac{d\ln P}{d\ln[x]}
\end{aligned}
\tag{13.31}
$$

As long as you can remember the formula for the binding polynomial, Equation 13.31 provides a speedy method to calculate fractional saturation curves. Although the logarithmic form is most compact, the penultimate derivative is easier to calculate. Also, you should recognize the similarity of Equation 13.31 to analogous expressions we derived for the average energy and fraction helix by differentiating q with respect to the inverse temperature (T^{-1}, Chapter 10) and the helix propagation equilibrium constant (κ, Chapter 12), respectively.

Populations from the binding polynomial

In addition to providing an easy route to calculating fractional saturation, the binding polynomial provides a simple means to calculate populations of different partly ligated states (as well as fully-ligated and unligated states). Recall that the binding polynomial can be regarded as a partition function describing macromolecular species; each $\beta_i x^i$ term can be thought of as a Boltzmann factor for the macrostate species Mx_i.[†] The probability of the macromolecule having j ligands bound is

$$
p_j = \frac{\beta_j[x]^j}{1+\displaystyle\sum_{i=1}^{s} \beta_i[x]^i} = \frac{\beta_j[x]^j}{P}
\tag{13.32}
$$

The formula above can be used to calculate the concentrations of various partly ligated states, which is useful to see if they are rarified (positive cooperativity) or enhanced (negative cooperativity or heterogeneous sites; see below and

---

[†] Recall that the Boltzmann factor for M_0 is represented by the leading term in the binding polynomial.

Problem 13.4), and to identify ligand concentrations where various partly ligated states are maximally populated (Problem 13.15). Extracting such information from Equation 13.32 is analogous to calculating the population of partly helical states in the helix–coil transition.

The binding capacity from the binding polynomial

As described above, the binding capacity can be calculated as the derivative of the fractional saturation with respect to log ligand concentration (Equation 13.9). Inserting the penultimate expression in Equation 13.31 (relating the fractional saturation to P) into Equation 13.9 (the binding capacity equation) gives

$$
\begin{aligned}
C_x &= 2.303 \frac{d\langle f_b \rangle}{d\ln[x]} = 2.303 \frac{d}{d\ln[x]} \left(\frac{1}{sP} \frac{dP}{d\ln[x]} \right) \\
&= \frac{2.303}{s} \left(\frac{dP^{-1}}{d\ln[x]} \frac{dP}{d\ln[x]} + \frac{1}{P} \frac{d}{d\ln[x]} \left\{ \frac{dP}{d\ln[x]} \right\} \right) \\
&= \frac{2.303}{s} \left(-\frac{1}{P^2} \frac{dP}{d\ln[x]} \frac{dP}{d\ln[x]} + \frac{1}{P} \frac{d^2P}{d\ln[x]^2} \right) \\
&= \frac{2.303}{sP} \left(\frac{d^2P}{d\ln[x]^2} - \frac{1}{P} \left\{ \frac{dP}{d\ln[x]} \right\}^2 \right) \\
&= \frac{2.303}{sP} \left(P'' - \frac{\{P'\}^2}{P} \right)
\end{aligned}
\tag{13.33}
$$

where P' and P'' are shorthand for the derivative of P with respect to $\ln[x]$. The advantage of the expression used here is that the steepness of the curve more directly reflects the sharpness of the transition (and thus the underlying cooperativity). The connection between the binding capacity and cooperativity can be seen by deriving an analytical expression for the Hill coefficient (see below).

AN EXAMPLE—THE MACROSCOPIC BINDING OF TWO LIGANDS

The binding-polynomial-based formulas above provide a simple means to think about and analyze the binding of two ligands to a macromolecule. Here we will use stepwise equilibrium constants to take advantage of their intuitive connection to affinity. Although this scheme (the first two lines of Figure 13.9A) may be thought of as resulting from two separate binding sites, it should be emphasized that the macroscopic description does not deal with site-specific binding, which generally requires more complex models (see Chapter 8).

We begin by writing the binding polynomial. For $s = 2$, the binding polynomial is

$$
P = 1 + K_1[x] + K_1 K_2 [x]^2
\tag{13.34}
$$

The fractional saturation is then obtained by differentiation of P with respect to $[x]$ (Equation 13.31)

$$
\langle f_b \rangle = \frac{[x]}{sP} \frac{dP}{d[x]}
$$

$$
\frac{[x]}{s(1 + K_1[x] + K_1 K_2[x]^2)} \frac{d}{d[x]} (1 + K_1[x] + K_1 K_2[x]^2)
\tag{13.35}
$$

$$
= \frac{K_1[x] + 2K_1 K_2[x]^2}{s(1 + K_1[x] + K_1 K_2[x]^2)}
$$

You can check that this is the same formula you would have arrived at if you started with the different ligand states times their ligation number, and divided by the sum of the macromolecular species. The derivative approach used in Equation 13.35 has fewer steps. The expression for average ligation number $\langle x_b \rangle$ can be obtained from fractional binding by multiplying by the total number of binding sites ($s = 2$). This leads to

$$\langle x_b \rangle = \frac{K_1[x] + 2K_1K_2[x]^2}{1 + K_1[x] + K_1K_2[x]^2} = \frac{[x]}{P}\frac{dP}{dx} \tag{13.36}$$

By inserting numerical values for K_1 and K_2, the fractional saturation of binding can be evaluated graphically. As will be demonstrated, the relative value of K_1 and K_2, has a profound effect on the shape of the saturation curve, and is connected to the physiological role of the macromolecule in its ligand interactions.

The binding capacity can be calculated for the two-step binding process using either the expression developed in terms of P (Equation 13.33), or by directly differentiating the expression for $\langle f_b \rangle$ (Equation 13.35) with respect to $\log_{10}[x]$ using Equation 13.12. Here we will do the latter:

$$C_x = 2.303[x] \times \frac{df_{bound}}{d[x]} = 2.303[x] \times \frac{d}{d[x]}\left\{\frac{K_1[x] + 2K_1K_2[x]^2}{2(1 + K_1[x] + K_1K_2[x]^2)}\right\}$$

$$= 1.1515\frac{(K_1[x] + 4K_1K_2[x]^2 + K_1^2K_2[x]^3)}{(1 + K_1[x] + K_1K_2[x]^2)^2} \tag{13.37}$$

Several intermediate steps are left out in generating the last line of Equation 13.37 (Problem 13.22).

$K_2 \gg K_1$: Positive cooperativity

In the two-site model, when the second binding step has a significantly larger stepwise constant than the first step, the binding curve shows features of cooperativity. Here we will examine this situation using the values $K_1 = 0.1$ M^{-1}, $K_2 = 10$ M^{-1} (the absolute values do not really matter for this example—what is relevant is that the second constant is 100-fold larger than the first). Inserting these two values into the above expression for $\langle f_b \rangle$ (Equation 13.35) gives

$$f_{bound} = \frac{1}{2} \times \frac{K_1[x] + 2K_1K_2[x]^2}{1 + K_1[x] + K_1K_2[x]^2}$$

$$= \frac{0.1[x] + (2 \times 0.1 \times 10 \times [x]^2)}{2 + 0.2[x] + (2 \times 0.1 \times 10 \times [x]^2)} \tag{13.38}$$

$$= \frac{0.1[x] + 2[x]^2}{2 + 0.2[x] + 2[x]^2}$$

The fractional saturation is plotted versus linear and logarithmic ligand concentration in **Figure 13.10**.

Unlike the saturation curves in the previous section, the two-site curve is sigmoidal when plotted as a function of *linear* ligand concentration. The sigmoidal shape is easiest to see by comparing to a single-site noncooperative reference curve (red, Figure 13.10). A nice reference is one that has the same midpoint; for the two-site model, midpoints match when K (the binding constant for the reference curve) is the geometric average of K_1 and K_2 (i.e., $K = \sqrt{K_1K_2} = \sqrt{0.1 \times 10} = 1\text{M}^{-1}$; Problem 13.16). For the two site curve with $K_1 \ll K_2$, saturation is lower than the reference curve at low ligand concentration, but is higher at high ligand concentrations. This is because at low ligand concentrations, most ligand binding produces Mx_1 from M_0, and is dominated by the low affinity K_1 constant, whereas at high ligand

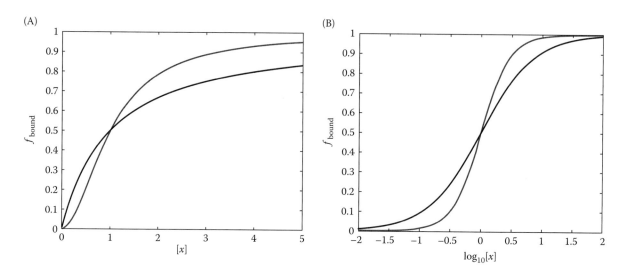

Figure 13.10 Binding of ligand to a two-site macromolecule with enhanced affinity in the second step. The gray fractional saturation curve is for $K_1 = 0.1$ M^{-1}, $K_2 = 10$ M^{-1}. The red curve provides a noncooperative reference with the same midpoint (1 M). By comparing to this reference, it is clear that enhancement of the second binding step results in a sigmoidal fractional saturation curve when plotted as a linear function of ligand concentration (A). Plotted as a function of the log of the ligand concentration, the curve enhanced in the second step has a steeper slope than the reference curve at its midpoint (B).

concentration, most ligand binding converts Mx_1 to Mx_2, and is dominated by the high affinity K_2 constant. In short, affinity appears to increase as saturation progress, a hallmark of positive cooperativity.

This progressive enhancement in affinity can also be seen when saturation is plotted as the logarithm of ligand concentration (**Figure 13.10B**). In logarithmic form both curves have similar sigmoidal shapes, but differ in steepness. The two-site curve appears to be steeper in general. This is particularly pronounced at the midpoint (whereas at low and high ligand concentration, the single site reference is actually steeper vs. log[x]). This steepness means that the positively cooperative macromolecule behaves as a "molecular switch," converting from unbound (M_0) to fully bound (Mx_s) over a very narrow range of ligand concentration. Graphical and analytical methods to represent cooperativity are developed below, and are applied to both this two-step scheme and for a binding scheme with values of the binding constants reversed.

$K_1 \gg K_2$: Negative cooperativity (or heterogeneous sites)

When the second binding step in the two-site model has a significantly smaller binding constant than the first step, the binding curve shows features of *negative* cooperativity. To illustrate this, we will evaluate the two-site expression (Equation 13.36) with the values of the stepwise constants flipped from those in the example above, that is, $K_1 = 10$ M^{-1}, $K_2 = 0.1$ M^{-1}. As before, the two binding constants differ by 100-fold, and their geometric mean is the same ($\sqrt{K_1 K_2} = \sqrt{10 \times 0.1} = 1$M^{-1}), but now ligand binds tighter in the first step than in the second. This results in a significant difference in the expressions for P (Problem 13.20) and for $\langle f_b \rangle$:

$$\langle f_b \rangle = \frac{1}{2} \times \frac{K_1[x] + 2K_1 K_2[x]^2}{1 + K_1[x] + K_1 K_2[x]^2}$$

$$= \frac{10[x] + (2 \times 0.1 \times 10 \times [x]^2)}{2 + 20[x] + (2 \times 0.1 \times 10 \times [x]^2)} \qquad (13.39)$$

$$= \frac{10[x] + 2[x]^2}{2 + 20[x] + 2[x]^2}$$

The fractional saturation is plotted versus linear (A) and logarithmic (B) ligand concentration in **Figure 13.11**.

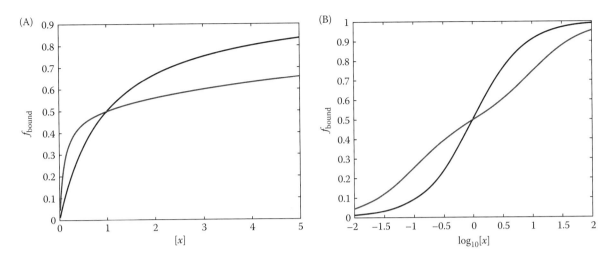

Figure 13.11 Binding of ligand to a two-site macromolecule with enhanced affinity in the first step. The gray fractional saturation curve is for $K_1 = 10$ M^{-1}, $K_2 = 0.1$ M^{-1}. The red curve is a single-site reference with the same midpoint (1 M). Weakening the second binding step results in a biphasic fractional saturation curve as a function of ligand concentration (A). Plotted as a function of log ligand concentration, the two-site curve has a shallower slope at its midpoint than the reference curve (B). Although this behavior is consistent with anticooperativity, it can also result from two different types of noninteracting binding sites, that is, site heterogeneity.

As a function of linear ligand concentration, the two-site $K_1 \gg K_2$ curve looks biphasic. At low ligand concentration, binding is quite facile, reflecting high affinity. However, as saturation passes the midpoint, further binding is difficult, requiring large increases in ligand concentration. It is as if the macromolecule has a strong site and a weak site. As a function of log ligand concentration, this biphasic saturation produces a markedly *nonsigmoidal* shape. In contrast to Figure 13.10, the binding curve looks like the sum of two binding curves of half amplitude, one with low affinity and one with high affinity. Between these two, the curve becomes flat compared to the reference, reflecting the fact that not much ligand is getting bound—the first ligand is on, and the second won't go until higher concentrations.

Binding capacity representation of two-step binding

As would be expected from the differences in the shapes of the fractional saturation curves for the two-step mechanisms, the binding capacity C_x depends on the relative values of K_1 and K_2, both in its magnitude and its overall shape (**Figure 13.12**).

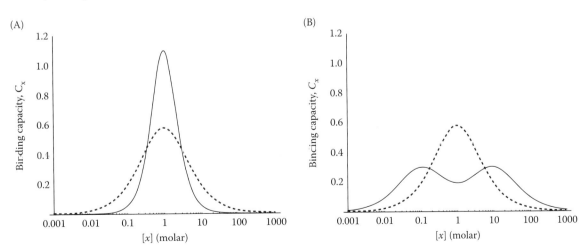

Figure 13.12 Binding capacity for two-step mechanisms. (A) When $K_1 \ll K_2$, corresponding to positive cooperativity, binding capacity is very high near the midpoint, but falls sharply away from the midpoint (black curve; red dashed curves represent a noncooperative reference, where $K_1 = 2$ M^{-1}, $K_2 = 0.5$ M^{-1}; see Chapter 14). (B) When $K_1 \gg K_2$, binding capacity is quite low (black), but is spread over a broad range of ligand concentration. The two peaks reflect the high and low affinity steps.

Again, the single-step mechanism provides a convenient point of reference, showing a single, symmetrical peak in C_x when plotted versus $\log_{10}[x]$. For the two-step mechanism with $K_2 \gg K_1$, C_x is very strongly peaked, reaching a higher value than the single site reference curve at the midpoint, but has lower values away from the midpoint (Figure 13.12A). This reflects the steepness of the binding curve, which saturates over a narrow range (Figure 13.10A). In contrast, when $K_1 \gg K_2$, C_x is broadly distributed, showing two peaks at ligand concentrations that are roughly equal to the dissociation constants for the first ($K_{d,1}=1/K_1=0.1$ M) and second ($K_{d,2}=1/K_2=10$ M) steps, and a low binding capacity in between. This reflects the biphasic nature of the fractional binding curve.

The binding capacity profiles are closely related to ligand binding cooperativity, which can be analyzed using graphical constructs such as the "Hill" plot.

Hill plots as a graphical representation of cooperativity

Another way to represent the steepness of the binding curve, and to quantify cooperativity is a transformation called the "Hill" plot.[†] In the Hill plot, the log of the ratio of $\langle f_b \rangle$ to the average fraction unbound ($\langle f_u \rangle = 1 - \langle f_b \rangle$) is plotted as a function of $\log[x]$ (Figure 13.11). The slope of the Hill plot, called the "Hill coefficient" (n_H), is interpreted as a measure of cooperativity. In its simplest interpretation (see more on this below), a value of n_H greater than one is considered to reflect positive cooperativity, whereas $n_H = 1$ is interpreted as reflecting no cooperativity (as for the single site reference curve). A value $0 < n_H < 1$ is interpreted as either a reflection of negative cooperativity or of site heterogeneity.

However, for most multisite binding reactions (except for macromolecules that have multiple identical and independent sites; see Chapter 8), there is no single value for n_H that describes the entire curve. Rather, the Hill coefficient varies with the extent of saturation. Because binding equilibrium is most easily studied around half-saturation ($\langle f_b \rangle = 1/2$), and because macromolecules often function at ligand concentrations near to half-saturation, Hill coefficients are often reported at half-saturation ($n_{H,\frac{1}{2}}$). The simple interpretations of n_H values given in the preceding paragraph ($n_H > 1$ for positive cooperativity, $n_H < 1$ for negative cooperativity or site heterogeneity) hold true at the half-saturation values. However, far from half-saturation, n_H values converge to unity. As illustrated below, these limiting regions of the binding curve provide additional information about the binding energies in multistep mechanisms, although accurate measurement of equilibrium at these extremes of the binding curve is very difficult.

To illustrate the complex features of Hill plots for multistep binding, we will consider our two examples where $K_1 = 0.1$ M^{-1} and $K_2 = 10$ M^{-1} (positive cooperativity, Figure 13.10) and where $K_1 = 10$ M^{-1} and $K_2 = 0.1$ M^{-1} (either negative cooperativity or site heterogeneity, Figure 13.11). The Hill plots for these two binding schemes are shown in **Figure 13.13**. For the scheme where $K_2 \gg K_1$, the slope of the Hill plot at the midpoint is higher than for the single-site reference (Figure 13.13, left). The Hill coefficient at the midpoint ($n_{H\frac{1}{2}}$) approaches 2.0 (1.904 by linear fitting to points near the transition, Problem 13.17; 1.9048 by differentiation, Problem 13.18). In contrast, for the scheme where $K_1 \gg K_2$, the slope of the Hill plot is lower than the single-site reference (Figure 13.13, right). Linear fitting near the midpoint gives an $n_{H\frac{1}{2}}$ value of around 1.837 (Problem 13.17); differentiation gives $n_{H\frac{1}{2}} = 0.3333$ (Problem 13.18).

As described above, at high and low ligand concentrations, the slopes of both Hill plots become the same as for the reference curve ($n_H = 1$). This is because at very low and very high ligand concentrations, binding is dominated by the first and last steps. Thus, the ligand concentrations at which this limiting behavior is obtained

---

[†] The Hill plot is named after Archibald Hill, who applied a simple version of the equation that bears his name to oxygen binding to human hemoglobin in 1910.

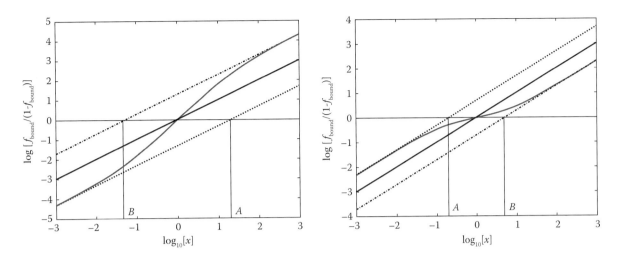

Figure 13.13 Hill plots for two-site binding schemes. Left, when $K_1 = 0.1\ M^{-1}$, $K_2 = 10\ M^{-1}$, the Hill plot (gray) is steeper at the midpoint than the reference curve (red), with a Hill coefficient at half-saturation ($n_{H1/2}$) approaching 2.0, reflecting positive cooperativity. Right, when $K_1 = 0.1\ M^{-1}$, $K_2 = 10\ M^{-1}$, the Hill plot is shallower at the midpoint, with a Hill coefficient at half-saturation ($n_{H1/2}$) less than one. For both curves, the slope converges to the same value as the reference curve, that is, $n_H = 1$. At these limiting ligand concentrations, binding appears as if to single-site macromolecules with affinities $K_1/2$ and $2K_2$ (lower and upper dotted lines). Numerical estimates of these quantities are obtained from x-intercepts of these limiting curves (A and B, respectively).

reflect the corresponding binding affinities (K_1 and K_2). By extrapolating the low and high ligand limits of the Hill plot to half-saturation ($\log_{10}\langle f_b\rangle/(1 - \langle f_b\rangle)=0$), values for the limiting binding constants can be determined from the x-intercepts. In Figure 13.13, values of K_1 and K_2 are determined by the inverse ligand concentrations at points A and B, respectively (see Problem 13.19).[†]

An analytical formula for the Hill coefficient

Since the Hill coefficient is the slope of the Hill plot, it can be calculated analytically by taking the first derivative:

$$n_H = \frac{d\log_{10}\{\langle f_b\rangle/(1-\langle f_b\rangle)\}}{d\log_{10}[x]} = \frac{d\ln\{\langle f_b\rangle/\langle f_u\rangle\}}{d\ln[x]}$$
$$= \frac{\langle f_u\rangle}{\langle f_b\rangle}\frac{d\{\langle f_b\rangle/\langle f_u\rangle\}}{d\ln[x]} \tag{13.40}$$

The second equality in the top line of Equation 13.40 results because the relationship $2.303 \times \log(x) \approx \ln(x)$ can be applied to both the numerator and denominator, and cancels. As before, $\langle f_u\rangle$ is the average fraction unbound. By applying the quotient rule and substituting Equation 13.31 to introduce the binding polynomial (Problem 13.23), the bottom line of Equation 13.40 can be rearranged to

$$n_H - \frac{1}{sP\langle f_b\rangle\langle f_u\rangle}\left(P'' - \frac{1}{P}\{P'\}^a\right) \tag{13.41}$$

where, as in Equation 13.33, P' and P'' represent derivatives of P with respect to $\ln[x]$. As can be seen from the slope of the Hill plots in Figure 13.13, the two-step scheme shows a peak in n_H when $K_1 = 0.1\ M^{-1}$, $K_2 = 10\ M^{-1}$ (positive cooperativity), and a trough when $K_1 = 10\ M^{-1}$, $K_2 = 0.1\ M^{-1}$ (**Figure 13.14**).

[†] Although there are statistical factors required to extract these binding constants (see Problem 13.19). The origin of these statistical factors can be best understood by using a microscopic model for binding (as opposed to a macroscopic model; see Chapter 8).

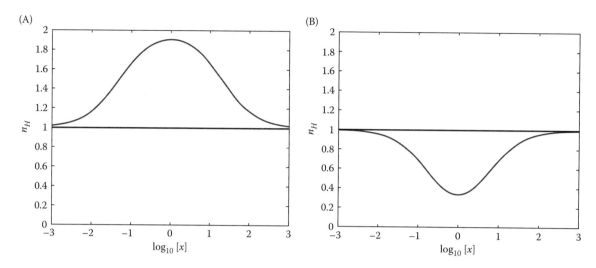

Figure 13.14 Hill coefficients for two-site binding schemes. When $K_1 = 0.1$ M^{-1}, $K_2 = 10$ M^{-1} gray (A), the Hill coefficient (Equation 13.41) shows a maximum near the midpoint, and converges to the value of the reference curve ($n_H = 1$, red) at high and low ligand concentrations. In contrast, when $K_1 = 10$ M^{-1}, $K_2 = 0.1$ M^{-1} (right B), the Hill coefficient (Equation 13.41) shows a minimum near the midpoint.

A close correspondence between the Hill coefficient and binding capacity, C_x, can be seen by comparing Equations 13.33 and 13.41:

$$C_x = 2.303 \langle f_b \rangle \langle f_u \rangle n_H \tag{13.42}$$

C_x can be thought of as being sensitive to both the fluctuation product that results at the midpoint of any binding transition, and also to the enhanced (or diminished) steepness due to cooperativity (or anticooperativity), represented by n_H. For the single-site model, where n_H is 1 at all ligand concentrations, C_x results solely from fluctuation, as was obtained in Equation 13.11. When there is positive cooperativity, as for the two-step scheme with $K_2 \gg K_1$, the peak in n_H (Figure 13.14A) further sharpens the peak in the binding capacity (Figure 13.12B). In contrast, when $K_1 \gg K_2$, the minimum in n_H (Figure 13.14B) suppresses C_x at the midpoint, and instead produces peaks at the extreme ligand concentrations (Figure 13.12B, right).

A simple limiting model for the Hill coefficient

Sometimes n_H is interpreted using a simple all-or-none scheme in which ligand binding occurs in a single step:

$$M_0 + jx \underset{}{\overset{\beta_j}{\rightleftharpoons}} Mx_j \tag{13.43}$$

where the only bound state considered is that with j ligands bound, with $0 < j \leq s$. For this simplified binding mechanism, the saturation function is simply

$$\langle f_b \rangle = \frac{\beta_j [x]^j}{1 + \beta_j [x]^j} \tag{13.44}$$

(Problem 13.24). Unlike the stepwise and overall schemes described above (Figure 13.9), the Hill plot for this type of-or-none scheme has the same slope (i.e., n_H) at all ligand concentrations. This is because partly ligated states are left out of the model. Moreover, for this scheme, the Hill coefficient is simply equal to j, the apparent number of ligands bound (Problem 13.25).

An advantage of the all-or-none scheme is its simplicity. However, it is a rather unrealistic binding scheme, especially for large values of j. The inadequacy of this model in describing oxygen binding to human hemoglobin is reflected in the observation that the n_H rarely exceeds three, even though there are four sites that can (and do) bind oxygen. Moreover, fitted Hill coefficients often have noninteger values, which in molecular terms, make little sense for macromolecules that remain in the same state of association in the bound and unbound states. Nonetheless, for many macromolecular binding reactions (hemoglobin being an exception), the average fractional saturation can only be determined accurately over a limited range (e.g., 5%–95%). For such cases, the Hill plot is approximately linear, and even a modest amount of error precludes accurate fitting of multiple binding constants (Problem 13.23). In such cases, using Equation 13.43 as an imperfect model to quantify cooperativity may be preferable to a more realistic model that is overparameterized.

Strengths and limitations of the macroscopic approach

As described above, the macroscopic approach is a simple way to describe the populations and energies of partly and fully ligated macromolecules relative to the unligated state. The binding polynomial functions as a partition function, and the magnitude of the different terms can be used to directly get insight into populations and their response to the ligand. For example, compare the expressions for P for our two-step examples (see Equation 13.34).

$$\text{For } K_1 = 0.1\,\text{M}^{-1}, \quad K_2 = 10\,\text{M}^{-1}: \quad P = 1 + 0.1[x] + [x]^2 \tag{13.45}$$

$$\text{For } K_1 = 10\,\text{M}^{-1}, \quad K_2 = 0.1\,\text{M}^{-1}: \quad P = 1 + 10[x] + [x]^2 \tag{13.46}$$

The difference here is in the middle terms. The middle term in Equation 13.45, which corresponds to the singly-ligated state, makes a small contribution to P, reflecting positive cooperativity, and thus, a rarification of intermediate forms. In contrast, the middle term in Equation 13.46 makes a large contribution to P, reflecting a high concentration of the singly-ligated state at intermediate ligand concentrations. In both expressions, the different powers of $[x]$ associated with each term ensures that population shifts from low-to-high states of ligation as $[x]$ increases.

However, the macroscopic approach falls short in providing details, such as what is going on *within* each ligation state. A clear example of how this can be a shortcoming comes from our fuzzy description of the two-step mechanism with stepwise constants given in Equation 13.46. The broadened binding curve is *consistent* with negative cooperativity—binding at one site interacts thermodynamically with the other site, lowering its affinity—but it is also consistent with simple differences in affinity between the two sites. These two molecular mechanisms look identical at the molecular level. To represent one or the other mechanisms requires a microscopic model that treats the individual sites and their potential coupling. Another example of where the macroscopic picture falls short is in ignoring the statistical factors associated with different states of ligation. For the two step scheme, there are two ways that a ligand can be bound in the singly ligated Mx state (to site 1 or to site 2) but only one way that the empty M_0 or fully ligated (Mx_2) state can be formed. The macroscopic treatment lumps these statistical factors into each stepwise (and overall) constant, partly obscuring the fundamental affinity differences at different sites. As an example, the two-site scheme that displays neither positive nor negative cooperativity (i.e., $n_H = 1$ at all ligand concentrations, matching the reference curve) has stepwise constants $K_1 = 2\,\text{M}^{-1}$ and $K_2 = 0.5\,\text{M}^{-1}$, although equal values of $1\,\text{M}^{-1}$ might be expected intuitively. By developing models using a microscopic (i.e., molecular) vantage point, these factors are represented explicitly.

However, before we turn to microscopic models in the next chapter, we will complete our discussion of macroscopic binding by treating the interaction of multiple, distinguishable ligands. The simple framework of the macroscopic expressions is particularly welcomed, given the added complexity brought about by including different ligand types. Moreover, this macroscopic treatment leads to a very powerful set of general linkage relationships.

BINDING TO MULTIPLE *DIFFERENT* LIGANDS: "HETEROTROPIC" BINDING

Many macromolecules can interact with more than one type of ligand. For example, transcription factors bind to DNA, but they often bind to other "effector" molecules (either small molecules, such as metabolites, or other protein factors). Often the binding of these effector molecules influences the affinity of transcription factors for their sites on DNA, forming the basis for regulation of gene expression. Other examples include the binding of enzymes by inducer and repressor molecules, and the binding of hydrogen ions and organic phosphates to oxygen transport proteins such as hemoglobin.

The approach we will use to build a general framework for binding of multiple different ligands (referred to as "heterotropic binding"), is similar to the treatment of binding of a single ligand ("homotropic binding") above. Consider a macromolecule M that can bind to multiple different ligands, x and y. As before, we will use s to represent the number of binding sites for ligand x. In addition, we will use u to represent the number of binding sites for ligand y. In the most general treatment, the binding sites for x and y differ, both in location and in number (that is, $s \neq u$). This scheme can be represented as

$$
\begin{array}{ccccccc}
M_0 & \rightleftharpoons & Mx & \rightleftharpoons & Mx_2 & \rightleftharpoons \; \ldots & Mx_s \\
\updownarrow & & \updownarrow & & \updownarrow & & \updownarrow \\
My & \rightleftharpoons & Mxy & \rightleftharpoons & Mx_2y & \rightleftharpoons \; \ldots & Mx_sy \\
\updownarrow & & \updownarrow & & \updownarrow & & \updownarrow \\
My_2 & \rightleftharpoons & Mxy_2 & \rightleftharpoons & Mx_2y_2 & \rightleftharpoons \; \ldots & Mx_sy_2 \\
\updownarrow & & \updownarrow & & \updownarrow & & \updownarrow \\
\vdots & & \vdots & & \vdots & & \vdots \\
My_u & \rightleftharpoons & Mxy_u & \rightleftharpoons & Mx_2y_u & \rightleftharpoons \; \ldots & Mx_sy_u
\end{array}
\tag{13.47}
$$

Although the macroscopic treatment in this chapter does not emphasize specific molecular models, scheme 13.47 clearly invokes distinct binding sites for x and y, since all s x ligands can bind even if all u y ligands are bound (and vice versa).

If instead, ligands x and y compete for the same sites, the maximum total number of ligands that can be bound is fixed. In the fully competitive case (that is, $u = s$), the scheme simplifies to

$$
\begin{array}{ccccccccc}
M_0 & \rightleftharpoons & Mx & \rightleftharpoons & Mx_2 & \rightleftharpoons \; \ldots & Mx_{s-1} & \rightleftharpoons & Mx_s \\
\updownarrow & & \updownarrow & & \updownarrow & & \updownarrow & & \\
My & \rightleftharpoons & Mxy & \rightleftharpoons & Mx_2y & \rightleftharpoons \; \ldots & Mx_{s-1}y & & \\
\updownarrow & & \updownarrow & & \updownarrow & & & & \\
My_2 & \rightleftharpoons & Mxy_2 & \rightleftharpoons & Mx_2y_2 & \rightleftharpoons & & & \\
\updownarrow & & \updownarrow & & \updownarrow & & & & \\
\vdots & & \vdots & & & & & & \\
My_{s-1} & \rightleftharpoons & Mxy_{s-1} & & & & & & \\
\updownarrow & & & & & & & & \\
My_s & & & & & & & &
\end{array}
\tag{13.48}
$$

Not only are the number of columns and rows in Equation 13.48 equal (to $s + 1$), the scheme has the shape of an upper triangular matrix. All the terms below the anti-diagonal (with elements Mx_iy_{s-i}) are omitted, because the total number of ligands would exceed the total number (s) of sites. An example of the competitive scheme (Equation 13.48) would be the binding of molecular oxygen (O_2) and carbon monoxide (CO) to the four heme sites in hemoglobin. An example of the non-competitive scheme (Equation 13.47) would be the binding of O_2 to the four heme sites and the binding of hydrogen ions (H^+) to the titratable groups hemoglobin (Asp, Glu, His, Tyr, Lys side chains, and the N- and C-termini).

As above, we will construct a binding polynomial based on macroscopic binding constants. However, the binding constants will represent ligation with both x and y ligands, and the resulting binding polynomial ($P(x,y)$) will depend on the concentrations of both x and y. Each term in $P(x,y)$ represents the thermodynamic weight of a macromolecule bound with i ligands of type x and j ligands of type y (Mx_iy_j). As before, thermodynamic weights are taken relative to the state with no ligand bound (M_0, the reference state):

$$
\begin{aligned}
P(x,y) = & \frac{[M_0]}{[M_0]} + \frac{[Mx]}{[M_0]} + \frac{[Mx_2]}{[M_0]} + \cdots + \frac{[Mx_s]}{[M_0]} \\
& + \frac{[My]}{[M_0]} + \frac{[Mxy]}{[M_0]} + \frac{[Mx_2y]}{[M_0]} + \cdots + \frac{[Mx_sy]}{[M_0]} \\
& + \frac{[My_2]}{[M_0]} + \frac{[Mxy_2]}{[M_0]} + \frac{[Mx_2y_2]}{[M_0]} + \cdots + \frac{[Mx_sy_2]}{[M_0]} \\
& \vdots \\
& + \frac{[My_u]}{[M_0]} + \frac{[Mxy_u]}{[M_0]} + \frac{[Mx_2y_u]}{[M_0]} + \cdots + \frac{[Mx_sy_u]}{[M_0]}
\end{aligned}
\qquad (13.49)
$$

Each term in $P(x,y)$ can be written as an overall binding constant $\beta_{i,j}$, representing the reaction[†]

$$
M_0 + ix + jy \underset{\phantom{\beta_{i,j}}}{\overset{\beta_{i,j}}{\rightleftharpoons}} Mx_iy_j
\qquad (13.50)
$$

Writing $\beta_{i,j}$ in terms of concentrations and rearranging gives

$$
\frac{[Mx_iy_j]}{[M_0]} = \beta_{i,j}[x]^i[y]^j
\qquad (13.51)
$$

Substituting Equation 13.51 into 13.49 gives

$$
\begin{aligned}
P(x,y) = & \quad \beta_{0,0} + \beta_{1,0}[x]^1 + \beta_{2,0}[x]^2 + \cdots + \beta_{s,0}[x]^s \\
& + \beta_{0,1}[y]^1 + \beta_{1,1}[x]^1[y]^1 + \beta_{2,1}[x]^2[y]^1 + \cdots + \beta_{s,1}[x]^s[y]^1 \\
& + \beta_{0,2}[y]^2 + \beta_{1,2}[x]^1[y]^2 + \beta_{2,2}[x]^2[y]^2 + \cdots + \beta_{s,2}[x]^s[y]^2 \\
& \vdots \\
& + \beta_{0,u}[y]^u + \beta_{1,u}[x]^1[y]^u + \beta_{2,u}[x]^2[y]^u + \cdots + \beta_{s,u}[x]^s[y]^u
\end{aligned}
\qquad (13.52)
$$

Here, $\beta_{0,0}$ is set to 1, so that the lead term is gives the Boltzmann factor for the unligated state, M_0. Note that Equation 13.52 can be condensed as a summation over both ligands:

[†] Alternatively, stepwise binding constants can be used (e.g., see Problem 13.26). As with homotropic binding, stepwise binding constants are more intuitive but less compact. However, as will be developed below, the full set of stepwise constants is "overdetermined," and as a result, carries redundant information.

$$P(x,y) = \sum_{j=0}^{u} \sum_{i=0}^{s} \beta_{i,j} [x]^i [y]^j \tag{13.53}$$

As with homotropic ligand binding, the primary means to connect the analytical framework (and the binding constants) for heterotropic binding to experimental data is through fractional saturation plots. In general, we could vary the concentration of both x and y, and evaluate the total saturation. However, this is a messy experiment to do, the results are difficult to represent, and the total saturation lacks ligand-specific information. Instead, we will plot fractional saturation with a single type of ligand (e.g., x) as a function of that ligand at different fixed concentrations of the other ligand (e.g., y). We will represent the average fractional saturation with ligand x as $\langle f_x \rangle$. Shifts in the binding curve for ligand x (that is, $\langle f_x \rangle$ versus either $[x]$ or $\log_{10}[x]$) at different y concentration[†] indicates coupling.

To demonstrate this coupling analytically, we need expressions for $\langle f_x \rangle$ and $\langle f_y \rangle$ in terms of $P(x,y)$. Since we are interested saturation functions at fixed concentration of the other ligands, these expressions take the same form as in Equation 13.31, but the derivative becomes partial to the ligand being varied. Specifically, we can express $\langle f_x \rangle$ and $\langle f_y \rangle$ as

$$\langle f_x \rangle = \frac{\langle x_b \rangle}{s} = \frac{[x]}{sP}\left(\frac{\partial P}{\partial [x]}\right)_y = \frac{1}{s}\left(\frac{\partial \ln P}{\partial \ln[x]}\right)_y \tag{13.54}$$

$$\langle f_y \rangle = \frac{\langle y_b \rangle}{u} = \frac{[y]}{uP}\left(\frac{\partial P}{\partial [y]}\right)_x = \frac{1}{u}\left(\frac{\partial \ln P}{\partial \ln[y]}\right)_x \tag{13.55}$$

Although we will not derive Equation 13.54 and 13.55 here (see Problem 13.27), they can be justified by recognizing that if we vary only one ligand (e.g., x for $\langle f_x \rangle$), the binding polynomial (Equation 13.52) can be represented as a set of constant (but rather cumbersome) terms, each multiplying a different power of the variable ligand x. For example, the terms in $P(x,y)$ that are first power in x can be written

$$\beta_{1,0}[x] + \beta_{1,1}[x][y] + \beta_{1,2}[x][y]^2 + \cdots + \beta_{1,u}[x][y]^u = (\beta_{1,0} + \beta_{1,1}[y] + \cdots + \beta_{1,u}[y]^u)[x]$$
$$= P(x_1, y)[x]$$
$$\tag{13.56}$$

The terms in parentheses on the right-hand side are constant if y is held constant. And terms for each power of x in $P(x,y)$ can be collected and factored into the same form as Equation 13.56. Thus, we can regard $P(x,y) = P'(x)$ at constant y, where the prime indicates that the constants will in general differ from the $P(x)$ expression in the absence of y. Thus, Equation 7.38 follows in direct analogy to Equation 13.31.

Effects of thermodynamic cycles on the stepwise constants

Unlike the schemes we developed to describe the binding of a single type of ligand (Equation 13.22), the binding of different ligands to a macromolecular species creates thermodynamic cycles between species. One of the important consequences of these cycles is that although a stepwise equilibrium constant *can* be written for each ligand binding step, not all stepwise constants are *needed* to fully describe the populations of the various ligation states. Rather, some of the binding constants are fixed by the values of the other constants; that is, they are not independent variables.[‡] This redundancy forms the basis of heterotropic coupling, which will be expanded on in the next section.

[†] And, as will be shown below, sensitivity of f_x to y mandates that f_y equally sensitive to x.

[‡] A related consequence of chemical reaction cycles is that the rate constants for each forward and reverse step are not independent, but must satisfy a condition called "detailed balance."

To illustrate this redundancy, consider a simple example where a macromolecule binds a single x and a single y ligand ($s = u = 1$) according to the noncompetitive scheme:

$$
\begin{array}{ccc}
M_0 & \xrightleftharpoons{K_{1,0}^x} & Mx \\[4pt]
K_{0,1}^y \updownarrow & & K_{1,1}^y \updownarrow \\[4pt]
My & \xrightleftharpoons{K_{1,1}^x} & Mxy
\end{array}
\tag{13.57}
$$

Here, stepwise binding constants are used because they correspond to a single edge in the reaction cycle. The superscripts on the stepwise constants in Equation 13.57 indicate which ligand is binding. They are needed to describe the formation of Mxy, which can be produced either by the binding of y to Mx or by the binding of x to My. The relationships between the overall and stepwise constants are

$$
\begin{aligned}
\beta_{1,0} &= K_{1,0}^x \\
\beta_{0,1} &= K_{0,1}^y \\
\beta_{1,1} &= K_{1,0}^x K_{1,1}^y = K_{0,1}^y K_{1,1}^x
\end{aligned}
\tag{13.58}
$$

Note that only three overall constants need to be specified to describe this scheme, since there are only three species that can be formed from M_0. This fact implies that there are only three independent stepwise constants. Indeed, the relation in the last line of Equation 13.58 states that if the value of three constants is specified, the fourth is fixed (e.g., $K_{1,1}^y = K_{0,1}^y K_{1,1}^x / K_{1,0}^x$).

Though the constraint on the fourth stepwise constant can be deduced by comparing the stepwise and overall formalisms, there is a more fundamental thermodynamic concept at work here. The four stepwise constants arise because we are specifying two separate paths (each with two steps) to Mxy. However, as described in Chapter 1, thermodynamic state functions are independent of path. Equilibrium constants, which are monotonically related to free energy changes, are state functions. The products of the stepwise equilibrium constants in the lower right-hand equality in Equation 13.58 reflect equilibrium constants along two different paths connecting M_0 to Mxy. These "product constants" must be path-independent, hence the equality. Another way to look at this is to convert the equilibrium constants to reaction free energies:

$$
\begin{aligned}
-RT \ln\left(K_{1,0}^x K_{1,1}^y\right) &= -RT \ln\left(K_{0,1}^y K_{1,1}^x\right) \\
-RT \ln K_{1,0}^x - RT \ln K_{1,1}^y &= -RT \ln K_{0,1}^y - RT \ln K_{1,1}^x \\
\Delta G_{1,0}^x + \Delta G_{1,1}^y &= \Delta G_{0,1}^y + \Delta G_{1,1}^x
\end{aligned}
\tag{13.59}
$$

The last line of Equation 13.59 states that the free energy difference between M_0 and $M_{1,1}$ is the same when x binds first and then y (left-hand side) as it is if y binds first and then x (right-hand side); in other words, free energy is independent of path.

To illustrate the types of linkage that can be obtained in the simple cycle in scheme 13.57, binding curves for x in increasing concentrations of y are graphed in **Figure 13.15**. Two cases are considered. In panels A and B, sites are separate, allowing both x and y to bind simultaneously (forming Mxy). In A, binding of x enhances binding of y by 10-fold, whereas in B, binding of x inhibits binding of y by 10-fold. In panel C, ligands compete for a single site. Binding constants for x in the absence of y, and y in the absence of x, are the same for all three panels ($K_{1,0}^x = 10^5\,\mathrm{M}^{-1}$ and $K_{0,1}^y = 10^6\,\mathrm{M}^{-1}$, respectively).

The binding curve for x is left shifted in (A), reflecting positive cooperativity between the sites, whereas it is left-shifted in (B), reflecting negative cooperativity. In both cases, the concentration of y must be approximately $1/K_{0,1}^y = 10^{-6}\,\mathrm{M}$ or higher to produce a measurable shift. Below this concentration, no y is bound, so x binds as if y was absent from solution. At very high y concentrations, the binding curves for

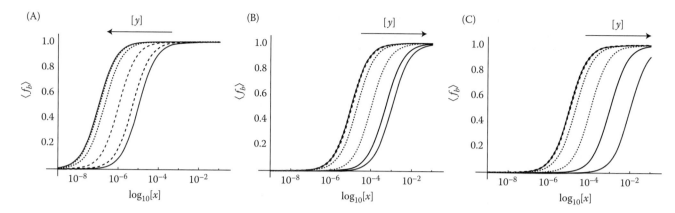

Figure 13.15 Heterotropic linkage of single binding sites for ligands *x* and *y*. For each of the three sets of curves a maximum of one *x* and one *y* ligand can be bound (see Equation 13.57). Binding constants for *x* in the absence of *y* ($K^x_{1,0}$), and *y* in the absence of *x* ($K^y_{0,1}$) are the same in all three cases (10^5 and 10^6 M^{-1}, respectively). In (A) and (B) ligands *x* and *y* can be bound simultaneously, whereas in (C), binding is mutually exclusive. (A) Positive coupling, in which binding of each ligand increases the binding constant of the other ligand by a factor of ten (i.e., $K^x_{1,1} = 10^6$ M^{-1}, $K^y_{1,1} = 10^7$ M^{-1}). Increasing the concentration of ligand *y* shifts the binding curve for *x* to the left, but the effect saturates at high concentrations of *y*. (B) Negative coupling, where binding of each ligand decreases the binding constant of the other ligand by a factor of ten (i.e., $K^x_{1,1} = 10^4$ M^{-1}, $K^y_{1,1} = 10^5$ M^{-1}). Increasing the concentration of ligand *y* shifts the binding curve for *x* to the right; as with positive coupling, the effect saturates at high concentrations of *y*. (C) Competitive binding, where *x* and *y* cannot simultaneously bind. As with negative coupling (B), increasing the concentration of ligand *y* shifts the binding curve for *x* to the right. However, the effect does not saturate at high concentrations of *y*.

x become unresponsive to further increase in *y*, because *M* is fully saturated with *y*. Like the negative cooperativity example, the binding curves for *x* are shifted to the right (lower affinity) in a competitive mechanism (panel C).[†] However, unlike the negative cooperativity example, the curve continues to shift with *y* at high concentrations, that is, it never saturates.

A GENERAL FRAMEWORK TO REPRESENT THERMODYNAMIC LINKAGE BETWEEN MULTIPLE INDEPENDENT LIGANDS

As illustrated above the simplest imaginable heterotropic binding scheme (a single binding site for ligand *x* and ligand *y*, Equation 13.57) can display functionally important coupling between sites (Figure 13.15). This coupling can be thought of as a result of the cycle in Equation 13.57. The more general mechanism shown in Equation 13.47 ($s, u > 1$) has multiple cycles, and can be expected to have extensive coupling. Here we will develop a general analytical framework to express this coupling. The resulting linkage relationships give mechanistic insight into heterotropic cooperativity, and through Maxwell relations, can be used to relate experimentally accessible heterotropic effects to hidden ones.

As seen in Figure 13.15, coupling between different types of ligands has a simple experimentally observable effect: addition of one ligand (e.g., *y*) results in a change in the fraction of the other ligand bound ($\langle f_x \rangle$). This can be quantified using "linkage coefficients", Λ:

$$\Lambda_{\langle f_x \rangle, y} = s \left(\frac{\partial \langle f_x \rangle}{\partial \ln[y]} \right)_{[x]} = \left(\frac{\partial \langle x_b \rangle}{\partial \ln[y]} \right)_{[x]}$$

$$\Lambda_{\langle f_y \rangle, x} = u \left(\frac{\partial \langle f_y \rangle}{\partial \ln[x]} \right)_{[y]} = \left(\frac{\partial \langle y_b \rangle}{\partial \ln[x]} \right)_{[y]}$$

(13.60)

[†] It is a general feature of competitive binding mechanisms to show what appears to be negative heterotropic cooperativity. Positive heterotropic cooperativity requires favorable coupling between non-overlapping sites.

Through a relationship analogous to the Maxwell relationships introduced in Chapter 5, these two linkage relationships can be shown to be identical. One way to show this is to recognize that $\ln P$ is a state function that depends on ligand concentration.[†] Thus, the differential of $\ln P$ can be written as

$$d\ln P = \left(\frac{\partial \ln P}{\partial \ln[x]}\right)_y d\ln[x] + \left(\frac{\partial \ln P}{\partial \ln[y]}\right)_x d\ln[y] \tag{13.61}$$

Substituting Equations 13.54 and 13.55 into Equation 13.61 gives

$$\begin{aligned} d\ln P &= s\langle f_x\rangle d\ln[x] + u\langle f_y\rangle d\ln[y] \\ &= \langle x_b\rangle d\ln[x] + \langle y_b\rangle d\ln[y] \end{aligned} \tag{13.62}$$

Because $d\ln P$ is an exact differential, the cross-derivatives of the sensitivity coefficients in Equation 13.61 are equal[‡]:

$$\left(\frac{\partial s\langle f_x\rangle}{\partial[y]}\right)_x = \left(\frac{\partial u\langle f_y\rangle}{\partial[x]}\right)_y \tag{13.63}$$

But the left and right sides of Equation 13.63 are simply the linkage coefficients, that is,

$$\Lambda_{\langle f_x\rangle,y} = \Lambda_{\langle f_y\rangle,x} \tag{13.64}$$

Equation 13.63 states that if the binding of one ligand is modulated by a second, then the binding of the second must be modulated by the first. Although this makes intuitive sense at a qualitative level, Equation 13.64 asserts that the linkages (as represented by the Λ coefficients) are *identical*.

This relationship can be useful in experimental studies of linkage. It may be difficult to measure the binding of one type of ligand and thus evaluate its sensitivity to a second type of ligand. Equation 13.59 states that the linkage coefficient can be obtained by measuring binding of the second type of ligand as a function of the first ligand. For example, the binding of hydrogen ions (H^+) to hemoglobin is linked to the binding of dioxygen (O_2) at distinct sites. Direct measurement of hydrogen binding is difficult, because there is no direct readout. Nonetheless, $\Lambda_{H+,O2}$ can be determined simply by measuring the how the extent of O_2 binding (which produces a large absorbance change) responds to change in H^+ concentration (pH).

Linkage coefficients for the simple two-site heterotropic model

The simple relationships between the linkage coefficients (Equation 13.59) are general, and can be applied to schemes of arbitrary complexity. Here we will calculate the linkage for our simple scheme involving a single binding site for x and for y (Figure 13.15) to gain insight into the magnitude of the coupling and how it relates to mechanistic features of the model. Linkage coefficients can be obtained by writing the binding polynomial and cross differentiating with respect to $\ln x$ and $\ln y$.[§] For competitive binding,

$$P = 1 + K_{1,0}^x[x] + K_{0,1}^y[y] \tag{13.65}$$

[†] $\ln P$ depends on concentrations of the ligands and on the binding constants. Both of these types of quantities are uniquely determined by the state of the system (concentrations are given unique values when the state is specified). Moreover, as described above, $\ln P$ is proportional to the free energy of ligand binding (with proportionality constant $-RT$), which is a state function.

[‡] This is another example of a Maxwell relation (see Chapter 5).

[§] In either order.

and

$$\langle f_x \rangle = \frac{1}{s}\left(\frac{\partial \ln P}{\partial \ln[x]}\right)_y = \frac{K_{1,0}^x[x]}{\left(1 + K_{1,0}^x[x] + K_{0,1}^y[y]\right)} \tag{13.66}$$

The linkage coefficients are obtained from differentiating $s\langle f_x \rangle$ from Equation 13.66 with respect to $\ln[y]$ (Equation 13.59):

$$\begin{aligned}
\Lambda_{\langle f_x \rangle, y} &= s\left(\frac{\partial \langle f_x \rangle}{\partial \ln[y]}\right)_x \\
&= s[y]\left(\frac{\partial \langle f_x \rangle}{\partial [y]}\right)_x \\
&= -\frac{K_{1,0}^x K_{0,1}^y[x][y]}{P^2}
\end{aligned} \tag{13.67}$$

(see Problem 13.28 for the derivative in the second line). Equation 13.67 can be rearranged to give

$$\Lambda_{\langle f_x \rangle, y} = -\langle f_x \rangle\langle f_y \rangle \tag{13.68}$$

Because the fractional saturation functions are always positive, the linkage coefficients for the competitive mechanism are always negative. This reflects the fact that the competitive scheme cannot generate positive cooperativity. The linkage coefficients have the largest (negative) value at ligand high concentration (such that the unbound ligand population goes to zero), where the ratio of x to y concentration is equal to the ratio of the binding constants, $K_{0,1}^y$ and $K_{1,0}^x$, respectively (Problem 13.29).

When a single x and a single y ligand can bind simultaneously (see Equation 13.57), the binding polynomial and fractional saturation are

$$P = 1 + K_{1,0}^x[x] + K_{0,1}^y[y] + K_{1,0}^x[x]K_{1,1}^y[y] \tag{13.69}$$

$$\langle f_x \rangle = \frac{1}{s}\left(\frac{\partial \ln P}{\partial \ln x}\right)_y = \frac{K_{1,0}^x[x] + K_{1,0}^x K_{1,1}^y[x][y]}{1 + K_{1,0}^x[x] + K_{0,1}^y[y] + K_{1,0}^x[x]K_{1,1}^y[y]} \tag{13.70}$$

The linkage coefficients are obtained from differentiating $s\langle f_x \rangle$ from Equation 13.69 with respect to $\ln[y]$ (see Problem 13.29):

$$\Lambda_{\langle f_x \rangle, y} = s\left(\frac{\partial \langle f_x \rangle}{\partial \ln[y]}\right)_x = -\frac{K_{1,0}^x\left(K_{0,1}^y - K_{1,1}^y\right)[x][y]}{P^2} \tag{13.71}$$

Here, the sign of the linkage coefficient can either be positive or negative, depending on the values of the binding constants in parentheses. If $K_{0,1}^y$ is larger than $K_{1,1}^y$, the linkage coefficient is negative, as described for competitive binding. However, in this case, the negative linkage coefficient results from negative cooperativity between distinct sites, rather than exclusive binding to a single site. If instead $K_{1,1}^y$ is larger than $K_{0,1}^y$, the linkage coefficient is positive, corresponding to positive cooperativity between the binding sites for x and y. It should be clear from Equation 13.71 that if $K_{0,1}^y = K_{1,1}^y$, the linkage coefficient is zero, corresponding to independent binding of ligands x and y.

Problems

13.1 For the fractional saturation data below, where the total macromolecule concentration is 10^{-6} M^{-1}, fit a single-site binding model. For this problem, you can assume that $[x]$ is the *free* ligand concentration (why is this case?).

$[x]$	$\langle f_b \rangle$	$[x]$	$\langle f_b \rangle$	$[x]$	$\langle f_b \rangle$
0.00	0.0005	0.57	0.3551	9.05	0.8988
0.05	0.0538	0.80	0.4522	12.80	0.9273
0.07	0.0714	1.13	0.5313	18.10	0.9441
0.10	0.0918	1.60	0.6177	25.60	0.9637
0.14	0.1193	2.26	0.6939	36.20	0.9685
0.20	0.1653	3.20	0.7582	51.20	0.9807
0.28	0.2281	4.53	0.8170		
0.40	0.2776	6.40	0.8661		

13.2 A single-site binding reaction has a binding constant of 10^{-7} M^{-1} at 25°C, and a molar enthalpy of binding of $\Delta \bar{H}° = -25 \, kJ/mol$. What is the molar entropy of binding at 25°C?

13.3 Assuming the enthalpy of binding is independent of temperature, make a plot of fractional saturation ($\langle f_b \rangle$) at 10°C, 25°C, and 40°C, on linear and $\log_{10}$ ligand concentration scales.

13.4 The three fractional saturation data sets (A, B, and C) below are for a single-site scheme, where the total macromolecule concentration $[M]_{tot} = 10^{-6}$ M^{-1}, and the ligand concentrations are *total* ligand rather than free ligand. Try to fit binding constants for these three data sets. What are the uncertainties on your three fitted binding constants?

$[x]_{tot}$	$\langle f_b \rangle$, set A	$\langle f_b \rangle$, set B	$\langle f_b \rangle$, set C
1.00×10^{-7}	0.089	0.118	0.148
1.41×10^{-7}	0.111	0.152	0.194
2.00×10^{-7}	0.103	0.158	0.220
2.82×10^{-7}	0.108	0.185	0.274
4.00×10^{-7}	0.103	0.208	0.336
5.65×10^{-7}	0.109	0.251	0.437
8.00×10^{-7}	0.102	0.291	0.557
1.13×10^{-6}	0.102	0.346	0.664
1.60×10^{-6}	0.101	0.404	0.682
2.26×10^{-6}	0.102	0.462	0.685
3.20×10^{-6}	0.112	0.521	0.691
4.52×10^{-6}	0.126	0.573	0.698
6.40×10^{-6}	0.139	0.612	0.703

Continued

$[x]_{tot}$	$\langle f_b \rangle$, set A	$\langle f_b \rangle$, set B	$\langle f_b \rangle$, set C
9.05×10^{-6}	0.149	0.635	0.699
1.28×10^{-5}	0.168	0.654	0.700
1.81×10^{-5}	0.190	0.666	0.699
2.56×10^{-5}	0.221	0.676	0.699
3.62×10^{-5}	0.258	0.683	0.699
5.12×10^{-5}	0.306	0.692	0.704
7.24×10^{-5}	0.348	0.689	0.697
1.02×10^{-4}	0.403	0.695	0.700
1.44×10^{-4}	0.453	0.694	0.698
2.04×10^{-4}	0.500	0.695	0.697
2.89×10^{-4}	0.539	0.691	0.693
4.09×10^{-4}	0.582	0.699	0.700
5.79×10^{-4}	0.613	0.700	0.701
8.19×10^{-4}	0.632	0.696	0.697
1.15×10^{-3}	0.652	0.699	0.700
1.63×10^{-3}	0.670	0.704	0.705
2.31×10^{-3}	0.663	0.687	0.688
3.27×10^{-3}	0.673	0.691	0.691

13.5 One thing saturation binding data (i.e., $K_d \ll [M]_{tot}$) is good for is accurately determining the total macromolecular concentration. For the saturation binding data (set C) from Problem 13.4, assume you don't know the value of $[M]_{tot}$ and fit it from the data. How does it compare with the true value given in Problem 13.4?

13.6 In deriving the equation for fractional saturation in the saturation-binding limit (Equation 13.19), we took the negative root of the quadratic equation. Demonstrate that the negative root is the correct one, by evaluating the concentration of Mx in the limits of low and high ligand concentration ($[x]_{tot}$).

13.7 In a myoglobin oxygen titration (Figure 13.6), a single-ligand binding constant of 1.06×10^{-6} M^{-1} is measured. Convert this constant to units of partial pressure, in mm Hg.

13.8 For the set of K_i's below, calculate β_i's.

Stepwise macroscopic constant	Stepwise microscopic constant
$K_1 = 100$ M^{-1}	$\beta_1 = ?$
$K_2 = 100$ M^{-1}	$\beta_2 = ?$
$K_3 = 100$ M^{-1}	$\beta_3 = ?$
$K_4 = 100$ M^{-1}	$\beta_4 = ?$

13.9 For the set of K_i's below, calculate β_i's.

Stepwise macroscopic constant	Stepwise microscopic constant
$\beta_1 = 10^4$ M^{-1}	$K_1 = ?$
$\beta_2 = 10^6$ M^{-2}	$K_2 = ?$
$\beta_3 = 10^7$ M^{-3}	$K_3 = ?$

13.10 From the values of the constants in Problems 13.8 and 13.9, do the systems possess positive, negative, or no cooperativity?

13.11 Using the overall binding version of the binding polynomial, substitute the $[x]$ terms in P with chemical potentials.

13.12 With the result from Problem 13.11, substitute the relationship for β_i in terms of standard state free energies of the overall reaction steps, and write in terms of chemical potentials (Equation 7.44). Interpret the resulting terms in the sum in words.

13.13 As given in Equations 13.29 and 13.30, P and q differ in their response to high ligand concentration and temperature, respectively, yet under these limits, a single ligand binding state is produced for systems described using P, whereas systems described using q spread out evenly among all available states. For a macromolecule with four sites, and the overall binding constants $\beta_1 = \beta_2 = \beta_3 = \beta_4 = 1$ M^{-i}, plot the thermodynamic weights (i.e., Boltzmann factors $\beta_i[x]^i$) of the four Mx_i states and, on a separate plot, the molar free energies of the four Mx_i states (relative to the M_0 state) as a function of ligand concentration. How does this differ from the behavior of the Boltzmann factors and relative free (in this case, Helmholtz) energies from a five level energy ladder (energy values $\bar{\varepsilon}_1 = 0$, $\bar{\varepsilon}_2 = 1$, $\bar{\varepsilon}_3 = 2$, $\bar{\varepsilon}_4 = 3$, $\bar{\varepsilon}_5 = 4$ kJmol^{-1}) as a function of temperature?

13.14 Derive a scheme and a binding polynomial for $s = 3$ binding sites. For the following sets of binding constants (set A and set B below), plot the fractional binding $<f_b>$ and the populations of each species (unligated, partly ligated, and fully ligated) as a function of free ligand concentration. Set A: $K_1 = 20$, $K_2 = 1$, $K_3 = 0.05$ M^{-1}; Set B: $K_1 = 0.05$, $K_2 = 1$, $K_3 = 20$ M^{-1}. Which set shows positive cooperativity?

13.15 Using the numerical values of the four stepwise binding constants for human hemoglobin in Table 13.1, find the O$_2$ concentration where the intermediate state Hb(O$_2$)^2 is maximally populated.

13.16 Starting with expressions for $\langle f_b \rangle$, show that the same midpoints are obtained for the single-site and the two site binding schemes, that is,

$$M_0 + X \xrightleftharpoons{\quad K \quad} MX$$

and

$$M_0 + 2X \xrightleftharpoons{\quad K_1 \quad} MX + X \xrightleftharpoons{\quad K_2 \quad} MX_2$$

when $K = \sqrt{K_1 K_2}$. Does this work for a three-site case, and if so, why does this work?

13.17 Using the fractional saturation data below, which was generated from the two-site binding scheme (second equation in Problem 13.16) with binding constants $K_1 = 0.1$ M^{-1}, $K_2 = 10$ M^{-1} (and includes 0.5% Gaussian error), generate a Hill plot and fit the Hill coefficient n_H using a linear approximation. Note, you will not want to use the extreme low and high ligand limits for your Hill plot (or fitting), due to the nonlinear effects of the Hill transformation on the error.

$[x]$	$\langle f_b \rangle$	$[x]$	$\langle f_b \rangle$	$[x]$	$\langle f_b \rangle$	$[x]$	$\langle f_b \rangle$
1.00×10^{-03}	-0.002	3.20×10^{-02}	-0.008	1.02×10^{00}	0.518	3.28×10^{01}	1.001
1.41×10^{-03}	0.008	4.53×10^{-02}	0.004	1.45×10^{00}	0.674	4.63×10^{01}	1.002
2.00×10^{-03}	0.005	6.40×10^{-02}	0.014	2.05×10^{00}	0.788	6.55×10^{01}	0.990
2.83×10^{-03}	-0.004	9.05×10^{-02}	0.017	2.90×10^{00}	0.881	9.27×10^{01}	1.000
4.00×10^{-03}	-0.001	1.28×10^{-01}	0.026	4.10×10^{00}	0.929	1.31×10^{02}	0.990
5.66×10^{-03}	0.000	1.81×10^{-01}	0.046	5.79×10^{00}	0.965	1.85×10^{02}	0.991
8.00×10^{-03}	0.000	2.56×10^{-01}	0.071	8.19×10^{00}	0.981	2.62×10^{02}	0.998
1.13×10^{-02}	0.006	3.62×10^{-01}	0.131	1.16×10^{01}	0.987	3.71×10^{02}	0.999
1.60×10^{-02}	0.005	5.12×10^{-01}	0.224	1.64×10^{01}	0.997	5.24×10^{02}	1.010
2.26×10^{-02}	-0.002	7.24×10^{-01}	0.358	2.32×10^{01}	0.985	7.41×10^{02}	0.988

13.18 For the two-site binding scheme (the second scheme in Problem 13.16) with binding constants $K_1 = 0.1$ M^{-1}, $K_2 = 10$ M^{-1}, find the ligand concentration where the Hill coefficient n_H is maximal, and find its value at that point. How does this value compare to the value you determined by fitting in Problem 13.17?

(a) Is the data from Problem 13.17 "good enough," (i.e., is the error low enough) to see the limiting K_1 and K_2 regions of the Hill plot (see Figure 13.13)?

(b) Using the data from Problem 13.17, fit $\langle f_b \rangle$ as a function of $[x]$ using the two-site mechanism to determine K_1 and K_2. How do your fitted values compare with the true values used to generate the data ($K_1 = 0.1$ M^{-1}, $K_2 = 10$ M^{-1})? Comparing your results from this problem and Problem 13.18a should demonstrate the value of directly fitting experimental (rather than transformed) data.

13.19 Demonstrate that for a macromolecule with two binding sites, the low- and high-ligand limits of the Hill plot are $\log_{10}(K_1[x]/2)$ and $\log_{10}(2K_2[x])$, respectively.

13.20 Write the binding polynomials for the two site examples in the text ($K_1 = 0.1$ M^{-1}, $K_2 = 10$ M^{-1}, and $K_1 = 10$ M^{-1}, $K_2 = 0.1$ M^{-1}). Regarding P as a partition function for binding, interpret the relative values of the terms in the two partition functions. Use these polynomials to plot the species fractions (M_0, Mx, and Mx_2) for the sets of K values as a function of ligand concentration. How do the two species plots differ?

13.21 For a six-site binding model with $K_1 = K_2 = K_3 = K_4 = K_5 = K_6 = 10^6$ M^{-1}, plot the binding curve and compare it to single-site binding with $K = 10^6$ M^{-1}. Describe the overall shape of the curve.

13.22 Show that for the two-site binding mechanism, the binding capacity C_x is given by Equation 13.37.

13.23 Show that the Equation 13.41 for the Hill coefficient can be obtained from Equation 13.40.

13.24 Assuming the simple all-or-none stoichiometry shown in Equation 13.43, show that the fraction binding is given by Equation 13.44. One simple way to do this is to write the binding polynomial, and differentiate it with respect to the binding constant β_j.

13.25 Assuming the all-or-none stoichiometry shown in Equation 13.43, transform the binding curve into logarithmic "Hill" form. Based on this transformation, describe the dependence of $\log(f_{bound}/\langle f_u \rangle)$ on $\log[x]$, and give its slope.

13.26 For the competitive binding of O_2 and CO to human hemoglobin ($s = 4$), write an expression for the formation of the species $Hb(O_2)_2(CO)_2$ (i.e., $i = j = 2$) from deoxyHb (i.e., $\beta_{2,2}$) in terms of stepwise constants. Is your answer unique? That is, can you come up with another set of stepwise constants to represent $\beta_{2,2}$?

13.27 Derive Equations 13.54 and 13.55, that is,

$$\langle f_x \rangle = \frac{1}{s}\left(\frac{\partial \ln P}{\partial \ln[x]}\right)_y \quad \text{and} \quad \langle f_y \rangle = \frac{1}{u}\left(\frac{\partial \ln P}{\partial \ln[y]}\right)_x$$

13.28 Derive the Equation 13.67, that is,

$$s\left(\frac{\partial \langle f_x \rangle}{\partial \ln[y]}\right)_x = -\frac{K^x_{1,0}K^y_{0,1}[x][y]}{P^2}$$

13.29 Show that for a simple competitive mechanism where there is a single binding site on M that can bind either x or y (with binding constant K_x and K_y; Equation 13.57 with $Mxy = 0$), the linkage coefficient $\Lambda_{x,y}$ is maximized for $\bar{X} = \bar{Y} = 0.5$.

13.30 Demonstrate that for the mechanism where a single x and a single y ligand can bind simultaneously to a macromolecule (Equation 13.57), the linkage coefficient is given by Equation 13.71.

13.31 For the mechanism where a single x and a single y ligand can bind simultaneously to a macromolecule (mechanism 13.57), show that if $K^y_{0,1} = K^y_{1,1}$, the binding polynomial can be factored into a component dependent on x (but not y), and a component dependent on y (but not x). Rationalize your answer in terms of probabilities and how they combine.

CHAPTER 14

Ligand Binding Equilibria from a Microscopic Perspective

Goals and Summary

The goal of this chapter is to extend the analytical framework developed in the previous chapter to describe binding using a microscopic, site-oriented approach to ligand binding. We will develop microstate versions of step-wise and overall binding constants. By borrowing ideas from topology and graph theory, we will show how the number of stepwise and overall constants relate to each other and to the number of thermodynamic cycle associated with binding. For the most general binding schemes, the large number of parameters required for microscopic analysis can be a disadvantage; however, this can be offset by the clear connection between the microscopic models and molecular mechanisms of binding.

Following our presentation of general microscopic schemes, we will introduce a number of simplifications (such as identical classes of sites and uniform coupling) that greatly simplify the binding partition function and expressions for saturation. By incorporating molecular features such as symmetry and periodicity, these simplifications can be justified, producing models that have elegant, compact forms, yet retain the key molecular details. A particularly interesting class of simplifying models are those that link conformational changes with ligand binding. Such "allosteric" mechanisms efficiently link distant binding sites, providing an effective means of homotropic cooperativity and heterotropic regulation.

In the previous chapter, we developed a formalism for binding in which we kept track of the number of ligands bound to a macromolecule, but not how those ligands were arranged on the binding sites. The connection to thermodynamics was made using macroscopic binding constants (either stepwise, K_i, or overall, β_i). Though these macroscopic constants provide some sense of the relative affinities at each step of the reaction, they lump multiple different modes of ligand binding into a single constant. Thus, they do not capture all of the details of binding at the atomic level. Here we will develop a microscopic description of binding in which the constants describe ligand association to individual *sites* on a macromolecule. For most of our analysis *stepwise* microscopic constants (which will be represented using the symbol kappa, κ) will be used to describe microscopic equilibria, but in some cases, *overall* microscopic constants (which will be represented using the symbol Б)[†] will be discussed to clarify the number of unique microscopic binding constants that describe the system.

The microscopic formalism for binding has several advantages. First, the microscopic description provides a more fundamental way to model the binding of multiple ligands than does the macroscopic description. For example, a simple

---

† Б is the upper-case cyrillic (Russian) letter Be (lower case is б). It is used here because of its connection to the overall macroscopic binding constant β. It is from the font set "*PT sans*."

noncooperative rectangular hyperbolic binding curve is produced when all of the *microscopic* stepwise constants are identical. If all of the *macroscopic* stepwise constants are identical, the resulting binding curve shows (modest) positive cooperativity (see Figure 14.7 below). As another example, if there are two different classes of sites on a macromolecule (high-affinity sites and low-affinity sites), the microscopic constants will directly reflect the intrinsic affinity differences of the two types of sites, whereas K_1 and K_2 represent a weighted combination of direct affinities to both sites and statistical factors that account for the number of ways to ligand arrangements.

A second advantage of the microscopic formalism is that it captures a greater level of mechanistic detail than the macroscopic formalism. For mechanisms that are complex at the molecular level, the macroscopic approach averages out important molecular details. This difference in complexity is reflected in the following rules relating macroscopic and microscopic constants:

1. The microscopic binding constants uniquely specify the macroscopic constants. Given values for the microscopic constants, the macroscopic constants are uniquely specified.
2. The macroscopic binding constants *do not* uniquely specify the microscopic binding constants. There are many sets of microscopic constants consistent with the macroscopic constants.
3. The macroscopic binding constants restrict microscopic constants to a range of values. The values of a set of macroscopic binding constants eliminate values of microscopic binding constants that are incompatible.

To provide a concrete comparison of the microscopic and macroscopic approaches, consider in the following section a macromolecule with three distinct binding sites for ligand x. This example can easily be extended to macromolecules with more than three sites. Within the three-site mechanism, we will begin with a completely general microscopic description of binding, to illustrate the high level of mechanistic detail that can be represented by the microstate approach. One downside from this treatment is that, in its most general form, the complexity of microscopic models can easily exceed the amount of information that can be resolved from even the best experimental binding data. Thus, in the subsequent sections, restrictions will be introduced that retain the fundamental, intuitive nature of the microscopic description but involve fewer parameters.

AN EXAMPLE OF GENERAL MICROSCOPIC BINDING: THREE LIGAND BINDING SITES

Consider a macromolecule that can bind three x ligands:

$$M_0 + 3x \rightleftharpoons Mx + 2x \rightleftharpoons Mx_2 + x \rightleftharpoons Mx_3 \qquad (14.1)$$

In the macroscopic description represented in Equation 14.1, we lump different arrangements of the same number i of bound ligands into a single species, Mx_i. o develop a microscopic description, we will treat each arrangement of bound and free sites explicitly. To do this, we need "label" each binding site (conceptually, at least); for the three-site example, we label sites 1, 2, and 3 (**Figure 14.1**).

It is clear that such labeling is warranted if each of the three sites is distinct. However, even if the sites are identical (which often occurs for macromolecules with internal symmetry, as is *implied* in the trimer in Figure 14.1), the concept of labeling helps to establish the statistical effects of having multiple sites (here, three ways to "arrange" the ligand, even when the sites are not formally distinguishable by experiment when sites are identical).

Figure 14.1 shows two ways to represent microscopic ligand binding. In Figure 14.1A, each binding site (often a subunit, but not always) is represented by an open

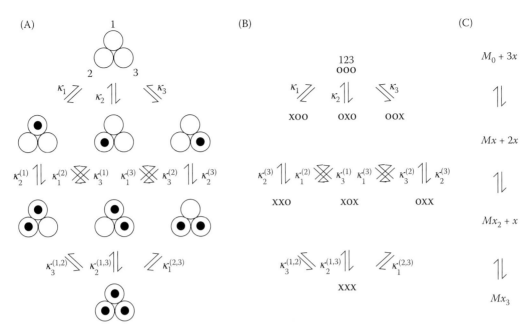

Figure 14.1 General microscopic binding scheme for a macromolecule with three binding sites. From top to bottom, ligand saturation increases from $i = 0$ to 3. (A) and (B) are microscopic depictions, (C) is a macroscopic description. In (A), macromolecular subunits are represented as large open circles, and bound ligands are represented as small filled circles (note, binding sites need not have a 1:1 correspondence to covalently separate subunits). In (B), free and bound sites with o's and x's, respectively, and distinguishes sites 1, 2, and 3 in the sequence from left to right (thus, xox has sites 1 and 3 bound, and site 2 empty). The model in (B) is the same as in (A), but the notation is more compact.

circle, and ligand occupancy is indicated by a smaller filled circle. In Figure 14.1B, site identity and ligand occupancy are represented using a sequence of o's and x's, which is more convenient for writing microstate mass action equations than cartoons showing subunits.

A stepwise microscopic description

By focusing on each binding site individually, we can define stepwise microscopic constants. For the three-site general model, these stepwise constants are shown in Figures 14.1 and 14.2. For binding of the first ligand, the constants are specified simply by identifying the site being bound, that is, κ_1, κ_2, and κ_3 for sites 1, 2, and 3. These constants can be defined using mass action relationships among the microstates:

$$\kappa_1 = \frac{[xoo]}{[ooo][x]} \quad \kappa_2 = \frac{[oxo]}{[ooo][x]} \quad \kappa_3 = \frac{[oox]}{[ooo][x]} \tag{14.2}$$

The three stepwise reactions represented by κ_1, κ_2, and κ_3 connect the top species (ooo in Figure 14.1; M_0 in Figure 14.1C) to the singly ligated states below it.

For the next binding step, each of these three singly-ligated states combines with a second ligand. In the most general (and perhaps most interesting) case, the occupancy of the first site influences the binding of the second ligand. Thus, microscopic binding constants for the second step must specify both the site being bound and the site already occupied. The site being bound is specified using a subscript, as for the first step. The already occupied site is specified with a superscripted parenthesis. For example, the microscopic binding constant for site 2, starting from the state with a single ligand at site 1 (xoo) is labeled $\kappa_2^{(1)}$. For the three-site example (Figure 14.1), there are six such binding reactions for the second step (three starting points, each with two empty sites; **Table 14.1**).

Table 14.1 Stepwise microscopic constants for the second step of binding in a three-site binding scheme

Site being bound	Site already bound		
	1	**2**	**3**
1	n.d.[a]	$\kappa_1^{(2)} = \dfrac{[xxo]}{[oxo][x]}$	$\kappa_1^{(3)} = \dfrac{[xox]}{[oox][x]}$
2	$\kappa_2^{(1)} = \dfrac{[xxo]}{[xoo][x]}$	n.d.[a]	$\kappa_2^{(3)} = \dfrac{[oxx]}{[oox][x]}$
3	$\kappa_3^{(1)} = \dfrac{[xox]}{[xoo][x]}$	$\kappa_3^{(2)} = \dfrac{[oxx]}{[oxo][x]}$	n.d.[a]

[a]These three entries do not need to be defined to describe microstate binding process. Technically, we could give them a value of one, consistent with our treatment of β_0 and $\beta_{0,0}$ in Chapter 13.

Similar constants can be written for the third step in binding, where the two sites occupied at the start of the reaction are included in the superscript. There are three such reactions (three starting points, only one site available for each), and three microscopic constants:

$$\kappa_1^{(23)} = \frac{[xxx]}{[oxx][x]} \quad \kappa_2^{(13)} = \frac{[xxx]}{[xox][x]} \quad \kappa_3^{(12)} = \frac{[xxx]}{[xxo][x]} \tag{14.3}$$

Together, Equations 14.2, 14.3, and Table 14.1 specify 12 microscopic binding constants, corresponding to the 12 equilibrium arrows in Figures 14.1A and 14.1B. This number of constants exceeds the number used in the macroscopic formalism for three sites (three total, either stepwise, K_1, K_2, and K_3 [Figure 14.1C] or overall, β_1, β_2, and β_3). Some of this excess is a result of the higher level of detail provided by the microscopic description. However, not all of the twelve $\kappa_i^{(jk)}$ values are independent. This can be appreciated by recognizing that the scheme in Figure 14.1 contains the same kinds of cycles that we encountered when treating the binding of distinct ligands (x and y) in Chapter 13. These cycles fix the values of some of the binding constants; thus, these constants are defined as combinations of the other constants, and are not free to vary.

So what is the total number of independent microscopic binding constants required to describe a particular binding scheme? In general, the number of unique constants is equal to the number of separate species, minus one ["minus one" because equilibrium constants compare species to a reference, which we take as the unligated state; we don't need a "constant" comparing the unligated state to itself (though sometimes this constant will make the math easier)]. In the three-site model (Figure 14.1) there are eight species, and thus there are seven independent stepwise microscopic binding constants. The other five can be written as combinations of these constants (Problem 14.5). In the next section, we will describe how parameterizing the microscopic binding scheme in terms of overall constants avoids this redundancy.

An overall microscopic description

A simple way to represent microscopic binding that provides a one-to-one correspondence between separate species and binding constants is to define *overall* (as opposed to stepwise) microscopic binding constants. Overall microscopic constants will be represented using the symbol Б. In analogy to the macrostate overall constants (β_i) described in Chapter 13, these constants relate the unligated state to partly (and fully) ligated states. Unlike the overall macrostate constants, each overall microstate constant describes a single ligand arrangement. For the general three-site model (Figure 14.1), the eight[†] overall microscopic binding constants are

$$\mathcal{B}_0 = \frac{[ooo]}{[ooo][x]^0} = 1 \quad \mathcal{B}_2 = \frac{[oxo]}{[ooo][x]^1} = \kappa_2 \quad \mathcal{B}_{13} = \frac{[xox]}{[ooo][x]^2} = \kappa_1 \kappa_3^{(1)} \quad \mathcal{B}_{123} = \frac{[xxx]}{[ooo][x]^3} = \kappa_1 \kappa_2^{(1)} \kappa_3^{(12)}$$

$$\mathcal{B}_1 = \frac{[xoo]}{[ooo][x]^1} = \kappa_1 \quad \mathcal{B}_{12} = \frac{[xxo]}{[ooo][x]^2} = \kappa_1 \kappa_2^{(1)}$$

$$\mathcal{B}_3 = \frac{[oox]}{[ooo][x]^1} = \kappa_3 \quad \mathcal{B}_{23} = \frac{[oxx]}{[ooo][x]^2} = \kappa_2 \kappa_3^{(2)}$$

$$(14.4)$$

The subscripts on each $\mathcal{B}$ specify which sites bind ligand in the corresponding reaction (see Figure 14.1B). Since the reaction starts with the unligated state, there is no need to include a superscript to specify the starting point, in contrast to the stepwise microscopic constants.

As indicated in Equation 14.4, the overall microscopic constants can be given as a product of stepwise microscopic constants. However, the fact that only seven overall microscopic constants are needed to define all of the species in the general three-site model means that, unlike stepwise microscopic constants, there is no redundancy in the overall microscopic approach. An implication of this is that many of the overall microscopic constants can be represented by more than one set of stepwise microscopic constants. Specifically, for a particular microscopic species, all possible sequences of steps that connect from the unligated state are valid, thermodynamically equivalent paths, and the product of stepwise constants along any of these paths can be multiplied together to give the overall microscopic constant. This relationship is probed further in Problem 14.6.

A geometric picture of the relationship between stepwise and overall microscopic constants

As described above, the number of overall microscopic constants needed to describe a general binding scheme is equal to the number of microstates minus one (in general, $2^s - 1$). The number of stepwise constants (which exceeds the number of overall constants) is equal to the number of equilibrium arrows that connecting species that are related by a single ligand binding step.[‡] The difference between the number of microscopic stepwise and overall constants is equal to the number of thermodynamic cycles embedded in the reaction scheme. Though these are important relations, they do not directly answer the following key question: for a given number of sites s, how many microscopic overall constants and how many stepwise constants (and thermodynamic cycles) are there?

A solution to these questions can be obtained by through an analogy between microscopic binding schemes and geometric figures referred to as "parallelotopes" (Coxeter, 1973). Parallelotopes can be thought of as versions of parallelograms expanded into higher dimensions.[§] The three-dimensional version of a parallelogram is a parallelepiped. The one dimensional version is a line segment. **Figure 14.2** shows binding schemes for $s = 1, 2,$ and 3 sites represented as parallelotopes.

[†] The eighth constant, $\mathcal{B}_0$, is included in Equation 14.4 for completeness. This trivial constant corresponds to the formation of the unbound state from the unbound state, by "binding" of zero ligands. Thus, $\mathcal{B}_0$ has a value of unity. In addition to providing symmetry, the apparently trivial constant $\mathcal{B}_0$ is useful for representing the statistical weight for the unligated state in a compact form of the binding polynomial (developed below).

[‡] Here, relation by a single ligand binding step means not only that the number of ligands bound to related species differs by one, but that related species are transformed by a single binding event (without ligand rearrangement among site). For example, species xxo can be formed directly by binding a ligand to xoo and oxo, but not to oox.

[§] Or, for a single site, compressed from two dimensions to one.

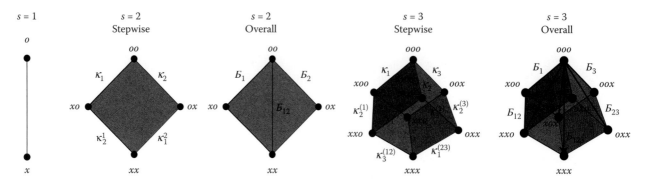

Figure 14.2 Geometric representation of ligand binding to multiple sites. Parallelotopes correspond to macromolecules binding to $s = 1$, 2, and 3 sites. As s is increased by one, a new figure is obtained by taking the previous figure (with s ligands), translating it in a dimension perpendicular to the other edges, and forming edges between the new (translated) and old vertices. In this process, the number of vertices double (from 2^s to 2^{s+1}, corresponding to a doubling of species), the number of edges increases by a factor of $2 + 2s$ (from $s \times 2^{s-1}$ to $(s + 1) \times 2^s$). In the progression shown, the figure transforms from a line (for a single site) to a square (two sites) to a cube (three sites). Adding a fourth site would produce a four-dimensional parallelotope (Figure 14.3 below; see Coxeter, 1973, for further details). In this representation, each vertex corresponds to one of the 2^s microstates. Each edge corresponds to an equilibrium arrow for ligand binding to an individual site. Thus, the number of edges is equal to the number of microscopic stepwise binding constants. Each face corresponds to a thermodynamic cycle, and provides a constraint on the microscopic stepwise constants.

For a single binding site, the scheme is represented by a line segment connecting the unbound (o) and bound (x) states. With two binding sites, the scheme is a square (obtained by dragging the $s = 1$ line segment in a perpendicular direction). With three binding sites, the scheme is a cube. Although extension to into four or more dimensions is conceptually challenging, we can adequately visualize the scheme for four binding sites by dragging the $s = 3$ cube in a new direction in three-dimensional space,[†] and connecting the new and old vertices with a new set of edges (**Figure 14.3**).

Figure 14.3 A parallelotope representing the binding of four ligands to a macromolecule. The occupancy of the fourth site is depicted by duplicating the three site cube (Figure 14.2), displacing the new copy (lower), and connecting displaced vertices by edges, relating to single-step binding reactions to the new site. As drawn, the resulting polyhedron looks like a box kite; alternatively, the direction of displacement can be thought of as perpendicular, into a fourth dimension (although this cannot be drawn easily).

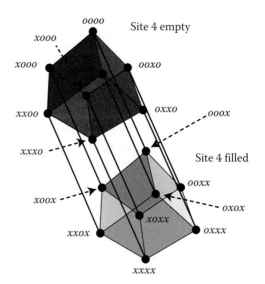

---

† "In a new direction" means that the translation does not go in the same direction as a previous translation, that is, it does not follow the same line. Doing so would make new features in the $s + 1$ figure (edges, faces, and vertices) difficult to distinguish from the old (s) features. For the parallelotopes in Figure 14.2, translations are perpendicular to previous translations, giving simple orthorhombic objects. However, extension past three binding sites would require us to imagine four- (or more-) dimensional space. In Figure 4.3, the $s = 4$ parallelotope remains in three-dimensions because translation was at an oblique angle to the $s = 3$ edges.

In addition to providing clear visualization, parallelotopes provide a good analytical framework for ligand binding, because their geometry is well understood. The geometry of parallelotopes (and other polyhedral) is uniquely defined by the number of vertices, edges, and faces (represented as V, F, and E). Though these three quantities are functions of the dimension of the parallelotope (1 for the line segment, 2 for the parallelogram, 3 for the parallelepiped, etc.), they are constrained by an important formula, namely,

$$V - E + F = 2 \tag{14.5}$$

This equation, which is called "Euler's polyhedron formula" holds no matter what the shape (or dimension) of the parallelotope, as long as it is convex.[†] Given any two of these three quantities, the third is uniquely determined by Equation 14.5.

As illustrated in Figures 14.2 and 14.3, there is a direct correspondence between thermodynamic quantities associated with various binding schemes and the defining geometric features of parallelotopes. The number of ligand binding sites (s) corresponds to the dimensions spanned by the parallelotope. The number of microstate species corresponds to the number of vertices (V). The number of stepwise microscopic binding constants (κ's) corresponds to number of edges (E). And the number of thermodynamic cycles corresponds to the number of faces (F). And the number of overall microscopic constants corresponds to the diagonals (referred to here as D) connecting the "unbound" vertex to the vertices.

With these correspondences in hand, we can approach the question posed above, namely "how many species, κ's, B's, and cycles are associated with an s-binding site scheme" by analysis of parallelotope geometry. It should be fairly obvious that the number of vertices (and thus binding microstates) is given by

$$V(s) = 2^s \tag{14.6}$$

A general formula for the number of edges (and thus κ's) as a function of s is not so obvious. It is a little easier is to relate the number of parallelotope edges for a given value of s [i.e., $E(s)$] with the number of edges for the $s - 1$ parallelotope:

$$E(s) = 2E(s-1) + 2^{s-1} \tag{14.7}$$

(Problem 14.7). Equations of the form of 14.8 are referred to as "recursion relationships," and can be used to build up step-by-step from known values at small s. This is done in **Table 14.2**, starting from $s = 1$.

Table 14.2 Geometric features of parallelotopes, and thermodynamic features of binding models

s (# of sites)	V (# of species)	E (# of κ's)	F (# of cycles)	D (# of B's)
1	2	1	0	1
2	4	4	1	3
3	8	12	6	7
4	16	32	18	15
5	32	80	50	31
...				
s	2^s	$s \times 2^{s-1}$	$s \times 2^{s-1} - 2^s + 2$	$2^s - 1$

[†] Convex means that any if a line segment is drawn between any two points inside the polytope, that line segment is also entirely inside the polytope.

By generating this list of E values using recursion formula 14.7, we can search for a pattern that leads to an explicit equation $E(s)$ as a function of s, rather than $E(s-1)$. You should confirm for yourself that the relationship is

$$E(s) = s \times 2^{s-1} \tag{14.8}$$

Combining Equations 14.7 and 14.9 with Euler's relation (Equation 14.5) gives an expression for the number of faces (thermodynamic cycles) in terms of the number of sites, s:

$$\begin{aligned} F(s) &= E(s) - V(s) + 2 \\ &= s \times 2^{s-1} - 2^s + 2 \end{aligned} \tag{14.9}$$

Finally, the number of diagonals (overall microscopic constants) is equal to the number of vertices minus one:

$$\begin{aligned} D(s) &= V(s) - 1 \\ &= 2^s - 1 \end{aligned} \tag{14.10}$$

Binding polynomials in terms of microscopic constants

As described in Chapter 7, the binding polynomial P plays a central role in describing the binding of multiple ligands to a macromolecule. P serves as a partition function for binding, giving populations of the partly ligated species, and providing a simple way to calculate (by differentiation) the extent of binding. As in Chapter 13, P is equal to the sum the concentrations of different species, relative to the unligated reference state:

$$\begin{aligned} P &= \frac{[ooo]}{[ooo]} + \frac{[xoo]}{[ooo]} + \frac{[oxo]}{[ooo]} + \frac{[oox]}{[ooo]} + \frac{[xxo]}{[ooo]} + \frac{[xox]}{[ooo]} + \frac{[oxx]}{[ooo]} + \frac{[xxx]}{[ooo]} \\ &= \frac{[M_0]}{[M_0]} + \frac{[M_x]}{[M_0]} + \frac{[Mx_2]}{[M_0]} + \frac{[Mx_3]}{[M_0]} \end{aligned} \tag{14.11}$$

The first line includes microscopic detail, whereas the second does not. The statement that these two are equal emphasizes the point that P has the same value regardless of the level of detail used to describe species. Given the role of the binding polynomial as a partition function for ligand binding, P has fundamental thermodynamic meaning, and should be insensitive to how we choose to represent binding. P varies, but only with changes in system variables such as with ligand concentration and external conditions (e.g., temperature, pressure, pH, etc.).

As with macroscopic constants, the concentration ratios in P (Equation 14.11) can be substituted with powers of ligand concentration and either stepwise ($\kappa_i^{(jk\ldots)}$) or overall (β_{ijk}) microscopic constants. As with the macroscopic representation, stepwise and overall microscopic constants have different advantages: both types of stepwise constants (K_i, κ_i^{jk}) are more intuitive, whereas both types of overall constants (β_i, β_{ijk}) lead to more compact partition functions. As described above, overall microscopic constants have the added advantage that they are unique, whereas some of the stepwise microscopic constants are redundant.

Generating *P* using stepwise microscopic stepwise constants

For the three-site model (Figure 14.1), we will use the combination of stepwise microscopic constants given on the right-hand side of Equation 14.4[†]:

---

† For no other reason than that we have already written them.

$$P = \frac{[ooo]}{[ooo]} + \frac{[xoo]}{[ooo]} + \frac{[oxo]}{[ooo]} + \frac{[oox]}{[ooo]} + \frac{[xxo]}{[ooo]} + \frac{[xox]}{[ooo]} + \frac{[oxx]}{[ooo]} + \frac{[xxx]}{[ooo]}$$

$$= 1 + \kappa_1 x + \kappa_2 x + \kappa_3 x + \kappa_1 \kappa_2^{(1)}[x]^2 + \kappa_1 \kappa_3^{(1)}[x]^2 + \kappa_2 \kappa_3^{(2)}[x]^2 + \kappa_1 \kappa_2^{(1)} \kappa_3^{(12)}[x]^3 \quad (14.12)$$

$$= 1 + (\kappa_1 + \kappa_2 + \kappa_3)[x] + (\kappa_1 \kappa_2^{(1)} + \kappa_1 \kappa_3^{(1)} + \kappa_1 \kappa_2^{(2)})[x]^2 + \kappa_1 \kappa_2^{(1)} \kappa_3^{(12)}[x]^3$$

$$= 1 + K_1[x] + K_1 K_2[x]^2 + K_1 K_2 K_3[x]^3$$

The last two lines compare microscopic and macroscopic versions of the binding polynomial (in terms of stepwise constants). The relationships among these constants can be built up explicitly starting from the lowest ligation states, which give the direct relationship

$$K_1 = \kappa_1 + \kappa_2 + \kappa_3 \quad (14.13)$$

Equation 14.13 emphasizes the point that microscopic constants provide more information than the macroscopic constants. There are an infinite number of combinations of κ_1, κ_2, and κ_3 that are consistent with a single K_1 (rule 2 above). One or two of the microscopic constants can be much smaller than K_1 (they can approach zero), as long as the third one has a value close to K_1. Nonetheless, Equation 14.13 significantly restricts the allowable values of κ_1, κ_2, and κ_3. This can be represented in a three-dimensional coordinate system with axes of κ_1, κ_2, and κ_3. In this coordinate system, Equation 14.13 restricts the allowable values of the microscopic constants to a region of a plane specified by the value of K_1 (**Figure 14.4**).

The relationship between K_2 and the microscopic binding constants can be obtained by again comparing the last two lines in Equation 14.12, which gives the relationship

$$K_1 K_2 = \kappa_1 \kappa_2^{(1)} + \kappa_1 \kappa_3^{(1)} + \kappa_2 \kappa_3^{(2)} \quad (14.14)$$

Combining this with Equation 14.13 leads to

$$K_2 = \frac{\kappa_1 \kappa_2^{(1)} + \kappa_1 \kappa_3^{(1)} + \kappa_2 \kappa_3^{(2)}}{K_1}$$

$$= \frac{\kappa_1 \kappa_2^{(1)} + \kappa_1 \kappa_3^{(1)} + \kappa_2 \kappa_3^{(2)}}{\kappa_1 + \kappa_2 + \kappa_3} \quad (14.15)$$

In the same way, K_3 can be expressed (Problem 14.4) as

$$K_3 = \frac{\kappa_1 \kappa_2^{(1)} \kappa_3^{(12)}}{\kappa_1 \kappa_2^{(1)} + \kappa_1 \kappa_3^{(1)} + \kappa_2 \kappa_3^{(2)}} \quad (14.16)$$

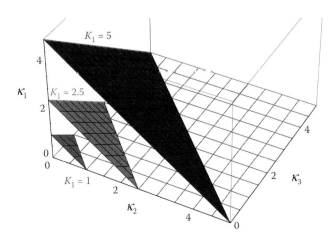

Figure 14.4 Relationship among the first stepwise microscopic constants κ_1, κ_2, and κ_3 for a three-site model. Given a value for the stepwise macroscopic binding constant K_1, the microscopic constants are restricted to a region of the plane defined by Equation 14.13, which can be rewritten as $\kappa_1 = K_1, -\kappa_2 - \kappa_3$. This plane intersects each of the κ_i axes at the value $\kappa_i = K_1$. All positive combinations of κ_1, κ_2, and κ_3 that are in the plane are consistent with a particular value of K_1. The three planes drawn are for K_1 values of 1, 2, and 4, from bottom to top.

This expression can be simplified to the following expression

$$K_3 = \left(\frac{1}{\kappa_3^{(12)}} + \frac{1}{\kappa_2^{(13)}} + \frac{1}{\kappa_1^{(23)}} \right)^{-1} \tag{14.17}$$

by taking advantage of the redundancy of the stepwise microscopic constants.

Generating *P* using overall microscopic constants

Using the overall microscopic constants, the binding polynomial can be written as

$$
\begin{aligned}
P &= \frac{[ooo]}{[ooo]} + \frac{[xoo]}{[ooo]} + \frac{[oxo]}{[ooo]} + \frac{[oox]}{[ooo]} + \frac{[xxo]}{[ooo]} + \frac{[xox]}{[ooo]} + \frac{[oxx]}{[ooo]} + \frac{[xxx]}{[ooo]} \\
&= 1 + Б_1[x] + Б_2[x] + Б_3[x] + Б_{12}[x]^2 + Б_{13}[x]^2 + Б_{23}[x]^2 + Б_{123}[x]^3 \\
&= 1 + (Б_1 + Б_2 + Б_3)[x] + (Б_{12} + Б_{13} + Б_{23})[x] + Б_{123}[x]^3 \\
&= 1 + \beta_1[x] + \beta_2[x] + \beta_3[x]^3
\end{aligned} \tag{14.18}
$$

The last two lines give relationships between the macroscopic and microscopic overall constants:

$$
\begin{aligned}
\beta_1 &= Б_1 + Б_2 + Б_3 \\
\beta_2 &= Б_{12} + Б_{13} + Б_{23} \\
\beta_3 &= Б_{123}
\end{aligned} \tag{14.19}
$$

Comparison with Equations 14.14 through 14.17 with highlights the relative simplicity of the overall representation of the general binding model.

SIMPLIFICATIONS TO MICROSCOPIC BINDING MODELS

Although the microscopic approach has the advantage of focusing on fundamental steps in binding, the derivations above demonstrate that regardless of the type of microscopic constant, there is usually too much detail in general models for microscopic binding to connect with experiment (i.e., there are too many constants). Remember that by "general model," we mean that every binding site is different and has unique energetic couplings to other sites. To maintain some of the mechanistic insight of the microscopic representation within an analytic framework that is tractable, we will simplify microscopic binding models.

One type of simplification would be that sites are **identical**, or that there are only a few classes of sites (fewer than *s*). A further simplification is that that there is only one type of coupling between sites. We can go even further and assume sites are **independent** (no coupling energies). These assumptions greatly simplify formulas for the binding polynomial and extent of binding, yet result in greater insight than the macroscopic approach. We will develop a few such models, starting with the simplest possible microscopic model: sites are both identical and independent. For this model, much of the complexity associated with the *stepwise* microscopic representation (e.g., Equations 14.16 through 14.18) is eliminated. Thus, because of their intuitive connections to mechanism and thermodynamics, the stepwise constants and will used to represent these simplified microscopic models.

Binding to three identical, independent sites

The most extreme simplification to a microscopic model for ligand binding is that all sites are identical *and* independent. For comparative purposes, we will return to

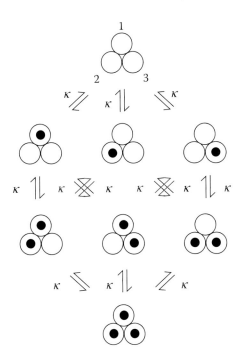

Figure 14.5 Microscopic binding scheme for a macromolecule with three binding sites, where each site is treated as identical and independent. When all sites are identical, the subscripts in the general scheme (Figure 14.1) can be omitted, e.g., $\kappa_1 = \kappa_2 = \kappa_3 = \kappa$. When all sites are independent (i.e., binding affinity at a given site does not depend on occupancy at the other sites), the superscripts used in the general scheme can safely be omitted, e.g., $\kappa_1 = \kappa_1^{(2)} = \kappa_1^{(23)} = \kappa$.

our three-site model (Figure 14.1), but unlike the general model above, this simplification allows us to generalize expressions for P, κ, and particularly, f_{bound}, to accommodate binding of ligand to an arbitrary number of sites quite easily.

If all sites independent and identical, we only need one microscopic constant, κ, to describe all microscopic binding events, regardless of ligation level (**Figure 14.5**).

Again, considering its general thermodynamic meaning, P should be independent of whether or not we label each site, and whether we represent binding with a macroscopic or microscopic description. We will use this generality to derive relationships between macroscopic and microscopic binding constants. Starting with the polynomial for the general three-site model with stepwise microstate constants (Equation 14.12), we can simply erase the subscripts and superscripts on the microscopic constants to get a representation for the independent, identical approximation shown in Figure 14.5:

$$
\begin{aligned}
P &= \frac{[ooo]}{[ooo]} + \frac{[xoo]}{[ooo]} + \frac{[oxo]}{[ooo]} + \frac{[oox]}{[ooo]} + \frac{[xxo]}{[ooo]} + \frac{[xox]}{[ooo]} + \frac{[oxx]}{[ooo]} + \frac{[xxx]}{[ooo]} \\
&= 1 + \kappa x + \kappa x + \kappa x + \kappa\kappa[x]^2 + \kappa\kappa[x]^2 + \kappa\kappa[x]^2 + \kappa\kappa\kappa[x]^3 \\
&= 1 + 3\kappa[x] + 3\kappa^2[x]^2 + \kappa^3[x]^3 \\
&= 1 + K_1[x] + K_1 K_2[x]^2 + K_1 K_2 K_3[x]^3 \\
&= 1 + \beta_1[x] + \beta_2[x]^2 + \beta_3[x]^3
\end{aligned}
\tag{14.20}
$$

The third line is considerably simpler than the corresponding expression in Equation 14.12 (which was the motivation for simplifying the microscopic model). Comparing the third line of Equation 14.20 with the analogous expression using stepwise macroscopic constants (line 4) gives a set of equations analogous to Equations 14.14 through 14.18, but with considerably simpler form:

$$
K_1 = 3\kappa
\tag{14.21}
$$

$$
K_2 = \kappa
\tag{14.22}
$$

$$
K_3 = \kappa/3
\tag{14.23}
$$

These three equations underscore a point made in Chapter 13 that even though sites have equal affinity, and as will be shown below, binding is noncooperative, the macroscopic stepwise binding constants are not equal. The first macroscopic constant is larger than the microscopic stepwise constant, and the last is smaller. This variation is *not* an expression of negative cooperativity; rather, it results from the statistics of ligand arrangements for different ligation levels. For the first step in binding (Mx from M_0), there are three ways that ligands can be arranged in the product state (*xoo*, *oxo*, and *oox*), but only one arrangement for the reactant state (*ooo*). Thus, the macroscopic stepwise constant K_1 is three times as large as the microscopic constant for single site binding. Conversely, at the last step in binding there is only a single product state (*xxx*) but three reactant states (*xxo*, *xox*, and *oxx*), so K_3 is 1/3 the value of κ. Using the same types of arguments you should be able to rationalize the relationship $K_2 = \kappa$ (Equation 14.22; see Problem 14.8). These relationships highlight the fundamental nature of microscopic binding constants: simple noncooperative binding is described by a single microscopic constant κ, but requires a collection of different values for macroscopic constants (e.g., K_1, K_2, K_3 for the three-site mechanism).

Relationships between the overall macroscopic constants and the microscopic stepwise constant κ can be obtained comparing lines three and five of Equation 14.20:

$$\beta_1 = 3\kappa \tag{14.24}$$

$$\beta_2 = 3\kappa^2 \tag{14.25}$$

$$\beta_3 = \kappa^3 \tag{14.26}$$

Again, β_i values differ for different ligation levels, reflecting both the number of ways to arrange the i ligands over the three sites, and the dimensional variation among the overall constants. Dimensional differences also affect the values of overall microscopic constants (given by b_1, b_2, and b_3 for the three-site model; see Problem 14.9).

Saturation curves for three independent identical sites

Given the generality of the binding polynomial, the fractional saturation and extent of binding can be calculated as before, by taking the derivative of the binding polynomial with respect to ligand concentration, and multiplying by x/P. This approach is not affected by the use of microscopic constants to represent affinities, other than to give a different (more fundamental) mechanistic representation. For the mechanism with three identical independent binding sites, the fractional saturation is given by

$$
\begin{aligned}
f_{bound} &= \frac{[x]}{3P}\frac{dP}{d[x]} \\
&= \frac{[x](3\kappa + 6\kappa^2[x] + 3\kappa^3[x]^2)}{3(1 + 3\kappa[x] + 3\kappa^2[x]^2 + \kappa^3[x]^3)} \\
&= \frac{\kappa[x] + 2\kappa^2[x]^2 + \kappa^3[x]^3}{1 + 3\kappa[x] + 3\kappa^2[x]^2 + \kappa^3[x]^3} \\
&= \frac{\kappa[x](1 + \kappa[x])^2}{(1 + \kappa[x])^3} \\
&= \frac{\kappa[x]}{1 + \kappa[x]}
\end{aligned} \tag{14.27}
$$

Notice that the fractional saturation function is a simple rectangular hyperbola, and is equivalent to the fractional saturation for the single-site scheme. This result makes intuitive sense, given the starting model that sites are independent and are

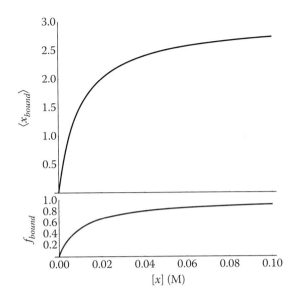

Figure 14.6 Ligand binding to a macromolecule with three identical independent sites. Each site has the same microscopic stepwise constant of $\kappa = 100$ M^{-1}. The top plot shows the average number of ligands bound per macromolecule, and the bottom plot shows the fractional saturation. The two curves have identical shapes, with a midpoint of 0.01 M, but they differ in their vertical scale: f_{bound} ranges from 0 to 1, whereas $\langle x_{bound} \rangle$ ranges from 0 to $s = 3$.

identical. Because the three sites are independent, we could (at least in principle) separate the three sites into three separate molecules, without any impact on the binding sites.[†] Clearly these separated sites would display simple rectangular hyperbolic binding curves.

For macromolecules that bind multiple ligands, it is sometimes useful to plot the average number of ligands bound per multi-site macromolecule ($\langle x_{bound} \rangle$) rather than the fractional saturation (f_{bound}). Both curves have the same shape (**Figure 14.6**), but the $\langle x_{bound} \rangle$ curve ranges from zero to s (three in this case), rather than zero to one.[‡] The half-saturation point occurs at $[x]_m = \kappa^{-1}$ (0.01 M in Figure 14.6), again highlighting the fundamental nature of the stepwise microscopic constant in this scheme.

BINDING TO S IDENTICAL, INDEPENDENT SITES

Although the independent, identical microscopic model has a rather featureless saturation curve, we will use this scheme as a building block for complicated binding schemes. For example, we will use identical independent site groupings to model macromolecules with multiple classes of sites, macromolecules that bind more than one type of ligand, macromolecules that show coupling between sites, and macromolecules that display allosteric ligand-induced conformational changes. But to do this, we need to generalize this model to macromolecules that can bind an arbitrary number of ligands.

As a starting place, we return to the binding polynomial for the three-site independent identical model (Equation 14.20). The coefficients multiplying powers of κx are binomial coefficients for $n = 3$, that is, 1:3:3:1. These coefficients reflect the number of ways to arrange the bound ligands (0, 1, 2, or 3 of them, respectively) over three sites (see Figure 14.1).[§] Thus, the binding polynomial for three independent identical sites can be written as

---

[†] Although in practice, such "separation" is probably not possible. Cutting macromolecules into pieces would likely perturb the overall structure of the macromolecule, and thus, the reactivity of the binding sites.

[‡] That is, $\langle x_{bound} \rangle = 3f_{bound}$.

[§] In our derivation, these coefficients came about because we first treated different arrangements of ligands as unique; when we simplified our model by treating each site as identical, we continued to count the contribution of each ligand arrangement (even though they cannot formally be distinguished for symmetric molecules).

$$P = \sum_{i=0}^{3} \frac{3!}{i!(3-i)!}(\kappa[x])^i$$

$$= (1 + \kappa[x])^3 \tag{14.28}$$

The second line of Equation 14.28, which results from the binomial theorem, is of particular importance: if sites are independent, the binding polynomial can be represented as a product of sub-polynomials[†] $(1 + \kappa[x])$ for each identical site (three in the example above). You should convince yourself that if you plug this factored form of the binding polynomial into Equation 14.27, you get the same expression for the fractional saturation (but with considerably less arithmetic; Problem 14.11).

At this point, it is fairly straightforward to generalize the relationships above for a macromolecule with three independent and identical binding sites to one with s sites. At each state of ligation, if there are i ligands bound, the number of ways to arrange the i ligands over s sites is given by the binomial coefficient

$$W(i;s) = \binom{s}{i} = \frac{s!}{(s-i)! \times i!} \tag{14.29}$$

These statistical factors combine with the microscopic binding constants to give the binding polynomial

$$P = \sum_{i=0}^{s} \frac{s!}{i!(s-i)!}(\kappa[x])^i$$

$$= (1 + \kappa[x])^s \tag{14.30}$$

As for the three-site independent identical model, P factors into a product of single-site binding polynomials $(1 + \kappa[x])$, in this case there are s of them in the product. The dependence of this simple form of the binding polynomial on ligand concentration is investigated in Problem 14.12.

The statistical representation of ligand binding arrangements to s sites (Equation 14.29) can be used to relate macroscopic binding constants to the microscopic constant κ. The relationship is given by using $W(i; s)$ to count the number of ligand arrangements in the product and the reactant. For both stepwise and overall binding macroscopic constants, the product is Mx_i, giving an arrangement number $W(i; s)$ in the numerator (Equations 14.31 and 14.32). For overall binding, the reactant is the macromolecule with zero ligands bound (M_0), and there is only one way to achieve this. As a result,

$$\beta_i = \frac{W(i;s)}{W(0;s)}\kappa^i = \frac{s!}{(s-i)! \times i!}\kappa^i \tag{14.31}$$

The binding constant is raised to the ith power because there are i binding events connecting M_0 to Mx_i. For stepwise binding, there is only a single power of the binding constant, because the reaction involves only one ligand binding step. However, the reactant (Mx_{i-1}) generally has multiple ligand arrangements, so the corresponding statistical factor is retained in the denominator:

$$K_i = \frac{W(i;s)}{W(i-1;s)}\kappa = \frac{s-i+1}{i}\kappa \tag{14.32}$$

[†] Each sub-polynomials are first-order ligand concentration; their product gives higher order terms.

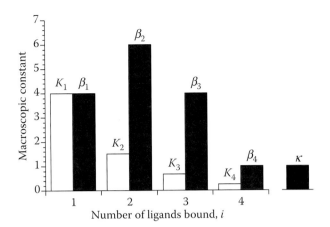

Figure 14.7 Variation in macroscopic stepwise (K_i) and overall (β_i) constants for four independent, identical sites. Values were calculated using a stepwise microscopic constant $\kappa = 1$ M^{-1} (black bar on the right), to avoid variations in the overall constants (β_i) from the different powers κ^i (Equation 14.31).

The simplification on the right-hand side of Equation 14.32 results from expanding the factorials (Problem 14.14). **Figure 14.7** shows how these statistical factors influence various overall binding constants for $s = 4$ sites, using a value of $\kappa = 1$ M^{-1} to sidestep dimensional variations in β_i that result from powers of κ.

Finally, as above, the fractional saturation and extent of binding remain simple rectangular hyperbolas:

$$f_{bound} = \frac{\langle x_{bound} \rangle}{s} = \frac{\kappa[x]}{1 + \kappa[x]} \tag{14.33}$$

This generalization (by differentiation) is left as a problem (Problem 14.10). Again, the saturation function has the same form as it would if the s sites were on separate molecules, emphasizing site independence—the fact that sites are on the same macromolecule does not affect their reactivity.

BINDING TO TWO CLASSES OF INDEPENDENT SITES

We will next consider a model in which there are two separate classes of binding sites for the same type of ligand x, numbering s_1 and s_2. Within each class, we will treat all sites as identical, with stepwise microscopic binding constants κ_1 and κ_2. We will treat all sites as independent. For each class of sites, we can write a sub-polynomial for binding, ignoring sites of the other class. Because sites within each class are identical, these sub-polynomials take the simple form

$$P_1 = (1 + \kappa_1[x])^{s_1} \tag{14.34}$$

$$P_2 = (1 + \kappa_2[x])^{s_2} \tag{14.35}$$

The full polynomial can be built simply by multiplying these two sub-polynomials:

$$P = P_1 \times P_2 = (1 + \kappa_1[x])^{s_1} \times (1 + \kappa_2[x])^{s_2} \tag{14.36}$$

This works because binding at the two classes of sites are independent of one another. Since partition functions are built from terms that represent the probabilities of each species, forming this product is equivalent to "and" combinations of probabilities for independent events. Based on this logic, it should be easy to see that for an arbitrary number of classes of independent sites (w), the overall binding polynomial can be built from the product of w sub-polynomials:

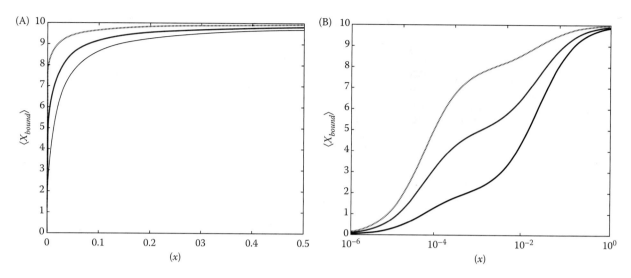

Figure 14.8 Binding to a macromolecule with two classes of independent sites. Within each class, sites have identical affinities, represented by stepwise microscopic constants $\kappa_1 = 50$ M^{-1} and $\kappa_2 = 20,000$ M^{-1}. All three curves have the same number of total sites ($s_1 + s_2 = 10$). The black dotted curve (left) has $s_1 = 2$, $s_2 = 8$; the red solid curve has $s_1 = 5$, $s_2 = 5$; the black solid curve (right) has $s_1 = 8$, $s_2 = 2$. The axis on the right shows the fraction of the total ($s_1 + s_2$) sites bound (see Problem 14.15). In linear scale (A), the high affinity binding phase is compressed against the y-axis. The logarithmic scale (B) allows both the high (κ_2) and low (κ_1) affinity phases to be resolved.

$$P = P_1 \times P_2 \times \cdots \times P_w = (1 + \kappa_1[x])^{s_1} \times (1 + \kappa_2[x])^{s_2} \times \cdots \times (1 + \kappa_w[x])^{s_w} \tag{14.37}$$

Returning to the model with two classes of sites, the extent of binding[†] can be obtained by differentiation with respect to $[x]$:

$$
\begin{aligned}
\langle x_{bound} \rangle &= (s_1 + s_2) f_{bound} = \frac{[x]}{P} \frac{dP}{d[x]} \\
&= \frac{[x]}{(1 + \kappa_1[x])^{s_1} (1 + \kappa_2[x])^{s_2}} \frac{d\{(1 + \kappa_1[x])^{s_1} (1 + \kappa_2[x])^{s_2}\}}{d[x]} \\
&= \frac{[x]\{s_1 \kappa_1 (1 + \kappa_1[x])^{s_1-1}(1 + \kappa_2[x])^{s_2} + s_2 \kappa_2 (1 + \kappa_1[x])^{s_1}(1 + \kappa_2[x])^{s_2-1}\}}{(1 + \kappa_1[x])^{s_1}(1 + \kappa_2[x])^{s_2}} \\
&= s_1 \frac{\kappa_1[x]}{1 + \kappa_1[x]} + s_2 \frac{\kappa_2[x]}{1 + \kappa_2[x]}
\end{aligned}
\tag{14.38}
$$

In other words, the extent of binding curve is given by the sum of two simple rectangular hyperbolic binding curves, each with saturation given by the corresponding microscopic binding constant (κ_1 and κ_2). The relative contributions of each of the two hyperbolic curves are given by the number of sites in each class (s_1 and s_2, **Figure 14.8**).

Following the logic in the previous section, the shape of the binding curve for two classes of binding independent sites can easily be rationalized. Since sites are independent, we could again separate the two classes of binding sites onto distinct macromolecules, at least in principle. Separated in this way, each class of macromolecule (e.g., type 1) would saturate following a rectangular hyperbola of the form of Equation 14.33. The binding curve would clearly be the sum of simple binding curves from each type of macromolecule, weighted by the relative number of sites (s_1 and s_2 in this example) in each type of macromolecule.

[†] Here the extent of biding is somewhat tidier and more intuitive than the fractional saturation, because the numbers of sites in the two classes are likely to be different ($s_1 \neq s_2$).

BINDING TO IDENTICAL COUPLED SITES

In the section above, we adapted the identical independent binding model to treat schemes in which sites are independent but have different affinities. Here we take our simplification in the other direction, developing models in which all sites have the same intrinsic affinities, but are influenced by occupancy at other sites. Compared to models in which sites are independent, models where site occupancy is coupled often include structural relationships among the sites. This is because coupling necessarily involves pairwise (or higher order) interactions among sites, whereas variation in intrinsic affinity involves interactions that are more closely localized to each binding site.

Even when affinities are treated as identical, models that include coupling between sites can become quite complicated. Some of this complexity can be avoided when the macromolecular binding sites have symmetrical relationships with each other. For example, discrete assemblies of protein subunits often possess cyclic (i.e., rotational) symmetry (**Figure 14.9A**) or tetrahedral symmetry (**Figure 14.9B**). Likewise, linear polymers such as nucleic acids and filamentous protein assemblies often possess translational symmetry in one dimension (**Figure 14.9C, D**).

To calculate fractional saturation of coupled ligand binding sites, we will use the same approach as for uncoupled sites, that is, we will build a binding polynomial

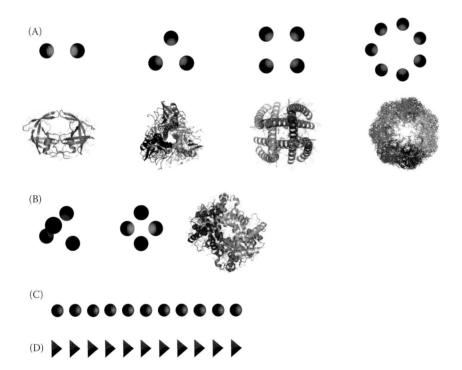

Figure 14.9 Some geometric representations of symmetrical oligomeric proteins.
(A) Geometric figures showing rotational or "cyclic" symmetries with a twofold axis (C_2, loft), threefold axis, fourfold axis, and sevenfold axis (C_7, right). Ribbon diagrams show proteins with cyclic symmetry, including HIV protease (C_2, pdb code 3hvp), influenza hemagglutinin (C_3, pdb code 2hmg), the Kcs-A transmembrane potassium channel (C_4, pdb code 1k4c), and GroEL/ES (C_7, pdb code 1aon). Note that although the simple geometric figures also show dihedral (mirror) symmetry, this is prevented by the asymmetric structure of most macromolecules.
(B) Tetrahedral symmetry (with to different views), along with a ribbon diagram of human hemoglobin (with approximate tetrahedral symmetry, (pdb code 2dn1). (C) One-dimensional translational lattice symmetries consistent with symmetric homopolymers such as double-stranded DNA. (D) One-dimensional translational lattice symmetries consistent with asymmetric homopolymers, such as single strand nucleic acids and cytoskeletal filaments (microtubules, filamentous actin). For more examples of symmetry in macromolecules and its consequences, see Goodsel and Olson (2000) and Marsh and Teichmann (2015).

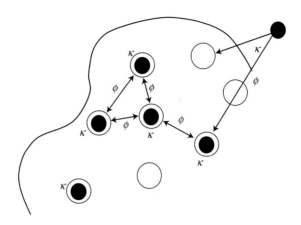

Figure 14.10 A general microscopic scheme for ligand binding. Ligands bind to sites with an intrinsic affinity κ (which applies to each bound ligand). Pairs of ligands couple with each other with an interaction constant ϕ in a way that depends on the structure of the macromolecule, and on the details of the model used to describe ligand binding. In the microstate shown, four of the five bound ligands are coupled to at least one other ligand, whereas the bound ligand on the lower left is not coupled. The Boltzmann factor for this $i = 5$ microstate is $\phi^4(\kappa x)^5$. Binding of the sixth ligand (upper right) is shown to make a single coupling interaction, which would contribute an additional multiple of $\phi\kappa x$ to the Boltzmann factor.

that sums the statistical weights of all the binding microstates. Each term will represent the intrinsic contribution to microstate binding as a product of i binding constants (κ^i) and i powers of ligand concentration ($[x]^i$). In addition, we must include coupling terms for each microstate, which will be represented using the symbol ϕ (**Figure 14.10**).[†] Like the intrinsic binding part of the statistical weight for each microstate ($\kappa^i[x]^i$), the coupling part will be given by ϕ^j, where j is the number of coupling interactions between ligands in a given microstate. The challenges in developing coupling models are to determine how many coupling terms to include for each microstate (i.e., what is j), and relatedly, how many microstates have a particular number of ligands and coupling terms (so that they can be combined with a corresponding statistical factor W).[‡] Before considering any specific geometric models, we will begin with two limiting models for coupling that lack structural detail. These "structure-blind" approaches have the advantage of generality, although they do a poor job of approximating the binding reactions of larger systems.

A simple model for identical coupling: One interaction per ligand

One analytically simple way to represent coupling among the binding sites of a multivalent macromolecule is to include a single interaction term (ϕ) between each ligand bound and the collection of other ligands. This model will be referred to as the OPL model (**o**ne interaction **p**er **l**igand). Because at a minimum, two bound ligands would be required for coupling, there is no coupling in the unligated and singly ligated states (M_0 and Mx_1). When the second ligand binds, a single ligand interaction forms, contributing a single power of ϕ to its Boltzmann factor. When a third ligand binds, it is modeled as making a single interaction with the other two, thus contributing another power of ϕ (for a total of ϕ^2) to Boltzmann factors with three or more ligands. In general, for i ligands bound, there will be $i - 1$ coupling terms contributed in the OPL model. This leads to a simple form for the OPL binding polynomial:

[†] Different types of couplings can be modeled using different coupling constants, that is, φ_1, φ_2, ... Note that the log will be related to the free energy of coupling, just as the log of an intrinsic stepwise constant is related to the free energy of binding to a given site.

[‡] Counting states here is similar to considerations that led to binomial coefficients in the independent identical site model, but can be complicated by structural relationships among binding sites.

$$P = 1 + W_1\kappa[x] + W_2\phi(\kappa[x])^2 + W_3\phi^2(\kappa[x])^3 + \cdots + W_s\phi^{s-1}(\kappa[x])^s \qquad (14.39)$$

The W_i terms in Equation 14.39 are statistical factors that describe the number of ways to arrange i ligands among s sites. Due to the lack of structural detail in the OPL model, the W_i terms are simply binomial coefficients:

$$P = 1 + \binom{s}{1}\kappa[x] + \binom{s}{2}\phi\kappa^2[x]^2 + \cdots + \binom{s}{s}\phi^{s-1}\kappa^s[x]^s \qquad (14.40)$$

Using binomial coefficients allows P to be represented in closed form, although slightly more manipulation is required to do so. Because there is one fewer power of ϕ than there is of κ and $[x]$ in each term of P (Equation 14.40), the binding polynomial cannot be collapsed directly into a simple product of linear $(1 + \phi\kappa[x])$ terms. However, this power mismatch between ϕ and $\kappa[x]$ can be fixed by multiplying and dividing the right-hand side of Equation 14.40 by a single power of ϕ, and by adding and subtracting $1/\phi$ terms. This rearrangement (Problem 14.16) leads to a closed form representation of the OPL binding polynomial:

$$P = 1 - \frac{1}{\phi} + \frac{(1+\phi\kappa[x])^s}{\phi} \qquad (14.41)$$

The fractional saturation is calculated by differentiation (Problem 14.16):

$$f_{bound} = \frac{[x]}{sP}\frac{dP}{d[x]} = \frac{\phi\kappa[x](1+\phi\kappa[x])^{s-1}}{\phi - 1 + (1+\phi\kappa[x])^s} \qquad (14.42)$$

Fractional saturation curves are shown for $s = 3$ sites with a range of ϕ values in **Figure 14.11**. With favorable coupling between sites ($\phi > 1$), the OPL binding curve shows sigmoidal behavior, reflecting positive cooperativity. This can be understood by recognizing that the first two terms in the denominator of Equation 14.42 (which together are greater than zero) suppress saturation at low ligand concentrations, but become overwhelmed at high ligand concentration. When there

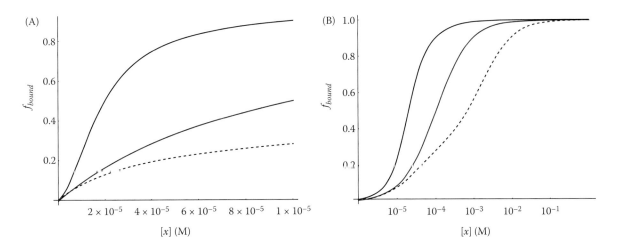

Figure 14.11 Fractional saturation for a model in which bound ligands make one interaction. Three identical sites are treated as having an intrinsic binding constant of $\kappa = 10^4\,\mathrm{M^{-1}}$, and having a single coupling (the OPL model). The black solid curve has a coupling constant of $\phi = 10$. The red curve has $\phi = 1$ (a noncooperative reference). The black dashed curve has a coupling constant of $\phi = 0.1$. Increasing the degree of coupling among sites enhances both positive and negative cooperativity. (A) Linear ligand concentration scale (note, the highest ligand concentration scale shown is $10^{-4}\,\mathrm{M}$). (B) Logarithmic ligand concentration scale, extended to one molar ligand.

is no coupling between sites ($\phi = 1$), a rectangular hyperbola is recovered with a midpoint of κ^{-1}.

With unfavorable coupling between sites ($\phi < 1$), the binding curve is spread out, reflecting negative cooperativity. At the extreme limit of unfavorable coupling ($\phi = 0$), both the numerator and denominator go to zero, and Equation 14.42 takes on an indeterminate form. However, by application of L'Hopital's rule (Problem 14.17), it can be shown that f_{bound} recovers a rectangular hyperbolic form ($s^{-1}\lambda[x]/(1 + \lambda[x])$, where $\lambda = s\kappa$), but with increased apparent affinity (with a "midpoint" of is $s^{-1}\kappa^{-1}$), and a decreased amplitude (with f_{bound} ranging from 0 to s^{-1}). The apparent affinity enhancement comes from the fact that there are s "first" sites to which the ligand can bind. The decreased amplitude comes from the restriction that after the first ligand binds, subsequent sites cannot be bound because of their extreme repulsion.

The strength of this model is that it leads to expressions for P and f_{bound} that show complex behavior (positive and negative cooperativity) yet are simple enough to understand at an intuitive level. The weakness of this model is that it is unlikely that each bound ligand only makes one other coupling interaction with a constellation of bound ligands (see the graphs in Problem 14.18). An alternative model (which also has the advantage of approximately binomial statistics) is developed in the next section.

A more involved model for identical coupling: Interactions among all sites

At the opposite extreme from the scheme in which each bound ligand couples to one other bound ligand, we can build a model where each bound ligand couples with *all* other bound ligands (we will refer to this as the "AS" model, for "**all sites**"). Like the polynomial for the OPL model, the binding polynomial for the AS model is a sum of statistical factors (W_i), intrinsic binding factors ($\kappa^i[x]^i$), and coupling factors ϕ. The first three terms of the AP polynomial (corresponding to zero, one, and two ligands bound) are identical to those of the OPL model.[†] However, higher powers of the coupling term ϕ are generated when more than two ligands are bound. With three ligands bound, there are now three pairwise couplings (ϕ^3), with four ligands bound there are six (ϕ^6), with five there are ten (ϕ^{10}, **Table 14.3**).

Table 14.3 The number of pairwise couplings for different ligation states in a macromolecule with five sites

# ligands bound (i)	1	2	3	4	5
Macrostate	MX_1	MX_2	MX_3	MX_4	MX_5
Number of pairwise couplings	0	.	3	6	10
Coupling contribution	1	ϕ^1	ϕ^3	ϕ^6	ϕ^{10}
Number of configurations	$\binom{5}{5}=5$	$\binom{5}{2}=10$	$\binom{5}{3}=10$	$\binom{5}{4}=5$	$\binom{5}{5}=1$
Graph representation of bound ligands[a]	•				

[a]Note that these graphs only depict the i bound sites (red vertices) and their couplings (edges), but omit the remaining ($s - i$) unbound sites.

[†] These three ligation states have zero, zero, and one coupling term, respectively.

The relationship between the number of bound ligands and the number of pairwise couplings in the AS model can be nicely illustrated using some simple ideas from graph theory.[†] Bound ligands are depicted as vertices of a "planar graph," and couplings are depicted as edges (Table 14.3). These graphs are similar to the parallelotopes described above, although their essential concepts can be represented in two dimensions (thus, these graphs are referred to as "planar"). Other differences are that vertices of the parallelotopes correspond to individual binding microstates, rather than individual bound ligands, and the edges of the parallelotopes depict adding a ligand, rather than interactions between ligands.

Each vertex of these graphs is connected to all the other vertices (this is the main assumption of the AS model). In graph theory, such fully connected graphs are referred to as "complete graphs" (every possible pairwise edge is drawn). To get a general formula for the number of couplings (c) when there are i ligands bound, we just need to count the number of edges in the complete graph with i vertices.[‡] There are two rather different-looking formulas that give this count:

$$c = \sum_{j=1}^{i-1} j \tag{14.43}$$

and

$$c = \frac{i(i-1)}{2} \tag{14.44}$$

These two equations are derived in Problems 14.18 and 14.19. The advantage of Equation 14.44 is that it allows the binding polynomial to be expressed in a simpler form than would be obtained with the summation for c:

$$P = 1 + \binom{s}{1}\kappa[x] + \binom{s}{2}\phi\kappa^2[x]^2 + \binom{s}{3}\phi^3\kappa^3[x]^3 + \cdots + \binom{s}{s}\phi^{s(s-1)/2}\kappa^s[x]^s$$

$$= 1 + \sum_{i=1}^{s} \binom{s}{s}\phi^{i(i-1)/2}\kappa^i[x]^i \tag{14.45}$$

Because ϕ is raised to different powers than $[x]$ and κ in Equation 14.45, and because this difference in powers is not constant from term to term (in contrast to the OPL model above, Equation 14.39), the binomial distribution cannot be used to get P into a closed form. Fractional saturation can still be calculated by differentiating P with respect to $[x]$, but it must be done term-by-term (Problem 14.22). **Figure 14.12** shows the fractional saturation of a three-site macromolecule, in comparison to the OPL model.

As with the OPL model, ϕ values greater than one lead to positive cooperativity for the AS model (steep, left-shifted binding curves), whereas ϕ values less than one lead to negative cooperativity (broad, right-shifted binding curves). However, close comparison of the two models (Problems 14.27 through 14.29) show that for given values of s and κ, the AS model shows greater cooperativity (both positive and negative) than the OPL model. This increase is a result of the larger number of couplings that contribute to saturated states.

[†] Here, the term "graph" is a mathematical object representing the connectivity between a set of points, rather than the x–y plots seen throughout this text. For an accessible description of graph theory, see "*Introduction to Graph Theory*" by Richard Trudeau.

[‡] Complete graphs with i vertices are often given the name "K_i."

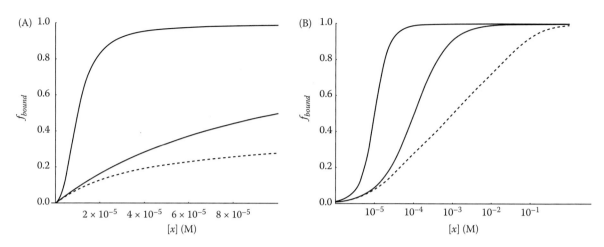

Figure 14.12 Fractional saturation for a model in which all sites are coupled. Three identical sites are treated as having an intrinsic binding constant of $\kappa = 10^4\,M^{-1}$, and having coupling between all bound ligands (the AS model). The black solid curve has a coupling constant of $\phi = 10$. The red curve has $\phi = 1$ (a noncooperative reference). The black dashed curve has a coupling constant of $\phi = 0.1$. Increasing the degree of coupling among sites enhances both positive and negative cooperativity. (A) Linear and (B) logarithmic ligand concentration scales, with the same limits as in Figure 14.11.

EXPLICIT STRUCTURAL MODELS FOR COUPLING AMONG BINDING SITES

The two coupling models above illustrate how general features of coupling between sites can be built into a partition function, and how such features influence binding curves. However, the lack of structural detail limits their applicability to specific macromolecular binding reactions. In this section, we will develop coupling models that are specific to particular geometric arrangements of sites, such as those that are shown in Figure 14.9. The advantage of this approach is that structural features of any given macromolecular binding reaction can be modeled (as long as these structural features are either known or imagined). A disadvantage is that the underlying equations (e.g., the binding polynomial, extent of binding) do not typically have closed-form solutions, but rather are built term-by-term.

As diagrammed in Figure 14.9, we will represent structural information using network diagrams from graph theory. Each binding site will be represented by a vertex, and thermodynamic interactions between sites will be represented as edges connecting the nodes (note that this differs from the graphs used to depict coupling in the AS model, where vertices represent the subset of sites that are bound with ligands). As with previous models, each binding site will have an intrinsic stepwise binding constant κ (defined as the binding affinity when all other sites are unoccupied). The simplest case is when all subunits are identical, and thus have the same intrinsic binding constants, although different types of subunits and binding constants can easily be accommodated.

An example of rotational symmetry: Binding to a hexagon

The primary feature that distinguishes different structural models (aside from the number of subunits present) is the pattern of pairwise coupling terms among the binding sites. Examples of different coupling patterns in a hexagonal subunit arrangement are given in **Figure 14.13**. Coupling can be modeled exclusively between adjacent subunits (a "nearest neighbor" model, Figure 14.13A), or it can involve interactions among more distant subunits. Local and nonlocal couplings can be modeled as the same, with a single coupling constant ϕ (Figure 14.8B), or can be given different values (Figure 14.8C).

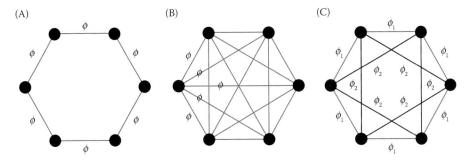

Figure 14.13 Structural models for coupled ligand binding in a six-site macromolecule. A hexagonal arrangement of ligand binding sites is depicted. Couplings are represented as edges connecting nodes (binding sites). (A) A model in which only nearest neighbors are coupled, with a single coupling constant ϕ. This model has some similarities to the to general models above (the OPL and all-sites models), except that here, bound ligands can have zero, one, or two couplings, depending on their arrangement. (B) A model where all subunits are coupled. This model is equivalent to the "all-the-sites" model above. For simplicity, only the couplings emanating from a single site are labeled, although all 15 are implied. (C) A model where nearest- and second-nearest neighbors (black and red, respectively) are coupled, but have different coupling constants, ϕ_1 and ϕ_2, respectively.

The binding polynomials for these coupling models have a similar form to those in the previous section, that is, they are a sum of terms that depend on ligand concentrations. Each term is a product of the factor $(\kappa[x])^i$, representing the intrinsic contribution binding of i ligands, and the factor ϕ^j, representing the number of coupling interactions between bound ligands. In addition, each term contains a statistical factor $(W_{i,j})$ describing the number of arrangements with i ligands and j coupling terms. These three factors can be represented as follows:

$$P = \sum_{i=0}^{s} \sum_{j=j_{\min}}^{j_{\max}} W_{i,j} (\kappa[x])^i \phi^j \qquad (14.46)$$

Equation 14.46 looks simpler than it is. Although the index i (the number of bound ligands) is straightforward, the index j (the number of couplings) is not: the limits $j_{\min}$ and $j_{\max}$ both depend on i, but not in a way that can be captured in a simple analytical expression. When using Equation 14.46, care must be taken to identify all of the ligand arrangements, the number of couplings for each arrangement, and the number of ways each arrangement can be achieved. Two checks can be applied to make sure all possible states are included. First, the total number of microstates should add up to 2^s. Second, for each ligation level i, the total number of microstates should equal the binomial coefficient for that level.

A good way to make sure to count all of the terms that go into the binding polynomial is to make a table that represents the possible ligand arrangements. **Table 14.4** shows the set of terms for a hexamer model with nearest-neighbor couplings (Figure 14.13A).

The entries in Table 14.4 are divided based on the number of gaps between ligands to help illustrate the connection between ligand arrangement and coupling (note there are two ways to get two gaps). Gaps are stretches of one or more empty sites in between full sites (Problem 14.30). Counting gaps is useful because the number of couplings is related to the difference between the number of ligands and the number of gaps (Problem 14.31).

Notice that as advertised above, for each state of ligation, the total number of microstates sums to give the corresponding binomial coefficient (bottom row, Table 14.4). In this regard, the *total* statistics are no different than when we considered binding to independent identical sites. However, due to differences in the number of coupling interactions within each ligation level, each binomial coefficient is distributed over different terms (the column entries in Table 14.4).

Table 14.4 Terms in the binding polynomial for a nearest-neighbor hexagon

Gaps between ligands	Number of ligands, i						
	0	1	2	3	4	5	6
0	1						$1(\kappa[x])^6\phi^6$
1		$6(\kappa[x])$	$6(\kappa[x])^2\phi$	$6(\kappa[x])^3\phi^2$	$6(\kappa[x])^4\phi^3$	$6(\kappa[x])^5\phi^4$	
2			$6(\kappa[x])^2$	$6(\kappa[x])^3\phi$	$6(\kappa[x])^4\phi^2$		
2			$3(\kappa[x])^2$	$6(\kappa[x])^3\phi$	$3(\kappa[x])^4\phi^2$		
3				$2(\kappa[x])^3$			
$\binom{6}{i}$	1	6	15	20	15	6	1

Note: The bottom row gives a summation of the different configurations with i ligands bound. The rows separate these configurations by the number of stretches of one or more empty sites between bound ligands. For two, three, and four bound ligands, there are two distinct configurations that result in two gaps (for three bound ligands, counting these two gap configurations to be distinct assumes asymmetry across the hexamer plane, a condition that is likely to be met for macromolecular subunits). For the state with no ligands ($i = 0$), we assign the number of gaps to be zero, although this assignment is difficult to justify on structural grounds.

The binding polynomial is obtained (after some factoring) as the sum of all the terms in Table 14.4:

$$P = 1 + 6(\kappa[x]) + (\kappa[x])^2(6\phi + 9) + (\kappa[x])^3(6\phi^2 + 12\phi + 2)$$
$$+ (\kappa[x])^4(6\phi^3 + 9\phi^2) + 6(\kappa[x])^5\phi^4 + (\kappa[x])^6\phi^6 \qquad (14.47)$$

Although there is no simple closed-form solution for Equation 14.47, obtaining fractional saturation by differentiating the binding polynomial is quite straightforward. A plot of saturation as a function of ligand concentration with a coupling constant $\phi = 10$ displays considerable cooperativity (**Figure 14.14**). Even though there are a large number of intermediates that can be formed, based on combinatorics, they have relatively low populations (Figure 14.14B). This is partly because the average number of coupling constants (j) for each state of ligation,

$$\langle j \rangle_i = \frac{1}{\binom{6}{i}} \sum_{j=j_{min}}^{j_{max}} j \times W_{i,j} \qquad (14.48)$$

increases nonlinearly with the number of ligands bound (Figure 14.14C).

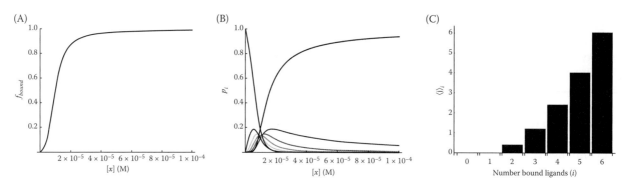

Figure 14.14 Ligand saturation, species populations, and degree of coupling for the nearest-neighbor hexagon model. Sites are assumed to be identical and to have an intrinsic binding constant of $\kappa = 10^4\,M^{-1}$. Nearest-neighbor coupling is represented using a single constant of $\phi = 10$. (A) Fractional saturation of the hexamer. (B) Species plots showing the population of macrostates with $i = 0$ to 6 bound ligands, calculated as the Boltzmann factor for each ligation state (each of the terms in Equation 14.47) divided by the binding polynomial (Equation 13.31). (C) The average number of coupling terms for each of the i ligation states (Equation 14.48).

ALLOSTERY IN LIGAND BINDING

So far, we have presented a general thermodynamic framework for how ligand binding at multiple sites can be energetically coupled. Both the microscopic (this chapter) and especially the macroscopic descriptions (Chapter 13) lack details on *how* these sites are coupled. This coupling is one of the most interesting aspects of macromolecular function. A description of this coupling would amount to a map (either mechanical or statistical) of how functional information flows between distant sites.

At the atomic level, direct forces between atoms and molecules tend to be short range. Thus, coupling among distant ligands must involve the intervening protein structure. Although it is possible that such coupling could involve forces along one (or a few) specific linked pathways between bonded (and nonbonded) groups, it would be very hard to evolve such mechanisms with high fidelity. A simpler way to link distant sites is through large-scale conformational change that affects the entire macromolecule (or large portions of it). Such changes can involve distinct folded structures, as has been described for cooperative oxygenation of hemoglobin, or can involve partial (or complete) unfolding transitions (see Motlagh et al. 2014).

Coupling of binding sites through macromolecular conformational change is often referred to as "allostery."[†] A critical component of allostery is that the affinities for ligands are different in different conformations. Thus, the overall ligand affinities change as conformation changes, and reciprocally, ligand binding induces conformational change. By having multiple ligand binding sites, the conformational changes linked to the early stages of the binding reaction can modify the affinity of later binding steps.

Here we will develop quantitative analytical models for this conformational linkage. We will begin with general *macrostate* approach. As we saw for ligand binding, although this macrostate approach to allostery is thermodynamically complete, it comes up short in terms of molecular insight, and in general, has too many parameters to be very useful. Thus, we will simplify our allosteric model, making assumptions that various conformations can either be omitted from the partition function or can be considered to be identical. Such assumptions are easiest to represent using a microstate approach.

A general (macrostate) allosteric model

Most generally, we will consider a macromolecule that can bind s ligands and can exist in unique conformational states $0, 1, 2, \ldots, n$. This scheme can be represented as an n by $s + 1$ matrix as follows:

$$
\begin{array}{ccccccc}
{}^{1}M_0 & \rightleftharpoons & {}^{1}Mx & \rightleftharpoons & {}^{1}Mx_2 & \rightleftharpoons & \cdots & {}^{1}Mx_s \\
\updownarrow & & \updownarrow & & \updownarrow & & & \updownarrow \\
{}^{2}M_0 & \rightleftharpoons & {}^{2}Mx & \rightleftharpoons & {}^{2}Mx_2 & \rightleftharpoons & \cdots & {}^{2}Mx_s \\
\updownarrow & & \updownarrow & & \updownarrow & & & \updownarrow \\
{}^{3}M_0 & \rightleftharpoons & {}^{3}Mx & \rightleftharpoons & {}^{3}Mx_2 & \rightleftharpoons & \cdots & {}^{3}Mx_s \\
\updownarrow & & \updownarrow & & \updownarrow & & & \updownarrow \\
\vdots & & \vdots & & \vdots & & & \vdots \\
{}^{n}M_0 & \rightleftharpoons & {}^{n}Mx & \rightleftharpoons & {}^{n}Mx_2 & \rightleftharpoons & \cdots & {}^{n}Mx_s
\end{array}
\tag{14.49}
$$

To derive an analytical form of the partition function with the least amount of work, we will use the overall binding constants ${}^{j}\beta_i$, where the left-superscript j denotes

[†] The term "allostery," which was first introduced by Monod and Jacob in the early 1960s, derives from the greek "allo-" meaning "other," and "stereo-," relating to solidness, three dimensionality , or structure.

the conformational state (from $j = 1$ to n), and as before, the subscript i reflects the number of ligands (x) bound (from 0 to s). In terms of concentrations of species,

$$^{j}\beta_i = \frac{\left[^{j}Mx_i\right]}{\left[^{j}M_0\right][x]^i}$$

(14.50)

We will take the unligated species in conformational state 1 (i.e., $^{1}M_0$) is taken as the starting point for each overall binding reaction, thus, $^{1}M_0$ serves as the reference state in the partition function:

$$\Pi = \sum_{j=1}^{n}\sum_{i=0}^{s} {}^{j}\beta_i[x]^i$$

(14.51)

Here, Π (Greek for P) is used to distinguish this partition function from the previous binding polynomials we have discussed, as Π sums over both conformational as well as ligation states. Equation 14.51 is nothing more than a sum of Boltzmann factors over all of the $s + 1$ ligand binding states in each of the n conformations.[†]

One way to simplify Equation 14.51 would be to express it as a sum of binding polynomials ^{j}P for each of the n conformational states.[‡] Although the "inside" (i.e., ligand-binding) sums in Equation 14.51 have the right form for such ^{j}P polynomials, they are referenced to the $^{1}M_0$ state rather than the $^{j}M_0$ state, that is,

$$\Pi = \sum_{j=1}^{n}\sum_{i=0}^{s} \frac{\left[^{j}M_i\right]}{\left[^{1}M_0\right]}$$

(14.52)

This mismatch can be corrected by moving the $^{1}M_0$ term outside the ligand-binding sum, and multiplying and dividing by $^{j}M_0$:

$$\Pi = \sum_{j=1}^{n} \frac{\left[^{j}M_0\right]}{\left[^{1}M_0\right]}\sum_{i=0}^{s} \frac{\left[^{j}M_i\right]}{\left[^{j}M_0\right]}$$
$$= \sum_{j=1}^{n} \frac{\left[^{j}M_0\right]}{\left[^{1}M_0\right]} \times {}^{j}P$$

(14.53)

The terms multiplying the ^{j}P binding polynomials can be regarded as conformational equilibrium constants for the n different unbound states, which we will represent using L:

$$^{j}L_0 = \frac{\left[^{j}M_0\right]}{\left[^{1}M_0\right]}$$

(14.54)

With this set of conformational constants, Π can be rewritten as

$$\Pi = \sum_{j=1}^{n} {}^{j}L_0 \times {}^{j}P$$

(14.55)

Note that the first allosteric constant, $^{1}L_0$, has a value of one (Equation 14.54). Thus, there are really only $n - 1$ L_0 values that need to be specified, but retaining the $^{1}L_0$ makes the summations such as Equation 14.55 easier to index.

[†] For a total of $ns + n$ terms.
[‡] Again, we are using a left-superscripted j to indicate the conformational state (in this case, the binding polynomial for the jth conformational state).

Before we derive the extent of binding from Π, some discussion of its form is in order. Equation 14.55 gives the total partition function Π as a sum of the binding polynomials of each of the n conformational states, weighted by the relative stability of each state.[†] The most stable conformational states make large contributions to Π, whereas the least stable conformational states make small contributions. One way to think of this is that each $^jL_0 \times {^jP}$ term is proportional to the relative probability of binding to each conformational state. Because the n conformational states are mutually exclusive, the relative probabilities add (see Chapter 1), consistent with the summation over conformational states in Equation 14.55.

The extent of ligand binding is calculated in the usual way, by differentiating the partition function with respect to ligand concentration. Starting with Equation 14.55, we can obtain a particularly intuitive form of the saturation function:

$$
\begin{aligned}
f_{bound} &= \frac{[x]}{s\Pi} \frac{d\Pi}{d[x]} \\[2mm]
&= \frac{[x]}{s\Pi} \frac{d}{d[x]} \left(\sum_{j=1}^{n} {^jL_0}\, {^jP} \right) \\[2mm]
&= \frac{[x]}{s\Pi} \sum_{j=1}^{n} {^jL_0}\, \frac{d\,{^jP}}{d[x]}
\end{aligned}
\tag{14.56}
$$

The third line in Equation 14.56 is valid because the conformational equilibrium constants $^jL_0$ are independent of ligand concentration. Each of the derivatives in the third line can be substituted with

$$
\frac{d\,{^jP}}{d[x]} = \frac{s \times {^jP} \times {^jf_{bound}}}{[x]}
\tag{14.57}
$$

which is a rearranged form of our standard equation for the fractional saturation of the jth conformational states (represented as $^jf_{bound}$; see Problem 14.29). Substituting into Equation 14.56 gives

$$
\begin{aligned}
f_{bound} &= \frac{[x]}{s\Pi} \sum_{j=1}^{n} {^jL_0}\, \frac{s \times {^jP} \times {^jf_{bound}}}{[x]} \\[2mm]
&= \sum_{j=1}^{n} \frac{{^jL_0} \times {^jP}}{\Pi}\, {^jf_{bound}}
\end{aligned}
\tag{14.58}
$$

The first term in the sum in the second line can be shown to be the fraction of M in the jth conformation $^j\alpha$ (Problem 14.34), that is,

$$
{^j\alpha} = \frac{{^jL_0} \times {^jP}}{\Pi}
\tag{14.59}
$$

The fractional population $^j\alpha$ includes all the $s + 1$ different states of ligation, and thus it depends on ligand concentration. Making this substitution in Equation 14.58 yields

$$
f_{bound} = \sum_{j=1}^{n} {^j\alpha} \times {^jf_{bound}}
\tag{14.60}
$$

Equation 14.60 is both its simple and intuitive. It says that the overall extent of binding is proportional to a weighted average of the extent of binding to each conformational state, where the weighting factors are simply the population of each conformational state. Here again, the population terms are summed, because the conformational states are mutually exclusive.

Overall, the general allosteric model provides a means through which complex ligand binding can be obtained. Conceptually, complexity results from the interplay between ligand binding and conformational change: binding of ligands to a subset of sites can shift conformational equilibrium, which in turn can change affinities at the remaining empty sites. A way to think about this in terms of equations is to focus on the ^jP terms. As in simpler applications (Chapter 13) the ^jP's increase with increasing ligand concentration. However, since each ^jP has its own set of $^j\beta_i$ constants, some ^jP's (those with large $^j\beta_i$ values) will dominate the others at high ligand concentrations, reflecting a shift in conformation toward those dominant j's, and a change in ligand affinity toward the dominant conformational state.

A two-conformation, two-site general allosteric model

Here we will illustrate the general allosteric model with the simplest possible scheme[†]: a macromolecule with two conformational states (referred to here $j = 1$ and $j = 2$), and two binding sites:

$$
\begin{array}{ccccc}
^1M_0 & \underset{}{\overset{^1K_1}{\rightleftharpoons}} & ^1Mx & \underset{}{\overset{^1K_2}{\rightleftharpoons}} & ^1Mx_2 \\[4pt]
^{1\to2}L_0 \updownarrow & & \updownarrow & & \updownarrow \\[4pt]
^2M_0 & \underset{}{\overset{^2K_1}{\rightleftharpoons}} & ^2Mx & \underset{}{\overset{^2K_2}{\rightleftharpoons}} & ^2Mx_2
\end{array}
\tag{14.61}
$$

Here, $^{1\to2}L_0$ is the conformational equilibrium constant between the two unligated states (conformations 1 and 2). Stepwise macroscopic binding constants (^jK_i) are used in Equation 14.61 because they are easy to map to the equilibrium arrows. These constants relate to the overall constants of Equation 14.50 using the relations described in Chapter 13.[‡]

Depending on the parameters, the two-conformation, two-site model can show cooperativity. Most notably, there must be a difference in stability between the two conformations combined with an opposing difference in affinity for the first ligand. Moreover, the magnitude of the difference in affinities must equal or exceed the difference in stability; if $^{1\to2}L_0 < 1$ (meaning the unligated $j = 1$ conformation is more stable than the $j = 2$ conformation), then

$$
\frac{^2K_1}{^1K_1} \ge \frac{1}{^{1\to2}L_0}
\tag{14.62}
$$

This requirement ensures that the difference in binding energy is enough to significantly shift the macromolecule from the $j = 1$ to the $j = 2$ conformation. In the example in **Figure 14.15**, $^2K_1/^1K_1 = 100$, whereas $1/^{1\to2}L_0 = 33$.

As long as condition 14.62 is satisfied (or its variant, see Problem 14.35), the two-conformation, two-site model undergoes a ligand induced conformational transition when the first ligand binds, and can display cooperativity through the second binding event. If affinity for the second ligand is higher in the new conformation[§] (i.e., $^2K_2 > ^1K_2$), positive cooperativity will result (Figure 14.15). If affinity is lower in the new conformation, negative cooperativity can result (Problem 14.37).

[†] If we had fewer conformational states there would be no possibility for switching between states. And if we had fewer ligands, we could not have changes in ligand affinity at unbound sites (though, surprisingly, we could still have conformational switching, see Problem 14.36).

[‡] Specifically, $^1\beta_1 = ^1K_1$ and $^1\beta_2 = ^1K_1{}^1K_2$; an analogous pair of relationships apply to the $j = 2$ conformation.

[§] More precisely, "new conformation" refers to the conformation that became populated in as a result of the first ligand binding event. For the example in Figure 14.15, this is the $j = 2$ state.

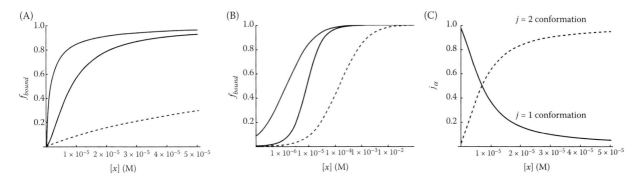

Figure 14.15 Ligand binding and conformational equilibria for a two-conformation, two-site allosteric model (Equation 14.61). (A, B) Fractional saturation curves (black lines) are given by Equation 14.56, using the conformational equilibrium constant $^{1\to2}L_0 = 0.03$, and the conformation-specific binding constants $^{1}K_1 = 2 \times 10^4$, $^{1}K_2 = 5 \times 10^3$, $^{2}K_1 = 2 \times 10^6$, and $^{2}K_2 = 5 \times 10^5$ (note, these values are consistent with independent binding within the $j = 1$ and $j = 2$ states, with microscopic binding constants $^{1}\kappa = 10^4$ and $^{2}\kappa = 10^6$ M^{-1}). Red curves show binding to $j = 1$ (dashed) and $j = 2$ (solid) conformations (Equation 14.57); compared to these rectangular hyperbolic references, the two-site allosteric model shows positive cooperativity. (C) The fraction of macromolecules shifts from the $j = 1$ conformation to the $j = 2$ conformation (Equation 14.59) with increasing ligand concentration, reflecting the high affinity (and low free energy) of the $j = 2$ state. This shift produces positive cooperativity in ligand binding (panels A and B).

Microscopic allosteric models and approximations

The general allosteric model above is a complete model, and it gives a conceptual sense for how conformational changes can be linked to changes in ligand affinity, producing either positive or negative cooperativity. However, as with the macroscopic binding polynomials discussed earlier, the general allosteric model suffers both from having *a lot* of parameters ($(s - 1) \times n$),[†] and from a somewhat awkward connection to molecular mechanism. As before, both of these afflictions can be cured, at least in part, by switching over to a microscopic description of ligand binding and by introducing some simplifying assumptions.

One way the picture can be simplified is by assuming that within each of the n conformational states, ligand binding to the s sites is identical and independent. With this assumption, we need only one microscopic binding site for each of the n conformational states, which we refer to as κ_1 through κ_n. As described above, within each conformational state, the binding polynomial ^{j}P will have a very simple form:

$$^{j}P = (1 + \kappa_j[x])^s \tag{14.63}$$

These assumptions can be combined with Equation 14.55 to obtain the total partition function polynomial that includes binding to each of the n conformational states:

$$\Pi = \sum_{j=1}^{n} {}^{j}L_0 \times {}^{j}P = \sum_{j=1}^{n} {}^{j}L_0 \times (1 + \kappa_j[x])^s \tag{14.64}$$

As usual, the extent of binding can be obtained by differentiation (Problem 14.38):

$$f_{bound} = \frac{\sum_{j=1}^{n} {}^{j}L_0 \times \kappa_j[x](1 + \kappa_j[x])^{s-1}}{\sum_{j=1}^{n} {}^{j}L_0 \times (1 + \kappa_j[x])^s} \tag{14.65}$$

[†] The number of parameters for the general allosteric model (each of the $^{j}\beta_i$) is equal to the number of distinct macrostates. There are n conformational states, and each conformational state contains some number of ligands bound from 0 to s, for a total of $s + 1$. Combining conformational and ligation states gives $(s + 1) \times n$ parameters. In practice, one of these parameters is fixed (i.e., $^{1}\beta_0 = 1$), and can be removed from the parameter count.

Although Equation 14.60 is more complex than the expression for fractional saturation for a single conformational state (Equation 14.33), it has fewer parameters $(2n - 1)^{\dagger}$ than the general allosteric model $((s - 1) \times n - 1)$.

Allosteric models involving subunits

Further simplification can be achieved by making restrictions on the numbers and types of macromolecular conformations, and how they relate to binding affinities. Such restrictions are particularly well-suited for proteins made of identical, symmetric subunits (Figure 14.9). Here we will invoke four related rules that apply to multi-subunit macromolecules and will evaluate their effects on binding equilibria. These rules form the basis for two popular models for allostery: the "MWC" and "KNF" models described below.

Rule 1: The macromolecule has the same number of subunits as it has binding sites (s).

Rule 2: Each subunit has two possible conformations, referred to as "T" and "R." In other words, it is a "two-state" model in terms of subunit conformation.

Rule 3: Binding affinity at any subunit depends only on the conformation of that subunit, and is otherwise independent of ligand occupancy at the other subunits.

Rule 4: T subunits have a single ligand affinity described by microscopic binding constant κ_T; likewise, R subunits have a single ligand affinity described by microscopic binding constant κ_R.

Models conforming to these four rules can be regarded as conformationally "two-state" at the subunit level. In terms of species, these rules reduce the general allosteric scheme (Equation 14.49) to a square matrix of $s + 1$ rows and $s + 1$ columns:

$$
\begin{array}{ccccccc}
R_sT_0 & \rightleftharpoons & R_sT_0x & \rightleftharpoons & R_sT_0x_2 & \rightleftharpoons & \cdots & R_sT_0x_s & {}^{R_sT_0}P=(1+\kappa_R[x])^s(1+\kappa_T[x])^0 \\
\updownarrow & & \updownarrow & & \updownarrow & & & \updownarrow & \\
R_{s-1}T_1 & \rightleftharpoons & R_{s-1}T_1x & \rightleftharpoons & R_{s-1}T_1x_2 & \rightleftharpoons & \cdots & R_{s-1}T_1x_s & {}^{R_{s-1}T_1}P=(1+\kappa_R[x])^{s-1}(1+\kappa_T[x])^1 \\
\updownarrow & & \updownarrow & & \updownarrow & & & \updownarrow & \\
R_{s-2}T_2 & \rightleftharpoons & R_{s-2}T_2x & \rightleftharpoons & R_{s-2}T_2x_2 & \rightleftharpoons & \cdots & R_{s-2}T_2x_s & {}^{R_{s-2}T_2}P=(1+\kappa_R[x])^{s-2}(1+\kappa_T[x])^2 \\
\updownarrow & & \updownarrow & & \updownarrow & & & \updownarrow & \\
\vdots & & \vdots & & \vdots & & & \vdots & \\
R_1T_{s-1} & \rightleftharpoons & R_1T_{s-1}x & \rightleftharpoons & R_1T_{s-1}x_2 & \rightleftharpoons & \cdots & R_1T_{s-1}x_s & {}^{R_1T_{s-1}}P=(1+\kappa_R[x])^1(1+\kappa_T[x])^{s-1} \\
\updownarrow & & \updownarrow & & \updownarrow & & & \updownarrow & \\
R_0T_s & \rightleftharpoons & R_0T_sx & \rightleftharpoons & R_0T_sx_2 & \rightleftharpoons & \cdots & R_0T_sx_s & {}^{R_0T_s}P=(1+\kappa_R[x])^0(1+\kappa_T[x])^s
\end{array}
\tag{14.66}
$$

For each conformational state (e.g., $R_{s-j}T_j$), the binding polynomial ${}^{R_{s-j}T_j}P$ can be generated as the product of binding sub-polynomials for the T- and R-state subunits $((1 + \kappa_T[x])^j$ and $(1 + \kappa_R[x])^{s-j}$, respectively). The T and R sub-polynomials are multiplied because, according to rules 3 and 4, binding to different subunits within a conformational state ($R_{s-j}T_j$) is independent. As described in Chapter 1, probabilities multiply for the "*and*" combination of outcomes of independent events.

To write the total conformation-dependent partition function Π, we need to weight these ${}^{R_{s-j}T_j}P$ using conformational equilibrium constants (Equation 14.55). We will

---

$\dagger$ The number of parameters for the allosteric model with identical independent binding sites for each n conformational state is equal to the number of conformational constants ($n \, {}^jL_0$ values) plus the stepwise microscopic binding constant κ_j for each of the n conformational states. In practice, one of the conformational constants is fixed (i.e., ${}^1L_0$).

use the conformation with all s subunits in the R state without ligand bound (upper left in Equation 14.66) as a reference conformation (analogous to the $j = 1$ conformation in the general allosteric model above), giving overall conformational constants

$$^0L_0 = \frac{[R_sT_0]}{[R_sT_0]} = 1, \quad ^1L_0 = \frac{[R_{s-1}T_1]}{[R_sT_0]}, \quad ^2L_0 = \frac{[R_{s-2}T_2]}{[R_sT_0]}, \quad \cdots, \quad ^sL_0 = \frac{[R_0T_s]}{[R_sT_0]} \quad (14.67)$$

The left-superscripts are a shorthand notation to describe conformational transitions from the R_sT_0 reference (i.e., $R_sT_0 \rightarrow R_{s-j}T_j$ for the transition j). Even though the first allosteric constant has a value of one, it is useful to define this constant in the same way as the other constants to simplify the derivations below. Note that the $^jL_0$ constants are macroscopic in that they lump all arrangements with $s-j$ R- and j T-subunits together. With these constants, the total partition function for the two-state subunit model is

$$\Pi = \sum_{j=0}^{s} {}^jL_0 \times {}^{R_{s-j}T_j}P = \sum_{j=0}^{s} {}^jL_0 \times (1 + \kappa_R[x])^{s-j}(1 + \kappa_T[x])^j \quad (14.68)$$

For $s > 2$ subunits, this partition function has fewer parameters (effectively $n + 2$, where $n = s$) than the allosteric model with identical independent binding constants for n different conformations (Equation 14.64; $2n - 1$ parameters).

The fractional saturation is calculated by differentiation with respect to ligand concentration in the usual way (Problem 14.39), leading to

$$f_{bound} = \frac{[x]\sum_{j=0}^{s} {}^jL_0 \times \left\{(s-j)\kappa_R(1 + \kappa_R[x])^{s-j-1}(1 + \kappa_T[x])^j + j\kappa_T(1 + \kappa_R[x])^{s-j}(1 + \kappa_T[x])^{j-1}\right\}}{s \times \sum_{j=0}^{s} {}^jL_0 \times (1 + \kappa_R[x])^{s-j}(1 + \kappa_T[x])^j}$$

$$(14.69)$$

Keep in mind that the sums in Equations 14.68 and 14.69 are over conformational states (the index j is the number of subunits in the T conformation, not the number of ligands bound). Positive cooperativity in ligand binding is achieved if the low-affinity conformations are most stable in the absence of ligand, and the high-energy conformations are close enough in energy that partial ligand saturation can shift the population to the high-affinity conformation. Historically, the T conformational state has been taken to be the low-affinity conformation; in this case positive cooperativity can be achieved when the conformational constants $^jL_0$ are greater than one.

Since ligand-linked conformational changes are essential for cooperativity in this model, it is useful to express conformational changes in terms of ligand concentration. The fraction of subunits in the jth conformational state ($^j\alpha$) can be expressed as the ratio weighted sub-polynomial for binding to the jth state divide by the total partition function:

$$^j\alpha = \frac{{}^jL_0 \times {}^jP}{\Pi} = \frac{{}^jL_0(1 + \kappa_R[x])^{s-j}(1 + \kappa_T[x])^j}{\sum_{j=0}^{s} {}^jL_0 \times (1 + \kappa_R[x])^{s-j}(1 + \kappa_T[x])^j} \quad (14.70)$$

We will finish our discussion of allostery in ligand binding by discussing two popular models for multisubunit proteins: the MWC model and the KNF model. These models have historical significance in the analysis of ligand binding in the last half-century, and are impressive in their elegance and economy of terms. These two models are worth studying together, in part as a study in contrasting styles.

The MWC model. This model, named after Monod, Wyman, and Changeaux (Monod et al. 1965), extends the two-state subunit behavior diagrammed in Equation 14.66 (i.e., two subunit structures T and R, with ligand affinities κ_T and κ_R) to the entire multisubunit macromolecule. In the MWC model, if one subunit switches from the T to the R conformation, all of the other subunits switch as well. This concerted switch preserves whatever symmetry the macromolecule possesses, by eliminating hybrid $R_{s-j}T_j$ conformations.[†]

In terms of species in Equation 14.66, the top and bottom row are retained, but all intermediate rows are eliminated. In terms of equilibrium constants, $^sL_0$ (the equilibrium constant of the all-T macromolecule relative to the all-R macromolecule, Equation 14.67) is retained, but all others are eliminated (having values of zero by assertion). As a result, the partition function Π simplifies from a weighted sum of $s + 1$ binding polynomials (Equation 14.68) to just two polynomials, ^0P and ^sP (for binding to the all-R and all T-quaternary structure):

$$\Pi = {}^0L_0(1+\kappa_R[x])^s + {}^sL_0(1+\kappa_T[x])^s$$
$$= (1+\kappa_R[x])^s + L_0(1+\kappa_T[x])^s \tag{14.71}$$

As can be seen in the second line, which makes use of the relationship $^0L_0 = 1$, the MWC model only has three thermodynamic parameters: L_0, κ_T, and κ_R.[‡]

Fractional saturation can be derived by differentiation (Problem 14.41):

$$f_{bound} = \frac{\kappa_R[x](1+\kappa_R[x])^{s-1} + {}^sL_0\kappa_R[x](1+\kappa_T[x])^{s-1}}{(1+\kappa_R[x])^s + {}^sL_0(1+\kappa_T[x])^s} \tag{14.72}$$

Conformational populations[§] can be derived using Equation 14.59:

$$^R\alpha = \frac{(1+\kappa_R[x])^s}{(1+\kappa_R[x])^s + L_0(1+\kappa_T[x])^s} \tag{14.73}$$

$$^T\alpha = \frac{L_0(1+\kappa_T[x])^s}{(1+\kappa_R[x])^s + L_0(1+\kappa_T[x])^s} \tag{14.74}$$

A specific example of the MWC scheme is shown in **Figure 14.16**. For this four subunit (and thus, four binding site) example, all 25 of the states given by the general allosteric model are shown, but only the top and bottom five, enclosed by the light gray rectangles are included in the MWC model. The 15 states with subunits in both T and R conformations are excluded.

For the right combinations of parameters, the MWC model results in positive cooperativity in ligand binding. The affinity of the T states and R states (T_4 and R_4 above) must differ (otherwise there is no cooperativity in ligand binding, and no ligand-induced conformational change; see Problem 14.43). In addition, the conformational equilibrium constant for the unbound state (L_0) must favor the low affinity state, so that ligand binding can shift the conformation to the high affinity state. However, this conformational bias cannot be so strong that it cannot be overcome by ligand binding (Problem 14.44). Note that affinity differences always shift toward the high-affinity quaternary structure, the MWC model cannot produce negative cooperativity in ligand binding.

The KNF model

The assumptions of this model, named after Koshland, Nemethy, and Filmer (Koshland et al., 1966) contrast quite sharply with those of the MWC model. Unlike

---

[†] Where $0 < j < s$.

[‡] Since there is only one relevant conformational equilibrium constant, we will omit the left-superscript, that is, $L_0 \equiv {}^sL_0 = [R_0T_s]/[R_sT_0]$.

[§] In this case there are just two conformational species: $^0\alpha \equiv {}^R\alpha$ and $^s\alpha \equiv {}^T\alpha = 1 - {}^R\alpha$.

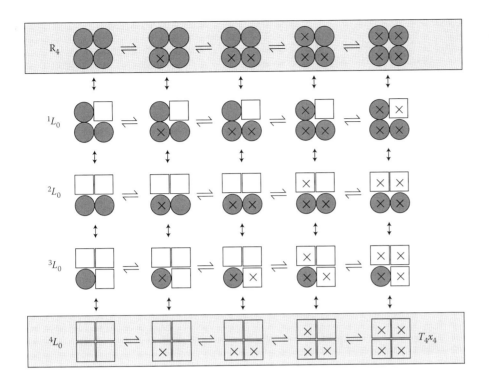

Figure 14.16 The MWC model for allosteric ligand binding. A macromolecule with four subunits is depicted as having two conformations per subunit (R, dark gray circles, top; T, white squares, bottom). Two conformations, R_4 (upper left) and T_4X_4 (lower right), are labeled as examples. Ligand binding reactions are horizontal, conformational transitions are vertical. Conformational equilibrium constants for unligated states (left) are taken relative to R_4. Only the all-R and all-T conformations (light gray rectangles) are included in the MWC model. This is equivalent to assuming that conformational constants $^1L_0$, $^2L_0$, and $^3L_0$ are negligibly small.

the MWC model, the KNF model allows for macromolecular structures that contain both T and R subunits. In the four-subunit example, quaternary structures from each row are allowed (**Figure 14.17**), not just the top and bottom rows. Thus, compared to the MWC model, the KNF model is more heterogeneous in terms of quaternary structure. However, the KNF model is more restrictive in terms of ligand binding, as it only allows ligand-bound subunits to be in the R conformation, and it restricts ligand-free subunits to the T conformation. Thus, only one state of ligation is allowed for each conformation, corresponding to i ligands bound to i R subunits, with $s - i$ T subunits unligated. Cooperativity in the KNF model results from differences in interaction energies among the subunits in different conformational states.

Returning to the four-subunit example, the only states included in the KNF model are the five states along the diagonal (in the light gray rectangle in Figure 14.17). The other 20 states are excluded, either because they have one or more unligated R subunits, or one or more ligated T subunits. Whereas quaternary structure symmetry is maintained in the MWC model, it is broken in the KNF model. Binding in the KNF model can be considered to be a sequence of concerted binding and local conformational transitions.

One way to generate the partition function for the KNF model is to begin with the macrostate partition function with two conformational states per subunit (Equation 14.68) and eliminate terms that correspond to excluded states. Since the species allowed for the ith ligation state is the conformation with all i R subunits bound and all T subunits empty, we can set the all T-state sub-polynomial in Equation 14.68 to one.[†] Moreover, the all R-state sub-polynomial is reduced to a single term: $(\kappa_R[x])^s$. Likewise, binding polynomials for hybrid $R_j T_{s-i}$ conformations are given by

$$^{R_iT_{s-i}}P = (\kappa_R[x])^i \tag{14.75}$$

[†] This is equivalent to setting κ_T to zero.

Figure 14.17 The KNF model for allosteric ligand binding to a tetrameric macromolecule. In the KNF model, subunits are in the R state if and only if they have a ligand bound. This corresponds to the diagonal species (light gray rectangle).

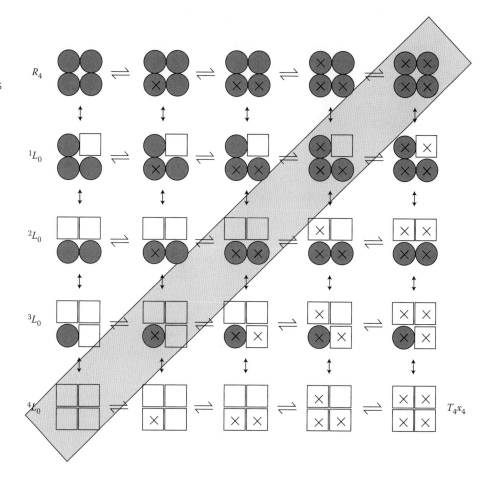

Combining Equations 14.68 and 14.75 gives the partition function for the KNF model:

$$\Pi = \sum_{i=0}^{s} {}^{i}L_0(\kappa_R[x])^i \qquad (14.76)$$

Fractional saturation f_{bound} and conformational populations ${}^{i}\alpha$ are calculated in the usual way (Problem 14.43).

Although the KNF partition function is more compact than the MWC model ($s + 1$ vs. $2s + 2$ terms), the number of parameters in the KNF partition function (Equation 14.76) exceeds the number for the MWC model (only three parameters). To further simplify the KNF partition function, we can treat the conformational transitions in a microscopic way,[†] resulting in a more mechanistic description of binding. However, this microscopic treatment requires us to explicitly enumerate the geometry of the subunit arrangements, along with the energetics of different types of subunit interactions. In a sense, this is the conformational equivalent to geometric models such as the hexagon models shown in Figure 14.13. As with those models, the simplest way to build a partition function for a specific arrangement is to make a table for each relevant microstate that keeps track of the extent of binding, conformational transitions, subunit interactions, and multiplicities.

Here we will develop a microscopic KNF model for a tetrameric macromolecule, with subunits arranged in a square (C_4) geometry. Starting with the unligated all T

---

† Recall that ${}^{j}L_0$ are microscopic in that they combine all subunit arrangements with j T conformations and $s - i$ R-conformations together.

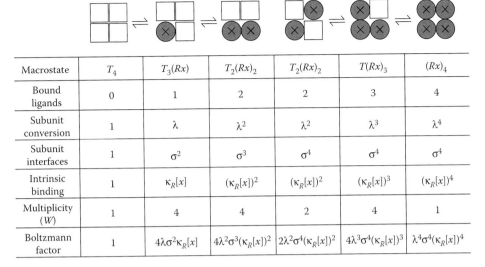

Macrostate	T_4	$T_3(Rx)$	$T_2(Rx)_2$	$T_2(Rx)_2$	$T(Rx)_3$	$(Rx)_4$
Bound ligands	0	1	2	2	3	4
Subunit conversion	1	λ	λ^2	λ^2	λ^3	λ^4
Subunit interfaces	1	σ^2	σ^3	σ^4	σ^4	σ^4
Intrinsic binding	1	$\kappa_R[x]$	$(\kappa_R[x])^2$	$(\kappa_R[x])^2$	$(\kappa_R[x])^3$	$(\kappa_R[x])^4$
Multiplicity (W)	1	4	4	2	4	1
Boltzmann factor	1	$4\lambda\sigma^2\kappa_R[x]$	$4\lambda^2\sigma^3(\kappa_R[x])^2$	$2\lambda^2\sigma^4(\kappa_R[x])^2$	$4\lambda^3\sigma^4(\kappa_R[x])^3$	$\lambda^4\sigma^4(\kappa_R[x])^4$

Figure 14.18 The species in the four-subunit KNF square, and factors influencing their Boltzmann factors. The quantity λ is an equilibrium constant for conversion of an unligated subunit from the T to the R conformation. The quantity σ is an equilibrium constant for disrupting the interface between two adjacent T subunits (here, subunits across the diagonal are assumed not to interact). The multiplicities are equal to the number of ways the T and Rx subunits can be arranged. Note that when two ligands are bound, there are two distinct arrangements of T and Rx subunits, with distinct subunit interactions and multiplicities. Boltzmann factors for each distinguishable $T_{4-i}R_i$ configuration are the product of subunit conversion, interaction between subunits, intrinsic binding to R subunits, and subunit arrangement multiplicities.

conformation, there are four components that determine the Boltzmann factors for the microstates: (1) conformational conversion of i unligated subunits from T to R, with an equilibrium constant λ for each subunit converted (λ^i total), (2) disruption of $T{:}T$ interfaces between subunits, with an equilibrium constant σ for each interface disrupted,[†] (3) the intrinsic binding of i ligands, with an equilibrium constant κ_R per ligand (and ligand concentration dependent weight of $\kappa_R[x]$), and (4) the number of ways (W) that i ligated R and $4-i$ unligated T subunits can be arranged on a square geometry. These four components are collected for each KNF microstate in **Figure 14.18**.

For the KNF square tetramer scheme in Figure 14.18, the binding partition can be written as the sum of the Boltzmann factors, that is,

$$\Pi = 1 + 4\sigma^2(\lambda\kappa_R[x]) + 4\sigma^3(\lambda\kappa_R[x])^2 + 2\sigma^4(\lambda\kappa_R[x])^2 + 4\sigma^4(\lambda\kappa_R[x])^3 + \sigma^4(\lambda\kappa_R[x])^4$$

$$(14.77)$$

Fractional saturation and conformational populations are calculated in the usual way. You should be able to convince yourself that the fraction of subunits in the R conformation is equal to the fractional saturation with ligand (see Problem 14.50).

Unlike some of the previous partition functions we have seen, where simplifications such as identical binding sites lead to closed-form solutions, binding partition functions for KNF-type models retain each Boltzmann factor as specified by a particular model. Details of these models depend on subunit number, subunit geometry, and the way subunit interactions are treated. Although KNF partition functions can be complex, the microscopic treatment does decrease the number

---

[†] Note there are other ways we could represent the subunit–subunit interaction (Problem 14.48). For example, we could use the number of $R{:}R$ interfaces formed rather than the number of $T{:}T$ interfaces broken. These two quantities are correlated (negatively), although the correlation is imperfect. Alternatively, we could use the number of $T{:}T$ interfaces formed or the number of $R{:}R$ interfaces disrupted. Another approach would be to use the number of "mismatched" $R{:}T$ interfaces.

of parameters required for the KNF model. For example, the tetramer KNF partition function has three parameters (σ, λ, and κ_R), the same number as for the MWC model. Moreover, λ and κ_R can be considered to be a single parameter, since the ligand binding and conformational transition are coincident at each stage of the KNF model (this is why they have the same exponent in each term of Equation 14.77). Combining λ and κ_R into a single parameter gives a cooperative model with a strong mechanistic foundation and just two independent parameters.

Problems

14.1   The amino acid glycine has two hydrogen ion binding sites, one on the amino group and one on the carboxylate group. In terms of macroscopic binding constants, this reaction can be written as

$$Gly^- + H^+ \xrightleftharpoons{K_1} Gly$$
$$Gly + H^+ \xrightleftharpoons{K_2} Gly^+$$

These hydrogen ion binding constants are often expressed as "pK_a"s, which are the base-ten log of the acid association constant. For example, the two reactions above,

$$pK_{a1} = \log_{10} K_1 = 9.60$$

$$pK_{a2} = \log_{10} K_2 = 2.34$$

(Note that such relationships are often given using acid *dissociation* constants (e.g., $K_{a1} = 1/K_1$), which introduce negative signs into the pK_a equations above). Using these macroscopic overall constants, calculate the fractional saturation as a function of hydrogen ion concentration.

Now using the relationship

$$pH = -\log[H^+]$$

plot the average number of hydrogen ions bound (ranging from 0 to 2) as function of pH.

14.2   Write a hydrogen ion binding scheme for glycine (see Problem 14.1 above) in terms of stepwise microscopic (i.e., site-resolved) reactions. What is the relationship between your stepwise microscopic binding constants and the overall constants in Problem 14.2?

14.3   Because the affinity of amino nitrogen for hydrogen ion is so much greater than the affinity of the carboxyl oxygen, there is a simpler relation between the macrostate constants K_1 and K_2 from Problem 14.1 and the microstate constants than you derived in Problem 14.2. What is this simpler relationship? If the sites do not interact (in fact, they do, through charge–charge interaction), what are the other two stepwise microstate constants?

14.4   Show that for the general three-site model, the macroscopic stepwise constant K_3 can be given as the combination of microscopic stepwise constants given in Equations 14.16 and 14.17.

14.5   For the general three-site model with different microscopic constants for each step, assume you are given unique parameters for the following parameters:

$$\kappa_1 = 1 \times 10^4 \qquad \kappa_1^{(2)} = 1 \times 10^5 \qquad \kappa_1^{(23)} = 1 \times 10^6$$
$$\kappa_2 = 0.5 \times 10^4 \qquad \kappa_2^{(3)} = 0.5 \times 10^5$$
$$\kappa_3 = 2 \times 10^4 \qquad \kappa_3^{(1)} = 2 \times 10^5$$

Give the other five stepwise parameters in terms of (1) nonnumerical formulas combining the microscopic constants and (2) numerical values based on the constants above.

14.6 For the general three-site model, are any seven microscopic stepwise constants sufficient to determine the values of the other five? If so, why? If not, what additional condition must be satisfied?

14.7 Use words and logic to rationalize the recursion Equation 14.8, which gives the number of edges in a parallelotope with n vertices in terms of the number of edges in a paralellotope with $n - 1$ vetices.

14.8 Rationalize Equation 14.22, based on the number of ways ligands can be arranged in the reactant and product states.

14.9 For the four-site identical independent binding scheme (analogous to that in Figure 14.5), give expressions for K_1 through K_4, β_1 through β_4, and $\mathit{Б}$ values in terms of the stepwise affinity constant κ (note there are multiple $\mathit{Б}$'s at most ligation levels, but they are all the same, so just give one example at each step). Next give numerical values for these constants for the value $\kappa = 10^5 \text{ M}^{-1}$.

	Generic in terms of κ			$\kappa = 10^5 \text{ M}^{-1}$		
	K_i	β_i	$\mathit{Б}$	K_i	β_i	$\mathit{Б}$
Step 1 (one ligand)						
Step 2 (two ligands)						
Step 3 (three ligands)						
Step 4 (four ligands)						

14.10 Derive the simple formula for fractional saturation for s independent identical sites given in Equation 14.33.

14.11 Give an expression for the binding polynomial for a hexagonal macromolecule with six independent identical sites. Differentiate this expression to obtain expressions for fractional saturation and extent of binding. For $\kappa = 500 \text{ M}^{-1}$, plot the binding polynomial and the fractional saturation over a concentration range that you think best illustrates the binding behavior (hint: for the binding polynomial, you might want to use a logarithmic scale).

14.12 Calculate, plot, and compare binding polynomials for identical, independent binding with a $\kappa = 10 \text{ M}^{-1}$, for $s = 1, 2, 3,$ and 5 sites. Do this for both linear and logarithmic scales, to try to maximize the information in the plots.

14.13 Using the parameters in Problem 14.12 to generate expressions and plot the ensemble average free energies of ligand interaction, using the expression

$$\bar{G} = -RT\ln P$$

In words, what does this free energy correspond to?

14.14 Demonstrate the equality given in Equation 14.32 using the properties of factorials.

14.15 Starting with the expression for $\langle X_{bound} \rangle$ for two classes of independent sites (Equation 14.38), demonstrate that the fractional saturation, f_{bound}, can be represented as

$$f_{bound} = f_1 \frac{\kappa_1 X}{(1 + \kappa_1 X)} + f_2 \frac{\kappa_2 X}{(1 + \kappa_2 X)}$$

where f_1 is the fraction of sites that are of type 1, and f_2 is the fraction that are of type 2.

14.16 Show that the binding polynomial for the singly coupled identical sites model (OPL, Equation 14.39) can be expressed as

$$P = 1 - \frac{1}{\phi} + \frac{(1 + \phi \kappa X)^s}{\phi}$$

and that the fractional saturation can be calculated as

$$f_{bound} = \frac{\phi \kappa X (1 + \phi \kappa X)^{s-1}}{\phi - 1 + (1 + \phi \kappa X)^s}$$

14.17 Evaluation of the saturation curve for the OPL coupling model (Equation 14.42) in the extreme negative coupling limit ($\phi = 0$) gives an expression of indeterminate form, namely

$$\lim_{\phi \to 0} f_{bound} = \lim_{\phi \to 0} \frac{\phi \kappa X (1 + \phi \kappa X)^{s-1}}{\phi - 1 + (1 + \phi \kappa X)^s}$$
$$= \frac{0 \kappa X (1 + 0 \kappa X)^{s-1}}{0 - 1 + (1 + 0 \kappa X)^s}$$
$$= \frac{0}{-1 + 1} = \frac{0}{0}$$

14.18 For the OPL model, fill in the table below (equivalent to Table 14.3 for the AS model) for the different states of ligation of a five-site macromolecule. Be sure to include a graph of each ligation state, where the vertices correspond to bound ligands, and the edges correspond to couplings between ligands.

Pairwise couplings, statistics, and graph representations in the "one-per-ligand" (OPL) model, for different ligation states in a macromolecule with five sites

# ligands bound (i)	1	2	3	4	5
Macrostate	MX_1	MX_2	MX_3	MX_4	MX_5
Number of pairwise couplings					
Coupling contribution					$\therefore$
Number of configurations					
Graph representation of bound ligands					

14.19 Demonstrate that the number of couplings, c, for the AS model is given by Equation 14.43, that is,

$$c = \sum_{j=1}^{i-1} j$$

where i is the number of ligands bound. To do this, think about the corresponding complete graph (see Table 14.3) with i vertices, and count the number of edges between each vertex, making sure not to count any edges twice.

14.20 Demonstrate that the number of couplings, c, for the AP model is given by Equation 14.44, that is,

$$c = \frac{i(i-1)}{2}$$

where i is the number of ligands bound. To do this, count the "edge endpoints" in the graph, and convert this to the number of edges.

14.21 For a macromolecule with four fully coupled ligand binding sites, show that the same number of coupling terms is obtained by the summation given in Table 14.3 as that obtained by counting edges (Equation 14.43), for each state of ligation. Arrange your work as in the table below:

	Number of bound ligands, i				
	0	1	2	3	4
$\#\phi = \sum_{j=1}^{i} (j-1)$					
$e = \dfrac{s(s-1)}{2}$					
$E = s \times 2^{s-1}$					

14.22 Show that Equations 14.43 and 14.44 are equal for arbitrary values of i.

14.23 Following up on your scheme for glycine in Problem 14.2, derive a two-site AS-like binding polynomial that accommodates different intrinsic affinities at the two binding sites (the amino and carboxyl groups—call them κ_N and κ_O). Is this model different than if you had used a two-site OPL-like model?

14.24 An alternative to the AS-like model for the glycine titration scheme (Problem 14.23) is to keep the two intrinsic binding constants for the amino and carboxyl sites κ_N and κ_O, but introduce a coupling term ψ that captures the favorable charge–charge interaction between the ammonium (+1 charge) and carboxylate (−1 charge). A partial scheme for this is shown below.

What would the value of the equilibrium constant for the fourth step be?

14.25 The pK_a of glycine methyl ester is 7.73.

Assuming this value represents the intrinsic binding of a hydrogen ion to the amino group κ_N (because the methyl ester cannot bind a proton, and it is uncharged), what are the values of the intrinsic binding constant for the carboxyl (κ_O) and for the charge–charge coupling term ψ from Problem 14.24? Use the pK_a's from Problem 14.1 (and some of your analysis from Problem 14.2).

14.26 Derive an expression for the fractional saturation for the general model with pairwise coupling among all bound ligands (the AS model, with the binding polynomial given by Equation 14.45). For $s = 5$ sites, plot the fractional saturation for $\kappa = 10^6\,\text{M}^{-1}$, and $\phi = 20$.

14.27 For a macromolecule with three sites, each with identical microscopic stepwise binding constants κ, compare binding polynomials (term by term) for the OPL and AS models. For comparison, arrange your answer in a table such as this:

	0	1	2	3
$P_{OPL} =$				
$P_{AS} =$				

14.28 Assuming microscopic stepwise binding constants $\kappa = 10^4\,\text{M}^{-1}$, plot and compare the fractional saturation curves for the $s = 3$ OPL and AS models, with $\phi = 10$, and with $\phi = 0.1$.

14.29 For the $s = 3$ OPL and AS models in Problem 14.28, calculate and plot the Hill coefficients as a function of ligand concentration. Based on these plots, which model (OPL or AS) results in greater cooperativity?

14.30 For the hexagon model, show the different gap configurations in Table 14.4, and give the statistical factors for each. Use hexagon pictures to illustrate.

14.31 Show that for nearest-neighbor cyclic n-gon (C_n; e.g., the hexagon on the left of Figure 14.13A for $n = 6$), the number of coupling c is given by the formula

$$c = i - g$$

where i is the number of ligands bound, and g is the number of gaps. Note, this formula does not work for $i = 0$ ligands bound.

14.32 Derive an expression for the binding polynomial for the hexagonal model in which nearest- and second-nearest neighbors are coupled with constants ϕ_1 and ϕ_2, respectively. Organize the terms in the binding polynomial as in Table 14.4. For the values $\phi_1 = 20$ and $\phi_2 = 4$ (i.e., stronger nearest-neighbor coupling), calculate the fractional saturation, and make a plot of fractional saturation versus linear ligand concentration.

14.33 Show that for the general allosteric model, the rearrangement in Equation 14.58 is consistent with our general expression for the fractional saturation (using differentiation of the binding polynomial). For the two-conformation, two-site model in Equation 14.61, write out the fractional saturation expression for the $j = 2$ conformational state as a ratio of concentrations.

14.34 Show that the first term in the sum in Equation 14.58 is equal to the fraction of macromolecules that are in the jth conformational state, that is, show that

$$\frac{{}^jL_0 \times {}^jP}{\Pi} = {}^j\alpha$$

For the two-conformation, two-site model in Equation 14.61, write out the fraction of macromolecules in the $j = 2$ conformational state as a ratio of concentrations.

14.35 For the two-conformation, two-site allosteric model (Equation 14.61), what is the condition for cooperativity if ${}^{1\to2}L_0 > 0$? In other words, how is Equation 14.62 modified?

14.36 Consider the general allosteric scheme with two conformational states and one binding site:

$$
\begin{array}{ccc}
{}^1M_0 & \xrightleftharpoons{\;{}^1K_1\;} & {}^1MX \\
{}^{1\to2}L_0 \updownarrow & & {}^{1\to2}L_1 \updownarrow \\
{}^2M_0 & \xrightleftharpoons{\;{}^2K_1\;} & {}^2MX
\end{array}
$$

Plot the fractional saturation, and the fractional saturation of conformational states 1 and 2 (that is, ${}^1f_{bound}$ and ${}^2f_{bound}$). Use the constants ${}^1K_1 = 10^4 \text{ M}^{-1}$, ${}^2K_1 = 10^6 \text{ M}^{-1}$, and ${}^{1\to2}L_0 > 0.05$). Also, plot the fraction of macromolecules in conformational states 1 and 2 as a function of ligand concentration (your plots should be the same format as in Figure 14.15).

Does the scheme show positive cooperativity? Does it show ligand-induced conformational switching? Reconcile these two answers.

14.37 Consider the equilibrium with two allosteric states and two ligand binding sites (Equation 14.61)

$$
\begin{array}{ccccc}
{}^1M_0 & \xrightleftharpoons{\;{}^1K_1\;} & {}^1MX & \xrightleftharpoons{\;{}^1K_2\;} & {}^1MX_2 \\
{}^{1\to2}L_0 \updownarrow & & \updownarrow & & \updownarrow \\
{}^2M_0 & \xrightleftharpoons{\;{}^2K_1\;} & {}^2MX & \xrightleftharpoons{\;{}^2K_2\;} & {}^2MX_2
\end{array}
$$

In Figure 14.15, we showed how positive cooperativity could result if the conformationally disfavored allosteric state (state 2, with ${}^{1\to2}L_0 = 0.03$) has higher affinity than the conformationally favored state (${}^1K_1 = 2 \times 10^4$, ${}^1K_2 = 5 \times 10^3$, ${}^2K_1 = 2 \times 10^6$, and ${}^2K_2 = 5 \times 10^5$), such that ligand binding switches the conformational state to the high affinity state. In the next two problems, we will consider some ways this scheme can lead to broad binding curves characteristic of negative cooperativity.

For this problem, plot the fractional saturation curves and the population states in conformation 1 and 2 (the same format as in Figure 14.15) using the constants ${}^{1\to2}L_0 = 0.03$, ${}^1K_1 = 2 \times 10^4$, ${}^1K_2 = 5 \times 10^3$, ${}^2K_1 = 2 \times 10^6$, and ${}^2K_2 = 5 \times 10^2$. Describe the shapes of the curves, and what is going on at the molecular level.

14.38 For the general allosteric model with s identical, independent sites in each of the s allosteric states, calculate the fractional saturation curve (Equation 14.65).

14.39 For the allosteric model with two-state subunits (Scheme 14.66), demonstrate that the fractional saturation is given by

$$f_{bound} = \frac{[X]\sum_{j=1}^{s} {}^jL_0 \times \left\{(s-j)\kappa_R(1+\kappa_R[X])^{s-j-1}(1+\kappa_T[X])^j + j\kappa_T(1+\kappa_R[X])^{s-j}(1+\kappa_T[X])^{j-1}\right\}}{s \times \sum_{j=1}^{s} {}^jL_0 \times (1+\kappa_R[X])^{s-j}(1+\kappa_T[X])^j}$$

14.40 For the allosteric model with two-state subunits (Scheme 14.66), derive a fractional saturation expression for an s subunit macromolecule for which the subunit equilibrium between the T and R tertiary structure is independent of the conformational states of the other subunits. In other words, there is a single microscopic stepwise constant,

$$^{R \to T}\kappa = \frac{[T]}{[R]}$$

that describes the equilibrium at each subunit. Does this system show cooperative ligand binding? Justify your answer with words (Note that $^{T \to R}\kappa$ is a *microscopic* equilibrium constant, and will generally be related to $^{j}L_0$ through statistical factors).

14.41 Derive an expression for fractional saturation (f_{bound}) for the MWC model.

14.42 For the MWC tetramer, plot the fraction binding and the populations of the T_4 and R_4 quaternary structures, using the values $\kappa_T = 10^4\,M^{-1}$, $\kappa_R = 10^6\,M^{-1}$, and $L_0 = 100$.

14.43 For an MWC tetramer model with identical affinities for the T and R subunits ($\kappa_T = \kappa_R$), derive expressions for fractional saturation and conformational populations ($^{j}\alpha$). Plot these quantities as a function of ligand concentration.

14.44 Consider an MWC tetramer model that includes the binding of an allosteric inhibitor (I). Derive a partition function (Π) that includes both ligand and inhibitor binding, subject to the following assumptions:

(a) I molecules bind to each subunit of the T state with identical, independent affinity, with microscopic stepwise binding constant κ_I.

(b) I does not bind to the R state.

(c) Binding sites for I are distant from sites for ligand x, so that the occupancy of the inhibitory site does not affect κ_R or κ_T.

14.45 Continuing the inhibited MWC scheme from Problem 14.44, derive an expression for the fractional saturation (f_{bound}) with ligand x in the presence of inhibitor. Plot f_{bound} with different concentrations of inhibitor ($[I] = 0$, $[I] = 2 \times 10^{-5}$, and $[I] = 10^{-3}\,M$), using the inhibitor binding constant $^{T}\kappa_I = 10^5\,M^{-1}$. Use the allosteric constants from Problem 14.42 ($\kappa_T = 10^4\,M^{-1}$, $\kappa_R = 10^6\,M^{-1}$, and $L_0 = 100$).

14.46 For the inhibited MWC scheme from Problem 14.44, make three-dimensional plots of the populations of the T and R quaternary structure as a function of ligand and inhibitor concentrations [x] and [I].

14.47 Derive the fractional saturation (f_{bound}) and conformational populations ($^{j}\alpha$) for the KNF model with s sites, using the partition function in terms of conformational constants $^{j}L_0$ (Equation 14.75).

14.48 Develop an expression for the binding partition function for the KNF tetramer in which subunit interactions are expressed in terms of the number of R:R interfaces formed. You can arrange your results as in Figure 14.18. How does this representation differ from that described in the text, where interactions are expressed in terms of the number of T:T interfaces broken?

14.49 For the KNF square, derive an expression for the fractional saturation, f_{bound}. Use the microscopic subunit description given in Figure 14.18.

14.50 For the KNF square, show that the population of subunits in the R-state is equal to the fractional saturation, f_{bound}.

14.51 For the four-subunit KNF square, can you come up with a set of parameters (σ and the combined parameter $\lambda\kappa_R$) that result in negative cooperativity in ligand binding?

References

Coxeter, H.S.M. 1973. *Regular Polytopes*. Dover Publications, Inc., New York.

Di Cera, E. 1996. *Thermodynamic Theory of Site-Specific Binding Processes in Biological Macromolecules*. Cambridge University Press, Cambridge, UK.

Trudeau, R. 1993. *Introduction to Graph Theory*. Dover Publications, Inc., New York.

Wyman, J. and Gill, S.J. 1990. *Binding and Linkage: Functional Chemistry of Biological Macromolecules*. University Science Books, Mill Valley, CA.

Symmetry in Macromolecular Structure

Goodsell, D.S. and Olson, A.J. 2000. Structural symmetry and protein function. *Annu. Rev. Biophys. Biomol. Struct.* 29, 105–153.

Marsh, J.A. and Teichmann, S.A. 2015. Structure, dynamics, assembly, and evolution of protein complexes. *Annu. Rev. Biochem.* 84, 551–574.

Allosteric Models

Koshland, D.E., Jr., Nemethy, G., and Filmer, D. 1966. Comparison of experimental binding data and theoretical models in proteins containing subunits. *Biochemistry* 5, 365–385.

Monod, J., Wyman, J., and Changeaux, J.P. 1965. On the nature of allosteric transitions: A plausible model. *J. Mol. Biol.* 12, 88–118.

Motlagh, H.N., Wrabl, J.O., Li, J., and Hilser, V.G. 2014. The ensemble nature of allostery. *Nature* 508, 331–339.

Appendix: How to Use Mathematica

Mathematica is a high-level mathematical computing software package that is available from Wolfram Research Incorporated. These instructions are written for the current Mathematica version (8) at the time of this writing. However, the online help is excellent, so if you want to take advantage of new functions, or if there are minor syntax changes, use the help commands. Also, Mathematica has a number of databases (stored online) that include information on chemistry, physics, social sciences, economics, etc., which can be readily imported for various types of analysis. One way to get started with Mathematica is to go through the introductory videos and work along with them. Here, specific tasks are presented that you will need to execute throughout this book (generating functions, plotting, numerical analysis, curve fitting), but there are other excellent features, including signal processing, graph theory, and topology.

One nice feature of Mathematica is that functions can be written on the fly instead of making a list (vector) of x values, and then operating on them to transport each x value to a y value. This is sometimes referred to as "symbolic math." Another nice feature is that the commands and results can be built into a single file (referred to as a "notebook"), rather than writing a script which executes a series of commands that put "results" (such as numbers, plots, and fits) into separate spreadsheets and windows. These Mathematica notebooks (designated with .nb extensions) can be used to prepare full reports (and homework assignments) in a single document. Finally, these notebooks can be prettied up much more than a simple MATLAB script file, with Titles, text, and images, and also with palates that make equations (fractions, exponents, square roots, sums, and integrals) look much more like we would write them, rather than as line-by-line computer script. And still, for the purists who think more stripped-down computer code is the way to go, all commands can be entered line by line using the Mathematica programming language. In my opinion, Mathematica is easiest for students to learn and for generating documents, but MATLAB may be easier to learn if you are already a fluent programmer. MATLAB and Scilab are quite similar, and although Scilab is much less developed, it is free, whereas Mathematica and MATLAB can be quite expensive (though many universities and laboratories have site licenses for them). Note also that MATLAB has "hidden costs," because many of the "toolbox" add-ons greatly expand the functionality, but they are not inexpensive.

As with the other appendices for mathematical analysis packages, in the instructions below, commands typed by the user in Mathematica are given in Courier font, with responses from Mathematica *italicized* when appropriate. Comments in the standard Helvetica font are sometimes given to the right.

GETTING STARTED IN MATHEMATICA

Once you install Mathematica on your machine (which works for PC and MAC), simply click the Mathematica icon to launch the program (again, you must be connected to the Internet). You will see a welcome window, which has some links to tutorials (these are useful), but for the purposes of this appendix, you should start a new notebook. In the upper left corner, there is a button that says, "Create new... Notebook." Click the notebook part and a new notebook will come up. Another way to do this is from the drag-down menu "file/new/notebook." Also, to the left of the welcome screen, it shows some notebooks you have worked on before, so if you want to continue work with one of them, just click that notebook.

Once you have created a new notebook, you are given a clean slate. You can put in titles, text, equations, calculations, plots, tables, etc. There are many ways you can do these things in Mathematica, depending on if you like pull-down menus, keyboard shortcuts, or formatting palettes. I will give some examples from each, and you can choose which one suits you. Imagine you are writing a homework assignment. The first thing you might want to do is give it a title, your name, and a date. Do this by adding text. One way is to go to the "Format" drag-down menu, and go to "style." You will see a lot of options for Titles, subsections, etc. Select "title" (Format/Style/Title) and enter the text for your problem set: "Problem set 1: The first law of thermodynamics." Then either hit a down arrow or click in the white space below what you have written. This gives you a new "cell" to type more information in. Make a subtitle (Format/Style/Subtitle) and type your name and the date. If you want, you can use a simple return to separate the name and date. Mathematica does not interpret a return as a command for execution, because it wants you to be able to use it as a text editor as well as a mathematics engine (to execute a command, you hold <Shift-enter>—more on that later). If you do not like the way your text is formatted, open the Palettes menu (top right) and select "Writing Palette." There you can center, change font size, type color, bold/italic, etc.

DOING ARITHMETIC AND SIMPLE CALCULATIONS WITH BUILT-IN FUNCTIONS

Clearly, being able to able to write pretty, formatted text is not the reason to use Mathematica. It is a bonus, but the point is to do math calculations. There are a few general rules you need to learn to do math in Mathematica:

1. Mathematica has built-in operators for arithmetic (see below) that can be used in line with numbers as you would in normal writing (3 + 2, for example).
2. Mathematica has lots of commands and built-in functions, and they all start with a capital letter (e.g., `Log`, not `log`). And when the command is a combination of two words (or word fragments), both words (or fragments) are capitalized. Examples include `PlotRange` or `ExpToTrig`.
3. Typically, when you use a command in Mathematica, it is followed by a square bracket [...] where the stuff inside the bracket is what the command is operating on.
4. Lists in Mathematica are made by curly brackets {xxx, yyy, ...} where the elements xxx, yyy, ... are treated as separate objects. Lists are very similar to vectors, and share some (but not all) characteristics of the vectors in MATLAB and Scilab (many of which make them easier to use). We will use lists a lot to compute and analyze data. Lists of lists (nested lists) have multiple curly braces (see below). These nested lists have the structure of matrices.
5. To group your operations together, use parenthesis, for example (4+6). This is in part because the other types of brackets have been used up with functions and lists (see above).
6. When you execute a Mathematica command, you must hit "shift" and "return" (or shift and enter) at the same time. Just hitting return makes Mathematica think you just want a new line within a cell. Here, I will designate this as <Shift-enter> for a while, so you get in the habit of it.
7. If you want to issue a set of similar commands within a single cell, must use a "return" between each. This will give you two multiple output cells, one for each input.
8. You can suppress the output of a command using semicolons (;) at the end of the command. This is particularly good when your command generates a very long list that will fill the screen. It is also nice for multiline (multicommand; see 7) input cells.

With these rules, we can get started with some simple examples of input and output in Mathematica notebooks. If you type 1+1 and then hit shift enter in your notebook, you get the following in your notebook[†]:

```
In[1]:= 1 + 1
Out[1]= 2
```

First, you get a blue In[1] := that tells you what your input was, and second you get blue Out[1]= that tells you what the output is (in this case, *2*). The lines on the right are "cells"—the rightmost one shows the first cell, and the left ones show input and output.

To make a second cell, just mouse-click below the lower line (the line with the plus tab on the left) or down-arrow until you see the line. You could next add 2 plus 3, and your notebook would look like this:

```
In[1]:= 1 + 1
Out[1]= 2

In[2]:= 2 + 3
Out[2]= 5
```

On the right, Mathematica designates another set of cells, and on the left, it increments the input and output numbers to 2. This would keep going as you added (and executed) more cells. You can save your notebook under file/save, but you should note that when you open it back up at a later date, none of the commands will be executed. Although variables were defined and commands were executed in the previous session, in the new session, they will not be there until you execute the newly opened notebook. You can do this line by line, but an easier way to do it is with the Evaluation/Evaluate notebook.

One feature of Mathematica is that you can refer to the output of the previous command using the percent (%) sign, and to earlier outputs using the % followed by the output number, and use these outputs to do further calculations:

```
In[1]:= 1 + 1
Out[1]= 2

In[2]:= 2 + 3
Out[2]= 5

In[3]:= %
Out[3]= 5

In[4].- 1%
Out[4]= 2

In[6]:= %2^%4
Out[6]= 25
```

[†] For the first few examples, I have included screenshots of the Mathematica notebook to give you a sense of how cells are structured. Thereafter, I will just switch to listing commands given and *output returned*.

A word of warning with this approach is in order. If you refer to results by line number (e.g., %4), save the notebook, and quit Mathematica, the next time you run the notebook in a new Mathematica session, the command numbers may change and you will get an incorrect result. These numbers are usually reliable within a single session, but not necessarily between sessions. As a result, the % feature should be avoided.

Other commands for arithmetic and calculations

Many of the calculations in Mathematic are intuitive, and you could probably guess at the symbols and syntax. Some common ones are listed here, but there are way too many to list, and you can find them quickly with the online help.

Arithmetic

Operation	Symbol or function	Example (and what it equals)
Addition	+	10+7 (17)
Subtraction	−	10−7 (3)
Multiplication	*	10*7 (70)
Division	/	10/7 (10/7)
Exponentiation	∧	10^7 (10,000,000)

Note that the output in the division example is the same as the input:

$$\text{In[1]:= } 10/7$$
$$\text{Out[1]= } \frac{10}{7}$$

This is because Mathematica gives exact answers unless you tell it otherwise. Sometimes you would rather see an approximate result, but in cases like 10/7, the value is a rational number that goes on forever. To get an approximate result, type //N after the input line (N for "numerical approximation"). This tells Mathematica to give you the result to six significant figures[†]:

$$\text{In[2]:= } 10/7//N$$
$$\text{Out[2]= } 1.42857$$

Transcendental functions

Operation	Symbol or function	Example (and what it equals)
Square root	Sqrt[]	Sqrt[9] (3)
Natural log (ln) of x	Log[x]	Log[15]//N (2.70805)
Base 10 log of x	Log[b,x]	Log[10,15] (1.17609)
e to the power of x	Exp[x]	E[5] (148.413)
Sine function of x	Sin[x]	Sin[Pi] (0)[‡]
Cosine function of x	Cos[x]	Cos[Pi] (−1)
Inverse sine	AcrSin[x]	Arcsin[1] ($\pi/2$)

[†] The //N is a shortcut to a function that can be run on its own. So, for example, the same could be achieved by saying N[Sqrt[10/7]]. Furthermore, the default of six significant figures can be changed by specifying the number desired after the argument of the N function. So if you want only three significant figures, you could enter N[Sqrt[10/7],3].

[‡] Note that the trigonometric functions calculate assuming the argument is in radians. Work with a degree value, you can use the function Degree[x], which converts to radians.

In addition, Mathematica has some *built-in constants* that are useful to know:

Constant	Symbol or function	Example (and what it equals)
π	Pi	Pi//N (3.14159)
e	E	E//N (2.71828)
∞	Infinity	Infinity//N (∞)

DEFINING VARIABLES OF DIFFERENT DATA TYPES

There are several types of objects that can be defined in Mathematica. These include scalar variables that are assigned a single numerical value (real or complex), string variables, lists (which are similar to vectors in MATLAB if they are lists of numbers, but they can also be lists of strings, graphics, functions, etc.), arrays (similar to matrices in MATLAB), and graphics of various types.

Scalar variables

The following example uses the assignment of variables to illustrate the simple calculation above, and can be run within a single cell with line breaks:

```
In[1]:= a = 1;
        b = 2;
        c = 3;
        d = a + a;
        e = b + c;
        f = c^d;
        d
        e
        f


Out[7]= 2

Out[8]= 5

Out[9]= 25
```

The variables a, b, and c are assigned the numerical values 1, 2, and 3 using an equals (=) sign. Like in MATLAB and Scilab, the use of semicolons is very useful for suppressing unwanted output. As you can see from the output numbers [7, 8, 9], the calculations still exist (e.g., in the above example, if you entered %5 as input, you would get out 5 (b + c)).

Lists

Lists are really important in Mathematica calculations. Lists are a series of numbers (real or complex), or other items, such as text strings, graphics, etc. In their simplest form (numbers), they have a lot of features in common with vectors. However, some of the features of vectors that can be a nuisance for data analysis do not apply to Mathematica lists. A little more on this later.

The simplest way to define a list is to put each object in the list in a set of curly braces, separated by commas:

```
{1,2,3,4,5}    <Shift-enter>    returns
Out[1]:=  {1,2,3,4,5}
```

Another very similar way to make a list is the command "`List`."

```
List[1,2,3,4,5]     <Shift-enter>          returns
Out[2]:= {1,2,3,4,5}
```

In other words, it is the same thing. These lists can be assigned to variables (again, starting with lowercase is a good idea). For example,

```
mylist={1,2,3,4,5} <Shift-enter>          returns
Out[3]:=  {1,2,3,4,5}
```

Again, it is the same thing. But now, you can call it back because you have created a variable that it is assigned to. If you say

```
mylist     <Shift-enter>          Mathematica returns
Out[4]:=  {1,2,3,4,5}
```

The advantage is that Mathematica can easily manipulate your list. If you want the square of the integers in `mylist`, type

```
mylist^2     <Shift-enter>          Mathematica returns
Out[5]:= {1, 4, 9, 16, 25}
```

Similarly, the factorial of the numbers in your list can be had by typing

```
Factorial[mylist] <Shift-enter>  or
mylist! <Shift-enter>            In both cases, Mathematica returns
Out[6]:= {1, 2, 6, 24, 120}
```

And unlike in MATLAB and Scilab, the elements of lists (of the same list) can be directly multiplied without any special characters to indicate "element-by-element" multiplication, and not a vector product. For example,

```
lista={1,2,3,4,5};
listb={5, 4, 3, 2, 1};
listc=lista+listb;
listc               <Shift-enter>          returns

Out[7]:= {5, 8, 9, 8, 5}
```

Automated lists

Lists are handy because they can be used as a range of x values for a function that operates on each element of the list. For example, we can create a list of volumes and then calculate pressures, given a temperature value, the number of molecules (or moles), and an equation of state (e.g., an ideal gas law, a van der Waals). We then might want to plot the results, or even add some noise to the data, and see how well we can extract constants (like van der Waals a and b constants). In short, we might want to simulate some data, and see what can be done with it. Rather than tediously typing in each of the many volumes we want to evaluate pressure, there are more efficient list generators that can simplify the task. One is called the "Range" function, and it is good for integers. If you enter `Range[n]`, Mathematica will return a list of integers starting at one and going up to n.

For example, if you type

```
Range[12] <Shift-enter>          and Mathematica returns
Out[1]:={1, 2, 3, 4, 5, 6, 7, 8, 9, 10, 11, 12}
```

If you want to start the list at an integer i other than 1, and end it at integer n, you can type 20, you can type `Range[i,n]`

For example, if you want a list that goes from 3 to 17, type

```
Range[3,17] <Shift-enter>          and Mathematica returns

Out[2]:= {3, 4, 5, 6, 7, 8, 9, 10, 11, 12, 13, 14, 15, 16, 17}
```

And if you want a list that does not go in increments of 1, you can specify smaller (or larger) increments for successive elements within the list. This is done with the command `Range[i,n,j]`, where the list starts at i, goes to n, and the numbers in the list change by increment j. For example, if you want a list that goes from 3 to 7 in steps of 0.2, type

```
Range[3,7,0.2]        <Shift-enter>        and Mathematica returns
```

```
Out[3]:={3., 3.2, 3.4, 3.6, 3.8, 4., 4.2, 4.4, 4.6, 4.8, 5.,
5.2, 5.4, 5.6, 5.8, 6., 6.2, 6.4, 6.6, 6.8, 7.}
```

Importantly, the starting and ending values do not need to be integers, but it is good that you specify an increment that actually gets you to the ending value. Here is an example where the list starts at 0.01, and increments by units of 0.01 to a final value of 1.0:

```
Range[0.01,1,0.01]   <Shift-enter>        and Mathematica returns
```

```
Out[4]:={0.01, 0.02, 0.03, 0.04, 0.05, 0.06, 0.07, 0.08, 0.09, 0.10,
   0.11, 0.12, 0.13, 0.14, 0.15, 0.16, 0.17, 0.18, 0.19, 0.20, 0.21,
   0.22, 0.23, 0.24, 0.25, 0.26, 0.27, 0.28, 0.29, 0.30, 0.31, 0.32,
   0.33, 0.34, 0.35, 0.36, 0.37, 0.38, 0.39, 0.40, 0.41, 0.42, 0.43,
   0.44, 0.45, 0.46, 0.47, 0.48, 0.49, 0.50, 0.51, 0.52, 0.53, 0.54,
   0.55, 0.56, 0.57, 0.58, 0.59, 0.60, 0.61, 0.62, 0.63, 0.64, 0.65,
   0.66, 0.67, 0.68, 0.69, 0.70, 0.71, 0.72, 0.73, 0.74, 0.75, 0.76,
   0.77, 0.78, 0.79, 0.80, 0.81, 0.82, 0.83, 0.84, 0.85, 0.86, 0.87,
   0.88, 0.89, 0.90, 0.91, 0.92, 0.93, 0.94, 0.95, 0.96, 0.97, 0.98,
   0.99, 1.00}
```

It is not very pretty, but one thing you can do is assign it to a variable to use later, and you can suppress the output with a semicolon.

Here is a little notebook which does just that, calculating the volume for an expansion.

```
In[1]:= V = Range[0.01, 1, 0.01];
```

```
In[2]:= V
Out[2]= {0.01, 0.02, 0.03, 0.04, 0.05, 0.06, 0.07, 0.08, 0.09, 0.1, 0.11, 0.12, 0.13, 0.14, 0.15,
        0.16, 0.17, 0.18, 0.19, 0.2, 0.21, 0.22, 0.23, 0.24, 0.25, 0.26, 0.27, 0.28, 0.29,
        0.3, 0.31, 0.32, 0.33, 0.34, 0.35, 0.36, 0.37, 0.38, 0.39, 0.4, 0.41, 0.42, 0.43,
        0.44, 0.45, 0.46, 0.47, 0.48, 0.49, 0.5, 0.51, 0.52, 0.53, 0.54, 0.55, 0.56, 0.57,
        0.58, 0.59, 0.6, 0.61, 0.62, 0.63, 0.64, 0.65, 0.66, 0.67, 0.68, 0.69, 0.7, 0.71,
        0.72, 0.73, 0.74, 0.75, 0.76, 0.77, 0.78, 0.79, 0.8, 0.81, 0.82, 0.83, 0.84, 0.85,
        0.86, 0.87, 0.88, 0.89, 0.9, 0.91, 0.92, 0.93, 0.94, 0.95, 0.96, 0.97, 0.98, 0.99, 1.}
```

In the second line, I typed

```
V               <Shift-enter>
```

so you can see that indeed, a variable is associated with the list that has the name "v." A handy command that you will use in calculations that involve combining different lists in various ways is

```
Length[{list}]              <Shift-enter>
```

which gives the number of elements in the list. For example, the number of volumes in the list V above can be calculated by saying

```
In[15]:= Length [V]
Out[15]= 100
```

Now we can manipulate the list to calculate the pressure during a reversible isothermal transformation over this volume. If the gas is ideal, we can use the equation

$$p = \frac{nRT}{V}$$

to calculate the isotherm. To do so, we need to pick values for T and n, and use the known value for R (about 8.314 in units of J mol^{-1} K^{-1}). The following notebook does this[†]:

```
In[1]:= V = Range[0.01, 1, 0.01];

In[4]:= R = 8.314;

In[5]:= T = 300;

In[7]:= n = 0.01;

In[9]:= p = n*R*T/V;

In[10]:= p

Out[10]= {2494.2, 1247.1, 831.4, 623.55, 498.84, 415.7, 356.314, 311.775, 277.133, 249.42,
          226.745, 207.85, 191.862, 178.157, 166.28, 155.888, 146.718, 138.567, 131.274,
          124.71, 118.771, 113.373, 108.443, 103.925, 99.768, 95.9308, 92.3778, 89.0786,
          86.0069, 83.14, 80.4581, 77.9438, 75.5818, 73.3588, 71.2629, 69.2833, 67.4108,
          65.6368, 63.9538, 62.355, 60.8341, 59.3857, 58.0047, 56.6864, 55.4267, 54.2217,
          53.0681, 51.9625, 50.902, 49.884, 48.9059, 47.9654, 47.0604, 46.1889, 45.3491,
          44.5393, 43.7579, 43.0034, 42.2746, 41.57, 40.8885, 40.229, 39.5095, 38.9719,
          38.3723, 37.7909, 37.2269, 36.6794, 36.1478, 35.6314, 35.1296, 34.6417, 34.1671,
          33.7054, 33.256, 32.8184, 32.3922, 31.9769, 31.5722, 31.1775, 30.7926, 30.4171,
          30.0506, 29.6929, 29.3435, 29.0023, 28.669, 28.3432, 28.0247, 27.7133, 27.4088,
          27.1109, 26.8194, 26.534, 26.2547, 25.9812, 25.7134, 25.451, 25.1939, 24.942}
```

The last line is to simply show that the product of the calculation is a list, which is a good thing, since our goal was to get pressure values at a bunch of different volumes. And it is clearly decreasing with increasing V, which matches our intuition. It is good practice when doing work with Mathematica to look at your output to check that is what you think it should be, and then you can suppress it with a semicolon to save space and simplify your code. A better way to check it is with a plot, which we will describe below.

Another way to make an automated list is with the command "Table." Table is extremely versatile, and can generate lists of all sorts of objects, including graphics arrays where one parameter is changed among the different list elements. Table works like Range, but it uses a function or an expression to generate the list. In addition to specifying starting and ending values and a step-size, an index (or counter) is defined, and it is used in an expression. In general, the format is

```
Table[expression, {i, imin, imax, step}]
```

where expression depends on the index i. As an example, the pressure data can be generated above without the vector V, using the command

```
Table[n*R*T/V,{V, 0.01, 1, 0.01}]
```

Manipulating elements in lists

There are many times when you will want to be able to obtain values for specific elements, or a range of elements, from one or more lists. This can either be done using the command "Part," or by using multiple square brackets. For a simple flat list like

---

[†] Note if you are wondering why the numbers in the "In[]" and "Out[]" statements do not increment as 1, 2, 3, 4, etc., it is because I messed up on some of them when I typed them the first time, and had to rerun a cell a second time. This is why you need to be careful calling a cell output by number (e.g., %23), which may evaluate a different cell the next time you run it. Assigning results like lists to variables is a really good way to get around this.

```
mylist=2*Range[12]
{2, 4, 6, 8, 10, 12, 14, 16, 18, 20, 22, 24}
```

`Part[mylist,3]` returns the third element in the list (6). The same result is obtained by the shortcut `mylist[[3]]`. We will use this double-bracket shortcut for the examples below.

If you want a range of elements from the list, for example, the fourth through the eighth, you can either `mylist[[4;;8]]`, which returns the list {8, 10, 12, 14, 16}.

Often, one has a nested list, and wants a single column or a single row. Working from the outside in, the first element of a nested list is the first row of data. For example, in the list

```
littlelist={{a, b}, {c, d}}
```

the first element is {a, b}. Thus, saying `littlelist[[2]]` returns the row {c, d}.

To get the first element of the second row, say `littlelist[[2,1]]`, which returns c. Notice that the command works from left to right, saying "take the second element (a row), and from that, take the first element" (a number). This works in the same way as elements of a matrix are specified (e.g., a_{ij} is the ith row, jth column element).

To get a *column*, use the "`all,j`" identifier to indicate that you will be selecting an element from the jth column. For the medium-sized list,

```
mediumlist={{a, b, c}, {d, e, f}, {g, h, i}}
```

you can get the entire second column by saying `mediumlist[[All,2]` or `Part[mediumlist,All,2]`, which returns {b, e, h}. If you want the second and third column, you can say `mediumlist[[All, 2;;3]`, which returns {{b, c}, {e, f}, {h, i}}.

The above command could also be done with a comma and curly braces to specify columns 2 and 3, that is, `mediumlist[[All, {2,3}]`.

This last approach is useful if you want nonadjacent columns, for example, columns 1 and 3:

```
mediumlist[[All, {1,3}]
```

Finally, there are times when you want to delete a row or a column from a nested list (for many data files, the first row is a text string of what the variables are—good to know, but bad for plotting and calculations). If you want to delete only the first row (or first element in a nonnested list), you can use the "`Rest`" command:

```
Rest[mediumlist] returns
{{d, e, f}, {g, h, i}}
```

The command "`Take`" can be used similarly, but allows several rows to be skipped (either from the beginning or end of a nested list).

Separating lists into nested lists can be done with the command "Partition [`list,n`]," where n is the number of elements you want in each sublist. In its simplest form, the number of elements in the list must be an integer multiple of n. `Partition` keeps the elements in order. For example, to break `mylist` into a nested list with three sublists, four elements each,

```
Partition[mylist,4]        returns
{{2, 4, 6, 8}, {10, 12, 14, 16}, {18, 20, 22, 24}}
```

The command "`Flatten`" undoes what `Partition` does.

There are many other ways to manipulate *single* lists, for example, deleting elements, removing elements. See the Mathematica documentation center for more information.

It is often useful to *combine lists together*. For example, let us combine `mylist` above with another list containing odd integers:

```
anotherlist=mylist-1
{1, 3, 5, 7, 9, 11, 13, 15, 17, 19, 21, 23}
```

To take lists of the same length and interleave the elements, use the command "Riffle":

```
Riffle[anotherlist,mylist]
{1, 2, 3, 4, 5, 6, 7, 8, 9, 10, 11, 12, 13, 14, 15, 16, 17, 18,
19, 20, 21, 22, 23, 24}
```

(Note that Riffle is sensitive to the order that you specify your lists.) `Riffle[mylist,anotherlist]` gives a different result. If you then Partition the riffled list above into sublists with two elements, the result is that the two lists are combined into sublists containing the first, second, third...element from each list:

```
Partition[Riffle[anotherlist,mylist],2]
{{1, 2}, {3, 4}, {5, 6}, {7, 8}, {9, 10}, {11, 12}, {13, 14},
{15, 16}, {17, 18}, {19, 20}, {21, 22}, {23, 24}}
```

(Note that the inner command, `Riffle`, is done first, followed by the second, `Partition`.) This is handy when you want to combine an x list and a y list for fitting.

A simpler way to get a nested list of x and y pairs from a list of x values and a list of y values is to first make a list of lists of the two values and then use the `Transpose [ ]` command. For example, for the following two lists:

```
xlist={1,2,3,4,5}
ylist={1,4,9,26,15}
```

the command

```
Transpose[{xlist, ylist}]
```

gives a list of lists of x, y pairs (note that without the `Transpose` command, we would just have a list of x values followed by a list of y values).

GENERATING LISTS OF RANDOM NUMBERS

One useful feature of Mathematica is its ability to generate random numbers from a variety of distributions. These include discrete distributions such as the binomial distribution, and continuous distributions such as the Gaussian distribution. In addition, there are many other distributions, including Poisson, geometric, chi-squared distribution, Student *t*, and *f*-distribution, to name a few.

Before describing how to generate random numbers, there is an important command you should put in all your notebooks with random number generation, and that is

```
SeedRandom[integer]
```

This command tells the random number generator in Mathematica where to begin. The integer is any number you choose to put in, and it will determine the sequence of random numbers generated. If this makes it sound like the "random numbers" generated by the computer are not really random, that is because random numbers generated by the computer are not really random. The computer needs to be told where to start generating random numbers. Using the SeedRandom command will ensure that you get the same random numbers if you run your notebook again. This reproducibility is more valuable than it sounds.

To get a random number from a particular distribution, use the command `RandomVariate[...]`. The argument of the RandomVariate command is whatever distribution you want random numbers from. For example, to get a random number from a Gaussian distribution centered on zero with a standard deviation of 0.01 (which might be useful for simulating 1 percent error), you say

```
RandomVariate[NormalDistribution[0, 0.01]]
```

If you want to generate a list of 1000 random numbers from a normal distribution with a mean of 4 and a standard deviation of 2, type

```
RandomVariate[NormalDistribution[4, 2], 1000]
```

Note that this can always be directed to a variable by writing `variable=` in front of the command above. Plotting this list using the `Histogram[...]` function gives

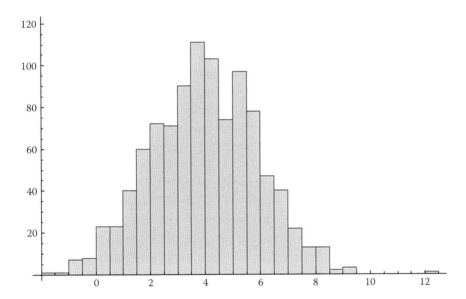

In addition to generating random numbers from built-in Mathematica distributions, there are commands to get other types of random numbers. Useful commands include

`RandomInteger[{min, max}]`	Generates a random integer
`RandomReal[{min, max}]`	Generates a random real number
`RandomChoice[list]`	Selects a random element from the list.

READING FROM AND WRITING TO FILES

One of the important applications for Mathematica is analysis of data sets (sometimes large data sets) collected outside Mathematica. Another is generation of data in Mathematica for analysis by other programs. In both cases, external files are required. The two operations used for this are "`Import`" and "`Export`." These commands will read (and write) files from (and to) your home directory (e.g., /Users/doug). Following the description of these commands is a section for how to change the "working" directory so that you do not have to move files to and from their final destination.

Importing data from files. The command used to read data sets into Mathematica from external files is "`Import`." For example, if you want to read a file from your computer that is named "filename.ext" that has a single column of numbers, and put it into a Mathematica list, you say

```
Import["filename.ext","List"]
```

The extension describes how the elements of the list are separated. Tab-separated elements are imported from `.txt` files. Comma-separated values are imported from `.csv` files. Space-separated values are imported from `.prn` files.

When you import a file with data arranged in multiple columns and you want to preserve the column structure, you say

```
Import["mybigfile.txt","Table"]          which returns
{{1, 1}, {2, 4}, {3, 9}, {4, 16}}
```

This generates a nested list. You will use this kind of "`Table`" importing a lot when you bring experimental measurements into Mathematica for analysis, especially for displaying data as a set of points using "`ListPlot`," and curve fitting. If you do not include the "`Table`" command, Mathematica imports the contents of the file, but it just treats it as a string of numbers.

A full list of allowable extensions can be seen by entering `$ImportFormats` in Mathematica. The safest bet for importing and exporting is a simple text (`txt`) file, which can be read by all other programs, and does not have any extra formatting (that can produce unintended, and thus undesirable, consequences) when importing back into Mathematica. Tab-delimited data imports as separate columns, as does comma-separated data.

Exporting data to files. Exporting works much the same way as importing. If you have a list, for example,

```
mylist={1,2,3,4,5,6}
```

and you want to save it to a text file, you can say

```
Export["myfile.txt", mylist]
```

This will create a file named `mylist.txt`. Note, the file does not need to exist beforehand, so you can call it anything you want, but if it does exist, it will overwrite (without asking permission, so be careful!). The file created by the `Export` command above puts each element of the list on a separate line, that is, it will be a column of data. If instead, you have a list of lists, for example,

```
mybiglist={{1,1},{2,4},{3,9},{4,16}}
```

and you want to save it as a file with a table format, enter

```
Export["mybigfile.txt",mybiglist,"Table"]
```

which creates (or overwrites) a file named `mybigfile.txt` in your current director with the contents

```
1. 1
2. 4
3. 9
4. 16
```

Allowable extensions for export can be seen by entering `$ExportFormats` in Mathematica. Note that export can also be used to create graphics files, with formats such as JPEG, PNG, EPS, PDF, and TIFF. This is one way to save plot output for use in other applications. See more on the Export command for graphics (and Image resolution!) in the plotting section below.

Changing the directory for import and export. Typically, your work is being done in a subdirectory of your file system. It is rather cumbersome to move files you need to import to Mathematica up to the default (home) directory, and put exported files back down when you are done (and if you forget, you get a very cluttered home directory very fast!). A good solution is to change the directory. This is done with the `SetDirectory["dir"]` command, for example,

```
SetDirectory["/Users/doug/book/Chapter 2/Figures"]
```

The format of the directory path can vary a little from one operating system to another (the path syntax above is for a Mac). To make sure you have really set the directory you want, you can say

```
Directory[]                which returns
/Users/doug/book/Chapter 2/Figures
```

You can list the contents of your current (working) directory with the command

```
Filenames[]
```

PLOTTING IN MATHEMATICA

The simplest and most common plot that can be made in Mathematica is the *two-dimensional plot* (2D) (an x–y plot). There are two basic plots that can be made. One is of the type one would generate with discrete data points, like the pressure-volume lists we generated above for an ideal gas. The other plots functions directly.

Plotting discrete sets of data. This is done with the function "`ListPlot[data]`." If the argument of `ListPlot` is simply a list of numbers (e.g., $\{1,4,9,16,25\}$), the numbers will just be evenly spaced with x values of $\{1,2,3,4,5\}$. For our *p-V* relationship above, the volume values, which we would like plotted on the x-axis, do not cover this range, but go from 0.01 to 1 in steps of 0.01. This can be specified as the option `DataRange->{`x_{min}`,`x_{max}`}` in `ListPlot`. For example, for our *p-V* relationship above, we would type

```
In[16]:= Listplot [p, Datarange → {0.01, 1}]
```

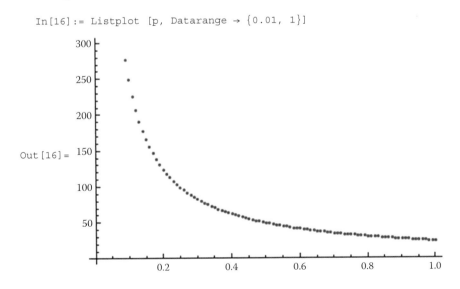

If the points seem a little too small (the default is 0.008, which is the fraction of the overall width of the graph [i.e., 0.8%]), you can change by modifying the "`PlotStyle`":

In[21]:= Listplot [p, Datarange → {0.01, 1}, Plotstyle → Pointsize[0.02]]

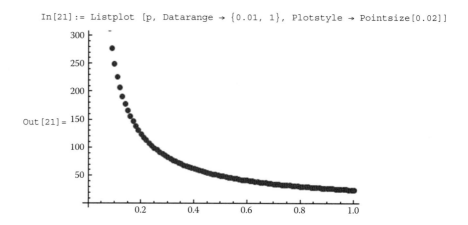

Out[21] =

Plotting functions directly

Plotting is also useful for directly exploring how certain functions behave, and to compare different functions. One (not very good) way to do this would be to generate a bunch of data values through lists (like our volume list), transform it to a dependent variable (like we did to get pressure), and "connect the dots." This is how a function might be represented in Excel or MATLAB. But this is a lot of work. One nice feature of Mathematica is that you can directly plot "functions" (e.g., $y = 3 + x^{0.5} + \sin(x)$). This is sometimes referred to as "symbolic math." The general syntax to plot a function f(x) is `Plot [f(x), {x, xmin, xmax}]`. Here is an example for the function given above[†]:

```
Plot [3 + x^0.5 + Sin[x], {x, 0, 100}]
```

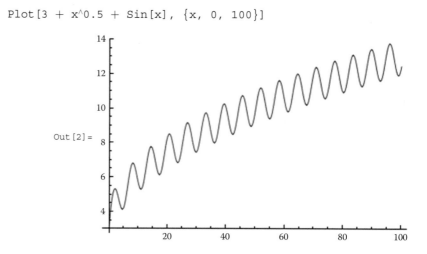

Out[2] =

In the example above, it is pretty clear what the parts of the function do: the sin(x) part gives the wiggle, the square root of x (i.e., x^0.5) gives the gradual rise, and the 3 gives an offset to the whole function. But in many cases, it is not so easy to see how parts of a function combine to give the whole. In this case, it is instructive to plot the parts along with the whole. As is often the case, there are several ways to do this in Mathematica. One way is that the command "Plot" allows you to plot a *list* of functions at the same time, in a single command. As always, lists are put in curly braces {} and are separated by commas. So, for example, the function could be plotted above, along with the three individual terms in the sum using the command

```
Plot [{3 + x^0.5 + Sin[x], 3, x^0.5, Sin[x]}, {x, 0, 100}]
```

[†] Note that within the plot command, which requires square brackets, there are additional square brackets following the Sin command, since it is also a function. In a sense, these are nested functions, and can be nested as deeply as one likes.

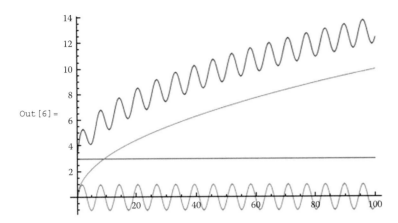

Out [6] =

This plot gives a clear picture of how the sin, the square root, and the offset combine to give the whole function. Mathematica uses a default set of colors to draw each function to a different color, which can help distinguish when things get complicated. It also chooses the y-axis scale to avoid a lot of white space, and show what it thinks are the most interesting aspects of the function. Sometimes you will disagree about what the most interesting aspects of the function are, and you will want to change the scale. In this case, use the command `PlotRange -> {ymin, ymax}`. Note, the x-axis scale is set by the Plot argument `{x,0,100}`, but can be narrowed in the `PlotRange` command as well.

The method above for plotting multiple different functions can get a little cumbersome if you want to include a lot of functions. There is a different way to plot multiple functions that is a little bit simpler, though it takes multiple lines. This is done by plotting each function individually, naming each plot. For example, for a bimolecular association reaction, which has the form (Chapter 13)

$$f_{bound} = \frac{K[x]}{1 + K[x]}$$

where $[x]$ is free ligand concentration and K is the binding constant, three different plots with three different binding constants can be generated as follows:

```
p1=Plot[0.3*x/(1+0.3*x),{x,0,10},PlotRange->{0,1}]
p2=Plot[1.0*x/(1+1.0*x),{x,0,10},PlotRange->{0,1}]
p3=Plot[3.0*x/(1+3.0*x),{x,0,10},PlotRange->{0,1}]
```

The command `Show [...]` can then be used to show multiple curves on the same set of axes as follows:

```
Show[p1, p2, p3, PlotRange -> {0, 1}]
```

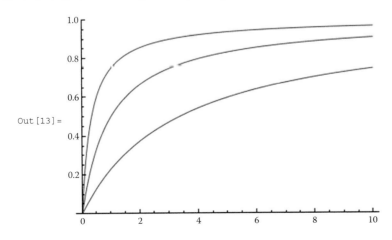

Out [13] =

In all of the plotting commands in this example, the option `PlotRange->{0,1}` was inserted so that when the plots were combined with the `Show` command, they span the entire x-axis. You should see what happens when this option is deleted from the four lines of code above.

One of the down sides of the `Show[{plot list}]` example above is that all three curves are the same color. This makes it hard to tell which curve corresponds to which value of the binding constant *K*. In addition, there are other minor problems with the plot. First, the numbers on the axes are kind of small, and hard to see. Second, there are no labels on the axes. It is bad form to present a plot without labeling the axes. Fortunately, there are options that can be included in plot commands (or even at the start of a notebook, if you know in advance that you would like to change some variables (font, font size, axis style) for all of your plots. The next section describes ways to improve the clarity and look of Mathematica plots.

Making 2D plots beautiful

The basic command for plotting above is "`Plot[f(x),{x,xmin,xmax}]`." Additional options can be added before the last square bracket to change the appearance of the plot. That is,

```
"Plot[f(x),{x,xmin,xmax},option1,option2...]"
```

To get a full list of options that can be modified in plots, you should look under the "options" for "graphics" in the Wolfram documentation center.

Changing the appearance of a plot. "`PlotStyle->`" can be used to change the contents of what is plotted (i.e., the color, dashing, thickness, reflectance, etc. of points, lines, and surfaces). This is done with "directives" that `PlotStyle` points to. These directives include

`Color`	can be one of Mathematica's default colors,[†] or it can be defined by various color schemes such as rgb, hsb, cmyk.
`Thickness[r]`	changes the thickness of lines, where r is the fraction of the width of the total plot.
`Pointsize[d]`	changes the size of markers, for example, the ListPlot points, where d is the fraction of the width of the total plot.
`Dashing[{r1,r2}]`	makes lines dashed. The values of r_i are, as above, the fraction of the total width of the plot. When two r values are given, they specify the length of the dash and the white space connecting to the next dash. To avoid messing with these values, can simply be given as `Tiny`, `Small`, `Medium`, or `Large`.

So for example, the simple curve

```
Plot[Sinc[x],{x,-20,20}]
```

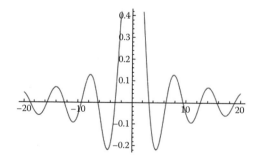

[†] Mathematica's default colors are Red, Green, Blue, Black, White, Gray, Cyan, Magenta, Yellow, Brown, Orange, Pink, Purple, LightRed, LightGreen, LightBlue, LightGray, LightCyan, LightMagenta, LightYellow, LightBrown, LightOrange, LightPink, and LightPurple.

can be made fancier by adding the PlotStyle option:

```
Plot[Sinc[x],{x,-20,20},
PlotStyle-> {Purple,Thickness[0.007],Dashing->{0.01,.01}}]
```

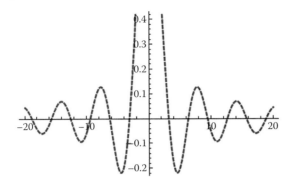

Changing the region that is plotted. Mathematica tries to show what it things will be the most informative y region in a plot. However, sometimes its choice cuts off part of the plot that you are interested in. You can change the axis limits of your plot with the PlotRange command. If just one pair of numbers is given as a list, it will change the dependent variable range only (y_{min} and y_{max} for a 2D plot, z_{min} and z_{max} for a 3D plot). If instead two ranges are given for a 2D plot will change both x and y; three ranges will change x, y, and z for a 3D plot. If you simply want to be sure all your data is showing, say PlotRange->All.

Here is an example where the range of the plot above is changed so that you can see the top of the Sinc(0)=1 peak:

```
Plot[Sinc[x], {x, -20, 20},
PlotStyle -> {Purple, Thickness[0.007], Dashing->{0.01, 0.01}},
PlotRange->{-0.3, 1.5}]
```

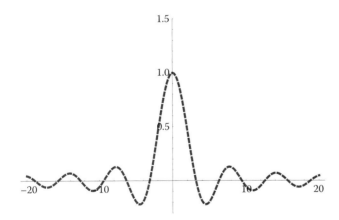

Adding labels to plots. In addition to showing the lines, points, or surfaces of a plot, there is important information that should be included in a plot, so that you and other readers can tell what they are looking at. For example, a 2D plot should indicate what quantities are being plotted, what the units are, and what different lines mean. In addition, it is often helpful to give an overall title to the plot. These commands are given below.

To put a title across the center of the plot using the command

```
PlotLabel->"Text"
```

If you want to put labels on the axes of your plot, there are two ways to do it.

```
AxesLabel->{xlabel, ylabel}
```

puts labels on the x- and y-axes, with the label to the right of the x-axis and directly on top of the y-axis. Unfortunately with this positioning, the y-label can look crowded, and the x-label gets lost. A way to get labels in the middle of the axes is to first draw a frame, and then use the command `FrameLabel`. The syntax for these commands is

```
Frame->True
FrameLabel->{xlabel, ylabel}
```

Here are some labeling commands that add to the sinc(x) plot from above:

```
Plot[Sinc[x],{x,-20,20},
PlotStyle -> {Purple, Thickness[0.007], Dashing -> {0.01, 0.01}},
PlotLabel -> "Sinc(x) versus x plot", Frame -> True, FrameLabel ->
{x, Sinc(x)}]
```

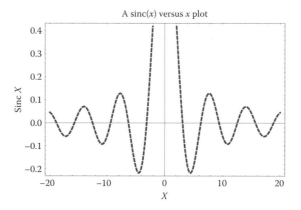

How to set a Mathematica notebook so all your plots have the same features. It is often the case that you will generate multiple plots in a single notebook. In such cases, you will want to change fonts, line thicknesses, etc., the same way for each plot. Rather than entering these commands for each plot, you can set them once at the start of your session. This is done with the command `SetOptions`. For example, you can set the font type and the size for all of the labels using the command.

```
SetOptions[Plot, BaseStyle -> {FontFamily -> "Helvetica",
FontSize -> 14}];
```

Now, subsequent text will be in 14-point Helvetica font. But for other graphical output, such as histograms and 3D plots, analogous "SetOptions" commands will need to be given in which `Plot` is changed to `Histogram` or `Plot3d`. It is often handy to adjust other features using a SetOptions command, for example, line thickness, PlotStyle, marker size (see above), and the "aspect ratio" of the plot (the height to the width). The default aspect ratio in Mathematica is the inverse of the number "phi" (about 1:1.68). I find this aspect ratio a bit squashed, and prefer a 3:4 aspect ratio instead, or about 0.75.

Exporting plots with high resolution. Many of the plots above were generated simply by selected by clicking on the plot, and writing it out as a file with specific graphic format (.png is nice, but .jpg is fairly universal too). These graphics files can then be imported as a picture into a standard document (like a Word or PowerPoint file). However, the default resolution on Mathematica graphics is rather low (72 dpi). This is adequate for "screen" resolution, but for printing or projection, such bitmapped formats end up looking rather pixelated (an alternative to this is to export a vector format like .eps).

To create a graphics file with higher resolution, the command `Export[]` should be used in combination with the command `ImageResolution`. The following command

```
Export["filename.ext", graphic, ImageResolution -> 300]
```

will create a file in your current directory of the appropriate format (specified by .ext; e.g., "`filemane.png`," "`filename.jpg`") at a resolution of 300 dpi, which should be adequate for most types of production. The three images below were exported at 30, 72, and 300 dots per inch (dpi) (from left to right).

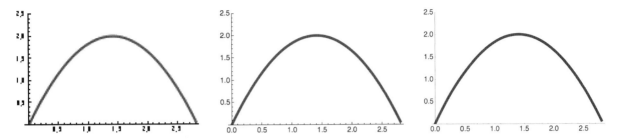

3D plots

One nice feature of Mathematica is the ability to generate and manipulate 3D plots. This allows functions of two independent variables, $z = f(x,y)$, to be plotted. Given that most thermodynamic systems depend on two (or more) independent variables, such plots are a useful way to visualize thermodynamic quantities. 3D plots are generated in much the same way that 2D plots are. The general syntax is

```
Plot3D[f(x,y),{x,x_min,x_max},{y,y_min,y_max}]
```

`f(x,y)` is an expression that can either be given inside the `Plot3D` command or can be defined as a separate function. The second and third lists define the independent variables to plot against and their limits (they do not need to be called x and y). For example, to plot the function

$$f(x,y) = \mathrm{sinc}(x)\mathrm{sinc}(y)$$

where sinc(x) equals $x^{-1}*\sin(x)$, the command is

```
Plot3D[Sinc[x]*Sinc[y], {x, -10, 10}, {y, -10, 10}]
```

This generates the plot shown in panel A below (with font changes).

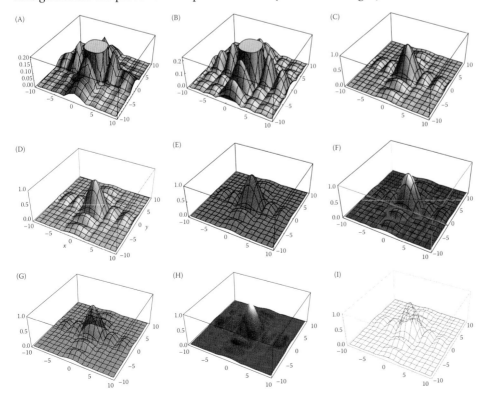

Though the default plot A is striking, it has some flaws. It appears much more jagged than the actual surface, because too few points are used to render the plotted surface. In panel B, more points are included with the command `PlotPoints -> 100`. In panel C, the z-axis is changed so that the peak is not clipped, using the command `PlotRange -> All`. In panel D, labels are added with the command `AxesLabel -> {x,y,}`. In panel E, the color is changed to red with the command `PlotStyle -> Red`. In panel F, a color gradient (Cherrytones) is used with the command `ColorFunction -> "CherryTones"`. In panel G, the plot is made transparent with the command `PlotStyle -> Opacity[0.5]`. In panel H (probably my favorite), the mesh lines are removed with the command `Mesh -> None`. And in panel H, only the mesh lines are shown with the command `PlotStyle -> None`, which gives a rather ghostly appearance!

As the 3D plots above show, a nearly endless variety of images can be generated, many of which are aesthetically striking. In addition, Mathematica can generate a number of other types of plots, many of which are used in this book, including contour plots, vector plots, bar charts, and 3D histograms. The Wolfram Documentation in the help pull-down provides syntax and examples of all these plots.

NONLINEAR LEAST SQUARES FITTING

One of the most important tools we will use in Mathematica is the nonlinear least-squares fitting tool, which allows a model $f(x; p_1, p_2, ...)$ to be fitted to observed data $(y_{obs} = y_1, y_2, ..., y_n)$ by adjusting parameters $(p_1, p_2, ...)$ to match the model to the data. The approach of least-squares optimization is described in detail in Chapter 2. Here we will simply re-state the problem and describe how to use Mathematica to do nonlinear least-squares optimization. The method we will use is a method called "least squares." The "least" refers to a minimization, the "squares" refers to the sum of the squared deviation between the model and the data. This deviation can be written as

$$SSR = \sum_{i=1}^{n} \left\{ y_i - f\left(x_i; p_1, p_2, ...\right) \right\}^2 \tag{A.1}$$

The algorithm goes through a number of iterations where it adjusts the parameters $(p_1, p_2, ...)$, and calculates the residual (the term in the curly brackets), and it returns the parameters that minimize SSR, that is, those that result in the "least squares." There are a number of important considerations relating to parameter optimization using least-squares methods; many of these have been discussed extensively in the literature (e.g., Johnson, 2008).

If you are familiar with linear regression, then you have some experience with "least-squares" approaches to determining parameter values (in that case, the parameters are the slope and intercept of a line). For linear regression, there is an exact analytical formula that can be used to find the slope and intercept that minimize the residual, and thus, SSR. However, unlike the equation for a line, most of the functions we will fit to data are not linear in their unknown parameters $(p_1, p_2, ...)$. In general, for such "nonlinear" least-squares problems, there is no direct, analytical solution.

Instead, nonlinear least-squares methods work by an iterative search mechanism where the parameters are adjusted a little bit, and if χ^2 decreases as a result of the adjustment, the new parameters are kept, and used in another optimization cycle. The process repeats until the parameters do not get better. This process can be considered a search for an optimum (minimum) in n-dimensional parameter space. There are a number of problems that can occur in such a search: sets of parameters can be found that provide a local minimum in χ^2, but not a global minimum. That is, another set of parameters result in a lower χ^2 value, but they are not identified because they are far away in parameter space. This is a problem of searching space efficiently. However, a very efficient search of parameter space may be very costly in terms of computing time. A number of different "optimization methods" have been developed, some with speed in mind, and others with thorough searching in mind. These include direct search methods (including Nelder–Mead "Simplex" search, Monte Carlo searching), "gradient" methods, Gauss–Newton, and other methods (see Bevington & Robinson, 2003; Hughes & Has, 2010). One of the most popular methods, which we encourage you to use for this book, is called the Levenberg–Marquardt method. This method combines features of the different optimization approaches, and as a result, it is both reasonably fast and does a good job of searching broadly (in the language of optimization, it has a "good radius of convergence").

Regardless of the specific method, iterative nonlinear optimization method, most nonlinear least-squares methods require a set of "initial guess" parameters to evaluate *SSR*. Although these guesses only need to be approximate, they must be close enough that the value of the residual is sensitive to modest adjustments to the parameter values (i.e., they need to be within the radius of convergence). Thus, there is sometimes a bit of adjusting that needs to be made to get the initial parameters to where the fitting program can work with them. Otherwise, the fitting algorithm will "diverge" (i.e., fail). Usually, you can just look at your experimental data and come up with decent initial guesses that will converge (e.g., by eyeballing midpoints, slopes, plateaus in the data). However, sometimes it is helpful to "simulate" the data with the fitting function, and try various combinations of parameter values that make the fitting function look like the data.

The command we will use in Mathematica to fit a model to data using nonlinear least squares is "NonlinearModelFit."[†] This function is called using the command

```
NonlinearModelFit[data, {function, constraints}, {fitting params
and guesses},x, options]
```

Format of the data. For a data set with *n* observations, the expression *data* above is a nested list of depth 2, containing *n* sublists. Each sublist corresponds to the values of the independent and dependent variable of the first "point" in the data set, the second to the second point, and so on, up through the *n*th (i.e., last) point. For example, for a simple fit of a function of a single variable and parameters $y = f(x, p_1, p_2, ...)$, the data would have the form

```
{{x1,y1},{x2,y2},{x3,y3},…,{xn,yn}}
```

This format is typically encountered when data is imported using the "Table" option. However, the data may either need to be arranged before (with a spreadsheet or editor) or after (using list manipulation) import to Mathematica.

Format of the function and constraints. The function is what you are going to fit to the data. It has the form given in Equation A.1. You can either enter it directly, or you can define it on a previous line. For example, if the data is fractional saturation of a macromolecule containing multiple sites with ligand at concentration x, and you wish to fit a Hill equation (Chapter 13, Equation 13.44) to the data, fitting for both the binding constant *K* and the Hill coefficient *n* (parameters p_1 and p_2), then you would enter:

```
(K*x)^n/(1+(K*x)^n)
```

If you wish to constrain the fitted parameters to a limited range of values (for instance, equilibrium constants are never negative), you can provide constraints. Constraining a parameter can also be useful when you have strong parameter correlation, especially if you have information on one of the parameter values from

---

[†] Another closely related function "FindFit" works almost the same way. However, Nonlinear ModelFit provides a more sophisticated analysis of the results of the fit, such as determination of confidence intervals on parameters, analysis of parameter correlation analysis, and analysis of variance (ANOVA).

another experiment. If you do not wish to use the variable x as your independent variable, you need to be sure to enter your chosen variable following the {Fitting params} instead of x.

Format of the fitting parameters and initial guesses. These are given as a nested list, with each sublist a two-element list consisting of the parameter name and the initial guess. For the example described here with the Hill model, the format would be

```
{{K,200},{n,2}}
```

I would use these initial values if the midpoint of the transition had a midpoint of around 0.005 concentration units, and the curve displayed some cooperativity.

Options. To see a full list of options that can be entered, go to the NonlinearModelFit entry in the Documentation center of Mathematica, and open the "more information" section. This is the place where you can set the type of algorithm used for optimization. I would always start with the option

```
Method->LevenbergMarquardt
```

Other useful things can be set using "options." For example, weights can be introduced if some y values have more error than others.[†] In addition, the maximum number of iterations to do before quitting can be set, and the "convergence criterion" can also be set (the point at which the fitter decides it is done, which is chosen when the decrease in *SSR* is smaller than a specified value).

Combining all these commands for the Hill example above would look like this:

```
NonlinearModelFit[data, (K*x)^n/(1+(K*x)^n), {{K,200},{n,2}},
x, Method->LevenbergMarquardt]
```

In addition, you might want to assign the fit to a variable name so that you can analyze various aspects of the fit (compare the fit to the data, check parameter values, plot residuals):

```
myfit=NonlinearModelFit[data, (K*x)^n/(1+(K*x)^n),
{{K,200},{n,2}},x, Method->LevenbergMarquardt]
```

[†] Nonlinear least squares assume that each y value has the same level of error, the error is uncorrelated from point to point, and that the error is Gaussian. See Chapter 2 for a discussion of weighting during fits.

Bibliography

Adkins, C.J. 1983. *Equilibrium Thermodynamics*, 3rd Edition. Cambridge University Press, Cambridge, UK.

Adkins, C.J. 1984. *Equilibrium Thermodynamics*. Cambridge University Press, Cambridge, United Kingdom. *An exceptionally clear treatment of classical thermodynamics, with some nice non-equilibrium problems.*

Aksel, T. and Barrick, D. 2009. Analysis of repeat-protein folding using nearest-neighbor statistical mechanical models. *Meth. Enzymol.*, 455, 95–125.

Antonini, E. and Brunori, M. 1971. *Hemoglobin and Myoglobin in their Reactions with Ligands*. North Holland Publishing Company, Amsterdam.

Atkins, P. 2014. *Physical Chemistry: Thermodynamics, Structure, and Change*. 10th edition. W.H. Freeman & Co., New York. *A classic text for teaching undergraduate physical chemistry.*

Atkins, P.W. 1984. The Second Law. Scientific American Books—W. H. Freeman and Company.

Babu, C.R., Hilser, V.J., and Wand, A.J. 2004. *Nat. Struct. Mol. Biol.* 11, 352.

Beard, D.A. and Qian, H. 2008. *Chemical Biophysics: Quantitative Analysis of Chemical Systems*. Cambridge University Press, UK.

Ben-Naim, A. 2008. *Entropy Demystified. The Second Law Reduced to Plain Common Sense with Seven Simulated Games*. World Scientific, Hackensack, NJ.

Blom, G. 1989. Probability and Statistics: Theory and Applications. Springer Texts in Statistics, Springer-Verlag, New York.

Bevington, P.R. 1969. *Data Reduction and Error Analysis for the Physical Sciences*. McGraw Hill, New York.

Bevington, P.R. and Robinson, B.K. 2003. Data Reduction and Error Analysis. McGraw-Hill, New York.

Callen, H. B. 1985. *Thermodynamics and Introduction to Thermostatistics*, 2nd Edition. John Wiley & Sons, New York.

Cantor, C.R. and Schimmel, P.R. 1080. *Biophysical Chemistry Part III: The Behavior of Biological Macromolecules*. W. H. Freeman & Co., New York, NY.

Chakrabartty, A., Kortemme, T., and Baldwin, R.L. 1994. Helix propensities of the amino acids measured in alanine-based peptides without helix-stabilizing side-chain interactions. *Prot. Sci.*, 3, 843–852.

Coxeter, H.S.M. 1973. *Regular Polytopes*. Dover Publications, Inc., New York.

Di Cera, E. 1996. *Thermodynamic Theory of Site-Specific Binding Processes in Biological Macromolecules*. Cambridge University Press, Cambridge, UK.

Fermi, E. 1936. *Thermodynamics*. Dover Publications, Inc., Mineola, NY.

Fermi, E. 1956. *Thermodynamics. Dover Books*, Mineola, NY. *This spanky little treatise cuts right through the crap!*

Gibbs, J.W. 1878. On the equilibrium of heterogeneous substances. *Trans. Connecticut Academy.* III, 108–248 and 343–524.

Gill, S.J. 1962. The chemical potential. *J. Chem. Edu.* 39, 506–510.

Goodsell, D.S. and Olson, A.J. 2000. Structural symmetry and protein function. *Annu. Rev. Biophys. Biomol. Struct.* 29, 105–153.

Guggenheim, E.A. 1933. *Modern Thermodynamics by the Methods of Willard Gibbs*. Methuen & Co., London.

Guggenheim, E.A. 1967. *Thermodynamics: An Advanced Treatment for Chemists and Physicists*. North Holland Publishing, Amsterdam.

Hill, T.G. 1987. *An Introduction to Statistical Thermodynamics*. Dover Press, New York.

Hill, T. 2005. *Free Energy Transduction and Biochemical Cycle Analysis*. Dover Books, Mineola, NY.

Hughes, I.G. and Hase, T.P.A. 2010. Measurements and their uncertainties: A practical guide to modern error analysis. Oxford University Press, Inc., New York.

Johnson, M.L. 2008. Nonlinear least-squares fitting methods. *Meth. Cell. Biol.*, 84, 781–805.

Johnson, M.L. 1992. Why, when, and how biochemists should use nonlinear least-squares. *Anal. Biochem.*, 202, 215–225.

Koshland, D.E., Jr., Nemethy, G., and Filmer, D. 1966. Comparison of experimental binding data and theoretical models in proteins containing subunits. *Biochemistry* 5, 365–385.

Lewis, G.N. and Randall, M. 1961. *Thermodynamics and the Free Energy of Substances*, 2nd edition. McGraw-Hill, New York.

Lifson, S. and Roig, A. 1961. On the theory of helix–coil transitions in polypeptides. *J. Chem Phys.*, 34, 1963–1974.

Lindley, D. 2001. *Boltzmann's Atom: The Great Debate that Launched a Revolution in Physics*. The Free Press, New York.

Lindley, D. 2004. *Degrees Kelvin: A Tale of Genius, Invention, and Tragedy*. Joseph Henry Press, Washington DC.

Marsh, J.A. and Teichmann, S.A. 2015. Structure, dynamics, assem and evolution of protein complexes. *Annu. Rev. Biochem.* 84, 551–574.

Marsh, K. N. and Richards, A. E. 1980. Excess volumes of ethanol + water mixtures at 10-K intervals from 278.15 to 338.15 K. *Aust. J. Chem.*, 33, 2121–2132.

Monod, J., Wyman, J., and Changeaux, J.P. 1965. On the nature of allosteric transitions: A plausible model. *J. Mol. Biol.* 12, 88–118.

Motlagh, H.N., Wrabl, J.O., Li, J., and Hilser, V.G. 2014. The ensemble nature of allostery. *Nature* 508, 331–339.

Poland, D. and Scheraga, H.A. 1970. *Theory of the Helix-Coil Transition in Biopolymers*. Academic Press, New York, NY.

Privalov, P. L. 2012. *Microcalorimetry of Macromolecules: The Physical Basis of Biological Structures*. John Wiley & Sons, Hoboken, NJ.

Pugh, E.M. and Winslow, G.H. 1966. *The Analysis of Physical Measurements*. Addison Wesley, Palo Alto.

Scholtz, J.M., Qian, H., York, E., Stewart, J.M., and Baldwin, R.L. 1991. Parameters for helix–coil transition theory for alanine-based peptides of varying chain-lengths in water. *Biopolymers*, 31(13), 1463–1470.

Shannon, C.E. 1948. A mathematical theory of communication. *Bell Syst. Tech. J.* 7(3), 379–423.

Silver, N. 2012. *The Signal and the Noise. Why So Many Predictions Fail—But Some Don't*. The Penguin Press, New York.

Strang, G. 2006. *Linear Algebra and its Applications*, 4th Edition. Cengage Publishers, Boston.

Tellinghuisin, J. 2008. Stupid statistics! *Meth. Cell. Biol.*, 84, 739–780.

Trudeau, R. 1993. *Introduction to Graph Theory*. Dover Publications, Inc., New York.

Weast, R.C. 1979. *The Handbook of Chemistry and Physics*, 60th Edition. CRC Press, Boca Raton, FL.

Wyman, J. and Gill, S.J. 1990. *Binding and Linkage: Functional Chemistry of Biological Macromolecules*. University Science Books, Mill Valley, CA.

Zemansky, M.W. 1964. *Temperatures Very Low and Very High*. Dover Books, Mineola, NY.

Zimm, B.H. and Bragg, J.K. 1959. Theory of the phase transition between helix and random coil in polypeptide chains. *J. Chem. Phys.*, 31, 526–535.

Index

Note: Page numbers followed by "*fn*" indicates footnotes.